VIROIDS
and
SATELLITES

Cover photographs

Front cover

Electron micrograph of small circular coconut cadang-cadang viroid molecules. (Courtesy John W. Randles.)

Coconut cadang-cadang viroid induced non-necrotic yellow spotting of a coconut palm leaflet (right) compared with an asymptomatic leaflet (left). (Courtesy John W. Randles.)

Potato spindle tuber viroid affected potato tuber showing typical spindle shape (right) compared with a normal potato tuber (left). (Courtesy Theodor O. Diener.)

Extreme leaf chlorosis (peach calico) symptoms induced by certain variants of peach latent mosaic viroid. (Courtesy Francesco Di Serio and Ricardo Flores.)

Symptoms of citrus exocortis viroid infection on a citron fruit (right) compared with a normal fruit (left). (Courtesy Joseph S. Semancik.)

Severe symptoms of a variant of cucumber mosaic virus Tfn-satellite RNA on tomato fruit cv. Italpeel. (Courtesy Donato Gallitelli and the late Makis Vovlas.)

Symptoms of the dapple apple variant of apple scar skin viroid on apple fruit cv. Starkrimson. (Courtesy Jean-Claude Desvignes.)

Back cover

Symptoms of chrysanthemum stunt viroid on white chrysanthemum flower (right) showing reduced size and quality compared with a normal flower (left). (Courtesy Teruo Sano.)

Avocado sunblotch viroid induced depression and distortion symptoms on an avocado fruit. (Courtesy David N. Kuhn.)

VIROIDS and SATELLITES

Edited by

AHMED HADIDI
U.S. Department of Agriculture, Beltsville, MD, United States

RICARDO FLORES
Polytechnic University of Valencia-CSIC, Valencia, Spain

JOHN W. RANDLES
The University of Adelaide, Waite Campus, Glen Osmond, Australia

PETER PALUKAITIS
Seoul Women's University, Seoul, South Korea

Academic Press is an imprint of Elsevier
125 London Wall, London EC2Y 5AS, United Kingdom
525 B Street, Suite 1800, San Diego, CA 92101-4495, United States
50 Hampshire Street, 5th Floor, Cambridge, MA 02139, United States
The Boulevard, Langford Lane, Kidlington, Oxford OX5 1GB, United Kingdom

British Library Cataloguing-in-Publication Data
A catalogue record for this book is available from the British Library

Library of Congress Cataloging-in-Publication Data
A catalog record for this book is available from the Library of Congress

ISBN: 978-0-12-801498-1

For Information on all Academic Press publications
visit our website at https://www.elsevier.com/books-and-journals

Publisher: Sara Tenney
Acquisition Editor: Linda Versteeg-Buschman
Editorial Project Manager: Fenton Coulthurst
Production Project Manager: Chris Wortley
Cover Designer: Miles Hitchen

Typeset by MPS Limited, Chennai, India

Contents

SECTION I
VIROIDS

PART I
VIROIDS: ECONOMIC SIGNIFICANCE

10. Changes in the Host Proteome and Transcriptome Induced by Viroid Infection

ROBERT A. OWENS, GUSTAVO GÓMEZ, PURIFICACIÓN LISÓN AND VICENTE CONEJERO

11. Viroids and RNA Silencing

ELENA DADAMI, ATHANASIOS DALAKOURAS AND MICHAEL WASSENEGGER

12. Origin and Evolution of Viroids

FRANCESCO DI SERIO, BEATRIZ NAVARRO AND RICARDO FLORES

13. Viroid Taxonomy

FRANCESCO DI SERIO, SHI-FANG LI, VICENTE PALLÁS, ROBERT A. OWENS, JOHN W. RANDLES, TERUO SANO, JACOBUS TH.J. VERHOEVEN, GEORGIOS VIDALAKIS AND RICARDO FLORES

PART III

VIROID DISEASES

21. Apple Scar Skin Viroid

AHMED HADIDI, MARINA BARBA, NI HONG AND VIPIN HALLAN

22. Other Apscaviroids Infecting Pome Fruit Trees

FRANCESCO DI SERIO, ENZA M. TORCHETTI, RICARDO FLORES AND TERUO SANO

29. Peach Latent Mosaic Viroid in Infected Peach

RICARDO FLORES, BEATRIZ NAVARRO, SONIA DELGADO, CARMEN HERNÁNDEZ, WEN-XING XU, MARINA BARBA, AHMED HADIDI AND FRANCESCO DI SERIO

30. Peach Latent Mosaic Viroid in Temperate Fruit Trees Other Than Peach

PANAYOTA E. KYRIAKOPOULOU, LUCIANO GIUNCHEDI, MARINA BARBA, IRAKLIS N. BOUBOURAKAS, MARIA S. KAPONI AND AHMED HADIDI

31. Chrysanthemum Chlorotic Mottle Viroid

RICARDO FLORES, SELMA GAGO-ZACHERT, PEDRO SERRA, MARCOS DE LA PEÑA AND BEATRIZ NAVARRO

PART VI

GEOGRAPHICAL DISTRIBUTION OF VIROIDS AND VIROID DISEASES

43. Geographical Distribution of Viroids in the Americas

EDWARD V. PODLECKIS

44. Geographical Distribution of Viroids in Europe

FRANCESCO FAGGIOLI, NURIA DURAN-VILA, MINA TSAGRIS AND VICENTE PALLÁS

45. Geographical Distribution of Viroids in Africa and the Middle East

KHALED A. EL-DOUGDOUG, KADRIYE ÇAĞLAYAN, AMINE ELLEUCH, HANI Z. AL-TUWARIQI, EBENEZER A. GYAMERA AND AHMED HADIDI

49. Genome Editing by CRISPR-Based Technology: Potential Applications for Viroids

AHMED HADIDI AND RICARDO FLORES

SECTION II

SATELLITES

PART VIII

INTRODUCTION

50. Satellite Viruses and Satellite Nucleic Acids

PETER PALUKAITIS

PART IX

IMPACT OF SATELLITES

51. Economic Significance of Satellites

TIZIANA MASCIA AND DONATO GALLITELLI

PART X

SATELLITE CHARACTERISTICS

55. Development and Application of Satellite-Based Vectors

MING-RU LIOU, CHUNG-CHI HU, NA-SHENG LIN AND YAU-HEIU HSU

56. Origin and Evolution of Satellites

FERNANDO GARCÍA-ARENAL AND AURORA FRAILE

57. Satellite Taxonomy

PETER PALUKAITIS

PART XI

TYPES OF SATELLITES

A

SATELLITE VIRUSES

58. Biology and Pathogenesis of Satellite Viruses

JESSE D. PYLE AND KAREN-BETH G. SCHOLTHOF

B

SATELLITE NUCLEIC ACIDS

59. Large Satellite RNAs

MAZEN ALAZEM AND NA-SHENG LIN

60. Small Linear Satellite RNAs

MIKYEONG KIM AND MARILYN J. ROOSSINCK

List of Contributors

Mazen Alazem Institute of Plant and Microbial Biology, Academia Sinica, Taipei, Taiwan

Hani Z. Al-Tuwariqi Ministry of Agriculture, Riyadh, Saudi Arabia

Marina Barba CREA-Research Centre for Plant Protection and Certification, Rome, Italy

Iraklis N. Boubourakas Directorate of Rural Economy and Veterinary of Piraeus, Attica Prefecture, Attica, Greece

Kadriye Çağlayan Mustafa Kemal University, Antakya, Turkey

Thierry Candresse INRA and University of Bordeaux, Bordeaux, France

Vicente Conejero Polytechnic University of Valencia-CSIC, Valencia, Spain

Elena Dadami RLP AgroScience, AlPlanta-Institute for Plant Research, Neustadt-Mußbach, Germany

Athanasios Dalakouras RLP AgroScience, AlPlanta-Institute for Plant Research, Neustadt-Mußbach, Germany

José-Antonio Daròs Polytechnic University of Valencia-CSIC, Valencia, Spain

Marcos De la Peña Polytechnic University of Valencia-CSIC, Valencia, Spain

Sonia Delgado Polytechnic University of Valencia-CSIC, Valencia, Spain

Francesco Di Serio National Research Council, Bari, Italy

Jonathan Dixon Seeka Kiwifruit Industries Limited, Te Puke, New Zealand

Nuria Duran-Vila Valencian Institute of Agricutural Research-IVIA, Valencia, Spain

Khaled A. El-Dougdoug Ain Shams University, Cairo, Egypt

Amine Elleuch University of Sfax, Sfax, Tunisia

Francesco Faggioli CREA-Research Centre for Plant Protection and Certification, Rome, Italy

Ricardo Flores Polytechnic University of Valencia-CSIC, Valencia, Spain

Aurora Fraile Center for Plant Biotechnology and Genomics, Polytechnic University of Madrid-INIA, Madrid, Spain

Selma Gago-Zachert Leibniz Institute of Plant Biochemistry, Halle, Germany

Donato Gallitelli University of Bari Aldo Moro, Bari, Italy

Fernando García-Arenal Center for Plant Biotechnology and Genomics, Polytechnic University of Madrid-INIA, Madrid, Spain

Andrew D.W. Geering The University of Queensland, St Lucia, QLD, Australia

Luciano Giunchedi University of Bologna, Bologna, Italy

Gustavo Gómez Polytechnic University of Valencia-CSIC, Valencia, Spain

Ebenezer A. Gyamera University of Cambridge, Cambridge, United Kingdom

Nuredin Habili The University of Adelaide, Waite Campus, Glen Osmond, SA, Australia

Ahmed Hadidi U.S. Department of Agriculture, Beltsville, MD, United States

Vipin Hallan CSIR-Institute of Himalayan Bioresource Technology, Palampur, Himachal Pradesh, India

Rosemarie W. Hammond U.S. Department of Agriculture, Beltsville, MD, United States

Dagmar Hanold The University of Adelaide, Waite Campus, Glen Osmond, SA, Australia

Tatsuji Hataya Hokkaido University, Sapporo, Japan

Carmen Hernández Polytechnic University of Valencia-CSIC, Valencia, Spain

Ni Hong Huazhong Agricultural University, Wuhan, China

Munetaka Hosakawa Kyoto University, Kyoto, Japan

Yau-Heiu Hsu National Chung Hsing University, Taichung, Taiwan

Chung-Chi Hu National Chung Hsing University, Taichung, Taiwan

Ying-Wen Huang National Chung Hsing University, Taichung, Taiwan

Delano James Centre for Plant Health-Canadian Food Inspection Agency, North Saanich, BC, Canada

Hendry Joseph Universiti Teknologi MARA (Sabah), Sabah, Malaysia

Kriton Kalantidis University of Crete, Heraklion, Greece

Maria S. Kaponi Formerly, Hirosaki University, Hirosaki, Japan

Mikyeong Kim National Institute of Agricultural Sciences, RDA, Wanju-gun, South Korea

Gary R. Kinard U.S. Department of Agriculture, Beltsville, MD, United States

Lih L. Kong Universiti Putra Malaysia, Selangor, Malaysia

Natalia Kovalskaya U.S. Department of Agriculture, Beltsville, MD, United States

David N. Kuhn U.S. Department of Agriculture, Miami, FL, United States

Panayota E. Kyriakopoulou Agricultural University of Athens, Athens, Greece

Irene Lavagi University of California, Riverside, CA, United States

Shi-Fang Li Chinese Academy of Agricultural Sciences, Beijing, China

Na-Sheng Lin Institute of Plant and Microbial Biology, Academia Sinica, Taipei, Taiwan

Kai-Shu Ling U.S. Department of Agriculture, Charleston, SC, United States

Ming-Ru Liou National Chung Hsing University, Taichung, Taiwan

Purificación Lisón Polytechnic University of Valencia-CSIC, Valencia, Spain

Amparo López-Carrasco Polytechnic University of Valencia-CSIC, Valencia, Spain

Marta Luigi CREA-Research Centre for Plant Protection and Certification, Rome, Italy

Ángel-Emilio Martínez de Alba Polytechnic University of Valencia-CSIC, Valencia, Spain

Tiziana Mascia University of Bari Aldo Moro, Bari, Italy

Chikara Masuta Hokkaido University, Sapporo, Japan

Jaroslav Matoušek Academy of Sciences of the Czech Republic (ASCR), České Budějovice, Czech Republic

Marie-Christine Maurel Sorbonne University, Paris, France

Ellis T.M. Meekes Naktuinbouw, Roelofarendsveen, The Netherlands

Sofia Minoia Polytechnic University of Valencia-CSIC, Valencia, Spain

Beatriz Navarro National Research Council, Bari, Italy

Xianzhou Nie Agriculture and Agri-Food Canada, Fredericton, NB, Canada

Robert A. Owens U.S. Department of Agriculture, Beltsville, MD, United States

Vicente Pallás Polytechnic University of Valencia-CSIC, Valencia, Spain

Peter Palukaitis Seoul Women's University, Seoul, South Korea

Dattaraj B. Parakh ICAR-National Bureau of Plant Genetic Resources, New Delhi, India

Jean-Pierre Perreault University of Sherbrooke, Sherbrooke, QC, Canada

Edward V. Podleckis U.S. Department of Agriculture, Riverdale, MD, United States

Jesse D. Pyle Harvard Medical School, Boston, MA, United States

John W. Randles The University of Adelaide, Waite Campus, Glen Osmond, SA, Australia

Detlev Riesner Heinrich Heine University, Düsseldorf, Germany

M. Judith B. Rodriguez Philippine Coconut Authority, Albay, Philippines

Johanna W. Roenhorst National Plant Protection Organization, Wageningen, The Netherlands

Marilyn J. Roossinck Pennsylvania State University, University Park, PA, United States

Luisa Rubino National Research Council, Bari, Italy

Jesus A. Sánchez-Navarro Polytechnic University of Valencia-CSIC, Valencia, Spain

Teruo Sano Hirosaki University, Hirosaki, Japan

Karen-Beth G. Scholthof Texas A&M University, College Station, TX, United States

Pedro Serra Polytechnic University of Valencia-CSIC, Valencia, Spain

Hanako Shimura Hokkaido University, Sapporo, Japan

Rudra P. Singh Agriculture and Agri-Food Canada, Fredericton, NB, Canada

Dijana Škorić University of Zagreb, Zagreb, Croatia

Giuseppe Stancanelli European Food Safety Authority, Parma, Italy

Gerhard Steger Heinrich Heine University Düsseldorf, Germany

Anna Taglienti CREA-Research Centre for Plant Protection and Certification, Rome, Italy

Matilde Tessitori University of Catania, Catania, Italy

Sathis S. Thanarajoo Universiti Putra Malaysia, Selangor, Malaysia

Antonio Tiberini Mediterranea University of Reggio Calabria, Reggio Calabria, Italy

Enza M. Torchetti National Research Council, Bari, Italy

Mina Tsagris University of Crete, Crete, Greece

Taro Tsushima Hirosaki University, Hirosaki, Japan

Ganesan Vadamalai Universiti Putra Malaysia, Selangor, Malaysia

Jacobus Th.J. Verhoeven National Plant Protection Organization, Wageningen, The Netherlands

Georgios Vidalakis University of California, Riverside, CA, United States

Qiao-Chun Wang Northwest A&F University, Xianyang, China

Michael Wassenegger RLP AgroScience, AlPlanta-Institute for Plant Research, Neustadt-Mußbach, Germany

Stephan Winter Leibniz Institute, DSMZ- German Collection of Microorganisms and Cell Cultures, Braunschweig, Germany

Wen-Xing Xu Huazhong Agricultural University, Wuhan, China

Xiuling Yang Chinese Academy of Agricultural Sciences, Beijing, China

Yongjiang Zhang Chinese Academy of Inspection and Quarantine, Beijing, China

Zhibo Zhang Northwest A&F University, Xianyang, China

Xueping Zhou Chinese Academy of Agricultural Sciences, Beijing, China; Zhejiang University, Hangzhou, China

Shuifang Zhu Chinese Academy of Inspection and Quarantine, Beijing, China

Foreword

I am delighted to write a Foreword for this well-conceived book, which, with the prominent authors chosen, is assured to take its place as an up-to-date, complete, and dependable textbook on the plant-infecting subviral viroids and satellites for years to come.

The third major extension in the history of the biosphere was ushered in by the discovery of viroids in 1971, following the earlier discoveries of the "subvisible" microorganisms by Antonie van Leeuwenhoek in 1675 and of the "submicroscopic" viruses by Dmitri Iosifovich Ivanovsky in 1892. The status of viroids is recognized by the International Committee for Virus Taxonomy (ICTV) with the creation of a new order of *subviral agents*, which currently includes two families, eight genera, and 32 species of viroids.

A great interest in viroids among plant virologists occurred immediately after their discovery and the important biological, biochemical, and evolutionary significance of subviral agents has been acknowledged by many other scientists who confirmed the fundamental nature of viroids (and, implicitly, of other subviral agents), which, like viruses, are replicated in infected cells.

It is now accepted that viroids, like viruses, share the most characteristic property of living beings: in an appropriate environment, they are able to generate copies of themselves; in other words, they are endowed with autonomous replication (and evolution). It is in this framework that viroids represent the frontier of life (infectious agents consisting of only 246–401 ribonucleotides), an aspect that should attract the attention of anybody interested in biology.

Theodor O. Diener

Preface

Many plant diseases currently known to be caused by viroids were first recognized because of their damaging economic impact many years ago. Potato spindle tuber disease was first described in the United States in the early 1920s and its infectious agent was identified by Theodor O. Diener of the US Department of Agriculture, Beltsville, Maryland, in 1971. He described it as a free RNA of 25,000–110,000 Daltons, much smaller than a viral genome, and named it as "viroid" in 1972. Similarly, the "exocortis" disease of citrus was first described in the late 1940s in the United States and in 1972, Joseph S. Semancik of the University of California, Riverside, showed that its causal agent is also a viroid of 125,000 Daltons. Cadang-cadang disease of coconut palms in the Philippines, so named because it leads to the premature decline and death of trees, was first described early in the 20th century but it was not until 1975 that John W. Randles of the University of Adelaide, South Australia, demonstrated that the disease is caused by a viroid. By 1977, it was found that viroids cause economically important diseases in vegetable crops, fruit trees, ornamentals, and palm species. The host range of viroids has expanded over the years to include grapevine and other cultivated and wild plant species. Viroid molecules are very small circular RNAs, ranging from 246 to 401 nucleotides in length and with a high degree of internal base-pairing. Their discovery has challenged the central dogma of genetic inheritance because unlike viruses they are naked and lack protein-coding ability. Although found only in plants, their discovery paved the way for other subviral agents such as prions and hepatitis delta virus to be recognized as pathogens of eukaryotic organisms. Viroids are the smallest known infectious RNAs.

Basil Kassanis of the Rothamsted Experiment Station, Harpenden, Hertfordshire, UK, coined the name "satellite" in 1962 for very small virus particles he found in some cultures of tobacco necrosis virus (TNV). He demonstrated that these 17-nm viral satellite particles were unable to replicate in the absence of the larger 30-nm TNV particles. Similarly, in 1972 Irving R. Schneider of the US Department of Agriculture in Beltsville found nonessential but biologically active satellite RNA molecules in preparations of various other viruses. In 1972, a devastating disease epidemic swept through tomato fields of the French Alsace province destroying the crop. The causal agent was not revealed until 1977, when Jacobus M. Kaper of the US Department of Agriculture in Beltsville clearly demonstrated that RNA 5 of cucumber mosaic virus (CMV) was required to cause the Alsace disease. Because RNA 5 alone cannot cause the disease, but needed other CMV RNAs for replication, this CMV RNA was recognized as the first RNA satellite to cause a plant disease.

Viroids and Satellites was initiated in mid-2013 as we recognized the need for a high quality book that advances knowledge on the biological, molecular, and genomic properties of these pathogens. The Editors have known each other professionally for decades, have been members of editorial boards of several virology and plant pathology journals, and have accumulated scientific research experience of many years. This book presents the result of the Editors' endeavor, providing up-to-date information in a comprehensive, scientific, and systematic way in a total of 63 chapters arranged into 12 parts as follows:

VIROIDS
- Part I Viroids: Economic Significance (Chapters 1–4)
- Part II Viroid Characteristics (Chapters 5–13)
- Part III Viroid Diseases (Chapters 14–32)
- Part IV Detection and Identification Methods (Chapters 33–38)
- Part V Control Measures for Viroids and Viroid Diseases (Chapters 39–42)
- Part VI Geographical Distribution of Viroids and Viroid Diseases (Chapters 43–47)
- Part VII Special Topics (Chapters 48–49)

SATELLITES
- Part VIII Introduction (Chapter 50)
- Part IX Impact of Satellites (Chapter 51)
- Part X Satellite Characteristics (Chapters 52–57)
- Part XI Types of Satellites (Chapters 58–62)
- Part XII Application to Control of Viruses (Chapter 63)

We would like to thank our colleagues from 24 countries in North America, Europe, Africa, the Middle East, Asia, and Oceania for their valuable contributions to this book, their patience, understanding, cooperation, and collaboration and, above all, for sharing their vast and excellent scientific expertise on viroids and/or satellites. We are also indebted to Elsevier staff in both the United States and Europe for their complete and unwavering support for publishing this book. It is our hope that this book serves as a new and preeminent resource for researchers, teachers, and students in plant virology, plant pathology, microbiology, biotechnology, horticulture, and agriculture, as well as for diagnosticians, regulators, and others worldwide.

Ahmed Hadidi, Ricardo Flores,
John W. Randles and Peter Palukaitis

Introduction

I appreciate the thoughtfulness of the editors in asking me to offer a few words of introduction to this volume, *Viroids and Satellites*. This production is a timely update to the book entitled *Viroids* (CSIRO Press, 2003) which it was my pleasure to coedit with Ahmed, John, and Ricardo over 13 years ago. I believe the addition of Peter to the new editorial team specifically for addressing plant viral satellites will expand the scope of this 2017 presentation on small RNAs causing disease in plants. I applaud the knowledge, insight, and dedication of the editors in producing this treatise.

The most recent findings and perspectives can now be found here in *Viroids and Satellites*. This current perspective is edited by four internationally acclaimed scientists in plant virology which includes viroids and satellites. The book contributors consisting of more than 100 authors from 24 countries are experts in their subjects. During the past 13 years, significant new information has been published on viroids, necessitating the publication of a new book. Moreover, with no book currently available on the subject of satellites, the combined discussion of viroids and satellites will expand the horizon of readers to small RNAs in plants.

Viroids and Satellites describes and discusses viroid and viral satellite diseases of plants as well as the resultant economic impact. It also describes the various protocols employed in the detection and control of pathological expression. This book offers a comprehensive resource as a textbook for students as well as a guide for research laboratories and commercial applications. Moreover, it is an excellent guide for regulatory applications of certification and quarantine of many plant species affected by the agents described here as well as strategies for production of pathogen-free plants.

In these few words, I offer no sage theories or novel interpretation. What follows are personal reflections and musings of past and present that may be questioned as any review or book chapter might. I hope it will encourage the turning of the pages with enhanced curiosity in a discovery that began about 50 years ago and remains a topical subject for investigation in the world of biology today.

My interest in plant viruses and pathology was extended to viroids with an introduction to exocortis disease of citrus by my colleague Lewis G. Weathers at the University of California, Riverside. Our reports of "infectious RNA existing as free nucleic acid" in 1968–70 was followed by "the first association of 'pathogenic activity' with a physically identifiable new RNA species" in *Nature* in 1972, a small pathogenic RNA, a "viroid" RNA. This finding was made possible when samples from physicochemical experiments in my laboratory at the University of Nebraska were assayed at the University of California, Riverside, to provide evidence supporting biological activity. What might be viewed as a whimsical curiosity is that the causal viroid (citrus exocortis viroid) of a disease of citrus was first seen in a laboratory in Nebraska! From that introduction, citrus was later found to harbor a variety of pathogenic and nonpathogenic viroid species.

It has been my fortune to be part of the viroid story for 30 years shared with talented colleagues associated or collaborating with my laboratory and introducing views from various perspectives and fields of interest. The focus of our laboratory has been not only the identification and characterization of viroids as pathogens but also on the biological interaction of viroids with host genome expression as "elements" altering growth and development. As an example, citrus dwarfing viroid and hop stunt viroid-citrus nonpathogenic variant are such acting as Transmissible small nuclear RNAs inducing citrus dwarfing for application to high-density plantings (see Chapter 2: Economic Significance of Fruit Tree and Grapevine Viroids, of this volume). Reports on similar subjects made in the 1970s and 1980s may be found in the chapter "Viroid Pathogenesis and Expressions of Biological Activity" coauthored with V. Conejero-Tomas in the book Viroids and Viroid-Like Pathogens (CRC Press, 1986) and subsequent research publications.

The research data of Martinus Beijerinck in describing a *"contagium vivum fluidum"* that he later termed a "virus" coupled with the findings of his contemporary Dmitri Ivanovsky brought into focus a new field of research, "Virology." Similarly, contributing an invaluable stimulus promoting research on the small pathogenic RNAs identified in the early 1970s was the reintroduction of the term "viroid" which Edgar Altenburg in 1946 coined to describe possible "... symbionts akin to viruses ..." in animal cells.

It is interesting that Martinus Beijerinck and Dmitri Ivanovsky were botanists who might now be grouped with plant pathologists. Both respected the value of the bioassay for the description of the virus. Similarities in physical properties may facilitate the definition of molecular relationships. Biological expressions of pathology and developmental alterations from interactions of transmissible molecules with cells may require more time and observation. In any taxonomic scheme, both the physical and biological properties need to be evaluated equally.

A focus on the physical properties as well as pathology and biological activities of viroids as a novel molecular species became of multinational interest in the United States, Germany, Canada, China, Japan, and Australia. In 1975, I was honored to share with Theodor O. Diener the Alexander Von Humboldt Award for recognition of viroids as "a previously unknown group of pathogens." I feel it fitting to also acknowledge the historic extensive contributions of the laboratories of Heinz L. Sänger (Germany), Rudra P. Singh (Canada), and Robert H. Symons (Australia) who though no longer with us, should not be forgotten.

Is the principal position of the plant kingdom in identifying both viruses and viroids significant or a curiosity of biology? The scope of viruses in the animal kingdom expanded interest and recognition greatly with the concerns for human health. In the search for commonality in the world of biology, what became of viroids in the animal world? In the absence of biological expressions of disease as displayed by plants, animal "viroids" may be more difficult to identify. Might these not be agents of disease but controlling elements of gene expression or as discussed in contributions to this volume, perhaps only relics preserved within the plant cell? Has the survey of the animal kingdom been exhausted or are the essential biological probes for the search of viroids in animal cells not yet available? Will the global initiatives to decipher the plant and animal microbiomes provide any answers to these questions?

I have not been attracted to theories and speculation, perhaps my failing. I am a victim of an imagination, intuition, and a curiosity that urges one to the bench to produce data. These data may confirm or reject the original motivating idea, however, all data have a story of their own to tell. It is only for each individual to read the story and perhaps its variations. Data can be seen as a group of words scrambled on the table as dominoes. Different sentences of note might be arranged by the eye as controlled by the person either producing or interpreting the words. Sentences of meaning may be fashioned from both.

Reflect on the wealth of information contained in this new "resource" volume *Viroids and Satellites*, but also be mindful that the content is formulated from "sources." These may speak with another perspective or simply be responding to a different question. It is for the reader to investigate these original sources as motivated. And the authors listed in sources are at times included simply as "et al." Their contributions should be acknowledged and their names not be diminished or forgotten.

No single person, professor, postdoctorate, technician, or student is without the capacity to generate an idea to be awakened to action and data that perhaps no other has been witness to. There are benefits of age in the fond memories of those bright eyes with whom I have interacted, some of whom have contributed to this volume, and produced data as a reward to honor their intellect and industry. Laboratory research might be viewed much in the words of Shunryu Suzuki "… a garden is never finished…" I once had a laboratory, now I have a garden. Both offer rewards of "growth and vision" to be nurtured and viewed with enjoyment and satisfaction.

Joseph S. Semancik

SECTION I

VIROIDS

PART I

VIROIDS: ECONOMIC SIGNIFICANCE

CHAPTER

1

Economic Significance of Viroids in Vegetable and Field Crops

Rosemarie W. Hammond

U.S. Department of Agriculture, Beltsville, MD, United States

INTRODUCTION

Potato spindle tuber viroid (PSTVd) (Diener, 1971) was shown to be the causal agent of the potato spindle tuber disease described in the early 1920s in Irish Cobbler potato in North America (Martin, 1922). Members of the family *Pospiviroidae* cause disease in several other vegetable and field crops and the economic significance of viroid infection in these crops is a reflection of the symptoms they cause in their host that may affect crop quality and/or yield, their mode of transmission, their distribution and their ability to cause an epidemic (Randles, 2003). Although citrus exocortis viroid (CEVd) was detected in asymptomatic carrot, eggplant, and turnip in a survey of vegetable crops in Spain (Fagoaga and Duran-Vila, 1996), and PSTVd was detected in pepino (*Solanum muricatum*) (Mertelík et al., 2010; Puchta et al., 1990; Shamloul et al., 1997), the agronomic significance of those infections is currently unknown.

POTATO

Potato (*Solanum tuberosum* L., family *Solanaceae*), the fourth most important food crop grown worldwide, after rice, wheat, and maize, is traditionally propagated from seed potatoes or from true seed harvested from potato fruits. In potato, the principal viroid pathogen is PSTVd;

Viroids and Satellites.
DOI: http://dx.doi.org/10.1016/B978-0-12-801498-1.00001-2

tomato chlorotic dwarf viroid has also been found in natural infections. Potato can be experimentally infected with Columnea latent viroid (Verhoeven et al., 2004), CEVd (Semancik et al., 1973), and pepper chat fruit viroid (PCFVd) (Verhoeven et al., 2009) resulting in tuber distortion and tuber growth reduction.

PSTVd is present worldwide in potato with multiple strains causing mild to severe symptoms (http://www.cabi.org/isc/datasheet/43659). The primary means of spread is mechanical transmission through contaminated cutting knives and farming equipment (Bonde and Merriam, 1951; Manzer and Merriam, 1961). PSTVd is also transmitted through true seed and pollen (Fernow et al., 1970; Singh et al., 1992). Transmission of PSTVd by aphids has been reported from plants coinfected with potato leafroll virus where PSTVd was transencapsidated in the virus particle (Querci et al., 1997; Syller et al., 1997).

Symptoms and yield loss from PSTVd infection in potato vary depending on the potato cultivar and environmental conditions, with little or no foliage symptoms under some environmental conditions, while tubers are elongated and may have severe growth cracks; yield loss may reach 64% due to reduced potato production (Diener, 1987; Pfannenstiel and Slack, 1980). In the Saco variety, a mild strain of PSTVd could result in a 17%–24% loss, while a severe strain would result in a 64% loss (Singh et al., 1971). Environmental conditions influence the yield loss as foliage symptoms of PSTVd in potato can become more severe if tubers are planted late in the season when temperatures are warm; high temperatures can also increase the viroid titer (Singh, 1983). PSTVd also reduces pollen viability and seed set in some potato cultivars (Grasmick and Slack, 1986).

Control of PSTVd in potato has been successful through strict quarantine measures with a zero tolerance for the presence of PSTVd and trade of certified, viroid-free seed potatoes. With these measures, PSTVd has essentially been eradicated from potato crops in North America and Western Europe (De Boer and DeHaan, 2005; Singh, 2014; Sun et al., 2004). PSTVd still poses significant problems for seed potato production in Russia and elsewhere (Owens et al., 2009). An economic impact assessment of the potential reintroduction of PSTVd into the European Union concluded that phytosanitary measures are economically justifiable to control the disease (EFSA Panel on Plant Health, 2011; Soliman et al., 2012).

TOMATO

Tomato (*Solanum lycopersicum* L., family *Solanaceae*) is the second most important vegetable crop next to potato. The present world production is approximately 100 million tons of fresh fruit produced on 3.7 million hectares and in 144 countries (http://faostat3.fao.org/home/). Tomato is grown in both open fields and greenhouses for fresh market and processing. Viroids that naturally infect tomato include PSTVd, tomato apical stunt viroid, tomato planta macho viroid, Mexican papita viroid (now a strain of tomato planta macho viroid), tomato chlorotic dwarf viroid, Columnea latent viroid, and Indian tomato bunchy top viroid (a distinct strain of CEVd) (Singh et al., 2003). Common symptoms of viroid infection in tomato, which are dependent on viroid species and strain, cultivar, temperature, and light conditions, include chlorosis, bronzing, leaf distortion, reduced growth, heavy yield loss, and unmarketable fruit (Singh et al., 2003).

In general, pospiviroids are seed-transmitted in tomato and infection reduces pollen viability and seed germination rates in some cultivars (Benson and Singh, 1964; Hooker et al., 1978). Secondary mechanical spread of tomato apical stunt viroid in tomato greenhouse crops in Israel was found to be a result of the pollination activity of bumblebees (Antignus et al., 2007). Viroid infections in commercial tomato fields and in greenhouse-grown plants have been linked to imported seed or ornamentals (Batuman and Gilbertson, 2013; Van Brunschot et al., 2014; Verhoeven et al., 2012). As viroids are transmitted through tomato seed to varying extents and surface disinfection of seeds does not prevent transmission, recent seed testing and certification protocols for pospiviroids have been proposed for pre- and post-entry of tomato seed shipments. With the increased detection of tomato-infecting viroids in asymptomatic ornamentals (Chapter 3: Economic Significance of Viroids in Ornamental Crops), and since many growers may propagate tomatoes and ornamentals in the same greenhouses, increased grower vigilance should be practiced to avoid chance infections of cultivated crops.

HOP

Hop (*Humulus lupulus* L., family *Cannabaceae*) is a dioecious perennial climbing plant grown commercially worldwide for its use in the pharmaceutical and brewing industries. Traditionally the annual stems (vines/bines) are pruned continuously during the season to promote growth of selected vines and to control disease. After harvest, shoots are trimmed to soil level for the winter; in spring, new shoots arise from the rootstock. The female plant produces hop flowers, or cones, which contain resin glands that synthesize lupulin. Lupulin contains the essential oils and resins (alpha and beta acids) that impart flavor.

Hop stunt viroid (HSVd), the causal agent of hop stunt disease (Sano, 2003), and hop latent viroid (HLVd) (Barbara and Adams, 2003; Puchta et al., 1988) are significant pathogens in commercial hop fields (Pethybridge et al., 2008). Two recently reported viroids, apple fruit crinkle viroid (AFCVd) (Sano et al., 2004) and citrus bark cracking viroid (CBCVd) (Jakse et al., 2015), pose additional threats to the hop industry.

Hop stunt disease, first observed in the Fukushima province of Japan in the 1950s and 1960s (Shikata, 1987), has since been found in North America, South Korea, China, and Europe (Eastwell and Nelson, 2007; Eastwell and Sano, 2009; Guo et al., 2008; Radisek et al., 2012; Sano, 2003). The disease is characterized by a delay in emergence and early growth, shortened internodes, leaf curl and yellowing, small cone formation, and dry root rot; the severity of symptoms is dependent on the hop cultivar, may take years to appear, and may be more severe in warmer climates (Sasaki and Shikata, 1977).

The effect of HSVd on cone yield and brewing alpha acids can result in losses of 50%–70% (Sano, 2003). Losses of 50%–80% in hop production areas in the Northwest Yakima Valley, the United States, were observed in the hop cultivar Willamette Glacer, with a reduction in brewing alpha acid levels of 50%–70% (Eastwell and Nelson, 2007).

HLVd is widespread in all hop-growing regions worldwide. The impact of HLVd infection on yield is relatively minor as there are no severe symptoms associated with its presence (Barbara and Adams, 2003). The effects of HLVd on yield and quality appear to be cultivar dependent as cone yield was reduced by 27% and alpha acids were reduced by 31% in the hop cultivar "Omega" in the United Kingdom (Barbara et al., 1990).

In the Akita Prefecture of Japan, Sano et al. (2004) detected a strain of the AFCVd in symptomatic hops, with vine stunting and severe leaf curl and considerably lower alpha acid content in dried cones; the plants were negative for HSVd. The viroid appears to have spread over the major hop-producing regions of Japan and AFCVd-infected hops may have been introduced in infected mother stocks (Sano et al., 2004). It is not known if infection by AFCVd alone causes symptoms, as all hops examined were also infected with HLVd and most hops cultivated in Japan also were infected with hop latent virus and apple mosaic virus (Kanno et al., 1993).

In 2007, a severe stunting of hop was observed in several hop gardens in Slovenia, with symptoms similar to those caused by HSVd. The disease spread rapidly to other hop farms, and several hectares of hops were eradicated. In addition to hop mosaic virus, HLVd, and HSVd (Radisek et al., 2012), some manifestations of the disease symptoms were not characteristic of HSVd infection. Next-generation sequencing identified the putative causal agent of the disease as a novel strain of CBCVd, and was confirmed as such by biolistic inoculation of cloned cDNAs onto hop and the development of severe leaf malformations and stunting (Jakse et al., 2015). To date, this is the first and only report of CBCVd in hop.

As infected hop plants cannot be cured, control of viroid diseases includes the use of viroid-free planting stock and clean tools, cultural practices that limit bine pruning, and the adoption of dwarf cultivars that have been reported to exhibit some resistance to viroid infection (Barbara and Adams, 2003; Pethybridge et al., 2008; Sano, 2003; Takahashi, 1979).

CUCUMBER

Cucumber (*Cucumus sativus* L., family *Cucurbitaceae*) is a warm season vining plant primarily seed propagated and grown for its fruit. Pale fruit disease, with its distinctive symptom of pale green fruit was first reported in 1963 in The Netherlands and was later discovered in natural infections of cucumber elsewhere in the country (Van Dorst and Peters, 1974). Cucumber pale fruit disease was shown to be caused by a viroid, later designated a cucumber isolate of HSVd (Sano et al., 1984). In 2009, HSVd was discovered in Finland in greenhouse-grown cucumber exhibiting symptoms of yellow, bottle-shaped fruits and crumpled flowers (Lemmetty et al., 2011) and estimated yield losses to the grower were 2%–3%.

PEPPER

Sweet (bell) pepper is a cultivar group in the species *Capsicum annuum* L., family *Solanaceae*, and is grown for its multicolored fruits for fresh market and processing. Sweet peppers are seed propagated, and both hybrid and open-pollinated varieties are grown. Experimental transmission of PSTVd to pepper was demonstrated by O'Brien and Raymer (1964); however, little or no resulting disease symptoms were observed. The first report of PSTVd naturally infecting pepper was made in New Zealand following evaluation of a diseased capsicum plant displaying subtle symptoms of wavy margins on the upper leaves with no significant effects on fruit production (Lebas et al., 2005). The occurrence of PSTVd in greenhouse-grown pepper was limited, however, due to the symptomless phenotype; infected pepper may serve as a source of inoculum where pepper and tomato are grown in the same greenhouse.

A new disease was observed in greenhouse-grown sweet pepper (cv. Jaguar) in The Netherlands and the pathogen, PCFVd, was proposed as a new pospiviroid species (Verhoeven et al., 2009). Symptoms of PCFVd in pepper included delayed fruit set, reduced fruit size and number, and reduced plant growth; it was also shown to be seed-transmitted. Mechanical inoculation revealed that PCFVd caused stem and petiole necrosis and stunting in tomato and small, distorted tubers in potato (Verhoeven et al., 2009). PCFVd was subsequently reported in pepper in Canada (Verhoeven et al., 2011), in a natural infection of field-grown tomato in Thailand (Reanwarakorn et al., 2011), and was intercepted in Australia in imported tomato seed from Thailand and Israel (Chambers et al., 2013). The crop losses associated with PCFVd in pepper are not yet known, however the destruction of PCFVd-infected tomato seed shipments impacts seed trade (Chambers et al., 2013).

DISEASE MANAGEMENT

No genetic resistance to viroids has been identified in these crops, and control measures include the use of clean planting stock, sanitary cultural practices, certification (Chapter 39: Quarantine and Certification for Viroids and Viroid Diseases), thermotherapy (Chapter 40: Viroid Elimination by Thermotherapy, Cold Therapy, Tissue Culture, In Vitro Micrografting or Cryotherapy), decontamination (Chapter 41: Decontamination Measures to Prevent Mechanical Transmission of Viroids), and engineered resistance (Chapter 42: Strategies to Introduce Resistance to Viroids).

References

Antignus, Y., Lachman, O., Pearlsman, M., 2007. The spread of tomato apical stunt viroid (TASVd) in greenhouse crops is associated with seed transmission and bumble bee activity. Plant Dis. 91, 47–50.

Barbara, D.J., Adams, A.N., 2003. Hop latent viroid. In: Hadidi, A., Flores, R., Randles, J.W., Semancik, J.S. (Eds.), Viroids. CSIRO Publishing, Collingwood, VIC, pp. 213–217.

Barbara, D.J., Morton, A., Adams, A.N., Green, C.P., 1990. Some effects of hop latent viroid on two cultivars of hop (*Humulus lupulus*) in the UK. Ann. Appl. Biol. 117, 359–366.

Batuman, O., Gilbertson, R.L., 2013. First report of columnea latent viroid (CLVd) in tomato in Mali. Plant Dis. 97, 692.

Benson, A.P., Singh, R.P., 1964. Seed transmission of potato spindle tuber virus in tomato. Am. Potato J. 41, 294.

Bonde, R., Merriam, D., 1951. Studies on the dissemination of the potato spindle tuber virus by mechanical inoculation. Am. Potato J. 28, 558–560.

Chambers, G.A., Seyb, A.M., Mackie, J., Constable, F.E., Rodoni, B.C., Letham, D., et al., 2013. First report of pepper chat fruit viroid in traded tomato seed, an interception by Australian biosecurity. Plant Dis. 97, 1386.

De Boer, S.H., DeHaan, T.L., 2005. Absence of potato spindle tuber viroid within the Canadian potato industry. Plant Dis. 89, 910.

Diener, T.O., 1971. Potato spindle tuber "virus": IV. A replicating, low molecular weight RNA. Virology 45, 411–428.

Diener, T.O. (Ed.), 1987. The Viroids. Plenum Press, New York, NY.

Eastwell, K.C., Nelson, M.E., 2007. Occurrence of viroids in commercial hop (*Humulus lupulus* L.) production areas of Washington State. Online. Plant Health Progress. Available from: http://dx.doi.org/10.1094/PHP-2007-1127-01-RS.

Eastwell, K.C., Sano, T., 2009. Hop stunt. In: Mahaffee, W., Pethybridge, S., Gent, D.H. (Eds.), Compendium of Hop Diseases and Pests. The American Phytopathological Society, St. Paul, MN, p. 48.

EFSA Panel on Plant Health, 2011. Scientific opinion on the assessment of the risk of solanaceous pospiviroids for the EU territory and the identification and evaluation of risk management options. EFSA J. 9, 2330.

Fagoaga, C., Duran-Vila, N., 1996. Naturally occurring variants of citrus exocortis viroid in vegetable crops. Plant Pathol. 45, 45–53.

Fernow, K.H., Peterson, L.C., Plaisted, R.L., 1970. Spindle tuber virus in seeds and pollen of infected potato plants. Am. Potato J. 47, 75–80.

Grasmick, M.E., Slack, S.A., 1986. Effect of potato spindle tuber viroid on sexual reproduction and viroid transmission in true potato seed. Can. J. Bot. 64, 336–340.

Guo, L., Liu, S., Wu, Z., Mu, L., Xiang, B., Li, S., 2008. Hop stunt viroid (HSVd) newly reported from hop in Xinjiang, China. Plant Pathol. 57, 764.

Hooker, W.J., Nimnoi, P.N., Tai, W., Young, T.C., 1978. Germination reduction in PSTV-infected tomato pollen. Am. Potato J. 55, 378.

Jakse, J., Radisek, S., Pokom, T., Matoušek, J., Javornik, B., 2015. Deep-sequencing revealed citrus bark cracking viroid (CBCVd) as a highly aggressive pathogen on hop. Plant Pathol. 64, 831–842.

Kanno, Y., Yoshikawa, N., Takahashi, T., 1993. Some properties of hop latent and apple mosaic viruses isolated from hop plants and their distributions in Japan. Ann. Phytopathol. Soc. Jpn. 59, 651–658.

Lebas, B.S.M., Clover, G., Ochoa Corona, F.M., Elliott, D.R., Tang, A., Alexander, B.J.R., 2005. Distribution of potato spindle tuber viroid in New Zealand glasshouse crops of capsicum and tomato. Australas. Plant Pathol. 34, 129–133.

Lemmetty, A., Werkman, A.W., Soukainen, M., 2011. First report of hop stunt viroid in greenhouse cucumber in Finland. Plant Dis. 95, 615.

Manzer, F.E., Merriam, D., 1961. Field transmission of the potato spindle tuber virus and virus X by cultivating and hilling equipment. Am. Potato J. 38, 346–352.

Martin, W.H., 1922. "Spindle tuber," a new potato trouble. Hints to potato growers. N. J. State Assoc. 3, 8.

Mertelík, J., Kloudová, K., Červená, G., Nečekalová, J., Mikulková, H., Levkaničová, Z., et al., 2010. First report of potato spindle tuber viroid (PSTVd) in *Brugmansia* spp., *Solanum jasminoides*, *Solanum muricatum* and *Petunia* spp. in the Czech Republic. J. Plant Pathol. 59, 392.

O'Brien, M.J., Raymer, W.B., 1964. Symptomless hosts of the potato spindle tuber virus. Phytopathology 54, 1045–1047.

Owens, R.A., Girsova, N.V., Kromina, K.A., Lee, I.M., Mozhaeva, K.A., Kastalyeva, T.B., 2009. Russian isolates of potato spindle tuber viroid exhibit low sequence diversity. Plant Dis. 93, 752–759.

Pethybridge, S.J., Hay, F.S., Barbara, D.J., Eastwell, K.C., Wilson, C.R., 2008. Viruses and viroids infecting hop: significance, epidemiology, and management. Plant Dis. 92, 324–338.

Pfannenstiel, M.A., Slack, S.A., 1980. Response of potato cultivars to infection by the potato spindle tuber viroid. Phytopathology 70, 922–926.

Puchta, H., Herold, T., Verhoeven, K., Roenhorst, A., Ramm, K., Schmidt-Puchta, W., et al., 1990. A new strain of potato spindle tuber viroid (PSTVd-N) exhibits major sequence differences as compared to all other PSTVd strains sequenced so far. Plant Mol. Biol. 15, 509–511.

Puchta, H.K., Ramm, H.L., Sänger, H.L., 1988. The molecular structure of hop latent viroid (HLV), a new viroid occurring worldwide in hops. Nucleic Acids Res. 16, 4197–4216.

Querci, M., Owens, R.A., Bartolini, I., Lazarte, V., Salazar, L.F., 1997. Evidence for heterologous encapsidation of potato spindle tuber viroid in particles of potato leafroll virus. J. Gen. Virol. 78, 1207–1211.

Radisek, S., Majer, A., Jakse, J., Javornik, B., Matousek, J., 2012. First report of hop stunt viroid infecting hop in Slovenia. Plant Dis. 96, 592.

Randles, J.W., 2003. Economic impact of viroid diseases. In: Hadidi, A., Flores, R., Randles, J.W., Semancik, J.S. (Eds.), Viroids. CSIRO Publishing, Collingwood, VIC, pp. 3–11.

Reanwarakorn, K., Klinkong, S., Porsoongnurn, J., 2011. First report of natural infection of pepper chat fruit viroid in tomato plants in Thailand. New Dis. Rep. 24, 6.

Sano, T., 2003. Hop stunt viroid. In: Hadidi, A., Flores, R., Randles, J.W., Semancik, J.S. (Eds.), Viroids. CSIRO Publishing, Collingwood, VIC, pp. 207–212.

Sano, T., Uyeda, I., Shikata, E., Ohno, T., Okada, Y., 1984. Nucleotide sequence of cucumber pale fruit viroid: homology to hop stunt viroid. Nucleic Acids Res. 12, 3427–3434.

Sano, T., Yoshida, H., Goshono, M., Monma, T., Kawasaki, H., Ishizaki, K., 2004. Characterization of a new viroid strain from hops: evidence for viroid speciation by isolation in different host species. J. Gen. Plant Pathol. 70, 181–187.

Sasaki, M., Shikata, E., 1977. On some properties of hop stunt disease agent, a viroid. Proc. Jpn. Acad. Ser. B. 53, 109–112.

Semancik, J.S., Magnuson, D.S., Weathers, L.G., 1973. Potato spindle tuber disease produced by pathogenic RNA from citrus exocortis disease: evidence for identity of the causal agents. Virology 52, 292–294.

Shamloul, A.M., Hadidi, A., Zhu, S.F., Singh, R.P., Sagredo, B., 1997. Sensitive detection of potato spindle tuber viroid using RT-PCR and identification of a viroid variant naturally infecting pepino plants. Can. J. Plant Pathol. 19, 89–96.

Shikata, E., 1987. Hop stunt. In: Diener, T.O. (Ed.), The Viroids. Plenum Press, New York, NY, pp. 279–290.

Singh, R.P., 1983. Viroids and their potential danger to potatoes in hot climates. Can. Plant Dis. Surv. 63, 13–18.

Singh, R.P., 2014. The discovery and eradication of potato spindle tuber viroid in Canada. Virus Dis. 25, 415–424.

Singh, R.P., Boucher, A., Somerville, T.H., 1992. Detection of potato spindle tuber viroid in the pollen and various parts of potato plant pollinated with viroid-infected pollen. Plant Dis. 76, 951–953.

Singh, R.P., Finnie, R.F., Bagnall, R.H., 1971. Losses due to the potato spindle tuber virus. Am. Potato J. 48, 262–267.

Singh, R.P., Ready, K.F.M., Nie, X., 2003. Viroids of Solanaceous species. In: Hadidi, A., Flores, R., Randles, J.W., Semancik, J.S. (Eds.), Viroids. CSIRO Publishing, Collingwood, VIC, pp. 125–126.

Soliman, T., Mourits, M.C.M., Oude Lansink, A.G.J.M., van der Wert, W., 2012. Quantitative economic impact assessment of an invasive plant disease under uncertainty-A case study for potato spindle tuber viroid (PSTVd) invasion into the European Union. Crop Prot. 40, 28–35.

Sun, M., Siemsen, S., Campbell, W., Guzman, P., Davidson, R., Whitworth, J.L., et al., 2004. Survey of potato spindle tuber viroid in seed potato growing areas of the United States. Am. J. Potato Res. 81, 227–231.

Syller, J., Marczewski, W., Pawlowicz, J., 1997. Transmission by aphids of potato spindle tuber viroid encapsidated by potato leafroll luteovirus particles. Eur. J. Plant Pathol. 103, 285–289.

Takahashi, T., 1979. Diagnosis and control of hop stunt disease. Agric. Hort. 54, 1031–1034.

Van Brunschot, S.L., Verhoeven, J.Th.J., Persley, D.M., Geering, A.D.W., Drenth, A., et al., 2014. An outbreak of potato spindle tuber viroid in tomato is linked to imported seed. Eur. J. Plant Pathol. 139, 1–7.

Van Dorst, H.J.M., Peters, D., 1974. Some biological observations on pale fruit, a viroid-incited disease of cucumber. Neth. J. Plant Pathol. 80, 85–96.

Verhoeven, J.Th.J., Botermans, M., Jansen, C.C.C., Roenhorst, J.W., 2011. First report of pepper chat fruit viroid in capsicum pepper in Canada. New Dis. Rep. 23, 15.

Verhoeven, J.Th.J., Botermans, M., Meekes, E.T.M., Roenhorst, J.W., 2012. Tomato apical stunt viroid in The Netherlands: most prevalent pospiviroid in ornamentals and first outbreak in tomatoes. Eur. J. Plant Pathol. 133, 803–810.

Verhoeven, J.Th.J., Jansen, C.C.C., Roenhorst, J.W., Flores, R., de la Peña, M., 2009. Pepper chat fruit viroid: biological and molecular properties of a proposed new species of the genus *Pospiviroid*. Virus Res. 144, 209–214.

Verhoeven, J.Th.J., Jansen, C.C.C., Willemen, T.M., Kox, L.F.F., Owens, R.A., et al., 2004. Natural infections of tomato by citrus exocortis viroid, columnea latent viroid, potato spindle tuber viroid, and tomato chlorotic dwarf viroid. Eur. J. Plant Pathol. 110, 823–831.

CHAPTER

2

Economic Significance of Fruit Tree and Grapevine Viroids

Ahmed Hadidi[1], *Georgios Vidalakis*[2] *and Teruo Sano*[3]

[1]U.S. Department of Agriculture, Beltsville, MD, United States
[2]University of California, Riverside, CA, United States
[3]Hirosaki University, Hirosaki, Japan

INTRODUCTION

Fruit trees and grapevine are important horticultural crops, responsible for tens of billions of dollars per year to the global economy, and comprising a major source of income for growers and allied businesses worldwide. Losses from viroid infections in these crops are difficult to measure unless infected hosts are visibly damaged. Viroids may be latent in some hosts yet have adverse effects that often go unnoticed. Whether viroid-infected fruit trees or grapevine show obvious detrimental effects or do not induce noticeable disease, viroid infections are responsible for significant economic losses to these crops. Table 2.1 lists viroids of fruit trees and grapevine and their alternative hosts. Specific variants of three citrus viroids may induce beneficial effects on citrus trees. Most grapevine viroids are latent and distributed worldwide. Only grapevine yellow speckle viroid 1 (GYSVd-1) and grapevine yellow speckle viroid 2 (GYSVd-2) can induce a yellow speckle symptom on grapevine leaves (Koltunow et al., 1989), which varies in intensity and may be absent under certain weather conditions.

Viroids and Satellites.
DOI: http://dx.doi.org/10.1016/B978-0-12-801498-1.00002-4

TABLE 2.1 Fruit Tree and Grapevine Viroids and Their Natural Hosts

Genus	Species	Length (nt)[a]	Major host (s)[b]	Minor host (s)[b]	Remarks
***FAMILY* POSPIVIROIDAE**					
Apscaviroid	*Apple scar skin viroid*	329–333	Apple, Pear	Wild Pear	Mainly in Asian countries. Also found in Greece and Argentina
	Apple fruit crinkle viroid[c]	368–372	Apple, Hop	Persimmon	Reported only from Japan. Symptomatic (stunt, leaf curl) in hop. Symptomless in persimmon
	Apple dimple fruit viroid	303–311	Apple	Fig	Reported from Italy, Lebanon, China, and Japan. The incidence of the disease is very limited. Symptomless in fig
	Australian grapevine viroid	367–372	Grapevine		Worldwide distribution. Latent
	Citrus bent leaf viroid	315–329	Citrus		
	Citrus dwarfing viroid	291–297	Citrus		
	Citrus viroid V	293–294	Citrus		Oman, California, Spain, and Nepal
	Citrus viroid VI	325–333	Citrus	Persimmon	
	Grapevine latent viroid[c]	328	Grapevine		Reported from Xinjiang province, China. Latent
	Grapevine yellow speckle viroid 1	365–368	Grapevine		Worldwide distribution

	Grapevine yellow speckle viroid 2	363	Grapevine		Worldwide distribution
	Pear blister canker viroid	314–316	Pear	Quince, Wild Pear	Minor, symptomless fruits, hosts tolerant
	Persimmon latent viroid[c]	396–402	Japanese persimmon		Syn. Persimmon viroid
	Persimmon viroid 2[c]	358	American persimmon		Reported from Japan
Cocadviroid	*Citrus bark cracking viroid*	284–286	Citrus	Hop	Associated with severe stunt disease in hops
Hostuviroid	*Hop stunt viroid*	294–303	Hop, Cucumber, *Prunus* spp., Citrus, Grapevine	Mulberry, Fig, Pomegranate, Jujube, Hibiscus	Fruits with disease symptoms are reduced in yield and marketable value. CVd-IIa variant in citrus does not cause cachexia disease
Pospiviroid	*Citrus exocortis viroid*	366–475	Citrus	Grapevine	Worldwide distribution. Causes citrus exocortis disease. The viroid is latent in grapevine
Pospiviroid	*Potato spindle tuber viroid*	341–364	Avocado		Reported only in Peru
***FAMILY* AVSUNVIROIDAE**					
Avsunviroid	*Avocado sunblotch viroid*	246–251	Avocado		Fruit yield and quality are reduced
Pelamoviroid	*Peach latent mosaic viroid*	335–351	Peach	Other *Prunus* spp., *Pyrus* spp.	Worldwide distribution With the exception of peach, other hosts are symptomless

[a] *Sequences available online: http://www.ncbi.nlm.nih.gov/nuccore/; http://subviral.med.uottawa.ca*
[b] *Selected economically important hosts, with emphasis on fruit trees and grapevine, are listed.*
[c] *Unassigned members of new species.*

POME FRUIT VIROIDS

Apple scar skin viroid (ASSVd) causes scar skin or dapple symptoms in apple fruits, especially in China, Japan, South Korea, and India. ASSVd causes serious yield losses in apple and affected fruits are significantly downgraded or unmarketable. ASSVd may be latent in many pear cultivars, which may be the source of viroid infection to adjacent apple trees. The viroid may also cause pear rusty skin and pear fruit crinkle diseases in China, pear fruit dimple disease in Japan, and variable fruit symptoms on cultivated and wild pear in Greece. Blemished pear fruits are reduced in quality and market value (Hadidi and Barba, 2011; Chapter 21: Apple Scar Skin Viroid).

Apple fruit crinkle viroid (AFCVd) causes crinkle and dappling symptoms on affected apple fruits. Moreover, it may cause brown necrotic areas in the flesh, fruit may drop early and/or fruits are smaller, and blister bark symptoms may appear in sensitive varieties (Koganezawa and Ito, 2011). AFCVd infection causes reduced yield and unmarketable fruits. The disease has been reported only from Japan, where the viroid also infects hop and persimmon. It causes major economic losses on hop plants by inciting severe stunting and reduction of alpha acids contents in the cones (see Chapter 22: Other Apscaviroids Infecting Pome Fruit Trees) but is symptomless in persimmon (Nakaune and Nakano, 2008).

Apple dimple fruit viroid (ADFVd) naturally infects apple (Di Serio et al., 2011) and fig (Chiumenti et al., 2014) with very limited incidence (Malfitano et al., 2004). Dimple fruit symptoms are observed only in red apple, in which the viroid may cause reduced yield and unmarketable fruits. Several apple cultivars such as Golden Delicious, Granny Smith, and others are symptomless. Pear seedlings infected with ADFVd are symptomless (see Chapter 22: Other Apscaviroids Infecting Pome Fruit Trees).

Pear blister canker viroid infects pear under natural conditions and the majority of pear cultivars are tolerant to infection and do not show bark symptoms (Flores et al., 2011). Sensitive infected trees show bark canker symptoms and quickly decline. The viroid also naturally infects wild pear and quince.

STONE FRUIT VIROIDS

Peach latent mosaic viroid (PLMVd) causes disease in peach (see Chapter 29: Peach Latent Mosaic Viroid in Infected Peach), but it is latent in several stone fruit species: almond, pear, and wild pear (see

Chapter 30: Peach Latent Mosaic Viroid in Temperate Fruit Trees Other Than Peach). Infected peach fruits are misshapen, discolored with cracked sutures and swollen roundish stones. Most PLMVd variants do not cause leaf symptoms in peach, however, some variants induce chlorotic or yellow mosaics or albino-variegated patterns, pink broken lines on petals, and possibly stem pitting. PLMVd infection causes yield reduction in peach fruits and fruits may become unmarketable. PLMVd in Japanese plum cv. Angeleno may cause "spotted fruit disease" (see Chapter 30: Peach Latent Mosaic Viroid in Temperate Fruit Trees Other Than Peach). The effect of PLMVd infection on yield and fruit quality has not yet been demonstrated in plum and other stone fruits, almond, and pears. PLMVd is very widely distributed.

Hop stunt viroid (HSVd) infects plum, peach, apricot, almond, and sweet cherry, and causes symptoms restricted to the fruit: dapple on plum and peach and yellow spots on apricot. Diseased sensitive plum cultivars were reported from Japan, South Korea, and China; some Japanese cultivars may produce fruits with yellowish red color and they may harden. Sensitive peach cultivars were reported from Japan. European *Prunus* cultivars are generally tolerant and may not show fruit symptoms. In Italy, however, some diseased plum cultivars may produce fruits with wine-red dappling or irregular reddish lines. Apricot cultivars in Spain produce fruits characterized by changes in their external appearance, which involves rugosity and loss of organoleptic characteristics (Amari et al., 2007). Yield, quality and marketable value of stone fruits with HSVd symptoms are reduced.

MULBERRY, POMEGRANATE, AND FIG VIROIDS

HSVd causes leaf vein clearing, yellow speckle, and deformation of mulberry leaves in Iran (Mazhar et al., 2014). The viroid is symptomless in pomegranate (Gorsane et al., 2010). HSVd and citrus exocortis viroid (CEVd) were identified in fig trees showing symptoms of fig mosaic disease (Yakoubi et al., 2007). Similarly, ADFVd was detected in fig accessions showing symptoms of fig viruses (Chiumenti et al., 2014).

PERSIMMON VIROIDS

Persimmon viroid 2 was reported from Japan by deep sequencing analysis of an American persimmon tree grafted onto a Japanese persimmon rootstock showing poor growth (Ito et al., 2013). However, the relationship between viroid infection and symptom expression has not yet been clarified, because not only persimmon viroid 2 but also multiple variants of a new unassigned closterovirus species, as well as AFCVd, citrus viroid VI, and persimmon latent viroid, were detected in the same tree.

Persimmon is also a natural host of three other apscaviroids in Japan: AFCVd, citrus viroid VI, and persimmon latent viroid (syn. persimmon viroid) (Nakaune and Nakano, 2008; Ito et al., 2013). They were identified in Japanese persimmons symptomatic and asymptomatic for fruit apex disorder. These viroids are graft transmissible from persimmon to persimmon.

AVOCADO VIROIDS

Avocado sunblotch viroid (ASBVd) causes a wide variety of symptoms on avocado fruits, twigs, and bark; and infected trees may be stunted. Trees with visible sunblotch symptoms often produce avocado fruits with reduced quality and yield (18%–30%) (Semancik, 2003a). A significant and dramatic reduction (95%) in fruit yield may also occur in some avocado trees in which ASBVd is latent (symptomless carriers) (Desjardins, 1987).

Potato spindle tuber viroid (PSTVd) was also reported to naturally infect avocado trees in Peru. PSTVd infection was latent, however, trees coinfected with ASBVd showed symptoms that included bunchy inflorescence, decrease in fruit size and number, and eventual decline and death of infected plants (Querci et al., 1995). It was suggested that PSTVd was introduced with the original avocado trees, about a century ago, mainly from Riverside, California (Bartolini and Salazar, 2003).

CITRUS VIROIDS

Citrus viroids infect a large number of citrus species and citrus relatives; however, most of the infections have no noted effect on the host. Citrus viroids induce disease only when a specific viroid (or variant) infects a sensitive citrus host. On the other hand, specific variants of three citrus viroid species (i.e., citrus bent leaf viroid, HSVd, and citrus dwarfing viroid) may induce desirable economic effects (Duran-Vila and Semancik, 2003; Semancik, 2003b; Semacik et al., 1997; Vidalakis et al., 2010, 2011).

From the seven known citrus viroids only two, namely, CEVd and HSVd, have been associated with citrus diseases that can result in economic losses (Gumpf et al., 2014; Timmer et al., 2000). Recently, it has been reported that citrus bark cracking viroid, which does not cause any important citrus disease, is a highly aggressive pathogen on hop plants and causes major economic losses Jakse et al. (2015).

Exocortis and cachexia diseases, induced by CEVd and certain HSVd variants, respectively, are posing significant economic risks to global citrus production. Both diseases have been reported from all major citrus producing areas of the world, as well as in early citrus budwood registration programs, since their original descriptions in 1948 and 1950, respectively (Duran-Vila and Semancik, 2003). In addition, both exocortis and cachexia diseases can reduce tree performance of commercially popular and economically important citrus, such as trifoliate orange (*Poncirus trifoliata* (L.) Raf.) and mandarins (*Citrus reticulata* Blanco).

Exocortis is a bark scaling, stunting, and yield reducing disease of trees propagated on trifoliate and trifoliate hybrid rootstocks. These rootstocks, however, are tolerant to citrus tristeza virus and have replaced the tristeza-sensitive rootstock sour orange (*Citrus aurantium* L.) in many citrus producing areas of the world (Moreno et al., 2008). Roistacher et al. (1996) elegantly presented the economic consequences of citrus viroids in the transition from sour to the trifoliate orange rootstock in Belize. Viroid-infected trees showing decline and severe bark cracking generally yielded one-fourth box (10.2 kg) or less per tree compared to healthy trees, which yielded 40.8 kg. In addition, rootstocks showing severe bark cracking were predisposed to *Phytophthora* infections, as the pathogen could readily enter through the exposed vascular tissues and many trees died before the eighth year in the field. Overall, it was estimated that the economic loss in Belize in 1996 was $5678/ha, which corresponds to approximately $8685/ha in 2014. Cachexia infected mandarin, tangerine, or clementine trees are stunted, chlorotic with gum impregnations in the phloem, and reduced yield (Hashemian et al., 2009; Murcia et al., 2014; Timmer et al., 2000; Vernière et al., 2004, 2006).

Some citrus viroids are unique among plant pathogens in regard to the question of economic significance because they can be considered as genetic elements for the modification of host genome responses that result in desirable horticultural, production, environmental, labor, and ultimately economic benefits. Originally proposed by Cohen (1968) and Mendel (1968), the use of viroids as means of producing dwarfed citrus trees has been under continuous investigation and has been reviewed by Hutton et al. (2000) and Semancik (2003b) in detail. Specific variants of three citrus viroids are approved for commercial use in California, where experiments with navel orange trees (*Citrus sinensis* (L.) Osb.) on trifoliate rootstock and navel orange and clementine mandarin trees on Carrizo citrange (*C. sinensis* × *P. trifoliata*) rootstock in standard (6 × 6.7 m) and high density (3 × 6.7 m) plantings reduced the canopy volume by 33%–53.5%, and almost doubled the yield per land surface unit (Vidalakis et al., 2010, 2011).

While increased fruit production per land surface unit is an obvious economic benefit of dwarfed citrus trees in high density plantings, the cost of harvesting, pest management, and irrigation are also important economic factors (Fridley, 1977; Phillips, 1978). A high density planting of dwarfed citrus trees with small canopies will also be advantageous for Huanglongbing disease management programs. Visual inspection of dwarfed trees will be simpler without the need for tractors and platforms (Belasque et al., 2010). The effects of tree removal on fruit production are mitigated, as shown by the analysis of citrus groves at two density plantings (494 and 988 trees/ha). Even at the extreme tree loss rate of 30% per year, the higher density grove approached 35.5 tons/ha in year six while the lower density grove reached a maximum of 17.2 tons/ha (Stover, 2008).

In the face of serious challenges threatening citrus production globally, and in the absence of a true dwarf citrus this technology presents an interesting alternative: it can be used immediately for the development of a simple, flexible, and consumer acceptable citrus management system since they have low cost (natural agents and grafting) and low biological and commercial risk with no natural transmission vectors and no genetically engineered materials involved. Finally, the noncachexia variant of HSVd, is present in the 143-year-old Parent Navel tree in Riverside, California, suggesting that this RNA species does not have a deleterious effect on tree health (University of California, Riverside, routine testing since the 1950s). Is it possible that the superior performance of the Parent Navel tree since 1873 perhaps is associated to a certain degree with the presence of a nucleic acid element such as a viroid RNA?

GRAPEVINE VIROIDS

There are at least six viroids reported to infect grapevines in nature (Table 2.1). Four of these, Australian grapevine viroid, grapevine latent viroid, HSVd (grapevine strain), and CEVd (grapevine strain), are latent and do not cause diseases in grapevine. Thus, they may be considered as economically unimportant. However, since HSVd can cause severe hop stunt disease on cultivated hops, their potential threats cannot be ignored (Kawaguchi-Ito et al., 2009). GYSVd-1 and GYSVd-2, however, have been implicated in the vein banding disease of grapevine, which is caused by coinfection with grapevine fanleaf virus or by the yellow speckle viroids alone (Martelli, 2014). With the exception of grapevine latent viroid, which was discovered recently in Xinjiang province, China in 2014 (Zhang et al., 2014), other grapevine viroids are distributed worldwide (Martelli, 2014). The incidences of HSVd and GYSVd-1 are extremely high in any of the grapevine producing areas. On the other hand, regional disparity can be seen in the incidence of Australian grapevine viroid, GYSVd-2 and CEVd; e.g., infection rate was relatively high, at 26%, 18.7%, and 9.3% in Tunisia, Iran, and India, respectively, but negligible in the other regions of the world (Adkar-Purushothama et al., 2014; Jiang et al., 2012). No information is available on the effect of any of the grapevine viroids on grape yield, vine vigor, and wine quality.

Acknowledgments

The senior author would like to thank Jon Hadidi BS, MBA, JD for his computer-related expertise.

References

Adkar-Purushothama, C.R., Kanchepalli, P.R., Sreenivasa, M.Y., Zhang, Z.-X., Sano, T., 2014. Detection, distribution, and genetic diversity of Australian grapevine viroid in grapevines in India. Virus Genes. 49, 304–311.

Amari, K., Ruiz, D., Gómez, G., Sánchez-Pina, M.A., Pallás, V., Egea, J., 2007. An important new apricot disease in Spain is associated with hop stunt viroid infection. Eur. J. Plant Pathol. 118, 173–181.

Bartolini, I., Salazar, L.F., 2003. Viroids in South America. In: Hadidi, A., Flores, R., Randles, J.W., Semancik, J.S. (Eds.), Viroids. CSIRO Publishing, Collingwood, VIC, pp. 265–267.

Belasque Jr, J., Bassanezi, R., Yamamoto, P., Ayres, A., Tachibana, A., Violante, A., et al., 2010. Lesson from huanglongbing management in São Paulo State, Brazil. J. Plant Pathol. 92, 285–302.

Chiumenti, M., Torchetti, E.M., Di Serio, F., Minafra, A., 2014. Identification and characterization of a viroid resembling apple dimple fruit viroid in fig (*Ficus carica* L.) by next generation sequencing of small RNAs. Virus Res. 188, 54–59.

Cohen, M., 1968. Exocortis virus as a possible factor in producing dwarf citrus trees. Proc. FL State Hortic. Soc. 81, 115–119.

Desjardins, P.R., 1987. Avocado sunblotch. In: Diener, T.O. (Ed.), The Viroids. Plenum Publishing, New York, NY, pp. 299–313.

Di Serio, F., Malfitano, M., Alioto, D., Ragozzino, A., Flores, R., 2011. Apple dimple fruit viroid. In: Hadidi, A., Barba, M., Candresse, T., Jelkmann, W. (Eds.), Virus and Virus-Like Diseases of Pome and Stone Fruits. APS Press, St. Paul, MN, pp. 49–52.

Duran-Vila, N., Semancik, J.S., 2003. Citrus viroids. In: Hadidi, A., Flores, R., Randles, J.W., Semancik, J.S. (Eds.), Viroids. CSIRO Publishing, Collingwood, VIC, pp. 178–194.

Flores, R., Ambrós, S., Llácer, G., Hernández, C., 2011. Pear blister canker viroid. In: Hadidi, A., Barba, M., Candresse, T., Jelkmann, W. (Eds.), Virus and Virus-Like Diseases of Pome and Stone Fruits. APS Press, St. Paul, MN, pp. 63–66.

Fridley, R., 1977. High density orchards facilitate harvest. California Agri. 31, 12–13.

Gorsane, F., Elleuch, A., Hamdi, I., Salhi-Hannachi, A., Fakhfakh, H., 2010. Molecular detection and characterization of hop stunt viroid sequence variants from naturally infected pomegranate (*Punica granatum* L.) in Tunisia. Phytopathol. Mediterr. 49, 152–162.

Gumpf, D.J., Polek, M., Vidalakis, G., 2014. Virus and viroid diseases. In: Ferguson, L., Grafton Cardwell, B. (Eds.), Citrus Production Manual. University of California Agricultural and Natural Resources (UC ANR), Oakland, CA, pp. 337–346.

Hadidi, A., Barba, M., 2011. Apple scar skin viroid. In: Hadidi, A., Barba, M., Candresse, T., Jelkmann, W. (Eds.), Virus and Virus-Like Diseases of Pome and Stone Fruits. APS Press, St. Paul, MN, pp. 57–62.

Hashemian, S.M.B., Serra, P., Barbosa, C.J., Juárez, J., Aleza, P., Corvera, J.M., et al., 2009. Effect of a field-source mixture of citrus viroids on the performance of "Nules" clementine and "Navelina" sweet orange trees grafted on carrizo citrange. Plant Dis. 93, 699–707.

Hutton, R.J., Broadbent, P., Bevington, K.B., 2000. Viroid dwarfing for high density citrus planting. Hortic. Rev. 24, 277–317.

Ito, T., Suzaki, K., Nakano, M., Sato, A., 2013. Characterization of a new apscaviroid from American persimmon. Arch. Virol. 158, 2629–2631.

Jakse, J., Radisek, S., Pokorn, T., Matousek, J., Javornik, B., 2015. Deep-sequencing revealed citrus bark cracking viroid (CBCVd) as a highly aggressive pathogen on hop. Plant Pathol. 64, 831–842.

Jiang, D., Sano, T., Tsuji, M., Araki, H., Sagawa, K., Adkar Purushothama, C.R., et al., 2012. Comprehensive diversity analysis of viroids infecting grapevine in China and Japan. Virus Res. 169, 237–245.

Kawaguchi-Ito, Y., Li, S.-F., Tagawa, M., Araki, H., Goshono, M., Yamamoto, S., et al., 2009. Cultivated grapevines represent a symptomless reservoir for the transmission of hop stunt viroid to hop crops: 15 years of evolutionary analysis. PLoS One 4, e8386.

Koganezawa, H., Ito, T., 2011. Apple fruit crinkle viroid. In: Hadidi, A., Barba, M., Candresse, T., Jelkmann, W. (Eds.), Virus and Virus-Like Diseases of Pome and Stone Fruits. APS Press, St. Paul, MN, pp. 53–55.

Koltunow, A.M., Krake, L.R., Johnson, S.D., Razian, M.A., 1989. Two related viroids cause grapevine yellow speckle disease independently. J. Gen. Virol. 70, 3411–3419.

Malfitano, M., Alioto, D., Ragozzino, A., Flores, R., Di Serio, F., 2004. Experimental evidence that apple dimple fruit viroid does not spread naturally. Acta Hortic. 657, 357–360.

Martelli, G.P., 2014. Directory of virus and virus-like diseases of the grapevine and their agents. J. Plant Pathol. 96, 1–136, Viroids.

Mazhar, M.A., Bagherian, S.A.A., Izadpanah, K., 2014. Variants of hop stunt viroid associated with mulberry vein clearing in Iran. J. Phytopathol. 162, 269–271.

Mendel, K., 1968. Interrelations between tree performance and some virus diseases. In: Childs, J.F.L (Ed.), Proceedings of the 4th Conference of the IOCV (International Organization of Citrus Virologists). University of Florida Press, Gainesville, FL, pp. 310–313.

Moreno, P., Ambros, S., Albiach-Martí, M.R., Guerri, J., Pena, L., 2008. Citrus tristeza virus: a pathogen that changed the course of the citrus industry. Mol. Plant Pathol. 9, 251–268.

Murcia, N., Hashemian, S.M.B., Serra, P., Pina, J.A., Duran-Vila, N., 2014. Citrus viroids: symptom expression and performance of Washington navel sweet orange trees grafted on carrizo citrange. Plant Dis. 99, 125–136.

Nakaune, R., Nakano, M., 2008. Identification of a new apscaviroid from Japanese persimmon. Arch. Virol. 153, 969–972.

Phillips, R.L., 1978. Citrus tree spacing and size control. International Citrus Congress (3rd: 1978: Sydney, Australia), Griffith, Australia, International Society of Citriculture. 1, 319–324.

Querci, M., Owens, R.A., Vargas, C., Salazar, L.F., 1995. Detection of potato spindle tuber viroid in avocado growing in Peru. Plant Dis. 79, 196–202.

Roistacher, C.N., Canton, H., Reddy, P.S., 1996. The economics of living with citrus diseases: exocortis in Belize. In: Proc. 13th Conf. IOCV., IOCV, Riverside, pp. 370–375.

Semancik, J.S., 2003a. Avocado sunblotch viroid. In: Hadidi, A., Flores, R., Randles, J.W., Semancik, J.S. (Eds.), Viroids. CSIRO Publishing, Collingwood, VIC, pp. 171–177.

Semancik, J.S., 2003b. Considerations for the introduction of viroids for economic advantage. In: Hadidi, A., Flores, R., Randles, J.W., Semancik, J.S. (Eds.), Viroids. CSIRO Publishing, Collingwood, VIC, pp. 357–362.

Semacik, J.S., Rakowski, A.G., Bash, J.A., Gumpf, D.J., 1997. Applications of selected viroids for dwarfing and enhancement of production of Valencia orange. J. Hort. Sci. 72, 563–570.

Stover, E., 2008. The citrus grove of the future and its implications for huanglongbing management. Proc. Fla. State Hort. Soc., 121, 155–159.

Timmer, L.W., Garnsey, S.M., Graham, J.H., 2000. Compendium of Citrus Diseases. APSPress, St. Paul, MN.

Vernière, C., Perrier, X., Dubois, C., Dubois, A., Botella, L., Chabrier, C., et al., 2004. Citrus viroids: symptom expression and effect on vegetative growth and yield of clementine trees grafted on trifoliate orange. Plant Dis. 88, 1189–1197.

Vernière, C., Perrier, X., Dubois, C., Dubois, A., Botella, L., Chabrier, C., et al., 2006. Interactions between citrus viroids affect symptom expression and field performance of clementine trees grafted on trifoliate orange. Phytopathology 96, 356–368.

Vidalakis, G., Pagliaccia, D., Bash, J.A., Afunian, M., Semacik, J.S., 2011. Citrus dwarfing viroid: effects on tree size and scion performance specific to *Poncirus trifoliata* rootstock for high density planting. Ann. Appl. Biol. 158, 2014–2217.

Vidalakis, G., Pagliaccia, D., Bash, J.A., Semacik, J.S., 2010. Effects of mixtures of citrus viroids as transmissible small nuclear RNA on tree dwarfing and commercial scion performance on Carrizo citrange rootstock. Ann. Appl. Biol. 157, 415–423.

Yakoubi, S., Elleuch, A., Besaies, N., Marrakchi, M., Fakhfakh, H., 2007. First report of hop stunt viroid and citrus exocortis viroid on fig with symptoms of fig mosaic disease. J. Phytopathol. 155, 125–128.

Zhang, Z., Shuishui, Qi, S., Tang, N., Zhang, X., Chen, S., Zhu, P., et al., 2014. Discovery of replicating circular RNAs by RNA-seq and computational algorithms. PLoS Pathog. 10, e1004553.

CHAPTER

3

Economic Significance of Viroids in Ornamental Crops

Jacobus Th.J. Verhoeven[1], *Rosemarie W. Hammond*[2] *and Giuseppe Stancanelli*[3]

[1]National Plant Protection Organization, Wageningen, The Netherlands
[2]U.S. Department of Agriculture, Beltsville, MD, United States
[3]European Food Safety Authority, Parma, Italy

INTRODUCTION

Similarly to plant viruses, viroids are problematic for their economic impact on agricultural crops. Actually, the reason for their discovery was their causing of serious diseases in main food crops, e.g., potato spindle tuber viroid (PSTVd) in potato (Diener, 1987; Martin, 1922), citrus exocortis viroid (CEVd) in citrus (Benton et al., 1949, 1950; Fawcett and Klotz, 1948), and coconut cadang-cadang viroid (CCCVd) in coconut (Kent, 1953; Randles and Rodriguez, 2003). In ornamental crops, especially chrysanthemum, diseases caused by viroids have also been reported (Brierley and Smith, 1949; Dimock and Geissinger, 1969).

Economic significance can be divided into direct and indirect impact. Direct impact is specifically related to the production process and includes losses due to reduction in growth, vitality, and quality as well as additional costs of production (e.g., extra hygienic measures), whereas indirect impact is related to changes in prices and effects on international trade (Bos, 1999; Soliman et al., 2012). Both types of impact will be considered for the viroids infecting ornamental crops. In addition, viroid infections in ornamentals may be economically significant as they may act as sources of viroid inoculum for food crops that may suffer severe diseases when infected. When discussing the significance of

Viroids and Satellites.
DOI: http://dx.doi.org/10.1016/B978-0-12-801498-1.00003-6

viroids in ornamental crops, those infecting fruit tree genera, such as ornamental *Malus* spp. and *Prunus* spp., are excluded.

VIROIDS INFECTING ORNAMENTAL CROPS

Since their discovery, viroid infections have been reported from food crops. The economic significance of viroids in these crops was generally obvious primarily due to severe symptoms. However, viroid infections in ornamentals were far less frequently reported. The first viroid reported in ornamentals was chrysanthemum stunt viroid (CSVd) in chrysanthemum. Symptoms were first reported by Dimock (1947), but the viroid etiology of the disease was only established in 1973 by Diener and Lawson. Subsequently, another type of symptom (Dimock and Geissinger, 1969) was reported in chrysanthemum, which appeared to be caused by another viroid; i.e., chrysanthemum chlorotic mottle viroid (CChMVd) (Navarro and Flores, 1997). Later, symptomless infections of CChMVd (Horst, 1975), coleus blumei viroid 1 (CbVd-1) in *Coleus blumei* (Fonseca et al., 1989, 1994) and columnea latent viroid (CLVd) in *Columnea erythrophae* (Hammond et al., 1989) were reported. More symptomless infections were reported for CbVd-1 and related variants or species (e.g., Spieker et al., 1990, 1996a) and, similarly, for CLVd in *Brunfelsia undulata* (Spieker, 1996a) and *Nematanthus wettsteinii* (Singh et al., 1992). With the aid of newly developed primers for the detection of a broad range of pospiviroids, many more infections have been reported in ornamental crops (Bostan et al., 2004; Verhoeven et al., 2004). The availability of these generic primers and the finding of PSTVd in various ornamental crops were an impetus to search for more pospiviroid infections in ornamentals, the results of which are listed in Table 3.1. It is obvious that the vast majority of the newly discovered viroid infections in ornamental crops are caused by members of the genus *Pospiviroid*. In this genus most natural infections are recorded for PSTVd and CSVd, i.e., 11 and 10 plant species, respectively.

In comparison with natural infections, a much wider host range has been identified by artificial inoculation. In particular, pospiviroids have been successfully inoculated to many plant species. Various pospiviroid–ornamental host combinations after artificial inoculation have been reported: citrus exocortis viroid (CEVd) (Niblett et al., 1980; Weathers et al., 1967; Weathers and Greer, 1968), CLVd (Matsushita and Tsuda, 2015), CSVd (Brierley, 1950, 1953), pepper chat fruit viroid (PCFVd) (Verhoeven et al., 2009), PSTVd (Matousek et al., 2014; Matsushita and Tsuda, 2015; Niblett et al., 1980; O'Brien, 1972; Singh, 1973), tomato apical stunt viroid (TASVd) (Matsushita and Tsuda, 2015), and tomato chlorotic

TABLE 3.1 Natural Viroid Infections in Ornamental Plant Species

	Pospiviroid								*Hostuviroid*		*Coleviroid*	*Pelamoviroid*	
Plant species/genus	**CEVd**[a]	**CLVd**	**CSVd**	**IrVd-1**	**PoLVd**	**PSTVd**	**TASVd**	**TCDVd**	**DLVd**	**HSVd**	**CbVd**	**CChMVd**	**References**[b]
Ageratum			+										1
Argyranthemum frutescens			+										2, 3, 4
Brugmansia sp.						+	+	+					5, 6, 7
Brunfelsia undulata		+											8
Calibrachoa sp.						+							9
Celosia cristata				+									10
Celosia plumosa				+									11
Cestrum sp.	+					+	+						5, 12
Chrysanthemum spp.			+			+						+	13, 14, 15, 16
Codieaum										+			17
Coleus blumei											+		18, 19
Columnea erythrophae		+											20
Dahlia sp.			+			+			+				21, 22, 23
Datura sp.						+							24
Glandularia pulchella	+												25
Gloxinia spp.		+											26
Hibiscus sp.										+			17, 27
Impatiens sp.	+												28, 29

(Continued)

TABLE 3.1 (Continued)

	Pospiviroid								*Hostuviroid*		*Coleviroid*	*Pelamoviroid*	
Plant species/genus	**CEVd**[a]	**CLVd**	**CSVd**	**IrVd-1**	**PoLVd**	**PSTVd**	**TASVd**	**TCDVd**	**DLVd**	**HSVd**	**CbVd**	**CChMVd**	**References**[b]
Iresine herbstii				+									30
Lycianthes rantonnetii	+					+	+						31, 32, 33
Nematanthus sp.		+											34
Pericallis sp.			+										9, 35
Petunia sp.	+		+			+		+					36, 37, 38, 39
Petunia × *Calibrachoa*	+												36
Physalis alkekengi			+										40
Pittosporum tobira								+					9
Portulaca sp.				+	+								9, 41, 42
Solanum jasminoides	+		+			+	+						5, 24, 32, 43, 44
Solanum pseudocapsicum						+	+						15, 45
Streptosolen jamesonii						+	+						33, 46
Verbena sp.	+		+	+			+	+					5, 7, 25, 28, 29
Vinca sp.			+	+				+					7, 28, 29, 47

[a] CEVd, *citrus exocortis viroid;* CLVd, *columnea latent viroid;* CSVd, *chrysanthemum stunt viroid;* IrVd-1, *Iresine viroid 1;* PoLVd, *portulaca latent viroid (tentative species);* PSTVd, *potato spindle tuber viroid;* TASVd, *tomato apical stunt viroid;* TCDVd, *tomato chlorotic dwarf viroid;* DLVd, *dahlia latent viroid;* HSVd, *hop stunt viroid;* CbVd, *Coleus blumei viroids;* CChMVd, *chrysanthemum chlorotic dwarf viroid.*

[b] *References: 1, GenBank Accession No. Z68201-Henkel and Sänger (1995); 2, Menzel and Maiss (2000); 3, Marais et al. (2011); 4, Torchetti et al. (2012); 5, Verhoeven et al. (2008a); 6, Olivier et al. (2011); 7, Verhoeven et al. (2010b); 8, Spieker (1996a); 9, Verhoeven (2010); 10, Sorrentino et al. (2015); 11, Verhoeven et al. (2010a); 12, Luigi et al. (2011a); 13, Brierley (1950); 14, Diener and Lawson (1973); 15, Lemmetty et al. (2010); 16, Dimock and Geissinger (1969); 17, Sänger (1988); 18, Fonseca et al. (1989); 19, Spieker et al. (1990); 20, Hammond et al. (1989); 21, Nakashima et al. (2007); 22, Tsushima et al. (2011); 23, Verhoeven et al. (2013); 24, Verhoeven et al. (2010b); 25, Singh et al. (2006); 26, Nielsen and Nicolaisen (2010); 27, Luigi et al. (2013); 28, Bostan et al. (2004); 29, Nie et al. (2005); 30, Spieker (1996b); 31, Luigi et al. (2011b); 32, Di Serio (2007); 33, Verhoeven et al. (2010c); 34, Singh et al. (1992); 35, GenBank Accession No. GQ174501-Daly et al. (2009); 36, Van Brunschot et al. (2014); 37, Verhoeven et al. (1998); 38, Mertelik et al. (2010); 39, Verhoeven et al. (2007); 40, Meekes, E.T.M. (personal communication); 41, Virscek Marn and Mavric Plasko (2012); 42, Verhoeven et al. (2015); 43, Verhoeven et al. (2008b); 44, Verhoeven et al. (2006); 45, Spieker et al. (1996b); 46, Verhoeven et al. (2008c); 47, Singh and Dilworth (2009).*

dwarf viroid (TCDVd) (Matsushita and Tsuda, 2015; Matsushita et al., 2009).

SYMPTOMS OF VIROID INFECTIONS IN ORNAMENTAL CROPS

Despite the many established and potential ornamental host–viroid combinations, symptoms have only been reported occasionally. The most severe symptoms are known for CSVd in chrysanthemum (*Chrysanthemum morifolium*). In this species viroid infection may cause growth reduction of up to 65%. In addition, flowers may show bleaching and leaves of some cultivars may show bright yellow spots of various sizes. However, variation in the severity of symptoms is high and infections may even be symptomless. Therefore, symptoms are often best observed in a crop with both infected and noninfected plants (Brierley and Smith, 1951; Horst et al., 1977). CSVd has also been reported in other ornamental crops. In *Argyranthemum frutescens*, growth reduction, flower distortion, and leaf necrosis were observed by Menzel and Maiss (2000) and Marais et al. (2011), whereas no symptoms were observed by Torchetti et al. (2012). In dahlia, symptoms similar to those in chrysanthemum have been reported, even including plant death. However, the role of CSVd in the symptomatology in this ornamental crop is unclear as the plants were coinfected with common viruses of dahlia (Nakashima et al., 2007). In a symptomatic plant of *Petunia* sp., CSVd infection was accompanied by both tobacco mosaic virus and potato virus Y (Verhoeven et al., 2008b); both viruses are known to cause severe symptoms in petunia. Symptoms were not reported for *Ageratum* sp. (GenBank Accession No. Z68201-Henkel and Sänger, 1995), *Pericallis* sp. (GenBank Accession No. GQ174501-Daly et al., 2009), *Solanum jasminoides* (Verhoeven et al., 2006), *Verbena* sp. or *Vinca major* (Bostan et al., 2004).

CChMVd may also cause symptoms in chrysanthemum. At elevated temperatures the viroid may cause mild mottling or mosaic of young leaves, which may be followed by general chlorosis, some growth reduction of leaves and a delay in blossom development (Dimock et al., 1971). The host range of CChMVd is restricted to some cultivars of *C. morifolium* and *Chrysanthemum zawadskii* (Horst, 1987).

Hop stunt viroid (HSVd) was identified in plants of two cultivars of *Hibiscus rosa-sinensis* showing both severe growth reduction and upward curling and deformation of leaves. Some common viruses of *Hibiscus* were not detected in these plants, suggesting that the viroid might be the cause of the symptoms (Luigi et al., 2013).

Viroid symptoms in naturally infected ornamental crops other than chrysanthemum and *Hibiscus* have not been reported. Although most successfully inoculated ornamental plant species were asymptomatic, symptoms have occasionally been reported in such plants. For example, Niblett et al. (1980) reported symptoms of both CEVd and PSTVd in *Gynura aurantiaca* and *C. morifolium*. In addition, Matsushita and Tsuda (2015) observed symptoms in *Calendula officinalis* for both CLVd and TASVd. Similarly, these authors observed symptoms in *Petunia* × *hybrida* cv. Mitchell after inoculation with PSTVd, TASVd, and TCDVd, but not in other cultivars of *Petunia* × *hybrida*. These examples indicate that, although most viroid infections in ornamentals are asymptomatic, symptoms may occur in several species and cultivars.

SIGNIFICANCE OF VIROIDS IN ORNAMENTAL CROPS: DIRECT IMPACT

CSVd may cause significant yield losses in chrysanthemum and *A. frutescens* due to severe reduction of growth and quality (Brierley and Smith, 1951; Horst et al., 1977; Marais et al., 2011; Menzel and Maiss, 2000). However, exact figures of yield loss have not been published. In some varieties, mild symptoms or no symptoms at all are evident and consequently, yield losses will be substantially lower. Losses can be prevented or reduced by using high quality propagation material from industry certification systems (EFSA Panel on Plant Health, 2012). In hosts other than chrysanthemum and *A. frutescens*, yield losses are assumed to be marginal as well, although uncertainty exists for dahlia.

In comparison with CSVd, the direct impact of CChMVd is assumed to be considerably less. Symptoms, the main reason for reduction in value, only appear at higher temperatures in a limited number of chrysanthemum cultivars.

Little can be concluded for the direct impact of HSVd on *H. rosa-sinensis*, as more data are needed.

SIGNIFICANCE OF VIROIDS IN ORNAMENTAL CROPS: INDIRECT IMPACT

Ornamentals may also suffer from indirect impacts of viroid infection. Viroids spread primarily through vegetative propagules (cuttings, ramets, tubers, and bulbs). Certification of source materials is an effective means to control this mode of spread. However, initiating a certification program is a costly activity. It requires inspection and testing of propagation source materials and the use of hygienic measures to prevent (re)infection. Costs will be less where a certification program is already in place for other pathogens, such as viruses. Established chrysanthemum certification schemes include protocols for sampling and testing for CSVd (EPPO, 2002).

For nonindigenous viroids, governments may enact phytosanitary measures to prevent the introduction and establishment of viroids. On the one hand, these measures result in extra costs for growers in need of regulated planting material; whereas, on the other hand they prevent severe financial costs arising from the introduction of quarantinable pathogens. For example, all plants of *Petunia* and *Calibrachoa* originating from outside the EU may only enter The Netherlands via a postentry quarantine program at a quarantine station. The program, paid for by the importers, includes testing for pospiviroids. Introduction and establishment of PSTVd in these crops would result in high eradication costs, as occurred with PSTVd infections of *Brugmansia* spp. and *S. jasminoides* discovered in Dutch commercial greenhouses in 2006 (Verhoeven et al., 2008, 2010b). The total cost for the eradication of these PSTVd infections was estimated at €3–5 million for growers, and €700,000 for governmental costs for inspection and testing (De Hoop et al., 2008). Quarantine regulations may also be applied to eradicating pathogens already present and for which the spread should be prevented, e.g., CSVd, in the EU. Control in such cases may be incorporated into the certification programs.

ORNAMENTAL CROPS AS A SOURCE OF VIROID INFECTIONS

In The Netherlands, the origin of occasional viroid infections in tomato from 1988 to 2001 could not be linked to either seed-borne or seedling-borne infection (Verhoeven et al., 2004). Surveys in ornamental plants, however, revealed that many were infected with various pospiviroids (Verhoeven et al., 2008a, 2010b, 2012). Subsequently, Seigner et al. (2008) and Verhoeven et al. (2010d, 2012) showed that infected ornamental crops of *Brugmansia*, *Lycianthes rantonnetii* and *S. jasminoides* could indeed function as a source of inoculum for tomato. Analysis of the genome sequences of the viroid isolates found in ornamental crops and tomato provided further evidence that infected ornamental plants had been sources of inoculum for tomato crops. Navarro et al. (2009) and Verhoeven et al. (2010b) correlated various outbreaks of PSTVd in tomato with infected plants of *S. jasminoides* (Fig. 3.1). Similar results were obtained for TASVd and *Brugmansia*, *Cestrum*, *L. rantonnetii*, and

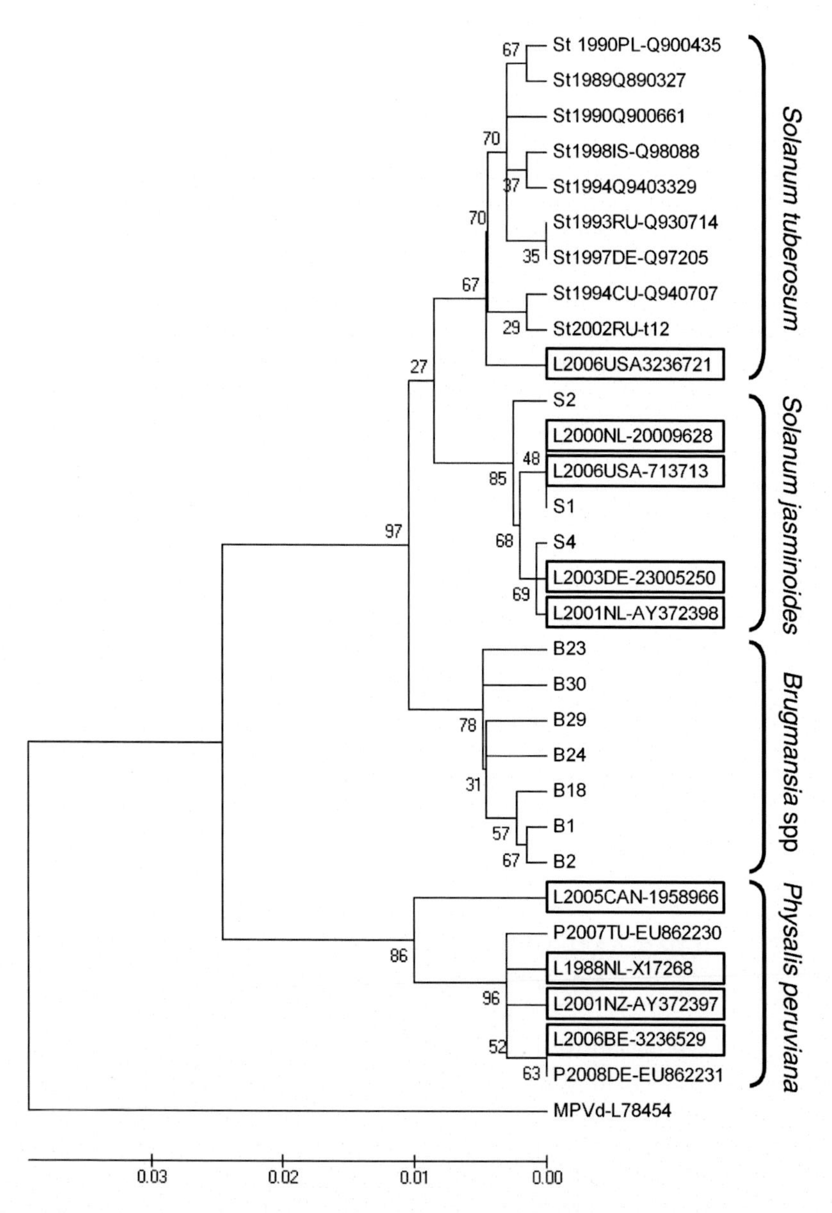

FIGURE 3.1 Evolutionary relationships of 31 potato spindle tuber viroid (PSTVd) genotypes from *Brugmansia* spp., *Solanum jasminoides*, *Physalis peruviana*, potato, and tomato. Other phylogenetic methods produced similar trees. PSTVd isolates from *Brugmansia* spp., *S. lycopersicum*, *P. peruviana*, *S. jasminoides*, and *S. tuberosum* start with B, L, P, S, and St, respectively. The letters B and S are followed by their genotype number; L, P, and St are followed by the year of sample submission, the country of the origin (when known), a hyphen, and letter/figures for genotype identification. PSTVd genome clusters from vegetatively propagated plant species are braced; individual PSTVd genomes from tomato are boxed. *Source: From Verhoeven, J.Th.J., Jansen, C.C.C., Botermans, M., Roenhorst, J.W., 2010b. Epidemiological evidence that vegetatively propagated, solanaceous plant species act as sources of Potato spindle tuber viroid inoculum for tomato. Plant Pathol. 59, 3–12.*

S. jasminoides (Parrella and Numitone, 2014; Verhoeven et al., 2012) and for TCDVd and *Petunia* (Shiraishi et al., 2013). All these data strongly support a significant role for infected ornamental crops providing viroid inoculum for tomato.

References

Benton, R.J., Bowman, F.T., Fraser, L., Kebby, R.G., 1949. Selection of citrus budwood to control scaly butt in trifoliata rootstock. Agr. Gaz. N. S. Wales 60, 31–34.

Benton, R.J., Bowman, F.T., Fraser, L., Kebby, R.G., 1950. Stunting and scaly butt associated with *Poncirus trifoliata* rootstock. N. S. Wales, Dept. Agr., Sci. Bull. 70, 1–20.

Bos, L., 1999. Economic impact of viruses. In: Bos, L. (Ed.), Plant Viruses, Unique and Intriguing Pathogens. Backhuys Publishers, Leiden, pp. 247–260.

Bostan, H., Nie, X., Singh, R.P., 2004. An RT-PCR primer pair for the detection of pospiviroid and its application in surveying ornamental plants for viroids. J. Virol. Methods 166, 189–193.

Brierley, P., 1950. Some host plants of chrysanthemum stunt virus. Phytopathology 40, 869.

Brierley, P., 1953. Some experimental hosts of the chrysanthemum stunt virus. Plant Dis. Rep. 37, 343–345.

Brierley, P., Smith, F.F., 1949. Chrysanthemum stunt. Phytopathology 39, 501.

Brierley, P., Smith, F.F., 1951. Chrysanthemum stunt. Control measures effective against virus in florists' crops. Florists' Rev. 107, 27–30.

De Hoop, M.B., Verhoeven, J.Th.J., Roenhorst, J.W., 2008. Phytosanitary measures in the European Union: a call for more dynamic risk management allowing more focus on real pest risks. Case study: *Potato spindle tuber viroid* (PSTVd) on ornamental *Solanaceae* in Europe. EPPO Bull. 38, 510–515.

Di Serio, F., 2007. Identification and characterization of *Potato spindle tuber viroid* infecting *Solanum jasminoides* and *S. rantonnetii* in Italy. J. Plant Pathol. 89, 297–300.

Diener, T.O., 1987. Potato spindle tuber. In: Diener, T.O. (Ed.), The Viroids. Plenum Press, New York, NY, pp. 221–233.

Diener, T.O., Lawson, R.H., 1973. Chrysanthemum stunt: a viroid disease. Virology 51, 94–101.

Dimock, A.W., 1947. Chrysanthemum stunt. N.Y. State Flower Grow. Bull. 26, 2.

Dimock, A.W., Geissinger, C.M., 1969. A newly recognized disease of chrysanthemums caused by a graft-transmissible agent. Phytopathology 59, 1024.

Dimock, A.W., Geissinger, C.M., Horst, R.K., 1971. Chlorotic mottle: a newly recognized disease of chrysanthemum. Phytopathology 61, 415–419.

EFSA Panel on Plant Health, 2012. Scientific opinion on the risk to plant health posed by Chrysanthemum stunt viroid for the EU territory, with identification and evaluation of risk reduction options. EFSA Journal. 10 (3027), [87 pp.]. Available online: www.efsa.europa.eu/efsajournal.

EPPO/OEPP (European and Mediterranean Plant Protection Organization), 2002. Certification scheme for chrysanthemum. EPPO Bull. 32, 105–114.

Fawcett, H.S., Klotz, L.J., 1948. Exocortis on trifoliate orange. Citrus Leaves 28, 8.

Fonseca, M.E.N., Boiteux, L.S., Singh, R.P., Kitajima, E.W., 1989. A small viroid in Coleus species from Brazil. Fitopatol. Bras. 14, 94–96.

Fonseca, M.E.N., Marcellino, L.H., Kitajima, E.W., Boiteux, L.S., 1994. Nucleotide sequence of the original Brazilian isolate of coleus yellow viroid from *Solenostemon scutellarioides* and infectivity of its complementary DNA. J. Gen. Virol. 75, 1447–1449.

Hammond, R., Smith, D.R., Diener, T.O., 1989. Nucleotide sequence and proposed secondary structure of *Columnea* latent viroid: a natural mosaic of viroid sequences. Nucleic Acids Res. 17, 10083–10094.

Horst, R.K., 1975. Detection of a latent infectious agent that protects against infection by chrysanthemum chlorotic mottle viroid. Phytopathology 65, 1000–1003.

Horst, R.K., 1987. Chrysanthemum chlorotic mottle. In: Diener, T.O. (Ed.), The Viroids. Plenum, New York, NY, pp. 291–295.

Horst, R.K., Langhans, R.W., Smith, S.H., 1977. Effects of chrysanthemum stunt, chlorotic mottle, aspermy and mosaic on flowering and rooting of chrysanthemums. Phytopathology 67, 9–14.

Kent, G.C., 1953. Cadang-cadang of coconut. Philipp. Agric. Sci. 37, 228–240.

Lemmetty, A., Laamanen, J., Soukained, M., Tegel, J., 2011. Emerging virus and viroid pathogen species identified for the first time in horticultural plants in Finland in 1997–2010. Agric. Food Sci. 20, 29–41.

Luigi, M., Luison, D., Tomassoli, L., Faggioli, F., 2011a. First report of *Potato spindle tuber* and *Citrus exocortis viroids* in *Cestrum* spp. in Italy. New Dis. Rep. 23, 4.

Luigi, M., Luison, D., Tomassoli, L., Faggioli, F., 2011b. Natural spread and molecular analysis of pospiviroids infecting ornamentals in Italy. J. Plant Pathol. 93, 491–495.

Luigi, M., Manglli, A., Tomassoli, L., Faggioli, F., 2013. First report of *Hop stunt viroid* in *Hibiscus rosa-sinensis* in Italy. New Dis. Rep. 27, 14.

Marais, A., Faure, C., Deogratias, J.M., Candresse, T., 2011. First report of *Chrysanthemum stunt viroid* in various cultivars of *Argyranthemum frutescens* in France. Plant Dis. 95, 1196.

Martin, W.H., 1922. "Spindle tuber," a new potato trouble. Hints to potato growers. N. J. State Potato Assoc. 8, 3.

Matousek, J., Piernikarczyk, R.J.J., Dědič, P., Mertelík, J., Uhlířová, K., Duraisamy, G.S., et al., 2014. Characterization of *Potato spindle tuber viroid* (PSTVd) incidence and new variants from ornamentals. Eur. J. Plant Pathol. 138, 93–101.

Matsushita, Y., Tsuda, S., 2015. Host ranges of *Potato spindle tuber viroid, Tomato chlorotic dwarf viroid, Tomato apical stunt viroid,* and *Columnea latent viroid* in horticultural plants. Eur. J. Plant Pathol. 141, 193–197.

Matsushita, Y., Usugi, T., Tsuda, S., 2009. Host range and properties of *Tomato chlorotic dwarf viroid*. Eur. J. Plant Pathol. 124, 349–352.

Menzel, W., Maiss, E., 2000. Detection of *Chrysanthemum stunt viroid* (CSVd) in cultivars of *Argyranthemum frutescens* by RT-PCR-ELISA. J. Plant Dis. Prot. 107, 548–552.

Mertelik, J., Kloudova, K., Cervena, G., Necekalova, J., Mikulkova, H., Levkanicova, Z., et al., 2010. First report of *Potato spindle tuber viroid* (PSTVd) in *Brugmansia* spp., *Solanum jasminoides, Solanum muricatum* and *Petunia* spp. in the Czech Republic. Plant Pathol. 59, 392.

Nakashima, A., Hosokawa, M., Maeda, S., Yazawa, S., 2007. Natural infection of *Chrysanthemum stunt viroid* in dahlia plants. J. Gen. Plant Pathol. 73, 225–227.

Navarro, B., Flores, R., 1997. Chrysanthemum chlorotic mottle viroid: unusual structural properties of a subgroup of self-cleaving viroids with hammerhead ribozymes. Proc. Natl. Acad. Sci. USA. 94, 11262–11267.

Navarro, B., Silletti, M.R., Trisciuzzi, V.N., Di Serio, F., 2009. Identification and characterization of *Potato spindle tuber viroid* infecting tomato in Italy. J. Plant Pathol. 91, 723–726.

Niblett, C.L., Dickson, E., Horst, R.K., Romaine, C.P., 1980. Additional host and an efficient purification procedure for four viroids. Phytopathology 70, 610–615.

Nie, X., Singh, R.P., Bostan, H., 2005. Molecular cloning, secondary structure, and phylogeny of three pospiviroids from ornamental plants. Can. J. Plant Pathol. 27, 592–602.

Nielsen, S.L., Nicolaisen, M., 2010. First report of *Columnea latent viroid* (CLVd) in *Gloxinia gymnostoma, G. nematanthodes* and *G. purpurascens* in a botanical garden in Denmark. New Dis. Rep. 22, 4.

O'Brien, M.J., 1972. Hosts of potato spindle tuber virus in suborder *Solanineae*. Am. Potato J. 49, 70–72.

Olivier, T., Demonty, E., Govers, J., Belkheir, K., Steyer, S., 2011. First report of a *Brugmansia* sp. infected by *Tomato apical stunt viroid* in Belgium. Plant Dis. 95, 495.

Parrella, G., Numitone, G., 2014. First report of *Tomato apical stunt viroid* in tomato in Italy. Plant Dis. 98, 1164.

Randles, J.W., Rodriguez, M.J.B., 2003. Coconut cadang-cadang viroid. In: Hadidi, A., Flores, R., Randles, J.W., Semancik, J.S. (Eds.), Viroids. CSIRO Publishing, Collingwood, VIC, pp. 233–241.

Sänger, H.L., 1988. Viroids and viroid diseases. Acta Hortic. 234, 79–87.

Seigner, L., Kappen, M., Huber, C., Kistler, M., Köhler, D., 2008. First trials for transmission of Potato spindle tuber viroid from ornamental *Solanaceae* to tomato using RT-PCR and an mRNA based internal positive control for detection. J. Plant Dis. Prot. 115, 97–101.

Shiraishi, T., Maejima, K., Komatsu, K., Hashimoto, M., Okano, Y., Kitazawa, Y., et al., 2013. First report of tomato chlorotic dwarf viroid isolated from symptomless petunia plants (*Petunia* spp.) in Japan. J. Gen. Plant. Pathol. 79, 214–216.

Singh, R.P., 1973. Experimental host range of the potato spindle tuber "virus." Am. Potato J. 50, 111–123.

Singh, R.P., Dilworth, A.D., 2009. *Tomato chlorotic dwarf viroid* in the ornamental plant *Vinca minor* and its transmission through tomato seed. Eur. J. Plant Pathol. 123, 111–116.

Singh, R.P., Dilworth, A.D., Baranwal, V.K., Gupta, K.N., 2006. Detection of Citrus exocortis viroid, Iresine viroid, and Tomato chlorotic dwarf viroid in new ornamental host plants in India. Plant Dis. 90, 1457.

Singh, R.P., Lakshman, D.K., Boucher, A., Tavantzis, S.M., 1992. A viroid from *Nematanthus wettsteinii* plants closely related to the *Columnea* latent viroid. J. Gen. Virol. 73, 2769–2774.

Soliman, T., Mourits, M.C.M., Oude Lansink, A.G.J.M., Van der Werf, W., 2012. Quantitative economic impact assessment of an invasive plant disease under uncertainty – a case study for potato spindle tuber viroid (PSTVd) invasion into the European Union. Crop Prot. 40, 28–35.

Spieker, R.L., 1996a. A viroid from *Brunfelsia undulata* closely related to the *Columnea* latent viroid. Arch. Virol. 141, 1823–1832.

Spieker, R.L., 1996b. The molecular structure of *Iresine* viroid, a new viroid species from *Iresine herbstii* ("beefsteak plant"). J. Gen. Virol. 77, 2631–2635.

Spieker, R.L., Haas, B., Charng, Y.C., Freimuller, K., Sänger, H.L., 1990. Primary and secondary structure of a new viroid "species" (CbVd 1) present in the *Coleus blumei* cultivar "Bienvenue". Nucleic Acids Res. 18, 3998.

Sorrentino, R., Minutolo, M., Alioto, D., Torchetti, E.M., Di Serio, F., 2015. First report of Iresine viroid 1 in ornamental plants in Italy and of *Celosia cristata* as a novel natural host. Plant Dis. 99, 1655.

Spieker, R.L., Marinkovic, S., Sänger, H.L., 1996a. A new sequence variant of Coleus blumei viroid 3 from the Coleus blumei cultivar "Fairway Ruby". Arch. Virol. 141, 1377–1386, Please, start the reference Spieker et al 2996a at a new line.

Spieker, R.L., Marinkovic, S., Sänger, H.L., 1996b. A viroid from *Solanum pseudocapsicum* closely related to the tomato apical stunt viroid. Arch. Virol. 141, 1387–1395.

Torchetti, E.M., Navarro, B., Trisciuzzi, V.N., Nuccitelli, L., Silletti, M.R., Di Serio, F., 2012. First report of Chrysanthemum stunt viroid in *Argyranthemum frutescens* in Italy. J. Plant Pathol. 94, 451–454.

Tsushima, T., Murakami, S., Ito, H., He, Y.H., Raj, A.P.C., Sano, T., 2011. Molecular characterization of *Potato spindle tuber viroid* in dahlia. J. Gen. Plant Pathol. 77, 253–256.

Van Brunschot, S.L., Persley, D.M., Roberts, A., Thomas, J.E., 2014. First report of pospiviroids infecting ornamental plants in Australia: *Potato spindle tuber viroid* in *Solanum laxum* (synonym *S. jasminoides*) and *Citrus exocortis viroid* in *Petunia* spp. New Dis. Rep. 29, 3.

Verhoeven, J.Th.J., 2010. Identification and epidemiology of pospiviroids. Thesis. Wageningen University, Wageningen. The Netherlands. ISBN 978-90-8585-623-8, 136 pp.

Verhoeven, J.Th.J., Arts, M.S.J., Owens, R.A., Roenhorst, J.W., 1998. Natural infection of petunia by chrysanthemum stunt viroid. Eur. J. Plant Pathol. 104, 383–386.

Verhoeven, J.Th.J., Botermans, M., Jansen, C.C.C., Roenhorst, J.W., 2010c. First report of *Tomato apical stunt viroid* in the symptomless hosts *Lycianthes rantonnetii* and *Streptosolen jamesonii* in the Netherlands. Plant Dis. 94, 791.

Verhoeven, J.Th.J., Botermans, M., Meekes, E.T.M., Roenhorst, J.W., 2012. *Tomato apical stunt viroid* in the Netherlands: most prevalent pospiviroid in ornamentals and first outbreak in tomatoes. Eur. J. Plant Pathol. 133, 803–810.

Verhoeven, J.Th.J., Huner, L., Virscek Marn, M., Mavric Plesko, I., Roenhorst, J.W., 2010d. Mechanical transmission of *Potato spindle tuber viroid* between plants of *Brugmansia suaveolens*, *Solanum jasminoides* and potatoes and tomatoes. Eur. J. Plant Pathol. 128, 417–421.

Verhoeven, J.Th.J., Jansen, C.C.C., Botermans, M., Roenhorst, J.W., 2010a. First report of *Iresine viroid 1* in *Celosia plumosa* in the Netherlands. Plant Dis. 94, 920.

Verhoeven, J.Th.J., Jansen, C.C.C., Botermans, M., Roenhorst, J.W., 2010b. Epidemiological evidence that vegetatively propagated, solanaceous plant species act as sources of *Potato spindle tuber viroid* inoculum for tomato. Plant Pathol. 59, 3–12.

Verhoeven, J.Th.J., Jansen, C.C.C., Roenhorst, J.W., 2006. First report of *Potato virus M* and *Chrysanthemum stunt viroid* in *Solanum jasminoides*. Plant Dis. 90, 1359.

Verhoeven, J.Th.J., Jansen, C.C.C., Roenhorst, J.W., 2008a. First report of pospiviroids infecting ornamentals in the Netherlands: *Citrus exocortis viroid* in *Verbena* sp., *Potato spindle tuber viroid* in *Brugmansia suaveolens* and *Solanum jasminoides*, and *Tomato apical stunt viroid* in *Cestrum* sp. Plant Pathol. 57, 399.

Verhoeven, J.Th.J., Jansen, C.C.C., Roenhorst, J.W., 2008c. *Streptosolen jamesonii* "Yellow," a new host plant of Potato spindle tuber viroid. Plant Pathol. 57, 399.

Verhoeven, J.Th.J., Jansen, C.C.C., Roenhorst, J.W., Flores, R., De la Peña, M., 2009. *Pepper chat fruit viroid*: biological and molecular properties of a proposed new species of the genus *Pospiviroid*. Virus Res. 144, 209–214.

Verhoeven, J.Th.J., Jansen, C.C.C., Roenhorst, J.W., Steyer, S., Schwind, N., et al., 2008b. First report of *Solanum jasminoides* infected by *Citrus exocortis viroid* in Germany and the Netherlands and *Tomato apical stunt viroid* in Belgium and Germany. Plant Dis. 92, 973.

Verhoeven, J.Th.J., Jansen, C.C.C., Werkman, A.W., Roenhorst, J.W., 2007. First report of *Tomato chlorotic dwarf viroid* in *Petunia hybrida* from the United States of America. Plant Dis. 91, 324.

Verhoeven, J.Th.J., Jansen, C.C.C., Willemen, T.M., Kox, L.F.F., Owens, R.A., et al., 2004. Natural infections of tomato by *Citrus exocortis viroid*, *Columnea latent viroid*, *Potato spindle tuber viroid* and *Tomato chlorotic dwarf viroid*. Eur. J. Plant Pathol. 110, 823–831.

Verhoeven, J.Th.J., Meekes, E.T.M., Roenhorst, J.W., Flores, R., Serra, P., 2013. Dahlia latent viroid: a recombinant new species of the family *Pospiviroidae* posing intriguing questions about its origin and classification. J. Gen. Virol. 94, 711–719.

Verhoeven, J.Th.J., Roenhorst, J.W., Hooftman, M., Meekes, E.T.M., Flores, R., et al., 2015. A pospiviroid from symptomless portulaca plants closely related to iresine viroid 1. Virus Res. 205, 22–26.

Virscek Marn, M., Mavric Plasko, I., 2012. First report of *Iresine viroid 1* in *Portulaca* sp. in Slovenia. J. Plant Pathol. 94 (S4), 97.

Weathers, L.G., Greer, F.C., 1968. Additional herbaceous hosts of the exocortis virus of citrus. Phytopathology 58, 1071.

Weathers, L.G., Greer, F.C., Harjung, M.K., 1967. Transmission of exocortis virus of citrus to herbaceous plants. Plant Dis. Rep. 51, 868–871.

CHAPTER

4

Economic Significance of Palm Tree Viroids

M. Judith B. Rodriguez[1], *Ganesan Vadamalai*[2] *and John W. Randles*[3]

[1]Philippine Coconut Authority, Albay, Philippines
[2]Universiti Putra Malaysia, Selangor, Malaysia
[3]The University of Adelaide, Waite Campus, Glen Osmond, SA, Australia

INTRODUCTION

Coconut cadang-cadang viroid (CCCVd) continues to threaten the coconut (*Cocos nucifera* L.) industry in the Philippines and is a newly recognized factor in the global oil palm (*Elaeis guineensis* Jacq.) industry. The cadang-cadang disease that it causes has killed an estimated 40 million (Mn) coconut palms in the Philippines, affected trade, and continues to spread (Randles and Rodriguez, 2003; Philippine Coconut Authority, 2014). Severe CCCVd variants such as those associated with the lamina depletion and more rapid lethal decline of coconut palms known as "brooming" (Rodriguez and Randles, 1993) occur naturally. CCCVd is also associated with the widespread orange leaf-spotting (OS) disorder of oil palm (Hanold and Randles, 1991, Vadamalai et al., 2006; Wu et al., 2013), so the viroid can be assumed to occur outside the Philippines wherever oil palms are grown. In addition, uncharacterized CCCVd-related RNAs occur in tropical monocotyledonous species in SE Asia and Oceania (Hanold, 1998; Hanold and Randles, 1998; Rodriguez, 1993). Research is needed to determine their role in the origins and epidemiology of CCCVd and other viroids, such as coconut tinangaja viroid (CTiVd), which has curtailed the coconut industry in Guam (Wall and Randles, 2003). A historical summary of cadang-cadang disease is presented in Table 4.1. In this chapter we first consider the economic impact

Viroids and Satellites.
DOI: http://dx.doi.org/10.1016/B978-0-12-801498-1.00004-8

TABLE 4.1 The History of CCCVd and CTiVd; Recognition of Disease, Viroid Characterization, Diagnosis, Surveillance, and a Forecast of Potential Impact

Year	Milestone	Notes	References
1914	Coconut dieback disease reported	Camarines Sur, Luzon, Philippines	Kent (1953)
1973	Viroid-like RNA associated	Polyacrylamide gel electrophoresis (PAGE) assay	Randles (1975)
1975	Viroid properties confirmed	Physico-chemistry and structure	Randles et al. (1976)
1975	Infectivity demonstrated	Mechanical transmission	Randles et al. (1977)
1975	Ecology of disease	Quantitative survey	Zelazny (1980)
1978	Diagnosis by hybridization assay	cDNA probe developed	Randles and Palukaitis (1979)
1978	Cadang-cadang survey	Total loss of 30 million palms	Zelazny et al. (1982)
1980	Additional palm hosts identified	Oil palm infected	Randles et al. (1980)
1980	CTiVd discovered	Tinangaja disease, Guam	Boccardo et al. (1981)
1981	CCCVd sequenced	246–574 nt forms described	Haseloff et al. (1982)
1981	Viroid diagnosis by circular structure	2-dimensional PAGE	Schumacher et al. (1983)
1985–91	No resistance in coconut palms	82 populations tested	Bonaobra et al. (1998)
1987	CTiVd sequenced	254 nt, 64% similarity to CCCVd	Keese et al. (1988)
1989	CCCVd in a commercial oil palm plantation	Orange leaf spotting, Solomon Islands	Hanold and Randles (1991)
1989	CCCVd -related RNA in other monocotyledons	Survey of 29 countries in SE Asia, Oceania	Hanold and Randles (1998)
1991	Natural variants of CCCVd occur	Severe "brooming" variants sequenced	Rodriguez and Randles (1993)
2002	Launch of disease containment program	Identification of disease free areas	Carpio (2011)
2005	Oil palm variants sequenced	98% sequence identity to CCCVd	Vadamalai et al. (2006)

(Continued)

TABLE 4.1 (Continued)

Year	Milestone	Notes	References
2007	Diagnosis of molecular variants	Ribonuclease protection assay	Vadamalai et al. (2009)
2012–13	Disease survey	Distribution, economic impact	Philippine Coconut Authority (2014)
2032	~3 million new infections predicted	Assumes control strategy is not effective	Philippine Coconut Authority (2014)

of viroids on coconut production and then describe the implications of the recent discovery of CCCVd in commercial oil palm.

COCONUT PALM

The coconut palm ("tree of life") grows throughout the wet tropics. The Philippines is the world's largest producer and exporter of coconut products, including copra, oil, fresh nuts, desiccated coconut, coconut water, lumber, and processed foods. Annual production from the 340 Mn bearing trees on 3.6 Mn ha is around 15 billion (Bn) nuts, yielding around 2.6 metric tons (mt) of copra worth about $2 Bn, and earning about $1.3 Bn in export income (Forbes, 2013). Cadang-cadang disease (from "gadan-gadan," a Bicol word meaning dead or dying; Rillo and Rillo, 1981) stops nut production and then slowly kills palms, and is considered to be a threat to other coconut producing countries. International trade restrictions on the export of Philippine coconut products prompted a research and development program (Rodriguez, 2011) directed at restricting the disease to its present geographical distribution in the Philippines (Carpio, 2011). A major survey in 2011–13 identified disease-free areas from which coconut germplasm and products could be sourced (Philippine Coconut Authority, 2014). Visual inspections and molecular diagnostic methods were used to monitor disease boundaries, establish buffer zones, and identify national quarantine checkpoints.

Fig. 4.1 shows the outcome of the survey where the incidence of cadang-cadang was mapped village by village. The survey confirmed that the disease still remains in the central eastern part of the country, the northernmost point being Polillo Island (14.83°N), and the southernmost Guiuan (11.03°N). Disease hotspots were the municipalities of Guiuan (14.2%) and Mercedes (10.8%) in Eastern Samar, the municipality of Jomalig in Quezon I (14.1%), and the municipality of Milagros in Masbate (11.4%). The disease disappeared from a region in Northern Samar.

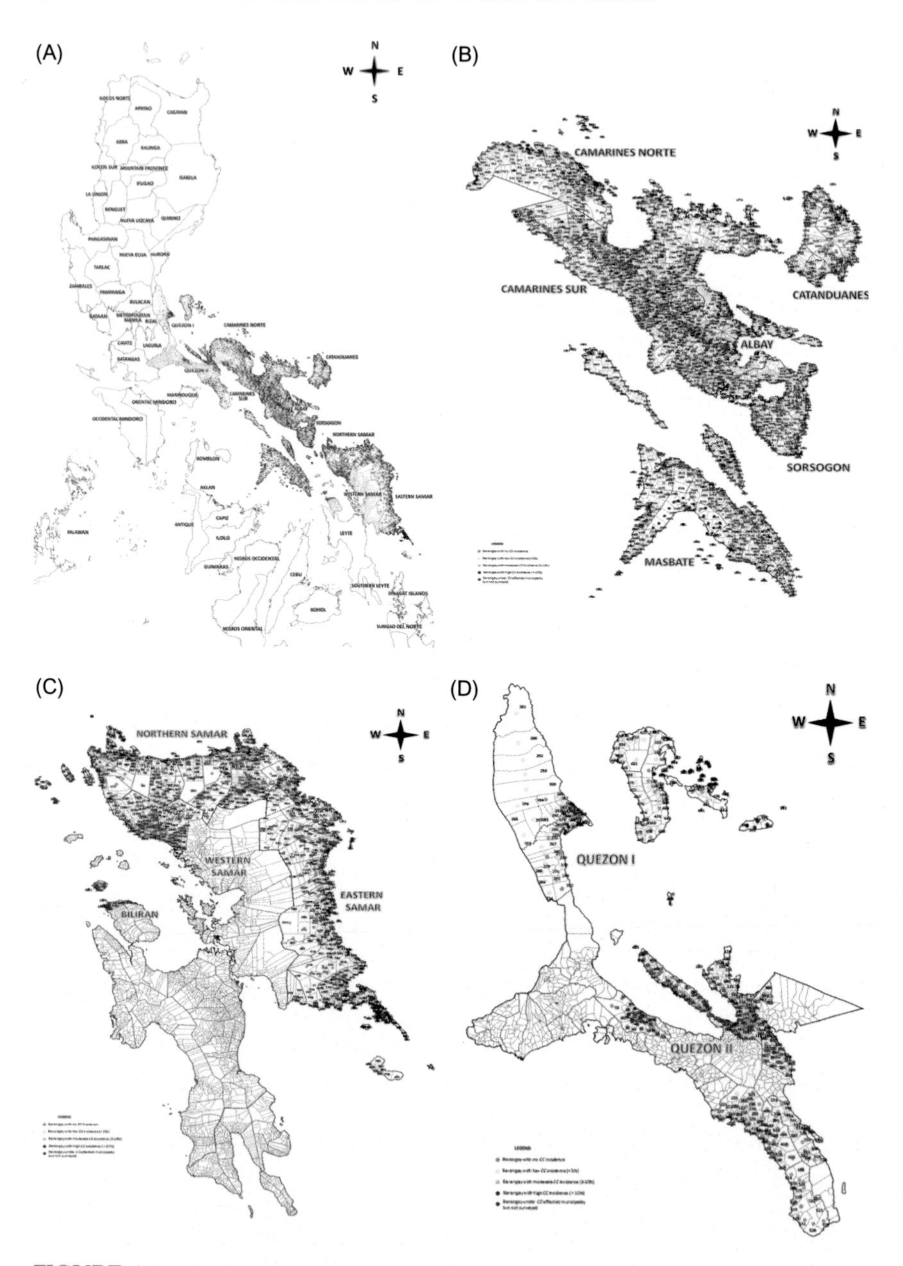

FIGURE 4.1 Maps showing the distribution and incidence of cadang-cadang disease in the Philippines. (A) Countrywide distribution; (B) Central region; (C) Southern boundary region; (D) Northern boundary region. Incidence in each district indicated as: nil (*green*); 3% (*yellow*); 3%–10% (*orange*); over 10% (*red*); not estimated (*mauve*). *Source: From Philippine Coconut Authority, 2014. Terminal report. Coconut cadang-cadang disease surveillance survey 2012–2013. Philippine Coconut Authority, Quezon City, Philippines, 70 pp, with kind permission.*

The sporadic distribution of cadang-cadang has not been explained, nor has the period of rapid increase in incidence between 1951 and 1959 in the central Philippines followed by a significant decline in 1978–80 (Zelazny and Pacumbaba, 1982; Zelazny et al., 1982). More recently, overall disease incidence was estimated at 2.5 Mn palms in 2006, declining to 691,068 in 2012–13 (Philippine Coconut Authority, 2014).

Affected areas are being rehabilitated by removal of infected palms and replacement with high yielding populations. In 2001–07, over 51,355 infected palms were cut, and nurseries were established to provide 180,000 healthy seedling replants (Carpio, 2011). This rehabilitation program is possibly linked to the decline in disease incidence since 2006, but ongoing surveys will be required to evaluate the success of this approach to control.

The total economic loss due to cadang-cadang was estimated in 2003 at $4 Bn, with further research costs of $10.8 Mn (Randles, 2003). The cost of the recent country survey was $240,000 (10 Mn pesos), but additional inputs such as public awareness campaigns, staff training, implementation of quarantine regulations, and laboratory diagnostics have not been quantified. A necessary future cost will be for the eradication of approximately 700,000 currently infected palms over an area of about 650,000 ha (Philippine Coconut Authority, 2014). Failure to control the disease is predicted to lead to an incidence of over 3 Mn infected palms in 15 years (Table 4.1). Based on a value of $100 per bearing palm ($5 per annum for 20 years) the cost of doing nothing can therefore be estimated at $300 Mn.

Coconut tinangaja disease is lethal but distinct from cadang-cadang. The causal CTiVd is distinct from CCCVd and occurs in Guam and Anatahan, in the Marianas archipelago (Wall and Randles, 2003; see Chapter 25: Coconut Cadang-Cadang Viroid and Coconut Tinangaja Viroid). Coconut is not a commercially important crop there, although it has cultural significance for the native people. The economic impact of tinangaja is minor (Randles, 2003).

OIL PALM

E. guineensis originates from equatorial West Africa (Hartley, 1988). Its development as a plantation crop started in South East Asia following the introduction of four seedlings from Mauritius and Amsterdam into the Bogor Botanic Garden (Indonesia) in 1848. Commercial oil palm plantations were first established in Indonesia (Sumatra), then in Malaysia in 1917 (Tennamaran Estate, Selangor) (MPOC, 2015). All modern commercial plantations comprise Tenera seedlings, a hybrid between the thick shelled Dura and shell-less Pisifera (D × P). Tissue-cultured or "clonal" palms (ramets), which provide "true copies" of selected high-yielding D × P palms, are being developed by the industry (MPOC, 2015).

The fresh fruit bunch yield in Malaysia in 2014 was 18.63 mt/ha, totaling approximately 87.2 Mn mt from 4.68 Mn ha (MPOB, 2015). Crude palm oil (CPO) production was 3.84 mt/ha, totaling approximately 18 Mn mt. CPO returned $578 (2144 ringgit) per mt.

Palm oil comes from the mesocarp and kernel of the fruit, and is used widely in the food industry, the oleochemical industry, and as a feedstock for biofuel production. As with the coconut palm, all other parts of oil palms are utilized, such as for animal feed, fuel, activated charcoal, and fiber board (RSPO, 2015). Indonesia is the largest producer, while Malaysia, which produces 39% of the world's palm oil, exported 25.1 Mn mt of oil palm derived products in 2014, earning $17.24 Bn (MPOB, 2015). This includes approximately 18 Mn mt of CPO and 7.1 Mn mt of palm kernel oil (PKO), palm kernel cake (PKC), oleochemicals, biodiesel, and other oil palm based products. From 1962 to 1982 global exports of palm oil increased from around 0.5 to 2.4 Mn mt per annum, then rose to 48 Mn mt by 2008, and global demand is forecast to double by 2020 and triple by 2050 (Prokurat, 2013).

Orange spotting of leaves is a well-known disorder of oil palm (Forde and Leyritz, 1968; Turner, 1981; see Chapter 25: Coconut Cadang-Cadang Viroid and Coconut Tinangaja Viroid). A distinction has been made between the symptoms characteristic of "confluent" orange spotting, which is associated with nutrient deficiencies, and "genetic" orange spotting (GOS) (Fig. 4.2). The documented association of GOS with particular parental lines, the lack of association with major nutrient deficiencies, and the absence of necrosis in the spots led to the hypothesis that the disorder is inherited. Because a Mendelian distribution could not be observed among the progeny (Forde and Leyritz, 1968; Gascon and Meunier, 1979), the term OS has replaced GOS.

Oil palms in the Philippines with OS and a decline resembling cadang-cadang disease in adjacent coconut palms were shown to contain CCCVd by gel electrophoresis and molecular hybridization assays (Randles et al., 1980). These symptoms could be reproduced in oil palm seedlings by mechanical inoculation with nucleic acid extracts from infected coconut palms, and CCCVd was reisolated from them (Imperial et al., 1985). OS and CCCVd were then detected in a Solomon Islands plantation (Hanold and Randles, 1991). Recently, CCCVd variants were characterized from commercial oil palms in Malaysian plantations showing OS symptoms. They also occurred in asymptomatic oil palms. All of the oil palm CCCVd variants showed greater than 90% sequence identity with CCCVd from coconut (Cheong, 2012; Joseph, 2012; Randles et al., 2009; Vadamalai, 2005; Vadamalai et al., 2006; Wu et al., 2013; Chapter 25: Coconut Cadang-Cadang Viroid and Coconut Tinangaja Viroid).

FIGURE 4.2 Symptoms of OS on a 6-year-old oil palm (refer also to Fig. 25.3).

OS has been observed in commercial oil palm plantations in West Africa, Indonesia, Malaysia, Thailand, Papua New Guinea, Solomon Islands, and Central America (Cheong, 2012; Hanold and Randles, 1991; Randles, unpublished; Selvaraja et al., 2012). Because oil palm has a very narrow genetic base and most of the planting materials for commercial oil palm plantations are derived from only four parents (Randles, 1998), the widespread distribution of OS could possibly be explained by the vertical transmission of a viroid agent from one or more parents and its distribution with seed (Hanold, 1998; Hanold and Randles, 1991). The incidence and severity of OS expression varies within plantations. Selvaraja et al. (2012), using a remote sensing method in a survey of 587 geo-referenced palms, found that 74.3% of the palms had OS symptoms. OS severity was low in 73% of the symptomatic palms, severity was moderate in 25%, and 2% had high severity. The relationship between severity of OS and presence and sequence of viroid is yet to be determined.

OS is associated with a yield reduction. The average yield obtained from OS-affected palms was 25%–50% lower than neighboring healthy palms (Forde and Leyritz, 1968; Hanold and Randles, 1991; Randles, 1998). Height measurements of 93 palms in an 8-year-old commercial plantation in the Solomon Islands showed that palms with both CCCVd-like sequences and OS were significantly stunted (Randles, 1998).

TABLE 4.2 Predicted Losses in Mature Palm Oil Plantations in Malaysia, Estimated for OS Incidences of 1%, 2%, or 5%. Calculations Assume a Yield Loss of 25% per OS Affected Palm; a Population of 655 Million Palms; and a Mean Annual Yield of 133 kg of Fruit and 27 kg of Oil per Palm (MPOB, 2015)

Assumed OS incidence (%)	Annual yield loss of fresh fruit (metric tons)	Annual loss of revenue from crude palm oil ($)
1	0.22×10^6	25.6×10^6
2	0.44×10^6	51.3×10^6
5	1.1×10^6	128.1×10^6

Due to a lack of information on overall economic losses from OS, we have prepared several hypothetical scenarios (Table 4.2). For example, if we assume a yield loss of 25% for each infected palm, and that the incidence of infection is either 1%, 2%, or 5%, we estimate that annual losses in revenue would range from $25.6 Mn to $128.1 Mn. This shows that the estimated annual loss per palm would be around $4.00 and this would double if the yield loss is 50%. While detailed field studies have yet to be done in Malaysia to determine actual yield depression due to OS and its overall incidence, these estimates show that the control of OS would have a high economic benefit.

Priorities for research include determining the disease cycle of the oil palm CCCVd variants in oil palm, whether production of viroid-free oil palm seedlings and ramets can be achieved, and whether secondary spread of CCCVd occurs in plantations. Thus, while it should be feasible to identify viroid-free parent palms for the production of seedlings and cloned ramets using established viroid diagnostic methods, it is not known whether the viroid spreads via pollen, seed, mechanical harvesting, or vectors, and whether it has a plant reservoir other than oil palm.

The CCCVd oil palm variants can be transmitted via mechanical inoculation into oil palm seedlings and ramets (Joseph, 2012; Thanarajoo, 2014) and there is a low rate of seed transmission of OS in oil palm (Hanold and Randles, 1991; Vadamalai, unpublished).

CONCLUSIONS

Oil palm is a newly recognized economically important host of CCCVd. It will be a useful species for research on the disease cycle of CCCVd, and the impact of mutations on disease phenotype. Mutations, especially in the pathogenicity domain, modulate symptom expression by some viroids (Schnölzer et al., 1985; Visvader and Symons, 1985; Wassenegger et al., 1996). Although a characteristic single base substitution has been reported in the pathogenicity domain of the oil palm CCCVd variant compared to that from coconut, evaluation of its role in OS symptom expression and modulation of pathogenicity is required (Joseph, 2012; Vadamalai et al., 2006).

Acknowledgments

The authors acknowledge the contributions of members of the Cadang-cadang Containment Program and the Cadang-cadang Technical Working Group of the Philippine Coconut Authority, survey data encoders, coconut farmers, landowners, and others, in the preparation of the maps depicted in Fig. 4.1.

References

Boccardo, G., Beaver, R.G., Randles, J.W., Imperial, J.S., 1981. Tinangaja and bristle top, coconut diseases of uncertain etiology in Guam, and their relationship to cadang-cadang disease of coconut in the Philippines. Phytopathology 71, 1104–1107.

Bonaobra III, Z.S., Rodriguez, M.J.B., Estioko, L.P., Baylon, G.B., Cueto, C.A., Namia, M.T. I., 1998. Screening of coconut populations for resistance to CCCVd using coconut seedlings. In: Hanold, D., Randles, J.W. (Eds.), Report on ACIAR-Funded Research on Viroids and Viruses of Coconut Palm and Other Tropical Monocotyledons 1985–1993. ACIAR Working Paper No. 51, Canberra, Australia, pp. 69–75.

Carpio, C.B., 2011. Practical strategies and regulatory measures adopted in the control, management and containment of the cadang-cadang (CCRNA) disease in the Philippines. Report, APCC/MCD & JED/CRI consultative meeting on phytoplasma/ wilt diseases in coconut. Lunuwila, Sri Lanka, 15–17 June 2011, pp. 160–170.

Cheong, L.C., 2012. Incidence of orange spotting and characterization of coconut cadang-cadang viroid variants in Selangor and Sabah oil palm plantations, Malaysia. MSc Thesis, University Putra Malaysia.

Forbes, E.G., 2013. Outlook for the coconut industry. Philippine Coconut Authority, Quezon City.

Forde, S.C.M., Leyritz, M.J.P., 1968. A study of confluent orange spotting of the oil palm in Nigeria. J. Nigerian Inst. Oil Palm Res. 4, 371–380.

Gascon, J.P., Meunier, J., 1979. Anomalies of genetic origin in the oil palm, *Elaeis*. Description and results. Oleagineux. 34, 437–447.

Hanold, D., 1998. Investigation of the characteristics of the CCCVd-related molecules. In: Hanold, D., Randles, J.W. (Eds.), Report on ACIAR-Funded Research on Viroids and Viruses of Coconut Palm and Other Tropical Monocotyledons 1985–1993. ACIAR Working Paper No. 51, Canberra, Australia, pp. 160–171.

Hanold, D., Randles, J.W., 1991. Detection of coconut cadang-cadang viroid-like sequences in oil and coconut palm and other monocotyledons in the South-West Pacific. Ann. Appl. Biol. 118, 139–151.

Hanold D., Randles J.W. (Eds.), Report on ACIAR-funded research on viroids and viruses of coconut palm and other tropical monocotyledons 1985–1993, ACIAR Working Paper No. 51. 1998. Canberra, Australia, pp. 222.

Hartley, C.W.S., 1988. The Oil Palm (*Elaeis guineensis* Jacq). 3rd ed. Longman Scientific and Technical, UK.

Haseloff, J., Mohamed, N.A., Symons, R.H., 1982. Viroid RNAs of cadang-cadang disease of coconuts. Nature 299, 316–320.

Imperial, J.S., Bautista, R.M., Randles, J.W., 1985. Transmission of the coconut cadang-cadang viroid to six species of palm by inoculation with nucleic acid extracts. Plant Pathol. 34, 391–401.

Joseph, H., 2012. Characterization and pathogenicity of coconut cadang-cadang viroid variants in oil palm (*Elaeis guineensis* Jacq.) seedlings. PhD Thesis, University Putra Malaysia.

Keese, P., Osorio-Keese, M.E., Symons, R.H., 1988. Coconut tinangaja viroid sequence homology with coconut cadang-cadang viroid and other potato spindle tuber related RNAs. Virology 162, 508–510.

Kent, G.C., 1953. Cadang-cadang of coconut. Philippine Agri. 37, 228–240.

MPOB, 2015. Malaysian Palm Oil Board. Statistics. http://bepi.mpob.gov.my.

MPOC, 2015. Malaysian Palm Oil Promotion Council. www.mpopc.org.my.

Philippine Coconut Authority, 2014. Terminal report. Coconut cadang-cadang disease surveillance survey 2012–2013. Philippine Coconut Authority, Quezon City, Philippines, 70 pp.

Prokurat, S., 2013. Palm oil—strategic source of renewable energy in Indonesia and Malaysia. J. Mod. Sci. 3, 425–443.

Randles, J.W., 1975. Association of two ribonucleic acid species with cadang-cadang disease of coconut palm. Phytopathology 65, 163–167.

Randles, J.W., 1998. CCCVd-related sequences in species other than coconut. In: Hanold, D., Randles, J.W. (Eds.), Report on ACIAR-Funded Research on Viroids and Viruses of Coconut Palm and Other Tropical Monocotyledons 1985–1993. ACIAR Working Paper No. 51, Canberra, Australia, pp. 144–152.

Randles, J.W., 2003. Economic impact of viroid diseases. In: Hadidi, A., Flores, R., Randles, J.W., Semancik, J.S. (Eds.), Viroids. CSIRO Publishing, Collingwood, VIC, pp. 3–11.

Randles, J.W., Boccardo, G., Imperial, J.S., 1980. Detection of the cadang-cadang associated RNA in African oil palm and buri palm. Phytopathology 70, 185–189.

Randles, J.W., Boccardo, G., Retuerma, M.L., Rillo, E.P., 1977. Transmission of the RNA species associated with cadang-cadang of coconut palm and insensitivity of the disease to antibiotics. Phytopathology 67, 1211–1216.

Randles, J.W., Palukaitis, P., 1979. *In vitro* synthesis and characterization of DNA complementary to cadang-cadang-associated RNA. J. Gen. Virol. 43, 649–662.

Randles, J.W., Rillo, E.P., Diener, T.O., 1976. The viroid-like structure and cellular location of anomalous RNA associated with the cadang-cadang disease. Virology 74, 128–139.

Randles, J.W., Rodriguez, M.J.B., 2003. Coconut cadang-cadang viroid. In: Hadidi, A., Flores, R., Randles., J.W., Semancik, J.S. (Eds.), Viroids. CSIRO Publishing, Collingwood, VIC, pp. 233–241.

Randles, J.W., Rodriguez, J.M.B., Hanold, D., Vadamalai, G., 2009. Coconut cadang-cadang viroid infection of African oil palm. Planter 85, 93–101.

Rillo, E.P., Rillo, A.R. 1981. Abstracts on the cadang-cadang disease of coconut 1937–1980. Philippine Coconut Authority, Quezon City, Philippines, 50pp.

Rodriguez, M.J.B., 1993. Molecular variation in coconut cadang-cadang viroid (CCCVd). PhD Thesis, The University of Adelaide.

Rodriguez, M.J.B., 2011. The nature of the cadang-cadang disease of coconut in the Philippines and review of the R&D programme: strategies and accomplishments. Report, APCC/MCD & JED/CRI consultative meeting on phytoplasma/wilt diseases in coconut. Lunuwila, Sri Lanka 15–17 June 2011, pp. 140–159.

Rodriguez, M.J.B., Randles, J.W., 1993. Coconut cadang-cadang viroid (CCCVd) mutants associated with severe disease vary in both the pathogenicity domain and the central conserved region. Nucleic Acids Res. 21, 2771.

RSPO, 2015. Roundtable on Sustainable Palm Oil. www.rspo.org/files/pdf/Factsheet-RSPO-AboutPalmOil.pdf.

Schnölzer, M., Haas, B., Ramm, K., Hofmann, H., Sänger, H.L., 1985. Correlation between structure and pathogenicity of potato spindle tuber viroid (PSTV). EMBO J. 4, 2181–2190.

Schumacher, J., Randles, J.W., Riesner, D., 1983. A two-dimensional electrophoretic technique for the detection of circular viroids and virusoids. Anal. Biochem. 135, 288–295.

Selvaraja, S., Balasundram, S.K., Vadamalai, G., Husni, M.H.A., 2012. Spatial variability of orange spotting disease in oil palm. J. Biol. Sci. 12, 232–238.

Thanarajoo, S.S., 2014. Rapid detection, accumulation and translocation of coconut cadang-cadang viroid (CCCVd) variants in oil palm. PhD Thesis, University Putra Malaysia.

Turner, P.D., 1981. Oil Palm Diseases and Disorders. Oxford University Press, Kuala Lumpur.

Vadamalai, G., 2005. An investigation of oil palm orange spotting disorder. PhD Thesis, The University of Adelaide.

Vadamalai, G., Hanold, D., Rezaian, M.A., Randles, J.W., 2006. Variants of the coconut cadang-cadang viroid isolated from an African oil palm (*Elaeis guineensis* Jacq.) in Malaysia. Arch. Virol. 151, 1447–1456.

Vadamalai, G., Perera, A.A.F.L.K., Hanold, D., Rezaian, M.A., Randles, J.W., 2009. Detection of coconut cadang-cadang viroid sequences in oil and coconut palm by ribonuclease protection assay. Ann. Appl. Biol. 154, 117–125.

Visvader, J.E., Symons, R.H., 1985. Eleven new sequence variants of citrus exocortis viroid and the correlation of sequence with pathogenicity. Nucleic Acids Res. 13, 2907–2920.

Wall, G.C., Randles, J.W., 2003. Coconut tinangaja viroid. In: Hadidi, A., Flores, R., Randles., J.W., Semancik, J.S. (Eds.), Viroids. CSIRO Publishing, Collingwood, VIC, pp. 242–245.

Wassenegger, M., Spieker, R.L., Thalmeir, S., Gast, F.U., Riedel, L., Sänger, H.L., 1996. A single nucleotide substitution converts potato spindle tuber viroid (PSTVd) from a non-infectious to an infectious RNA for *Nicotiana tabacum*. Virology 226, 191–197.

Wu, Y.H., Cheong, L.C., Meon, S., Lau, W.H., Kong, L.L., Joseph, H., et al., 2013. Characterization of coconut cadang-cadang viroid variants from oil palm affected by orange spotting disease in Malaysia. Arch. Virol. 158, 1407–1410.

Zelazny, B., 1980. Ecology of cadang-cadang disease of coconut palm in the Philippines. Phytopathology 70, 700–703.

Zelazny, B., Pacumbaba, E., 1982. Incidence of cadang-cadang disease of coconut palm in the Philippines. Plant Dis. 66, 547–549.

Zelazny, B., Randles, J.W., Boccardo, G., Imperial, J.S., 1982. The viroid nature of the cadang-cadang disease of coconut palm. Scientia Filipinas. 2, 46–63.

PART II

VIROID CHARACTERISTICS

CHAPTER

5

Viroid Biology

Dijana Škorić
University of Zagreb, Zagreb, Croatia

INTRODUCTION

Viroids were the first small circular RNA molecules recognized in the history of biology as noncoding independently replicating genomes (Diener, 1971). With the increasing set of circular RNAs discovered in all domains of life, they are no longer such an oddity (Lasda and Parker, 2014) but still remain unique in several aspects including autonomous replication in plant hosts and pathogenicity. Although viroid discovery is linked to plant disturbances previously supposed to be of viral etiology (Diener, 1971), our knowledge of viroids latently infecting plants has increased considerably over the last few decades (Flores et al., 2011, 2015; Singh et al., 2003). The application of next-generation sequencing technologies looking into the global RNA content and the expansion of the circular RNA research field will inevitably identify new viroid species and hosts (Barba et al., 2014; Barba and Hadidi, 2015; Hadidi et al., 2016; Zhang et al., 2014). Nevertheless, proving the infectious nature of a putative viroid, through its transmissibility to a certain plant host range, and investigating symptoms and other features demonstrating the ability of this minimal RNA replicon to exert biological functions, will remain important in viroid research (Di Serio et al., 2014). As other chapters of this book deal with some aspects of viroid biology in more detail, this chapter attempts to briefly introduce the main concepts, and highlights some of the latest developments, relying extensively on previous reviews (Diener, 1987; Flores et al., 2011, 2015; Singh et al., 2003).

Viroids and Satellites.
DOI: http://dx.doi.org/10.1016/B978-0-12-801498-1.00005-X

HOST RANGE

Apart from the *Saccharomyces cerevisiae* experimental system supporting avocado sunblotch viroid (ASBVd) replication (Delan-Forino et al., 2011) and the ability of *Chlamydomonas reinhardtii* chloroplasts to process dimeric RNAs of ASBVd and two other members of the family *Avsunviroidae* (Molina-Serrano et al., 2007), angiosperms are the only confirmed natural hosts of viroids (Singh et al., 2003). A notable biological difference between *Avsunviroidae* and *Pospiviroidae,* the two existing viroid families (Di Serio et al., 2014; see also Chapter 13: Viroid Taxonomy), is the host range reflecting diverse structural and functional features of their members (reviewed in Flores et al., 2015). Interestingly, plastid-replicating members of the family *Avsunviroidae* infect only dicotyledonous plants within a range limited to the natural host and a few related species, while nuclear-replicating members of the family *Pospiviroidae* can infect dicots and monocots. Members in the genera *Pospiviroid* and *Hostuviroid* have particularly broad host ranges infecting woody and herbaceous hosts with new hosts often being found (Matsushita and Tsuda, 2015; Singh and Dilworth, 2009; Singh et al., 2003). Members of the genus *Apscaviroid* naturally infect a broad range of woody plants; however, experimental infection of herbaceous hosts has been demonstrated by mechanical inoculation of purified viroid preparations for pear blister canker viroid in cucumber (Flores et al., 1991) and Australian grapevine viroid in cucumber and tomato (Rezaian et al., 1988). Recently, the use of an apple scar skin viroid (ASSVd) dimeric infectious clone for agroinoculation, and mechanical inoculation of dimeric plasmids, transcripts, or sap, resulted in finding nine new herbaceous hosts (Walia et al., 2014). Conversely, coleviroids and most cocadviroids exemplify constituents of the family *Pospiviroidae* with narrow host ranges. Only a small number of viroids have been found in monocots, specifically in palms (Hadidi et al., 2003). Whether this uneven distribution of plant host types is a matter of research bias imposed by agricultural practices or for biological reasons remains to be determined.

Viroid host definition is a consequence of a viroid's ability to replicate, accumulate, and systemically invade a plant (Flores et al., 2011, 2015). This can be radically affected by a single nucleotide change as demonstrated by the ability of potato spindle tuber viroid (PSTVd) to infect *Nicotiana tabacum* (Wasseneger et al., 1996). As with RNA viruses, viroids exist as quasispecies of closely related sequence variants with differential properties. Proving the host's role in selecting out particular sequence variants and thus shaping the viroid population structure up to a point of having biologically different host and tissue specific variants was first demonstrated for citrus exocortis viroid and ASBVd (Semancik and Szychowski, 1994; Semancik et al., 1993). Clonally propagated hosts, especially long-living citrus and grapevine plants, often harbor several viroids belonging to different genera of the family *Pospiviroidae*, or even combinations of viruses and viroids (Serra et al., 2014; Singh et al., 2003; Vernière et al., 2006). This provides a system for interactions that can affect the evolution of viroid species (e.g., due to recombination between different viroids) and symptom modifications ranging from fading or disappearance of symptoms in the case of preinoculation with a mild viroid (cross-protection) to exacerbation (synergism) as described for viruses (Flores et al., 2011).

SYMPTOMS

Viroid infection may or may not induce symptoms, the result of complex viroid–host–environment interactions. Plant regulatory networks and defense systems, where RNA silencing probably plays a major role in symptom development, combine with viroid structural elements in the expression of differential plant responses (Di Serio et al., 2013; Flores et al., 2015). Viroid and virus symptoms are similar in respect to the plant structures, organelles, and metabolic steps affected, but they are induced quite differently because the viroids do not code for proteins and their effects depend on the structure of their genomic RNA or some derivative thereof. Stunting is typically the most conspicuous viroid symptom at the whole plant level but dramatic alterations can be observed on the leaves (distortion, epinasty, rugosity, chloroses of various types, necrosis), stems (shortening, thickening), bark (cracking, scaling, gumming, pitting, and pegging), tubers (malformation), reproductive organs like flowers, fruits (color and shape alterations, reduction in numbers), and seeds (abortion, enlarged stones). The symptom intensity may vary with the host and the sequence of the viroid (Hadidi et al., 2003). Symptomless viroid hosts are apparently more common than previously suggested (Matsushita and Tsuda, 2015; Singh and Dilworth, 2009; Singh et al., 2006; Zhang et al., 2014). This may turn into a problem if mechanical transmission occurs from symptomless viroid reservoirs to sensitive hosts as in greenhouse growing of tomatoes or ornamentals. Symptomatic plants are often termed sensitive and some are suitable as bioassay hosts, especially herbaceous ones. Their cultivation is easier and the bioassay usually lasts for weeks, as opposed to months or years for woody hosts. Even for viroids infecting woody plants, such as ASSVd, which are considered to be difficult or not transmissible mechanically to herbaceous hosts, continuing the search for new herbaceous hosts would be advantageous. Besides the host and its cultivar, the growing conditions, especially light intensity and temperature, may affect symptom development (Walia et al., 2014). High temperatures and light intensity have been considered favorable for severe symptom expression and possibly explain why viroids are important in (sub)tropical and greenhouse production systems (Diener, 1987; Flores et al., 2011).

CYTOPATHOLOGY

The pathogenicity of viroids can also be observed at the subcellular level. Cytopathic effects of members of the family *Pospiviroidae* include invaginations of the plasmalemma (paramural bodies or plasmalemmasomes), irregularities and thickening of cell walls; however, the role of the former in pathogenesis has not been clarified (Diener, 1987; Di Serio et al., 2013). Chloroplast malformations were demonstrated in infections by members of both viroid families, but have been well studied only for peach latent mosaic viroid (PLMVd) leading to a model where extreme peach albinism symptoms are associated with chloroplast defects and linked to specific events in the RNA silencing process and maturation of plastid ribosomal RNAs (Rodio et al., 2007). This study combined biochemical, molecular, and ultrastructural approaches to understand the links between cellular and tissue disturbances and the onset of macroscopic symptoms. Further studies using the ever developing set of tools and enabling the integration of knowledge on cytopathic effects, plant regulatory networks, and biochemical pathways represent a promising approach for understanding somewhat neglected viroid cytopathic effects and the complexity of plant–viroid interactions in general (Di Serio et al., 2013).

TRANSMISSION

Horizontal transmission is the most important dispersal mode for the majority of viroids. Contaminated farming implements, machinery, tools (e.g., pruning scissors, grafting knives), even hands, gloves, and clothes can be instrumental in viroid mechanical transmission. Contemporary agriculture relies on large monoculture production areas wherein the high density of genetically uniform plants can facilitate viroid transmission from infected source plants. These agents may be introduced from a distant infection focus by international trade. If the first reports of viroid diseases from the beginning of the 20th century are viewed in light of modern agriculture practices, they leave little doubt about their anthropogenic origin (Diener, 1987; Flores et al., 2011). The inoculum sources are frequently leaves and stems. Tubers, flowers, fruits, and seeds can also be important, depending on a viroid–host combination as viroids are found in virtually all plant organs and tissue types (Di Serio et al., 2010; Singh et al., 2003). Consequently, grafting, vegetative, or in vitro plant propagation are effective modes of viroid transmission also used in experimental procedures. However, mechanical inoculations by stem slashing, stem puncturing, or leaf rubbing are more frequently used (Hadidi et al., 2003).

Recently, a few reports support the possibility of viroid transmission through roots as implied long ago for hop stunt viroid and ASSVd (Hadidi et al., 2003). Transmission could be achieved directly by root anastomoses or indirectly by acquiring a viroid via root system from a substrate. Direct transmission of PSTVd from tomato root phloem to a holoparasitic plant *Phelipanche ramosa* (syn. *Orobanche ramosa*) corroborates the root connection as a possible transmission mode (Vachev et al., 2010). However, its unidirectionality, probably due to the lack of viroids in *P. ramosa* tubercles, still provides no evidence for suggesting its more significant impact on PSTVd epidemiology. The overwintering of tomato chlorotic dwarf viroid (TCDVd) in infected *Vinca minor* roots, and of chrysanthemum stunt viroid in infected *V. major* roots, could result in mechanical transmission of viroid from groundcover to neighboring plants during the spring growth flush (Singh and Dilworth, 2009).

The uptake of PSTVd by roots from hydroponic systems has been addressed in tomato and tomato/potato experimental systems (Mehle et al., 2014). PSTVd was released into the nutrient solution from injured roots and remained infectious for up to 7 weeks. Water-mediated PSTVd transmission to tomato plants cultivated hydroponically was possible, although with low efficiency. Nevertheless, more efficient mechanical transmission of PSTVd (Singh et al., 2003) could be envisaged from these source plants. The increase in hydroponic and greenhouse production, the use of recycled water, and the demonstrated longevity of viroids therein (Mehle et al., 2014) should be reasons enough to start considering the aqueous route as an alternative transmission route, especially in viroid disease control schemes.

Reports of insect transmission are still inconsistent and limited to PSTVd and a few other pospiviroids. Tomato planta macho viroid was the first viroid confirmed to be persistently transmitted from wild reservoirs (e.g., *Physalis foetens*) to tomato by *Myzus persicae*. On the other hand, PSTVd transmission to potato and tomato by the same aphid was found to be dependent on transencapsidation with potato leafroll virus, while similar systems potentially important in PSTVd epidemiology are yet to be addressed (Diener, 1987; Singh et al., 2003). Tomato apical stunt viroid (TASVd) and TCDVd were recently transmitted in greenhouse tomatoes by the pollinating activity of the bumblebees, *Bombus terrestris* and *Bombus ignites*, respectively (Antignus et al., 2007; Matsuura et al., 2010). These two viroids could have been transmitted either mechanically in crude sap on bumblebee mouthparts, or, more likely, via viroid contaminated pollen carried by the insects. Later work reporting the absence of *B. terrestris* PSTVd transmission from *Petunia* sp. to tomato (Nielsen et al., 2012) suggests that effective pollination needs to occur for viroid transmission in this system. The same report demonstrates that there was no PSTVd transmission to tomato either from *Brugmansia* sp. by western honey bee (*Apis mellifera*) or from *Solanum jasminoides* by the polyphagous thrips species, *Thrips tabaci* and *Frankliniella occidentalis*, even though the insects carried small quantities of PSTVd. Conversely, PSTVd and TASVd were detected in the stylets and foreguts of 20%–40% of *M. persicae* adults while no viroids were localized in the aphid embryos (Van Bogaert et al., 2015). Regardless, the preliminary transmission experiments in tomato, *Physalis* sp., and *Nicotiana benthamiana* have not established the aphid's ability to transmit these viroids. These new lines of investigation should help in determining the exact mode(s) of viroid insect transmission and their role in the spread of viroid diseases.

The vertical transmission of viroids has been well documented historically with the seed and pollen transmission reports available for members of both viroid families (Flores et al., 2011; Hadidi et al., 2003). Within the family *Avsunviroidae*, only ASBVd and PLMVd are pollen-borne and -transmissible. Both are seed-borne (detected in seed structures) but only ASBVd is seed-transmissible (detected in seedlings produced from infected seeds). The inability of PLMVd to be transmitted in seed is assumed to be due to its absence from the embryo (Barba et al., 2007).

The continuing research described below has provided some clues into the largely unknown mechanisms underlying pollen and seed transmission. Whereas RNA silencing blocks the PSTVd invasion of developing *N. benthamiana* flower and meristem structures (Di Serio et al., 2010), this situation is avoided in the PLMVd-infected peach, due to bypassing the RNA surveillance system and is possibly correlated with the plastid localization of PLMVd where no RNA silencing has been reported yet (Rodio et al., 2007). The dissimilarity observed between PSTVd and PLMVd as members of different families was observed even within the family *Pospiviroidae* in different pospiviroid-host combinations. The ubiquitous PSTVd distribution in tomato floral organs was demonstrated and its seed transmissibility confirmed (Matsushita et al., 2011). Conversely, the closely related TCDVd could neither be localized in tomato ovules and placenta, nor be seed-transmitted (Matsushita et al., 2011) as previously reported (Singh and Dilworth, 2009), suggesting that there is a complex interplay between viroid variants and a host cultivar's defense mechanisms determining the patterns of reproductive organ invasion and seed transmissibility. Histochemical data from the PSTVd–petunia system showed indirect PSTVd delivery to the embryo through ovule or pollen before embryogenesis (Matsushita and Tsuda, 2014). These data indicate that the route and time of embryo invasion could be critical determinants in successful seed transmission, but how viroids evade plant defense mechanisms to achieve seed and pollen transmission remains to be investigated.

The seed transmission efficiency of different viroids varies considerably (Flores et al., 2011; Singh et al., 2003), but even for a highly seed transmissible viroid like coleus blumei viroid 1 the 0%–100% range has been reported to depend on the host cultivar (Chung and Choi, 2008). Moreover, seed transmission of multiple viroids is possible in one host, as shown for five viroids in grapevine by the application of the most sensitive detection methods available at the time (Wan Chow Wah and Symons, 1999). This approach has also enabled linking a PSTVd outbreak in tomato to imported seeds (Van Brunschot et al., 2014). The latter may become an increasingly important issue in the epidemiology of quarantinable pospiviroids infecting annual hosts whose cultivation relies heavily on seed propagules.

Acknowledgments

This work is partially supported by the University of Zagreb grant no. 2028124.

References

Antignus, Y., Lachman, O., Pearlsman, M., 2007. Spread of tomato apical stunt viroid (TASVd) in greenhouse tomato crops is associated with seed transmission and bumble bee activity. Plant Dis. 91, 47–50.

Barba, M., Czosnek, H., Hadidi, A., 2014. Historical perspective, development and applications of next-generation sequencing in plant virology. Viruses 6, 106–136.

Barba, M., Hadidi, A., 2015. An overview of plant pathology and application of next-generation sequencing technologies. CAB Rev. 10, 1–21.

Barba, M., Ragozzino, E., Faggioli, F., 2007. Pollen transmission of peach latent mosaic viroid. J. Plant Pathol. 89, 287–289.

Chung, B.N., Choi, G.S., 2008. Incidence of coleus blumei viroid 1 in seeds of commercial *Coleus* in Korea. Plant Pathol. J. 24, 305–308.

Delan-Forino, C., Maurel, M.C., Torcet, C., 2011. Replication of avocado sunblotch viroid in the yeast *Saccharomyces cerevisiae*. J. Virol. 85, 3229–3238.

Di Serio, F., De Stradis, A., Delgado, D., Flores, R., Navarro, B., 2013. Cytopathic effects incited by viroid RNAs and putative underlying mechanisms. Front. Plant Sci. 3, 288.

Di Serio, F., Flores, R., Verhoeven, J.Th.J., Li, S.-F., Pallás, V., et al., 2014. Current status of viroid taxonomy. Arch. Virol. 159, 3467–3478.

Di Serio, F., Martínez de Alba, A.-E., Navarro, B., Gisel, A., Flores, R., 2010. RNA-dependent RNA polymerase 6 delays accumulation and precludes meristem invasion of a viroid that replicates in the nucleus. J. Virol. 84, 2477–2489.

Diener, T.O., 1971. Potato spindle tuber "virus": IV. A replicating, low molecular weight RNA. Virology 45, 411–428.

Diener, T.O., 1987. Biological properties. In: Diener, T.O. (Ed.), The Viroids. Plenum Press, New York, NY, pp. 9–35.

Flores, R., Di Serio, F., Navarro, B., Duran-Vila, N., Owens, R.A., 2011. Viroids and viroid diseases of plants. In: Hurst, C.J. (Ed.), Studies in Viral Ecology, Volume One: Microbial and Botanical Host Systems. Wiley-Blackwell, Chichester, England, UK, pp. 311–346.

Flores, R., Hernández, C., Llácer, G., Desvignes, J.C., 1991. Identification of a new viroid as the putative causal agent of pear blister canker disease. J. Gen. Virol. 72, 1199–1204.

Flores, R., Minoia, S., Carbonell, A., Gisel, A., Delgado, S., López-Carrasco, A., et al., 2015. Viroids, the simplest RNA replicons: how they manipulate their hosts for being propagated and how their hosts react for containing the infection. Virus Res. 209, 136–145.

Hadidi, A., Flores, R., Candresse, T., Barba, M., 2016. Next-generation sequencing and genome editing in plant virology. Front. Microbiol. 7, 1325.

Hadidi, A., Flores, R., Randles, J.W., Semancik, J.S. (Eds.), 2003. Viroids. CSIRO Publishing, Collingwood, VIC.

Lasda, E., Parker, R., 2014. Circular RNAs: diversity of form and function. RNA. 20, 1829–1842.

Matsushita, Y., Tsuda, S., 2014. Distribution of potato spindle tuber viroid in reproductive organs of petunia during its developmental stages. Phytopathology 104, 964–969.

Matsushita, Y., Tsuda, S., 2015. Host ranges of potato spindle tuber viroid, tomato chlorotic dwarf viroid, tomato apical stunt viroid, and columnea latent viroid in horticultural plants. Eur. J. Plant Pathol. 141, 193–197.

Matsushita, Y., Usugi, T., Tsuda, S., 2011. Distribution of tomato chlorotic dwarf viroid in floral organs of tomato. Eur. J. Plant Pathol. 130, 441–447.

Matsuura, S., Matsushita, Y., Kozuka, R., Shimizu, S., Tsuda, S., 2010. Transmission of tomato chlorotic dwarf viroid by bumblebees (*Bombus ignitus*) in tomato plants. Eur. J. Plant Pathol. 126, 111–115.

Mehle, N., Gutiérrez-Aguirre, I., Prezelj, N., Delić, D., Vidic, U., Ravnikar, M., 2014. Survival and transmission of potato virus Y, pepino mosaic virus, and potato spindle tuber viroid in water. Appl. Environ. Microbiol. 80, 1455–1462.

Molina-Serrano, D., Suay, L., Salvador, M.L., Flores, R., Daròs, J.A., 2007. Processing of RNAs of the family *Avsunviroidae* in *Chlamydomonas reinhardtii* chloroplasts. J. Virol. 81, 4363–4366.

Nielsen, S.L., Enkegaard, A., Nicolaisen, M., Kryger, P., Virščeк Marn, M., Mavrič Pleško, I., et al., 2012. No transmission of potato spindle tuber viroid shown in experiments with thrips (*Frankliniella occidentalis, Thrips tabaci*), honey bees (*Apis mellifera*) and bumblebees (*Bombus terrestris*). Eur. J. Plant Pathol. 133, 505–509.

Rezaian, M.A., Koltunow, A.M., Krake, L.R., 1988. Isolation of three viroids and a circular RNA from grapevines. J. Gen. Virol. 69, 413–422.

Rodio, M.E., Delgado, S., De Stradis, A.E., Gómez, M.D., Flores, R., Di Serio, F., 2007. A viroid RNA with a specific structural motif inhibits chloroplast development. Plant Cell. 19, 3610–3626.

Semancik, J.S., Szychowski, J.A., 1994. Avocado sunblotch disease: a persistent viroid infection in which variants are associated with differential symptoms. J. Gen. Virol. 75, 1543–1549.

Semancik, J.S., Szychowski, J.A., Rakowski, A.G., Symons, R.H., 1993. Isolates of citrus exocortis viroid recovered by host and tissue selection. J. Gen. Virol. 74, 2427–2436.

Serra, P., Hashemian, S.M.B., Fagoaga, C., Romero, J., Ruiz-Ruiz, S., Gorris, M.T., et al., 2014. Virus-viroid interactions: citrus tristeza virus enhances the accumulation of citrus dwarfing viroid in Mexican lime via virus-encoded silencing suppressors. J. Virol. 88, 1394–1397.

Singh, R.P., Dilworth, A.D., 2009. Tomato chlorotic dwarf viroid in the ornamental plant *Vinca minor* and its transmission through tomato seed. Eur. J. Plant Pathol. 123, 111–116.

Singh, R.P., Dilworth, A.D., Baranwal, V.K., Baranwal, V.K., Gupta, K.N., 2006. Detection of citrus exocortis viroid, Iresine viroid, and tomato chlorotic dwarf viroid in new ornamental host plants in India. Plant Dis. 90, 1457.

Singh, R.P., Ready, K.F.M., Nie, X., 2003. Biology. In: Hadidi, A., Flores, R., Randles, J.W., Semancik, J.S. (Eds.), Viroids. CSIRO Publishing, Collingwood, VIC, pp. 30–48.

Vachev, T., Ivanova, D., Minkov, I., Tsagris, M., Gozmanova, M., 2010. Trafficking of the potato spindle tuber viroid between tomato and *Orobanche ramosa*. Virology 399, 187–193.

Van Bogaert, N., De Jonghe, K., Van Damme, E.J.M., Maes, M., Smagghe, G., 2015. Quantitation and localization of pospiviroids in aphids. J. Virol. Meth. 211, 51–54.

Van Brunschot, S.L., Verhoeven, J.Th.J., Persley, D.M., Geering, A.D.W., Drenth, A., et al., 2014. An outbreak of potato spindle tuber viroid in tomato is linked to imported seed. Eur. J. Plant Pathol. 139, 1–7.

Vernière, C., Perrier, X., Dubois, C., Dubois, A., Botella, L., Chabrier, C., et al., 2006. Interactions between citrus viroids affect symptom expression and field performance of clementine trees grafted on trifoliate orange. Phytopathology 96, 356–368.

Walia, Y., Dhir, S., Ram, R., Zaidi, A.A., Hallan, V., 2014. Identification of the herbaceous host range of apple scar skin viroid and analysis of its progeny variants. Plant Pathol. 63, 684–690.

Wan Chow Wah, Y.F., Symons, R.H., 1999. Transmission of viroids via grape seeds. J. Phytopathol. 147, 285–291.

Wasseneger, M., Spieker, R.L., Thalmeir, S., Gast, F.U., Riedel, L., Sänger, H.L., 1996. A single nucleotide substitution converts potato spindle tuber viroid (PSTVd) from noninfectious to an infectious RNA for *Nicotiana tabacum*. Virology 226, 191–197.

Zhang, Z., Qi, S., Tang, N., Zhang, X., Chen, S., Zhu, P., et al., 2014. Discovery of replicating circular RNAs by RNA-Seq and computational algorithms. PLoS Pathog. 10, e1004553.

CHAPTER

6

Viroid Structure

Gerhard Steger[1,*], *Detlev Riesner*[1], *Marie-Christine Maurel*[2] *and Jean-Pierre Perreault*[3,*]

[1]Heinrich Heine University Düsseldorf, Germany
[2]Sorbonne University, Paris, France
[3]University of Sherbrooke, Sherbrooke, QC, Canada

INTRODUCTION

The elucidation of the structures adopted by viroids is paramount for understanding the different mechanisms involved in their replication, pathogenesis, and transport. In general, the secondary structures of viroids have been predicted using computer software. However, the characterization of biological structures in solution (in vitro) and in the cell (in vivo) is obviously more important for the elucidation of the structure–function relationships of a viroid. Recent advances have now revealed the structure in solution of several viroid species (Steger and Perreault, 2016).

STRUCTURE OF THE *POSPIVIROIDAE*

The native structure of circular forms of members of the family *Pospiviroidae* is generally described as rod-like without any bifurcation (Fig. 6.1A). This structure consists of five domains that are described below:

The terminal left (TL) domain of most pospi- and apscaviroids (type members are potato spindle tuber viroid, PSTVd, and apple scar skin viroid, respectively) contains an imperfect repeat that could form either a rod-like or a Y-shaped (Fig. 6.1B) structure. However, the rod-like

*Both authors contributed equally.

Viroids and Satellites.
DOI: http://dx.doi.org/10.1016/B978-0-12-801498-1.00006-1

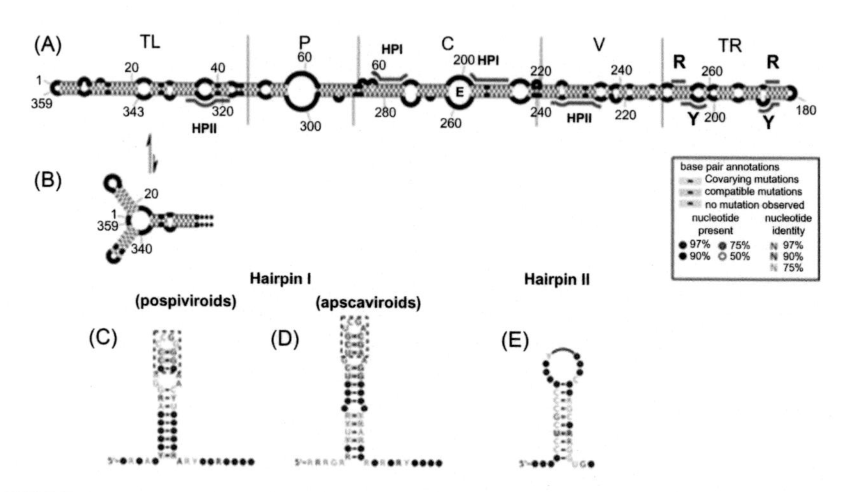

FIGURE 6.1 Secondary structure of the *Pospiviroidae*. (A) Consensus secondary structure scheme of potato spindle tuber viroid. The color code used for the annotation of base-pairs is given in the box. Borders of the five domains (see text) are marked by lines. The loop E is labeled with the letter E. The nucleotides involved in formation of the extrastable hairpins I and II (HPI, HPII) are underlined. RY motifs in the TR critical for binding of the viroid RNA-binding protein VirP1 are outlined and marked by R and Y (Gozmanova et al., 2003). (B) The suboptimal Y-shaped alternative of the TL domain is shown. (C–E) Nucleotides forming the extrastable HPI and HPII in thermodynamically metastable structures (Y = U or C; R = A or G).

structure is thermodynamically preferred (Dingley et al., 2003). PSTVd is transcribed by the host-encoded DNA-dependent RNA polymerase II redirected to accept RNA templates. A single start site for (-) strands was identified in the TL hairpin loop, either at position U_{359} or at C_1, in an in vitro transcription assay (Kolonko et al., 2006).

The pathogenicity (P) domain contains an oligopurine stretch in the upper strand and the corresponding oligopyrimidine stretch in the lower strand of most pospiviroids (Fig. 6.1A). The resulting region has a relatively low thermodynamic stability; hence the name "premelting" region. Furthermore, after sequencing the first severe and mild variants of PSTVd (Gross et al., 1981), it was realized that slight sequence variations in this region influence the pathogenicity of the viroid. It should be noted that pathogenicity-modulating mutations and/or regions are not restricted to this domain.

The central (C) domain of members of the family *Pospiviroidae* is a highly conserved region (Fig. 6.1A). In addition to the rod-like structure with base-pairs between the upper strand (upper central conserved region, UCCR) and the lower strand (lower central conserved region), both strands can form hairpins. In pospiviroids, hostuviroids (type member hop stunt viroid) and cocadviroids (type member coconut cadang-cadang viroid, CCCVd) the UCCR strand can form a thermodynamically stable hairpin I (HPI) with a basal stem of nine base-pairs during thermal denaturation (Fig. 6.1C). In apscaviroids, the basal stem of HPI consists of two helices separated by a mismatch (Fig. 6.1D). In the center of the C domain of many pospiviroids lies a particular internal loop (positions 98–102 and 255–260 in PSTVd; Fig. 6.1A) similar to loop E of eukaryotic 5S RNA, in which all nucleotides are involved in non-Watson-Crick base-pairings.

Longer-than-unit-length (+)-strand-transcripts of PSTVd that contain the UCCR twice, i.e., at the 5′- and 3′-termini, can fold into a metastable multihelix junction. This junction can be cleaved in a potato nuclear extract between G_{95} and G_{96} of (+) PSTVd, within a stem with a GNRA tetraloop. A local conformational change switches the tetraloop motif into a loop E motif; after the second cleavage between G_{95} and G_{96}, close to the 3′-terminus, both terminal nucleotides are base-paired in an optimal juxtaposition for ligation (Baumstark et al., 1997; Schrader et al., 2003). From experiments in vivo with transgenic lines of *Arabidopsis thaliana*, Gas et al. (2007) proposed an alternative model wherein the replication intermediate in a trihelical structure of two UCCRs (Steger et al., 1986) is cleaved into monomers, presumably by a class III RNase (Gas et al., 2008). The activity of the trihelical structure however, could not be found in in vitro experiments.

The variable (V) domain is the most varying one, showing low sequence similarity between otherwise closely related viroids (Fig. 6.1A). The boundaries of the V domain have been defined by a change from low sequence similarity to the adjacent C and terminal right (TR) domains (Keese and Symons, 1985).

By sequence duplications of the TR domain, the smallest CCCVd, termed CCCVd-1fast or -1small (246/247 nt), gives rise to longer molecules, termed CCCVd-1slow or -1long (287–301 nt). PSTVd moves from cell to cell via plasmodesmata (Ding et al., 1997); distinct structural motifs that presumably interact with specific cellular factors are required for movement across various cellular boundaries to achieve systemic trafficking (Takeda et al., 2011). The structure or stability of the TR stem-loop is essential for cell-to-cell and/or long-distance movement (Hammond, 1994).

ALTERNATIVE STRUCTURES WITHIN THE *POSPIVIROIDAE*

From optical melting curves and kinetic studies it was concluded that members of this family denature in a highly cooperative transition, in which all native base-pairs are dissociated and stable new hairpins are formed. This cooperativity is based on the formation of stable hairpins that are not part of the rod-like structure. Common to all members of the family *Pospiviroidae*, HPI is formed from sequences located in the UCCR of the rod-like conformation. Hairpin II (HPII, Fig. 6.1E), unique to PSTVd and closely related viroids (pospiviroids), is a GC-rich hairpin of up to 12 base-pairs (Riesner et al., 1979). The formation of a thermodynamically metastable structure including HPII is critical for infectivity. In particular HPII acts as a functional element of the (−)-strand replication intermediate. Its presence in vivo in the (−)-strand could be demonstrated (Schröder and Riesner, 2002).

STRUCTURE OF THE *AVSUNVIROIDAE*

Peach Latent Mosaic Viroid (PLMVd): A Model for Structure Determination

The stable secondary structure of PLMVd in solution was determined over the years for several sequence variants and using various probing techniques including more recently RNA-selective 2′-hydroxyl acylation analyzed by primer extension (e.g., Bussière et al., 2000; Dubé et al, 2011). All these structures are in good agreement and indicate that PLMVd folds into a complex branched secondary structure with differences associated with sequence variations. One of these structures, containing 14 helices and stem-loops, P1−P14, is illustrated in Fig. 6.2. The left domain, including the P1 and P11 to P14 stem-loop structures, is characterized by significant differences depending on the sequence variants. This region includes a cruciform structure as well as the second pseudoknot P14 in some variants that are absent in others (Ambrós et al., 1998). The cruciform structure corresponds to the formation of the stem II of the hammerhead self-cleaving structure of the lower strand, and to its counterpart in the upper strand (i.e., hairpins P12 and P13). The pseudoknot P14 is formed by base-pairing nucleotides from both the hairpin loops of P1 and P11. These two motifs are not essential and are independent of each other. Conversely, the right-hand domain, which is composed of the P2−P10 stem-loop structures, is virtually identical regardless of the sequence variant probed. This domain includes the P8 pseudoknot that is conserved in all PLMVd variants and seems to allow folding into a very compact form. This pseudoknot has been demonstrated to be essential for PLMVd accumulation *in planta* (Dubé et al., 2010).

Some important motifs within PLMVd have been revealed. For example, positions A_{50} and especially C_{51} were identified as the initiation site in the (+) strand, and U_{284} in the (−) strand as the most important initiation sites of the rolling circle replication (Delgado et al., 2005; Motard

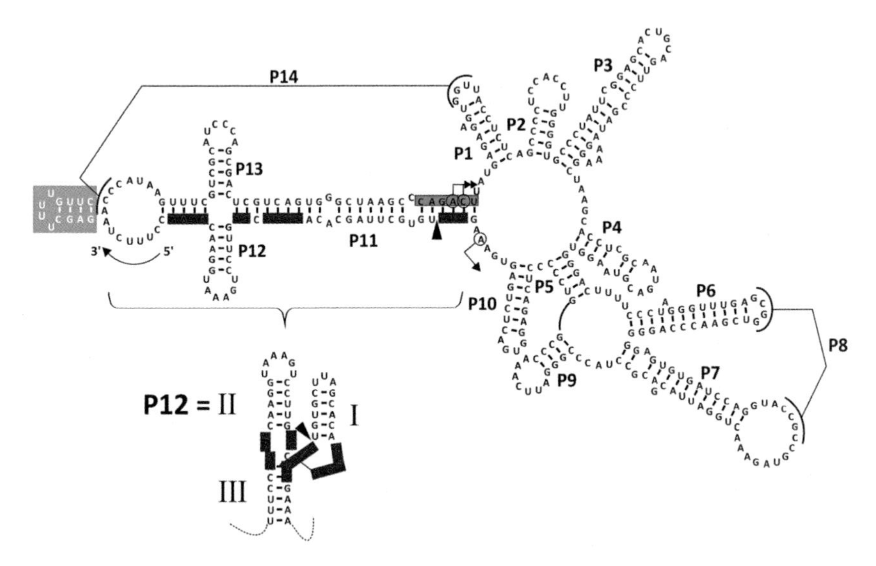

FIGURE 6.2 Nucleotide sequence and secondary structure of a peach latent mosaic viroid variant (+) strand. The arrowhead indicates the hammerhead self-cleavage site and the red boxes identify the core nucleotides of the hammerhead structure. The hammerhead structure can adopt the catalytically active folding with stems I–III shown below. The insertion associated with peach calico and consisting of a short stem-loop found in a few variants is shown in blue. The green boxes highlight a highly conserved CAGAC box that is reminiscent of the sequence found in the vicinity of the replication initiation sites for each polarity (Motard et al., 2008). Finally, the circled nucleotides (A_{50} and especially C_{51} of (+) and U_{284} of (−) strands, respectively) with arrows indicate the replication initiation sites.

et al., 2008). Moreover, the catalytic hammerhead structures, which are alternative, biologically active, conformations of the P11 stem-loop, are responsible for the self-cleavage of the longer-than-unit conformers resulting from polymerization. Folded into its active structure, the hammerhead structure is composed of three stems flanking an apparently unpaired central core that includes highly conserved nucleotides (Fig. 6.2). Finally, an additional stem-loop insertion of 12 or 13 nt has been reported in the left terminal loop of the P11 stem-loop (Fig. 6.2; Malfitano et al., 2003). Its presence during PLMVd infection is responsible for an extensive chlorosis known as peach calico. Viroid small RNAs containing the peach calico insertion direct cleavage by an RNA-induced silencing complex of a specific host mRNA (Navarro et al., 2012).

Secondary Structure of All Members of the Family *Avsunviroidae*

Secondary structures for both polarities of all members of the *Avsunviroidae* have been reported (Fig. 6.3; Giguère et al., 2014). The

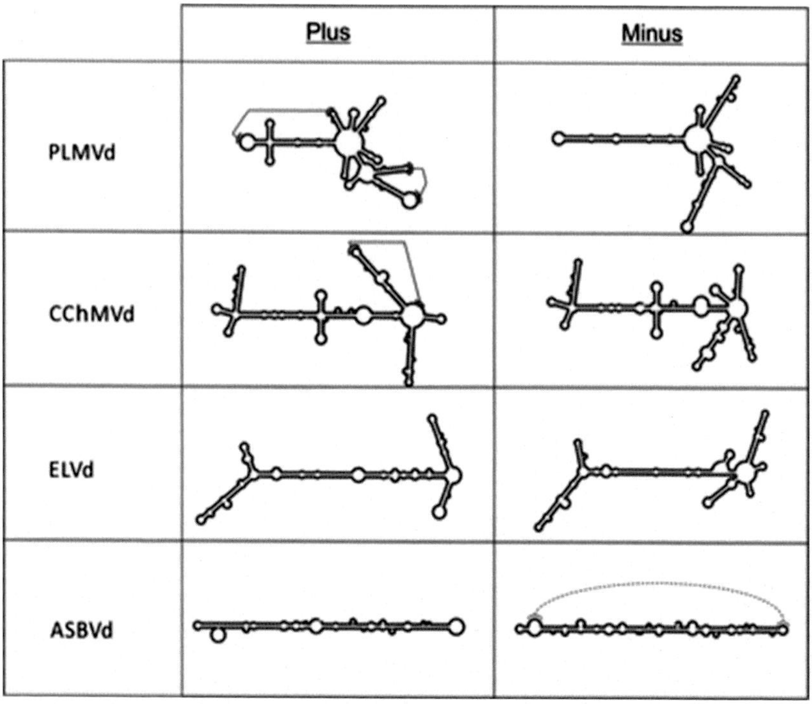

	Plus	Minus
PLMVd		
CChMVd		
ELVd		
ASBVd		

FIGURE 6.3 Structures of viroids of the family *Avsunviroidae*. Schematic representation of the most stable structures of both polarities of each member of the family *Avsunviroidae*. PLMVd, peach latent mosaic viroid; CChMVd, chrysanthemum chlorotic mottle viroid; ELVd, eggplant latent viroid; ASBVd, avocado sunblotch viroid. The gray and dotted lines mark proposed pseudoknots.

PLMVd (−) strand contains only eight stems (P1−P8), versus 14 for the (+) strand, several of them being very long (e.g., P2, P6, and P8). Moreover, the P8 and P14 pseudoknots of the (+) strand have not been detected in the (−) strand, resulting in a more linear (rod-like) structure. Chrysanthemum chlorotic mottle viroid (CChMVd), the longest known member of this family (i.e., 398−401 nt), adopted a complex, branched, secondary structure like PLMVd. CChMVd (+) strand includes a pseudoknot reminiscent of P8 in PLMVd. The 2 four-way junctions of the left domains, as well as the larger internal loop are conserved. The right domain is also a multibranched region. The CChMVd (−) strand is similar to that of its (+) counterpart. Avocado sunblotch viroid is characterized by its relatively high content of both adenosine and uridine residues. Probing one sequence variant in the presence of $MgCl_2$ by two independent groups did not permit the detection of any distinct structures in the (+) strand (Fig. 6.3). But controversy exists regarding the (−)

strand. Probing led to a proposal for the presence of a kissing-loop interaction between nucleotides located near both terminal loops in one study (Delan-Forino et al., 2014), which was not detected in another study (Giguère et al., 2014). Finally, both strands of eggplant latent viroid have been shown to fold into a relatively long rod-like central domain that ends with a three-way junction in the left domain (Fig. 6.3). The (+) strand also possesses a three-way junction with three stems that forms its right domain, as initially proposed by free energy minimization and by analysis of covariation in natural variants of this viroid (Fadda et al., 2003), whereas the right domain on the (−) strand ends with five stems. In this case, the presence of $MgCl_2$ did not lead to the detection of any significant differences, suggesting that these structures lack a pseudoknot.

References

Ambrós, S., Hernández, C., Desvignes, J.C., Flores, R., 1998. Genomic structure of three phenotypically different isolates of peach latent mosaic viroid: implications of the existence of constraints limiting the heterogeneity of viroid quasi-species. J. Virol. 72, 7397–7406.

Baumstark, T., Schröder, A., Riesner, D., 1997. Viroid processing: switch from cleavage to ligation is driven by a change from a tetraloop to a loop E conformation. EMBO J. 16, 599–610.

Bussière, F., Ouellet, J., Côté, F., Lévesque, D., Perreault, J., 2000. Mapping in solution shows the peach latent mosaic viroid to possess a new pseudoknot in a complex, branched secondary structure. J. Virol. 74, 2647–2654.

Delan-Forino, C., Deforges, J., Benard, L., Sargueil, B., Maurel, M., Torchet, C., 2014. Structural analyses of avocado sunblotch viroid reveal differences in the folding of plus and minus RNA strands. Viruses 6, 489–506.

Delgado, S., Martínez de Alba, A.E., Hernández, C., Flores, R., 2005. A short double-stranded RNA motif of peach latent mosaic viroid contains the initiation and the self-cleavage sites of both polarity strands. J. Virol. 79, 12934–12943.

Ding, B., Kwon, M., Hammond, R., Owens, R., 1997. Cell-to-cell movement of potato spindle tuber viroid. Plant J. 12, 931–936.

Dingley, A., Steger, G., Esters, B., Riesner, D., Grzesiek, S., 2003. Structural characterization of the 69 nucleotide potato spindle tuber viroid left-terminal domain by NMR and thermodynamic analysis. J. Mol. Biol. 334, 751–767.

Dubé, A., Baumstark, T., Bisaillon, M., Perreault, J., 2010. The RNA strands of the plus and minus polarities of peach latent mosaic viroid fold into different structures. RNA 16, 463–473.

Dubé, A., Bolduc, F., Bisaillon, M., Perreault, J., 2011. Mapping studies of the peach latent mosaic viroid reveal novel structural features. Mol. Plant Pathol. 12, 688–701.

Fadda, Z., Darós, J., Fagoaga, C., Flores, R., Duran-Vila, N., 2003. Eggplant latent viroid, the candidate type species for a new genus within the family *Avsunviroidae* (hammerhead viroids). J. Virol. 77, 6528–6532.

Gas, M., Hernández, C., Flores, R., Daròs, J., 2007. Processing of nuclear viroids *in vivo*: an interplay between RNA conformations. PLoS Pathog. 3, e182.

Gas, M., Molina-Serrano, D., Hernández, C., Flores, R., Daròs, J., 2008. Monomeric linear RNA of citrus exocortis viroid resulting from processing *in vivo* has

5′-phosphomonoester and 3′-hydroxyl termini: implications for the RNase and RNA ligase involved in replication. J. Virol. 82, 10321–10325.

Giguère, T., Adkar-Purushothama, C., Bolduc, F., Perreault, J., 2014. Elucidation of the structures of all members of the *Avsunviroidae* family. Mol. Plant Pathol. 15, 767–779.

Gozmanova, M., Denti, M., Minkov, I., Tsagris, M., Tabler, M., 2003. Characterization of the RNA motif responsible for the specific interaction of potato spindle tuber viroid RNA (PSTVd) and the tomato protein Virp1. Nucleic Acids Res. 31, 5534–5543.

Gross, H.J., Liebl, U., Alberty, H., Krupp, G., Domdey, H., Ramm, K., et al., 1981. A severe and a mild potato spindle tuber viroid isolate differ in three nucleotide exchanges only. Biosci. Rep. 1, 235–241.

Hammond, R., 1994. Agrobacterium-mediated inoculation of PSTVd cDNAs onto tomato reveals the biological effect of apparently lethal mutations. Virology 201, 36–45.

Keese, P., Symons, R., 1985. Domains in viroids: evidence of intermolecular RNA rearrangement and their contribution to viroid evolution. Proc. Natl. Acad. Sci. USA 82, 4582–4586.

Kolonko, N., Bannach, O., Aschermann, K., Hu, K.H., Moors, M., Schmitz, M., et al., 2006. Transcription of potato spindle tuber viroid by RNA polymerase II starts in the left terminal loop. Virology 347, 392–404.

Malfitano, M., Di Serio, F., Covelli, L., Ragozzino, A., Hernández, C., Flores, R., 2003. Peach latent mosaic viroid variants inducing peach calico (extreme chlorosis) contain a characteristic insertion that is responsible for this symptomatology. Virology 313, 492–501.

Motard, J., Bolduc, F., Thompson, D., Perreault, J., 2008. The peach latent mosaic viroid replication initiation site is located at a universal position that appears to be defined by a conserved sequence. Virology 373, 362–375.

Navarro, B., Gisel, A., Rodio, M., Delgado, S., Flores, R., Di Serio, F., 2012. Small RNAs containing the pathogenic determinant of a chloroplast-replicating viroid guide the degradation of a host mRNA as predicted by RNA silencing. Plant J. 70, 991–1003.

Riesner, D., Henco, K., Rokohl, U., Klotz, G., Kleinschmidt, A.K., Domdey, H., et al., 1979. Structure and structure formation of viroids. J. Mol. Biol. 133, 85–115.

Schrader, O., Baumstark, T., Riesner, D., 2003. A mini-RNA containing the tetraloop, wobble-pair and loop E motifs of the central conserved region of potato spindle tuber viroid is processed into a minicircle. Nucleic Acids Res. 31, 988–998.

Schröder, A., Riesner, D., 2002. Detection and analysis of hairpin II, an essential metastable structural element in viroid replication intermediates. Nucleic Acids Res. 30, 3349–3359.

Steger, G., Perreault, J.-P., 2016. Structure and associated biological functions of viroids. Adv. Virus Res. 94, 141–172.

Steger, G., Tabler, M., Brüggemann, W., Colpan, M., Klotz, G., Sänger, H., et al., 1986. Structure of viroid replicative intermediates: physico-chemical studies on SP6 transcripts of cloned oligomeric potato spindle tuber viroid. Nucleic Acids Res. 14, 9613–9630.

Takeda, R., Petrov, A., Leontis, N., Ding, B., 2011. A three-dimensional RNA motif in potato spindle tuber viroid mediates trafficking from palisade mesophyll to spongy mesophyll in *Nicotiana benthamiana*. Plant Cell 23, 258–272.

CHAPTER

7

Viroid Replication

Ricardo Flores[1], *Sofia Minoia*[1], *Amparo López-Carrasco*[1], *Sonia Delgado*[1], *Ángel-Emilio Martínez de Alba*[1] *and Kriton Kalantidis*[2]

[1]Polytechnic University of Valencia-CSIC, Valencia, Spain
[2]University of Crete, Heraklion, Greece

INTRODUCTION

Viroids differ from viruses, particularly plant RNA viruses, in fundamental aspects that include structure, function, and evolutionary origin. Although for clarity these aspects are treated independently (as in this book), they are deeply intermingled. This is vividly illustrated in viroid replication. Plant RNA viruses code for proteins, which with others from their hosts, form a complex (RNA-dependent RNA polymerase or RNA replicase) that catalyzes initiation and elongation of viral strands; this process takes place in membranous vesicles induced, at least in part, by viral proteins (den Boon and Ahlquist, 2010). However, since viroids are nonprotein-coding RNAs—a likely consequence of their structure and limited size—they must be replicated by preexisting cellular RNA polymerases and processing enzymes that they manipulate. The accumulation of potato spindle tuber viroid (PSTVd), the type member of the family *Pospiviroidae*, in nuclei of infected tomato (Diener, 1971; Harders et al., 1989; Qi and Ding, 2003), and of avocado sunblotch viroid (ASBVd), the type member of the family *Avsunviroidae*, in chloroplasts of infected avocado (Bonfiglioli et al., 1994; Lima et al., 1994; Mohamed and Thomas, 1980), is directly associated with the presence of such enzymes in these two organelles. Moreover, one step of the replication cycle in the family *Avsunviroidae* is catalyzed by hammerhead ribozymes embedded in viroid strands (see below), a likely evolutionary

Viroids and Satellites.
DOI: http://dx.doi.org/10.1016/B978-0-12-801498-1.00007-3

relic of the emergence of protoviroids in the RNA world (Diener, 1989; Flores et al., 2014).

SETTING THE STAGE: ALTERNATIVE PATHWAYS OF A REPLICATION MECHANISM WITH ONE OR TWO ROLLING CIRCLES

The identification of multimeric minus (−) viroid RNAs (complementary to the predominant infectious (+) strand) in PSTVd-infected tomato (Branch et al., 1981), and of double-stranded PSTVd complexes containing unit- and longer-than-unit length (+) and (−) strands (Hadidi et al., 1982), as well as multimeric (+) viroid RNAs in ASBVd-infected avocado (Bruening et al., 1982), led to the proposal that one or both of these longer-than-unit length strands resulted from the reiterative transcription of circular templates through a rolling-circle mechanism (Branch and Robertson, 1984) with only RNA intermediates (Grill and Semancik, 1978; Hadidi et al., 1982). Moreover, the localization of viroid multimeric strands supported the conclusion that PSTVd not only accumulated in, but also replicated in, the nucleus (Spiesmacher et al., 1983), and that ASBVd replicated in the chloroplast (Bonfiglioli et al., 1994; Navarro et al., 1999). The evidence available supports the view that this differential behavior is a typical feature of each family.

These and other observations were interpreted as indicative of a mechanism with two alternative pathways (Branch and Robertson, 1984), which with modifications is the model prevailing today. In the first pathway, named asymmetric because it operates through a single rolling circle, the incoming circular (+) viroid RNA is repeatedly transcribed into linear multimeric (−) strands, which subsequently serve as templates for generating linear multimeric (+) strands. In the second pathway, named symmetric because it functions through two rolling circles, the linear multimeric (−) strands are first cleaved and ligated into the monomeric circular (−) RNAs, then serving as the template for synthesis of the linear multimeric (+) RNAs. In both instances, the multimeric (+) strands synthesized in the second RNA-RNA transcription round are cleaved into unit-length strands that are ligated to produce the mature circular (+) RNAs (Fig. 7.1). The absence of the monomeric circular PSTVd (−) RNA in naturally infected plants supports the conclusion that PSTVd replication proceeds through the asymmetric pathway (Branch et al., 1988; Feldstein et al., 1998). In contrast, the observation that (+) and (−) dimeric RNAs of ASBVd and other members of the family *Avsunviroidae* self-cleave in vitro (Flores et al., 2000; Hutchins et al., 1986) provides indirect evidence against multimeric RNAs of either polarity acting as templates for these

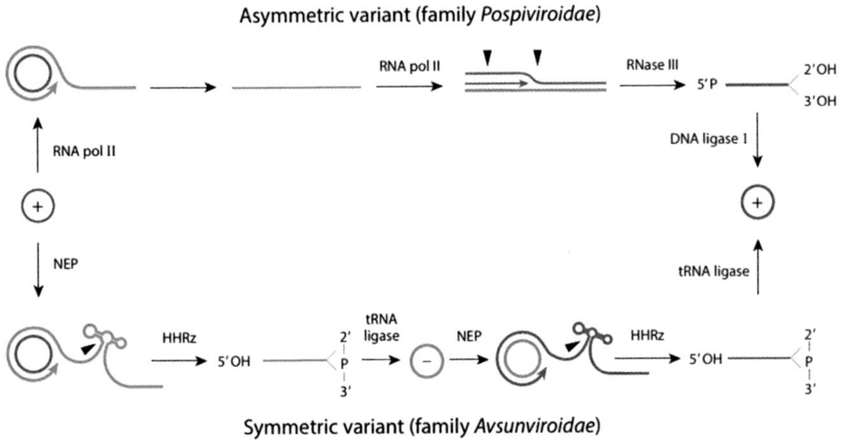

FIGURE 7.1 Asymmetric and symmetric variants of the rolling-circle mechanism proposed for replication of members of the families *Pospiviroidae* and *Avsunviroidae*, respectively. Orange and blue colors refer to plus and minus polarities, respectively, with cleavage sites denoted by arrowheads. The enzymes and ribozymes that presumably catalyze the replication steps are indicated. Notice that RNA polymerase II (and NEP) is redirected to transcribe RNA templates and DNA ligase 1 to circularize RNA substrates. *HHRz*, hammerhead ribozyme; *NEP*, nuclear-encoded polymerase. *Source: Reproduced from Flores, R., Gago-Zachert, S., Serra, P., Sanjuán, R., Elena, S.F., 2014. Viroids, survivors from the RNA world? Annu. Rev. Microbiol. 68, 395–414.*

viroids. This finding, together with the presence of monomeric circular ASBVd (+) and (−) RNAs in multistranded complexes isolated from infected avocado (Navarro et al., 1999), support the view that replication of ASBVd occurs through a symmetric pathway with two rolling circles (Daròs et al., 1994).

Replication of RNA viruses requires an RNA replicase. A rolling-circle mechanism, however, demands three catalytic activities: RNA polymerase, ribonuclease (RNase), and RNA ligase. The enzymes, and ribozymes, on which these activities reside are described below. To exert their function, they most likely recruit additional host proteins. For instance, a tomato bromodomain-containing protein with a nuclear localization signal, Virp1, has been isolated by its specific interaction with PSTVd (+) RNA in vitro and in vivo (Martínez de Alba et al., 2003). Protoplast transfection has revealed that Virp1-suppressed cells cannot sustain viroid replication (Kalantidis et al., 2007). Interestingly, Virp1 is necessary for the nuclear import of a satellite RNA of cucumber mosaic virus (Chaturvedi et al., 2014), indicating that viroids and satellite RNAs may share specific aspects in their replication (Rao and Kalantidis, 2015). Nevertheless, the replication stage at which Virp1 participates remains unknown.

INITIATION AND ELONGATION OF RNA STRANDS: DNA-DEPENDENT RNA POLYMERASES TRANSCRIBE RNA TEMPLATES

Experiments in vivo with cucumber protoplasts from plants infected by hop stunt viroid (HSVd, family *Pospiviroidae)* showed that its synthesis is blocked by nanomolar concentrations of α-amanitin that typically inhibit RNA polymerase II (Pol II) (Mühlbach and Sänger, 1979). This result was surprising because Pol II transcribes DNA under physiological conditions, thus implying that the viroid could alter the template specificity of the enzyme. Experiments in vitro with other members of the family *Pospiviroidae*, PSTVd and citrus exocortis viroid (CEVd), confirmed that such low levels of α-amanitin inhibit the synthesis of one or both polarity strands (Flores and Semancik, 1982; Schindler and Mühlbach, 1992). Moreover, other results showed that a monoclonal antibody against the largest subunit of Pol II, when added to a nuclear extract of CEVd-infected tomato, coimmunoprecipitated both viroid strands (Warrilow and Symons, 1999). Furthermore, in vitro transcription with a nuclear extract from potato cells primed with the monomeric (+) circular PSTVd RNA, mapped the initiation of PSTVd (−) strands at the hairpin loop of the left terminal domain in the rod-like secondary structure of the former, with this proposed start site being also supported by the infectivity *in planta* and replication ability of a group of site-directed mutants (Kolonko et al., 2006). Two additional pertinent points are: (1) the finding by in situ hybridization of both polarity strands of PSTVd in the nucleoplasm (Qi and Ding, 2003), wherein Pol II resides, is consistent with RNA polymerization occurring in this compartment and (2) given that transcription factor IIIA (TFIIIA) from *Arabidopsis thaliana* binds PSTVd (+) RNA in vitro with similar affinity as its physiological ligand 5S rRNA, and that this protein also binds the 5S DNA gene, the dual ability of TFIIIA could allow it to act as a bridge between the viroid RNA template and Pol II (Eiras et al., 2011). Recent data support this notion, because PSTVd interacts in vitro and in vivo with TFIIIA from *Nicotiana benthamiana*, and its downregulation and overexpression results in reduced and increased accumulation of PSTVd, respectively (Wang et al., 2016).

Regarding the family *Avsunviroidae*, experiments in vitro showed that ASBVd transcription is insensitive to α-amanitin and tagetitoxin. Considering that among the two chloroplastic RNA polymerases tagetitoxin inhibits the plastid-encoded polymerase (PEP) but not the nuclear-encoded polymerase (NEP), the latter is the most likely candidate to mediate replication of ASBVd and possibly other members of this family (Navarro et al., 2000). Consistent with this view, sequence variants of peach latent mosaic viroid (PLMVd) that incite an extreme albinism (calico) in peach leaves replicate actively in these leaves despite their lack of PEP (Rodio et al., 2007). On the other hand, NEP starts synthesis of both polarity strands of ASBVd and PLMVd in specific sequence/structure motifs (Delgado et al., 2005; Navarro and Flores, 2000), as revealed by experiments that take advantage of a characteristic property of chloroplastic transcripts: their 5′ termini have a triphosphorylated group that can be labeled in vitro and subsequently mapped. In this context it is also worth noting that the mutation rate of chrysanthemum chlorotic mottle viroid (family *Avsunviroidae*), is the highest reported for any biological entity. This extreme rate, which presumably is the consequence of the errors introduced by NEP when transcribing nonphysiological RNA templates, and its lack of proofreading activity, is most likely tolerated because of the small size of the viroid genomes (Gago et al., 2009).

CLEAVAGE OF OLIGOMERIC STRANDS: HOST-ENCODED RNASES AND VIROID-EMBEDDED RIBOZYMES

Correct processing to monomeric circular forms has been observed in vitro by mixing potato nuclear extracts and longer-than-unit PSTVd (+) RNAs (Baumstark et al., 1997; Tsagris et al., 1987), and in vivo by expressing transgenically in *A. thaliana* dimeric (+) transcripts of three representative members of the family *Pospiviroidae:* CEVd, HSVd, and apple scar skin viroid (ASSVd) (Gas et al., 2007). However, the corresponding dimeric (−) transcripts expressed in the same way were not processed as expected for the asymmetric rolling-circle replication pathway (Gas et al., 2007). This result, together with the finding that the PSTVd (−) strands accumulate exclusively in the nucleoplasm while those of (+) polarity are split between the nucleolus and the nucleoplasm (Qi and Ding, 2003), suggest that PSTVd (+) strands contain motifs mediating their trafficking to the nucleolus, wherein they are processed.

Besides its most stable rod-like conformation, PSTVd can adopt metastable folding with hairpins, prominent among which is hairpin I (HP I) formed by the upper strand of the central conserved region (CCR) and flanking nucleotides (Riesner et al., 1979). The strict conservation in all members of the family *Pospiviroidae* of this element of secondary structure, despite changes in sequence, supports its key role in vivo. In consonance with this idea, the cleavage sites of the dimeric (+) transcripts of CEVd, HSVd, and ASSVd—mimicking replicative intermediates—expressed in the transgenic lines of *A. thaliana*, have been mapped at similar positions of the capping loop of HP I (Gas et al., 2007). Moreover, as a consequence of the particular properties of HP I, the corresponding sequences of the dimeric or oligomeric RNAs would facilitate the adoption of an alternative long double-stranded region proposed previously (Diener, 1986; Visvader et al., 1985), the actual substrate for cleavage. Cleavage at both strands of this double-stranded region leaves two 3′ protruding nucleotides, the fingerprint of RNases of class III, which act preferentially upon double-stranded substrates to generate products with 5′-phosphomonoester and 3′-hydroxyl groups. In support of this notion, the monomeric viroid linear (+) strands that accumulate in transgenic *A. thaliana* expressing the corresponding dimeric transcripts contain such terminal groups (Gas et al., 2008).

Remarkably, both (+) and (−) multimeric strands of members of the family *Avsunviroidae* self-cleave in vitro and in vivo through hammerhead ribozymes (Flores et al., 2000; Hutchins et al., 1986). These ribozymes act in *cis* upon the RNAs in which they are contained, and their activity is facilitated by tertiary contacts between apical loops (De la Peña et al., 2003; Khvorova et al., 2003), and by interactions with host proteins (Daròs and Flores, 2002) that promote the adoption of the catalytically active conformation during transcription. Hammerhead ribozymes produce 5′-hydroxyl and 2′,3′-cyclic phosphodiester termini, an aspect relevant when considering the subsequent circularization step.

CIRCULARIZATION OF MONOMERIC LINEAR STRANDS: RNA LIGASES AND DNA LIGASES OPERATING ON RNA SUBSTRATES

Initial in vitro data suggested the participation in PSTVd circularization of an enzyme with properties similar to the wheat germ tRNA ligase (Branch et al., 1982), which requires 5′-hydroxyl and 2′,3′-cyclic phosphodiester termini (Konarska et al., 1981). However, the subsequent finding that the CEVd monomeric linear RNA processed in vivo has 5′-phosphomonoester and 3′-hydroxyl termini (Gas et al., 2008), prompted the search for an alternative RNA ligase with specificity for such termini and, hence, similarity to that of phage T4 (Ho and Shuman, 2002; Wang et al., 2006). The search, surprisingly, did not result in the expected outcome but in the identification of the nuclear DNA ligase 1 (Nohales et al., 2012a). This enzyme is able to accept RNA substrates, thus illustrating the ability of viroids to usurp not only the transcription but also the processing machinery of their hosts. The circularization is further promoted by the rod-like secondary structure that in its CCR contains a local element of tertiary structure (the loop E). This element, conserved in members of the genus *Pospiviroid*, to which PSTVd belongs, positions the termini to be ligated in close proximity and proper orientation (Gas et al., 2007).

Therefore, while cleavage is determined by the upper CCR strand, ligation involves both the upper and lower CCR strands.

Because the hammerhead-mediated self-cleavage operating in the family *Avsunviroidae* generates monomeric linear forms with the 5′-hydroxyl and 2′,3′-cyclic phosphodiester termini specific for the wheat germ tRNA ligase, an enzyme of this class appeared as the best candidate for catalyzing circularization in this family. The wheat germ tRNA ligase, however, is targeted in most organisms to the nucleus, wherein it facilitates tRNA splicing (Abelson et al., 1998). The observation that in plants there is an additional isoform targeted to chloroplasts (Englert et al., 2007), added further credence to its involvement in replication of chloroplastic viroids. Experiments with a recombinant version of the chloroplastic isoform of the tRNA ligase from eggplant substantiated this view by showing the efficient circularization in vitro of the (+) and (−) monomeric linear replication intermediates from the four members of the family *Avsunviroidae*. Moreover, while this tRNA ligase specifically recognizes the genuine monomeric linear (+) eggplant latent viroid (ELVd) replication intermediate, it does not do so with five other monomeric linear (+) ELVd RNAs with their ends mapping at different sites along the molecule, despite containing the same 5′-hydroxyl and 2′,3′-phosphodiester terminal groups (Nohales et al., 2012b). Although detection in infected peach of circular PLMVd RNA forms locked through a 2′,5′-phosphodiester linkage has been interpreted as a result of nonenzymatic self-ligation (Côté et al., 2001), several arguments do not favor the view that this could be the circularization mechanism in vivo (Flores et al., 2011; Nohales et al., 2012b).

THE OTHER SIDE OF THE COIN: DEGRADATION OF VIROID STRANDS

The final titers that viroids reach in vivo result from the balance between replication (transcription and processing) and decay. In contrast to replication, decay of viroid RNAs has only been studied recently. As described in other chapters of this book, viroids from both families are recognized by their hosts as foreign RNAs and targeted for degradation by the RNA silencing machinery, whose two key components are dicer-like and argonaute proteins (see Chapter 11: Viroids and RNA Silencing). Besides, examination by northern-blot hybridization of PSTVd infecting solanaceous hosts has disclosed in eggplant, and to a lesser extent in *N. benthamiana* and tomato, a series of defined PSTVd (+) subgenomic RNAs (sgRNAs) accompanying the monomeric circular and linear forms. Further analyses by primer extension and 5′- and 3′-RACE (rapid amplification of cDNA ends) have failed to detect a common 5′ terminus in the (+) sgRNAs, dismissing the idea that they could result from aborted transcriptions initiated at one specific site, and showed that they have 5′-hydroxyl and 3′-phophomonoester termini (Minoia et al., 2015). The same methodology has revealed the presence of PSTVd (−) sgRNAs with similar properties (Minoia et al., 2015). Detection of PSTVd sgRNAs of both polarities suggests that they are generated in the nucleoplasm during elongation of viroid strands, possibly because these nascent strands are particularly vulnerable to endonucleolytic cleavage. In agreement with this view, cell fractionation studies have shown that the PSTVd (+) sgRNAs accumulate preferentially in the nuclei. These results provide a novel insight into viroid decay in vivo, indicating that synthesis and decay of PSTVd strands might be linked (Minoia et al., 2015). Whether decay routes, apart from RNA silencing, operate also in members of the family *Avsunviroidae* remains to be explored.

CONCLUDING REMARKS

Research on replication of viroids, apart from disclosing the unique manner in which these minimal noncoding RNAs have evolved to parasitize their host transcriptional and processing machineries, has led to discovering the hammerhead ribozyme. This ribozyme, one of the simplest reported, has attracted much interest for working out the fine details of RNA-mediated catalysis and for manipulating it as a therapeutic tool. Moreover, studies on viroid replication illuminated subsequent research on how the replication of certain plant satellite RNAs and, particularly, of the hepatitis delta virus (HDV, an important human pathogen) proceeds (Flores et al., 2011). It is amazing to realize how viroids and HDV RNA have developed similar rolling-circle replication strategies mediated by DNA-dependent RNA polymerases—redirected to accept RNA templates—and by RNA-embedded ribozymes. This parallelism illustrates that, when it comes to fundamental processes, frontiers between plant and animal systems become blurred.

Acknowledgments

Research in R.F. laboratory is currently funded by grant BFU2014−56812-P from the Spanish Ministerio de Economía y Competitividad (MINECO) and in K.K laboratory by grant KA 4499 (AristeiaII, ESPA 2007−2013) from the Greek Ministry of Culture, Education and Religious Affairs.

References

Abelson, J., Trotta, C.R., Li, H., 1998. tRNA splicing. J. Biol. Chem. 273, 12685−12688.

Baumstark, T., Schröder, A.R.W., Riesner, D., 1997. Viroid processing: switch from cleavage to ligation is driven by a change from a tetraloop to a loop E conformation. EMBO J. 16, 599−610.

Bonfiglioli, R.G., McFadden, G.I., Symons, R.H., 1994. In situ hybridization localizes avocado sunblotch viroid on chloroplast thylakoid membranes and coconut cadang cadang viroid in the nucleus. Plant J. 6, 99–103.

Branch, A.D., Benenfeld, B.J., Robertson, H.D., 1988. Evidence for a single rolling circle in the replication of potato spindle tuber viroid. Proc. Natl. Acad. Sci. USA 85, 9128–9132.

Branch, A.D., Robertson, H.D., 1984. A replication cycle for viroids and other small infectious RNAs. Science 223, 450–455.

Branch, A.D., Robertson, H.D., Dickson, E., 1981. Longer-than-unit-length viroid minus strands are present in RNA from infected plants. Proc. Natl. Acad. Sci. USA 78, 6381–6385.

Branch, A.D., Robertson, H.D., Greer, C., Gegenheimer, P., Peebles, C., Abelson, J., 1982. Cell-free circularization of viroid progeny RNA by an RNA ligase from wheat germ. Science 217, 1147–1149.

Bruening, G., Gould, A.R., Murphy, P.J., Symons, R.H., 1982. Oligomers of avocado sunblotch viroid are found in infected avocado leaves. FEBS Lett. 148, 71–78.

Chaturvedi, S., Kalantidis, K., Rao, A.L.N., 2014. A bromodomain-containing host protein mediates the nuclear importation of a satellite RNA of cucumber mosaic virus. J. Virol. 88, 1890–1896.

Côté, F., Lévesque, D., Perreault, J.P., 2001. Natural 2′,5′-phosphodiester bonds found at the ligation sites of peach latent mosaic viroid. J. Virol. 75, 19–25.

Daròs, J.A., Flores, R., 2002. A chloroplast protein binds a viroid RNA *in vivo* and facilitates its hammerhead-mediated self-cleavage. EMBO J. 21, 749–759.

Daròs, J.A., Marcos, J.F., Hernández, C., Flores, R., 1994. Replication of avocado sunblotch viroid: evidence for a symmetric pathway with two rolling circles and hammerhead ribozyme processing. Proc. Natl. Acad. Sci. USA 91, 12813–12817.

De la Peña, M., Gago, S., Flores, R., 2003. Peripheral regions of natural hammerhead ribozymes greatly increase their self-cleavage activity. EMBO J. 22, 5561–5570.

Delgado, S., Martínez de Alba, E., Hernández, C., Flores, R., 2005. A short double-stranded RNA motif of peach latent mosaic viroid contains the initiation and the self-cleavage sites of both polarity strands. J. Virol. 79, 12934–12943.

den Boon, J.A., Ahlquist, P., 2010. Organelle-like membrane compartmentalization of positive-strand RNA virus replication factories. Annu. Rev. Microbiol. 64, 241–256.

Diener, T.O., 1971. Potato spindle tuber "virus": a plant virus with properties of a free nucleic acid. III. Subcellular location of PSTV-RNA and the question of whether virions exist in extracts or in situ. Virology 43, 75–89.

Diener, T.O., 1986. Viroid processing: a model involving the central conserved region and hairpin I. Proc. Natl. Acad. Sci. USA 83, 58–62.

Diener, T.O., 1989. Circular RNAs: relics of precellular evolution? Proc. Natl. Acad. Sci. USA 86, 9370–9374.

Eiras, M., Nohales, M.A., Kitajima, E.W., Flores, R., Daròs, J.A., 2011. Ribosomal protein L5 and transcription factor IIIA from *Arabidopsis thaliana* bind in vitro specifically potato spindle tuber viroid RNA. Arch. Virol. 156, 529–533.

Englert, M., Latz, A., Becker, D., Gimple, O., Beier, H., Akama, K., 2007. Plant pre-tRNA splicing enzymes are targeted to multiple cellular compartments. Biochimie 89, 1351–1365.

Feldstein, P.A., Hu, Y., Owens, R.A., 1998. Precisely full length, circularizable, complementary RNA: an infectious form of potato spindle tuber viroid. Proc. Natl. Acad. Sci. USA 95, 6560–6565.

Flores, R., Daròs, J.A., Hernández, C., 2000. The Avsunviroidae family: viroids with hammerhead ribozymes. Adv. Virus Res. 55, 271–323.

Flores, R., Gago-Zachert, S., Serra, P., Sanjuán, R., Elena, S.F., 2014. Viroids, survivors from the RNA world? Annu. Rev. Microbiol. 68, 395–414.

Flores, R., Grubb, D., Elleuch, A., Nohales, M.A., Delgado, S., Gago, S., 2011. Rolling-circle replication of viroids, viroid-like satellite RNAs and hepatitis delta virus: variations on a theme. RNA Biol. 8, 200–206.

Flores, R., Semancik, J.S., 1982. Properties of a cell-free system for synthesis of citrus exocortis viroid. Proc. Natl. Acad. Sci. USA 79, 6285–6288.

Gago, S., Elena, S.F., Flores, R., Sanjuán, R., 2009. Extremely high variability of a hammerhead viroid. Science 323, 1308.

Gas, M.E., Hernández, C., Flores, R., Daròs, J.A., 2007. Processing of nuclear viroids *in vivo*: an interplay between RNA conformations. PLoS Pathog. 3, 1813–1826.

Gas, M.E., Molina-Serrano, D., Hernández, C., Flores, R., Daròs, J.A., 2008. Monomeric linear RNA of citrus exocortis viroid resulting from processing *in vivo* has 5′-phosphomonoester and 3′-hydroxyl termini: implications for the ribonuclease and RNA ligase involved in replication. J. Virol. 82, 10321–10325.

Grill, L.K., Semancik, J.S., 1978. RNA sequences complementary to citrus exocortis viroid in nucleic acid preparations from infected *Gynura aurantiaca*. Proc. Natl. Acad. Sci. USA 75, 896–900.

Hadidi, A., Hashimoto, J., Diener, T.O., 1982. Potato spindle tuber viroid-specific double-stranded RNA in extracts from infected tomato leaves. Ann. Inst. Pasteur Virol. 133, 15–31.

Harders, J., Lukacs, N., Robert-Nicoud, M., Jovin, J.M., Riesner, D., 1989. Imaging of viroids in nuclei from tomato leaf tissue by *in situ* hybridization and confocal laser scanning microscopy. EMBO J. 8, 3941–3949.

Ho, C.K., Shuman, S., 2002. Bacteriophage T4 RNA ligase 2 (gp24.1) exemplifies a family of RNA ligases found in all phylogenetic domains. Proc. Natl. Acad. Sci. USA 99, 12709–12714.

Hutchins, C., Rathjen, P.D., Forster, A.C., Symons, R.H., 1986. Self-cleavage of plus and minus RNA transcripts of avocado sunblotch viroid. Nucleic Acids Res. 14, 3627–3640.

Kalantidis, K., Denti, M.A., Tzortzakaki, S., Marinou, E., Tabler, M., Tsagris, M., 2007. Virp1 is a host protein with a major role in potato spindle tuber viroid infection in Nicotiana plants. J. Virol. 81, 12872–12880.

Khvorova, A., Lescoute, A., Westhof, E., Jayasena, S.D., 2003. Sequence elements outside the hammerhead ribozyme catalytic core enable intracellular activity. Nat. Struct. Biol. 10, 708–712.

Kolonko, N., Bannach, O., Aschermann, K., Hu, K.H., Moors, M., Schmitz, M., et al., 2006. Transcription of potato spindle tuber viroid by RNA polymerase II starts in the left terminal loop. Virology 347, 392–404.

Konarska, M., Filipowicz, W., Domdey, H., Gross, H.J., 1981. Formation of a 2′-phosphomonoester, 3′,5′-phosphodiester linkage by a novel RNA ligase in wheat germ. Nature 293, 112–116.

Lima, M.I., Fonseca, M.E.N., Flores, R., Kitajima, E.W., 1994. Detection of avocado sunblotch viroid in chloroplasts of avocado leaves by in situ hybridization. Arch. Virol. 138, 385–390.

Martínez de Alba, A.E., Sägesser, R., Tabler, M., Tsagris, M., 2003. A bromodomain-containing protein from tomato specifically binds potato spindle tuber viroid RNA *in vitro* and *in vivo*. J. Virol. 77, 9685–9694.

Minoia, S., Navarro, B., Delgado, S., Di Serio, F., Flores, R., 2015. Viroid RNA turnover: characterization of the subgenomic RNAs of potato spindle tuber viroid accumulating in infected tissues provides insights into decay pathways operating in vivo. Nucleic Acids Res. 43, 2313–2325.

Mohamed, N.A., Thomas, W., 1980. Viroid-like properties of an RNA species associated with the sunblotch disease of avocados. J. Gen. Virol. 46, 157–167.

Mühlbach, H.P., Sänger, H.L., 1979. Viroid replication is inhibited by α-amanitin. Nature 278, 185–188.

Navarro, J.A., Daròs, J.A., Flores, R., 1999. Complexes containing both polarity strands of avocado sunblotch viroid: identification in chloroplasts and characterization. Virology 253, 77–85.

Navarro, J.A., Flores, R., 2000. Characterization of the initiation sites of both polarity strands of a viroid RNA reveals a motif conserved in sequence and structure. EMBO J. 19, 2662–2670.

Navarro, J.A., Vera, A., Flores, R., 2000. A chloroplastic RNA polymerase resistant to tagetitoxin is involved in replication of avocado sunblotch viroid. Virology 268, 218–225.

Nohales, M.A., Flores, R., Daròs, J.A., 2012a. A viroid RNA redirects host DNA ligase 1 to act as an RNA ligase. Proc. Natl. Acad. Sci. USA 109, 13805–13810.

Nohales, M.A., Molina-Serrano, D., Flores, R., Daròs, J.A., 2012b. Involvement of the chloroplastic isoform of tRNA ligase in the replication of the viroids belonging to the family *Avsunviroidae*. J. Virol. 86, 8269–8276.

Qi, Y., Ding, B., 2003. Differential subnuclear localization of RNA strands of opposite polarity derived from an autonomously replicating viroid. Plant Cell 15, 2566–2577.

Rao, A.L.N., Kalantidis, K., 2015. Virus-associated small satellite RNAs and viroids display similarities in their replication strategies. Virology 479–480, 627–636.

Riesner, D., Henco, K., Rokohl, U., Klotz, G., Kleinschmidt, A.K., Domdey, H., et al., 1979. Structure and structure formation of viroids. J. Mol. Biol. 133, 85–115.

Rodio, M.E., Delgado, S., De Stradis, A.E., Gómez, M.D., Flores, R., Di Serio, F., 2007. A viroid RNA with a specific structural motif inhibits chloroplast development. Plant Cell 19, 3610–3626.

Schindler, I.M., Mühlbach, H.P., 1992. Involvement of nuclear DNA-dependent RNA polymerases in potato spindle tuber viroid replication: a reevaluation. Plant Sci. 84, 221–229.

Spiesmacher, E., Mühlbach, H.P., Schnölzer, M., Haas, B., Sänger, H.L., 1983. Oligomeric forms of potato spindle tuber viroid (PSTV) and of its complementary RNA are present in nuclei isolated from viroid-infected potato cells. Biosci. Rep. 3, 767–774.

Tsagris, M., Tabler, M., Mühlbach, H.P., Sänger, H.L., 1987. Linear oligomeric potato spindle tuber viroid (PSTV) RNAs are accurately processed *in vitro* to the monomeric circular viroid proper when incubated with a nuclear extract from healthy potato cells. EMBO J. 6, 2173–2183.

Visvader, J.E., Forster, A.C., Symons, R.H., 1985. Infectivity and *in vitro* mutagenesis of monomeric cDNA clones of citrus exocortis viroid indicates the site of processing of viroid precursors. Nucleic Acids Res. 13, 5843–5856.

Wang, L.K., Schwer, B., Shuman, S., 2006. Structure-guided mutational analysis of T4 RNA ligase 1. RNA 12, 2126–2134.

Wang, Y., Qu, J., Ji, S., Wallace, A.J., Wu, J., Li, Y., et al., 2016. A land plant-specific transcription factor directly enhances transcription of a pathogenic noncoding RNA template by DNA-dependent RNA polymerase II. Plant Cell 28, 1094–1107.

Warrilow, D., Symons, R.H., 1999. Citrus exocortis viroid RNA is associated with the largest subunit of RNA polymerase II in tomato in vivo. Arch. Virol. 144, 2367–2375.

8

Viroid Movement

Vicente Pallás and Gustavo Gómez
Polytechnic University of Valencia-CSIC, Valencia, Spain

INTRODUCTION

Eukaryotic cells have evolved subcellular structures (organelles) specialized in diverse biological processes. Hence, sorting RNAs into specific destinations within a cell is essential for controlling and/or coordinating diverse aspects of cellular homeostasis. An increasing number of studies have begun to provide molecular insights into how mammalian cells select RNAs for transport to specific destinations (Blower, 2013). Yet, the mechanism that controls the specific movement and compartmentalization of RNAs in the plant cell has still not been entirely elucidated. Viroids have recently emerged as ideal model systems to study RNA transport within and between cells (Wang and Ding, 2010). Conventional viroid infection of a host plant comprises a series of coordinated steps that involve: (1) intracellular movement, including subcellular compartmentalization for replication and subsequent exit; (2) export of the viroid progeny to neighboring cells; (3) entry to vascular tissue for long-distance trafficking to distant plant organs; and (4) systemic infection development (Ding and Wang, 2009). Although the critical importance of this set of events, commonly defined as movement, is evident, our knowledge about their molecular and functional basis is limited. In this chapter we summarize the more relevant results obtained about the viroid movement study, and how this study has contributed to understanding noncoding RNAs (ncRNAs) traffic in plants.

INTRACELLULAR MOVEMENT

Viroids possess two well-determined replication sites in the infected cell: the nucleus for members of the family *Pospiviroidae* and chloroplasts

Viroids and Satellites.
DOI: http://dx.doi.org/10.1016/B978-0-12-801498-1.00008-5

for members of the family *Avsunviroidae*. With no apparent protein coding ability, these RNAs are compelled to interact directly with host-cell factors to achieve proper subcellular compartmentalization and efficient replication. Consequently, subverting the plant-endogenous and poorly described pathways that regulate ncRNA compartmentalization into both cellular organelles is the first obstacle that viroids face after host-cell invasion.

Nuclear Transport

The use of labeled and/or chimeric transcripts provided evidence that the nuclear import of potato spindle tuber viroid (PSTVd), family *Pospiviroidae*, is mediated by *cis*-acting RNA sequences or structural motifs via a saturable host receptor that is unrelated to the cytoskeleton route (Woo et al., 1999). The import of labeled PSTVd RNA was specifically inhibited by nonlabeled homologous transcripts (Woo et al., 1999). Zhao et al. (2001) used a vector derived from the plant cytoplasmic virus potato virus X to study and ultimately identify the nuclear targeting signals of PSTVd in a whole plant system. The import of this viroid to the nucleus was proposed to occur through a precise process controlled by a sequence or structural motif mapping at the upper strand of the central conserved region (Abraitiene et al., 2008).

The specific subnuclear localization of the pospiviroids is another critical step in their infection cycle. By in situ hybridization, it has been demonstrated that (minus)-PSTVd-RNAs are localized in the nucleoplasm, but not in the nucleolus (Qi and Ding, 2003). In contrast, (plus)-PSTVd-RNAs have been detected in both the nucleolus and the nucleoplasm. These findings support a replication model in which only (plus)-strand PSTVd RNAs are selectively imported into the nucleolus, suggesting that they may contain a specific domain capable of recognizing the host factors commonly involved in the specific localization of endogenous RNAs in the plant-cell nucleolus.

Despite the host factors directly involved in the recognition of these potential specific domains remaining unidentified, a bromodomain-containing host protein (VirP1) with a nuclear localization signal, which showed in vitro and in vivo viroid-binding activity, has been proposed to mediate the nuclear compartmentalization of PSTVd-RNA in infected cells (Gozmanova et al., 2003; Martínez de Alba et al., 2003) (Fig. 8.1). Subsequent studies, which employed transgenic plants with suppressed VirP1 expression, have shown a direct correlation between this protein and infection by PSTVd and citrus exocortis viroid (Kalantidis et al., 2007). This finding reinforces the role of VirP1 in the replication of at least some pospiviroids. Recent studies have shown that the nuclear import of the satellite RNA of cucumber mosaic virus strain Q is also

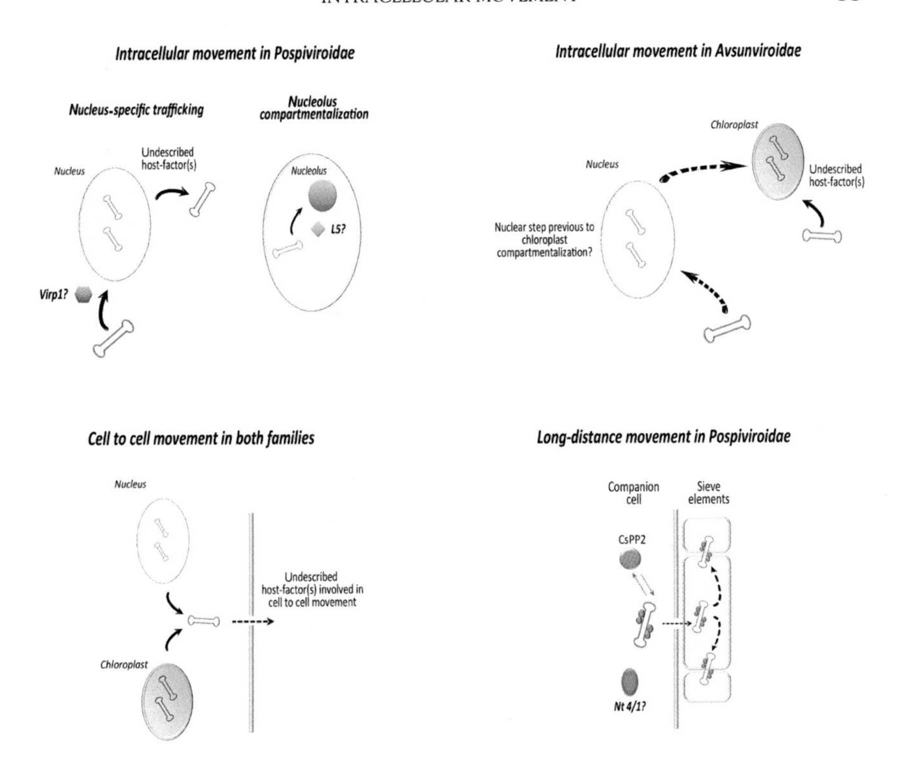

FIGURE 8.1 Simplified general overview of *viroid movement* in infected plants. The different aspects regarding *intracellular, cell-to-cell,* and *long-distance* trafficking are explained in the text. *Virp1,* Bromodomain-containing protein; *L5,* ribosomal protein L5; *CsPP2, Cucumis sativus* Phloem Protein 2; *Nt 4/1, Nicotiana tabacum* 4/1 protein.

mediated by VirP1 (Chaturvedi et al., 2014). Finally, it has also been demonstrated that (plus)-PSTVd RNA binds in vitro the ribosomal protein L5 and transcription factor IIIA (TFIIIA) from *Arabidopsis thaliana* (Eiras et al., 2011). In this work, the authors suggested that L5 could mediate the selective trafficking of the (plus)-PSTVd forms into the nucleolus (Fig. 8.1). The mechanisms and viroid sequences and/or structures involved in the nuclear export of the pospiviroids have not been elucidated (Ding and Owens, 2003).

Transport to Chloroplasts

Although the specific import of RNAs into chloroplasts has been suggested for both nuclear-encoded tRNAs (Bungard, 2004) and the mRNA that codes for translation factor eIF4E (Nicolai et al., 2007), members of the family *Avsunviroidae* are the only pathogenic RNAs known to selectively traffic into this organelle. Consequently, chloroplast targeting is a critical step that permits access to the host machinery responsible for replication of these viroids.

By cellular approaches, it has been demonstrated that a partial-length RNA sequence derived from eggplant latent viroid (ELVd) acting as a 5′UTR was able to mediate specific trafficking and accumulation of functional GFP mRNA from the nucleus into chloroplasts (Gómez and Pallás, 2012a,b). This finding supports the existence of an unknown plant signaling mechanism mediated by noncoding RNAs capable of regulating *in cis* the selective import of nuclear transcripts into chloroplasts (Fig. 8.1). This pathway may prove to be a novel mechanism capable of regulating the accumulation of some nuclear-encoded proteins in chloroplasts as an alternative to that based on signal-peptides (Gómez and Pallás, 2012a,b). Interestingly, it was suggested later that this long noncoding RNA-directed mechanism may also mediate the protein import into the cyanobacterial endosymbiont/plastids of the rhizarian amoeba *Paulinella chromatophora* (Mackiewicz et al., 2012).

More recent results, obtained by combined cytoplasmic and nuclear expression assays, have shown that ELVd transcripts traffic from the cytoplasm to the nucleus, and subsequently from the nucleus to chloroplasts (Gómez and Pallás, 2012a). According to this pathway, once the ELVd invades the cell cytoplasm, it is imported into the nucleus. Next the viroid uses this organelle as a subcellular port to be launched into the chloroplast, where replication occurs (Gómez and Pállas, 2012b). Further studies are necessary to identify and characterize the host factor involved in the regulation and coordination of these events.

INTERCELLULAR AND VASCULAR MOVEMENT

Although it had been suspected that, like plant viruses (reviewed in Pallás et al., 2011; Vuorinen et al., 2011), viroids might use the phloem for their long-distance movement, evidence for this assumption had to wait until the seminal work of Palukaitis (1987). Later, using fluorescently labeled PSTVd transcripts, Ding et al. (1997) showed that viroid RNA moved rapidly from cell to cell when injected into symplasmically connected mesophyll cells. Remarkably, a 1400 nt RNA with vector sequences was able to move out of the injected mesophyll cells only when fused to the PSTVd sequence. These results strongly suggest that PSTVd moves from cell to cell via plasmodesmata, and that this movement might be mediated by a specific sequence or structural motif. More importantly, these results together with structural data of the genomic viroid molecule make viroids appropriate models to study the intercellular and vascular trafficking of RNAs (Ding and Wang, 2009).

More specifically, an RNA motif in PSTVd has been shown to be required for trafficking from the palisade to the spongy mesophyll in *Nicotiana benthamiana* leaves (Takeda et al., 2011). This motif, called loop 6, contains the sequence 5′-CGA-3′... 5′-GAC-3′ flanked on both sides by *cis* Watson-Crick G/C and G/U wobble-base pairs. The authors propose that this motif could be recognized by distinct cellular factors, which are components of cell-specific RNA trafficking machinery.

The phloem is a plant-specific conduit (Dinant et al., 2010) that delivers not only sugars, amino acids, mineral nutrients, and hormones, but also peptides, proteins, and RNAs from autotrophic to heterotrophic tissues (Atkins et al., 2011). The long-distance transport of certain host non-cell autonomously acting RNAs has been shown to be a key regulator of essential processes like gene silencing, pathogen defense, and development (Banerjee et al., 2009; Dunoyer et al., 2010; Haywood et al., 2005; Kim et al., 2001; Melnyk et al., 2011; Yoo et al., 2004). How these RNAs are transported through the phloem is still an unsolved question, and studies on phloem viroid transport can help to shed some light on this relevant issue. It is commonly accepted that without the protection of specific RNA-binding proteins (RBPs), naked cellular RNA molecules quickly fall prey to degradative processes (Shyu et al., 2008). Thus, viroids most likely interact with some RBP to be transported through the phloem. The phloem of angiosperms is rich in a myriad of RBPs (Pallás and Gómez, 2013). Owens et al. (2001) and Gómez and Pallás (2001) independently reported that one of the most abundant RBPs of the phloem of cucurbits, phloem protein 2 (CsPP2), strongly interacts in vitro with hop stunt viroid, suggesting the involvement of this protein in viroid phloem transport (Fig. 8.1). Remarkably, immunoprecipitation experiments together with intergeneric graft assays have demonstrated that CsPP2 interacts in vivo with hop stunt viroid, and that the corresponding ribonucleoprotein complex is translocated to the scion (Gómez and Pallás, 2004). A double-stranded RNA-binding motif identified in the predicted structure of CsPP2 has been proposed to be responsible for the RNA-binding properties of this protein (Gómez and Pallás, 2004). Other phloem proteins with similar characteristics to CsPP2 have been identified in melon and shown to bind and translocate RNAs through intergeneric grafts (Gómez et al., 2005). The tomato protein VirP1 (see above) also has been suggested to be involved in systemic viroid spread (Maniataki et al., 2003). More recently, the *Arabidopsis thaliana* 4/1 (At-4/1) protein has been proposed to mediate signaling in the vasculature, which includes the mobility of pathogen-related and cellular RNAs (Morozov et al., 2014; Solovyev et al., 2013). When PSTVd was inoculated onto *N. benthamiana* plants in which 4/1 mRNA expression was downregulated by virus-induced gene silencing, long-distance movement of the viroid into developing young leaves above the inoculated leaf was much more efficient than in comparable unsilenced control plants (Solovyev et al., 2013) (Fig. 8.1).

A loss-of-function genetic analysis identified an RNA motif (nucleotides U43/C318) forming a small loop (loop 7), which is required for PSTVd to traffic from non-vascular to vascular phloem tissue (Zhong et al., 2007). By using two PSTVd variants that differed in their intercellular trafficking capacities, Qi et al. (2004) identified a bipartite motif that is both necessary and sufficient to mediate the trafficking of viroid RNA from the bundle sheath to the mesophyll in young leaves; one part of the motif is located in the right terminal domain and the other in the pathogenicity domain. Remarkably, this motif is not necessary for trafficking in the reverse direction (i.e., from the mesophyll to bundle sheath). This result further supports the hypothesis that the bundle sheath plays a critical role in regulating the macromolecular trafficking between the phloem and nonvascular tissues. A genome-wide mutational analysis of the PSTVd RNA molecule has identified multiple loops/bulges as critical functional motifs for replication and/or systemic RNA trafficking, and has provided a framework to enable high-throughput studies on the tertiary structures and functional mechanisms of the RNA motifs that regulate viroid replication and trafficking (Ding and Wang, 2009).

FUTURE PROSPECTS

Although the cell factors and functional mechanisms that direct the overall aspects of viroid movement in infected plants remain to be entirely elucidated, considerable progress has been made in answering this intriguing question. In recent years viroids have become uniquely simple and tractable models to elucidate the regulation of the cell-to-cell trafficking of RNAs (Ding and Wang, 2009). Lack of coding capacity and compact structure make viroid RNAs ideal models to not only study structure–function relationships, but to also search for the host partners involved in the formation of translocatable ribonucleotide complexes. Although the structural features of the viroid RNA molecule have been elucidated with a high a degree of accuracy (Flores et al., 2005; Kovalskaya and Hammond, 2014; Palukaitis, 2014), a genetic model system that could help in deciphering how and where these features act within the plant host-cell is still lacking. Remarkably, avocado sunblotch viroid has been shown to replicate in yeast cells (Delan-Forino et al., 2011). Although this system could facilitate the study of the replication, intracellular transport, and subcellular compartmentalization of viroids, it would not be useful for studying intercellular RNA trafficking. Thus genetic models that can provide insights into the poorly understood mechanism of intercellular trafficking and subcellular compartmentalization of long noncoding RNAs in plant cells are one of the main challenges for the future.

Acknowledgments

This chapter is dedicated to the memory of Dr. Biao Ding whose contributions on the study of viroid movement have been essential. Work in the laboratories of G. Gómez and V. Pallás was supported by grants AGL2013–47886-R (GG) and BIO2014–54362-R (VP) from the Spanish Granting Agency (Direccion General Investigacion Cientifica) and PROMETEO program 2015/010 (Gen. Valenciana).

References

Abraitiene, A., Zhao, Y., Hammond, R., 2008. Nuclear targeting by fragmentation of the potato spindle tuber viroid genome. Biochem. Biophys. Res. Commun. 368, 470–475.
Atkins, C.A., Smith, P.M., Rodriguez-Medina, C., 2011. Macromolecules in phloem exudates—a review. Protoplasma 248, 165–172.
Banerjee, A.K., Lin, T., Hannapel, D.J., 2009. Untranslated regions of a mobile transcript mediate RNA metabolism. Plant Physiol. 151, 1831–1843.
Blower, M.D., 2013. Molecular insights into intracellular RNA localization. Inter. Rev. Cell Mol. Biol. 302, 1–39.
Bungard, R.A., 2004. Photosynthetic evolution in parasitic plants: insight from the chloroplast genome. Bioessays. 26, 235–247.
Chaturvedi, S., Kalantidis, K., Rao, A.L.N., 2014. A bromodomain-containing host protein mediates the nuclear importation of a satellite RNA of cucumber mosaic virus. J. Virol. 88, 1890–1896.
Delan-Forino, C., Maurel, M.C., Torchet, C., 2011. Replication of avocado sunblotch viroid in the yeast *Saccharomyces cerevisiae.* J. Virol. 85, 3229–3238.
Dinant, S., Bonnemain, J.L., Girousse, C., Kehr, J., 2010. Phloem sap intricacy and interplay with aphid feeding. Compt. Rend. Biol. 333, 504–515.
Ding, B., Kwon, M.O., Hammond, R., Owens, R., 1997. Cell-to-cell movement of potato spindle tuber viroid. Plant J. 12, 931–936.
Ding, B., Owens, R., 2003. Movement. In: Hadidi, A., Flores, R., Randles, J.W., Semancik, J.S. (Eds.), Viroids. CSIRO Publishing, Collingwood, pp. 49–54.
Ding, B., Wang, Y., 2009. Viroids: uniquely simple and tractable models to elucidate regulation of cell-to-cell trafficking of RNAs. DNA Cell Biol. 28, 51–56.
Dunoyer, P., Schott, G., Himber, C., Meyer, D., Takeda, A., Carrington, J.C., et al., 2010. Small RNA duplexes function as mobile silencing signals between plant cells. Science 328, 912–916.
Eiras, M., Nohales, M.A., Kitajima, E.W., Flores, R., Daròs, J.A., 2011. Ribosomal protein L5 and transcription factor IIIA from *Arabidopsis thaliana* bind in vitro specifically potato spindle tuber viroid RNA. Arch. Virol. 156, 529–533.
Flores, R., Hernández, C., Martínez de Alba, A.E., Darós, J.A., Di Serio, F., 2005. Viroids and viroid–host interactions. Annu. Rev. Phytopathol. 43, 117–139.
Gómez, G., Pallás, V., 2001. Identification of an in vitro ribonucleoprotein complex between a viroid RNA and a phloem protein from cucumber. Mol Plant Microbe Interact. 14, 910–913.

Gómez, G., Pallás, V., 2004. A long distance translocatable phloem protein from cucumber forms a ribonucleoprotein complex in vivo with hop stunt viroid RNA. J. Virol. 78, 10104–10110.

Gómez, G., Pallás, V., 2012a. Studies on subcellular compartmentalization of plant pathogenic non coding RNAs give new insights into the intracellular RNA-traffic mechanisms. Plant Physiol. 159, 558–564.

Gómez, G., Pallás, V., 2012b. A pathogenic non coding RNA that replicates and accumulates in chloroplasts traffics to this organelle through a nuclear-dependent step. Plant Sig. Behav. 7, 882–884.

Gómez, G., Torres, H., Pallás, V., 2005. Identification of translocatable RNA-binding phloem proteins from melon, potential components of the long distance RNA transport system. Plant J. 41, 107–116.

Gozmanova, M., Denti, M.A., Minkov, I.N., Tsagris, M., Tabler, M., 2003. Characterization of the RNA motif responsible for the specific interaction of potato spindle tuber viroid RNA (PSTVd) and the tomato protein Virp1. Nucleic Acids Res. 31, 5534–5543.

Haywood, V., Yu, T.S., Huang, N.C., Lucas, W.J., 2005. Phloem long-distance trafficking of gibberellic acid-insensitive RNA regulates leaf development. Plant J. 42, 49–68.

Kalantidis, K., Denti, M., Tzortzakaki, S., Marinou, E., Tabler, M., Tsagris, M., 2007. Viroid binding protein 1 is necessary for the infection of potato spindle tuber viroid (PSTVd). J. Virol. 81, 12872–12880.

Kim, M., Canio, W., Kessler, S., Sinha, N., 2001. Developmental changes due to long-distance movement of a homeobox fusion transcript in tomato. Science 293, 280–289.

Kovalskaya, N., Hammond, R.W., 2014. Molecular biology of viroid–host interactions and disease control strategies. Plant Sci. 228, 48–60.

Mackiewicz, P., Bodył, A., Gagat, P., 2012. Possible import routes of proteins into the cyanobacterial endosymbionts/plastids of *Paulinella chromatophora*. Theory Biosci. 131, 1–18.

Maniataki, E., Tabler, M., Tsagris., M., 2003. Viroid RNA systemic spread may depend on the interaction of a 71-nucleotide bulged hairpin with the host protein VirP1. RNA 9, 346–354.

Martínez de Alba, A.E., Sägesser, R., Tabler, M., Tsagris, M., 2003. A bromodomain-containing protein from tomato specifically binds potato spindle tuber viroid RNA in vitro and in vivo. J. Virol. 77, 9685–9694.

Melnyk, C.W., Molnar, A., Bassett, A., Baulcombe, D.C., 2011. Mobile 24 nt small RNAs direct transcriptional gene silencing in the root meristems of Arabidopsis thaliana. Curr. Biol. 21, 1678–1683.

Morozov, S.Y., Makarova, S.S., Erokhina, T.N., Kopertekh, L., Schiemann, J., Owens, R.A., et al., 2014. Plant 4/1 protein: potential player in intracellular, cell-to-cell and long-distance signaling. Front. Plant Sci. 5, 1–7.

Nicolai, M., Duprat, A., Sormani, R., Rodriguez, C., Roncato, M.A., Rolland, N., et al., 2007. Higher plant chloroplasts import the mRNA coding for the eukaryotic translation initiation factor 4E. FEBS Lett. 581, 3921–3926.

Owens, R.A., Blackburn, M., Ding, B., 2001. Possible involvement of a phloem lectin in long distance viroid movement. Mol. Plant Microbe Interact. 14, 905–909.

Pallás, V., Genovés, A., Sanchez-Pina, M.A., Navarro, J.A., 2011. Systemic movement of viruses via the plant phloem. In: Caranta, C., Aranda, M.-A., Tepfer, M., Lopez Moya, J.-J. (Eds.), Recent Advances in Plant Virology. Caister Academic Press, Norfolk, pp. 75–102.

Pallás, V., Gómez, G., 2013. Phloem RNA-binding proteins as potential components of the long-distance RNA transport system. Front. Plant Sci. 4, 130.

Palukaitis, P., 1987. Potato spindle tuber viroid: investigation of the long-distance, intra-plant transport route. Virology 158, 239–241.

Palukaitis, P., 2014. What has been happening with viroids? Virus Genes 49, 175–184.

Qi, Y., Ding, B., 2003. Differential subnuclear localization of RNA strands of opposite polarity derived from an autonomously replicating viroid. Plant Cell. 15, 2566–2577.
Qi, Y., Pélissier, T., Itaya, A., Hunt, E., Wassenegger, M., Ding, B., 2004. Direct role of a viroid RNA motif in mediating directional RNA trafficking across a specific cellular boundary. Plant Cell. 16, 741–1752.
Shyu, A.B., Wilkinson, M.F., van Hoof, A., 2008. Messenger RNA regulation: to translate or to degrade. EMBO J. 27, 471–481.
Solovyev, A.G., Makarova, S.S., Remizowa, M.V., Lim, H.S., Hammond, J., Owens, R.A., et al., 2013. Possible role of the Nt-4/1 protein in macromolecular transport in vascular tissue. Plant Sig. Behav. 8, e25784.
Takeda, R., Petrov, A.I., Leontis, N.B., Ding, B., 2011. A three-dimensional RNA motif in potato spindle tuber viroid mediates trafficking from palisade mesophyll to spongy mesophyll in *Nicotiana benthamiana*. Plant Cell. 23, 258–272.
Vuorinen, A.L., Kelloniemia, J., Valkonen, J.P.T., 2011. Why do viruses need phloem for systemic invasion of plants? Plant Sci. 181, 355–363.
Wang, Y., Ding, B., 2010. Viroids: small probes for exploring the vast universe of RNA trafficking in plants. J. Intergr. Plant Biol. 52, 17–27.
Woo, Y.M., Itaya, A., Owens, R.A., Tang, L., Hammond, R.W., Chou, H.C., et al., 1999. Characterization of nuclear import of potato spindle tuber viroid RNA in permeabilized protoplasts. Plant J. 17, 627–635.
Yoo, B.C., Kragler, F., Varkonyi-Gasic, E., Haywood, V., Archer-Evans, S., Lee, Y.M., et al., 2004. A systemic small RNA signaling system in plants. Plant Cell. 16, 1979–2000.
Zhao, Y., Owens, R.A., Hammond, R.W., 2001. Use of a vector based on potato virus X in a whole plant assay to demonstrate nuclear targeting of potato spindle tuber viroid. J. Gen. Virol. 82, 1491–1497.
Zhong, X., Tao, X., Stombaugh, J., Leontis, N., Ding, B., 2007. Tertiary structure and function of an RNA motif required for plant vascular entry to initiate systemic trafficking. EMBO J. 26, 3836–3846.

CHAPTER

9

Viroid Pathogenesis

Ricardo Flores[1], *Francesco Di Serio*[2], *Beatriz Navarro*[2] *and Robert A. Owens*[3]

[1]Polytechnic University of Valencia-CSIC, Valencia, Spain
[2]National Research Council, Bari, Italy
[3]U.S. Department of Agriculture, Beltsville, MD, United States

INTRODUCTION

From early research on viroids, "How do viroids make plants sick?" has been a question of interest. Most symptoms associated with viroid diseases can also be seen in plants infected with viruses, and the small size and noncoding nature of the viroid genome requires that viroid-induced disease be an essentially host-generated process. Previous reviews by three viroid pioneers (e.g., Diener, 1987; Sänger, 1982; Semancik, 2003) summarized information from the early, observational phases of work on viroid pathogenicity. Soon after the first complete viroid sequence was reported (Gross et al., 1978), speculation describing possible molecular mechanism(s) of disease induction appeared. In the absence of viroid-encoded polypeptides, disease was assumed to result from the *direct interaction* of the viroid (or its complement) with cellular constituents. Only later were *indirect interactions* between viroid and host via newly discovered regulatory pathways like RNA silencing considered.

Initially, very little experimental data was available to test these hypotheses. More recently, the situation has changed, and in this chapter we summarize evidence indicating the involvement of multiple molecular mechanisms in viroid disease induction. Promising strategies to expand our understanding of the linkages among different aspects of the host response are also described.

Viroids and Satellites.
DOI: http://dx.doi.org/10.1016/B978-0-12-801498-1.00009-7

MACROSCOPIC SYMPTOMS AND CYTOPATHIC EFFECTS

Although many viroids were discovered due to their ability to cause disease, not all viroid infections are symptomatic. Stunting is often considered a characteristic of viroid disease, but additional symptoms may include epinasty, distortions affecting leaves and flowers, veinal necrosis, bark disorders, changes in fruit shape/color, sterility, and rapid aging (see Chapter 5: Viroid Biology). The root systems of infected plants are often reduced in size and reserve organs, like tubers, may also be misshapen. For many viroids, symptom expression can be enhanced by high temperatures—in contrast to most plant viruses. Only in coconut cadang-cadang and Tinangaja diseases does infection result in death of the whole plant.

At the microscopic level, reported cytopathic effects of viroid infection include: (1) proliferation of cytoplasmic membranes to form "plasmalemmasomes"; (2) distortion of cell walls; (3) chloroplast abnormalities associated with members of both families; and (4) formation of electron-dense deposits (in the cytoplasm and chloroplasts) (Di Serio et al., 2013). How these effects are related to symptoms at the whole plant level remains unclear, but disruption of chloroplast function is also a prominent feature of many plant virus infections (see Chapter 10: Changes in the Host Proteome and Transcriptome Induced by Viroid Infection).

BIOCHEMICAL CHANGES ACCOMPANYING VIROID INFECTION

Many symptoms associated with viroid infection (e.g., stunting) are consistent with disturbances in hormone metabolism. Evidence of hormonal changes has been presented for tomato infected with citrus exocortis viroid (CEVd) (Duran-Vila and Semancik, 1982) and cucumber infected with hop stunt viroid (HSVd) (Yaguchi and Takahashi, 1985). Other changes in the host metabolome include (1) increased levels of gentisic acid (a signaling molecule related to salicylic acid) in CEVd-infected tomato (Bellés et al., 1999) and (2) reduced alpha acid content of hop cones harvested from HSVd-infected vines (see Chapter 19: Hop Stunt Viroid).

Other alterations affect host macromolecules. Among the first reported were changes in two low molecular weight pathogenesis-related (PR) proteins in CEVd-infected *Gynura aurantiaca* (Conejero and Semancik, 1977). PR protein synthesis is one component of *systemic acquired resistance,* a term describing a coordinated series of plant responses to pathogen attack. Additional information concerning PR protein accumulation in viroid-infected plants can be found in Chapter 10, Changes in the Host Proteome and Transcriptome Induced by Viroid Infection. Other proteins are modified (e.g., phosphorylated; Hiddinga et al., 1988) in response to viroid infection. Overexpression in tobacco of a signal-transducing tomato protein kinase (PKV), resulted in stunting, modified vascular development, reduced root formation, and male sterility (Hammond and Zhao, 2009). PKV may regulate plant development by modulating critical signaling pathways involved in gibberellic acid metabolism.

MOLECULAR MECHANISMS

Determination of the complete nucleotide sequence of potato spindle tuber viroid (PSTVd) (Gross et al., 1978) was a landmark in plant biology. Certain similarities with U1 small nuclear RNA and eukaryotic introns suggested a possible mechanism for pathogenicity (Diener, 1981). Two further developments—the finding that minor sequence differences distinguish PSTVd strains inducing very different symptoms (Gross et al., 1981) and construction of infectious PSTVd-cDNA clones (Cress et al., 1983)—caused the emphasis in pathogenicity research to shift, turning away from the host to focus almost exclusively on the viroid. Site-directed mutagenesis allowed experimental testing of mechanisms proposed for viroid pathogenesis.

As sequences of other members of the family *Pospiviroidae* became available, Keese and Symons (1985) proposed that their rod-like genomes contain five structural/functional domains, one of which is the pathogenicity (P) domain. For some viroids, sequences outside the P domain also play an important role in symptom expression (e.g., Sano et al., 1992). Even for PSTVd the situation has proved to be more complex than originally believed; thus, a single nucleotide substitution in the loop E motif of the central domain can either (1) broaden the host range to include tobacco (Wassenegger et al., 1996) or (2) modify the tomato symptoms to produce a "flat top" phenotype (Qi and Ding, 2003).

Schnölzer et al. (1985) proposed that a "virulence-modulating" region within the P domain of PSTVd modulates symptoms by interacting with unidentified host constituents. Diener et al. (1993) described a differential activation of the interferon-induced, dsRNA-activated, mammalian protein kinase by PSTVd strains of varying pathogenicity, but efforts to clone a plant homolog have been unsuccessful.

THE HOST CONTRIBUTION TO PATHOGENICITY REEXAMINED

Skoric et al. (2001), using two stable, naturally occurring sequence variants of CEVd, observed that increasing the daytime temperature from 35° to 40°C caused symptoms of the mild (but not the severe) variant to intensify; furthermore, this effect was (1) reversible and (2) only observed when the host used was *G. aurantiaca*. Structural calculations failed to reveal any linkage with predicted CEVd secondary structures.

Modern genomic techniques such as microarray or proteomic analysis have been used to examine changes in host gene expression associated with viroid infection (see Chapter 10: Changes in the Host Proteome and Transcriptome Induced by Viroid Infection). In tomato (where analysis is facilitated by the availability of a complete genomic sequence), PSTVd induces numerous changes in mRNAs encoding enzymes involved in hormone biosynthesis and components of the corresponding signaling pathways (Owens et al., 2012). Using a combination of proteomic analysis and qRT-PCR, Lisón et al. (2013) have shown that control of the accumulation of some proteins resides at the translational rather than transcriptional level.

RNA SILENCING LIMITS THE SPREAD OF VIROID INFECTION

Discovery of RNA silencing, first in plants and subsequently in most other eukaryotes, revealed novel regulatory roles for RNA (apart from those involved in protein translation) and integrated diverse observations into a single interpretative scheme. This topic is discussed extensively in Chapter 11, Viroids and RNA Silencing. In brief, RNA silencing most likely emerged as a defense mechanism against invading agents and then was coopted for other roles including regulation of gene expression. Transcriptional and posttranscriptional gene silencing are initiated by dsRNA, which in plants are cleaved by Dicer-like (DCL) enzymes into small RNAs (sRNAs): microRNAs (miRNAs, 21–22 nt) of endogenous origin and small interfering RNAs (siRNAs, 21, 22 or 24 nt) of both endogenous and exogenous origin (Axtell, 2013). One strand of the sRNA duplex guides the RNA inducing silencing complex (RISC), to inactivate either complementary RNA or DNA. Plant RNA and DNA viruses are restrained by the combined action of DCLs and RISC.

RNA silencing also restrains viroid infection, as shown by the detection (Itaya et al., 2001; Martínez de Alba et al., 2002; Papaefthimiou et al., 2001) and characterization (Di Serio et al., 2009; Itaya et al., 2007;

Machida et al., 2007; Martín et al., 2007; Navarro et al., 2009) of viroid-derived sRNAs (vd-sRNAs) in tissues infected by members of both viroid families. These vd-sRNAs reduce, in a sequence-specific manner, the expression of a reporter gene (Itaya et al., 2007; Vogt et al., 2004), and the viroid titer in infected plants (Carbonell et al., 2008). Moreover, one specific RNA-directed RNA polymerase (RDR6) acts to reduce the titer of PSTVd and block its entry into meristems (Di Serio et al., 2010).

These data, particularly the sequence-specific effects of vd-sRNAs, suggest that (1) a "dicing-only" mechanism is unlikely and (2) one or more Argonaute proteins (AGOs) (Mallory and Vaucheret, 2010), at the core of RISC, are also involved. Indeed, pioneering work revealed that PSTVd-cDNA sequences in transgenic tobacco become methylated de novo in an RNA-directed sequence-specific manner following infection and replication of PSTVd (Wassenegger et al., 1994), thus providing the first example of transcriptional gene silencing, subsequently shown to be mediated by specific DCL-generated siRNAs and AGOs.

There is direct evidence that AGOs recruit vd-sRNAs (Minoia et al., 2014). In PSTVd-infected *Nicotiana benthamiana*, both endogenous AGO1 and epitope-tagged *Arabidopsis thaliana* AGO1, AGO2, AGO4, and AGO5 associate with vd-sRNAs displaying the same properties (5′-terminal nucleotide and size) as endogenous and viral sRNAs (Mi et al., 2008). Furthermore, overexpression of these four AGOs attenuated viroid accumulation, supporting the involvement of RISC in antiviroid defense and consistent with other in vitro results showing that RISC mediates cleavage of viral RNAs with a compact conformation—like those of viroids—by targeting bulged regions within this conformation (Schuck et al., 2013).

RNA SILENCING ALSO MEDIATES VIROID PATHOGENESIS

In addition to guiding RISC against viroid RNAs, vd-sRNAs could also target specific host mRNAs, whose inactivation would ultimately result in symptom development through a signal transduction cascade (Gómez et al., 2009; Papaefthimiou et al., 2001; Wang et al., 2004). This intriguing hypothesis has gathered increasing experimental support.

For viroids that replicate in plastids (family *Avsunviroidae*), certain variants of peach latent mosaic viroid (PLMVd) containing a specific 12–14 nt insertion elicit an extreme albinism known as "peach calico" (PC) (Chapter 29: Peach Latent Mosaic Viroid in Infected Peach; Malfitano et al., 2003; Rodio et al., 2007). This insertion contains the determinant for PC, and the initial defect is most likely triggered by two 21-nt PLMVd-sRNAs that include the pathogenic determinant and differ in one position. The two PLMVd-sRNAs are complementary to a 21-nt sequence in the peach mRNA coding for the chloroplastic heat-shock protein 90 (cHSP90), and target it for cleavage at the expected site (between positions 10 and 11 of the PLMVd-sRNAs).

cHSP90 is a chaperone that mediates chloroplast biogenesis (a process impaired in PC-affected tissue); furthermore, a single-amino acid mutation in cHSP90 induces a yellow phenotype in *A. thaliana*. Hence, these results provide a plausible mechanism for the initiation of PC pathogenesis (Navarro et al., 2012). An account of the properties of the two vd-sRNA that support their involvement in RNA silencing-mediated cleavage of cHSP90 mRNA has been reported elsewhere (Flores et al., 2015). Prominent among these properties is the presence of a 5′-terminal U, indicating that these viroid sRNAs most likely bind and guide AGO1 (Mi et al., 2008), which plays a key role in RNA silencing *via* cleavage or translation arrest of endogenous and invading RNA targets (Baumberger and Baulcombe, 2005; Morel et al., 2002).

For viroids that replicate in the nucleus (family *Pospiviroidae)*, a similar RNA silencing mechanism has been proposed for PSTVd and tomato planta macho viroid (TPMVd). For PSTVd, constitutive expression of artificial miRNAs derived from one of its pathogenicity determinants induces abnormal phenotypes in two *Nicotiana* species (Eamens et al., 2014), and vd-sRNAs from this determinant silence callose synthase genes in tomato (Adkar-Purushothama et al., 2015). Results for TPMVd are based on in silico prediction and experimental validation of the targeting of a tomato mRNA by vd-sRNA from the same determinant (Avina-Padilla et al., 2015). How inactivation of the mRNAs targeted—coding for a soluble inorganic pyrophosphatase, callose synthases, and a WD40-repeat protein, respectively—leads to disease induction is not yet clear. Also unclear is the AGO involved, considering that most of the vd-sRNAs lack a 5′-terminal U. Finally, infection of cucumber by HSVd results in dynamic changes in the methylation status of ribosomal RNA promotor sequences (Martínez et al., 2014). The potential role of these changes in pathogenesis remains to be examined.

VIROID PATHOGENESIS ENGAGES MULTIPLE REGULATORY NETWORKS

The biochemical and molecular mechanisms used by plants to defend themselves against attack by diverse pathogens (i.e., fungi, bacteria, viruses, and viroids) exhibit many common features (Hammond-Kosack and Jones, 2015). For example, photosynthetic activity almost always decreases following infection, probably as part of a host effort to conserve energy resources needed to mount the defense response. The studies described above have identified single host genes that respond to signals originating from the viroid genome, either directly (via interaction with host proteins) or indirectly (via RNA silencing). How the effects of viroid infection spread from the limited number of initial interactions characterized thus far to affect diverse metabolic pathways is currently unknown.

As discussed in Chapter 10, Changes in the Host Proteome and Transcriptome Induced by Viroid Infection, microarray analysis of PSTVd-infected "Rutgers" tomato has revealed that infection-related changes in transcript levels involve more than half of the approximately 10,000 genes present on the array. Among the genes affected were those encoding enzymes involved in biosynthesis of gibberellin and other hormones as well as components of their respective signaling pathways. Proteomic analysis of CEVd-infected tomato has shown that posttranscriptional changes may involve translational arrest/inhibition as well as sRNA-mediated mRNA degradation. Large-scale sRNA sequence analyses have also revealed changes in host miRNA levels following viroid infection. Direct evidence for cleavage of the target genes is limited; nevertheless, the list of potential target genes includes a number of transcription factors and other regulatory proteins likely to occupy hub positions in the still-to-be-determined host interactome. The picture of viroid pathogenicity that has begun to emerge is clearly complex, with regulatory controls operating at multiple levels.

To date, genome-wide studies involving viroids have focused on the host transcriptome. However, integration of transcriptional and proteomic data sets may yield novel perspectives (Payne, 2015). Until recently, transcriptional and protein–protein interaction networks were available only for *A. thaliana*, but similar attempts with other plant species are expanding rapidly (Braun et al., 2013). A recent meta-analysis comparing the effects of eight plant viruses on *A. thaliana* gene expression (Rodrigo et al., 2012) illustrates the benefits of such an approach and is discussed in more detail in Chapter 10, Changes in the Host Proteome and Transcriptome Induced by Viroid Infection. The data sets and tools required to compare the effects of virus and viroid infections in tomato should soon become available, and the expected results should provide a much more detailed picture of viroid pathogenicity.

Acknowledgments

Research in R.F. laboratory is currently funded by grant BFU2014–56812-P from the Spanish Ministerio de Economía y Competitividad (MINECO). Research in F.D.S. and B.N laboratory has been partially supported by a dedicated grant (CISIA) of the Ministero dell'Economia e Finanze Italiano to the CNR (Legge n. 191/2009).

References

Adkar-Purushothama, C.R., Brosseau, C., Giguère, T., Sano, T., Moffett, P., Perreault, J.P., 2015. Small RNA derived from the virulence modulating region of the potato spindle tuber viroid silences callose synthase genes of tomato plants. Plant Cell 27, 2178–2194.

Avina-Padilla, K., Martinez de la Vega, O., Rivera-Bustamante, R., Martínez-Soriano, J.P., Owens, R.A., et al., 2015. *In silico* prediction and validation of potential gene targets for pospiviroid-derived small RNAs during tomato infection. Gene 564, 197–205.

Axtell, M.J., 2013. Classification and comparison of small RNAs from plants. Annu. Rev. Plant Biol. 64, 137–159.

Baumberger, N., Baulcombe, D.C., 2005. Arabidopsis ARGONAUTE1 is an RNA slicer that selectively recruits microRNAs and short interfering RNAs. Proc. Natl. Acad. Sci. USA 102, 11928–11933.

Bellés, J.M., Garro, R., Fayos, J., Navarro, P., Primo, J., Conejero, V., 1999. Gentisic acid as a pathogen-inducible signal, additional to salicylic acid for activation of plant defenses in tomato. Mol. Plant Microbe Interact. 12, 227–235.

Braun, P., Aubourg, S., Van Leene, J., De Jaeger, G., Lurin, C., 2013. Plant protein interactomes. Annu. Rev. Plant Biol. 64, 161–187.

Carbonell, A., Martínez de Alba, A.E., Flores, R., Gago, S., 2008. Double-stranded RNA interferes in a sequence-specific manner with infection of representative members of the two viroid families. Virology 371, 44–53.

Conejero, V., Semancik, J.S., 1977. Exocortis viroid: alteration in the proteins of *Gynura aurantiaca* accompanying viroid infection. Virology 77, 221–232.

Cress, D.E., Kiefer, M.C., Owens, R.A., 1983. Construction of infectious potato spindle tuber viroid cDNA clones. Nucleic Acids Res. 11, 6821–6835.

Di Serio, F., De Stradis, A., Delgado, S., Flores, R., Navarro, B., 2013. Cytopathic effects incited by viroid RNAs and putative underlying mechanisms. Front. Plant Sci. 3, 288.

Di Serio, F., Gisel, A., Navarro, B., Delgado, S., Martínez de Alba, A.E., Donvito, G., et al., 2009. Deep sequencing of the small RNAs derived from two symptomatic variants of a chloroplastic viroid: implications for their genesis and for pathogenesis. PLoS One 4, e7539.

Di Serio, F., Martínez de Alba, A.E., Navarro, B., Gisel, A., Flores, R., 2010. RNA-dependent RNA polymerase 6 delays accumulation and precludes meristem invasion of a nuclear-replicating viroid. J. Virol. 84, 2477–2489.

Diener, T.O., 1981. Are viroids escaped introns? Proc. Natl. Acad. Sci. USA. 78, 5014–5015.

Diener, T.O., 1987. Biological properties. In: Diener, T.O. (Ed.), The Viroids. Plenum Press, New York, NY, pp. 9–35.

Diener, T.O., Hammond, R.W., Black, T., Katze, M.G., 1993. Mechanism of viroid pathogenesis: differential activation of the interferon-induced, double-stranded RNA-activated, M_r 68,000 protein kinase by viroid strains of varying pathogenicity. Biochimie 75, 533–538.

Duran-Vila, N., Semancik, J.S., 1982. Effects of exogenous auxins on tomato tissue infected with the citrus exocortis viroid. Phytopathology 72, 777–781.

Eamens, A.L., Smith, N.A., Dennis, E.S., Wassenegger, M., Wang, M.B., 2014. In Nicotiana species, an artificial microRNA corresponding to the virulence modulating region of potato spindle tuber viroid directs RNA silencing of a soluble inorganic pyrophosphatase gene and the development of abnormal phenotypes. Virology 450–451, 266–277.

Flores, R., Minoia, S., Carbonell, A., Gisel, A., Delgado, S., López-Carrasco, A., et al., 2015. Viroids, the simplest RNA replicons: How they manipulate their hosts for being propagated and how their hosts react for containing the infection. Virus Res. 209, 136–145.

Gómez, G., Martínez, G., Pallás, V., 2009. Interplay between viroid-induced pathogenesis and RNA silencing pathways. Trends Plant Sci. 14, 264–269.

Gross, H.J., Domdey, H., Lossow, C., Jank, P., Raba, M., Alberty, H., et al., 1978. Nucleotide sequence and secondary structure of potato spindle tuber viroid. Nature 273, 203–208.

Gross, H.J., Liebl, U., Alberty, H., Krupp, G., Domdey, H., Ramm, K., et al., 1981. A severe and a mild potato spindle tuber viroid isolate differ in three nucleotide exchanges only. Biosci. Rep. 1, 235–241.

Hammond, R.W., Zhao, Y., 2009. Modification of tobacco plant development by sense and antisense expression of the tomato viroid-induced AGC VIIIa protein kinase PKV suggests involvement in gibberellin signaling. BMC Plant Biol. 9, 108.

Hammond-Kosack, K.E., Jones, J.D.G., 2015. Responses to plant pathogens. In: Buchanan, B.B., Gruissem, W., Jones, R.L. (Eds.), Biochemistry and Molecular Biology of Plants, 2nd ed. Am. Soc. Plant Physiol., Rockville, MD, pp. 984–1050.

Hiddinga, H.J., Crum, C.J., Hu, J., Roth, D.A., 1988. Viroid-induced phosphorylation of a host protein related to a dsRNA-dependent protein kinase. Science 241, 451–453.

Itaya, A., Folimonov, A., Matsuda, Y., Nelson, R.S., Ding, B., 2001. Potato spindle tuber viroid as inducer of RNA silencing in infected tomato. Mol. Plant Microbe Interact. 14, 1332–1334.

Itaya, A., Zhong, X., Bundschuh, R., Qi, Y., Wang, Y., Takeda, R., et al., 2007. A structured viroid RNA is substrate for dicer-like cleavage to produce biologically active small RNAs but is resistant to RISC-mediated degradation. J. Virol. 81, 2980–2994.

Keese, P., Symons, R.H., 1985. Domains in viroids: evidence of intermolecular RNA rearrangements and their contribution to viroid evolution. Proc. Natl. Acad. Sci. USA 82, 4582–4586.

Lisón, P., Tárraga, S., López-Gresa, P., Saurí, A., Torres, C., Campos, L., et al., 2013. A noncoding plant pathogen provokes both transcriptional and posttranscriptional alterations in tomato. Proteomics 13, 833–844.

Machida, S., Yamahata, N., Watanuki, H., Owens, R.A., Sano, T., 2007. Successive accumulation of two size classes of viroid-specific small RNA in potato spindle tuber viroid-infected tomato plants. J. Gen. Virol. 88, 3452–3457.

Malfitano, M., Di Serio, F., Covelli, L., Ragozzino, A., Hernández, C., Flores, R., 2003. Peach latent mosaic viroid variants inducing peach calico (extreme chlorosis) contain a characteristic insertion that is responsible for this symptomatology. Virology 313, 492–501.

Mallory, A., Vaucheret, H., 2010. Form, function, and regulation of ARGONAUTE proteins. Plant Cell 22, 3879–3889.

Martín, R., Arenas, C., Daròs, J.A., Covarrubias, A., Reyes, J.L., Chua, N.H., 2007. Characterization of small RNAs derived from citrus exocortis viroid (CEVd) in infected tomato plants. Virology 367, 135–146.

Martínez de Alba, A.E., Flores, R., Hernández, C., 2002. Two chloroplastic viroids induce the accumulation of the small RNAs associated with post-transcriptional gene silencing. J. Virol. 76, 13094–13096.

Martínez, G., Castellano, M., Tortosa, M., Pallás, V., Gómez, G., 2014. A pathogenic noncoding RNA induces changes in dynamic DNA methylation of ribosomal RNA genes in host plants. Nucleic Acids Res. 42, 1553–1562.

Mi, S., Cai, T., Hu, Y., Chen, Y., Hodges, E., Ni, F., et al., 2008. Sorting of small RNAs into Arabidopsis argonaute complexes is directed by the 5′ terminal nucleotide. Cell 133, 116–127.

Minoia, S., Carbonell, A., Di Serio, F., Gisel, A., Carrington, J.C., Navarro, B., et al., 2014. Specific ARGONAUTES bind selectively small RNAs derived from potato spindle tuber viroid and attenuate viroid accumulation in vivo. J. Virol. 88, 11933–11945.

Morel, J.B., Godon, C., Mourrain, P., Béclin, C., Boutet, S., Feuerbach, F., et al., 2002. Fertile hypomorphic ARGONAUTE (ago1) mutants impaired in post-transcriptional gene silencing and virus resistance. Plant Cell 14, 629–639.

Navarro, B., Gisel, A., Rodio, M.E., Delgado, S., Flores, R., Di Serio, F., 2012. Small RNAs containing the pathogenic determinant of a chloroplast-replicating viroid guide the degradation of a host mRNA as predicted by RNA silencing. Plant J. 70, 991–1003.

Navarro, B., Pantaleo, V., Gisel, A., Moxon, S., Dalmay, T., Bisztray, G., et al., 2009. Deep sequencing of viroid-derived small RNAs from grapevine provides new insights on the role of RNA silencing in plant-viroid interaction. PLoS One 4, e7686.

Owens, R.A., Tech, K.B., Shao, J.Y., Sano, T., Baker, C.J., 2012. Global analysis of tomato gene expression during potato spindle tuber viroid infection reveals a complex array of changes affecting hormone signaling. Mol. Plant Microbe Interact. 25, 582–598.

Papaefthimiou, I., Hamilton, A.J., Denti, M.A., Baulcombe, D.C., Tsagris, M., Tabler, M., 2001. Replicating potato spindle tuber viroid RNA is accompanied by short RNA fragments that are characteristic of post-transcriptional gene silencing. Nucleic Acids Res. 29, 2395–2400.

Payne, J.L., 2015. The utility of protein and mRNA correlation. Trends Biochem. Sci. 40, 1–3.

Qi, Y., Ding, B., 2003. Inhibition of cell growth and shoot development by a specific nucleotide sequence in a noncoding viroid RNA. Plant Cell 15, 1360–1374.

Rodio, M.E., Delgado, S., De Stradis, A.E., Gómez, M.D., Flores, R., Di Serio, F., 2007. A viroid RNA with a specific structural motif inhibits chloroplast development. Plant Cell. 19, 3610–3626.

Rodrigo, G., Carrera, J., Ruiz-Ferrer, V., del Toro, F.J., Llave, C., Voinnet, O., et al., 2012. A meta-analysis reveals the commonalities and differences in *Arabidopsis thaliana* response to different viral pathogens. PLoS One 7, e40526.

Sänger, H.L., 1982. Biology, structure, functions and possible origin of viroids. In: Parthier, B., Boulter, D. (Eds.), Encyclopedia of Plant Physiology New Series Vol. 14B. Springer-Verlag, Berlin, pp. 368–454.

Sano, T., Candresse, T., Hammond, R.W., Diener, T.O., Owens, R.A., 1992. Identification of multiple structural domains regulating viroid pathogenicity. Proc. Natl. Acad. Sci. USA 89, 10104–10108.

Schnölzer, M., Haas, B., Ramm, K., Hofmann, H., Sänger, H.L., 1985. Correlation between structure and pathogenicity of potato spindle tuber viroid (PSTV). EMBO J. 4, 2181–2190.

Schuck, J., Gursinsky, T., Pantaleo, V., Burgyan, J., Behrens, S.E., 2013. AGO/RISC-mediated antiviral RNA silencing in a plant in vitro system. Nucleic Acids Res. 41, 5090–5103.

Semancik, J.S., 2003. Pathogenesis. In: Hadidi, A., Flores, R., Randles, J.W., Semancik, J.S. (Eds.), Viroids. CSIRO Publishing, Collingwood, VIC, pp. 61–66.

Skoric, D., Conerly, M., Szychowski, J.A., Semancik, J.S., 2001. CEVd-induced symptom modification as a response to a host-specific temperature sensitive reaction. Virology 280, 115–123.

Vogt, U., Pelissier, T., Putz, A., Razvi, F., Fischer, R., Wassenegger, M., 2004. Viroid-induced RNA silencing of GFP-viroid fusion transgenes does not induce extensive spreading of methylation or transitive silencing. Plant J. 1, 107–118.

Wang, M.B., Bian, X.Y., Wu, L.M., Liu, L.X., Smith, N.A., Isenegger, D., et al., 2004. On the role of RNA silencing in the pathogenicity and evolution of viroids and viral satellites. Proc. Natl. Acad. Sci. USA 101, 3275–3280.

Wassenegger, M., Heimes, S., Riedel, L., Sänger, H.L., 1994. RNA-directed *de novo* methylation of genomic sequences in plants. Cell 76, 567–576.

Wassenegger, M., Spieker, R.L., Thalmeir, S., Gast, F.U., Riedel, L., Sänger, H.L., 1996. A single nucleotide substitution converts potato spindle tuber viroid (PSTVd) from a noninfectious to an infectious RNA for *Nicotiana tabacum*. Virology 226, 191–197.

Yaguchi, S., Takahashi, T., 1985. Studies on pathogenesis in viroid infections. 13. syndrome characteristics and endogenous indoleacetic-acid levels in cucumber plants incited by hop stunt viroid. Zeit. Pflanzen Krank Pflanzenschutz-J. Plant Dis. Protec. 92, 263–269.

CHAPTER

10

Changes in the Host Proteome and Transcriptome Induced by Viroid Infection

Robert A. Owens[1], *Gustavo Gómez*[2], *Purificación Lisón*[2] *and Vicente Conejero*[2]

[1]**U.S. Department of Agriculture, Beltsville, MD, United States**
[2]**Polytechnic University of Valencia-CSIC, Valencia, Spain**

INTRODUCTION

Their noncoding nature implies that the symptoms associated with viroid infection result from alterations in host gene expression. Over the years many different hypotheses to explain such changes have been put forward; only recently, however, has a series of technological advances together with an increasingly detailed understanding of information flow in plant cells (Fig. 10.1) allowed these hypotheses to be tested experimentally.

The study of protein changes accompanying infection by citrus exocortis viroid (CEVd) or potato spindle tuber viroid (PSTVd) led to the discovery of viroid-induced "pathogenesis-related" (PR) proteins as components of a general host response to pathogen attack (Conejero et al., 1990). Recent proteomic studies have led to the identification of additional pathogenesis- and defense-related proteins that are differentially expressed during viroid infection.

The Arabidopsis Genome Initiative (http://arabidopsis.org) and Tomato Genome Sequencing Project (http://solgenomics.net), have allowed thousands of individual genes and dozens of metabolic and regulatory pathways to be screened for changes in transcript levels following viroid infection. RNA silencing has emerged as an important mediator of viroid–host interaction (see Chapter 9: Viroid Pathogenesis

Viroids and Satellites.
DOI: http://dx.doi.org/10.1016/B978-0-12-801498-1.00010-3

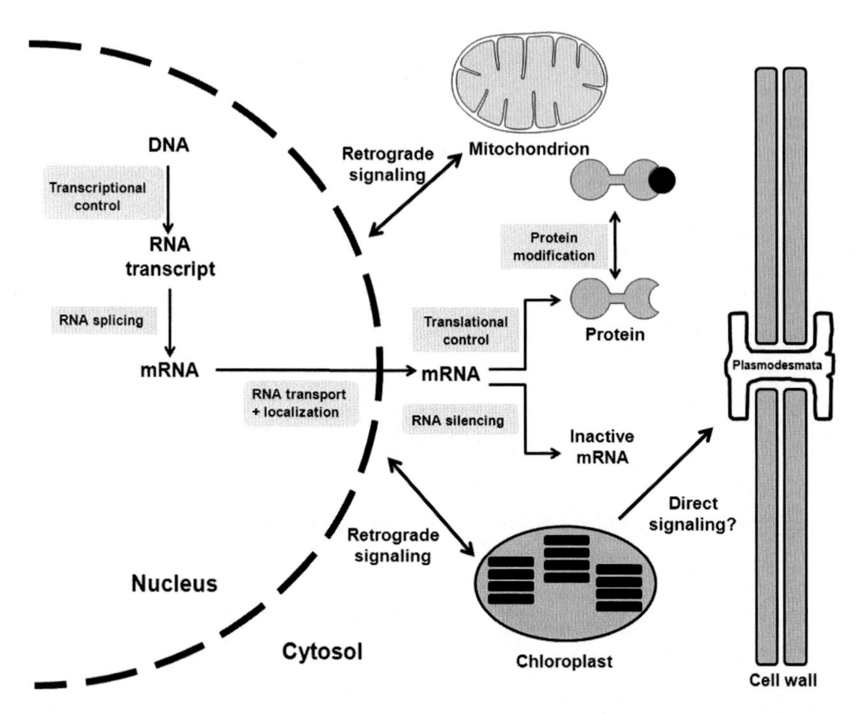

FIGURE 10.1 Regulation of plant gene expression. In addition to controls at the transcriptional, posttranscriptional, and translational levels, mRNA levels in plant cells are also sensitive to (1) retrograde signaling between plastids and nucleus (Chi et al., 2013) and (2) "regulation from without" involving the plasmodesmata (Burch-Smith and Zambryski, 2012).

and Chapter 11: Viroids and RNA Silencing), and genomics-based approaches provide an efficient means to compare expression levels of gene products containing potential binding sites for small viroid-derived RNAs.

Physical interaction between proteins and nucleic acids results in formation of molecular interaction networks; e.g., protein–protein interaction networks. Identification of such *interactomes* allows integration of data obtained by transcriptomic and proteomic approaches. The final section of this chapter describes interactions between viroids and certain host proteins and outlines some promising directions for future systems-based research.

PROTEIN CHANGES ASSOCIATED WITH VIROID INFECTION

Because viroids encode no proteins, changes in protein profiles during infection must involve only host-encoded proteins. Early studies used SDS-PAGE, density gradient centrifugation and in vivo labeling of plant proteins to demonstrate changes in the abundance of tomato PR proteins after viroid infection (Conejero et al., 1979; Hadidi, 1988; Henriquez and Sänger, 1982). Additional proteins that accumulate as a consequence of viroid infection were identified, but their roles in pathogenesis and defense reactions remain largely unknown.

Proteomics is a holistic approach to the study of cellular proteins. Separation methods exploit differences in protein size, charge, and/or hydrophobicity, and proteomic studies have been carried out to investigate plant responses to a wide variety of pathogens (see reviews by Delaunois et al., 2014; Jayaraman et al., 2012; Rampitsch and Bykova, 2012). In accordance with earlier proposals (e.g., Conejero et al., 1990), many common features have been observed (Fig. 10.2). Photosynthetic activity, for example, was reported to be reduced in almost every study, probably to conserve energy resources required for the defense response (Delaunois et al., 2014). Respiration and secondary metabolism also increase in many pathosystems. Additional proteins accumulating upon pathogen attack include PR proteins, antioxidant enzymes, and certain protein components of the translation machinery.

Using two-dimensional differential gel electrophoresis coupled with mass spectrometry to study the response in tomato to CEVd infection, Lisón et al. (2013) characterized 45 proteins. A direct correlation between changes in protein and mRNA levels (as determined by RT-PCR) was observed for defense-related proteins. Interestingly, changes in levels of a second group of proteins occurred without any alterations

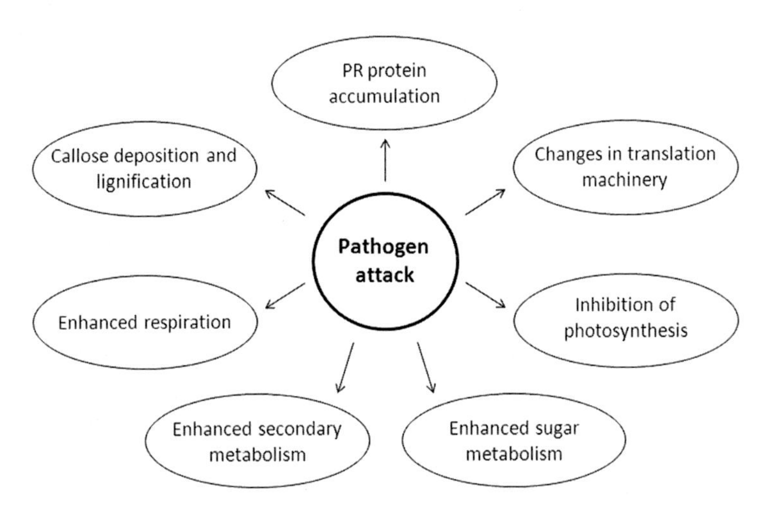

FIGURE 10.2 Global view of the host response to viroid infection illustrating the widespread nature of the resulting proteomic changes and targeting of multiple organelles.

in levels of their respective mRNAs. Members of this group include ribosomal proteins S3, S5, and L10 as well as eukaryotic translation factors eEF1A, eEF2, and eIF5A. These results illustrate the potential synergy between proteome and transcriptome analyses.

INTERACTION OF HOST PROTEINS WITH VIROIDS

Besides several proteins isolated from the nucleus (Wolff et al., 1985), many other cellular proteins have been shown to interact with PSTVd and other viroids. An interaction between hop stunt viroid and phloem lectin 2 apparently involved in long-distance movement is discussed in Chapter 8: Viroid Movement.

A tomato protein known as Virp1 (Viroid-binding protein 1), which contains a C-terminal RNA-binding domain, a nuclear localization signal, and a bromodomain (a structural feature commonly found in many chromatin remodeling factors) (Martínez de Alba et al., 2003), has been proposed to play different roles in the infection cycle. Virp1 may act to transfer PSTVd and related viroids to the nucleus and bring them into contact with the chromatin. Virp1 binding requires the presence of an "RY motif" that is located in the terminal right domain of PSTVd and related viroids. In *Nicotiana benthamiana* the orthologous protein (BRP1) mediates nuclear importation of highly structured cucumber mosaic virus Q-satRNA (Chaturvedi et al., 2014). Interestingly, a PSTVd mutant defective for Virp1 binding motif is unable to spread systemically (Hammond, 1994).

Additional host proteins interact with other PSTVd structural domains. RNA polymerase II interacts in vitro with the terminal left domain of PSTVd, and sequence changes affecting certain bulge loops reduce/abolish this interaction (Bojić et al., 2012). PSTVd also interacts with transcription factor IIIA and ribosomal protein L5, two *Arabidopsis* proteins that bind to the loop E motif of 5S rRNA (Eiras et al., 2011). Other proteins that can be inferred to interact with viroids include: (1) Dicer-like 1–4 and the corresponding Argonaute proteins (cleavage of viroid replicative intermediates into small interfering RNA); (2) nuclear DNA ligase 1 (pospiviroid circularization in the nucleus); (3) chloroplast tRNA ligase (circularization of chloroplast-replicating viroids); and (4) RNA-directed RNA polymerase 6 (inhibition of PSTVd movement into meristematic tissues).

Arabidopsis protein 4/1 and its orthologs from various *Nicotiana* spp. contain as many as five coiled–coil (CC) domains, and the tertiary structure of its C-terminal CC domain is strikingly similar to that of She2p, a yeast protein involved in actin-dependent RNA transport. Like She2p, 4/1 exhibits RNA-binding activity in vitro, and virus-induced silencing of 4/1 expression in *N. benthamiana* leads to altered patterns of PSTVd accumulation and movement. Localization studies indicate that this protein may play a key role in RNA-mediated signaling in the vasculature (Morozov et al., 2014).

Less is known about the interaction of cellular proteins with viroids that replicate in the chloroplast. Characterization of host protein adducts formed by UV irradiation of avocado leaf tissue infected by avocado sunblotch viroid revealed interactions involving two closely-related chloroplast RNA-binding proteins (Daròs and Flores, 2002). PARBP33 was shown to act as an RNA chaperone and facilitates the ribozyme-mediated self-cleavage of dimeric avocado sunblotch viroid transcripts. In a separate study (Dubé et al., 2009), northwestern analysis revealed interactions between peach latent mosaic viroid (PLMVd) and six proteins isolated from peach leaf tissue, one of which, eEF1A, also interacts with both PSTVd and CEVd.

TRANSCRIPTIONAL CHANGES ASSOCIATED WITH VIROID INFECTION

Microarray-based functional genomics has been widely used to monitor global changes in host transcription during plant–virus interaction (e.g., Wise et al., 2007). To date, four studies have examined the genome-wide effects of viroid infection on host gene expression.

The first study used a custom-designed macroarray containing 1,156 differentially expressed tomato cDNAs to compare gene expression patterns in Rutgers plants infected with mild or severe strains of PSTVd (Itaya et al., 2002). More genes, including many involved in defense/stress response, cell wall structure, chloroplast function, and protein metabolism, were induced than suppressed.

Next, Owens et al. (2012) used a combination of microarray and large-scale RNA sequence analysis to compare changes in gene expression and microRNA levels associated with PSTVd infection in two tomato cultivars. Transformed plants expressing PSTVd small interfering RNAs in the absence of viroid replication were also examined. For the sensitive cultivar "Rutgers" infection-related changes in transcript levels were extensive, involving more than half of the approximately 10,000 genes present on the Affymetrix array. Chloroplast biogenesis was downregulated in both sensitive and tolerant cultivars, and effects on mRNAs encoding enzymes involved in biosynthesis of gibberellin and other hormones were accompanied by numerous changes affecting their respective signaling pathways.

Using their Gene Ontology descriptions, tomato genes affected by viroid infection can be grouped according to three different criteria; i.e.,

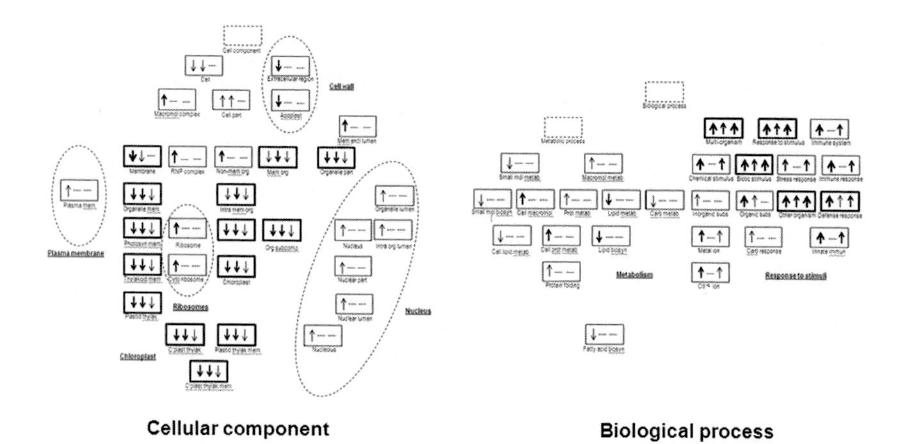

FIGURE 10.3 Effect of PSTVd infection on tomato gene expression. Groups of overrepresented genes associated with specific cellular components (left) or different biological functions (right) were identified for PSTVd-infected "Rutgers" plants showing strong symptoms. Dark outlines indicate groupings also present in the other two treatments, and arrows denote the relative degree of overrepresentation for, from left to right, infected "Rutgers," infected "Moneymaker," and uninfected "Moneymaker" plants expressing PSTVd small RNAs. *Source: Modified from Owens, R.A., Tech, K.B., Shao, J.Y., Sano, T., Baker, C.J., 2012. Global analysis of tomato gene expression during potato spindle tuber viroid infection reveals a complex array of changes affecting hormone signaling. Mol. Plant Microbe Interact. 25, 582–598.*

cellular component, biological process, or molecular function. Fig. 10.3 (left) illustrates the association of >5000 affected genes with different cellular components. A consistent downregulation of chloroplast function is immediately apparent across all three treatments. In contrast, many genes encoding proteins associated with other cellular compartments—nucleus, plasma membrane, ribosomes, cell wall, and apoplast—were upregulated in symptomatic "Rutgers" plants. Regrouping affected genes by biological function (Fig. 10.3, right) reveals additional differences among treatments. The dwarf tomato cultivar "MicroTom" contains a defect in brassinosteroid synthesis, and genes involved in response to stress and other stimuli were upregulated only when exogenous brassinosteroid was applied to infected plants, thereby providing the first evidence for the involvement of brassinosteroid-mediated signaling in viroid disease induction.

To determine how gene expression changes with time following viroid infection, Rizza et al. (2012) compared gene expression at early (presymptomatic) and late (postsymptomatic) stages of CEVd infection in Etrog citron using a citrus cDNA microarray. Only 132 putative citrus unigenes showed evidence of differential expression over time, and as observed for PSTVd-infected tomato, CEVd infection was shown to trigger changes in chloroplast, cell wall, peroxidase, and symporter activities.

Finally, microarray analysis has also been used to examine synergistic interactions between viroids and viruses. Herranz et al. (2013) used a specialized microarray to compare gene expression in fruits harvested from trees infected with PLMVd and/or prunus necrotic ringspot virus. Singly infected trees exhibited relatively few statistically significant changes. Fruits from doubly infected trees, in contrast, exhibited a large number of differentially expressed (mostly downregulated) genes. Functional categorization of the gene expression changes accompanying double PLMVd and prunus necrotic ringspot virus infection revealed protein modification and degradation as the category with the highest proportion of repressed genes. Other induced genes encoded proteins related to phosphate, C-compound, and carbohydrate metabolism. The changes observed are consistent with an active host counter-defense mechanism following pathogen infection.

Studies with other economically important species are currently frustrated by a lack of community-based genomic and postgenomic tools. One such host is chrysanthemum where infection by chrysanthemum stunt viroid is responsible for serious economic losses. Jo et al. (2014) used high-throughput next generation sequencing methods to characterize a total of 11,600 expressed sequence tags from a cDNA library derived from chrysanthemum stunt viroid-infected chrysanthemum leaf tissue that can be used to create a microarray suitable for transcriptome studies.

MOLECULAR MECHANISMS UNDERLYING TRANSCRIPTIONAL CHANGES

As discussed in detail elsewhere (see Chapter 9: Viroid Pathogenesis and Chapter 11: Viroids and RNA Silencing), small RNAs derived from viroids can act as elicitors of symptom expression by directing posttranscriptional cleavage of host mRNAs, or methylation of promoter regions. Several groups (e.g., Diermann et al., 2010; Ivanova et al., 2014; Tsushima et al., 2015) have reported changes in host microRNA levels following viroid infection. The extent of these changes and even their direction were quite variable, and no direct evidence for cleavage of the corresponding target gene is yet available. Nevertheless, the list of potential target genes includes a number of transcription factors and other regulatory proteins likely to occupy hub positions in the still-to-be-determined host interactomes (see below).

A GLIMPSE INTO THE FUTURE

To date, genome-wide studies of viroid pathogenesis have focused on effects on the host transcriptome. A meta-analysis comparing the effects of eight plant viruses on *Arabidopsis* gene expression (Rodrigo et al., 2012) illustrates the potential benefits of a broader *network-based approach* to future studies of viroid–host interaction. Lists of virus-responsive genes were mapped onto *Arabidopsis* transcriptional (Carrera et al., 2009) and protein–protein interaction (Geisler-Lee et al., 2007) networks, thereby revealing significant reprogramming of the host cell transcriptome during infection, possibly as a central requirement to activate host defenses. Affected genes were highly connected, located in the center of networks, and organized in modules.

Transcriptional and protein–protein interaction networks are based on information about steady-state transcript abundance, and attempts to compare molecular networks from different plant species are expanding rapidly (Braun et al., 2013). As valuable as such information can be, understanding the spatiotemporal aspects of viroid–host interaction will require additional *microgenomics technologies* that allow genome-wide quantitation of cell-type-specific gene expression (Bailey-Serres, 2013). Nearly 50 years after the discovery of viroids, we may finally be beginning to understand how these small, non-coding RNAs cause disease.

References

Bailey-Serres, J., 2013. Microgenomics: genome-scale, cell-specific monitoring of multiple gene regulation tiers. Annu. Rev. Plant Biol. 64, 293–325.

Bojić, T., Beeharry, Y., Zhang, D.J., Pelchat, M., 2012. Tomato RNA polymerase II interacts with the rod-like conformation of the left terminal domain of the potato spindle tuber viroid positive RNA genome. J. Gen. Virol. 93, 1591–1600.

Braun, P., Aubourg, S., Van Leene, J., De Jaeger, G., Lurin, C., 2013. Plant protein interactomes. Annu. Rev. Plant Biol. 64, 161–187.

Burch-Smith, T.M., Zambryski, P.C., 2012. Plasmodesmata paradigm shift: regulation from without versus within. Annu. Rev. Plant Biol. 63, 239–260.

Carrera, J., Rodrigo, G., Jaramillo, A., Elena, S.F., 2009. Reverse-engineering the *Arabidopsis thaliana* transcriptional network under changing environmental conditions. Genome Biol. 10, R96.

Chaturvedi, S., Kalantidis, K., Rao, A.L., 2014. A bromodomain-containing host protein mediates the nuclear importation of a satellite RNA of cucumber mosaic virus. J. Virol. 88, 1890–1896.

Chi, W., Sun, X.W., Zhang, L.X., 2013. Intracellular signaling from plastid to nucleus. Annu. Rev. Plant Biol. 64, 559–582.

Conejero, V., Bellés, J.M., García-Breijo, F., Garro, R., Hernández-Yago, J., Rodrigo, I., et al., 1990. Signaling in viroid pathogenesis. In: Fraser, R.S.S. (Ed.), Recognition and Response in Plant-Virus Interactions. Springer-Verlag, Berlin, pp. 233–261.

Conejero, V., Picazo, I., Segado, P., 1979. Citrus exocortis viroid (CEV): protein alterations in different hosts following viroid infection. Virology 97, 454–456.

Daròs, J.A., Flores, R., 2002. A chloroplast protein binds a viroid RNA in vivo and facilitates its hammerhead-mediated self-cleavage. EMBO J. 21, 749–759.

Delaunois, B., Jeandet, P., Clément, C., Baillieul, F., Dorey, S., Cordelier, S., 2014. Uncovering plant-pathogen crosstalk through apoplastic proteomic studies. Front. Plant Sci. 5, 249.

Diermann, N., Matoušek, J., Junge, M., Riesner, D., Steger, G., 2010. Characterization of plant miRNAs and small RNAs derived from potato spindle tuber viroid (PSTVd) in infected tomato. Biol. Chem. 391, 1379–1390.

Dubé, A., Bisaillon, M., Perreault, J.P., 2009. Identification of proteins from *Prunus persica* that interact with peach latent mosaic viroid. J. Virol. 83, 12057–12067.

Eiras, M., Nohales, M.A., Kitajima, E.W., Flores, R., Daròs, J.A., 2011. Ribosomal protein L5 and transcription factor IIIA from *Arabidopsis thaliana* bind *in vitro* specifically potato spindle tuber viroid RNA. Arch. Virol. 156, 529–533.

Geisler-Lee, J., O'Toole, N., Ammar, R., Provart, N.J., Harvey Millar, A., Geisler, M., 2007. A predicted interactome for *Arabidopsis*. Plant Physiol. 145, 317–329.

Hadidi, A., 1988. Synthesis of disease associated proteins in viroid-infected tomato leaves and binding of viroid to host proteins. Phytopathology 78, 575–578.

Hammond, R.W., 1994. *Agrobacterium*-mediated inoculation of PSTVd cDNAs onto tomato reveals the biological effect of apparently lethal mutations. Virology 201, 36–45.

Henriquez, C.A., Sänger, H.L., 1982. Analysis of acid-extractable tomato leaf proteins after infection with a viroid, two viruses, and a fungus and partial purification of the "pathogenesis-related" protein p14. Arch. Virol. 74, 181–196.

Herranz, M.C., Niehl, A., Rosales, M., Fiore, N., Zamorano, A., Granell, A., et al., 2013. A remarkable synergistic effect at the transcriptomic level in peach fruits doubly infected by prunus necrotic ringspot virus and peach latent mosaic viroid. Virol. J. 10, 164.

Itaya, A., Matsuda, Y., Gonzales, R.A., Nelson, R.S., Ding, B., 2002. Potato spindle tuber viroid strains of different pathogenicity induces and suppresses expression of common and unique genes in infected tomato. Mol. Plant Microbe Interact. 15, 990–999.

Ivanova, D., Milev, I., Vachev, T., Baev, V., Yahubyan, G., Minkov, G., et al., 2014. Small RNA analysis of potato spindle tuber viroid infected *Phelipanche ramosa*. Plant Physiol. Biochem. 74, 276–282.

Jayaraman, D., Forshey, K.L., Grimsrud, P.A., Ané, J.M., 2012. Leveraging proteomics to understand plant-microbe interactions. Front. Plant Sci. 3, 44.

Jo, Y.-H., Jo, K.-M., Park, S.-H., Kim, K.-H., Cho, W.-K., 2014. Transcriptomic landscape of chrysanthemums infected by chrysanthemum stunt viroid. Plant Omics J. 7, 1–11.

Lisón, P., Tárraga, S., Lápez-Gresa, P., Saurı́, A., Torres, C., Campos, L., et al., 2013. A non-coding plant pathogen provokes both transcriptional and posttranscriptional alterations in tomato. Proteomics 13, 833–844.

Martínez de Alba, A.E., Sägesser, R., Tabler, M., Tsagris, M., 2003. A bromodomain-containing protein from tomato specifically binds potato spindle tuber viroid RNA *in vitro* and *in vivo*. J. Virol. 77, 9685–9694.

Morozov, S.Y., Makarova, S.S., Erokhina, T.N., Kopertekh, L., Schiemann, J., Owens, R.A., et al., 2014. Plant 4/1 protein: potential player in intracellular, cell-to-cell and long-distance signaling. Front. Plant Sci. 5, 26.

Owens, R.A., Tech, K.B., Shao, J.Y., Sano, T., Baker, C.J., 2012. Global analysis of tomato gene expression during potato spindle tuber viroid infection reveals a complex array of changes affecting hormone signaling. Mol. Plant Microbe Interact. 25, 582–598.

Rampitsch, C., Bykova, N.V., 2012. The beginnings of crop phosphoproteomics: exploring early warning systems of stress. Front. Plant Sci. 3, 144.

Rizza, S., Conesa, A., Juarez, J., Catara, A., Navarro, L., Duran-Vila, N., et al., 2012. Microarray analysis of Etrog citron (*Citrus medica* L.) reveals changes in chloroplast, cell wall, peroxidase and symporter activities in response to viroid infection. Mol. Plant Pathol. 13, 852–864.

Rodrigo, G., Carrera, J., Ruiz-Ferrer, V., del Toro, F.J., Llave, C., Voinnet, O., et al., 2012. A meta-analysis reveals the commonalities and differences in *Arabidopsis thaliana* response to different viral pathogens. PLoS One 7, e40526.

Tsushima, D., Adkar-Purushothama, C.R., Taneda, A., Sano, T., 2015. Changes in relative expression levels of viroid-specific small RNAs and microRNAs in tomato plants infected with severe and mild symptom-inducing isolates of potato spindle tuber viroid. J. Gen. Plant Pathol. 81, 49–62.

Wise, R.P., Moscou, M.J., Bogdanove, A.J., Whitham, S.A., 2007. Transcript profiling in host-pathogen interactions. Annu. Rev. Phytopathol. 45, 329–369.

Wolff, P., Gilz, R., Schumacher, J., Riesner, D., 1985. Complexes of viroids with histones and other proteins. Nucleic Acids Res. 13, 355–367.

CHAPTER

11

Viroids and RNA Silencing

Elena Dadami, Athanasios Dalakouras and Michael Wassenegger

RLP AgroScience, AlPlanta-Institute for Plant Research, Neustadt-Mußbach, Germany

RNA SILENCING IN PLANTS

RNA silencing is induced by double-stranded RNA (dsRNA). Sources of dsRNA include replication intermediates of viruses/viroids, transcription of inverted repeats, stress-induced overlapping antisense transcripts, and RNA-directed RNA polymerase (RDR) transcription of aberrant transcripts (Wassenegger and Krczal, 2006). The Dicer-like (DCL) endonucleases DCL1/DCL4, DCL2, and DCL3 process dsRNA into 21-, 22-, and 24-nucleotide (nt) siRNAs, respectively (Blevins et al., 2006; Bouche et al., 2006; Deleris et al., 2006; Fusaro et al., 2006). Mature siRNAs exhibit 3′ 2-nt overhangs and become stabilized through 3′ end methylation by Hue enhancer 1 (HEN1) protein (Yang et al., 2006). Depending on their 5′ terminal nucleotide, siRNAs are loaded onto specific Argonaute (AGO) proteins. The formation of functional RNA-induced silencing complexes involves retention of one, the siRNA guide strand, and release of the other, the siRNA passenger strand (Kim, 2008). In general, 21-nt siRNAs with a 5′ terminal U are loaded onto AGO1 and mediate cleavage of complementary RNA in a process termed posttranscriptional gene silencing (PTGS) (Hamilton and Baulcombe, 1999; Martínez de Alba et al., 2013). Importantly, besides transcript degradation, siRNAs can also mediate translational arrest (Brodersen et al., 2008). In the nucleus, AGO4-loaded 24-nt siRNAs are suggested to guide Domains Rearranged Methyltransferase 2 to cognate DNA leading to de novo methylation (Matzke and Mosher, 2014; Matzke et al., 2015) in a process termed RNA-directed DNA methylation (RdDM) (Wassenegger et al., 1994).

Viroids and Satellites.
DOI: http://dx.doi.org/10.1016/B978-0-12-801498-1.00011-5

VIROIDS TRIGGER RNA SILENCING

The family *Pospiviroidae* contains more than 24 species, including *Potato spindle tuber viroid* (type species), *Citrus exocortis viroid, Hop latent viroid*, and *Hop stunt viroid*, while the family *Avsunviroidae* includes *Avocado sunblotch viroid* (type species), *Chrysanthemum chlorotic mottle viroid, Eggplant latent viroid*, and *Peach latent mosaic viroid* (Flores et al., 2000, 2004, 2014; Tabler and Tsagris, 2004; Tsagris et al., 2008).

Similarly to viruses and viral satellite RNAs, replicating viroids elicit the plant RNA silencing machinery (Itaya et al., 2001; Papaefthimiou et al., 2001; Wang et al., 2004). To defend themselves, plants have developed strategies to combat viroid infection. All four DCLs seem to process viroid RNAs (Dadami et al., 2013; Navarro et al., 2009). Mature viroid RNA molecules are directly targeted by DCLs mediating their endonucleolytic degradation into viroid-derived siRNAs (vd-siRNAs) (Itaya et al., 2007). The finding that transcripts of potato spindle tuber viroid (PSTVd) and peach latent mosaic viroid (PLMVd) were processed into vd-siRNAs when incubated with DCL-containing in vitro extracts supports the processing of mature viroid RNA (Itaya et al., 2007; Landry and Perreault, 2005). In addition, vd-siRNAs could be processed from dsRNA viroid replication intermediates.

In plants, aberrant transcripts, such as RNAs lacking a 5′ cap and/or 3′ polyadenylated tail, are copied into dsRNA by RDR6 in the cytoplasm and most probably also in the nucleus (Gazzani et al., 2004; Luo and Chen, 2007). It is reasonable to assume that viroids or their replication intermediates resemble such aberrant RNAs. The notion that RDR6 indeed transcribes PSTVd RNA was indirectly validated by studying viroid infection in an *rdr6* knockdown mutant of *Nicotiana benthamiana* (Di Serio et al., 2010). It was observed that RDR6 is associated with a delay of PSTVd accumulation and with its exclusion from the meristem. In addition to RDR6, Cajal body-located RDR2 may also copy oligomeric (+)-transcripts of viroids in the family *Pospiviroidae* that accumulate in the nucleus (Tabler and Tsagris, 2004). Whether other RDRs are also involved and whether members of the family *Avsunviroidae* are RDR-processed, is not clear. Interestingly, RDR1 was shown to be significantly upregulated in PSTVd-infected tomato plants (Schiebel et al., 1998). Irrespective of their origin, vd-siRNAs are loaded onto AGOs (Minoia et al., 2014) to target the mature viroid and/or the replication intermediates (cis-PTGS). In addition to cis-PTGS, vd-siRNAs may also target host genes for PTGS (trans-PTGS) and RdDM (see below).

DCL-PROCESSING OF VIROIDS INTO VD-SIRNAS

Immunofluorescence analysis revealed that the functionally redundant *Arabidopsis* DCL1, DCL2, DCL3, and DCL4 colocalize in the nuclear periphery (Pontes et al., 2013). Members of both the *Pospiviroidae* and *Avsunviroidae* are processed by DCLs (Martínez de Alba et al., 2002; Papaefthimiou et al., 2001). In members of the family *Pospiviroidae*, nuclear DCL-processing could take place during their replication, while in the members of the family *Avsunviroidae* this could conceivably occur during their nuclear trafficking prior to their movement into chloroplasts (Gómez and Pallás, 2012). Importantly, no 24-nt siRNAs of chloroplast-replicating viroids accumulate in infected plants indicating that DCL3 has no access to these RNA molecules or that it is not able to process them. In addition, it cannot be excluded that viroids are targeted by DCLs in the cytoplasm during their cell-to-cell and long distance movement.

The finding that vd-siRNAs map at hotspot regions not covering the entire viroid sequence reinforced the idea that viroid single-stranded RNAs (ssRNAs), rather than dsRNAs, are processed by DCLs (Itaya et al., 2007). However, it should be noted that DCL-processing of even perfect dsRNAs generally does not result in homogeneous small RNA (sRNA) populations mapping along the entire dsRNA length, but also exhibits hotspot regions (Sasaki et al., 2014). In transgenic tomato plants expressing a hairpin-PSTVd transgene, hairpin-derived siRNAs showed hotspot distribution patterns similar to those that were detected in natural infections (Schwind et al., 2009). Whether hotspots reflect differential DCL-processing and/or AGO loading/stability, or just simply sRNA cloning/sequencing bias is not clear. In any case, siRNA hotspots do not necessarily indicate the processing of ssRNA substrates. It is important to note that, due to mismatches, perfectly paired regions within mature viroid RNA molecules are not longer than 14 base pairs (bp) (Wang et al., 2004). A 14-bp dsRNA stretch is below the minimum DCL processivity requirement of 19 bp of perfect duplex dsRNA (Yu et al., 2002). On the other hand, the observation that both polarities of vd-siRNAs have been detected by deep sequencing and northern blot analysis (Itaya et al., 2007; Machida et al., 2007; Wang et al., 2011), does not necessarily suggest "dsRNA processing." One may hypothesize that, besides dsRNAs, ssRNA viroid replication intermediates of (−) polarity can also be processed, although their accumulation level is considerably lower than that of (+) polarity. In summary, the currently available data do not suggest that DCL processes only either dsRNA or ssRNA of viroids. Most probably, both processes take place.

Vd-siRNAs from the following members of the families *Pospiviroidae* and *Avsunviroidae* have been detected by northern blot analysis and/or by sRNA deep sequencing (sRNA-seq): PSTVd (Dadami et al., 2013; Di Serio et al., 2010; Diermann et al., 2010; Itaya et al., 2001; Machida et al., 2007; Matousek et al., 2007; Papaefthimiou et al., 2001; Wang et al., 2011), avocado sunblotch viroid (Markarian et al., 2004), citrus exocortis viroid (Martin et al., 2007), hop stunt viroid (Gómez et al., 2008; Martínez et al., 2010; Navarro et al., 2009), PLMVd (Bolduc et al., 2010; Di Serio et al., 2009; Martínez de Alba et al., 2002), chrysanthemum chlorotic mottle viroid (Martínez de Alba et al., 2002), grapevine yellow speckle viroid 1 (Navarro et al., 2009), apple dimple fruit viroid (Chiumenti et al., 2014), and coconut cadang-cadang viroid (Vadamalai, 2005; Chapter 34, Gel Electrophoresis). They were found to be phosphorylated at their 5′-end, methylated at their 3′end by HEN1 (Martin et al., 2007) and in all cases, both polarities have been detected. Vd-siRNAs(+) map to hotspot regions along the viroid genome, whereas vd-siRNAs(−) are distributed more homogeneously along the viroid genome (Itaya et al., 2007; Wang et al., 2011). Whether this finding is of biological relevance or simply reflects sequencing bias and/or differential vd-siRNA stability is not clear. It is also not clear whether vd-siRNAs are perfect duplexes (processed from dsRNAs) or have asymmetric bulges (processed from mature viroid RNA molecules or single-stranded replication intermediates). However, this could be of particular biological significance since asymmetric bulge-containing sRNAs may be implicated in transitive silencing through initiation of conformational changes of AGO1 (see below) (Chen et al., 2010; Cuperus et al., 2010; Manavella et al., 2012; McHale et al., 2013).

Coimmunoprecipitation analyses of AGO proteins from *Arabidopsis* transiently expressed in PSTVd-infected *N. benthamiana* revealed that AGO1, AGO2, and AGO3 preferentially bind 21- and 22-nt vd-siRNAs, whereas AGO4, AGO5, and AGO9 additionally coimmunoprecipitated with 24-nt vd-siRNAs (Minoia et al., 2014). Whether AGO-loaded vd-siRNAs are present in both the cytoplasm and nucleus is not clear. At least in tomato, PSTVd derived-siRNAs were reported to only accumulate in the cytoplasm (Denti et al., 2004). Accordingly, PSTVd siRNAs triggered PTGS in the cytoplasm while target RNA cleavage was not detected in the nucleus of tobacco cells (Dalakouras et al., 2015).

VIROIDS AS TARGETS FOR PTGS

Due to their extensive secondary structure, viroids are suggested to be resistant to PTGS (Itaya et al., 2007; Wang et al., 2004). Obviously, during natural infections, the vd-siRNAs produced fail to eliminate the corresponding viroid. However, targeting of viroids by PTGS has been demonstrated as follows. (1) Transgenic tomato plants highly expressing a hairpin-PSTVd transgene were resistant to viroid infection (Schwind et al., 2009). (2) Transgenic *N. benthamiana* rootstocks expressing a hairpin-PSTVd transgene under a strong companion cell-specific promoter attenuated symptoms of a viroid-infected scion (Kasai et al., 2013). (3) Viroid infection was impaired, in tomato and chrysanthemum, by mechanical coinoculation of PSTVd and chrysanthemum chlorotic mottle viroid with dsRNAs sharing homology with the corresponding viroid (Carbonell et al., 2008). (4) Transient expression of AGO1, AGO2, AGO4 and AGO5 through agroinfiltration in PSTVd-infected tissue of *N. benthamiana* attenuated the level of viroid RNAs, suggesting that they, or their precursors, are siRNA-targeted (Minoia et al., 2014). (5) Coinfection of Mexican lime with citrus dwarfing viroid and citrus tristeza virus resulted in enhanced viroid titers due to silencing by one of the virus suppressors, indicating that viroids do not fully evade cis-PTGS (Serra et al., 2014). Vd-siRNAs seemingly fail to trigger nuclear PTGS (Dalakouras et al., 2015). Thus, it is reasonable to assume that viroids are only targeted in the cytoplasm, probably when they are moving from the nucleus (*Pospiviroidae*) and chloroplast (*Avsunviroidae*) into neighboring cells, respectively.

VIROID-INDUCED TRANS-PTGS OF ENDOGENES

Vd-siRNAs may target homologous host transcripts for trans-PTGS. Thus, viroid-induced symptoms could be based on silencing of host genes (Hammann and Steger, 2012; Papaefthimiou et al., 2001; Wang et al., 2004). Support for this assumption is derived from several observations. (1) In tomato, expression of a hairpin-RNA transgene consisting of a nearly full-length PSTVd sequence led to phenotypes resembling PSTVd disease symptoms (Wang et al., 2004). (2) In *Nicotiana* species, an artificial miRNA corresponding to the virulence modulating region of PSTVd was predicted to direct PTGS of a soluble inorganic pyrophosphatase mRNA and led to the development of phenotypes (Eamens et al., 2014). (3) In PSTVd-infected tomato, the accumulation of several endogenous miRNAs is suppressed, suggesting vd-siRNA-induced downregulation of the respective pre-miRNAs (Diermann et al., 2010). (4) Two tomato genes that are involved in gibberellin and jasmonic acid biosynthesis contain binding sites for vd-siRNAs in their ORFs and are downregulated early in PSTVd infection (Wang et al., 2011). (5) In PLMVd-infected peach (*Prunus persica*), vd-siRNAs derived from the pathogenic determinant targeted a host mRNA for cleavage as predicted by RNA silencing (Navarro et al., 2012). (6) In tomato planta macho viroid-infected tomato plants, a host mRNA was predicted and validated to be cleaved by vd-siRNAs (Avina-Padilla et al., 2015). (7) In PSTVd-infected tomato, a vd-siRNA derived from the virulence modulating region of PSTVd silenced the callose synthase genes CalS11-like and CalS12-like by targeting them for PTGS (Adkar-Purushothama et al., 2015).

Vd-siRNAs are not necessarily perfect duplexes but may contain asymmetric bulges. Such siRNAs are suggested to change AGO1 into a structure that enables the initiation of transitive silencing by recruiting RDR6 (Manavella et al., 2012). One may speculate that asymmetric bulge-containing vd-siRNAs are targeting endogenous transcripts not only for trans-PTGS but also for the induction of transitivity. Transitivity involves the production of secondary siRNAs that could have the potential to target additional host genes not sharing homology with the viroid (off-target effect). It should be noted that in *N. benthamiana*, HSVd-induced symptoms are RDR6-dependent (Gómez et al., 2008), although PSTVd-induced symptoms in *N. benthamiana* are RDR6-independent (Di Serio et al., 2010). Importantly, it was recently shown that activation of antiviral RNA silencing in *Arabidopsis* is accompanied by RDR1-dependent production of virus-activated 21-nt siRNAs (vasiRNAs). These vasiRNAs are transcribed from host genes and trigger the silencing of hundreds of host genes (Cao et al., 2014). Whether viroid infection triggers a similar response is currently not known.

VIROID-INDUCED RDDM

In addition to the activation of PTGS, members of the *Pospiviroidae* have the potential to trigger RdDM of homologous sequences (Wassenegger et al., 1994; Wassenegger, 2000). Chloroplast DNA is not methylated, but, so far, it has not been determined whether members of the *Avsunviroidae* can also trigger RdDM of cognate nuclear DNA sequences. RdDM was discovered in PSTVd-infected tobacco plants that contained multimeric genome-integrated PSTVd cDNA copies (Wassenegger et al., 1994). Viroid-induced RdDM is highly sequence-specific, requires a minimal target size of about 30 bp (Pelissier and Wassenegger, 2000) but could be inhibited by flanking DNA sequences (Dalakouras et al., 2010). It results in dense cytosine methylation in CG, CHG, and CHH contexts (Pelissier et al., 1999) of which only CG methylation is transgenerationally maintained in viroid-free progeny derived from viroid-infected plants (Dalakouras et al., 2012). The overall biological significance of viroid-induced DNA methylation during infection is not clear. In cucumber, HSVd infection induced dynamic changes in the DNA methylation of ribosomal RNA genes (Martínez et al., 2014) indicating that epigenetics may have an impact on viroid symptom initiation. How viroids induce RdDM is not clear. The current RdDM models suggest that 24-nt siRNAs guide the RdDM machinery to their targets (Matzke and Mosher, 2014; Matzke et al., 2015). However, it seems likely that siRNAs are not the actual RdDM-guide molecules but, instead, are involved in an intermediate step producing longer dsRNA molecules which would guide the RdDM machinery (Dalakouras et al., 2013; Dalakouras and Wassenegger, 2013). Further studies are needed to illuminate if viroid infection is indeed associated with *de novo* methylation of host genes that are involved in symptom development.

References

Adkar-Purushothama, C.R., Brosseau, C., Giguere, T., Sano, T., Moffett, P., Perreault, J.P., 2015. Small RNA derived from the virulence modulating region of the potato spindle tuber viroid silences callose synthase genes of tomato plants. Plant Cell 27, 2178–2194.

Avina-Padilla, K., Martínez de la Vega, O., Rivera-Bustamante, R., Martínez-Soriano, J.P., Owens, R.A., et al., 2015. In silico prediction and validation of potential gene targets for pospiviroid-derived small RNAs during tomato infection. Gene. 564, 197–205.

Blevins, T., Rajeswaran, R., Shivaprasad, P.V., Beknazariants, D., Si-Ammour, A., Park, H.S., et al., 2006. Four plant Dicers mediate viral small RNA biogenesis and DNA virus induced silencing. Nucleic Acids Res. 34, 6233–6246.

Bolduc, F., Hoareau, C., St-Pierre, P., Perreault, J.P., 2010. In-depth sequencing of the siRNAs associated with peach latent mosaic viroid infection. BMC Mol. Biol. 11, 16.

Bouche, N., Lauressergues, D., Gasciolli, V., Vaucheret, H., 2006. An antagonistic function for Arabidopsis DCL2 in development and a new function for DCL4 in generating viral siRNAs. EMBO J. 25, 3347–3356.

Brodersen, P., Sakvarelidze-Achard, L., Bruun-Rasmussen, M., Dunoyer, P., Yamamoto, Y.Y., Sieburth, L., et al., 2008. Widespread translational inhibition by plant miRNAs and siRNAs. Science. 320, 1185–1190.

Cao, M., Du, P., Wang, X., Yu, Y.Q., Qiu, Y.H., Li, W., et al., 2014. Virus infection triggers widespread silencing of host genes by a distinct class of endogenous siRNAs in Arabidopsis. Proc. Natl. Acad. Sci. USA 111, 14613–14618.

Carbonell, A., Martínez de Alba, A.E., Flores, R., Gago, S., 2008. Double-stranded RNA interferes in a sequence-specific manner with the infection of representative members of the two viroid families. Virology 371, 44–53.

Chen, H.M., Chen, L.T., Patel, K., Li, Y.H., Baulcombe, D.C., Wu, S.H., 2010. 22-Nucleotide RNAs trigger secondary siRNA biogenesis in plants. Proc. Natl. Acad. Sci. USA 107, 15269–15274.

Chiumenti, M., Torchetti, E.M., Di Serio, F., Minafra, A., 2014. Identification and characterization of a viroid resembling apple dimple fruit viroid in fig (Ficus carica L.) by next generation sequencing of small RNAs. Virus Res. 188, 54–59.

Cuperus, J.T., Carbonell, A., Fahlgren, N., Garcia-Ruiz, H., Burke, R.T., Takeda, A., et al., 2010. Unique functionality of 22-nt miRNAs in triggering RDR6-dependent siRNA biogenesis from target transcripts in Arabidopsis. Nat. Struct. Mol. Biol. 17, 997–1003.

Dadami, E., Boutla, A., Vrettos, N., Tzortzakaki, S., Karakasilioti, I., Kalantidis, K., 2013. DICER-LIKE 4 but not DICER- LIKE 2 may have a positive effect on potato spindle tuber viroid accumulation in *Nicotiana benthamiana*. Mol. Plant 6, 232–234.

Dalakouras, A., Dadami, E., Bassler, A., Zwiebel, M., Krczal, G., Wassenegger, M., 2015. Replicating potato spindle tuber viroid mediates de novo methylation of an intronic viroid sequence but no cleavage of the corresponding pre-mRNA. RNA Biol. 12, 268–275.

Dalakouras, A., Dadami, E., Wassenegger, M., 2013. Viroid-induced DNA methylation in plants. BioMol. Concepts 4, 557–565.

Dalakouras, A., Dadami, E., Zwiebel, M., Krczal, G., Wassenegger, M., 2012. Transgenerational maintenance of transgene body CG but not CHG and CHH methylation. Epigenetics 7, 1071–1078.

Dalakouras, A., Moser, M., Krczal, G., Wassenegger, M., 2010. A chimeric satellite transgene sequence is inefficiently targeted by viroid-induced DNA methylation in tobacco. Plant Mol. Biol. 73, 439–447.

Dalakouras, A., Wassenegger, M., 2013. Revisiting RNA-directed DNA methylation. RNA Biol. 10, 453–455.

Deleris, A., Gallego-Bartolome, J., Bao, J., Kasschau, K.D., Carrington, J.C., Voinnet, O., 2006. Hierarchical action and inhibition of plant dicer-like proteins in antiviral defense. Science 313, 68–71.

Denti, M.A., Boutla, A., Tsagris, M., Tabler, M., 2004. Short interfering RNAs specific for potato spindle tuber viroid are found in the cytoplasm but not in the nucleus. Plant J. 37, 762–769.

Di Serio, F., Gisel, A., Navarro, B., Delgado, S., Martínez de Alba, A.E., Donvito, G., et al., 2009. Deep sequencing of the small RNAs derived from two symptomatic variants of a chloroplastic viroid: implications for their genesis and for pathogenesis. PLoS One 4, e7539.

Di Serio, F., Martínez de Alba, A.E., Navarro, B., Gisel, A., Flores, R., 2010. RNA-dependent RNA polymerase 6 delays accumulation and precludes meristem invasion of a viroid that replicates in the nucleus. J. Virol. 84, 2477–2489.

Diermann, N., Matousek, J., Junge, M., Riesner, D., Steger, G., 2010. Characterization of plant miRNAs and small RNAs derived from potato spindle tuber viroid (PSTVd) in infected tomato. Biol. Chem. 391, 1379–1390.

Eamens, A.L., Smith, N.A., Dennis, E.S., Wassenegger, M., Wang, M.B., 2014. In Nicotiana species, an artificial microRNA corresponding to the virulence modulating region of potato spindle tuber viroid directs RNA silencing of a soluble inorganic pyrophosphatase gene and the development of abnormal phenotypes. Virology 450–451, 266–277.

Flores, R., Daros, J.A., Hernández, C., 2000. Avsunviroidae family: viroids containing hammerhead ribozymes. Adv. Virus Res. 55, 271–323.

Flores, R., Delgado, S., Gas, M.E., Carbonell, A., Molina, D., Gago, S., De la Pena, M., 2004. Viroids: the minimal non-coding RNAs with autonomous replication. FEBS Lett. 567, 42–48.

Flores, R., Gago-Zachert, S., Serra, P., Sanjuán, R., Elena, S.F., 2014. Viroids: survivors from the RNA world. Annu. Rev. Microbiol. 68, 395–414.

Fusaro, A.F., Matthew, L., Smith, N.A., Curtin, S.J., Dedic-Hagan, J., Ellacott, G.A., et al., 2006. RNA interference-inducing hairpin RNAs in plants act through the viral defence pathway. EMBO Rep. 7, 1168–1175.

Gazzani, S., Lawrenson, T., Woodward, C., Headon, D., Sablowski, R., 2004. A link between mRNA turnover and RNA interference in Arabidopsis. Science 306, 1046–1048.

Gómez, G., Martínez, G., Pallás, V., 2008. Viroid-induced symptoms in *Nicotiana benthamiana* plants are dependent on RDR6 activity. Plant Physiol. 148, 414–423.

Gómez, G., Pallás, V., 2012. Studies on subcellular compartmentalization of plant pathogenic noncoding RNAs give new insights into the intracellular RNA-traffic mechanisms. Plant Physiol. 159, 558–564.

Hamilton, A.J., Baulcombe, D.C., 1999. A species of small antisense RNA in posttranscriptional gene silencing in plants. Science 286, 950–952.

Hammann, C., Steger, G., 2012. Viroid-specific small RNA in plant disease. RNA Biol. 9, 809–819.

Itaya, A., Folimonov, A., Matsuda, Y., Nelson, R.S., Ding, B., 2001. Potato spindle tuber viroid as inducer of RNA silencing in infected tomato. Mol. Plant Microbe Interact. 14, 1332–1334.

Itaya, A., Zhong, X., Bundschuh, R., Qi, Y., Wang, Y., Takeda, R., et al., 2007. A structured viroid RNA serves as a substrate for dicer-like cleavage to produce biologically active small RNAs but is resistant to RNA-induced silencing complex-mediated degradation. J. Virol. 81, 2980–2994.

Kasai, A., Sano, T., Harada, T., 2013. Scion on a stock producing siRNAs of potato spindle tuber viroid (PSTVd) attenuates accumulation of the viroid. PLoS One 8, e57736.

Kim, V.N., 2008. Sorting out small RNAs. Cell 133, 25–26.

Landry, P., Perreault, J.P., 2005. Identification of a peach latent mosaic viroid hairpin able to act as a dicer-like substrate. J. Virol. 79, 6540–6543.

Luo, Z., Chen, Z., 2007. Improperly terminated, unpolyadenylated mRNA of sense transgenes is targeted by RDR6-mediated RNA silencing in Arabidopsis. Plant Cell 19, 943–958.

Machida, S., Yamahata, N., Watanuki, H., Owens, R.A., Sano, T., 2007. Successive accumulation of two size classes of viroid-specific small RNA in potato spindle tuber viroid-infected tomato plants. J. Gen. Virol. 88, 3452–3457.

Manavella, P.A., Koenig, D., Weigel, D., 2012. Plant secondary siRNA production determined by microRNA-duplex structure. Proc. Natl. Acad. Sci. USA 109, 2461–2466.

Markarian, N., Li, H.W., Ding, S.W., Semancik, J.S., 2004. RNA silencing as related to viroid induced symptom expression. Arch. Virol. 149, 397–406.

Martin, R., Arenas, C., Daros, J.A., Covarrubias, A., Reyes, J.L., Chua, N.H., 2007. Characterization of small RNAs derived from Citrus exocortis viroid (CEVd) in infected tomato plants. Virology 367, 135–146.

Martínez de Alba, A.E., Elvira-Matelot, E., Vaucheret, H., 2013. Gene silencing in plants: A diversity of pathways. Biochim. Biophys. Acta. 1829, 1300–1308.

Martínez de Alba, A.E., Flores, R., Hernández, C., 2002. Two chloroplastic viroids induce the accumulation of small RNAs associated with posttranscriptional gene silencing. J. Virol. 76, 13094–13096.

Martínez, G., Castellano, M., Tortosa, M., Pallás, V., Gómez, G., 2014. A pathogenic non-coding RNA induces changes in dynamic DNA methylation of ribosomal RNA genes in host plants. Nucleic Acids Res. 42, 1553–1562.

Martínez, G., Donaire, L., Llave, C., Pallás, V., Gómez, G., 2010. High-throughput sequencing of hop stunt viroid-derived small RNAs from cucumber leaves and phloem. Mol. Plant Pathol. 11, 347–359.

Matousek, J., Kozlova, P., Orctova, L., Schmitz, A., Pesina, K., Bannach, O., et al., 2007. Accumulation of viroid-specific small RNAs and increase in nucleolytic activities linked to viroid-caused pathogenesis. Biol. Chem. 388, 1–13.

Matzke, M.A., Kanno, T., Matzke, A.J., 2015. RNA-directed DNA methylation: the evolution of a complex epigenetic pathway in flowering plants. Annu. Rev. Plant Biol. 66, 243–267.

Matzke, M.A., Mosher, R.A., 2014. RNA-directed DNA methylation: an epigenetic pathway of increasing complexity. Nat. Rev. Genet. 15, 394–408.

McHale, M., Eamens, A.L., Finnegan, E.J., Waterhouse, P.M., 2013. A 22-nt artificial microRNA mediates widespread RNA silencing in Arabidopsis. Plant J. 76, 519–529.

Minoia, S., Carbonell, A., Di Serio, F., Gisel, A., Carrington, J.C., Navarro, B., et al., 2014. Specific argonautes bind selectively small RNAs derived from potato spindle tuber viroid and attenuate viroid accumulation in vivo. J. Virol. 88, 11933–11945.

Navarro, B., Gisel, A., Rodio, M.E., Delgado, S., Flores, R., Di Serio, F., 2012. Small RNAs containing the pathogenic determinant of a chloroplast-replicating viroid guide the degradation of a host mRNA as predicted by RNA silencing. Plant J. 70, 991–1003.

Navarro, B., Pantaleo, V., Gisel, A., Moxon, S., Dalmay, T., Bisztray, G., et al., 2009. Deep sequencing of viroid-derived small RNAs from grapevine provides new insights on the role of RNA silencing in plant-viroid interaction. PLoS One 4, e7686.

Papaefthimiou, I., Hamilton, A., Denti, M., Baulcombe, D., Tsagris, M., Tabler, M., 2001. Replicating potato spindle tuber viroid RNA is accompanied by short RNA fragments that are characteristic of post-transcriptional gene silencing. Nucleic Acids Res. 29, 2395–2400.

Pelissier, T., Thalmeir, S., Kempe, D., Sänger, H.L., Wassenegger, M., 1999. Heavy de novo methylation at symmetrical and non-symmetrical sites is a hallmark of RNA-directed DNA methylation. Nucleic Acids Res. 27, 1625–1634.

Pelissier, T., Wassenegger, M., 2000. A DNA target of 30 bp is sufficient for RNA-directed DNA methylation. RNA. 6, 55–65.

Pontes, O., Vitins, A., Ream, T.S., Hong, E., Pikaard, C.S., Costa-Nunes, P., 2013. Intersection of small RNA pathways in Arabidopsis thaliana sub-nuclear domains. PLoS One 8, e65652.

Sasaki, T., Lee, T.F., Liao, W.W., Naumann, U., Liao, J.L., Eun, C., et al., 2014. Distinct and concurrent pathways of Pol II- and Pol IV-dependent siRNA biogenesis at a repetitive trans-silencer locus in Arabidopsis thaliana. Plant J. 79, 127–138.

Schiebel, W., Pélissier, T., Riedel, L., Thalmeir, S., Schiebel, R., Kempe, D., et al., 1998. Isolation of a RNA-directed RNA polymerase-specific cDNA clone from tomato leaf-tissue mRNA. Plant Cell 10, 2087–2101.

Schwind, N., Zwiebel, M., Itaya, A., Ding, B., Wang, M.B., Krczal, G., et al., 2009. RNAi-mediated resistance to potato spindle tuber viroid in transgenic tomato expressing a viroid hairpin RNA construct. Mol. Plant Pathol. 10, 459–469.

Serra, P., Bani Hashemian, S.M., Fagoaga, C., Romero, J., Ruiz-Ruiz, S., Gorris, M.T., et al., 2014. Virus-viroid interactions: citrus tristeza virus enhances the accumulation of citrus dwarfing viroid in Mexican lime via virus-encoded silencing suppressors. J. Virol. 88, 1394–1397.

Tabler, M., Tsagris, M., 2004. Viroids: petite RNA pathogens with distinguished talents. Trends Plant Sci. 9, 339–348.

Tsagris, E.M., Martínez de Alba, A.E., Gozmanova, M., Kalantidis, K., 2008. Viroids. Cell Microbiol. 10, 2168–2179.

Vadamalai, G., 2005. An Investigation of Oil Palm Orange Spotting Disorder (Ph.D. thesis). The University of Adelaide, South Australia. https://digital.library.adelaide.edu.au/dspace/handle/2440/37756.

Wang, M.B., Bian, X.Y., Wu, L.M., Liu, L.X., Smith, N.A., Isenegger, D., et al., 2004. On the role of RNA silencing in the pathogenicity and evolution of viroids and viral satellites. Proc. Natl. Acad. Sci. USA 101, 3275–3280.

Wang, Y., Shibuya, M., Taneda, A., Kurauchi, T., Senda, M., Owens, R.A., et al., 2011. Accumulation of potato spindle tuber viroid-specific small RNAs is accompanied by specific changes in gene expression in two tomato cultivars. Virology 413, 72–83.

Wassenegger, M., 2000. RNA-directed DNA methylation. Plant Mol. Biol. 43, 203–220.

Wassenegger, M., Heimes, S., Riedel, L., Sänger, H.L., 1994. RNA-directed de novo methylation of genomic sequences in plants. Cell 76, 567–576.

Wassenegger, M., Krczal, G., 2006. Nomenclature and functions of RNA-directed RNA polymerases. Trends Plant Sci. 11, 142–151.

Yang, Z., Ebright, Y.W., Yu, B., Chen, X., 2006. HEN1 recognizes 21-24 nt small RNA duplexes and deposits a methyl group onto the 2′ OH of the 3′ terminal nucleotide. Nucleic Acids Res. 34, 667–675.

Yu, J.Y., DeRuiter, S.L., Turner, D.L., 2002. RNA interference by expression of short-interfering RNAs and hairpin RNAs in mammalian cells. Proc. Natl. Acad. Sci. USA 99, 6047–6052.

CHAPTER

12

Origin and Evolution of Viroids

Francesco Di Serio[1], Beatriz Navarro[1] and Ricardo Flores[2]

[1]National Research Council, Bari, Italy
[2]Polytechnic University of Valencia-CSIC, Valencia, Spain

INTRODUCTION

Discovery of the atypical properties of viroids, including small size, circularity, self-complementarity, and inability to code for proteins (Diener, 1971; Gross et al., 1978) prompted intriguing evolutionary questions about their origin. In addition, the finding that the quasispecies model (Biebricher and Eigen, 2006) could be applied to viroid populations (Ambrós et al., 1999; Góra-Sochacka et al., 1997) opened additional queries on the constraints affecting the evolution of these minimal RNA replicons. This chapter addresses these evolutionary topics, briefly discussing early and recent hypotheses that were proposed or adapted to match discoveries on multiple facets of RNA, including structural, biochemical, and regulatory features of this nucleic acid, and its pivotal role in the origin of life. Most of the arguments summarized here have been presented and discussed previously in a seminal article (Diener, 1989) and in subsequent reviews (Diener, 1995; Elena et al., 2009; Flores et al., 2014).

WHERE DO VIROIDS COME FROM?

Even if there are important differences between RNA viruses and viroids, with the former having a genomic size at least 10 times larger and encoding protein(s) of their own, both are intracellular parasites that replicate through RNA intermediates. Therefore, RNA viruses remain the infectious agents most closely related to viroids, and early

Viroids and Satellites.
DOI: http://dx.doi.org/10.1016/B978-0-12-801498-1.00012-7

hypotheses on the origin of viroids considered that they could be primitive or degenerate RNA viruses (Diener, 1974). However, this and the alternative hypotheses that viroids might have originated from nuclear RNAs (Diener, 1974), escaped introns (Diener, 1981; Hadidi, 1986), transposable elements (Kiefer et al., 1983), and mitochondrial retroplasmids or retroplasmid satellites, were later dismissed based on different experimental evidence (Diener, 1995).

The discovery that RNA can store genetic information (Fraenkel-Conrat, 1956; Gierer and Schramm, 1956) and display catalytic activity (Kruger et al., 1982) consolidated a novel view on the origin of life: present cellular life, based on DNA and proteins as the vehicles for the storage and expression of genetic information, respectively, was preceded by an RNA world exclusively populated by RNA molecules able to catalyze their own synthesis (Gilbert, 1986). According to this view, the ribozyme activity of certain introns, particularly that of the *Tetrahymena thermophila* rRNA precursor, was considered a relict of the RNA world (Cech, 1986). Self-cleaving activity was soon discovered also in certain viroid-like satellite RNAs, small circular RNAs that rely on a helper virus for replication and encapsidation (see Chapter 61: Small Circular Satellite RNAs), and in avocado sunblotch viroid (ASBVd), the type member of chloroplast-replicating viroids (family *Avsunviroidae*) (see Chapter 28: Avocado Sunblotch Viroid). Based on these findings, Diener (1989), proposed that viroids and viroid-like RNAs could also be regarded as relics of an ancient RNA world because of other salient features: (1) minimal genome and (with the exception of ASBVd) high G + C content, which would provide higher replication fidelity of primitive RNAs; (2) circular structure, which would exclude the need of specific tags for initiating RNA replication and would ensure transcription of the entire genome; (3) structural periodicity consistent with the recombinant nature of some viroids (see below) and with evolutionary models based on modular assembly (Manrubia and Briones, 2007); and (4) absence of coding capacity, which is consistent with these RNAs having appeared before the emergence of the ribosome. Therefore, according to Diener's suggestion, viroids and viroid-like RNAs were even better candidates than introns to be "living fossils" of a precellular RNA world (Diener, 1989; for a review see Flores et al., 2014). This idea gained additional supporting evidence that included the identification of other self-cleaving viroids (Flores et al., 2000) and viroid-like satellite RNAs (see Chapter 61: Small Circular Satellite RNAs), and the finding that in vitro-selected ribozymes can catalyze RNA synthesis (Martin et al., 2015).

Phylogenetic reconstructions suggest a monophyletic origin for viroids and viroid-like satellite RNAs (Elena et al., 1991; Elena et al., 2001). Despite some controversy (Elena et al., 2001; Jenkins et al., 2000), these reconstructions group viroid and viroid-like satellite RNAs in different clades consistent with their specific biological properties. However, a polyphyletic viroid origin cannot be dismissed, particularly considering that ASBVd is A + U rich, in contrast to the other viroids (Flores et al., 2014). In the first phylogenetic analysis (Elena et al., 1991), a common origin was assumed not only for viroids and viroid-like satellite RNAs, but also for the viroid-like domain of the human hepatitis delta virus (HDV), a circular RNA requiring hepatitis B virus as a helper for packaging and transmission (for a review see Flores et al., 2012a). Since it was subsequently shown that HDV RNA self-cleaves through specific ribozymes (Been, 1994), different from those present in plant viroids and viroid-like satellite RNAs (hammerhead and hairpin structures), HDV was not considered in subsequent analyses (Elena et al., 2001). However, a common origin for the hammerhead, hairpin, and HDV ribozymes has been proposed (Harris and Elder, 2000).

HOW DID VIROIDS ADAPT TO A CELLULAR ENVIRONMENT?

In the monophyletic scenario proposed above, emergence of chloroplastic viroids (family *Avsunviroidae*) preceded that of nuclear viroids like potato spindle tuber viroid (family *Pospiviroidae*). This consideration is based on the proposal that plastids evolved from primitive cyanobacteria by symbiogenesis to generate the ancestor of eukaryotic photosynthetic cells (Douglas, 1998). Had the cyanobacteria hosted an ancestor viroid, this replicating RNA would have infected the resulting symbiotic cell. The ancestor viroid could have adapted to the protein-catalyzed environment of the primordial endosymbiont, with just the hammerhead-catalyzed self-cleavage being preserved over time. By invading other subcellular niches of the endosymbiotic cell, ancestral viroids might have found in the nucleus an environment sufficiently complex to favor their complete adaptation to a protein-based replication. This idea is consistent with the partial sequence similarity observed between a portion of the hammerhead structure mediating self-cleavage in chloroplastic viroids and a portion of the region directing host-mediated cleavage of nuclear viroids (see Chapter 7: Viroid Replication), a region that could have evolved by conversion of the ribozyme into structural elements recognized by host enzymes (Diener, 1995). A vestigial nonfunctional hammerhead-like structure identified in hop stunt viroid (Amari et al., 2001) may be taken as evidence supporting the loss of self-cleaving activity in members of family *Pospiviroidae*. However, the evolutionary implications of this finding, which seems limited to only this viroid, have not been further investigated.

It is unclear how the ancestral viroids, when adapting to a cellular environment, may have developed the ability to utilize host enzymes for their replication. In this context, the nuclear RNA polymerase II (Pol II) became involved in replication of nuclear viroids and the nuclear-encoded plastid RNA polymerase in the replication of chloroplastic viroids (see Chapter 7: Viroid Replication). Therefore, both types of viroid have adopted the same strategy of redirecting DNA-dependent RNA polymerases to transcribe RNA templates. In the absence of RNA-directed RNA polymerases in plastids, this strategy can easily be assumed for chloroplast-replicating viroids, but why nuclear-replicating viroids preferred Pol II instead of a nuclear RNA-directed RNA polymerase is less straightforward. Intriguingly, besides viroids, Pol II also transcribes HDV strands (Taylor, 2009) and certain mammalian noncoding RNAs (i.e., B2 RNA in mouse) (Wagner et al., 2013). Moreover, dissection of the crystal structure of the complete Pol II bound to a scaffold is consistent with the possibility that a Pol II ancestor could replicate primitive RNA genomes (Lehmann et al., 2007). A similar situation can be envisaged for another enzyme involved in the replication of nuclear viroids, namely the DNA ligase I, which functions as an RNA ligase in the circularization of the (+) monomeric viroid RNAs with 5′-phosphomonoester and 3′-hydroxyl termini (see Chapter 7: Viroid Replication). Interestingly, some RNA ligases, like T4 RNA ligase 1, demand the same termini and operate through a similar mechanism to that proposed for DNA ligase I (Ho and Shuman, 2002; Wang et al., 2006). Moreover, an ancestral catalytic module mediating RNA repair has been advocated as a possible common precursor for these enzymes (Shuman and Lima, 2004), thus supporting the view that, similarly to Pol II, the ability of DNA ligase 1 to act on viroid RNAs could also be reminiscent of the original template of its precursor. In contrast, for ligation, chloroplast-replicating viroids have evolved to recruit a chloroplastic tRNA ligase isoform with specificity for the 5′-hydroxyl and 2′, 3′ cyclic phosphodiester termini produced by hammerhead ribozymes (see Chapter 7: Viroid Replication).

HOW DID VIROIDS SUBSEQUENTLY EVOLVE?

Viroid populations infecting a single host assume the typical features of quasispecies (Codoñer et al., 2006), which are composed of closely related sequence variants. This situation has clearly been observed when, following inoculation with single infectious variants, the resulting progeny of members of the family *Pospiviroidae* (Gandía and Durán-Vila, 2004; Góra-Sochacka et al., 1997; Tessitori et al., 2013) and *Avsunviroidae* (Ambrós et al., 1999; De la Peña et al., 1999; Navarro et al., 2012) are heterogeneous, with the heterogeneity being considerably higher in the latter. The error-prone replication mediated by RNA polymerases transcribing RNA instead of their physiological DNA templates, is presumably the major cause for the sequence variability observed in viroids. Actually, the mutation rate of the chloroplast-replicating chrysanthemum chlorotic mottle viroid (Navarro and Flores, 1997), estimated from the frequency of lethal mutations, is the highest ever reported for any biological entity (Gago et al., 2009).

When using an RNA template, Pol II is slower and less processive than when displaying DNA-dependent activity (Lehmann et al., 2007). This lower processivity during viroid replication might promote stalling and subsequent jumping of the polymerase, which with the bound nascent transcript, might reinitiate transcription on another template and lead to recombinant variants of the same viroid (Fadda et al., 2003; Haseloff et al., 1982; Szychowski et al., 2005) or of different viroids coinfecting the same host (Hammond et al., 1989; Rezaian, 1990). Although chimeric viroids have not been reported in the family *Avsunviroidae,* a similar lower processivity may be presumed for the nuclear-encoded plastid RNA polymerase mediating replication of members of this family. Therefore, the 12–14 nt insertion identified in some variants of peach latent mosaic viroid, which contains the pathogenic determinant of peach calico disease and may be acquired de novo during infection (Rodio et al., 2006), might be the outcome of a recombination event between peach latent mosaic viroid RNA and another—still undetermined—host RNA.

Although both the high mutation rate and the recombination occurring during replication are the main causes of genetic diversity in viroids, the variants actually accumulating depend on selection pressures imposed by the host and environmental factors. The first evidence of a host selection pressure was the identification of specific mutations in variants of citrus exocortis viroid (family *Pospiviroidae*) recovered after serial transmission in tomato of a viroid population originally infecting citrus (Semancik et al., 1993). Additional studies confirmed the genetic variability imposed on citrus exocortis viroid by several herbaceous and woody hosts (Gandía et al., 2007; Szychowski et al., 2005), and revealed the ability of some of them to set evolutionary restrictions ultimately converging into the same host-specific variants independently of the divergent viroid populations inoculated (Bernad et al., 2009). Based on phylogenetic relationships among viroid populations isolated from different hosts, similar ideas were proposed for hop stunt viroid (Amari et al., 2001; Kofalvi et al., 1997). In another member of the family *Pospiviroidae* (hop latent viroid), it was shown that heat shock treatment of plants induced significant increase in the sequence polymorphism of the infecting viroid population, identifying temperature as a key environmental factor (Matoušek et al., 2001).

Preservation of a compact secondary structure has been identified as one of the major constraints for viroid evolution (reviewed by Elena et al., 2009). Indeed, compact secondary structures have been proposed to have evolved as a viroid response against the selection pressure imposed by inactivation mediated by Argonaute proteins (see Chapter 11: Viroids and RNA Silencing), or as a compromise between resistance to the latter and to Dicer-like enzymes (Carbonell et al., 2008; Minoia et al., 2014), which act preferentially on RNAs with relaxed and compact conformations, respectively. Interestingly, the rod-like and branched secondary structures proposed for viroids are composed of short loops interspersed between double-stranded regions (see Chapter 6: Viroid Structure), thus supporting an evolution in part driven by RNA silencing.

A major role of secondary structure in viroid evolution is further supported by algorithms predicting RNA secondary structure indicating that viroids have evolved toward increased robustness (resistance to deleterious mutations) and decreased antagonistic epistasis (interaction between deleterious mutations at different loci) (Sanjuán et al., 2006a,b). Hence, the need for preserving secondary and tertiary structural elements involved in replication (see Chapter 6: Viroid Structure) and in intracellular, intercellular, and long distance trafficking of viroids (see Chapter 8: Viroid Movement), must also be considered. In addition, the recent finding that viroids may modify host gene expression through RNA silencing may also have evolutionary implications (see Chapter 9: Viroid Pathogenesis).

CONCLUDING REMARKS

Viroids possess a genotype and express a phenotype without resorting to protein intermediation and, additionally, some display catalytic activity. These unique properties support the view that viroids are relics of an ancient RNA world (Diener, 1989; Flores et al., 2014). In contrast to RNA viruses, viroids have evolved without the constraints imposed by the need to code for proteins, although they must preserve specific structural elements required to interact with the transcription, processing, and trafficking machineries of their hosts, and to escape the defense strategies the latter mount. Because viroid genomes are very small, they are composed of regions with overlapping functional roles (Flores et al., 2012b), a situation contributing to the generation of strong robustness to mutations, with modular redundancy and recombination playing additional roles in viroid evolution (Elena et al., 2009). In this complex network of constraints delimiting the fitness landscape in which viroids must evolve, RNA silencing may have forced the adoption of compact secondary structure by genomic RNAs. However, the findings that viroid-derived small RNAs may target host mRNAs (Navarro et al., 2012) and viroid RNAs (Minoia et al., 2014) for inactivation indicate that the role of RNA silencing on viroid evolution possibly needs a reevaluation (for a review see Flores et al., 2015). Moreover, the recent identification of a viroid degradation pathway unrelated to RNA silencing (Minoia et al., 2015) poses the question as to whether this RNA decay mechanism also has evolutionary implications.

Acknowledgments

Work in the laboratories of B.N. and F.D.S. has been partially supported by a dedicated grant (CISIA) of the Ministero dell'Economia e Finanze Italiano to the CNR (Legge n. 191/2009). Research in R.F. laboratory is currently funded by grant BFU2014–56812-P from the Spanish Ministerio de Economía y Competitividad (MINECO).

References

Amari, K., Gómez, G., Myrta, A., Di Terlizzi, B., Pallás, V., 2001. The molecular characterization of 16 new sequence variants of hop stunt viroid reveals the existence of invariable regions and a conserved hammerhead-like structure on the viroid molecule. J. Gen. Virol. 82, 953–962.

Ambrós, S., Hernández, C., Flores, R., 1999. Rapid generation of genetic heterogeneity in progenies from individual cDNA clones of peach latent mosaic viroid in its natural host. J. Gen. Virol. 80, 2239–2252.

Been, M., 1994. Cis- and trans-acting ribozymes from a human pathogen, hepatitis delta virus. Trends Biochem. Sci. 19, 251–256.

Bernad, L., Durán-Vila, N., Elena, S.F., 2009. Effect of citrus hosts on the generation, maintenance and evolutionary fate of genetic variability of citrus exocortis viroid. J. Gen. Virol. 90, 2040–2049.

Biebricher, C.K., Eigen, M., 2006. What is a quasispecies? Curr. Top. Microbiol. Immunol. 299, 1–31.

Carbonell, A., Martínez de Alba, A.E., Flores, R., Gago, S., 2008. Double stranded RNA interferes in a sequence-specific manner with infection of representative members of the two viroid families. Virology 371, 44–53.

Cech, T.R., 1986. RNA as an enzyme. Sci. Am. 255, 64–75.

Codoñer, F.M., Darós, J.A., Solé, R.V., Elena, S.F., 2006. The fittest versus the flattest: experimental confirmation of the quasispecies effect with subviral pathogens. PLoS Pathog. 2, 1187–1193.

De la Peña, M., Navarro, B., Flores, R., 1999. Mapping the molecular determinant of pathogenicity in a hammerhead viroid: a tetraloop within the in vivo branched RNA conformation. Proc. Natl. Acad. Sci. USA 96, 9960–9965.

Diener, T.O., 1971. Potato spindle tuber "virus": IV. A replicating, low molecular weight RNA. Virology 45, 411–428.

Diener, T.O., 1974. Viroids as prototypes or degeneration products of viruses. In: Kurstak, E., Maramorosch, K. (Eds.), Viruses, Evolution and Cancer. Academic Press, New York, NY, pp. 757–783.

Diener, T.O., 1981. Are viroids escaped introns? Proc. Natl. Acad. Sci. USA 78, 5014–5015.

Diener, T.O., 1989. Circular RNAs: relics of precellular evolution? Proc. Natl. Acad. Sci. USA 86, 9370–9374.

Diener, T.O., 1995. Origin and evolution of viroids and viroid-like satellite RNAs. Virus Genes 11, 119–131.

Douglas, S.E., 1998. Plastid evolution: origins, diversity, trends. Curr. Opin. Genet. Dev. 8, 655–661.

Elena, S.F., Dopazo, J., de la Peña, M., Flores, R., Diener, T.O., Moya, A., 2001. Phylogenetic analysis of viroid and viroid-like satellite RNAs from plants: a reassessment. J. Mol. Evol. 53, 155–159.

Elena, S.F., Dopazo, J., Flores, R., Diener, T.O., Moya, A., 1991. Phylogeny of viroids, viroidlike satellite RNAs, and the viroidlike domain of hepatitis delta virus RNA. Proc. Natl. Acad. Sci. USA 88, 5631–5634.

Elena, S.F., Gómez, G., Daròs, J.A., 2009. Evolutionary constraints to viroid evolution. Viruses 1, 241–254.

Fadda, Z., Daròs, J.A., Fagoaga, C., Flores, R., Durán-Vila, N., 2003. Eggplant latent viroid, the candidate type species for a new genus within the family Avsunviroidae (hammerhead viroids). J. Virol. 77, 6528–6532.

Flores, R., Daròs, J.A., Hernández, C., 2000. Avsunviroidae family: viroids containing hammerhead ribozymes. Adv. Virus Res. 55, 271–323.

Flores, R., Gago-Zachert, S., Serra, P., Sanjuán, R., Elena, S.F., 2014. Viroids: survivors from the RNA world? Annu. Rev. Microbiol. 68, 395–414.

Flores, R., Minoia, S., Carbonell, A., Gisel, A., Delgado, S., López-Carrasco, A., et al., 2015. Viroids, the simplest RNA replicons: how they manipulate their hosts for being propagated and how their hosts react for containing the infection. Virus Res. 209, 136–145.

Flores, R., Ruiz-Ruiz, S., Serra, P., 2012a. Viroids and hepatitis delta virus. Semin. Liver Dis. 32, 201–210.

Flores, R., Serra, P., Minoia, S., Di Serio, F., Navarro, B., 2012b. Viroids: from genotype to phenotype just relying on RNA sequence and structural motifs. Front. Microbiol. 3, 217.

Fraenkel-Conrat, H., 1956. The role of the nucleic acid in the reconstitution of active tobacco mosaic virus. J. Am. Chem. Soc. 78, 882–883.

Gago, S., Elena, S.F., Flores, R., Sanjuán, R., 2009. Extremely high mutation rate of a hammerhead viroid. Science 323, 1308.

Gandía, M., Bernad, L., Rubio, L., Durán-Vila, N., 2007. Host effect on the molecular and biological properties of a citrus exocortis viroid isolate from *Vicia faba*. Phytopathology 97, 1004–1010.

Gandía, M., Durán-Vila, N., 2004. Variability of the progeny of a sequence variant citrus bent leaf viroid (CBLVd). Arch. Virol. 149, 407–416.

Gierer, A., Schramm, G., 1956. Infectivity of ribonucleic acid from tobacco mosaic virus. Nature 177, 702–703.

Gilbert, W., 1986. Origin of life: the RNA world. Nature 319, 618.

Góra-Sochacka, A., Kierzek, A., Candresse, T., Zagórski, W., 1997. The genetic stability of potato spindle tuber viroid (PSTVd) molecular variants. RNA. 3, 68–74.

Gross, H.J., Domdey, H., Lossow, C., Jank, P., Raba, M., Alberty, H., et al., 1978. Nucleotide sequence and secondary structure of potato spindle tuber viroid. Nature 273, 203–208.

Hadidi, A., 1986. Relationship of viroids and certain other plant pathogenic nucleic acids to group I and II introns. Plant Mol. Biol. 7, 129–142.

Hammond, R., Smith, D.R., Diener, T.O., 1989. Nucleotide sequence and proposed secondary structure of Columnea latent viroid: a natural mosaic of viroid sequences. Nucleic Acids Res. 17, 10083–10094.

Harris, R.J., Elder, D., 2000. Ribozyme relationships: the hammerhead, hepatitis delta and hairpin ribozymes have a common origin. J. Mol. Evol. 51, 182–184.

Haseloff, J., Mohamed, N.A., Symons, R.H., 1982. Viroid RNAs of cadang-cadang disease of coconuts. Nature 299, 316–321.

Ho, C.K., Shuman, S., 2002. Bacteriophage T4 RNA ligase 2 (gp24.1) exemplifies a family of RNA ligases found in all phylogenetic domains. Proc. Natl. Acad. Sci. USA 99, 12709–12714.

Jenkins, G.M., Woelk, C.H., Rambaut, A., Homes, E.C., 2000. Testing the extent of sequence similarity among viroids, satellite RNAs, and the hepatitis delta virus. J. Mol. Evol. 50, 98–102.

Kiefer, M.C., Owens, R.A., Diener, T.O., 1983. Structural similarities between viroids and transposable genetic elements. Proc. Natl. Acad. Sci. USA 80, 6234–6238.

Kofalvi, S.A., Marcos, J.F., Cañizares, M.C., Pallás, V., Candresse, T., 1997. Hop stunt viroid (HSVd) sequence variants from *Prunus* species: evidence for recombination between HSVd isolates. J. Gen. Virol. 78, 3177–3186.

Kruger, K., Grabowski, P.J., Zaug, A.J., Sands, J., Gottschling, D.E., Cech, T.R., 1982. Self-splicing RNA: autoexcision and autocyclization of the ribosomal RNA intervening sequence of Tetrahymena. Cell 31, 147–157.

Lehmann, E., Brueckner, F., Cramer, P., 2007. Molecular basis of RNA-dependent RNA polymerase II activity. Nature 450, 445–449.

Manrubia, S.C., Briones, C., 2007. Modular evolution and increase of functional complexity in replicating RNA molecules. RNA. 13, 97–107.

Martin, L.L., Unrau, P.J., Müller, F.U., 2015. RNA synthesis by *in vitro* selected ribozymes for recreating an RNA world. Life (Basel). 5 (1) 247–268.

Matoušek, J., Patzak, J., Orctová, L., Schubert, J., Vrba, L., Steger, G., et al., 2001. The variability of hop latent viroid as induced upon heat treatment. Virology 287, 349–358.

Minoia, S., Carbonell, A., Di Serio, F., Gisel, A., Carrington, J.C., Navarro, B., et al., 2014. Specific argonautes selectively bind small RNAs derived from potato spindle tuber viroid and attenuate viroid accumulation in vivo. J. Virol. 88, 11933–11945.

Minoia, S., Navarro, B., Delgado, S., Di Serio, F., Flores, R., 2015. Viroid RNA turnover: characterization of the subgenomic RNAs of potato spindle tuber viroid accumulating in infected tissues provides insights into decay pathways operating in vivo. Nucleic Acids Res. 43, 2313–2325.

Navarro, B., Flores, R., 1997. Chrysanthemum chlorotic mottle viroid: unusual structural properties of a subgroup of self-cleaving viroids with hammerhead ribozymes. Proc. Natl. Acad. Sci. USA 94, 11262–11267.

Navarro, B., Gisel, A., Rodio, M.E., Delgado, S., Flores, R., Di Serio, F., 2012. Small RNAs containing the pathogenic determinant of a chloroplast-replicating viroid guide the degradation of a host mRNA as predicted by RNA silencing. Plant J. 70, 991–1003.

Rezaian, M.A., 1990. Australian grapevine viroid—evidence for extensive recombination between viroids. Nucleic Acids Res. 18, 1813–1818.

Rodio, M.E., Delgado, S., Flores, R., Di Serio, F., 2006. Variants of peach latent mosaic viroid inducing peach calico: uneven distribution in infected plants and requirements of the insertion containing the pathogenicity determinant. J. Gen. Virol. 87, 231–240.

Sanjuán, R., Forment, J., Elena, S.F., 2006a. In silico predicted robustness of viroids RNA secondary structures. I. The effect of single mutations. Mol. Biol. Evol. 23, 1427–1436.

Sanjuán, R., Forment, J., Elena, S.F., 2006b. In silico predicted robustness of viroid RNA secondary structures. II. Interaction between mutation pairs. Mol. Biol. Evol. 23, 2123–2130.

Semancik, J.S., Szychowski, J.A., Rakowski, A.G., Symons, R.H., 1993. Isolates of citrus exocortis viroid recovered by host and tissue selection. J. Gen. Virol. 74, 2427–2436.

Shuman, S., Lima, C.D., 2004. The polynucleotide ligase and RNA capping enzyme superfamily of covalent nucleotidyltransferases. Curr. Opin. Struct. Biol. 14, 757–764.

Szychowski, J.A., Vidalakis, G., Semancik, J.S., 2005. Host-directed processing of citrus exocortis viroid. J. Gen. Virol. 86, 473–477.

Taylor, J.M., 2009. Replication of the hepatitis delta virus RNA genome. Adv. Virus Res. 74, 103–121.

Tessitori, M., Rizza, S., Reina, A., Causarano, G., Di Serio, F., 2013. The genetic diversity of citrus dwarfing viroid populations is mainly dependent on the infected host species. J. Gen. Virol. 94, 687–693.

Wagner, S.D., Yakovchuk, P., Gilman, B., Ponicsan, S.L., Drullinger, L.F., Kugel, J.F., et al., 2013. RNA polymerase II acts as an RNA-dependent RNA polymerase to extend and destabilize a non-coding RNA. EMBO J. 32, 781–790.

Wang, L.K., Schwer, B., Shuman, S., 2006. Structure-guided mutational analysis of T4 RNA ligase 1. RNA 12, 2126–2134.

CHAPTER

13

Viroid Taxonomy

Francesco Di Serio[1], *Shi-Fang Li*[2], *Vicente Pallás*[3], *Robert A. Owens*[4], *John W. Randles*[5], *Teruo Sano*[6], *Jacobus Th.J. Verhoeven*[7], *Georgios Vidalakis*[8] *and Ricardo Flores*[3]

[1]National Research Council, Bari, Italy [2]Chinese Academy of Agricultural Sciences, Beijing, China [3]Polytechnic University of Valencia-CSIC, Valencia, Spain [4]U.S. Department of Agriculture, Beltsville, MD, United States [5]The University of Adelaide, Waite Campus, Glen Osmond, SA, Australia [6]Hirosaki University, Hirosaki, Japan [7]National Plant Protection Organization, Wageningen, The Netherlands [8]University of California, Riverside, CA, United States

INTRODUCTION

The Code of Virus Classification and Nomenclature, established by the International Committee on Virus Taxonomy (ICTV), states that "rules concerned with the classification of viruses shall also apply to the classification of viroids" (King et al., 2012). Therefore, similarly to viruses, a viroid species can be defined as "a monophyletic group sharing properties that can be distinguished from those of other species by multiple criteria." These criteria, which are proposed and periodically revised by ICTV, take into consideration the structural, functional, and evolutionary diversity of viroids with respect to viruses.

In contrast to viruses, viroids: (1) have a small, circular and nonprotein-coding genomic RNA that is approximately ten times smaller than the smallest DNA and RNA viral genomes; (2) express, at least in some cases, ribozyme activity; and (3) have an independent evolutionary history that has been traced back to the RNA world proposed to have preceded the present cellular world (Diener, 1989; Flores et al., 2014; Chapter 12: Origin and Evolution of Viroids). Several issues

Viroids and Satellites.
DOI: http://dx.doi.org/10.1016/B978-0-12-801498-1.00013-9

regarding viroid taxonomy have been recently addressed by the authors, who are the members of the ICTV Viroid Study Group (Di Serio et al., 2014). This chapter reviews the current viroid classification scheme, putting particular emphasis on the ICTV guidelines for recognizing viroid species and for their inclusion in higher taxa (genus and family). The requirements for accepting as viroid species several novel viroid-like RNAs recently identified by next-generation sequencing (NGS) and bioinformatics tools are briefly discussed.

VIROID VARIANTS AND VIROID SPECIES

Similarly to other RNA replicons, the low fidelity of RNA polymerases and recombination events during replication generate sequence heterogeneity in viroid populations, which are composed of sequence variants that differ from each other to varying extents. Therefore, resembling RNA viruses, viroid populations infecting a single host adopt the typical features of a quasispecies (Biebricher and Eigen, 2006; Codoner et al., 2006), posing the key taxonomic question of when related viroid variants should be considered as a single species or as different species.

Early taxonomic guidelines proposed that an arbitrary limit of less than 90% sequence identity over the entire genome was sufficient for creating a new species, although biological aspects (e.g., distinct host range) were also considered for resolving conflicting situations (Flores, 1995; Flores et al., 1998, 2000a, 2005). Some years later, the ICTV required at least two independent criteria for establishing new virus and viroid species, and the ICTV Viroid Study Group decided to consider as mandatory, besides a sequence identity lower than 90% over the entire genome, the existence of some divergent biological features between different species (Owens et al., 2012).

The importance now attributed to biological properties in viroid taxonomy has already led to the taxonomic reassessment of two previously recognized species. Thus, careful experimental reevaluation of the biological features of variants of *Mexican papita viroid* and *Tomato planta macho viroid* did not confirm the existence of differences in their host range (as initially assumed), leading the ICTV to accept the proposal, also supported by phylogenetic analyses, of merging these two species into a single species named *Tomato planta macho viroid* (Verhoeven et al., 2011). This decision was ratified in 2015 (http://ictvonline.org/virusTaxonomy.asp?bhcp=1). The ICTV currently recognizes 32 viroid species, but proposals for other species are expected to be submitted in the near future (see below).

VIROID FAMILIES

Since the second taxonomic scheme for viroids was adopted by the ICTV (Flores et al., 2000a), viroid species have been classified into higher taxa, consisting of families and genera, according to certain rules (Owens et al., 2012). Structural elements, including the presence or absence of specific RNA motifs, and relevant biological features such as sites of replication and accumulation, are the main criteria used to allocate viroid species into one of the two recognized families: *Pospiviroidae* and *Avsunviroidae*. The first family (type species *Potato spindle tuber viroid*) includes viroids that adopt a rod-like or quasi rod-like conformation with a central conserved region (CCR) formed by two stretches of conserved nucleotides in the upper and lower strands (Fig. 13.1) (McInnes and Symons 1991). Nucleotides in the upper CCR strand are flanked by an imperfect inverted repeat that may assume an alternative conformation (hairpin I, see inset in Fig. 13.1) with a crucial role in viroid replication via an asymmetric rolling-circle mechanism (Gas et al., 2007; Riesner et al., 1979). Experimental evidence collected for potato spindle tuber viroid (PSTVd) and other members of the family *Pospiviroidae* indicates that this process takes place in the nucleus (Bonfiglioli et al., 1996; Diener, 1971; Gambino et al., 2011; Harders et al., 1989; Qi and Ding, 2003; Spiesmacher et al., 1983).

In contrast, members of the family *Avsunviroidae* (type species *Avocado sunblotch viroid*) replicate and accumulate in plastids (mostly chloroplasts) (Bonfiglioli et al., 1994; Bussière et al., 1999; Lima et al., 1994; Mohamed and Thomas, 1980; for a review see Flores et al., 2000b). Additional demarcation criteria include: (1) absence of a CCR; and (2) presence of hammerhead ribozymes in both the genomic and antigenomic RNAs involved in replication through a symmetric rolling-circle mechanism (Fig. 13.2).

VIROID GENERA

In the absence of a formal taxonomy, early proposals suggested sequence similarity (%) among viroids (Puchta et al., 1988) or—for those lacking hammerhead ribozymes—the type of CCR (Koltunow and Rezaian, 1989) as the demarcating criteria for their allocation into higher-order groups (later recognized as genera). Elena et al. (1991) proposed that phylogenetic relationships could also be considered for classification of viroids, with the advantage of taking their evolutionary history into account. For assigning names to these groups (now considered as genera), the same authors also suggested a nomenclature code resembling that used for plant viruses: a prefix, consisting of the name in abbreviated form of the type member of the group, followed by the suffix viroid (e.g., "pospiviroid" for viroids in the group having *Potato spindle tuber viroid* as the type species). This nomenclatural code was adopted by the ICTV, which now classifies viroid species belonging to families *Pospiviroidae* and *Avsunviroidae* into five and three genera, respectively (Fig. 13.1),

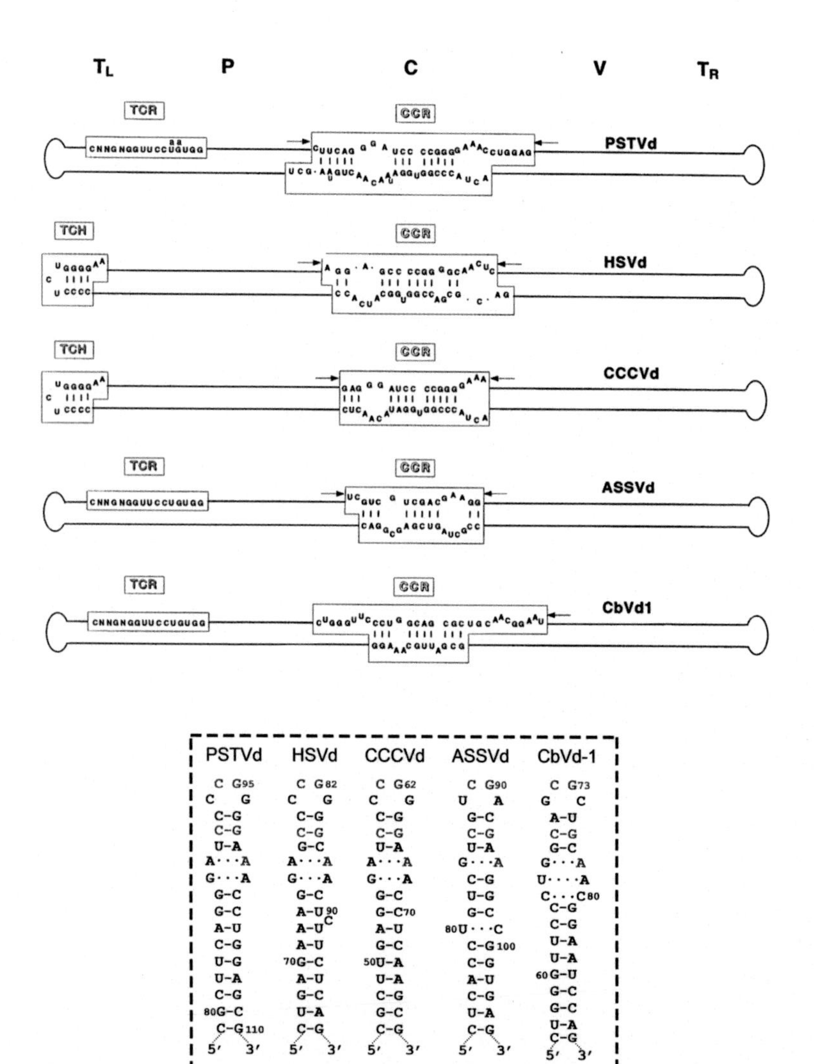

FIGURE 13.1 Schematic rod-like secondary structures proposed for the type members of the five genera in the family *Pospiviroidae*. Abbreviations for each viroid are indicated on the right (*PSTVd*, potato spindle tuber viroid; *HSVd*, hop stunt viroid; *CCCVd*, coconut cadang-cadang viroid; *ASSVd*, apple scar skin viroid; *CbVd-1*, Coleus blumei viroid 1). The terminal conserved region (TCR), terminal conserved hairpin (TCH), and central conserved region (CCR), with arrows representing the flanking sequences that together with the core nucleotides of the upper CCR strand form hairpin I, are indicated. Lower-case fonts refer to substitutions in the CCR and TCR, illustrated here for iresine viroid (IrVd-1), a member of the genus *Pospiviroid*. In the genus *Coleviroid*, the TCR exists only in the two largest members, CbVd-2 and CbVd-3 being replaced by the TCH in CbVd-1. *Source: Reproduced from Flores, R., Di Serio, F., Hernández, C., 1997. Viroids: the noncoding genomes. Semin. Virol. 8, 65–73 with permission from Elsevier. Inset.* Hairpin I formed by the upper CCR strand and flanking nucleotides of PSTVd, HSVd, CCCVd, ASSVd, and CbVd-1. Conserved nucleotides in structurally similar positions are shown in red. Continuous and broken lines represent Watson-Crick and noncanonical base pairs, respectively. Notice that the variability preserves the overall structure of hairpin I, including the terminal palindromic tetraloop, the adjacent 3-bp stem, and the long stem. *Source: Reproduced with modifications from Gas, M. E., Hernández, C., Flores, R., Daròs, J.A., 2007. Processing of nuclear viroids in vivo: an interplay between RNA conformations. PLoS Pathog. 3, e182.*

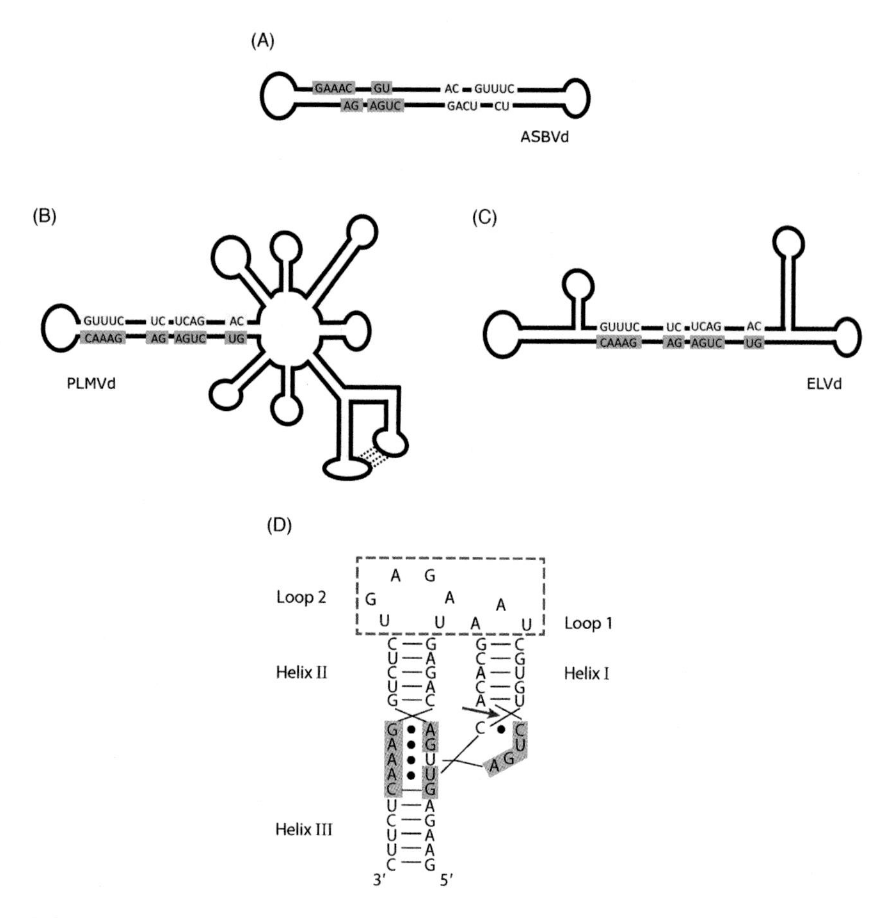

FIGURE 13.2 Schematic rod-like, branched, and partially branched secondary structures proposed for avocado sunblotch viroid (ASBVd, genus *Avsunviroid*), peach latent mosaic viroid (PLMVd, genus *Pelamoviroid*), and eggplant latent viroid (ELVd, genus *Elaviroid*), with nucleotides strictly or highly conserved in natural hammerhead structures shown on yellow and green backgrounds for plus and minus polarities, respectively. Lower panel, schematic representation of the plus polarity hammerhead ribozyme of PLMVd. Continuous lines and dots between nucleotides represent canonical (Watson-Crick) and noncanonical base pairs, respectively, with the red arrow marking the self-cleavage site. Dashed rectangles denote tertiary interactions between loops that either stabilize the global viroid conformation or promote the catalytically active ribozyme folding. *Source: Lower panel is reproduced with modification from Flores, R., Gago-Zachert, S., Serra, P., Sanjuan, R., Elena, S. F., 2014. Viroids: survivors from the RNA world? Annu. Rev. Microbiol. 68, 395–414.*

according to criteria mainly based on RNA structural features complemented by phylogenetic relationships among species.

The type of CCR and presence or absence of two other conserved regions, the so-called terminal conserved region (TCR) and terminal conserved hairpin (TCH) (Flores et al., 1997; Koltunow and Rezaian, 1988; Puchta et al., 1988), are used to divide the species in the family

Pospiviroidae into five genera: *Apscaviroid, Cocadviroid, Coleviroid, Hostuviroid,* and *Pospiviroid*. Each genus has a type species with a characteristic CCR, which with minimal modifications (see below) is also present in the genomes of the other members of the same genus (Fig. 13.1). In addition, members of genera *Pospiviroid* and *Apscaviroid* (vernacular names: pospiviroids and apscaviroids, respectively), also contain a TCR, while those of genera *Cocadviroid* and *Hostuviroid* (vernacular names cocadviroids and hostuviroids, respectively) contain a TCH. The presence of a TCR or TCH is not diagnostic for members of the genus *Coleviroid* (vernacular name coleviroids): a TCH is characteristic of the type species of this genus (*Coleus blumei viroid 1*), but is replaced by a TCR in the other two species of the same genus, *Coleus blumei viroid 2* and *Coleus blumei viroid 3* (Fig. 13.1).

In the family *Avsunviroidae,* the morphology of the hammerhead structures and other structural properties of the viroid RNA, including the minimal free energy conformation, G + C content and solubility in 2M LiCl, have been considered in grouping the species into three genera, e.g., *Avsunviroid, Pelamoviroid,* and *Elaviroid* (Navarro and Flores, 1997; Owens et al., 2012). Sequence variants of *Avocado sunblotch viroid,* the type species of the family and the only species in the genus *Avsunviroid,* are soluble in 2M LiCl, have a low G + C content, and assume a rod-like secondary structure of minimal free energy (Fig. 13.2A). In contrast, variants of species within the genus *Pelamoviroid* (*Peach latent mosaic viroid* and *Chrysanthemum chlorotic mottle viroid,* vernacular name pelamoviroids) have branched secondary structures (Fig. 13.2B), a high G + C content, and low solubility in 2M LiCl. The genus *Elaviroid* includes only a single species so far (*Eggplant latent viroid*), the variants of which have structural properties that are intermediate between those of avocado sunblotch viroid (ASBVd) and pelamoviroids (Fig. 13.2C). Moreover, while the monomeric (+) and (−) strands of pelamoviroids and eggplant latent viroid (ELVd) can form stable hammerhead structures (Fig. 13.2D), those of ASBVd are thermodynamically unstable, with a stable double hammerhead structure most likely mediating self-cleavage of the oligomeric ASBVd RNAs resulting from rolling-circle replication (Davies et al., 1991; Forster et al., 1988).

The continuous rather than discrete character of natural variation and the role played by recombination in viroid evolution may occasionally make strict application of these criteria difficult, thus demanding some flexibility. This is the case for the species *Columnea latent viroid, Dahlia latent viroid,* and *Citrus bark cracking viroid* in the family *Pospiviroidae.* The first of these species has been classified in the genus *Pospiviroid,* in spite of the existence of variants containing a CCR typical of members of the genus *Hostuviroid* (type species *Hop stunt viroid*). This decision is supported by: (1) phylogenetic analyses of the whole genome, which grouped columnea latent viroid (CLVd) together with other pospiviroids; (2) the presence of a TCR, a region shared by pospiviroids and apscaviroids; and (3) the ability of CLVd to infect solanaceous hosts, a biological feature common to most pospiviroids (Flores et al., 1998).

In contrast, the presence of a TCR in the genome of dahlia latent viroid (DLVd) has not been considered sufficient to prevent classification of the species *Dahlia latent viroid* within the genus *Hostuviroid*, although the presence of a TCH rather than a TCR is typical of the other member of this genus. This decision is further supported by phylogenetic analysis, which clusters DLVd and hop stunt viroid (HSVd) in the same group, and by the host range of DLVd, which so far appears restricted to dahlia and, in contrast to most pospiviroids, does not include solanaceous species (Verhoeven et al., 2013). Finally, phylogenetic relationships and structural features (e.g., type of CCR) were also considered pertinent for classifying *Citrus bark cracking viroid* (formerly *Citrus viroid IV*) within the genus *Cocadviroid* (Flores et al., 1998). The recent report of citrus bark cracking viroid (CBCVd) in stunted hop plants (Jakse et al., 2015), a natural host of another cocadviroid (hop latent viroid) (Puchta et al., 1988), supports this choice, even though certain CBCVd variants exhibit host range similarities with a pospiviroid (Semanicik and Vidalakis, 2005).

CURRENT VIROID CLASSIFICATION

Fig. 13.3 summarizes the current classification scheme for viroids. Recently endorsed by ICTV (http://ictvonline.org/virusTaxonomy.asp?bhcp=1), this scheme differs from the last ICTV report (Owens et al., 2012) in two respects: (1) the species *Mexican papita viroid* has been removed, with variants of this species now classified within the species *Tomato plant macho viroid* and (2) the species *Dahlia latent viroid* has been created within the genus *Hostuviroid*. Based on sequence properties, other viroids or viroid-like RNAs (the latter lacking evidence for autonomous replication) can be classified as unassigned members of new species to be eventually accepted by ICTV once proper biological requirements are fulfilled. Meanwhile these RNAs are included in a provisional list (Table 13.1).

Several viroids or viroid-like RNAs in this list, including persimmon viroid 2 (Ito et al., 2013), grapevine latent viroid (Zhang et al., 2014), grapevine hammerhead viroid-like RNA (Wu et al., 2012), and apple hammerhead viroid-like RNA (Zhang et al., 2014), have been identified by NGS and *in silico* assembly of viroid-derived small RNAs (vd-sRNAs) that accumulate in infected hosts. The vd-sRNAs are generated by host-encoded

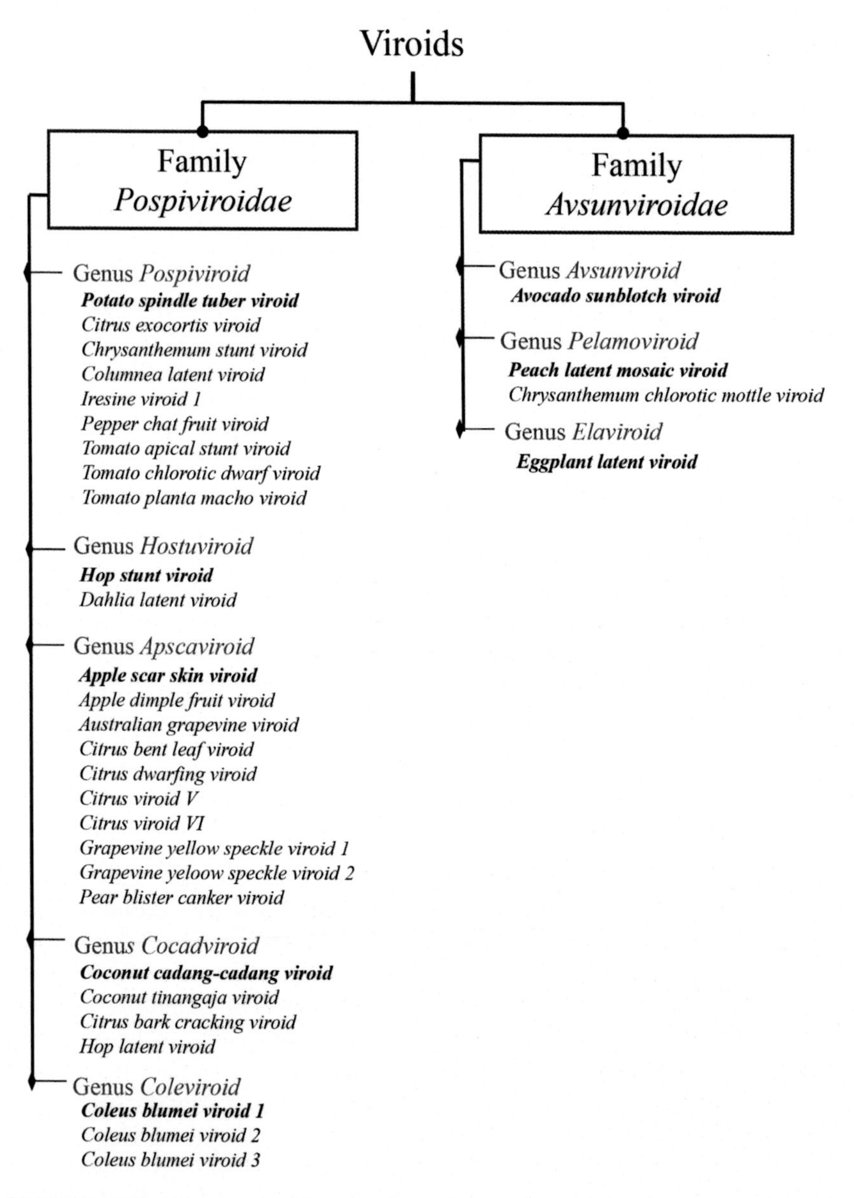

FIGURE 13.3 Current classification scheme for viroids.

RNase III-like enzymes (Dicer-like in plants) that target viroid RNAs as part of the plant antiviroid defense (reviewed by Gómez et al., 2009; Navarro et al., 2012; Chapter 11: Viroids and RNA Silencing). The bioinformatic tools developed for specifically identifying circular RNAs from NGS data (Wu et al., 2012; Zhang et al., 2014) will most likely lead to the

TABLE 13.1 List of Unassigned Viroids and Viroid-Like RNAs

Family[a]	Genus[a]	Viroid/viroid-like RNA[b]	ID[c]	Size (nt)	Reference
Pospiviroidae	*Apscaviroid*	Apple fruit crinkle viroid (AFCVd)	E29032	371	Sano et al. (2008)
Pospiviroidae	*Apscaviroid*	Grapevine yellow speckle viroid 3 (GYSVd-3)	DQ371462	366	Jiang et al. (2009)
Pospiviroidae	*Apscaviroid*	Persimmon latent viroid (PLVd)	AB366022	396	Nakaune and Nakano (2008)
Pospiviroidae	*Apscaviroid*	Persimmon viroid 2 (PVd-2)	AB817729	358	Ito et al. (2013)
Pospiviroidae	*Apscaviroid*	Grapevine latent viroid (GLVd)	KR605505	328	Zhang et al. (2014)
Pospiviroidae	*Coleviroid*	Coleus blumei viroid 5 (CbVd-5)	FJ151370	274	Hou et al. (2009a)
Pospiviroidae	*Coleviroid*	Coleus blumei viroid 6 (CVd-6)	FJ615418	342	Hou et al. (2009b)
Pospiviroidae	*Pospiviroid*	Portulaca latent viroid (PoLVd)	KR677387	351	Verhoeven et al. (2015)
Avsunviroidae		Grapevine hammerhead viroid-like RNA	KR736334	375	Wu et al. (2012)
Avsunviroidae		Apple hammerhead viroid-like RNA	KR605506	434	Zhang et al. (2014)

[a] *Classification in families and genera based exclusively on the presence of a certain type of CCR or hammerhead ribozymes in the respective circular RNAs.*
[b] *Viroid-like RNAs are those for which the proof of autonomous replication and transmissibility is still lacking.*
[c] *ID: GenBank accession number.*

discovery of additional new viroid-like RNAs in the near future. However, such RNAs should not be considered *bona fide* viroids until bioassays demonstrate their autonomous replication and transmissibility. Moreover, the structural and biological criteria discussed above must be met before proposing to ICTV the creation of any new viroid species.

References

Biebricher, C.K., Eigen, M., 2006. What is a quasispecies? Curr. Top. Microbiol. 299, 1–31.
Bonfiglioli, R.G., McFadden, G.I., Symons, R.H., 1994. *In situ* hybridization localizes avocado sunblotch viroid on chloroplast thylakoid membranes and coconut cadang-cadang viroid in the nucleus. Plant J. 6, 99–103.

Bonfiglioli, R.G., Webb, D.R., Symons, R.H., 1996. Tissue and intra-cellular distribution of coconut cadang-cadang viroid and citrus exocortis viroid determined by *in situ* hybridization and confocal laser scanning and transmission electron microscopy. Plant J. 9, 457–465.

Bussière, F., Lehoux, J., Thompson, D.A., Skrzeczkowski, L.J., Perreault, J.P., 1999. Subcellular localization and rolling circle replication of peach latent mosaic viroid: hallmarks of group A viroids. J. Virol. 73, 6353–6630.

Codoner, F., Daros, J., Sole, R., Elena, S., 2006. The fittest versus the flattest: experimental confirmation of the quasispecies effect with subviral pathogens. PLoS Pathog. 2, 1187–1193.

Davies, C., Sheldon, C.C., Symons, R.H., 1991. Alternative hammerhead structures in the self-cleavage of avocado sunblotch viroid RNAs. Nucleic Acids Res. 19, 1893–1898.

Di Serio, F., Flores, R., Verhoeven, J.Th.J., Li, S.F., Pallas, V., Randles, J.W., et al., 2014. Current status of viroid taxonomy. Arch. Virol. 159, 3467–3478.

Diener, T.O., 1971. Potato spindle tuber "virus": IV. A replicating, low molecular weight RNA. Virology 45, 411–428.

Diener, T.O., 1989. Circular RNAs: relics of precellular evolution? Proc. Natl. Acad. Sci. USA 86, 9370–9374.

Elena, S.F., Dopazo, J., Flores, R., Diener, T.O., Moya, A., 1991. Phylogeny of viroids, viroidlike satellite RNAs, and the viroidlike domain of hepatitis δ virus RNA. Proc. Natl. Acad. Sci. USA 88, 5631–5634.

Flores, R., 1995. Subviral agents: Viroids. In: Murphy, F.A., Fauquet, C.M., Bishop, D.H.L., Ghabrial, S.A., Jarvis, A.W., Martelli, G.P., Mayo, M.A., Summers, M.D. (Eds.), Virus Taxonomy, Sixth Report of the International Committee on Taxonomy of Viruses. Springer, Vienna, AT, pp. 495–497.

Flores, R., Daròs, J.A., Hernández, C., 2000b. The *Avsunviroidae* family: viroids with hammerhead ribozymes. Adv. Virus Res. 55, 271–323.

Flores, R., Di Serio, F., Hernández, C., 1997. Viroids: the noncoding genomes. Semin. Virol. 8, 65–73.

Flores, R., Gago-Zachert, S., Serra, P., Sanjuan, R., Elena, S.F., 2014. Viroids: survivors from the RNA world. Annu. Rev. Microbiol. 68, 395–414.

Flores, R., Randles, J.W., Bar-Joseph, M., Diener, T.O., 1998. A proposed scheme for viroid classification and nomenclature. Arch. Virol. 143, 623–629.

Flores, R., Randles, J.W., Owens, R.A., Bar-Joseph, M., Diener, T.O., 2000a. Subviral agents: Viroids. In: van Regenmortel., M.H.V., Fauquet, C.M., Bishop, D.H.L., Carstens, E.B., Estes, M.K., Lemon, S.M., Mc Geoch, D.J., Maniloff, J., Mayo, M.A., Pringle, C.R., Wickner, R.B. (Eds.), Virus Taxonomy, Seventh Report of the International Committee on Taxonomy of Viruses. Academic Press, San Diego, CA, pp. 1009–1024.

Flores, R., Randles, J.W., Owens, R.A., Bar-Joseph, M., Diener, T.O., 2005. Viroids. In: Fauquet, C.M., Mayo, M.A., Maniloff, J., Desselberger, U., Ball, A.L. (Eds.), Virus Taxonomy, Eighth Report of the International Committee on Taxonomy of Viruses. Elsevier/Academic Press, London, pp. 1145–1159.

Forster, A.C., Davies, C., Sheldon, C.C., Jeffries, A.C., Symons, R.H., 1988. Self-cleaving viroid and newt RNAs may only be active as dimers. Nature 334, 265–267.

Gambino, G., Navarro, B., Vallania, R., Gribaudo, I., Di Serio, F., 2011. Somatic embryogenesis efficiently eliminates viroid infections from grapevines. Eur. J. Plant Pathol. 130, 511–519.

Gas, M.E., Hernández, C., Flores, R., Daròs, J.A., 2007. Processing of nuclear viroids in vivo: an interplay between RNA conformations. PLoS Pathog. 3, e182.

Gómez, G., Martínez, G., Pallás, V., 2009. Interplay between viroid-induced pathogenesis and RNA silencing pathways. Trends Plant Sci. 14, 264–269.

Harders, J., Lukacs, N., Robert-Nicoud, M., Jovin, J.M., Riesner, D., 1989. Imaging of viroids in nuclei from tomato leaf tissue by in situ hybridization and confocal laser scanning microscopy. EMBO J. 8, 3941–3949.

Hou, W.Y., Li, S.F., Wu, Z.J., Jiang, D.M., Sano, T., 2009b. Coleus blumei viroid 6: a new tentative member of the genus *Coleviroid* derived from natural genome shuffling. Arch. Virol. 154, 993–997.

Hou, W.Y., Sano, T., Li, S.F., Li, F., Li, L., Wu, Z.J., 2009a. Identification and characterization of a new coleviroid (CbVd-5). Arch. Virol. 154, 315–320.

Ito, T., Suzaki, K., Nakano, M., Sato, A., 2013. Characterization of a new apscaviroid from American persimmon. Arch. Virol. 158, 2629–2631.

Jakse, J., Radisek, S., Pokorn, T., Matousek, J., Javornik, B., 2015. Deep-sequencing revealed citrus bark cracking viroid (CBCVd) as a highly aggressive pathogen on hop. Plant Pathol. 64, 831–842.

Jiang, D., Guo, R., Wu, Z., Wang, H., Li, S.F., 2009. Molecular characterization of a member of a new species of grapevine viroid. Arch. Virol. 154, 1563–1566.

King, A.M.K., Adams, M.J., Carstens, E.B., Lefkowitz, E.J., 2012. Virus Taxonomy: Ninth Report of the International Committee on Taxonomy of Viruses. Elsevier/Academic Press, London, UK.

Koltunow, A.M., Rezaian, M.A., 1989. A scheme for viroid classification. Intervirology 30, 194–201.

Koltunow, A.M., Rezaian, M.A., 1988. Grapevine yellow speckle viroid: structural features of a new viroid group. Nucleic Acids Res. 16, 849–864.

Lima, M.I., Fonseca, M.E.N., Flores, R., Kitajima, E.W., 1994. Detection of avocado sunblotch viroid in chloroplasts of avocado leaves by *in situ* hybridization. Arch. Virol. 138, 385–390.

McInnes, J.L., Symons, R.H., 1991. Comparative structure of viroids and their rapid detection using radioactive and nonradioactive nucleic acid probes. In: Maramorosch, K. (Ed.), Viroids and Satellites: Molecular Parasites at the Frontier of Life. CRC Press, Boca Raton, FL, pp. 21–58.

Mohamed, N.A., Thomas, W., 1980. Viroidlike properties of an RNA species associated with the sunblotch disease of avocados. J. Gen. Virol. 46, 157–167.

Nakaune, R., Nakano, M., 2008. Identification of a new apscaviroid from Japanese persimmon. Arch. Virol. 153, 969–972.

Navarro, B., Flores, R., 1997. Chrysanthemum chlorotic mottle viroid: unusual structural properties of a subgroup of self-cleaving viroids with hammerhead ribozymes. Proc. Natl. Acad. Sci. USA 94, 11262–11267.

Navarro, B., Gisel, A., Rodio, M.E., Delgado, S., Flores, R., Di Serio, F., 2012. Viroids: how to infect a host and cause disease without encoding proteins. Biochimie 94, 1474–1480.

Owens, R.A., Flores, R., Di Serio, F., Li, S.F., Pallás, V., Randles, J.W., et al., 2012. Viroids. In: King, A.M.Q., Adams, M.J., Carstens, E.B., Lefkowitz, E.J. (Eds.), Virus Taxonomy, Ninth Report of the International Committee on Taxonomy of Viruses. Elsevier/ Academic Press, London UK, pp. 1221–1234.

Puchta, H., Ramm, K., Sänger, H.L., 1988. The molecular structure of hop latent viroid (HLVd), a new viroid occurring worldwide in hops. Nucleic Acids Res. 16, 4197–4216.

Qi, Y., Ding, B., 2003. Differential subnuclear localization of RNA strands of opposite polarity derived from an autonomously replicating viroid. Plant Cell. 15, 2566–2577.

Riesner, D., Henco, K., Rokohl, U., Klotz, G., Kleinschmidt, A.K., Domdey, H., et al., 1979. Structure and structure formation of viroids. J. Mol. Biol. 133, 85–115.

Sano, T., Isono, S., Matsuki, K., Kawaguchi-Ito, Y., Tanaka, K., Kondo, K., et al., 2008. Vegetative propagation and its possible role as a genetic bottleneck in the shaping of the apple fruit crinkle viroid populations in apple and hop plants. Virus Genes 37, 298–303.

Semancik, J.S., Vidalakis, G., 2005. The question of citrus viroid IV as a cocadviroid. Arch. Virol. 150, 1059–1067.

Spiesmacher, E., Mühlbach, H.P., Schnölzer, M., Haas, B., Sänger, H.L., 1983. Oligomeric forms of potato spindle tuber viroid (PSTV) and of its complementary RNA are present in nuclei isolated from viroid infected potato cells. Biosci. Rep. 3, 767–774.

Verhoeven, J.Th.J., Meekes, E.T., Roenhorst, J.W., Flores, R., Serra, P., 2013. Dahlia latent viroid: a recombinant new species of the family *Pospiviroidae* posing intriguing questions about its origin and classification. J. Gen. Virol. 94, 711–719.

Verhoeven, J.Th.J., Roenhorst, J.W., Hooftman, M., Meekes, E.T.M., Flores, R., et al., 2015. A pospiviroid from symptomless portulaca plants closely related to iresine viroid 1. Virus Res. 205, 22–26.

Verhoeven, J.Th.J., Roenhorst, J.W., Owens, R.A., 2011. *Mexican papita viroid* and *tomato planta macho viroid* belong to a single species in the genus *Pospiviroid*. Arch. Virol. 156, 1433–1437.

Wu, Q., Wang, Y., Cao, M., Pantaleo, V., Burgyan, J., Li, W.X., et al., 2012. Homology-independent discovery of replicating pathogenic circular RNAs by deep sequencing and a new computational algorithm. Proc. Natl. Acad. Sci. USA 109, 3938–3943.

Zhang, Z., Qi, S., Tang, N., Zhang, X., Chen, S., Zhu, P., et al., 2014. Discovery of replicating circular RNAs by RNA-seq and computational algorithms. PLoS Pathog. 10, e1004553.

PART III

VIROID DISEASES

CHAPTER

14

Potato Spindle Tuber Viroid

Robert A. Owens[1] and Jacobus Th.J. Verhoeven[2]

[1]U.S. Department of Agriculture, Beltsville, MD, United States

[2]National Plant Protection Organization, Wageningen, The Netherlands

INTRODUCTION

Of the many diseases now caused by viroids, the spindle tuber disease of potatoes was the first to be recognized and studied by plant pathologists. Nearly 50 years elapsed between the initial description of spindle tuber disease in the early 1920s and identification of its causal agent, a small, highly structured, covalently closed circular RNA known as potato spindle tuber viroid (PSTVd) (Diener, 1971). PSTVd remains an important threat to potato and tomato production, and a recent increase in the number of reported latent infections of ornamental species is creating new challenges. PSTVd is also a favorite object of study for viroid molecular biologists, due in large part to its ability to replicate to high titers in tomato where certain strains rapidly induce the appearance of a characteristic disease syndrome that includes stunting and epinasty.

TAXONOMIC POSITION AND NUCLEOTIDE SEQUENCE

Potato spindle tuber viroid is the type species of the genus *Pospiviroid* and the family *Pospiviroidae* (Owens et al., 2012). Its single-stranded, circular RNA genome contains 356–364 nt and adopts a rod-like conformation with a central conserved region (CCR), formed by stretches of conserved nucleotides in the upper and lower strands, as well as a terminal conserved region. Nucleotides in the upper portion of the CCR are flanked by an imperfect, inverted repeat that may assume an alternative conformation and is involved in viroid replication (see Fig. 13.1).

Viroids and Satellites.
DOI: http://dx.doi.org/10.1016/B978-0-12-801498-1.00014-0

Comparative sequence analysis suggests that PSTVd contains five structural domains; from left to right, these are known as the terminal left, pathogenicity, central, variable, and terminal right domains.

The Viroid Genome Database (http://www.ncbi.nlm.nih.gov/) currently contains the complete nucleotide sequences of ca. 300 PSTVd variants. Phylogenetic analysis groups these variants into several clusters according to the host species from which they were isolated (e.g., Verhoeven, 2010a).

ECONOMIC SIGNIFICANCE

Potato spindle tuber disease was first described in the early 1920s (Martin, 1922; Schultz and Folsom, 1923). Serious outbreaks affecting both commercial production and potato breeding material were reported in several parts of the world over the next several decades. Estimated yield losses in North America were approximately 1%, low as a percentage but still significant in absolute terms. Beginning in the 1980s, extensive testing and seed certification programs in many countries began to eliminate (or at least substantially reduce) the impact of PSTVd on potato production. This success could only have been achieved because new PSTVd introductions into potato crops are rare.

PSTVd also causes occasional disease outbreaks in tomato (e.g., Puchta et al., 1990). Although the impact on individual growers may be severe, the yearly number of registered outbreaks is very low. In contrast, the number of latent PSTVd infections in ornamental species, especially those growing in greenhouses, has increased dramatically (EFSA Panel on Plant Health, 2011). These infections have little or no effect on crop production but may indirectly affect the grower's income by restricting export opportunities due to existing phytosanitary measures.

SYMPTOMS

PSTVd causes disease in *Solanum tuberosum* (potato), *S. lycopersicum* (tomato), and *Capsicum annuum* (pepper), where symptoms vary considerably depending on plant species, variety, viroid strain, and environmental conditions. In potato, growth of infected plants may be severely reduced or even cease entirely. As shown in Fig. 14.1, the vines of infected plants may be smaller, more upright, and produce smaller leaves than their healthy counterparts. Infected tubers may be small, elongated (from which the disease derives its name), misshapen, and cracked. Their eyes may be more pronounced than normal and may be borne on knob-like protuberances that may develop into small tubers.

JOURNAL OF AGRICULTURAL RESEARCH

VOL. XXV WASHINGTON, D. C., JULY 14, 1923 No. 2

TRANSMISSION, VARIATION, AND CONTROL OF CERTAIN DEGENERATION DISEASES OF IRISH POTATOES[1]

By E. S. Schultz, *Pathologist, Cotton, Truck, and Forage Crop Disease Investigations, Bureau of Plant Industry, United States Department of Agriculture*, and Donald Folsom, *Plant Pathologist, Maine Agricultural Experiment Station*

INTRODUCTION

Progress in solving the well-known problem of degeneration in the Irish potato, *Solanum tuberosum* L., has been comparatively rapid during the last decade. With this progress the apparent complexity of the problem has increased. Consequently the results of many investigators are needed and frequent reports from the various workers in this field are desirable.

FIGURE 14.1 Spindle tuber disease of potato. Left, title and introductory paragraph from Schultz and Folsom (1923) describing the biological properties of PSTVd; right, foliar and tuber symptoms in potato. (1) uninfected control; (2) PSTVd-infected plant.

The first symptoms of PSTVd infection in tomato are growth reduction and chlorosis in the top of the plant. Subsequently, chlorosis may become more severe, turning into reddening and/or purpling, and the plants may become stunted. At this stage, leaves may become brittle. Stunting is generally permanent; occasionally, however, plants may either die or partially recover. As stunting begins, flower and fruit development stop.

Peppers, in contrast, display very mild symptoms (Lebas et al., 2005). The only visible symptom of PSTVd infection is a certain "waviness" or distortion of the leaf margins near the top of infected plants. Infection in most of the other hosts listed in Table 14.1 (including many ornamental species) is symptomless.

HOST RANGE

The natural host range of PSTVd includes many solanaceous species (Table 14.1). In addition to potato and a diverse array of ornamental species (many native to Central and/or South America), PSTVd has also been isolated from several crop species native to the Americas; namely, tomato, pepper, and Cape gooseberry (*Physalis peruviana*). Reported woody hosts include avocado (*Persea americana*) and rubber (*Hevea brasiliensis*).

The experimental host range of PSTVd is considerably broader than its natural host range. As reported by Singh (1973) more than half of all plant species and varieties tested (i.e., 138 of 232) were susceptible to PSTVd infection. Susceptible selections were distributed among ten different families, but only 12 selections from the families *Solanaceae* and *Compositae* showed visible symptoms (see also Diener, 1979; Singh et al., 2003).

TABLE 14.1 Reported Natural Hosts of Potato Spindle Tuber Viroid (PSTVd)

Plant species	Reference
Atriplex semilunaris	Mackie et al. (2016)
Brugmansia spp.	Verhoeven et al. (2008a, 2010a)
Calibrachoa sp.	Verhoeven (2010)
Capsicum annuum	Lebas et al. (2005)
Cestrum spp.	Luigi et al. (2011)
Chrysanthemum sp.	Lemmetty et al. (2011)
Conyza bonariensis	Mackie et al. (2016)
Dahlia sp.	Tsushima et al. (2011)
Datura leichhardtii	Mackie et al. (2016)
Datura sp.	Verhoeven et al. (2010a)
Hevea brasiliensis	Ramachandran et al. (2000), Kumar et al. (2015)
Lycianthes rantonnetii	Di Serio (2007)
Nicandra physalodes	Mackie et al. (2016)
Persea americana	Querci et al. (1995)
Petunia × hybrida	Mertelik et al. (2009)
Physalis angulata	Mackie et al. (2016)
Physalis peruviana	Verhoeven et al. (2009)
Rhagodia eremaea	Mackie et al. (2016)
Solanum jasminoides	Di Serio (2007), Verhoeven et al. (2008a)
Solanum lycopersicum	Puchta et al. (1990)
Solanum nigrum	Mackie et al. (2016)
Solanum tuberosum	Martin (1922), Schultz and Folsom (1923)
Streptoglossa sp.	Mackie et al. (2016)
Streptosolen jamesonii	Verhoeven et al. (2008b)

TRANSMISSION

PSTVd may spread in four different ways:

Vegetative Propagation

Propagation by tubers, cuttings, and microplants provides a very efficient means of PSTVd transmission. Once established, infection is persistent; therefore, plants from infected lots act as a permanent source of inoculum for other lots and crops. Vegetative propagation is the major pathway for PSTVd transmission in potato and ornamentals such as *Brugmansia* spp. and *Solanum jasminoides*. The absence of symptoms in ornamental species increases the risk that infected plants will be used for propagation.

Mechanical Transmission

Under favorable conditions (e.g., warm temperatures), PSTVd is readily transmitted by normal cultivation activities (Verhoeven et al., 2010b). The viroid can be disseminated very readily in potato fields by contact of healthy vines with contaminated cultivating and hilling equipment (Manzer and Merriam, 1961).

Infected Seed and Pollen

PSTVd is assumed to have spread among potato germplasm collections all over the world via infected true seed (Diener, 1996). Once present, the viroid can be transmitted to uninfected plants either mechanically or by pollen transfer (Singh, 1970). Seed is also a potential source of infection for other crops, such as tomato and pepper that are propagated by seed.

Insect Transmission

Aphid transmission of PSTVd requires the source plant to be infected by both PSTVd and potato leafroll virus (PLRV), thereby limiting the number of potential infection sources (Querci et al., 1997; Syller et al., 1997). The viroid is assumed to be encapsidated by PLRV coat protein; such trans-encapsidation protects the viroid from digestion by micrococcal nuclease in vitro, suggesting that a similar protective effect may occur in vivo. Under greenhouse conditions, bumblebee activity can result in transmission of both tomato apical stunt viroid (Antignus et al., 2007) and tomato chlorotic dwarf viroid (Matsuura et al., 2010).

DETECTION

Detection of PSTVd is usually accomplished by conventional or real-time RT-PCR using primers designed for pospiviroid amplification (Boonham et al., 2004; Bostan et al., 2004; Botermans et al., 2013; Shamloul et al., 1997; Verhoeven et al., 2004). In all cases, additional tests are required to obtain a full-length sequence for definite identification. Other molecular methods, including dot-blot hybridization with PSTVd-specific probes (Owens and Diener, 1981), can also be used but require additional tests for definitive identification.

These rapid and sensitive molecular methods have almost completely displaced two older methods, bioassay on "Rutgers" or other sensitive tomato cultivar (Raymer and O'Brien, 1962) and polyacrylamide gel electrophoresis (e.g., Morris and Wright, 1975; Singh and Boucher, 1987). Although laborious and time-consuming (days/weeks vs hours), these older methods are still useful in certain situations. Both methods can be used to distinguish mild from severe strains of PSTVd, and two-dimensional or return-polyacrylamide gel electrophoresis analysis provides a powerful means to detect small circular RNA molecules that does not require prior sequence information (Hanold et al., 2003; Chapter 34: Gel Electrophoresis).

GEOGRAPHICAL DISTRIBUTION AND EPIDEMIOLOGY

PSTVd is widely distributed throughout the world (CABI/EPPO, 2014). A relatively wide host range, which includes both major crops and asymptomatically infected ornamental plant species (Table 14.1), and the ability to spread by both vegetative and generative propagation has facilitated its spread to many countries. Nevertheless, several successful eradications have been reported, especially for potato (e.g., De Boer and De Haan, 2005; Sun et al., 2004).

PSTVd epidemiology (its distribution, transmission, and control) is complicated because of the large number of natural hosts (Table 14.1) and potential transmission routes. For vegetatively propagated crops such as potato and many ornamentals, the main mode of spread is propagation by infected tubers and/or cuttings. Transmission by contaminated machinery (Manzer and Merriam, 1961) or aphids carrying both PLRV and PSTVd may result in further spread.

Contaminated seed is an important source of infection for crops, like tomato, that are grown from seed (Simmons et al., 2015; Singh, 1970), but other routes of transmission should also be considered. For example, comparison of the respective nucleotide sequences indicated connections between PSTVd infections in several lots of tomato and other infections in the ornamental species *S. jasminoides*. Because *S. jasminoides* (and its vegetatively propagated progeny) is kept in the greenhouse year-round, it can serve as a persistent source of inoculum for tomato crops, which are removed from the greenhouse after every production cycle. This indicates that *S. jasminoides* was the original source of infection in tomato and not vice versa (Verhoeven et al., 2010a).

PSTVd has never been isolated from any of the wild potato species native to the Andes, center of origin for the cultivated potato. Furthermore, no genetic resistance to PSTVd has been identified in *S. tuberosum*. These observations suggest that (1) PSTVd and its potato host did not coevolve and (2) spindle tuber disease originated by chance transfer of the viroid from a wild host, with further spread by propagation and breeding activities (Diener, 1987, 1996). How and when this presumably recent transfer occurred is unknown.

Some wild *Solanum cardiophyllum* plants growing in Mexico are latently infected with Mexican papita viroid, a pospiviroid now considered to be a variant of tomato planta macho viroid (Martinez-Soriano et al., 1996). These authors suggest that other wild solanaceous species brought to the United States from Mexico in the late 19th century may have contained the initial source of PSTVd later detected in potato. Also, PSTVd may have been introduced into South Africa in plants of *Physalis peruviana* imported from South America centuries ago as a vegetable. In South Africa, the viroid may have caused the "bunchy top" disease of tomato (McClean, 1931). Also, variants from the *P. peruviana* cluster of PSTVd sequences have been isolated from several plant species growing in Australia and New Zealand (Verhoeven et al., 2004, 2010a). Many plant introductions to Oceania from the Americas have passed through South Africa.

CONTROL

Successful control of PSTVd infection requires both prevention of infection and viroid eradication.

Prevention of infection includes all measures to prevent the introduction of PSTVd into a specific crop. It is very important to start a new cultivation with viroid-free planting material. Many countries regard PSTVd as a quarantine "organism," and governmental measures to prevent introduction of PSTVd with plants from other countries will often be applied. Certification schemes including testing may be required to provide further guarantees that tubers, seeds, or plants are free from PSTVd (EFSA Panel on Plant Health, 2011). It is also important to prevent viroid introduction via human activities. Because PSTVd is mechanically transmissible, it can be introduced into potential host plants via the hands, clothes, or equipment used by people working in or visiting the greenhouse.

Viroid eradication involves destruction of PSTVd-infected plants and thorough cleaning of equipment and greenhouses where infected plants have been grown. All infected plants together with those from an adequate buffer zone should be destroyed. For field-grown potatoes, crop rotations involving nonhost species help eliminate infected volunteer plants. In symptomless infections such as those commonly observed in ornamentals, all plants in the lot should be destroyed. Any rock wool or plastic used to cover the soil should also be removed and destroyed. When PSTVd is identified in a greenhouse-grown crop, all parts of the greenhouse should be thoroughly cleaned and disinfected.

References

Antignus, Y., Lachman, O., Pearlsman, M., 2007. The spread of tomato apical stunt viroid (TASVd) in greenhouse tomato crops is associated with seed transmission and bumble bee activity. Plant Dis. 91, 47–50.

Boonham, N., Pérez, L.G., Méndez, M.S., Peralta, E.L., Blockley, A., Walsh, K., et al., 2004. Development of a real-time RT-PCR assay for the detection of potato spindle tuber viroid. J. Virol. Methods 116, 139–146.

Bostan, H., Nie, X., Singh, R.P., 2004. An RT-PCR primer pair for the detection of pospiviroid and its application in surveying ornamental plants for viroids. J. Virol. Methods 116, 189–193.

Botermans, M., van de Vossenberg, B.T., Verhoeven, J.Th.J., Roenhorst, J.W., Hooftman, M., Dekter, R., et al., 2013. Development and validation of a real-time RT-PCR assay for generic detections of pospiviroids. J. Virol. Methods 187, 43–50.

CABI/EPPO, 2014. Potato Spindle Tuber Viroid. Distribution Maps of Plant Diseases No. 729. CABI Head Office, Wallingford, UK.

De Boer, S.H., De Haan, T.L., 2005. Absence of potato spindle tuber viroid within the Canadian potato industry. Plant Dis. 89, 910.

Di Serio, F., 2007. Identification and characterization of potato spindle tuber viroid infecting *Solanum jasminoides* and *S. rantonnetii* in Italy. J. Plant Pathol. 89, 297–300.

Diener, T.O., 1971. Potato spindle tuber "virus" IV. A replicating, low molecular weight RNA. Virology 45, 411–428.

Diener, T.O., 1979. Viroids and Viroid Diseases. Wiley-Interscience, New York, p. 252.

Diener, T.O., 1987. Potato spindle tuber. In: Diener, T.O. (Ed.), The Viroids. Plenum Press, New York, pp. 221–233.

Diener, T.O., 1996. Origin and evolution of viroids and viroid-like satellite RNAs. Virus Genes 11, 119–131.

EFSA Panel on Plant Health, 2011. Scientific opinion on the assessment of the risk of solanaceous pospiviroids for the EU territory and the identification and evaluation of risk management options. EFSA J. 9, 2330.

Hanold, D., Semancik, J.S., Owens, R.A., 2003. Polyacrylamide gel electrophoresis. In: Hadidi, A., Flores, R., Randles, J.W., Semancik, J.S. (Eds.), Viroids. CSIRO Publishing, Collingwood, VIC, pp. 95–102.

Kumar, A., Pandey, D.M., Abraham, T., Mathew, J., Jyothsna, P., Ramachandran, P., et al., 2015. Molecular characterization of viroid associated with tapping panel dryness syndrome of *Hevea brasiliensis* from India. Current Sci. 108, 1520–1527.

Lebas, B.S.M., Clover, G.R.G., Ochoa-Corona, F.M., Elliott, D.R., Tang, Z., Alexander, B.J.R., 2005. Distribution of potato spindle tuber viroid in New Zealand glasshouse crops of capsicum and tomato. Austral. Plant Pathol. 34, 129–133.

Lemmetty, A., Laamanen, J., Soukained, M., Tegel, J., 2011. Emerging virus and viroid pathogen species identified for the first time in horticultural plants in Finland in 1997–2010. Agric. Food Sci. 20, 29–41.

Luigi, M., Luison, D., Tomassoli, L., Faggioli, F., 2011. First report of potato spindle tuber and citrus exocortis viroids in Cestrum spp. in Italy. New Dis. Rep. 23, 4.

Mackie, A.E., Rodoni, B.C., Barbetti, M.J., McKirdy, S.J., Jones, R.A.C., 2016. Potato spindle tuber viroid: alternative host reservoirs and strain found in a remote subtropical irrigation area. Eur. J. Plant Pathol. 145, 433–446.

Manzer, F.E., Merriam, D., 1961. Field transmission of the potato spindle tuber virus and virus X by cultivating and hilling equipment. Am. Potato J. 38, 346–352.

Martin, W.H., 1922. "Spindle tuber," a new potato trouble. Hints to potato growers, New Jersey State Potato Association 3, 8.

Martinez-Soriano, J.P., Galindo-Alonso, J., Maroon, C.J.M., Yucel, I., Smith, D.R., Diener, T.O., 1996. Mexican papita viroid: putative ancestor of crop viroids. Proc. Nat. Acad. Sci. USA 93, 9397–9401.

Matsuura, S., Matsushita, Y., Kozuka, R., Shimizu, S., Tsuda, S., 2010. Transmission of tomato chlorotic dwarf viroid by bumblebees (*Bombus ignites*) in tomato plants. Eur. J. Plant Pathol. 126, 111–115.

McClean, A.P.D., 1931. Bunchy top disease of tomato. South African Department of Agriculture Science Bulletin 100, 36.

Mertelik, J., Kloudova, K., Cervena, G., Necekalova, J., Mikulkova, H., Levkanicova, Z., et al., 2009. First report of potato spindle tuber viroid (PSTVd) in *Brugmansia* spp., *Solanum jasminoides, Solanum muricatum* and *Petunia* spp. in the Czech Republic. New Dis. Rep. 19, 27.

Morris, T.J., Wright, N.S., 1975. Detection on polyacrylamide gel of a diagnostic nucleic acid from tissue infected with potato spindle tuber viroid. Am. Potato J. 52, 57–63.

Owens, R.A., Diener, T.O., 1981. Sensitive and rapid diagnosis of potato spindle tuber viroid disease by nucleic acid hybridization. Science 213, 670–672.

Owens, R.A., Flores, R., Di Serio, F., Li, S.F., Pallás, V., Randles, J.W., et al., 2012. Viroids. In: King, A.M.Q., Adams, M.J., Carstens, E.B., Lefkowitz, E.J. (Eds.), Virus Taxonomy, Ninth Report of the International Committee on Taxonomy of Viruses, Elsevier/ Academic Press, London, UK, pp. 1221–1234.

Puchta, H., Herold, T., Verhoeven, K., Roenhorst, A., Ramm, K., Schmidt-Puchta, W., et al., 1990. A new strain of potato spindle tuber viroid (PSTVd-N) exhibits major sequence differences as compared to all other PSTVd strains sequenced so far. Plant Mol. Biol. 15, 509–511.

Querci, M., Owens, R.A., Vargas, C., Salazar, L.F., 1995. Detection of potato spindle tuber viroid in avocado growing in Peru. Plant Dis. 79, 196–202.

Querci., M., Owens, R.A., Bartoli, I., Lazarte, V., Salazar, L.F., 1997. Evidence for heterologous encapsidation of potato spindle tuber viroid in particles of potato leafroll virus. J. Gen. Virol. 78, 1207–1211.

Ramachandran, P., Francis, L., Mathur, S., Varma, A., Mathew, J., Mathew, N.M., et al., 2000. Evidence for association of a low molecular weight RNA with tapping panel dryness syndrome of rubber. Plant Dis. 84, 1155.

Raymer, W.J., O'Brien, M.J., 1962. Transmission of potato spindle tuber virus to tomato. Am. Potato J. 39, 401–408.

Schultz, E.S., Folsom, D., 1923. Transmission, variation, and control of certain degeneration diseases of Irish potatoes. J. Agric. Res. 25, 43–118.

Shamloul, A.M., Hadidi, A.F., Zhu, S.F., Singh, R.P., Sagredo, B., 1997. Sensitive detection of potato spindle tuber viroid using RT-PCR and identification of a viroid variant naturally infecting pepino plants. Can. J. Plant Pathol. 19, 89–96.

Simmons, H.E., Ruchti, T.B., Munkvold, G.P., 2015. Frequencies of seed infection and transmission to seedlings by potato spindle tuber viroid (a pospiviroid) in tomato. J. Plant Pathol. Microbiol. 6, 275.

Singh, R.P., 1970. Seed transmission of potato spindle tuber virus in tomato and potato. Am. Potato J. 47, 225–227.

Singh, R.P., 1973. Experimental host range of potato spindle tuber "virus." Am. Potato J. 50, 111–123.

Singh, R.P., Boucher, A., 1987. Electrophoretic separation of a severe from mild strains of potato spindle tuber viroid. Phytopathology 77, 1588–1591.

Singh, R.P., Ready, K.F.M., Nie, X., 2003. Viroids of solanaceous species. In: Hadidi, A., Flores, R., Randles, J.W., Semancik, J.S. (Eds.), Viroids. CSIRO Publishing, Collingwood, VIC, pp. 125–133.

Sun, M., Siemsen, S., Campbell, W., 2004. Survey of potato spindle tuber viroid in seed potato growing areas of the United States. Amer. J. Potato Res. 81, 227–231.

Syller, J., Marczewski, W., Pawlowicz, J., 1997. Transmission by aphids of potato spindle tuber viroid encapsidated by potato leafroll luteovirus particles. Eur. J. Plant Pathol. 103, 285–289.

Tsushima, T., Murakami, S., Ito, H., He, Y.H., Raj, A.P.C., Sano, T., 2011. Molecular characterization of potato spindle tuber viroid in dahlia. J. Gen. Plant Pathol. 77, 253–256.

Verhoeven, J.Th.J., 2010a. Identification and Epidemiology of Pospiviroids. Thesis Wageningen University, Wageningen, The Netherlands, ISBN 978-90-8585-623-8. 136pp.

Verhoeven, J.Th.J., Botermans, M., Roenhorst, J.W., Westerhof, J., Meekes, E.T.M., 2009. First report of *Potato spindle tuber viroid* in Cape gooseberry (*Physalis peruviana*) from Turkey and Germany. Plant Dis. 93, 316.

Verhoeven, J.Th.J., Hüner, L., Virscek Marn, M., Mavric Plesko, I., Roenhorst, J.W., 2010b. Mechanical transmission of potato spindle tuber viroid between plants of *Brugmansia suaveoles, Solanum jasminoides* and potato and tomatoes. Eur. J. Plant Pathol. 128, 417–421.

Verhoeven, J.Th.J., Jansen, C.C.C., Botermans, M., Roenhorst, J.W., 2010a. Epidemiological evidence that vegetatively propagated, solanaceous plant species act as sources of potato spindle tuber viroid inoculum for tomato. Plant Pathol. 59, 3–12.

Verhoeven, J.Th.J., Jansen, C.C.C., Roenhorst, J.W., 2008a. First report of pospiviroids infecting ornamentals in the Netherlands: citrus exocortis viroid in *Verbena* sp., potato spindle tuber viroid in *Brugmansia suaveolens* and *Solanum jasminoides*, and tomato apical stunt viroid in *Cestrum* sp. Plant Pathol. 57, 399.

Verhoeven, J.Th.J., Jansen, C.C.C., Roenhorst, J.W., 2008b. *Streptosolen jamesonii* "Yellow," a new host plant of potato spindle tuber viroid. Plant Pathol. 57, 399.

Verhoeven, J.Th.J., Jansen, C.C.C., Willemen, T.M., Kox, L.F.F., Owens, R.A., et al., 2004. Natural infections of tomato by citrus exocortis viroid, columnea latent viroid, potato spindle tuber viroid and tomato chlorotic dwarf viroid. Eur. J. Plant Pathol. 110, 823–831.

CHAPTER

15

Other Pospiviroids Infecting Solanaceous Plants

Thierry Candresse[1], Jacobus Th.J. Verhoeven[2], Giuseppe Stancanelli[3], Rosemarie W. Hammond[4] and Stephan Winter[5]

[1]INRA and University of Bordeaux, Bordeaux, France [2]National Plant Protection Organization, Wageningen, The Netherlands [3]European Food Safety Authority, Parma, Italy [4]U.S. Department of Agriculture, Beltsville, MD, United States [5]Leibniz Institute, DSMZ- German Collection of Microorganisms and Cell Cultures, Braunschweig, Germany

INTRODUCTION

Of the nine recognized members of the genus *Pospiviroid*, eight are reported to naturally infect solanaceous crops (tomato, potato, and pepper) and ornamentals (*P. hybrida*, *Solanum* spp., *Brugmansia* spp.). The sole exception is Iresine viroid 1 (IrVd-1), which has never been found on solanaceous hosts and which could not experimentally infect tomato or potato (Spieker, 1996; Verhoeven et al., 2010c). While most pospiviroids are strongly associated with solanaceous hosts, two of them, citrus exocortis viroid (CEVd) and chrysanthemum stunt viroid (CSVd) have their major hosts in other families (*Citrus* spp. for CEVd, *Asteraceae* for CSVd). The present chapter focuses on the pospiviroids infecting solanaceous plants other than potato spindle tuber viroid (PSTVd), CEVd, and CSVd, which are addressed in separate chapters.

Viroids and Satellites.
DOI: http://dx.doi.org/10.1016/B978-0-12-801498-1.00015-2

TAXONOMIC POSITION AND NUCLEOTIDE SEQUENCE

According to the ICTV 9th report updated online in 2015 (http://www.ictvonline.org/virusTaxonomy.asp), the genus *Pospiviroid* comprises nine species: *Potato spindle tuber viroid, Citrus exocortis viroid, Columnea latent viroid, Iresine viroid 1, Tomato apical stunt viroid, Tomato chlorotic dwarf viroid, Tomato planta macho viroid, Chrysanthemum stunt viroid,* and *Pepper chat fruit viroid* (see Chapter 13, Viroid Taxonomy; Owens et al., 2012).

For all solanaceous pospiviroids, complete genome sequences are available, sometimes in large numbers, in international sequence databases and reference genome sequences have been defined in GenBank: Columnea latent viroid (CLVd, NC_003538), pepper chat fruit viroid (PCFVd, NC_011590), tomato apical stunt viroid (TASVd, NC_001553), tomato chlorotic dwarf viroid (TCDVd, NC_000885), tomato planta macho viroid (TPMVd, NC_001558).

HOST RANGE

The pospiviroids addressed here have narrow natural host ranges comprising from one to six natural hosts in the family *Solanaceae* (Table 15.1). However, new hosts are occasionally discovered and the experimental host range is usually significantly larger than the natural one, suggesting the latter is incompletely known.

SYMPTOMS AND ECONOMIC SIGNIFICANCE

Symptoms vary widely depending on the host. Tomato is the main host for the five pospiviroids considered here, which all incite symptoms similar to those caused by PSTVd. Symptom severity varies with viroid isolate and tomato variety. In addition, symptoms are more severe when plants are infected at early developmental stages or grown under conditions of high light intensity and temperature. Early infections result in growth reduction, stunted growth, and bunchy appearance. Leaves show chlorosis and epinasty and may become brittle, reddened, or necrotic. Flowering and fruit development are severely affected in stunted plants and yield losses of up to 100% can be observed in plants infected at an early stage. The impact is less in plants infected late since fruit already initiated may still develop. Verhoeven et al. (2006) also reported a delay in the ripening of TASVd-infected

TABLE 15.1 Natural Host Range of the Solanaceous Pospiviroids Addressed Here

Host plant	Viroid and reference
HOSTS IN THE SOLANACEAE	
Brugmansia sp.	TASVd (Olivier et al., 2011); TCDVd (Verhoeven et al., 2010d)
Brunfelsia undulata	CLVd (Spieker, 1996)
Capsicum annuum	PCFVd (Verhoeven et al., 2009)
Cestrum sp.	TASVd (Verhoeven et al., 2008a)
Petunia sp.	TCDVd (Verhoeven et al., 2007)
Solanum cardiophyllum	TPMVd (as MPVd, Martinez-Soriano et al., 1996)
Solanum jasminoides	TASVd (Verhoeven et al., 2008b)
Solanum lycopersicum	CLVd (Verhoeven et al., 2004), TPMVd (Galindo 1987), TASVd (Candresse et al., 1987; Walter, 1987), PCFVd (Reanwarakorn et al., 2011), TCDVd (Singh et al., 1999)
Solanum pseudocapsicum	TASVd (Spieker et al., 1996)
Solanum stramonifolium	CLVd (Genbank database)
Streptosolen jamesonii	TASVd (Verhoeven et al., 2010a)
Lycianthes rantonnetii	TASVd (Verhoeven et al., 2010a)
HOST PLANTS OUTSIDE THE SOLANACEAE	
Columnea erythrophae	CLVd (Hammond et al., 1989)
Gloxinia gymnostoma	CLVd (Nielsen and Nicolaisen, 2010)
Gloxinia nematanthodes	CLVd (Nielsen and Nicolaisen, 2010)
Gloxinia purpurascens	CLVd (Nielsen and Nicolaisen, 2010)
Nematanthus wettsteinii	CLVd (Singh et al., 1992)
Pittosporum tobira	TCDVd (Verhoeven, Th. 2010)
Verbena sp.	TCDVd (Nie et al., 2005)
Vinca sp.	TCDVd (Singh and Dilworth, 2009)

tomatoes and, irrespective of the viroid involved, a reduction of fruit storage life.

As with other mechanically transmissible agents, symptoms are generally observed along plant rows (Matsushita et al., 2008; Verhoeven et al.,

FIGURE 15.1 A healthy fruit of pepper cv. Lamborgini (left) and two fruit produced by a PCFVd naturally infected plant (right). Plant Protection Service, Wageningen, The Netherlands.

2004). Varying infection rates (from 0.1% up to 80%–90%) have been recorded (Candresse et al., 2010; Ling and Bledsoe, 2009; Verhoeven et al., 2004), probably as a consequence of different initial infection pressures but also due to variations in management practices and environmental conditions.

Of the pospiviroids addressed in the present chapter, in pepper only PCFVd infections are known in nature (Verhoeven et al., 2009, 2011). Plant growth is slightly affected but the fruit produced can be 50% smaller (Fig. 15.1). However, maximum levels reported for natural infections reached only up to 3% of the plants during PCFVd outbreaks so that the overall impact was limited.

No natural infection of potato has been reported for the pospiviroids considered here. However, upon artificial inoculation they all cause PSTVd-like symptoms in this crop. Yield reduction in cv. Nicola grown from infected progeny tubers reached up to 72%–82% for CLVd (Verhoeven et al., 2004).

Natural symptomless infections have also been reported for CLVd, TASVd, and TCDVd in a number of ornamental species within and outside the *Solanaceae* (Table 15.1). Similarly, PCFVd is infectious to these hosts and albeit not found naturally, experimental inoculation of *Brugmansia suaveolens*, *Lycianthes rantonnetii*, *Solanum jasminoides*, and eggplant also resulted in symptomless infection (Verhoeven et al., 2009).

TRANSMISSION AND EPIDEMIOLOGY

Human-assisted spread plays a significant role in pospiviroid dissemination. Over large distances, this occurs through the global transport of infected propagation material (Owens and Verhoeven, 2009). Locally, human-assisted spread is via mechanical transmission through crop handling (Matsushita et al., 2008; Verhoeven et al., 2004) and may involve contaminated hands or pruning tools (Verhoeven et al., 2010b). However, disinfection of pruning tools with sodium hypochlorite (2% solution) or commercial bleach is quite efficient (Matsuura et al., 2010a; Singh et al., 1989).

TASVd and TCDVd are seed-transmitted in tomato (Antignus et al., 2007; Candresse et al., 2010; Singh and Dilworth, 2009) and PCFVd is seed-transmitted in pepper (Verhoeven et al., 2009), with high rates of transmission reported for TASVd (80%) (Antignus et al., 2007) and PCFVd (19%) (Verhoeven et al., 2009). There is circumstantial but inconclusive evidence that CLVd and TPMVd could be similarly seed-transmitted in tomato. There is conflicting evidence as to whether pospiviroids can be transmitted to the pollinated mother plant in chrysanthemum (Chung and Pak, 2008) or in tomato (Kryczynski et al., 1988), indicating that this process may be host- and/or viroid-dependent or may involve additional factors such as the action of pollinating insects. TASVd is carried in internal tissues of contaminated tomato seeds and not only on external seed envelopes, and is not affected by conventional seed decontamination techniques (Antignus et al., 2007). The same applies to TCDVd (Singh and Dilworth, 2009).

There is circumstantial evidence that TASVd could be transmitted by aphids (Walter, 1987), and *Myzus persicae* has been reported to efficiently transmit TPMVd (Galindo, 1987). However, the possibility of aphid transmission of pospiviroids in the absence of a helper virus is widely discounted today. Bumblebees (*Bombus ignitus* Smith) have been shown to efficiently transmit TASVd (Antignus et al., 2007) and TCDVd (Matsuura et al., 2010b) within a tomato crop; however, the transmission mechanism involved has not been investigated.

The reservoir(s) and mechanism(s) leading to pospiviroid outbreaks are poorly understood. Although it has been speculated that wild hosts, often not showing any symptoms of infection, may constitute the ultimate reservoir of viroids, only a few reports substantiate this hypothesis. For TASVd and TCDVd, the observation that the same viroid genotypes were found in tomato and in ornamentals has led to the hypothesis that at least some outbreaks may have been caused by transfer from these ornamental reservoirs to tomato crops (Parrella and Numitone, 2014; Shiraishi et al., 2013; Verhoeven et al., 2012).

DETECTION

Pospiviroid detection does not pose any specific challenge and a wide range of efficient techniques is available. The most efficient approaches involve RT-PCR or real-time RT-PCR with generic primers, allowing the broad detection of genus members (Botermans et al., 2013; Olivier et al., 2014; Shamloul et al., 1997; Verhoeven et al., 2004), or with species-specific primers (Monger et al., 2010). Other valuable methods are available, such as northern blot hybridization with species-specific or polyprobes (Owens and Diener, 1981; Torchetti et al., 2012) or the use of microarrays or bead-based suspension arrays (Tiberini and Barba, 2012; van Brunschot et al., 2014; Zhang et al., 2013).

Suitable protocols and techniques have been developed for the detection of pospiviroids in seeds (Bakker et al., 2015). Because detection of a pospivirioid in a seed does not necessarily result in transmission and seedling infection (T. Candresse and S. Winter unpublished observations), the diagnostic results from seed testing are sometimes difficult to interpret.

GEOGRAPHICAL DISTRIBUTION

The pospiviroids discussed here have mainly been studied and reported from symptomatic crops, so that knowledge on their geographic distribution is largely based on viroid disease expression. Infections in symptomless hosts are likely underreported, raising the possibility that these viroids could have a wider distribution than currently known. TPMVd (including Mexican papita viroid) and PCFVd are only known from a few countries: Canada and Mexico for TPMVd (Galindo, 1987; Ling and Bledsoe, 2009); Canada, Thailand, and the Netherlands for PCFVd (Reanwarakorn et al., 2011; Verhoeven et al., 2009, 2011). PCFVd has also been intercepted in pepper seeds originating from Israel and Thailand (Chambers et al., 2013). TASVd, TCDVd and CLVd had, until recently, only been reported from a few countries: Costa Rica, USA, Canada, and Denmark (in a botanical garden) for CLVd (Hammond et al., 1989; Nielsen and Nicolaisen, 2010; Singh et al., 1992); Indonesia, Israel, Ivory Coast, Senegal, and Tunisia for TASVd (Antignus et al., 2002; Candresse et al., 1987; Walter, 1987). In Europe these viroids were found (and later eradicated in some cases) in a number of countries during recent surveys of ornamental plants and were also detected in tomato in a few isolated outbreaks (EFSA PLH Panel, 2011).

CONTROL

The pospiviroids addressed here do not significantly differ in their biology and epidemiology from other viroids and the strategies for their control are largely similar to those used for other viroids. In the absence of curative treatments or of resistant host varieties, control is best achieved by a combination of quarantine/eradication and prophylactic measures involving in particular the use of certified or quality-controlled plant propagation materials. These measures should be complemented by the use of sanitary-oriented best cropping practices, which provide both protection against outbreaks and efficient management in case of a crisis (EFSA PLH Panel, 2011).

References

Antignus, Y., Lachman, O., Pearlsman, M., 2007. Spread of tomato apical stunt viroid (TASVd) in greenhouse tomato crops is associated with seed transmission and bumble bee activity. Plant Dis. 91, 47–50.

Antignus, Y., Lachman, O., Pearlsman, M., Gofman, R., Bar-Joseph, M., 2002. A new disease of greenhouse tomatoes in Israel caused by a distinct strain of tomato apical stunt viroid (TASVd). Phytoparasitica 30, 502–510.

Bakker, D., Bruimsma, M., Dekter, R.W., Toonen, M.A.J., Verhoeven, J.Th.J., et al., 2015. Detection of PSTVd and TCDVd in seeds of tomato using real-time RT-PCR. Bulletin OEPP/EPPO Bulletin 45, 14–21.

Botermans, M., van de Vossenberg, B.T., Verhoeven, J.T.J., Roenhorst, J.W., Hooftman, M., Dekter, R., et al., 2013. Development and validation of a real-time RT-PCR assay for generic detection of pospiviroids. J. Virol. Methods 187, 43–50.

Candresse, T., Marais, A., Tassus, X., Suhard, P., Renaudin, I., Leguay, A., et al., 2010. First report of tomato chlorotic dwarf viroid in tomato in France. Plant Dis. 94, 633.

Candresse, T., Smith, D., Diener, T.O., 1987. Nucleotide-sequence of a full-length infectious clone of the Indonesian strain of tomato apical stunt viroid (TASV). Nucleic Acids Res. 15, 10597.

Chambers, G.A., Seyb, A.M., Mackie, J., Constable, F.E., Rodoni, B.C., Letham, D., et al., 2013. First report of Pepper chat fruit viroid in traded tomato Seed, an interception by Australian Biosecurity. Plant Dis. 97, 1386.

Chung, B.N., Pak, H.S., 2008. Seed transmission of chrysanthemum stunt viroid in chrysanthemum. Plant Pathol. J. 24, 31–35.

EFSA PLH (Plant health) Panel, 2011. Scientific opinion on the assessment of the risk of solanaceous pospiviroids for the EU territory and the identification and evaluation of risk management options. EFSA J. 9, 2330.

Galindo, J.A., 1987. Tomato planta macho. In: Diener, T.O. (Ed.), The Viroids. Plenum Press, New York, USA, pp. 315–320.

Hammond, R., Smith, D.R., Diener, T.O., 1989. Nucleotide sequence and proposed secondary structure of *Columnea latent viroid*: a natural mosaic of viroid sequences. Nucleic Acids Res. 17, 10083–10094.

Kryczynski, S., Paduch-Cichal, E., Skrzeczkowski, L.J., 1988. Transmission of three viroids through seed and pollen of tomato plants. J. Phytopathol. 121, 51–57.

Ling, K.S., Bledsoe, M.E., 2009. First report of Mexican papita viroid infecting greenhouse tomato in Canada. Plant Dis. 93, 839.

Martinez-Soriano, J.P., Galindo-Alonso, J., Maroon, C.J.M., Yucel, I., Smith, D.R., Diener, T.O., 1996. Mexican papita viroid: putative ancestor of crop viroids. Proc. Natl. Acad. Sci. USA 93, 9397–9401.

Matsushita, Y., Kanda, A., Usugi, T., Tsuda, S., 2008. First report of a tomato chlorotic dwarf viroid disease on tomato plants in Japan. J. Gen. Plant Pathol. 74, 182–184.

Matsuura, S., Matsushita, Y., Kozuka, R., Shimizu, S., Tsuda, S., 2010b. Transmission of tomato chlorotic dwarf viroid by bumblebees (*Bombus ignitus*) in tomato plants. Eur. J. Plant Pathol. 126, 111–115.

Matsuura, S., Matsushita, Y., Usugi, T., Tsuda, S., 2010a. Disinfection of tomato chlorotic dwarf viroid by chemical and biological agents. Crop Prot. 29, 1157–1161.

Monger, W., Tomlinson, J., Booonham, N., Marn, M.V., Plesko, I.M., Molinero-Demilly, V., et al., 2010. Development and inter-laboratory evaluation of real-time PCR assays for the detection of pospiviroids. J. Virol. Methods 169, 207–210.

Nie, X., Singh, R.P., Bostan, H., 2005. Molecular cloning, secondary structure, and phylogeny of three pospiviroids from ornamental plants. Can. J. Plant Pathol. 27, 592–602.

Nielsen, S.L., Nicolaisen, M., 2010. First report of Columnea latent viroid (CLVd) in *Gloxinia gymnostoma*, *G. nematanthodes* and *G. purpurascens* in a botanical garden in Denmark. New Dis. Rep. 22, 4.

Olivier, T., Demonty, E., Fauche, F., Steyer, S., 2014. Generic detection and identification of pospiviroids. Arch. Virol. 159, 2097–2102.

Olivier, T., Demonty, E., Govers, J., Belkheir, K., Steyer, S., 2011. First report of a *Brugmansia* sp. infected by tomato apical stunt viroid in Belgium. Plant Dis. 95, 495.

Owens, R.A., Diener, T.O., 1981. Sensitive and rapid diagnosis of potato spindle tuber viroid disease by nucleic-acid hybridization. Science 213, 670–672.

Owens, R.A., Flores, R., Di Serio, F., Li, S.F., Pallás, V., Randles, J.W., et al., 2012. Pospiviroid. In: King, A.M.Q., Adams, M.J., Carstens, E.B., Lefkowitz, E.J. (Eds.), Virus Taxonomy, Classification and Nomenclature of Viruses, Ninth Report of the International Committee on Taxonomy of Viruses. Elsevier-Academic Press, London, pp. 1229–1230.

Owens, R.A., Verhoeven, J.T.J., 2009. Potato spindle tuber. The Plant Health Instructor. Available from: http://dx.doi.org/10.1094/PHI-I-2009-0804-01.

Parrella, G., Numitone, G., 2014. First report of tomato apical stunt viroid in tomato in Italy. Plant Dis. 98, 1164.

Reanwarakorn, K., Klinkong, S., Porsoongnurn, J., 2011. First report of natural infection of pepper chat fruit viroid in tomato plants in Thailand. New Dis. Rep. 24, 6.

Shamloul, A.M., Hadidi, A., Zhu, S.F., Singh, R.P., Sagredo, B., 1997. Sensitive detection of potato spindle tuber viroid using RT-PCR and identification of a viroid variant naturally infecting pepino plants. Can. J. Plant Pathol. 19, 89–96.

Shiraishi, T., Maejima, K., Komatsu, K., Hashimoto, M., Okano, Y., Kitazawa, Y., et al., 2013. First report of tomato chlorotic dwarf viroid isolated from symptomless petunia plants (*Petunia* spp.) in Japan. J. Gen. Plant Pathol. 79, 214–216.

Singh, R.P., Boucher, A., Somerville, T.H., 1989. Evaluation of chemicals for disinfection of laboratory equipment exposed to potato spindle tuber viroid. Am. Potato J. 66, 239–246.

Singh, R.P., Dilworth, A.D., 2009. Tomato chlorotic dwarf viroid in the ornamental plant *Vinca minor* and its transmission through tomato seed. Eur. J. Plant Pathol. 123, 111–116.

Singh, R.P., Lakshman, D.K., Boucher, A., Tavantzis, S.M., 1992. A viroid from *Nematanthus wettsteinii* plants closely related to the Columnea latent viroid. J. Gen. Virol. 73, 2769–2774.

Singh, R.P., Nie, X., Singh, M., 1999. Tomato chlorotic dwarf viroid: an evolutionary link in the origin of pospiviroids. J. Gen. Virol. 80, 2823–2828.

Spieker, R.L., 1996. A viroid from *Brunfelsia undulata* closely related to the Columnea latent viroid. Arch. Virol. 141, 1823–1832.

Spieker, R.L., Marinkovic, S., Sänger, H.L., 1996. A viroid from *Solanum pseudocapsicum* closely related to the tomato apical stunt viroid. Arch. Virol. 141, 1387–1395.

Tiberini, A., Barba, M., 2012. Optimization and improvement of oligonucleotide microarray-based detection of tomato viruses and pospiviroids. J. Virol. Methods 185, 43–51.

Torchetti, E.M., Navarro, B., Di Serio, F., 2012. A single polyprobe for detecting simultaneously eight pospiviroids infecting ornamentals and vegetables. J. Virol. Methods 186, 141–146.

van Brunschot, S.L., Bergervoet, J.H., Pagendam, D.E., de Weerdt, M., Geering, A.D., Drenth, A., et al., 2014. Development of a multiplexed bead-based suspension array for the detection and discrimination of pospiviroid plant pathogens. PLoS One 9, e84743.

Verhoeven, J.Th.J., 2010. Identification and epidemiology of pospiviroids. Thesis Wageningen University, Wageningen, The Netherlands. ISBN 978-90-8585-623-8. 136 pp.

Verhoeven, J.Th.J., Botermans, M., Jansen, C.C.C., Roenhorst, J.W., 2010a. First report of tomato apical stunt viroid in the symptomless hosts *Lycianthes rantonnetii* and *Streptosolen jamesonii* in the Netherlands. Plant Dis. 94, 791.

Verhoeven, J.Th.J., Botermans, M., Jansen, C.C.C., Roenhorst, J.W., 2011. First report of pepper chat fruit viroid in capsicum pepper in Canada. New Dis. Rep. 23, 15.

Verhoeven, J.Th.J., Botermans, M., Meekes, E.T.M., Roenhorst, J.W., 2012. Tomato apical stunt viroid in the Netherlands: most prevalent pospiviroid in ornamentals and first outbreak in tomatoes. Eur. J. Plant Pathol. 133, 803–810.

Verhoeven, J.Th.J., Huner, L., Virscek Marn, M., Mavric Plesko, I., Roenhorst, J.W., 2010b. Mechanical transmission of potato spindle tuber viroid between plants of *Brugmansia suaveolens*, *Solanum jasminoides* and potatoes and tomatoes. Eur. J. Plant. Pathol. 128, 417–421.

Verhoeven, J.Th.J., Jansen, C.C.C., Botermans, M., Roenhorst, J.W., 2010c. First report of Iresine viroid 1 in *Celosia plumosa* in the Netherlands. Plant Dis. 94, 920.

Verhoeven, J.Th.J., Jansen, C.C.C., Botermans, M., Roenhorst, J.W., 2010d. Epidemiological evidence that vegetatively propagated, solanaceous plant species act as sources of potato spindle tuber viroid inoculum for tomato. Plant Pathol. 59, 3–12.

Verhoeven, J.Th.J., Jansen, C.C.C., Roenhorst, J.W., 2006. First report of tomato apical stunt viroid in tomato in Tunisia. Plant Dis. 90, 528.

Verhoeven, J.Th.J., Jansen, C.C.C., Roenhorst, J.W., 2008a. First report of pospiviroids infecting ornamentals in the Netherlands: citrus exocortis viroid in *Verbena* sp., potato spindle tuber viroid in *Brugmansia suaveolens* and *Solanum jasminoides*, and tomato apical stunt viroid in *Cestrum* sp. Plant Pathol. 57, 399.

Verhoeven, J.Th.J., Jansen, C.C.C., Roenhorst, J.W., Flores, R., de la Peña, M., 2009. Pepper chat fruit viroid: biological and molecular properties of a proposed new species of the genus *Pospiviroid*. Virus Res. 144, 209–214.

Verhoeven, J.Th.J., Jansen, C.C.C., Roenhorst, J.W., Steyer, S., Schwind, N., et al., 2008b. First report of *Solanum jasminoides* infected by citrus exocortis viroid in Germany and the Netherlands and tomato apical stunt viroid in Belgium and Germany. Plant Dis. 92, 973.

Verhoeven, J.Th.J., Jansen, C.C.C., Werkman, A.W., Roenhorst, J.W., 2007. First report of tomato chlorotic dwarf viroid in *Petunia hybrida* from the United States of America. Plant Dis. 91, 324.

Verhoeven, J.Th.J., Jansen, C.C.C., Willemen, T.M., Kox, L.F.F., Owens, R.A., et al., 2004. Natural infections of tomato by citrus exocortis viroid, Columnea latent viroid, potato spindle tuber viroid and tomato chlorotic dwarf viroid. Eur. J. Plant Pathol. 110, 823–831.

Walter, B., 1987. Tomato Apical Stunt. In: Diener, T.O. (Ed.), The Viroids. Plenum Press, New York, NY, pp. 321–326.

Zhang, Y.J., Yin, J., Jiang, D.M., Xin, Y.Y., Ding, F., Deng, Z.N., et al., 2013. A Universal oligonucleotide microarray with a minimal number of probes for the detection and identification of viroids at the genus level. PLoS One 8, e64474.

CHAPTER

16

Citrus Exocortis Viroid

Nuria Duran-Vila

Valencian Institute of Agricutural Research-IVIA, Valencia, Spain

INTRODUCTION

The "exocortis" disease was first described in 1948 as a bark scaling disorder affecting the rootstock of citrus trees grafted on trifoliate orange (*Poncirus trifoliata*) (Fawcett and Klotz, 1948). A similar disorder was also described as "scaly butt" of trifoliate orange in Australia (Benton et al., 1949) and as "Rangpur lime disease" affecting the rootstock of trees grafted on Rangpur lime (*Citrus limonia*) (Moreira, 1955). This disease, now termed exocortis, was demonstrated to be graft transmissible (Benton et al., 1950; Bitters, 1952) and the etiological agent was considered a virus until viroids were described as a new class of plant pathogens (Diener, 1971; Semancik and Weathers, 1972).

Initial biological indexing tests were conducted using trifoliate orange or Rangpur lime. The identification of citron (*Citrus medica*) as a highly sensitive species (Salibe, 1961) prompted the observation of symptoms ranging from mild to severe in inoculated seedlings. Subsequently, a selection (Arizona 861-S1) of Etrog citron has been widely used for indexing purposes (Roistacher et al., 1977) and the range of symptoms on this indicator was erroneously considered as evidence of the existence of many strains of the exocortis agent.

Citrus exocortis viroid (CEVd) was identified after its transmission to gynura (*Gynura aurantiaca*) an experimental host displaying stunting, epinasty, and leaf distortion and yielding high viroid titers. Unexpectedly, isolates inducing mild and moderate symptoms in Etrog citron were not transmissible to gynura and other herbaceous hosts (Fig. 16.1). This situation delayed the identification of citrus viroids, other than CEVd, until laboratory tools became available (Duran-Vila et al., 1986, 1988; Ito et al., 2001; Sano et al., 1986; Serra et al., 2008).

Viroids and Satellites.
DOI: http://dx.doi.org/10.1016/B978-0-12-801498-1.00016-4

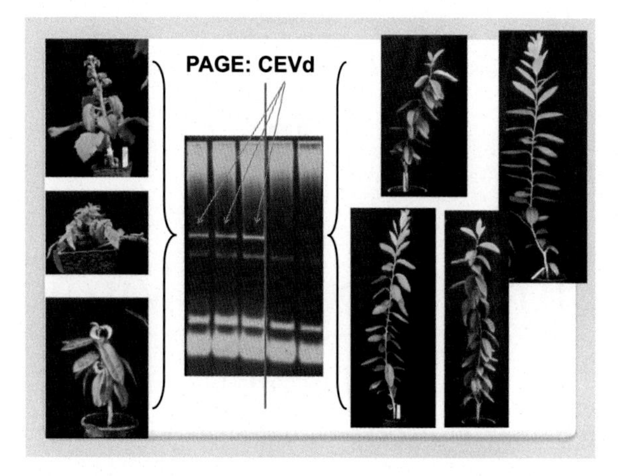

FIGURE 16.1 Detection of CEVd by 5% PAGE and ethidium bromide staining. CEVd is recovered from gynura (top left), tomato (center left), and Etrog citron (bottom left) displaying the characteristic symptoms of severe stunting and leaf epinasty, but not in Etrog citron plants showing mild and moderate symptoms (right).

CEVd was also identified in grapevines (Flores et al., 1985), tomatoes showing the "tomato bushy top syndrome" (Mishra et al., 1991), and several vegetable crop species (Fagoaga et al., 1995; Fagoaga and Duran-Vila, 1996). CEVd also naturally infects many ornamental species acting as symptomless carriers.

TAXONOMIC POSITION AND NUCLEOTIDE SEQUENCE

CEVd was first shown to have a 371-nucleotide sequence with a predicted rod-like secondary structure (Gross et al., 1982; Visvader et al., 1982). Following the ICTV criteria, an arbitrary level of less than 90% sequence similarity and distinct biological properties (host range and symptoms), this viroid has been allocated to the genus *Pospiviroid* within the family *Pospiviroidae*. It lacks RNA self-cleavage activity and contains a central conserved region formed by two sets of conserved nucleotides in the upper and lower strands of its rod-like secondary structure, and a terminal conserved region. The rod-like structure has five structural domains: terminal left (TL), pathogenic (P), central (C), variable (V), and terminal right (TR) (Keese and Symons, 1985). Using tomato as an experimental host, Visvader and Symons (1985, 1986) classified CEVd isolates into severe (Class A) and mild (Class B), differing by a minimum of 26 nucleotides affecting the PL and PR regions located, respectively, in the P and V domains. CEVd isolates of these two classes induce distinct symptoms in gynura (Chaffai et al., 2007), but only subtle differences in trifoliate orange used as a rootstock and a similar overall performance of the infected trees (Vernière et al., 2004). Sequencing of additional isolates showed that strains different from Class A and Class B existed, and that the relationship between sequence and pathogenicity was more complex than initially proposed.

Biological indexing identified an unusual strain (CEVdCOL) inducing severe symptoms in tomato but, in spite of clustering into Class A, it unexpectedly induced extremely mild symptoms in Etrog citron. Using site-directed mutagenesis, two nucleotide substitutions (314 A→G and 315U→A) in the lower strand of the P domain of CEVdCOL resulted in an artificial CEVdMCOL variant severe in Etrog citron. Infectivity assays using natural and mutated variants demonstrated that pathogenic determinants of CEVd are in fact host dependent (Murcia et al., 2011).

Visvader and Symons (1985) defined a CEVd isolate as the RNA from a single tree that might consist of one or more sequence variants. They reported isolates containing mixtures of variants and hypothesized that they were either due to a high copy error rate of the RNA polymerase involved in replication, or the result of several viroid transmissions. Later, it was shown that viroids display high genetic diversity due to the absence of proofreading activity of the RNA polymerases, thus generating populations of mutants (termed quasispecies) varying around a consensus sequence. Following inoculations with single variants, de novo populations are generated (Gandía et al., 2005) following the quasispecies model proposed by Eigen (1971). Differences in nucleotide sequence can also be the result of host and tissue selection (Semancik et al., 1993) and the genetic diversity being host dependent. For instance, the genetic diversity of CEVd in trifoliate orange and sour orange after graft-transmission from Etrog citron differed from that of the inoculum source, with sour orange (a tolerant host) containing populations more variable than those from trifoliate orange (a sensitive host) and the original source of Etrog citron (Bernad et al., 2009).

Several examples show that larger CEVd molecules resulting from specific duplications in the rod-like secondary structure can be generated. These enlarged CEVd molecules have been identified as stable CEVd variants containing a 92 nucleotide duplication (CEVd-D92) in a hybrid tomato (*Solanum lycopersicum* × *Solanum peruvianum*) (Semancik et al., 1994) or a 96 nucleotide duplication (CEVd-D96) in eggplant (Fadda et al., 2003), but preserving the rod-like secondary structure. The identification of additional enlarged as well as transient forms of CEVd variants (Szychowski et al., 2005), suggests that this phenomenon, first observed in coconut cadang-cadang viroid (Haseloff et al., 1982), is not as unusual as initially envisaged.

ECONOMIC SIGNIFICANCE

Vegetative propagation by rooting or grafting is the most efficient perpetuation route for viroids, especially in their spread by the exchange of propagation materials. Although most CEVd-host combinations are symptomless, economic significance is associated with symptoms in trifoliate orange, Rangpur lime, citranges (*P. trifoliata* × *Citrus sinensis*), and certain citrumelos (*Citrus paradisi* × *P. trifoliata*) used as rootstocks. The economic importance of exocortis depends on the response of commercial species grafted on these sensitive rootstocks, which may show reduction of tree size and fruit production. High temperatures and high light intensities favor symptom expression and viroid accumulation (Carbonell et al., 2008), illustrating that viroids mainly affect tropical, subtropical, and greenhouse crops.

The economic impact of CEVd by itself is hard to evaluate because natural sources also contain other viroids that either enhance or reduce symptom expression (Vernière et al., 2006). The effect of viroid interactions in the performance of trees grown on sensitive rootstocks indicates that the impact on fruit production and quality depends on coinfection with viroids other than CEVd (Bani Hashemian et al., 2009, 2010). In spite of the negative effects of CEVd infection, its widespread presence in citrus and grapevines suggests that it may be linked to desirable properties (Rossetti et al., 1980; Solel et al., 1995; Vera et al., 1993).

SYMPTOMS

CEVd was initially identified because of its ability to induce symptoms in trifoliate orange and Rangpur lime and was further characterized as a viroid after transmission to herbaceous hosts also displaying characteristic symptoms. Since commercial citrus and grapevines are coinfected with several viroids, symptom development caused by CEVd could not be studied until sources devoid of other viroids became available through the inoculation of herbaceous hosts.

The results of long-term field assays conducted with clementine trees grafted on trifoliate orange showed that CEVd induces bark scaling and cracking symptoms as well as conspicuous bumps in the wood of sensitive rootstocks (Fig. 16.2). Infected trees are smaller (rootstock and scion circumference) than viroid-free controls and show yield reduction, with such effects being strain dependent (Vernière et al., 2004). Similar but milder symptoms are also induced in trees grafted on Carrizo citrange (Murcia et al., 2015) (Fig. 16.3A), which develop bark cracking symptoms associated with lesions of the roots and poor development of the overall root system (Fig. 16.3B–D).

FIGURE 16.2 Effect of CEVd infection on trees grafted on trifoliate orange: (left) bark scaling symptoms; (right) bumps in the wood.

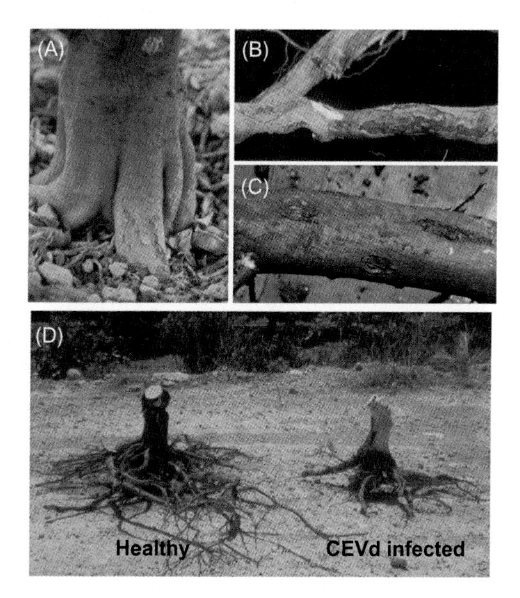

FIGURE 16.3 Effect of CEVd infection on trees grafted on Carrizo citrange: (A) mild bark scaling symptoms; (B) severe lesions in the root system; (C) characteristic lesions in the roots; (D) poor development of the overall root system.

An assay conducted with 22 species of *Citrus* and 10 species of related genera that had been graft-inoculated with a CEVd containing a mixture of citrus viroids showed that most species were symptomless carriers indicating that symptom expression is unusual, despite the ability of citrus and citrus relatives to become infected (Barbosa et al., 2002).

CEVd causes characteristic symptoms of stunting, epinasty, and leaf rugosity in experimental herbaceous hosts such as *Chrysanthemum morifolium, Cucurbita pepo, G. aurantiaca, G. sarmentosa, Petunia axilaris, P. hybrida, P. violacea, Physalis floridana, P. ixocarpa, P. peruviana, Solanum* spp., *Zinnia elegans,* and probably others.

Under field conditions, CEVd has also been reported to cause bunchy-top and/or leaf chlorosis in tomato (Mishra et al., 1991; Verhoeven et al., 2004), whereas other naturally infected plants such as grapevines and several vegetable species (broad bean, eggplant, turnip, and carrot) act as symptomless carriers (Fagoaga et al., 1995; Fagoaga and Duran-Vila, 1996).

HOST RANGE

CEVd, like other pospiviroids, has a wide host range. The experimental host range was initially established by transmission assays and observation of symptom expression of the inoculated hosts. The availability of laboratory detection tools allowed the identification of additional symptomless hosts.

CEVd has been shown to naturally infect, among others, *Verbena* spp., petunia, and *Solanum jasminoides* without eliciting symptoms (Singh et al., 2006; Verhoeven et al., 2008a,b; Van Brunschot et al., 2014).

TRANSMISSION

In vegetatively propagated woody species (citrus and grapevine) CEVd is transmitted through propagation by grafting and/or rooting. There is no evidence of seed transmission in citrus, but mechanical transmission can be quite efficient. Studies conducted before CEVd was identified, demonstrated that it could be transmitted from citron to citron with 100% efficiency simply by slashing with a knife blade (Garnsey and Jones, 1967). Mechanical transmission was further confirmed in commercial citrus (clementine and lemon) as well as in commercial plantations subjected to standard agronomic practices (Barbosa et al., 2005).

Even though there is no evidence of seed transmission in trifoliate orange and other citrus species, Singh et al. (2009) reported relatively high but variable seed transmission rates in *Impatiens* and *Verbena* species, a situation that explains the occurrence and prevalence of CEVd in these hosts.

DETECTION

The first detection methods were biological assays using trifoliate orange or Rangpur lime, until Etrog citron was characterized as an indicator displaying symptoms in a period 3–6 months after graft-inoculation. These assays were developed before CEVd was characterized. With the development of nucleic acid analysis procedures, biological indexing tests have been replaced by rapidly evolving molecular techniques.

PAGE and ethidium bromide staining was the first molecular technique used for detection of CEVd but sensitivity required an adequate accumulation level. Higher resolution was achieved using two-dimensional PAGE systems, which coupled with silver staining allowed the visualization of viroid circular forms. Technical details have been recently reviewed (Owens et al., 2012; Chapter 34: Gel Electrophoresis).

In the 1980s, dot-blot hybridization and hybridization of tissue imprints (Owens and Diener, 1981; Romero-Durban et al., 1995) gained popularity because they allowed the processing of large number of samples, but adequate sensitivity required the use of a viroid amplification host (Palacio et al., 2000). A northern hybridization protocol based on the analysis of preparations from bark tissues was found to be sensitive enough to detect CEVd and other citrus viroids from field grown plants of different species and cultivars (Murcia et al., 2009; Chapter 35: Molecular Hybridization Techniques for Detecting and Studying Viroids).

A number of RT-PCR approaches have been proposed (Hadidi and Yang, 1990; Levy et al., 1992; Yang et al., 1992) with ongoing modifications to simplify sample preparation and defining the most adequate primer pairs. A one-step RT-PCR approach has been proposed to avoid amplicon cross-contamination during routine diagnosis (Bernad and Duran-Vila, 2006). A number of extraction protocols and types of analytical approaches are presented in great detail in Owens et al. (2012).

GEOGRAPHICAL DISTRIBUTION AND EPIDEMIOLOGY

CEVd and other citrus viroids are widespread in all citrus-growing countries. Whenever indexing tests have been performed, viroids have been identified and viroid-free sources are only available where sanitation programs have been implemented. Viroid-free citrus plants can be recovered as nucellar seedlings (true-to-type plants resulting from the nucellar embryos of polyembryonic cultivars) or by the application of shoot-tip grafting in vitro techniques.

CONTROL

In citrus, the exocortis disease can only be controlled by prevention measures. Viroid-free budwood must be used for graft-propagation of commercial cultivars. Desirable clones can be freed of viroids by shoot-tip grafting in vitro (Navarro et al., 1975) but the sanitary status of the recovered plants has to be confirmed with appropriate viroid detection tools.

Acknowledgments

I would like to acknowledge Prof. J.S. Semancik who introduced me to the field of CEVd research. I would also like to acknowledge my students (Carmen Fagoaga, Juan Romero Durbán, Ana Palacio Bielsa, Ziad Ghaleb Nimer Fadda, Mónica Gandía, Cristiane de Jesús Barbosa, Karelia Velázquez Caballero, Lucía Bernad, Pedro Serra Alonso, Seyed Mehdi Bani Hashemian, and Nubia Murcia) who produced as part of their PhD thesis much of the information on CEVd presented in this chapter.

References

Bani Hashemian, S.M., Serra, P., Barbosa, C., Juárez, J., Aleza, P., Corvera, J.M., et al., 2009. The effect of a field-source mixture of citrus viroids on the performance of "Nules" clementine and "Navelina" sweet orange trees grafted on Carrizo citrange. Plant Dis. 93, 699–707.

Bani Hashemian, S.M., Murcia, N., Trenor, I., Duran-Vila, N., 2010. Low performance of citrus trees grafted on Carrizo citrange is associated with viroid infection. J. Plant Pathol. 92, 511–517.

Barbosa, C.J., Pina, J.A., Navarro, L., Duran-Vila, N., 2002. Replication/accumulation and symptom expression of citrus viroids on some species of Citrus and related genera. In: Duran-Vila, N., Milne, R.G., da Graça, J.V. (Eds.), Proc. 15th Conf. Int. Organ. Citrus Virol. University of California, Riverside, CA, pp. 264–271.

Barbosa, C.J., Pina, J.A., Bernad, L., Serra, P., Navarro, L., Duran-Vila, N., 2005. Mechanical transmission of citrus viroids. Plant Dis. 89, 749–754.

Benton, R.J., Bowman, F.T., Fraser, L., Kebby, R.G., 1949. Selection of citrus budwood to control scaly butt in trifoliata rootstock. Agr. Gaz. N. S. Wales 60, 31–34.

Benton, R.J., Bowman, F.T., Fraser, L., Kebby, R.G., 1950. Stunting and scaly butt associated with *Poncirus trifoliata* rootstock. N. S. Wales Dept. Agr. Sci. Bull. 70, 1–20.

Bernad, L., Duran-Vila, N., 2006. A novel RT-PCR approach for detection and characterization of citrus viroids. Mol. Cell. Probes 20, 105–113.

Bernad, L., Duran-Vila, N., Elena, S.F., 2009. Effect of citrus hosts on the generation, maintenance and evolutionary fate of genetic variability of citrus exocortis viroid. J. Gen. Virol. 90, 2040–2049.

Bitters, W.P., 1952. Exocortis disease of citrus. California Agric. 6, 5–6.

Carbonell, A., Martínez de Alba, A.E., Flores, R., Gago, S., 2008. Double-stranded RNA interferes in a sequence specific manner with infection of representative members of the two viroid families. Virology 371, 44–53.

Chaffai, M., Serra, P., Gandía, M., Hernández, C., Duran-Vila, N., 2007. Molecular characterization of CEVd strains that induce different phenotypes in *Gynura aurantiaca:* structure-pathogenicity relationships. Arch. Virol. 152, 1283–1294.

Diener, T.O., 1971. Potato spindle tuber virus: a plant virus with properties of a free nucleic acid III. Subcellular location of PSTV-RNA and the question of whether virions exist in extracts or *in situ*. Virology 43, 75–98.

Duran-Vila, N., Flores, R., Semancik, J.S., 1986. Characterization of viroid-like RNAs associated with the citrus exocortis syndrome. Virology 150, 75–84.

Duran-Vila, N., Roistacher, C.N., Rivera-Bustamante, R., Semancik, J.S., 1988. A definition of citrus viroid groups and their relationship to the exocortis disease. J. Gen. Virol. 69, 3069–3080.

Eigen, M., 1971. Self-organization of matter and the evolution of biological macromolecules. Naturwissenschaften. 58, 465–523.

Fadda, Z., Daròs, J.A., Flores, R., Duran-Vila, N., 2003. Identification in eggplant of a variant of citrus exocortis viroid (CEVd) with a 96 nucleotide duplication in the right terminal region of the rod-like secondary structure. Virus Res. 97, 145–149.

Fagoaga, C., Duran-Vila, N., 1996. Natural occurrence of variants of the citrus exocortis viroid in vegetable crop species. Plant Pathol. 45, 45–53.

Fagoaga, C., Semancik, J.S., Duran-Vila, N., 1995. A citrus exocortis viroid variant from broad bean (*Vicia faba* L.): infectivity and pathogenesis. J. Gen. Virol. 76, 2271–2277.

Fawcett, H.S., Klotz, L.J., 1948. Exocortis on trifoliate orange. Citrus Leaves 28, 8.

Flores, R., Duran-Vila, N., Pallás, V., Semancik, J.S., 1985. Detection of viroid and viroid-like RNAs from grapevine. J. Gen. Virol. 66, 2095–2102.

Gandía, M., Rubio, L., Palacio, A., Duran-Vila, N., 2005. Genetic variation and population structure of an isolate of citrus exocortis viroid (CEVd) and of the progenies of two infectious sequence variants. Arch. Virol. 150, 1945–1957.

Garnsey, S.M., Jones, J.W., 1967. Mechanical transmission of exocortis virus with contaminated budding tools. Plant Dis. Rep. 51, 410–413.

Gross, H.J., Krupp, G., Domdey, H., Raba, M., Jank, P., Lossow, C., et al., 1982. Nucleotide sequence and secondary structure of citrus exocortis and chrysanthemum stunt viroid. Eur. J. Biochem. 121, 249–257.

Hadidi, A., Yang, X., 1990. Detection of pome fruit viroids by enzymatic cDNA amplification. J. Virol. Methods 30, 261–270.

Haseloff, J., Mohamed, N.A., Symons, R.H., 1982. Viroid RNAs of cadang-cadang disease of coconuts. Nature 299, 316–321.

Ito, T., Ieki, H., Ozaki, K., Ito, T., 2001. Characterization of a new citrus viroid species tentatively termed citrus viroid OS. Arch. Virol. 146, 975–982.

Keese, P., Symons, R.H., 1985. Domains in viroids: evidence of intermolecular RNA rearrangements and their contribution to viroid evolution. Proc. Natl. Acad. Sci. USA 82, 4582–4586.

Levy, L., Hadidi, A., Garnsey, S.M., 1992. Reverse transcription-polymerase chain reaction assays for the rapid detection of citrus viroids using multiplex primer sets. Proc. Int. Soc. Citricul. 2, 800–803.

Mishra, M.D., Hammond, R.W., Owens, R.A., Smith, D.R., Diener, T.O., 1991. Indian bunchy top disease of tomato plants is caused by a distinct strain of citrus exocortis viroid. J. Gen. Virol. 71, 1781–1785.

Moreira, S., 1955. Sintomas de "exocortis" em limoneiro cravo. Bragantia 14, 19–21.

Murcia, N., Serra, P., Olmos, A., Duran-Vila, N., 2009. A novel hybridization approach for detection of citrus viroids. Mol. Cell. Probes 23, 95–102.

Murcia, N., Bernad, L., Duran-Vila, N., Serra, P., 2011. Two nucleotide positions in the citrus exocortis viroid RNA associated with symptom expression in Etrog citron but not in experimental herbaceous hosts. Mol. Plant Pathol. 12, 203–208.

Murcia, M., Bani Hashemian, V., Serra, P., Pina, V., Duran-Vila, N., 2015. Citrus viroids: symptom expression and performance of Washington navel sweet orange trees grafted on Carrizo citrange. Plant Dis. 99, 119–124.

Navarro, L., Roistacher, C.N., Murashige, T., 1975. Improvement of shoot tip grafting in vitro for virus-free citrus. J. Am. Soc. Hortic. Sci. 100, 471–479.

Owens, R.A., Diener, T.O., 1981. Sensitive and rapid diagnosis of potato spindle tuber viroid disease by nucleic acid hybridization. Science 213, 670–672.

Owens, R.A., Sano, T., Duran-Vila, N., 2012. Plant viroids: isolation, characterization/detection, and analysis. Methods Mol. Biol. 894, 253–271.

Palacio, A., Foissac, X., Duran-Vila, N., 2000. Indexing of citrus viroids by imprint hybridization. Eur. J. Plant Pathol. 105, 897–903.
Roistacher, C.N., Calavan, E.C., Blue, R.L., Navarro, L., Gonzales, R., 1977. A new more sensitive citron indicator for the detection of mild isolates of citrus exocortis viroid (CEV). Plant Dis. Rep. 61, 135–139.
Romero-Durban, J., Cambra, M., Duran-Vila, N., 1995. A simple imprint hybridization method for detection of viroids. J. Virol. Methods 55, 37–47.
Rossetti, V., Pompeu, J., Rodriguez, O., 1980. Reaction of exocortis-infected and healthy tees to experimental *Phytophthora* inoculations. In: Calavan, E.C., Garnsey, S.M., Timmer, L.W. (Eds.), Proc. 8th Conf. Int. Organ. Citrus Virol. University of California, Riverside, CA, pp. 209–214.
Salibe, A.A., 1961. Contribucao ao estudo da doença exocorte dos citros. Thesis, Escola Superior de Agricultura "Luiz de Queiroz." Universidade de Sao Paulo.
Sano, T., Hataya, T., Sasaki, A., Shikata, E., 1986. Etrog citron is latently infected with hop stunt viroid-like RNA. Proc. Japan Acad. 62, 325–328.
Semancik, J.S., Szychowski, J.A., Rakowski, A.G., Symons, R.H., 1993. Isolates of citrus exocortis viroid recovered by host and tissue selection. J. Gen. Virol. 74, 2427–2436.
Semancik, J.S., Szychowski, J.A., Rakowski, A.G., Symons, R.H., 1994. A stable 463 nucleotide variant of citrus exocortis viroid produced by terminal repeats. J. Gen. Virol. 75, 727–732.
Semancik, J.S., Weathers, L.G., 1972. Exocortis virus: an infectious free-nucleic acid plant virus with unusual properties. Virology 46, 456–466.
Serra, P., Barbosa, C.J., Daròs, J.A., Flores, R., Duran-Vila, N., 2008. Citrus viroid V: molecular characterization and synergistic interactions with other members of the genus Apscaviroid. Virology 370, 102–112.
Singh, R.P., Dilworth, A.D., Ao, X., Singh, M., 2009. Citrus exocortis viroid transmission through commercially-distributed seeds of Impatiens and Verbena plants. Eur. J. Plant Pathol. 124, 691–694.
Singh, R.P., Dilworth, A.D., Baranwal, V.K., Gupta, K.N., 2006. Detection of citrus exocortis viroid, Iresine viroid and tomato chlorotic dwarf viroid in new ornamental host plants in India. Plant Dis. 90, 1457.
Solel, Z., Mogilner, N., Gafni, R., Bar-Joseph, M., 1995. Induced tolerance to mal secco disease in Etrog citron and Rangpur lime by infection with the citrus exocortis viroid. Plant Dis. 79, 60–62.
Szychowski, J.A., Vidalakis, G., Semancik, J.S., 2005. Host-directed processing of citrus exocortis viroid. J. Gen. Virol. 86, 473–477.
Van Brunschot, S.L., Persley, D.M., Roberts, A., Thomas, J.E., 2014. First report of pospiviroids infecting ornamental plants in Australia: potato spindle tuber viroid in *Solanum laxum* (synonym *S. jasminoides*) and citrus exocortis viroid in *Petunia* spp. New Dis. Rep. 29, 3.
Vera, P., Tornero, P., Conejero, V., 1993. Cloning and expression analysis of a viroid-induced peroxidase from tomato plants. Mol. Plant-Microbe Interact. 6, 790–794.
Verhoeven, J., Jansen, C.C.C., Roenhorst, J.W., 2008a. First report of pospiviroids infecting ornamentals in the Netherlands: citrus exocortis viroid in *Verbena* sp., potato spindle tuber viroid in *Brugmansia suaveolens* and *Solanum jasminoides*, and tomato apical stunt viroid in *Cestrum* spp. Plant Pathol. 57, 399.
Verhoeven, J., Jansen, C.C.C., Roenhorst, J.W., Steyer, S., Schwind, N., et al., 2008b. First report of *Solanum jasminoides* infected by citrus exocortis viroid in Germany and the Netherlands and tomato apical stunt viroid in Belgium and Germany. Plant Dis. 92, 973.
Verhoeven, J., Jansen, C.C.C., Willemen, T.M., Kox, L.F.F., Owens, R.A., et al., 2004. Natural infections of tomato by citrus exocortis viroid, columnea latent viroid, potato spindle tuber viroid and tomato chlorotic dwarf viroid. Eur. J. Plant Pathol. 110, 823–831.

Vernière, C., Perrier, X., Dubois, C., Dubois, A., Botella, L., Chabrier, C., et al., 2004. Citrus viroids: symptom expression and effect on vegetative growth and yield on clementine trees grafted on trifoliate orange. Plant Dis. 88, 1189–1197.

Vernière, C., Perrier, X., Dubois, C., Dubois, A., Botella, L., Chabrier, C., et al., 2006. Interactions between citrus viroids affect symptom expression and field performance of clementine trees grafted on trifoliate orange. Phytopathology 96, 356–368.

Visvader, J.E., Gould, A.R., Bruening, G.E., Symons, R.H., 1982. Citrus exocortis viroid: nucleotide sequence and secondary structure of an Australian isolate. FEBS Lett. 137, 288–292.

Visvader, J.E., Symons, R.H., 1985. Eleven new sequence variants of citrus exocortis viroid and the correlation of sequence with pathogenicity. Nucleic Acids Res. 13, 2907–2920.

Visvader, J.E., Symons, R.H., 1986. Replication of in vitro constructed viroid mutants: location of the pathogenicity modulating domain of citrus exocortis viroid. EMBO J. 5, 2051–2055.

Yang, X., Hadidi, A., Garnsey, S.M., 1992. Enzymatic cDNA amplification of citrus exocortis and cachexia viroids from infected citrus hosts. Phytopathology 82, 279–285.

CHAPTER

17

Chrysanthemum Stunt Viroid

Peter Palukaitis

Seoul Women's University, Seoul, South Korea

INTRODUCTION

The stunting disease of chrysanthemums was first described by Dimock (1947) and Brierley and Smith (1949). Transmission experiments suggested that a virus was most likely the causal agent (Brierley, 1952, 1953; Keller, 1951; Olson, 1949). However, the failure to identify a virus causing the stunting disease (Lawson and Hearon, 1971) led to experiments showing a viroid etiology (Diener and Lawson, 1973; Hollings and Stone, 1973). The purification and molecular characterization of chrysanthemum stunt viroid (CSVd) verified its viroid nature (Palukaitis and Symons, 1980).

TAXONOMIC POSITION AND NUCLEOTIDE SEQUENCE

Chrysanthemum stunt viroid is a species in the genus *Pospiviroid* in the family *Pospiviroidae*. It is most closely related to *Tomato apical stunt viroid* and *Citrus exocortis viroid* among the nine species of the genus *Pospiviroid* (Owens et al., 2012). The extent of sequence homology between CSVd, potato spindle tuber viroid (PSTVd), and citrus exocortis viroid (67%–73%) (Gross et al., 1982) is sufficient to allow cross-protection in chrysanthemums among these viroids (Niblett et al., 1978), as well as between CSVd, PSTVd, and hop stunt viroid (Kryczyński and Paduch-Cichal, 1987).

The nucleotide sequences of CSVd isolates from 16 countries have been determined. Most isolates contain between 354 and 356 nt (Yoon and Palukaitis, 2013). The sequence of an Australian isolate (Haseloff and Symons, 1981) showed 10 nt differences (2.8% sequence variation)

Viroids and Satellites.
DOI: http://dx.doi.org/10.1016/B978-0-12-801498-1.00017-6

from a UK isolate (Gross et al., 1982). Among 117 global CSVd isolates and cDNA clones, the range of sequence variation within countries (0%–4%) is similar to the variation observed among isolates from different countries (0%–4.5%), and in some cases cDNA clones of an isolate can vary as much in sequence as for isolates between different countries (Yoon and Palukaitis, 2013). Sequences responsible for pathogenicity on various hosts have not been delineated.

ECONOMIC SIGNIFICANCE

Chrysanthemum is in the top three of the world's most important cutflowers. CSVd is the causal agent of a major disease of florist's chrysanthemum (known variously as *Chrysanthemum morifolium*, *C. indicum*, *Dendranthema morifolium*, *D. grandiflora*, and *D. grandiflorum*). The stunt disease was first discovered in the United States, where it had a serious effect on the chrysanthemum production industry (Horst et al., 1977; Lawson, 1987). Subsequently, the disease was found in all chrysanthemum-growing countries (Lawson, 1987). The incidence of infection varies considerably from country to country and over time, even within a single country; both with sampling locations and cultivars, with incidences up to 100% in South Korea (Chung et al., 2005), 90% in Japan (Matsushita et al., 2007), 70% in India (Singh et al., 2010), and an average of less than 12% in China (Zhang et al., 2011). Yield losses are difficult to quantify, due to mixed infections with various viruses and/or another viroid, but Hill et al. (1996) gave a figure of $3 million for losses in Australia due to a CSVd outbreak in chrysanthemum in 1987.

SYMPTOMS

CSVd-induced symptoms on chrysanthemums include effects on plant growth, the flower, and leaf development (Fig. 17.1). The plants are usually stunted in their growth with poor root development (Hollings and Stone, 1973; Horst et al., 1977; Matsushita, 2013), and reduction in height of 47% reported for the United Kingdom (Hollings and Stone, 1973), 4%–30% reported for the United States (Horst et al., 1977), and 32%–50% reported for South Korea (Chung et al., 2005), with variation depending on the cultivars and the flowering season. Effects on the flowers include reductions in number, size, and quality (color bleaching, color-breaking, malformation, and premature opening) (Chung et al., 2005; Hollings and Stone, 1973; Horst et al., 1977;

FIGURE 17.1 Chrysanthemum plants showing stunting (left panel) and a chrysanthemum flower showing reduced size and quality (right panel) as a consequence of infection by CSVd. *Source: Photos courtesy of Teruo Sano.*

Matsushita, 2013). Effects on leaf development include reduction in leaf size, general chlorosis, small or large chlorotic spots, large necrotic lesions, leaf crumpling, vein clearing, and vein banding (Chung et al., 2005; Hollings and Stone, 1973). Beside cultivar effects, symptoms also are affected by light intensity and temperature (Bachelier et al., 1976; Hollings and Stone, 1973). In general, CSVd causes symptomless infections in natural hosts other than cultivated chrysanthemum (Bachelier et al., 1976; Hollings and Stone, 1973; Matsushita, 2013). Some infected cultivated chrysanthemums are also symptomless (Bachelier et al., 1976; Doi and Kato, 2004; Hollings and Stone, 1973).

HOST RANGE

The main host of CSVd is chrysanthemum, although the viroid has been found naturally in *Ageratum* spp. (GenBank Accession No. Z68201), *Argyranthemum frutescens* (Marais et al., 2011; Menzel and Maiss, 2000; Torchetti et al., 2012), *Dahlia* spp. (Nakashima et al., 2007), *Petunia hybrida* (Verhoeven et al., 1998), *Solanum jasminoides* (Verhoeven et al., 2006), *Vinca major* (Nie et al., 2005), as well as in the following wild chrysanthemum species: *C. crassum*, *C. indicum*, *C. japonense*, *C. makinoi*, *C. wakasaense*, *C. weyrichii*, *C. yoshinaganthum*, and *C. zawadskii* (Matsushita et al., 2007).

The experimental host range was obtained largely from the work of Brierley (1953), Hollings and Stone (1973), and Runia and Peters (1980). In some cases there are contradictions between the results of these and/ or other studies. In those cases, the negative infection result is excluded below and citations are included for work demonstrating infection by CSVd in such plant species. Additional hosts given by other workers are similarly identified. Some hosts listed below were later identified as natural hosts (see previous paragraph). Plants listed as susceptible are as follows: *Achiellea millefolium*, *A. ptarmica*, *Ageratum houstonianum*, *Ambrosia trifida*, *Anthemis tinctoria*, *Benincasa cerifera*, *Capsicum annum*, *Centaurea cyanus*, *Chrysanthemum boreale* (Matsushita et al., 2012), *C. carinatum*, *C. cinerariaefolium*, *C. coccineum*, *C. coronarium*, *C. corymbosum*, *C. frutescens*, *C. indicum* (Matsushita et al., 2012), *C. lacustre*, *C. leucanthemum*, *C. majus*, *C. maximum*, *C. morifolium*, *C. myconis*, *C. nivellei*, *C. pacificum* (Matsushita and Penmetcha, 2009), *C. parthenium*, *C. praealtum*, *C. shiwogiku* (Matsushita et al., 2012), *C. viscosum*, *Cucumis sativus* (one cv.), *Dahlia pinnata*, *Echinacea purpurea*, *Emilia sagittata*, *Gynura aurantiaca* (Bachelier et al., 1976), *Heliopsis pitcheriana*, *Liatris pycnostachya*, *Nicandra physaloides*, *Nicotiana benthamiana* (R. Flores, pers. com.), *Petunia axilaris*, *P. hybrida*, *P. inflata*, *P. nyctaginiflora*, *Sanvitalia procumbens*, *Scopolia sinensis* (Kryczyński and Paduch-Cichal, 1987), *Senecio cruentus*, *S. glastifolius*, *S. mikanioides*, *Solanum lycopersicum* (formerly *Lycopersicon esculentum*; Kryczyński and Paduch-Cichal, 1987; Niblett et al., 1980), *S. melongena* (one cv.), *S. tuberosum*, *Sonchus asper*, *Tanacetum boreale*, *T. camphoratum*, *T. vulgare*, *Tithonia rotundifolia*, *Venidium fastuosum*, *Verbesina encelioides*, *Zinnea elegans*. Most of these hosts did not develop symptoms. CSVd infection induced symptoms in some cultivars of *S. lycopersicum* (tomato) but not others (Chung et al., 2001; Matsushita and Penmetcha, 2009; Niblett et al., 1980; Runia and Peters, 1980; Yoon et al., 2012, 2014).

TRANSMISSION

CSVd is sap transmissible, but infection is erratic in chrysanthemums after rubbing sap onto leaves (Hollings and Stone, 1973). The viroid is more readily transmitted by either stem slashing with a razor blade (Hollings and Stone, 1973; Runia and Peters, 1980) or needle pricking (Palukaitis and Symons, 1980), both through sap applied to the stem. CSVd can be transmitted by handling infected plants and on tools used for making cuttings or grafting (Brierley and Smith, 1949, 1951a; Chung et al., 2009; Hollings and Stone, 1973). Transmission in chrysanthemums by plant-to-plant contact also occurs (Brierley and Smith, 1949, 1951a; Hollings and Stone, 1973). Soil transmission does not appear to occur (Sugiura and Hanada, 1998), but transmission through root contact does (Matsushita, 2013). Insects and one mite species tested did not transmit the disease (Brierley and Smith, 1951b). Seed transmission occurs in chrysanthemum under some experimental conditions (Chung and Pak, 2008; Monsion et al., 1973), but not others (Hollings and Stone, 1973); the degree of seed transmission depended on the temperature at the time of fertilization and whether one or both parents were infected (Chung and Pak, 2008). Pollen transmission is low (Chung and Pak, 2008). Transmission in tomato plants also occurred, through both pollen and seed (Kryczyński et al., 1988).

DETECTION

The various bioassay techniques used previously for CSVd detection have been reviewed (Bouwen and van Zaayen, 2003; Lawson, 1987), but as they require weeks to show symptoms, they are used largely when economics prevents the use of more rapid tests (see Chapter 33: Viroid Detection and Identification by Bioassay). Other techniques such as polyacrylamide gel electrophoresis (see Chapter 34: Gel Electrophoresis) and nucleic acid hybridization (see Chapter 35: Molecular Hybridization Techniques for Detecting and Studying Viroids) are still used in some cases, but have been largely supplanted by the reverse transcription of the viroid RNA followed by the polymerase chain (RT-PCR) of the cDNA (see Chapter 36: Viroid Amplification Methods: RT-PCR, Real-Time RT-PCR, and RT-LAMP). Several such RT-PCR methods have been used with CSVd, with different experimental designs or goals (Gobatto et al., 2014; Hooftman et al., 1996; Nakahara et al., 1999; Ragozzino et al., 2004; Sugiura and Hanada, 1998). To eliminate the need for a thermocycler for the PCR step, RT combined with loop-mediated isothermal amplification can be used (Fukuta et al., 2005).

GEOGRAPHICAL DISTRIBUTION AND EPIDEMIOLOGY

CSVd has been reported from 26 countries (Yoon and Palukaitis, 2013). These countries include 15 in Europe (Austria, Belgium, Czech Republic, Denmark, France, Germany, Hungary, Italy, Latvia, Norway,

Poland, Spain, Sweden, The Netherlands, and United Kingdom), four in Asia (China, India, Japan, and South Korea), two in Africa (Egypt and South Africa), two in North America (Canada and United States), two in Oceania (Australia and New Zealand), and one in South America (Brazil).

It is likely that CSVd has been spread by the international trade in cuttings (Lawson, 1987). The presence of symptomless alternative hosts among ornamental plants (see above) may also lead to the contamination of mother plants used for cuttings and breeding lines (Marais et al., 2011; Matsushita, 2013; Nakashima et al., 2007; Verhoeven et al., 1998). Spread through chrysanthemum seed in breeding programs is also likely (Chung and Pak, 2008).

CONTROL

The control of CSVd consists mostly of obtaining viroid-free germplasm and removing sources of infectious material. Treating contaminated cutting tools was not very efficient as a means of control (Hollings and Stone, 1973). The use of meristem-tip culture (MTC) combined with heat therapy was not successful at eliminating CSVd (Hollings and Stone, 1970). MTC using leaf primordial-free shoots allowed some level of viroid-free plantlets to be obtained (Hosokawa, 2008; Nabeshima et al., 2014). The use of chemotherapy with amantadine and MTC showed mixed success (Horst and Cohen, 1980; Savitri et al., 2013). Psychrotherapy alone did not eliminate CSVd (Chung et al., 2006), while the combination of psychrotherapy and MTC showed better recovery of viroid-free plant materials (Paduch-Cichal and Kryczyński, 1987; Savitri et al., 2013), especially in combination with chemotherapy, using ribavirin (Savitri et al., 2013).

Genetic resistance to CSVd has been identified in several chrysanthemum cultivars. In two studies, tolerance to CSVd was obtained, with low CSVd accumulation in the inoculated plants (Omori et al., 2009; Nabeshima et al., 2012), while in two other studies heritable resistance was obtained with no detectable CSVd accumulation (Matsushita et al., 2012; Nabeshima et al., 2014). Transgenic resistance also has been obtained. Transgenic chrysanthemums expressing a double-stranded RNA nuclease, *pacI*, showed delayed and reduced frequency of infection with reduced CSVd accumulation (Ogawa et al., 2005), while transgenic chrysanthemums expressing a unique nuclease activity intrinsic to a single-chain variable fragment antibody showed systemic resistance to infection by CSVd (Tran et al., 2016).

Acknowledgments

Work in the author's lab is supported by a grant from the Korean Rural Development Administration (grant no. PJ011309).

References

Bachelier, J.C., Monsion, M., Dunez, J., 1976. Possibilities of improving detection of chrysanthemum stunt and obtaining viroid-free plants by meristem-tip culture. Acta Hortic. 59, 63–69.

Bouwen, I., van Zaayen, A., 2003. Chrysanthemum stunt viroid. In: Hadidi, A., Flores, R., Randles, J.W., Semancik, J.S. (Eds.), Viroids. CSIRO Publishing, Collingwood, VIC, pp. 218–223.

Brierley, P., 1952. Exceptional heat tolerance and some other properties of the chrysanthemum stunt virus. Plant Dis. Rep. 36, 243–244.

Brierley, P., 1953. Some experimental hosts of the chrysanthemum stunt virus. Plant Dis. Rep. 37, 343–345.

Brierley, P., Smith, F.F., 1949. Evidence points to virus as cause of chrysanthemum stunt. Florists' Rev. 103, 36–37.

Brierley, P., Smith, F.F., 1951a. Hardy varieties main target of mum stunt as five year war eases greenhouse threat. Florists Exch. 116, 11.

Brierley, P., Smith, F.F., 1951b. Chrysanthemum stunt. Control measures effective against virus in florists' crop. Florists' Rev. 107, 27–30.

Chung, B.N., Cho, J.D., Cho, I.S., Choi, G.S., 2009. Transmission of chrysanthemum stunt viroid in chrysanthemum by contaminated cutting tools. Hort. Environ. Biotechnol. 50, 536–538.

Chung, B.N., Choi, G.S., Kim, H.R., Kim, J.S., 2001. Chrysanthemum stunt viroid in *Dendranthema grandifolium*. Plant Pathol. J. 17, 194–200.

Chung, B.N., Huh, E.J., Kim, J.S., 2006. Effect of temperature on the concentration of chrysanthemum stunt viroid in CSVd-infected chrysanthemum. Plant Pathol. J. 22, 152–154.

Chung, B.N., Lim, J.H., Choi, S.Y., Kim, J.S., Lee, E.J., 2005. Occurrence of chrysanthemum stunt viroid in Korea. Plant Pathol. J. 21, 377–382.

Chung, B.N., Pak, H.S., 2008. Seed transmission of chrysanthemum stunt viroid in *Chrysanthemum*. Plant Pathol. J. 24, 31–35.

Diener, T.O., Lawson, R.H., 1973. Chrysanthemum stunt: a viroid disease. Virology 51, 94–101.

Dimock, A.W., 1947. Chrysanthemum stunt. N.Y. State Flower Growers' Bull. 26, 2.

Doi, M., Kato, K., 2004. Nucleotide sequence of chrysanthemum stunt viroid (CSVd) occurred in Shizuoka Prefecture and symptoms of chrysanthemum cultivar. Ann. Rept. Kansai Pl. Prot. 46, 11–14.

Fukuta, S., Nima, Y., Ohishi, K., Yoshimura, Y., Anai, N., Hotta, M., et al., 2005. Development of reverse transcription loop-mediated isothermal amplification (RT-LAMP) method for detection of two viruses and chrysanthemum stunt viroid. Ann. Rept. Kansai Pl. Prot. 47, 31–36.

Gobatto, D., Chaves, A.L.R., Harakava, R., Marque, J.M., Daròs, J.A., Eiras, M., 2014. Chrysanthemum stunt viroid in Brazil: survey, identification, biological and molecular characterization and detection methods. J. Plant Pathol. 96, 111–119.

Gross, H.J., Krupp, G., Domdey, H., Raba, M., Jank, P., Lossow, C., et al., 1982. Nucleotide sequence and secondary structure of citrus exocortis and chrysanthemum stunt viroid. Eur. J. Biochem. 121, 249–257.

Haseloff, J., Symons, R.H., 1981. Chrysanthemum stunt viroid: primary sequence and secondary structure. Nucleic Acids Res. 9, 2741–2752.
Hill, M.F., Giles, R.J., Moran, J.R., Hepworth, G., 1996. The incidence of chrysanthemum stunt viroid, chrysanthemum B carlavirus, tomato aspermy cucumovirus and tomato spotted wilt tospovirus in Australian chrysanthemum crops. Australas. Plant Path. 25, 174–178.
Hollings, M., Stone, O.M., 1970. Attempts to eliminate chrysanthemum stunt from chrysanthemum by meristem-tip culture after heat treatment. Ann. Appl. Biol. 65, 311–315.
Hollings, M., Stone, O.M., 1973. Some properties of chrysanthemum stunt, a virus with characteristics of an uncoated ribonucleic acid. Ann. Appl. Biol. 74, 333–348.
Hooftman, R., Arts, M.-J., Shamloul, A.M., Van Zaayen, A., Hadidi, A., 1996. Detection of chrysanthemum stunt viroid by reverse transcription-polymerase chain reaction and by tissue blot hybridization. Acta Hortic. 432, 120–128.
Horst, R.K., Cohen, D., 1980. Amantadine supplement tissue culture medium: a method for obtaining chrysanthemum free of chrysanthemum stunt viroid. Acta Hortic. 110, 311–315.
Horst, R.K., Langhans, R.W., Smith, S.H., 1977. Effects of chrysanthemum stunt, chlorotic mottle, aspermy and mosaic on flowering and rooting in chrysanthemums. Phytopathology 67, 9–14.
Hosokawa, M., 2008. Leaf primordial-free shoot apical meristem culture: a new method for production of viroid-free plants. J. Japan Soc. Hort. Sci. 77, 341–349.
Keller, J.R., 1951. Report on indicator plants for chrysanthemum stunt virus and on a previously unreported chrysanthemum virus. Phytopathology 41, 947–949.
Kryczyński, S., Paduch-Cichal, E., 1987. A comparative study of four viroids. J. Phytopathol 120, 121–129.
Kryczyński, S., Paduch-Cichal, E., Skrzeczkowski, L.J., 1988. Transmission of three viroids through seed and pollen of tomato plants. J. Phytopathol. 121, 51–57.
Lawson, R.H., 1987. Chrysanthemum stunt. In: Diener, T.O. (Ed.), The Viroids. Plenum Press, New York, pp. 247–259.
Lawson, R.H., Hearon, S.S., 1971. Ultrastructure of chrysanthemum stunt virus-infected and stunt-free mistletoe chrysanthemum. Phytopathology 61, 653–656.
Marais, A., Faure, C., Deogratias, J.-M., Candresse, T., 2011. First report of chrysanthemum stunt viroid in various cultivars of *Argyranthemum frutescens* in France. Plant Dis. 95, 1196.
Matsushita, Y., 2013. Chrysanthemum stunt viroid. Jpn. Agric. Res. Quart. 47, 237–242.
Matsushita, Y., Aoki, K., Sumitomo, K., 2012. Selection and inheritance of resistance to chrysanthemum stunt viroid. Crop Protect. 35, 1–4.
Matsushita, Y., Penmetcha, K.K.R., 2009. In vitro-transcribed chrysanthemum stunt viroid RNA is infectious to chrysanthemum and other plants. Phytopathology 99, 58–66.
Matsushita, Y., Tsukiboshi, T., Ito, Y., Chikuo, Y., 2007. Nucleotide sequence and distribution of chrysanthemum stunt viroid in Japan. J. Jpn. Soc. Hort. Sci. 76, 333–337.
Menzel., W., Maiss, E., 2000. Detection of chrysanthemum stunt viroid (CSVd) in cultivars of *Argyranthemum frutescens* by RT-PCR-ELISA. Z. Pflanzenk. Pflanzensch. 107, 548–552.
Monsion, M., Bachelier, J.C., Dunez, J., 1973. Quelques proprieties d'un viroïde: Le rabougrissement du chrysantheme. Ann. Phytopathol. 5, 467–469.
Nabeshima, T., Hosokawa, M., Yano, S., Ohishi, K., Doi, M., 2012. Screening of chrysanthemum cultivars with resistance to chrysanthemum stunt viroid. J. Jpn. Soc. Hort. Sci. 81, 285–294.
Nabeshima, T., Hosokawa, M., Yano, S., Ohishi, K., Doi, M., 2014. Evaluation of chrysanthemum stunt viroid (CSVd) infection in newly-expanded leaves from

CSVd-inoculated shoot apical meristems as a method of screening for CSVd-resistant chrysanthemum cultivars. J. Hort. Sci. Biotechnol. 89, 29–34.

Nakahara, K., Hataya, T., Uyeda, I., 1999. A simple, rapid method of nucleic acid extraction without tissue homogenization for detecting viroids by hybridization and RT-PCR. J. Virol. Methods. 77, 47–58.

Nakashima, A., Hosokawa, M., Maeda, S., Yazawa, C., 2007. Natural infection of chrysanthemum stunt viroid in dahlia plants. J. Gen. Plant Pathol. 73, 225–227.

Niblett, C.L., Dickson, E., Fernow, K.H., Horst, R.K., Zaitlin, M., 1978. Cross protection among four viroids. Virology 91, 198–203.

Niblett, C.L., Dickson, E., Horst, K.H., Romaine, C.P., 1980. Additional hosts and efficient purification procedure for four viroids. Phytopathology 70, 610–615.

Nie, X., Singh, R.P., Bostan, H., 2005. Molecular cloning, secondary structure, and phylogeny of three pospiviroids from ornamental plants. Can. J. Plant Pathol. 27, 592–602.

Ogawa, T., Toguri, T., Kudoh, H., Okamura, M., Momma, T., Yoshioka, M., et al., 2005. Double-stranded RNA-specific ribonuclease confers tolerance against chrysanthemum stunt viroid and tomato spotted wilt virus in transgenic chrysanthemum plants. Breeding Sci. 55, 49–55.

Olson, C.J., 1949. A preliminary report on transmission of chrysanthemum stunt. Bull. Chrysanthemum Soc. Amer. 17, 2–9.

Omori, H., Hosokawa, M., Shiba, H., Shitsukawa, N., Murai, K., Yazawa, S., 2009. Screening of chrysanthemum plants with strong resistance to chrysanthemum stunt viroid. J. Jpn. Soc. Hort. Sci. 78, 350–355.

Owens, R.A., Flores, R., Di Serio, F., Li, S.-F., Pallás, V., Randles, J.W., et al., 2012. Viroids. In: King, A.M.Q., Adams, M.J., Carstens, E.B., Lefkowitz, E.J. (Eds.), Virus Taxonomy. Classification and Nomenclature of Viruses. Ninth Report of the International Committee on Taxonomy of Viruses. Elsevier Academic Press, San Diego, CA, pp. 1221–1234.

Paduch-Cichal, E., Kryczyński, S., 1987. A low temperature therapy and meristem-tip culture for eliminating four viroids from infected plants. J. Phytopathol. 118, 341–346.

Palukaitis, P., Symons, R.H., 1980. Purification and characterization of the circular and linear forms of chrysanthemum stunt viroid. J. Gen. Virol. 46, 477–489.

Ragozzino, E., Faggioli, F., Barba, M., 2004. Development of a one tube–one step RT-PCR protocol for the detection of seven viroids in four genera: *Apscaviroid*, *Hostuviroid*, *Pelamoviroid* and *Pospiviroid*. J. Virol. Methods 121, 25–29.

Runia, W., Th., Peters, D., 1980. The response of plant species used in agriculture and horticulture to viroid infections. Neth. J. Plant Pathol. 86, 135–146.

Savitri, W.D., Park, K.I., Jeon, S.M., Chung, M.Y., Han, J.-S., Kim, C.K., 2013. Elimination of chrysanthemum stunt viroid (CSVd) from meristem tip culture combined with prolonged cold treatment. Hort. Environ. Biotechnol. 54, 177–182.

Singh, D., Pathania, M., Ram, R., Zaidi, A.A., Verma, N., 2010. Screening of chrysanthemum cultivars for chrysanthemum stunt viroid in an Indian scenario. Arch. Phytopathol. Plant Protect. 43, 1517–1523.

Sugiura, H., Hanada, K., 1998. Chrysanthemum stunt viroid, a viroid disease of large-flowered chrysanthemum in Niigata Prefecture. J. Jpn. Soc. Hort. Sci. 67, 432–438.

Torchetti, E.M., Navarro, B., Trisciuzzi, V.N., Nuccitelli, L., Silletti, M.R., Di Serio, F., 2012. First report of chrysanthemum stunt viroid in *Argyranthemum frutescens* in Italy. J. Plant Pathol. 94, 451–454.

Tran, D.H., Cho, S., Hoang, P.M., Kim, J., Kil, E.-J., Lee, T.-K., et al., 2016. A codon-optimized nucleic acid hydrolyzing single-chain antibody confers resistance to chrysanthemums against chrysanthemum stunt viroid. Plant Mol. Biol. Rep. 34, 221–232.

Verhoeven, J.Th.J., Arts, M.S.J., Owens, R.A., Roenhorst, J.W., 1998. Natural infection of petunia by chrysanthemum stunt viroid. Eur. J. Plant Pathol. 104, 383–386.

Verhoeven, J.Th.J., Jansen, C.C.C., Roenhorst, J.W., 2006. First report of potato virus M and chrysanthemum stunt viroid in *Solanum jasminoides*. Plant Dis. 90, 1359.

Yoon, J.-Y., Baek, E., Palukaitis, P., 2012. Are there strains of chrysanthemum stunt viroid? J. Plant Pathol. 94, 697–701.

Yoon, J.-Y., Cho, I.-S., Choi, G.-S., Choi, S.-K., 2014. Construction of infectious cDNA clone of a chrysanthemum stunt viroid Korean isolate. Plant Pathol. J. 30, 68–74.

Yoon, J.-Y., Palukaitis, P., 2013. Sequencing comparison of global chrysanthemum stunt viroid variants: multiple polymorphic positions scattered through the viroid genome. Virus Genes 46, 97–104.

Zhang, Z., Ge, B., Pan, S., Zhao, Z., Wang, H., Li, S., 2011. Molecular detection and sequence analysis of chrysanthemum stunt viroid. Acta Horticult. Sin. 38, 2349–2356.

CHAPTER

18

Iresine Viroid 1 and a Potential New Pospiviroid From Portulaca

Jacobus Th.J. Verhoeven[1], *Ricardo Flores*[2] *and Pedro Serra*[2]

[1]**National Plant Protection Organization, Wageningen, The Netherlands**
[2]**Polytechnic University of Valencia-CSIC, Valencia, Spain**

INTRODUCTION

Iresine viroid 1 (IrVd-1) was described for the first time in 1996. The viroid was isolated from a symptomless plant of *Iresine herbstii* (beefsteak plant) during a survey of ornamental plants (Spieker, 1996). Subsequently, the viroid was detected in a few more ornamental plant species. A potential new pospiviroid, portulaca latent viroid (PoLVd), was recently characterized from symptomless plants of *Portulaca* sp. that were coinfected with alternanthera mosaic virus (Verhoeven et al., 2015).

TAXONOMIC POSITION AND NUCLEOTIDE SEQUENCE

IrVd-1 has been assigned to the genus *Pospiviroid* of the family *Pospiviroidae* (Di Serio et al., 2014; Owens et al., 2012). The newly characterized PoLVd has similar features and, therefore, is also affiliated to these taxa (Verhoeven et al., 2015). The secondary structure of minimal free energy of both IrVd-1 and PoLVd is a rod-like conformation resembling that proposed for members of the family *Pospiviroidae* (Fig. 18.1). Embedded in the rod-like secondary structure of both viroids are the central conserved region (CCR) characteristic of the genus *Pospiviroid*, and the terminal conserved region (TCR) present in members of the genera *Apscaviroid* and *Pospiviroid* and in some members of the genus

Viroids and Satellites.
DOI: http://dx.doi.org/10.1016/B978-0-12-801498-1.00018-8

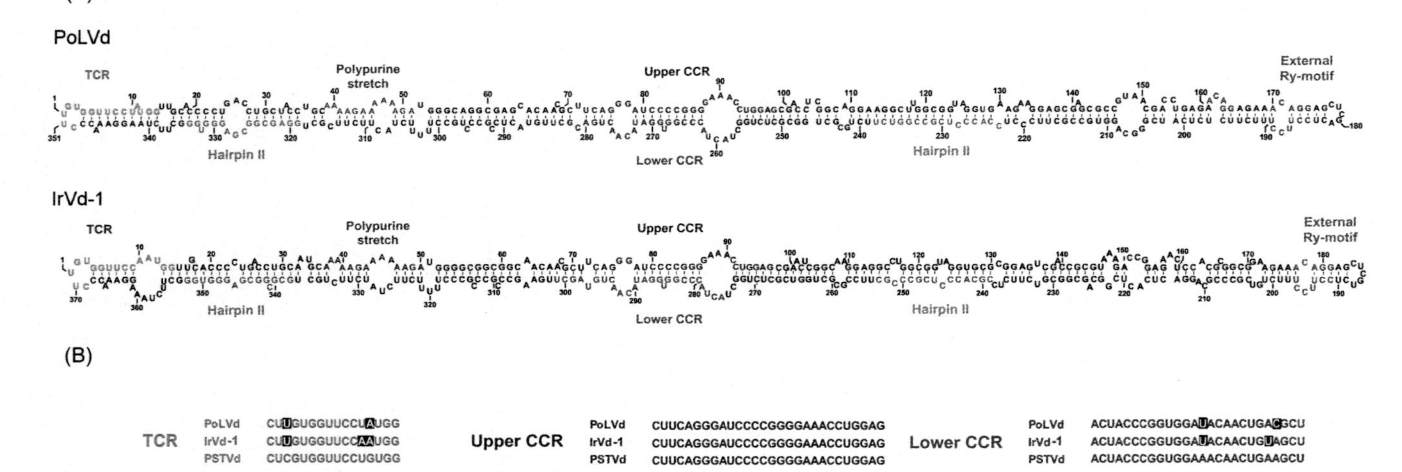

FIGURE 18.1 (A) Primary and secondary structure of minimum free energy predicted for PoLVd and IrVd-1. Nucleotides forming the terminal conserved region (TCR), polypurine stretch, upper and lower central conserved region (CCR), hairpin II, and the core of the RY motif are highlighted with different colors. (B) Comparison between the TCR and the upper and lower CCR of PoLVd, IrVd-1 and PSTVd as the representative member of the family *Pospiviroidae*. Minor differences in PoLVd and IrVd-1 are highlighted with a black background. *Source: From Verhoeven J.Th.J., Roenhorst J.W., Hooftman M., Meekes E.T.M., Flores R., et al., A pospiviroid from symptomless portulaca plants closely related to iresine viroid 1. Virus Res. 205, 2015, 22–26.*

Coleviroid (Flores et al., 1997). In addition, both viroids contain the polypurine tract conserved in the family *Pospiviroidae* (Keese and Symons, 1985), and the external RY-motif present in all members of the genus *Pospiviroid* (Gozmanova et al., 2003). The conservation of two structural elements, i.e., hairpin I and II, which can be adopted in alternative metastable conformations, further contributes to their classification within the genus *Pospiviroid*. Hairpin I —comprising the upper CCR strand and two flanking nucleotides at both sides forming an imperfect inverted repeat (Riesner et al., 1979), and involved in facilitating processing of the oligomeric (+) RNA replication intermediates (Gas et al., 2007)— is identical in PoLVd and IrVd (positions 69–100 and 68–99, respectively). Hairpin II —formed by sequences located in the lower strand of the rod-like conformation (Riesner et al., 1979), and crucial for infectivity (Candresse et al., 2001; Qu et al., 1993)— has the characteristic GC-rich stem of 10 bp for IrVd-1 and 16 bp for PoLVd.

IrVd-1 (370–371 nt) and PoLVd (351 nt) show circa 80% sequence identity, whereas identities with other members of the genus *Pospiviroid* are below 60%. Using the ICTV demarcation criterion of less than 90% sequence similarity (Owens et al., 2012), Irvd-1 and PoLVd should be considered as two distinct species. However, the demarcation criteria of the ICTV also prescribe distinct biological properties, particularly host range and symptoms, to separate viroid species within a genus (Owens et al., 2012). So far, *Portulaca* sp. is the only natural and experimental host of PoLVd, in which no symptoms are induced. However, this species may also be symptomlessly infected by IrVd-1 (Verhoeven, 2010; Virscek Marn and Mavric Plesko, 2012) and, therefore, distinguishing biological properties for PoLVd remain to be found. Consequently, PoLVd should currently be listed as an unassigned member in the genus *Pospiviroid*, and not classified as a species.

Interestingly, IrVd-1 and PoLVd are separated from the other members in the genus *Pospiviroid* by both biological and molecular features. Unlike other pospiviroids, IrVd-1 and PoLVd do not infect tomato, and part of the TCR of both viroids forms the terminal left loop of the rod-like conformation (Spieker, 1996; Verhoeven et al., 2015). Moreover, alignment of both sequences shows the deletion of two stretches of 9–10 nt located approximately opposite each other in the upper and lower strands of the right terminal domain in the rod-like conformation of PoLVd, suggesting that polymerase jumping during IrVd-1 replication and some other minor accompanying changes could have resulted in preservation of the PoLVd conformation.

Phylogenetic reconstructions (Fig. 18.2) following a multiple alignment also showed that PoLVd is more closely related to IrVd-1 than to any other pospiviroid. In addition, PoLVd is clearly separated from the two members of the genus *Hostuviroid* (hop stunt viroid, and dahlia latent viroid). Therefore, these data further support the inclusion of PoLVd in the genus *Pospiviroid*.

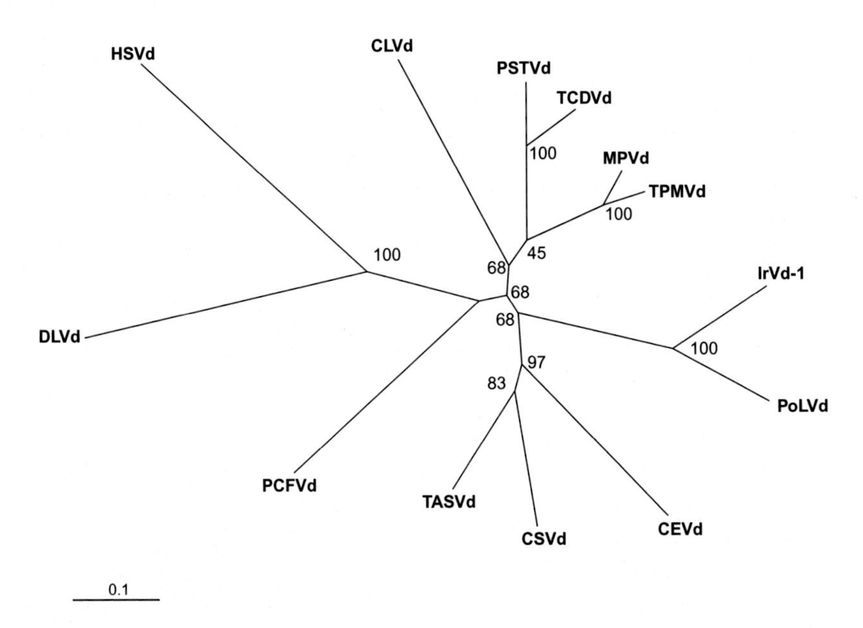

FIGURE 18.2 Evolutionary relationships of the pospiviroids citrus exocortis viroid (CEVd; M34917), chrysanthemum stunt viroid (CSVd; X16408), Columnea latent viroid (CLVd; X15663), IrVd-1 (X95734), Mexican papita viroid (MPVd; L78454; currently considered a strain of tomato planta macho viroid, TPMVd), pepper chat fruit viroid (PCFVd, FJ409044), PoLVd (KR677387), potato spindle tuber viroid (PSTVd; V01465), tomato apical stunt viroid (TASVd; NC001553), tomato chlorotic dwarf viroid (TCDVd; AF162131), and tomato planta macho viroid (TPMVd; K00817) and the hostuviroids, hop stunt viroid (HSVd; EU872277) and dahlia latent viroid (DLVd; JX263426). The phylogenetic reconstruction was performed with the Neighbor-joining method (Saitou and Nei, 1987). The fraction (%) of 5000 replicate trees in which the groups clustered together in the bootstrap test (Felsenstein, 1985) is shown next to the branches. Branch lengths, which were estimated with the Maximum Composite Likelihood method (Tamura et al., 2004) and expressed in base substitutions per site, are proportional to the evolutionary distances used to infer the phylogenetic tree. Phylogenetic analyses were conducted with MEGA4 (Tamura et al., 2007). *Source: From Verhoeven J.Th.J., Roenhorst J.W., Hooftman M., Meekes E. T.M., Flores R., et al., A pospiviroid from symptomless portulaca plants closely related to iresine viroid 1. Virus Res. 205, 2015, 22–26.* IrVd-1 and PoLVd are more related to each other than to any of the other viroids and both clearly better match with the other pospiviroids than with both hostuviroids.

SYMPTOMS, HOST RANGE, AND ECONOMIC SIGNIFICANCE

Infections by IrVd-1 have been reported for six ornamental plant species: *Celosia cristata* (Sorrentino et al., 2015), *Celosia plumosa* (Verhoeven et al., 2010), *I. herbstii* (Spieker, 1996), *Portulaca* sp. (Verhoeven, 2010; Virscek Marn and Mavric Plesko, 2012), *Verbena* spp., and *Vinca major*

(Bostan et al., 2004; Nie et al., 2005) and for the aquatic plant *Alternanthera sessilis* (Singh et al., 2006). No symptoms have been observed or specified.

The IrVd-1 isolate from *I. herbstii* was successfully inoculated mechanically to plants of cv. "Aureo-reticulata" of this species, but not to tomato (*Solanum lycopersicum*) cv. "Rentita" (Spieker, 1996). Similarly, the IrVd-1 isolate from *C. plumosa* was successfully inoculated mechanically to plants of *C. plumosa*, but not to tomato cv. "Moneymaker", chrysanthemum (*Chrysanthemum* × *morifolium*) cv. "White Delianne", or potato (*Solanum tuberosum*) cv. "Nicola" (Verhoeven et al., 2010). Consequently, the viroid may not infect major host plants of other members of the genus *Pospiviroid*.

The direct economic impact of IrVd-1 is assumed to be low because its host range seems limited, excluding the major pospiviroid crops tomato and potato, and because no symptoms have been reported for its known host plants. In addition, the indirect economic impact also seems low because no specific phytosanitary requirements apply to this viroid.

Infections by PoLVd have only been reported for two plants of *Portulaca*, both of which were coinfected with alternanthera mosaic virus and showed mild chlorotic mottling and some leaf deformation. Mechanical inoculation was successful for *Portulaca* (symptomless infections), but not for chrysanthemum cvs. "Bonnie Jean" and "Mistletoe", *Nicotiana benthamiana*, or tomato cv. "Moneymaker". Inoculated *Portulaca* plants, singly infected with PoLVd, did not show symptoms (Verhoeven et al., 2015). Consequently, similarly to IrVd-1, the economic significance of PoLVd is assumed to be low.

TRANSMISSION

Both IrVd-1 and PoLVd are transmitted by vegetative propagation of infected plants and mechanical inoculation. However, the efficiency of the mechanical inoculation of PoLVd may be low since only one out of eight *Portulaca* plants (resulting from cuttings from a viroid-free plant) appeared infected after rub-inoculating the viroid circular RNAs, separated from other nucleic acids by rPAGE, onto carborundum-dusted leaves. Dipping a razor blade in this inoculum and then slashing three stems of eight *Portulaca* plants did not result in viroid transmission. Most successful was mechanical inoculation by leaf rubbing of head-to-tail dimeric transcripts of PoLVd generated in vitro, as two out of six *Portulaca* plants became infected in this way (Verhoeven et al., 2015).

DETECTION

IrVd-1 was initially detected and isolated by bidirectional electrophoresis (Schumacher et al., 1983). IrVd-1 comigrated with citrus exocortis viroid (371 nt; Gross et al., 1982), a viroid of similar size. Various RT-PCR approaches are now available for the detection of IrVd-1 with either specific primers that are not expected to amplify other pospiviroids (Verhoeven et al., 2010), with generic primers for pospiviroids (Bostan et al., 2004; Verhoeven et al., 2004), or with real-time RT-PCR (Botermans et al., 2013). For further identification, the resulting amplicons of the RT-PCRs should be sequenced directly or after cloning.

For PoLVd, the same RT-PCRs for generic detection of pospiviroids may be used. For specific detection, RT-PCR primers have been developed (Verhoeven et al., 2015).

GEOGRAPHICAL DISTRIBUTION AND EPIDEMIOLOGY

IrVd-1 has been reported in Canada (Bostan et al., 2004), Germany (Spieker, 1996), India (Singh et al., 2006), Italy (Sorrentino et al., 2015), The Netherlands (Verhoeven, 2010), and Slovenia (Virscek Marn and Mavric Plesko, 2012). Since the reports originate from countries in three continents and most refer to symptomless ornamentals that are traded frequently, it is likely that the viroid will be present in more countries. PoLVd has been identified only in The Netherlands (Verhoeven et al., 2015).

CONTROL

The economic significance is low, since no symptoms have been observed. Nevertheless, in The Netherlands, *Celosia* spp., *Verbena* spp., and *Vinca* spp. can be included in the Naktuinbouw Elite certification system on a voluntary basis, which implies testing for pospiviroids.

Acknowledgments

Research in R.F. laboratory is currently funded by grant BFU2014–56812-P from the Spanish Ministerio de Economía y Competitividad (MINECO).

References

Bostan, H., Nie, X., Singh, R.P., 2004. An RT-PCR primer pair for the detection of pospiviroid and its application in surveying ornamental plants for viroids. J. Virol. Methods 166, 189–193.

Botermans, M., Van de Vossenberg, B.T.L.H., Verhoeven, J.Th.J., Roenhorst, J.W., Hooftman, M., et al., 2013. Development and validation of a real-time RT-PCR assay for generic detection of pospiviroids. J. Virol. Methods 187, 43–50.

Candresse, T., Góra-Sochacka, A., Zagórski, W., 2001. Restoration of secondary hairpin II is associated with restoration of infectivity of a non-viable recombinant viroid. Virus Res. 75, 29–34.

Di Serio, F., Flores, R., Verhoeven, J.Th.J., Li, S.-F., Pallás, V., et al., 2014. Current status of viroid taxonomy. Arch. Virol. 159, 3467–3478.

Felsenstein, J., 1985. Confidence limits on phylogenies: an approach using the boot-strap. Evolution 39, 783–791.

Flores, R., Di Serio, F., Hernández, C., 1997. Viroids: the non-coding genomes. Semin. Virol. 8, 65–73.

Gas, M.E., Hernández, C., Flores, R., Daròs, J.A., 2007. Processing of nuclear viroids in vivo: an interplay between RNA conformations. PLoS Pathog. 3, e182.

Gozmanova, M., Denti, M.A., Minkov, I.N., Tsagris, M., Tabler, M., 2003. Characterization of the RNA motif responsible for the specific interaction of potato spindle tuber viroid RNA (PSTVd) and the tomato protein Virp1. Nucleic Acids Res. 31, 5534–5543.

Gross, H.J., Krupp, G., Domdey, H., Raba, M., Alberty, H., Lossow, C.H., et al., 1982. Nucleotide sequence and secondary structure of citrus exocortis and chrysanthemum stunt viroid. Eur. J. Biochem. 121, 249–257.

Keese, P., Symons, R.H., 1985. Domains in viroids: evidence of intermolecular RNA rearrangements and their contribution to viroid evolution. Proc. Natl. Acad. Sci. USA 82, 4582–4586.

Nie, X., Singh, R.P., Bostan, H., 2005. Molecular cloning, secondary structure, and phylogeny of three pospiviroids from ornamental plants. Can. J. Plant Pathol. 27, 592–602.

Owens, R.A., Flores, R., Di Serio, F., Li, S.-F., Pallás, V., Randles, J.W., et al., 2012. Viroids. In: King, A.M.Q., Adams, M.J., Carstens, E.B., Lefkowitz, E.J. (Eds.), Virus Taxonomy, Classification and Nomenclature of Viruses, Ninth Report of the International Committee on Taxonomy of Viruses. Elsevier-Academic Press, London, pp. 1221–1234.

Qu, F., Heinrich, C., Loss, P., Steger, G., Tien, P., Riesner, D., 1993. Multiple pathways of reversion in viroids for conservation of structural elements. EMBO J. 12, 2129–2139.

Riesner, D., Henco, K., Rokohl, U., Klotz, G., Kleinschmidt, A.K., Domdey, H., et al., 1979. Structure and structure formation of viroids. J. Mol. Biol. 133, 85–115.

Saitou, N., Nei, M., 1987. The neighbor-joining method: a new method for reconstructing phylogenetic trees. Mol. Biol. Evol. 4, 406–425.

Schumacher, J., Randles, J.W., Riesner, D., 1983. A two-dimensional electrophoresis technique for the detection of circular viroids and virusoids. Anal. Biochem. 135, 288–295.

Singh, R.P., Dilworth, A.D., Baranwal, V.K., Gupta, K.N., 2006. Detection of citrus exocortis viroid, iresine viroid, and tomato chlorotic dwarf viroid in new ornamental host plants in India. Plant Dis. 90, 1457.

Sorrentino, R., Minutolo, M., Alioto, D., Torchetti, E.M., Di Serio, F., 2015. First report of iresine viroid 1 in ornamental plants in Italy and of *Celosia cristata* as a novel natural host. Plant Dis. 99, 1655.

Spieker, R.L., 1996. The molecular structure of iresine viroid, a new viroid species from *Iresine herbstii* ("beefsteak plant"). J. Gen. Virol. 77, 2631–2635.

Tamura, K., Dudley, J., Nei, M., Kumar, S., 2007. MEGA4: molecular evolutionary genetics analysis (MEGA) software version 4.0. Mol. Biol. Evol. 24, 1596–1599.

Tamura, K., Nei, M., Kumar, S., 2004. Prospects for inferring very large phylogenies by using the neighbor-joining method. Proc. Nat. Acad. Sci. 101, 11030–11035.

Verhoeven, J.Th.J., 2010. Identification and epidemiology of pospiviroids. Thesis. Wageningen University, Wageningen, The Netherlands. ISBN 978-90-8585-623-8, 136 pp.

Verhoeven, J.Th.J., Jansen, C.C.C., Botermans, M., Roenhorst, J.W., 2010. First report of iresine viroid 1 in *Celosia plumosa* in the Netherlands. Plant Dis. 94, 920.

Verhoeven, J.Th.J., Jansen, C.C.C., Willemen, T.M., Kox, L.F.F., Owens, R.A., et al., 2004. Natural infections of tomato by citrus exocortis viroid, columnea latent viroid, potato spindle tuber viroid and tomato chlorotic dwarf viroid. Eur. J. Plant Pathol. 110, 823–831.

Verhoeven, J.Th.J., Meekes, E.T.M., Roenhorst, J.W., Flores, R., Serra, P., 2013. Dahlia latent viroid: a recombinant new species of the family Pospiviroidae posing intriguing questions about its origin and classification. J. Gen. Virol. 94, 711–719.

Verhoeven, J.Th.J., Roenhorst, J.W., Hooftman, M., Meekes, E.T.M., Flores, R., et al., 2015. A pospiviroid from symptomless portulaca plants closely related to iresine viroid 1. Virus Res. 205, 22–26.

Virscek Marn, M., Mavric Plesko, I., 2012. First report of iresine viroid 1 in *Portulaca* sp. in Slovenia. J. Plant Pathol. 94, 97.

CHAPTER

19

Hop Stunt Viroid

Tatsuji Hataya[1], *Taro Tsushima*[2] *and Teruo Sano*[2]
[1]Hokkaido University, Sapporo, Japan [2]Hirosaki University, Hirosaki, Japan

INTRODUCTION

Hop stunt viroid (HSVd) was first identified in hop (*Humulus lupulus*) plants showing abnormal dwarfing of bines and low α-acids content in the cone and known as hop stunt disease (Sasaki and Shikata, 1977). The causal agent of cucumber pale fruit disease in The Netherlands was found to be an isolate of HSVd (Sano et al., 1984). Subsequently, variants were detected in grapevine and citrus trees worldwide (Sano et al., 1986, 1988), and in plum and peach showing dapple fruit or yellow fruit symptoms (Sano et al., 1989).

HSVd is now known to be distributed in fruit plants of the families *Moraceae* (fig and mulberry), *Rosaceae* (almond, apple, apricot, cherry, peach, pear, and plum), and *Rutaceae* (*Citrus* spp.) (Table 19.1). Most host plants are symptomless carriers but some sensitive hosts and/or specific variants show disease symptoms.

TAXONOMIC POSITION AND NUCLEOTIDE SEQUENCE

Hop stunt viroid is the type species of genus *Hostuviroid*, family *Pospiviroidae*. HSVd contains a genus specific central conserved region and a terminal conserved hairpin, but lacks a terminal conserved region (Flores et al., 1998; Chapter 13).

More than 600 HSVd sequences are deposited in the DDBJ/GenBank/EMBL sequence databases. Initial comparisons of overall sequence similarity among HSVd isolates from hop, cucumber, grapevine, citrus, plum, and peach revealed that they can be divided into three groups (i.e., plum-type, grape/hop-type, and citrus-type) (Shikata,

Viroids and Satellites.
DOI: http://dx.doi.org/10.1016/B978-0-12-801498-1.00019-X

TABLE 19.1 List of HSVd Variants and Diseases in the Natural Host Plants

Host species	Symptoms/diseases (Country)	Length (nt)	Sequence type	References
Almond	Latent (Spain, China)	296, 297	Plum	Astruc et al. (1996), Cañizares et al. (1999), Zhang et al. (2012)
Apple, wild apple	Probably latent (Greece, Iran)	296–299	Some similar to plum and P-C type, but the others similar to citrus type	Kaponi et al. (2012), DNA database
Apricot	Degeneración (Spain), latent (Mediteranean region, Serbia, China)	294–300	Plum, P-C, P-H/cit3	Amari et al. (2001, 2007), Astruc et al. (1996), El Maghraby et al. (2007), Mandic et al. (2008), Zhang et al. (2012)
Cherry	Latent (Greece)	297, 298	Similar to plum and P-C type, but one similar to citrus type	Kaponi et al. (2012), DNA database
Citrus	Latent (worldwide)	c. 302–303	Citrus (CVd-IIa)	Duran-Vila et al. (1988), Sano et al. (1988)
	Cachexia (worldwide)	c. 295–299	Citrus (CCaVd, CVd-IIb, CVd-IIc)	Palacio-Bielsa et al. (2004), Roistacher et al. (1983), Semancik et al. (1988)
	[a]Gummy bark (sweet orange/ Turkey, Egypt)	296–300	Citrus (similar to CVd-IIb, CVd-IIc)	Önelge et al. (2004), Sofy et al. (2010)
	[a]Yellow corky vein (Kagzi lime/India)	295	Citrus (CVd-IIc)	Roy and Ramachandran (2003)
	[a]Yellow corky vein (sweet orange/Iran)	302	Citrus (similar to CVd-IIa)	Bagherian and Izadpanah (2010)
	[a]Split bark (sweet lime/Iran)	299	Citrus (similar to CVd-IIc)	Bagherian and Izadpanah (2010)
Cucumber	Cucumber pale fruit (The Netherlands, Finland)	301, 303	Citrus	Lemmetty et al. (2011), Sano et al. (1984), Van Dorst and Peters (1974)
Fig		297–299, 303–306	Plum, citrus	

(*Continued*)

TABLE 19.1 (Continued)

Host species	Symptoms/diseases (Country)	Length (nt)	Sequence type	References
	Probably latent (Tunisia, Syria, Lebanon, Iran)			Elbeaino et al. (2012a, 2013), Yakoubi et al. (2007), DNA database
Grapevine	Latent (worldwide)	294–300	Grape/hop	Sano et al. (1986)
Hibiscus	Dwarfing and leaf curling (Italy)	296, 299	Grape/hop, P-C	Luigi et al. (2013)
Hop	Stunt (Japan, USA, China)	296–303	Grape/hop	Eastwell and Nelson (2007), Guo et al. (2008), Sasaki and Shikata (1977)
	Stunt (Slovenia)	301	Citrus	Radisek et al. (2012)
Jujube	Latent (China)	297	Plum	Zhang et al. (2009)
Mulberry	Latent (Lebanon, Italy)	303	Citrus	Elbeaino et al. (2012b)
	Vein clearing and yellow speckling (Iran)	302, 305	Plum	Mazhar et al. (2014)
Peach	Dapple fruit/latent (Japan, Spain, China)	297	Plum	Kofalvi et al. (1997), Sano et al. (1989), Zhou et al. (2006)
Pear	Latent (Tunisia)	c. 300	Plum	Hassen et al. (2004)
Pistachio	Probably latent but may cause yellowing (Tunisia)	294–298	Plum, grape/hop, P-C, P-H/cit3	Elleuch et al. (2013)
Plum	Dapple fruit (Japan, Italy, Korea)/yellow fruit (Japan)/latent (China, worldwide)	297, 299	Plum	Cho et al. (2011), Ragozzino et al. (2002), Sano et al. (1989), Zhang et al. (2012)
Pomegranate	Probably latent (Spain, Tunisia)	297–300	Citrus, P-H/cit3	Astruc et al. (1996), Gorsane et al. (2010)

[a] *Involvement of HSVd has not yet been shown clearly.*
CVd-IIb, -IIc, CCaVd are cachexia symptom-inducing variant.
CVd-IIa is noncachexia symptom-inducing variant.
P-C and P-H/cit3 are probably derived from recombination events between variants of the plum- and citrus-type or between variants of the plum- and hop-type or cit3 variant (Kofalvi et al., 1997; Amari et al., 2001).

1990). More detailed and comprehensive phylogenetic analysis revealed two new groups (i.e., plum-citrus-type and plum-hop/cit3-type) (Amari et al., 2001; Kofalvi et al., 1997).

ECONOMIC SIGNIFICANCE

Hop stunt disease emerged in Japan in the 1940–50s. The infected hops were characterized by their dwarf shape and stunted growth, and were clearly distinguished from healthy bines during growth and at the time of picking. The economic impact was so high because the average length of the bines, number of cones per bine, average cone weight, and alpha acids content in the infected cones were nearly 1/2 to 1/4 of those of the normal plants (Yamamoto et al., 1970). The disease spread during rapid expansion of hop cultivation areas and caused severe economic losses when it reached high incidences in hop gardens over the next three decades. The proportion of HSVd-infected gardens reached 10%–20% in the 1970s. The disease epidemic ended in the second half of 1980s and is now sporadic. However, new epidemics were reported in North America in 2004 and in China in 2007 (Eastwell and Sano, 2009; Guo et al., 2008; Sano, 2013). HSVd was also identified in Germany by large scale surveillance in 2008–12 (Seigner et al., 2013). More recently, a similar but much more severe hop disorder emerged in Slovenia, from which a variant of citrus bark cracking viroid was detected in addition to HSVd (Jakse et al., 2015; Radisek et al., 2012). HSVd-hop normally belongs to the grape/hop-type (Kawaguchi-Ito et al., 2009), however, the Slovenian isolate was composed of 301-nt, identical to those inducing cucumber pale fruit disease–citrus-type.

Cucumber pale fruit disease, caused by cucumber pale fruit viroid—an isolate of HSVd—was first reported in The Netherlands, then in Finland, causing economic losses in greenhouse cucumber, although the fraction of affected plants was usually less than 0.1% (Lemmetty et al., 2011; Van Dorst and Peters, 1974).

Cachexia disease of citrus mainly affects some mandarins, mandarin hybrids such as tangelos, and *Citrus macrophylla*. Most other commercial citrus species are apparently symptomless unless grafted on sensitive rootstocks. Sensitive citrus trees affected are chlorotic and stunted, and may decline and die. The viroid nature of cachexia was suggested in 1980s (Roistacher et al., 1983), and citrus viroid-IIb, a strain of HSVd, was identified as the causal agent and named citrus cachexia viroid (Semancik et al., 1988). Xyloporosis disease in "Palestine" sweet lime used as rootstock in Israel is also caused by cachexia-inducing variants (Reanwarakorn and Semancik, 1999).

Dapple fruit disease of plum and peach was first recognized in Japan, then in Korea and Italy, as a fruit disorder causing serious damage to commercial value (Cho et al., 2011; Ragozzino et al., 2002; Sano, et al., 1989). The disease incidence is sporadic but the economic impact is higher in sensitive varieties. Similar dapple fruit symptom is more commonly observed in the peach market in China (Zhou et al., 2006).

SYMPTOMS

Hop Stunt

Fig. 19.1 shows hop stunt disease. Typical symptoms reported include stunting, leaf curling, bent leaf, and small cones. Stunting appears several years after established plants become infected from observations of the cultivars in Japan (Sano, 2003) and a bit earlier in sensitive cultivars in the United States (Eastwell and Nelson, 2007). In the US commercial production, yellow–green leaves are prominent among the basal foliage that develops early in the growing season and

FIGURE 19.1 Hop stunt disease. Hop bines in the center show stunting and failure to reach the top-wire of the shelf.

the development of lateral branches is inhibited. On some sensitive cultivars, yellow speckling along the major leaf veins is also evident.

Citrus Disorders

Cachexia-sensitive mandarins and tangelos develop discoloration and gum impregnation of the bark. Xyloporosis in "Palestine" sweet lime is characterized by fine wood pitting and no discoloration or gumming reaction (Reanwarakorn and Semancik, 1999).

Although involvement of HSVd has not yet been shown clearly, a cachexia-inducing citrus viroid-IIc has been implicated in yellow corky vein disease of Kagzi lime (*Citrus aurantifolia*) from India (Roy and Ramachandran, 2003), which is characterized by yellow spots on leaf lamina and veins and corky tissues. Cachexia-inducing variants have also been associated with gummy bark disease of sweet orange in Turkey and Egypt (Önelge et al., 2004; Sofy et al., 2010) and split bark disorder of sweet lime in Iran (Bagherian and Izadpanah, 2010). Gummy bark is characterized by a reddish-brown line and gum-impregnated tissue under the bark around the scion circumference and near the bud union. Affected trees are stunted to varying degrees. Split bark is characterized by cracks in bark of the stem and branches. The affected tree shows retarded growth without any symptoms in leaves and fruits.

Stone Fruit Disorders

The diseased sensitive plum cvs. "Taiyo", "Oishi-wase sumomo", "Kiyo", and "Shiho" in Japan and Korea, and "Xiangshanhong" in China, produce fruits with irregular reddish blotches on the pericarp (Fig. 19.2). "Florentia" and "Sorriso di Primavera" in Italy produce wine-red

FIGURE 19.2 Plum dapple fruit disease (cv. "Shiho"). Dapple fruit with irregular reddish blotches on the pericarp (left) compared with healthy (right) plum cv. "Shiho".

FIGURE 19.3 Peach dapple fruit disease (cv. "Hakuhou"). Chlorotic blotches on the pericarp of dapple fruit of peach cv. "Hakuhou".

dappling and irregular reddish lines, respectively. The plum cv. "Soldam" in Japan produces fruits with yellowish red-colored flesh accompanied by a polished appearance of pericarp due to poor formation of the wax layer. Fruit maturation is delayed by 7–10 days. The sensitive peach cvs. "Asama-Hakutou", "Hakuhou", and "Hikawa-Hakuhou" in Japan produce chlorotic blotches or crinkling on the pericarp of mature fruits, respectively (Fig. 19.3). "*Degeneración*" of apricot is characterized by changes in their external appearance involving rugosity and a loss of organoleptic characteristics (Amari et al., 2007).

Cucumber Pale Fruit

Infected cucumber develops stunting of the whole plant, leaf curling, flower deformation, and pale colored fruits.

Others

Vein clearing, yellow speckle, and leaf deformation of mulberry in Iran (Mazhar et al., 2014), and severe reduction in plant growth and upward curling and deformation of leaves on hibiscus in Italy (Luigi et al., 2013), are suspected to be incited by HSVd infection.

HOST RANGE

HSVd has a wide natural host range which includes almond (*Prunus dulcis*), apple (*Malus domestica*), wild apple (*M. sylvestris*), apricot (*P. armeniaca*), *Citrus* spp., cucumber (*Cucumis sativus*), fig (*Ficus carica*), grapevine (*Vitis vinifera*), hibiscus (*Hibiscus rosa-sinensis*), hop, jujube

(*Ziziphus jujuba*), mulberry (*Morus alba*), peach (*P. persica*), pear (*Pyrus communis*), pistachio (*Pistacia vera*), plum (*P. domestica*), sweet cherry (*P. avium*), wild cherry (*P. undulata*), and pomegranate (*Punica granatum*). HSVd sequences from Korla fragrant pear (*Pyrus × sinkiangensis*) in China (KM278991), *Ixeridium dentatum* in South Korea (KT725429), and quince (*Cydonia oblonga*) in Iran (KP126943) are deposited in nucleotide sequence databases.

Experimental host range includes several species in the *Cucurbitaceae*, *Moraceae*, and *Solanaceae*, such as cucumber, melon (*C. melo*), winter melon (*Benincasa hispida*), *H. japonicus*, *H. lupulus* var. *cordifolius*, tomato (*Solanum lycopersicum*), *Nicotiana tabacum*, and *N. benthamiana*.

TRANSMISSION

The primary mode of transmission is through mechanical means in the case of hop stunt epidemics. Once HSVd becomes established in a hop planting, it is easily transmitted by workers, cutting tools, and equipment during cultural activities like trimming, thinning, and mechanical leaf stripping. In fruit trees, HSVd can be transmitted by grafting infected scions. HSVd is poorly transmitted through seeds, although this way of transmission may play a role for survival of the viroid in certain hosts such as grapevine (Wan Chow Wah and Symons, 1999).

DETECTION

The sensitive cucumber cv. "Suyo" was used as an indicator plant. Mechanical inoculation to "Suyo" by sap or nucleic acids extracts leads to severely stunted plants displaying leaf epinasty and curling when grown at 25–30°C for 3–4 weeks.

Detection and identification of HSVd has been performed by: (1) several formats of polyacryamide gel electrophoresis (PAGE), such as return-PAGE (Li et al., 1995), two-dimensional-PAGE, and sequential-PAGE; (2) dot-blot and RNA-gel blot hybridization; and (3) RT-PCR (Yang et al., 1992).

GEOGRAPHIC DISTRIBUTION AND EPIDEMIOLOGY

HSVd distributes unevenly in fruit trees. HSVd-grapevine, a major variant of 297 nt in length (accession E01844), occurs widely in the world, but distinct variants have been reported from German cvs.

"Riesling" and "5BB" (Puchta et al., 1989). HSVd-citrus is also distributed worldwide; the noncachexia-inducing type is more common but the cachexia-inducing type is also distributed widely. More details are summarized in Table 19.1.

CONTROL

The best strategy to manage HSVd-related diseases is to use HSVd-free planting or grafting material. Because of the long latency of infection, removal of symptomatic plants may allow infected plants to remain in the hop yard or fruit tree orchard that then serve as sources of inoculum. Plants adjacent to those infected should also be removed, paying special attention in the case of hops to remove as many roots as possible in preparation for replanting. Field operations should be performed in infected yards after those in HSVd-free yards or orchards. Exposure of equipment and tools for 10 minutes to a solution containing 2% formaldehyde and 2% sodium hydroxide is effective for disinfection but corrosive and potentially hazardous for workers.

References

Amari, K., Gómez, G., Myrta, A., Di Terlizzi, B., Pallás, V., 2001. The molecular characterization of 16 new sequence variants of hop stunt viroid reveals the existence of invariable regions and a conserved hammerhead-like structure on the viroid molecule. J. Gen. Virol. 82, 953–962.

Amari, K., Ruiz, D., Gómez, G., Sánchez-Pina, M.A., Pallás, V., Egea, J., 2007. An important new apricot disease in Spain is associated with hop stunt viroid infection. Eur. J. Plant Pathol. 118, 173–181.

Astruc, N., Marcos, J.F., Macquaire, G., Candresse, T., Pallás, V., 1996. Studies on the diagnosis of hop stunt viroid in fruit trees: identification of new hosts and application of a nucleic acid extraction procedure based on non-organic solvents. Eur. J. Plant Pathol. 102, 837–846.

Bagherian, S.A.A., Izadpanah, K., 2010. Two novel variants of hop stunt viroid associated with yellow corky vein disease of sweet orange and split bark disorder of sweet lime. 21st International Conference on Virus and Other Graft Transmissible Diseases of Fruit Crops, 427. Julius-Kühn-Archiv, pp. 105–113.

Cañizares, M.C., Marcos, J.F., Pallás, V., 1999. Molecular characterization of an almond isolate of hop stunt viroid (HSVd) and conditions for eliminating spurious hybridization in its diagnosis in almond samples. Eur. J. Plant Pathol. 105, 553–558.

Cho, I.S., Chung, B.N., Cho, J.D., Choi, S.K., Choi, G.S., Kim, J.S., 2011. Hop stunt viroid (HSVd) sequence variants from dapple fruits of plum (*Prunus salicina* L.) in Korea. Res. Plant Dis. 17, 358.

Duran-Vila, N., Roistacher, C., Rivera-Bustamante, R., Semancik, J.S., 1988. A definition of citrus viroid groups and their relationship to the exocortis disease. J. Gen. Virol. 69, 3069–3080.

Eastwell, K., Sano, T., 2009. Hop Stunt Disease (HSVd). Compendium of Hop Diseases, Arthropod Pests and Other Disorders. APS Press, pp. 48–51.

Eastwell, K.C., Nelson, M.E., 2007. Occurrence of viroids in commercial hop (*Humulus lupulus* L.) production areas of Washington State. Plant Health Prog. Available from: http://dx.doi.org/10.1094/PHP-2007-1127-01-RS.

El Maghraby, I., Matićc, S., Fahmy, H., Myrta, A., 2007. Viruses and viroids of stone fruit in Egypt. J. Plant Pathol. 89, 427–430.

Elbeaino, T., Abou Kubaa, R., Ismaeil, F., Mando, J., Digiaro, M., 2012a. Viruses and hop stunt viroid of fig trees in Syria. J. Plant Pathol. 94, 687–691.

Elbeaino, T., Choueiri, E., Digiaro, M., 2013. First report of hop stunt viroid in Lebanese fig trees. J. Plant Pathol. 95, 218.

Elbeaino, T., Kubaa, R.A., Choueiri, E., Michele Digiaro, M., Navarro, B., 2012b. Occurrence of hop stunt viroid in mulberry (*Morus alba*) in Lebanon and Italy. J. Phytopathol. 160, 48–51.

Elleuch, A., Hamdi, I., Ellouze, O., Ghrab, M., Fkahfakh, H., Drira, N., 2013. Pistachio (*Pistacia vera* L.) is a new natural host of hop stunt viroid. Virus Genes 47, 330–337.

Flores, R., Randles, J.W., Bar-Joseph, M., Diener, T.O., 1998. A proposed scheme for viroid classification and nomenclature. Arch. Virol. 143, 623–629.

Gorsane, F., Elleuch, A., Hamdi, I., Salhi-hannachi, A., Fakhfakh, H., 2010. Molecular detection and characterization of hop stunt viroid sequence variants from naturally infected pomegranate (*Punica granatum* L.) in Tunisia. Phytopathol. Mediterr. 49, 152–162.

Guo, L., Liu, S., Wu, Z., Mu, L., Xiang, B., Li, S., 2008. Hop stunt viroid (HSVd) newly reported from hop in Xinjiang, China. Plant Pathol. 57, 764.

Hassen, I.F., Kummert, J., Marbot, S., Fakhfakh, H., Marrakchi, M., Jijakli, M.H., 2004. First report of pear blister canker viroid, peach latent mosaic viroid, and hop stunt viroid infecting fruit trees in Tunisia. Plant Dis. 88, 1164.

Jakse, J., Radisek, S., Pokorn, T., Matousek, J., Javornik, B., 2015. Deep-sequencing revealed citrus bark cracking viroid (CBCVd) as a highly aggressive pathogen on hop. Plant Pathol. 64, 831–842.

Kaponi, M., Luigi, M., Kyriakopoulou, P.E., 2012. Mixed infections of pome and stone fruit viroids in cultivated and wild trees in Greece. New Dis. Rep. 26, 8.

Kawaguchi-Ito, Y., Li, S.F., Tagawa, M., Araki, H., Goshono, M., Yamamoto, S., et al., 2009. Cultivated grapevines represent a symptomless reservoir for the transmission of hop stunt viroid to hop crops: 15 years of evolutionary analysis. PLoS One 4, e8386. Available from: http://dx.doi.org/10.1371/journal.pone.0008386.

Kofalvi, S.A., Marcos, J.F., Cañizares, M.C., Pallás, V., Candresse, T., 1997. Hop stunt viroid (HSVd) sequence variants from *Prunus* species: evidence for recombination between HSV isolates. J. Gen. Virol. 78, 3177–3186.

Lemmetty, A., Werkman, A.W., Soukainen, M., 2011. First report of hop stunt viroid in greenhouse cucumber in Finland. Plant Dis. 95, 615.

Li, S., Onodera, S., Sano, T., Yoshida, K., Wang, G., Shikata, E., 1995. Gene diagnosis of viroids: comparisons of return-PAGE and hybridization using DIG-labeled DNA and RNA probes for practical diagnosis of hop stunt, citrus exocortis and apple scar skin viroids in their natural host plants. Ann. Phytopathol. Soc. Jpn. 61, 381–390.

Luigi, M., Manglli, A., Tomassoli, L., Faggioli, F., 2013. First report of hop stunt viroid in *Hibiscus rosa-sinensis* in Italy. New Dis. Rep. 27, 14.

Mandic, B., Rwahnih, M.A., Myrta, A., Gómez, G., Pallás, V., 2008. Incidence and genetic diversity of peach latent mosaic viroid and hop stunt viroid in stone fruits in Serbia. Eur. J. Plant Pathol. 120, 167–176.

Mazhar, M.A., Bagherian, S.A.A., Izadpanah, K., 2014. Variants of hop stunt viroid associated with mulberry vein clearing in Iran. J. Phytopathol. 162, 269–271.

Önelge, N., Cinar, A., Szychowski, J.A., Vidalakis, G., Semancik, J.S., 2004. Citrus viroid II variants associated with "Gummy Bark" disease. Eur. J. Plant Pathol. 110, 1047–1052.

Palacio-Bielsa, A., Romero-Durbán, J., Duran-Vila, N., 2004. Characterization of citrus HSVd isolates. Arch. Virol. 149, 537–552.

Puchta, H., Ramm, K., Luckinger, R., Freimüller, K., Sänger, H.L., 1989. Nucleotide sequence of a hop stunt viroid (HSVd) isolate from the German grapevine rootstock 5BB as determined by PCR-mediated sequence analysis. Nucleic Acids Res. 17, 5841.

Radisek, S., Majer, A., Jakse, A., Javornik, B., Matoušek, J., 2012. First report of hop stunt viroid infecting hop in Slovenia. Plant Dis. 96, 592.

Ragozzino, E., Faggioli, F., Amatruda, G., Barba, M., 2002. Occurrence of dapple fruit of plum in Italy. Phytopathol. Mediterr. 41, 72–75.

Reanwarakorn, K., Semancik, J.S., 1999. Correlation of hop stunt viroid variants to cachexia and xyloporosis diseases of citrus. Phytopathology 89, 568–574.

Roistacher, C.N., Gumpf, D.J., Nauer, E.M., Gonzales, R., 1983. Cachexia disease: virus or viroid. Citrograph 68, 111–113.

Roy, A., Ramachandran, P., 2003. Occurrence of a hop stunt viroid (HSVd) variant in yellow corky vein disease of citrus in India. Curr. Sci. India 85, 1608–1612.

Sano, T., 2003. Hop stunt viroid. In: Hadidi, A., Flores, R., Randles, J.W., Semancik, J.S. (Eds.), Viroids. CSIRO Publishing, Collingwood, VIC, pp. 207–212.

Sano, T., 2013. History, origin, and diversity of hop stunt disease and hop stunt viroid. Acta Hortic. 1010, 87–96.

Sano, T., Hataya, T., Shikata, E., 1988. Complete nucleotide sequence of a viroid isolated from Etrog citron, a new member of hop stunt viroid group. Nucleic Acids Res. 16, 347.

Sano, T., Hataya, T., Terai, Y., Shikata, E., 1989. Hop stunt viroid strains from dapple fruit disease of plum and peach in Japan. J. Gen. Virol. 70, 1311–1319.

Sano, T., Ohshima, K., Hataya, T., Uyeda, I., Shikata, E., Chou, T.G., et al., 1986. A viroid resembling hop stunt viroid in grapevines from Europe, the United States and Japan. J. Gen. Virol. 67, 1673–1678.

Sano, T., Uyeda, I., Shikata, E., Ohno, T., Okada, Y., 1984. Nucleotide sequence of cucumber pale fruit viroid: homology to hop stunt viroid. Nucleic Acids Res. 12, 3427–3434.

Sasaki, M., Shikata, E., 1977. On some properties of hop stunt disease agent, a viroid. Proc. Japan Acad. 53B, 109–112.

Seigner, L., Lutz, A., Seigner, E., 2013. Monitoring of hop stunt viroid and dangerous viruses in German hop gardens. In Abstracts of International Hop Growers. Convention, Kiev, Ukraine, p. 60.

Semancik, J.S., Roistacher, C.N., Rivera-Bustamante, R., Duran-Vila, N., 1988. Citrus cachexia viroid, a new viroid of citrus: relationship to viroids of the exocortis disease complex. J. Gen. Virol. 69, 3059–3068.

Shikata, E., 1990. New viroids from Japan. Semin. Virol. 1, 107–115.

Sofy, A.R., Soliman, A.M., Mousa, A.A., Ghazal, S.A., El-Dougdoug, K.A., 2010. First record of citrus viroid II (CVd-II) associated with gummy bark disease in sweet orange (*Citrus sinensis*) in Egypt. New Dis. Rep. 21, 24.

Van Dorst, H.J.M., Peters, D., 1974. Some biological observations on pale fruit, a viroid-incited disease of cucumber. Neth. J. Plant Pathol. 80, 85–96.

Wan Chow Wah, Y.F., Symons, R.H., 1999. Transmission of viroids via grape seeds. J. Phytopathol. 147, 285–291.

Yakoubi, S., Elleuch, A., Besaies, N., Marrakchi, M., Fakhfakh, H., 2007. First report of hop stunt viroid and citrus exocortis viroid on fig with symptoms of fig mosaic disease. J. Phytopathol. 155, 125–128.

Yamamoto, H., Kagami, Y., Kurokawa, M., Nishimura, S., Kubo, S., Inoue, M., et al., 1970. Studies on hop stunt disease I. Mem. Fac. Agr. Hokkaido Univ. 7, 491–512.

Yang, X., Hadidi, A., Garnsey, S.M., 1992. Enzymatic cDNA amplification of citrus exocortis and cachexia viroids from infected citrus hosts. Phytopathology 82, 279–285.

Zhang, B., Liu, G.Y., Liu, C., Wu, Z., Jiang, D., Li, S., 2009. Characterization of hop stunt viroid (HSVd) isolates from jujube trees (*Ziziphus jujuba*). Eur. J. Plant Pathol. 125, 665–669.

Zhang, Z., Zhou, Y., Guo, R., Mu, L., Yang, Y., Li, S., 2012. Molecular characterization of Chinese hop stunt viroid isolates reveals a new phylogenetic group and possible cross transmission between grapevine and stone fruits. Eur. J. Plant Pathol. 134, 217–225.

Zhou, Y., Guo, R., Cheng, Z., Sano, T., Li, S.F., 2006. First report of hop stunt viroid from *Prunus persica* with dapple fruit symptoms in China. Plant Pathol. 55, 564.

CHAPTER

20

Dahlia Latent Viroid

Jacobus Th.J. Verhoeven[1], *Ellis T.M. Meekes*[2], *Johanna W. Roenhorst*[1], *Ricardo Flores*[3] *and Pedro Serra*[3]

[1]National Plant Protection Organization, Wageningen, The Netherlands [2]Naktuinbouw, Roelofarendsveen, The Netherlands [3]Polytechnic University of Valencia-CSIC, Valencia, Spain

INTRODUCTION

Three viroids have been reported in dahlia in the last decade. Chrysanthemum stunt viroid and potato spindle tuber viroid were reported by Nakashima et al. (2007) and Tsushima et al. (2011), respectively. The third viroid was dahlia latent viroid (DLVd) (Verhoeven et al., 2013). This viroid was first detected by return polyacrylamide gel electrophoresis in pot-grown dahlias (*Dahlia* spp.) without discernible symptoms during a routine inspection of ornamental plants in The Netherlands in 2010.

TAXONOMIC POSITION AND NUCLEOTIDE SEQUENCE

The reference sequence variant of DLVd (NCBI GenBank accession JX263426), consists of 342 nt: 72A (21.1%), 69 U (20.2%), 105G (30.7%), and 96C (28.1%), making a G + C content of 58.8%. Sequencing of 10 clones of DLVd revealed minimal variations (Verhoeven et al., 2013). The predicted secondary structure of minimum free energy is a rod-like conformation similar to that proposed for most members of the family *Pospiviroidae* (Di Serio et al., 2014; Keese and Symons, 1985; Sänger et al., 1976), except for a short bifurcation in the right terminal domain,

Viroids and Satellites.
DOI: http://dx.doi.org/10.1016/B978-0-12-801498-1.00020-6

wherein 67.8% of the nucleotides are paired (65.5% G:C, 26.7% A:U, and 7.8% G:U pairs) (Fig. 20.1A).

DLVd has less than 56% sequence identity with other viroids. The viroid displays characteristic features of members the family *Pospiviroidae* that include a predicted rod-like secondary structure of minimum free-energy with a central conserved region identical to that of hop stunt viroid (HSVd) in the genus *Hostuviroid* (Ohno et al., 1983). Furthermore, DLVd has the terminal conserved region present in members of the genus *Pospiviroid* (Fig. 20.1A) and lacks the terminal conserved hairpin present in HSVd. Similar structural features have been reported for Columnea latent viroid (Hammond et al., 1989). Consequently, both viroids pose particular questions regarding classification at the genus level; these questions most likely derive from the recombinant nature of two viroids, as is shown for DLVd in Fig. 20.2. Finally, DLVd can form hairpins I and II that are typical of members of the genus *Pospiviroid*.

Phylogenetic reconstructions (Fig. 20.3) do not solve the classification problem because they indicate that DLVd is most closely related to a viroid species from each genus; i.e., HSVd and pepper chat fruit viroid. So far, the host range of DLVd is restricted to dahlia and, as such, the viroid differs from HSVd and pepper chat fruit viroid biologically. Hence, DLVd fulfills molecular and biological criteria to be considered a novel species in the family *Pospiviroidae*. Allocation of the viroid to a genus entails difficulties but it has been assigned to the genus

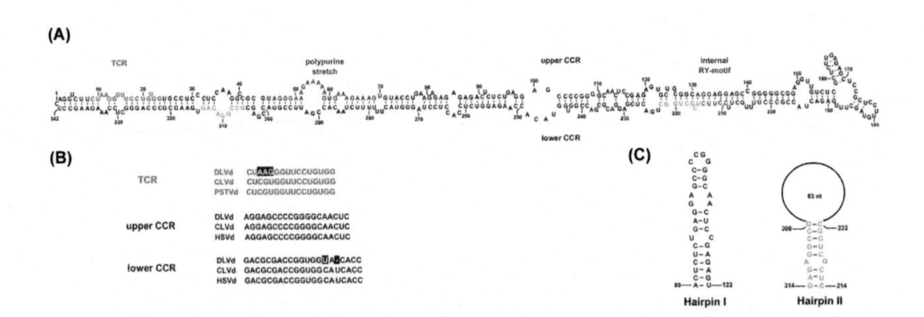

FIGURE 20.1 (A) Primary and minimum free energy secondary structure predicted for DLVd. Nucleotides forming the terminal conserved region (TCR), polypurine stretch, upper and lower central conserved region (CCR) strands, hairpins I and II, and the core of the RY motif are highlighted with different colors. (B) Comparison between the TCR and the upper and lower CCR strands of DLVd with other representative members of the family *Pospiviroidae*. Minor differences in DLVd are highlighted with a black background. (C) Schematic representation of hairpins I and II that DLVd can potentially form. *Source: From Verhoeven, J.Th.J., Meekes, E.T.M., Roenhorst, J.W., Flores, R., Serra, P., 2013. Dahlia latent viroid: a recombinant new species of the family Pospiviroidae posing intriguing questions about its origin and classification. J. Gen. Virol. 94, 711–719, with permission.*

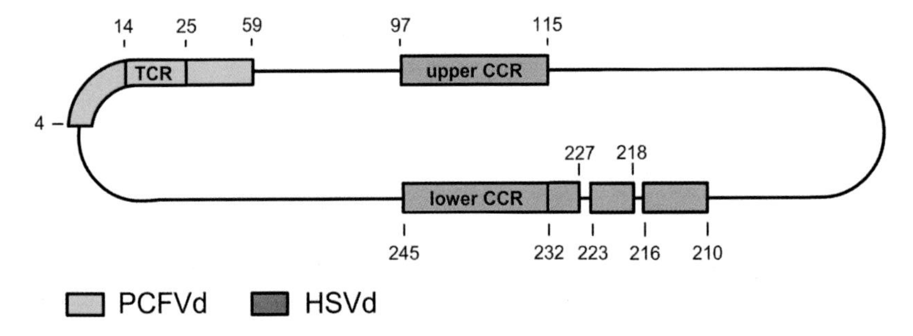

FIGURE 20.2 Schematic representation of DLVd illustrating regions of the sequence that are almost identical to hop stunt viroid and pepper fruit chat viroid. *Source: From Verhoeven, J.Th.J., Meekes, E.T.M., Roenhorst, J.W., Flores, R., Serra, P., 2013. Dahlia latent viroid: a recombinant new species of the family Pospiviroidae posing intriguing questions about its origin and classification. J. Gen. Virol. 94, 711–719, with permission.*

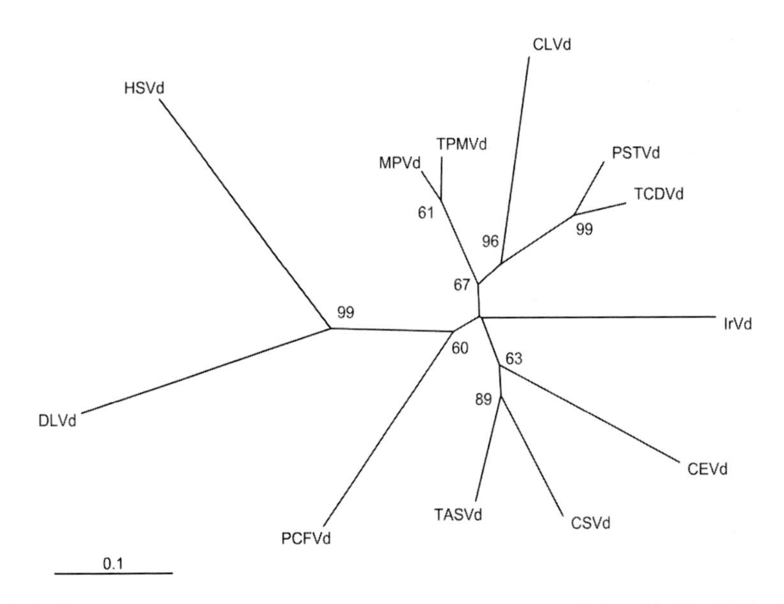

FIGURE 20.3 Evolutionary relationships of dahlia latent viroid (DLVd, JX263426) with hop stunt viroid (HSVd, EU872277) and all known pospiviroids: chrysanthemum stunt viroid (CSVd, X16408), citrus exocortis viroid (CEVd, M34917), columnea latent viroid (CLVd, X15663), iresine viroid 1 (IrVd-1, X95734), Mexican papita viroid (MPVd, L78454), pepper chat fruit viroid (PCFVd, FJ409044), potato spindle tuber viroid (PSTVd, V01465), tomato apical stunt viroid (TASVd, NC001553), tomato chlorotic dwarf viroid (TCDVd, AF162131), and tomato planta macho viroid (TPMVd, K00817). The phylogenetic reconstruction was performed with the neighbor-joining method (Saitou and Nei, 1987). The fraction (%) of 5000 replicate trees in which the groups clustered together in the bootstrap test (Felsenstein, 1985) is shown next to the branches. Branch lengths, which were estimated with the Maximum Composite Likelihood method (Tamura et al., 2004) and expressed in base substitutions per site, are proportional to the evolutionary distances used to infer the phylogenetic tree. Phylogenetic analyses were conducted with MEGA4 (Tamura et al., 2007). *Source: From Verhoeven, J.Th.J., Meekes, E.T.M., Roenhorst, J.W., Flores, R., Serra, P., 2013. Dahlia latent viroid: a recombinant new species of the family Pospiviroidae posing intriguing questions about its origin and classification. J. Gen. Virol. 94, 711–719, with permission.*

Hostuviroid based on the most important demarcating criterion established for this purpose; i.e., a common central conserved region with the type species of this genus.

SYMPTOMS, HOST RANGE, AND ECONOMIC SIGNIFICANCE

Dahlia is the only known host plant of DLVd. In both naturally and artificially infected plants, however, no symptoms have been observed. Moreover, agroinoculation and mechanical sap inoculation failed to transmit the viroid to chrysanthemum, potato, tomato, or *Nicotiana benthamiana*, suggesting that it has a narrow host range (Verhoeven et al., 2013). In addition, Tsushima et al. (2015) failed to transmit DLVd to *Chrysanthemum pacificum*, *Tithonia rotundifolia*, *Helianthus annuus*, and *Glebionis coronary*, and coinoculations of DLVd and PSTVd to dahlia and wild hop (*Humulus lupulus* var. *cordifolius*) were only successful for dahlia. Therefore, the economic significance of DLVd is low.

TRANSMISSION AND CONTROL

DLVd is transmitted by vegetative propagation of infected plants and by mechanical inoculation. Both ways of transmission, in combination with the asymptomatic appearance of the viroid, have probably contributed to the widespread occurrence of DLVd in dahlia in The Netherlands.

No control strategies have been implemented, since no symptoms have been observed and the economic significance is estimated to be low.

DETECTION

DLVd can be detected by return polyacylamide gel electrophoresis (Roenhorst et al., 2000; Schumacher et al., 1986). Furthermore, Verhoeven et al. (2013) developed a new set of primers (DLVd-P1 and DLVd-P2) for amplification of the complete viroid genome in RT-PCR. For further identification of the viroid, the PCR product may be sequenced, either directly or after cloning. The viroid is neither detected by RT-PCRs with generic primers for pospiviroids as reported by Verhoeven et al. (2004) and Botermans et al. (2013), nor with specific primers for HSVd (Osaki et al., 1998).

GEOGRAPHICAL DISTRIBUTION AND EPIDEMIOLOGY

DLVd has been reported in dahlia plants in The Netherlands and Japan. The viroid was widespread in both field- and pot-grown plants. However, occurrence of the viroid in dahlia plants in more countries is assumed for the following reasons: (1) infections in dahlia are symptomless, which allows unperceived spread of the viroid and (2) The Netherlands is a main producer and exporter of dahlia plants.

Acknowledgments

Research in R.F. laboratory is currently funded by grant BFU2014–56812-P from the Spanish Ministerio de Economía y Competitividad (MINECO).

References

Botermans, M., Van de Vossenberg, B.T.L.H., Verhoeven, J.Th.J., Roenhorst, J.W., et al., 2013. Development and validation of a real-time RT-PCR assay for generic detection of pospiviroids. J. Virol. Meth. 187, 43–50.

Di Serio, F., Flores, R., Verhoeven, J.Th.J., Li, S.-F., Pallás, V., et al., 2014. Current status of viroid taxonomy. Arch. Virol. 159, 3467–3478.

Felsenstein, J., 1985. Confidence limits on phylogenies: an approach using the boot-strap. Evolution 39, 783–791.

Hammond, R., Smith, D.R., Diener, T.O., 1989. Nucleotide sequence and proposed secondary structure of columnea latent viroid: a natural mosaic of viroid sequences. Nucleic Acids Res. 17, 10083–10094.

Keese, P., Symons, R.H., 1985. Domains in viroids: evidence of intermolecular RNA rearrangements and their contribution to viroid evolution. Proc. Natl. Acad. Sci. USA 82, 4582–4586.

Nakashima, A., Hosokawa, M., Maeda, S., Yazawa, S., 2007. Natural infection of chrysanthemum stunt viroid in dahlia plants. J. Gen. Plant Pathol. 73, 225–227.

Ohno, T., Takamatsu, N., Meshi, T., Okada, Y., 1983. Hop stunt viroid: molecular cloning and nucleotide sequence of the complete cDNA copy. Nucleic Acids Res. 11, 6185–6197.

Osaki, H., Ohtsu, Y., Kudo, A., 1998. Two rapid extraction methods to detect apple scar skin and hop stunt viroids by RT-PCR. Acta Hortic. 472, 603–611.

Roenhorst, J.W., Butôt, R.P.T., Van der Heijden, K.A., Hooftman, M., Van Zaayen, A., 2000. Detection of chrysanthemum stunt viroid and potato spindle tuber viroid by return-polyacrylamide gel electrophoresis. Bull OEPP/EPPO Bull. 30, 453–456.

Saitou, N., Nei, M., 1987. The neighbor-joining method: a new method for reconstructing phylogenetic trees. Mol. Biol. Evol. 4, 406–425.

Sänger, H.L., Klotz, G., Riesner, D., Gross, H.J., Kleinschmidt, A., 1976. Viroids are single-stranded covalently-closed circular RNA molecules existing as highly base-paired rod-like-structures. Proc. Natl. Acad. Sci. USA 73, 3852–3856.

Schumacher, J., Meyer, N., Riesner, D., Weidemann, H.L., 1986. Diagnostic procedure for detection of viroids and viruses with circular RNAs by return-gel electrophoresis. J. Phytopathol. 115, 332–343.

Tamura, K., Dudley, J., Nei, M., Kumar, S., 2007. MEGA4: molecular evolutionary genetics analysis (MEGA) software version 4.0. Mol. Biol. Evol. 24, 1596–1599.

Tamura, K., Nei, M., Kumar, S., 2004. Prospects for inferring very large phylogenies by using the neighbor-joining method. Proc. Natl. Acad. Sci. USA 101, 11030–11035.

Tsushima, T., Matsushita, Y., Fuji, S., Sano, T., 2015. First report of dahlia latent viroid and potato spindle tuber viroid mixed-infection in commercial ornamental dahlia in Japan. New Dis. Rep. 31, 11.

Tsushima, T., Murakami, S., Ito, H., He, Y.-H., Raj, A.P.C., Sano, T., 2011. Molecular characterization of potato spindle tuber viroid in dahlia. J. Gen. Plant Pathol. 77, 253–256.

Verhoeven, J.Th.J., Jansen, C.C.C., Willemen, T.M., Kox, L.F.F., Owens, R.A., et al., 2004. Natural infections of tomato by citrus exocortis viroid, columnea latent viroid, potato spindle tuber viroid and tomato chlorotic dwarf viroid. Eur. J. Plant. Pathol. 110, 823–831.

Verhoeven, J.Th.J., Meekes, E.T.M., Roenhorst, J.W., Flores, R., Serra, P., 2013. Dahlia latent viroid: a recombinant new species of the family *Pospiviroidae* posing intriguing questions about its origin and classification. J. Gen. Virol. 94, 711–719.

CHAPTER

21

Apple Scar Skin Viroid

Ahmed Hadidi[1], *Marina Barba*[2], *Ni Hong*[3] *and Vipin Hallan*[4]

[1]U.S. Department of Agriculture, Beltsville, MD, United States [2]CREA-Research Centre for Plant Protection and Certification, Rome, Italy [3]Huazhong Agricultural University, Wuhan, China [4]CSIR-Institute of Himalayan Bioresource Technology, Palampur, Himachal Pradesh, India

INTRODUCTION

Apple scar skin disease was first reported from China in the 1930s (Ohtsuka, 1935, 1938), then from Japan in 1953 (Ushirozawa et al., 1968). The viroid etiology of the disease was suggested in the 1980s (Koganezawa, 1985, 1986) and its causal agent, apple scar skin viroid (ASSVd) from Japan and China, was purified, cloned, and sequenced (Hashimoto and Koganezawa, 1987; Puchta et al., 1990).

TAXONOMIC POSITION AND NUCLEOTIDE SEQUENCE

Apple scar skin viroid (Hashimoto and Koganezawa, 1987) is the type species of the genus *Apscaviroid*, family *Pospiviroidae* (Owens et al., 2012; Chapter 13: Viroid Taxonomy). Table 21.1 describes nucleotide sequence information of known ASSVd variants that infect apple, pear, apricot, peach, sweet cherry, and Himalayan wild cherry. Most deletions, insertions, and changes among variants of the viroid occur in regions corresponding to the pathogenicity and left terminal domains of ASSVd (i.e., Puchta et al., 1990; Shamloul et al., 2004; Yang et al., 1992; Zhu et al., 1995).

Viroids and Satellites.
DOI: http://dx.doi.org/10.1016/B978-0-12-801498-1.00021-8

TABLE 21.1 Characteristics, Size, and Sequence of ASSVd Variants From Naturally Infected Pome and Stone Fruit Trees

Variant	Size (nt)	Remarks	Reference
Scar skin from Japan	330	From apple (reference variant)	Hashimoto and Koganezawa (1987)
Scar skin from China	329	From apple; differs from apple reference variant (Hashimoto and Koganezawa, 1987) at 4 nt positions	Puchta et al. (1990)
Scar skin from China	330	From apple; differs from another Chinese variant (Puchta et al., 1990) in one base insertion; differs from the apple reference variant at 3 nt positions	Yang et al. (1992)
Dapple apple from Canada	331	Shares 97% sequence identity with apple reference variant	Zhu et al. (1995)
Dapple apple from South Korea	331	Shares over 99% sequence identity with the apple reference variant; differs at one site, and 1 nt is inserted	Lee et al. (2001)
Dapple apple from India	330 or 331	Variants share 94%–100% sequence identity with each other. Two variants are identical to a Chinese variant; six variants share 99% sequence identity with a Korean variant. Two variants are similar to a Chinese and a Japanese variants	Walia et al. (2009)
Symptomless pear from China	330	Differs from apple reference variant at 4 nt positions	Zhu et al. (1995)
Pear rusty skin from China	334	Shares 92% sequence identity with apple reference variant and differs at 25 nt positions	Zhu et al. (1995)
Pear fruit dimple from Japan	330	Differs from apple reference variant at 3 nt positions	Osaki et al. (1996)
Pear fruit crinkle from China	333	Shares 95% sequence identity with apple reference variant and differs at 19 nt positions. Shares 92% sequence identity with pear rusty skin reference variant and differs at 27 nt positions. Differs from pear fruit crinkle reference variant at 6 nt sites	Shamloul et al. (2004)

(Continued)

TABLE 21.1 (Continued)

Variant	Size (nt)	Remarks	Reference
Variants from apple and pear, from Iran	329–334	Variants form a cluster distinct from other variants	Yazarlou et al. (2012)
Variants from apple, pear, apricot and peach from China	315–330 (apple) 250–327 (pear) 330 (apricot) 330–332 (peach)	From Xinjiang Province; variants from apple, pear, peach, and apricot share 98%, 89%, 98%, 98% sequence identity, respectively, with apple scar skin variant from China (Puchta et al., 1990)	Wang et al. (2012)
Variants from sweet cherry from Greece	327–340	Differ from apple reference variant at up to 29 positions	Kaponi et al. (2010, 2013)
Variants from Himalayan wild cherry from India	330	Variants share 91%–98% sequence identity with sweet cherry variants from Greece and differ at 10–30 positions; differ from the apple reference variant at up to 18 positions	Walia et al. (2012)

ECONOMIC SIGNIFICANCE

Apple scar skin disease in apple caused severe losses in China in the 1950s, especially in the Liadong peninsula area and several provinces (Liu et al., 1957), where thousands of trees were affected and apple fruits were rendered unmarketable. Recently, the disease has become problematic for the production of cultivar "Red Fuji" in China, which became the main apple cultivar after it was introduced from Japan (N. Hong, unpublished). In Japan in the 1970s, the disease was observed in major apple growing areas, however, it was sporadic; diseased fruits lost their market value (Koganezawa et al., 2003). The disease was also reported from several regions in South Korea (Lee et al., 2001; Kim et al., 2010), as well as from the Indian State of Himachal Pradesh (Walia et al., 2009). Affected fruits are significantly downgraded.

Many pear trees in China and Japan are infected latently with ASSVd (Hadidi and Barba, 2011). These trees may be the source of viroid infection to apple trees that eventually will develop scar skin and/or dapple symptoms (Liu et al., 1957, 1962, 1985; Ohtsuka, 1935). Also, latently infected, cultivated or wild pear trees in Greece may infect susceptible pear cultivars and induce fruit symptoms (Kyriakopoulou et al., 2003). ASSVd infection in pear trees causes pear rusty skin and pear fruit crinkle diseases in China (Chen et al., 1987; Shamloul et al., 2003, 2004; Zhu et al., 1995), pear fruit dimple disease in Japan (Osaki et al., 1996), and diseased pear in Greece (Kyriakopoulou and Hadidi, 1998; Kyriakopoulou et al., 2001, 2003). Blemished fruit are reduced in value and become unmarketable.

SYMPTOMS

The expression of apple scar skin and/or dapple apple fruit symptoms may be related to the apple cultivar (Koganezawa et al., 2003). Scar skin symptoms appear on cultivars such as "Ralls Janet," "Indo," and "Golden Delicious," while dapple symptoms generally appear on red-skinned cultivars such as "Jonathan," "Red Gold," and "Red Fuji." Both types of symptoms may appear on "Starking Delicious" and "Red Delicious." Fig. 21.1 shows dapple apple symptoms on "Red Fuji" during the fruit coloring stage. The dappling symptoms become less obvious every subsequent year while the scar skin symptoms become more

FIGURE 21.1 Dapple apple symptoms on apple cultivar "Red Fuji" fruit caused by ASSVd. *Source: Courtesy WenXing Xu.*

pronounced (Yamaguchi and Yanase, 1976). Fig. 21.2 shows scar skin symptoms on "Golden Delicious."

Leaf roll or leaf epinasty symptoms may develop, under certain conditions, on certain apple cultivars such as "Stark's Earliest," "Sugar Crab," and "Ralls Janet" (Chen et al., 1988a; Ito and Yoshida, 1993; Liu et al., 1957; Skrzeczkowski et al., 1993).

Pear rusty skin disease symptoms are restricted to the fruit of Chinese pears such as cultivar "Muoli." Similar disease symptoms were observed on fruit of the Italian cultivar "Passacrassana" grafted on quince rootstocks in central Italy (Kyriakopoulou et al., 2001). Pear fruit crinkle disease is characterized by crinkling symptoms on fruit of the Chinese pear cultivars "Xuehuali" and "Yali" in Hebei Province (Shamloul et al., 2004). Pear fruit dimple disease symptoms in Japan are restricted to fruits of the Japanese pear cultivars "Nitaka" and "Yoshino" (Ohtsu et al., 1990). Symptoms of ASSVd infection on Greek pear cultivars and wild pears are also observed on fruit and include russet, scarring, and cracking (Kyriakopoulou et al., 2001).

ASSVd-infected bean leaves develop interveinal chlorotic symptoms whether inoculated with ASSVd infectious transcripts or with viroidliferous whiteflies (Walia et al., 2015).

FIGURE 21.2 Scar skin symptoms on apple cultivar "Golden Delicious" fruit.

HOST RANGE

ASSVd naturally infects apple, wild apple (*Malus sylvestris*), pear, wild pear (*Pyrus amygdaliformis*), apricot, and sweet cherry trees (Hadidi and Barba, 2011), as well as peach (Wang et al., 2012) and Himalayan wild cherry (*Prunus cerasoides*) trees (Walia et al., 2012). Grafting experiments showed that the viroid also infects species of *Malus, Pyrus, Sorbus, Chaenomeles, Cydonia*, and *Pyronia* (Desvignes et al., 1999). Using genetic engineering approaches (see "Section Transmission") Walia et al. (2014) extended the host range of ASSVd to herbaceous hosts which include cucumber, tomato, garden pea, eggplant, *Nicotiana benthamiana, N. tabacum, N. glutinosa, Chenopodium quinoa*, and *C. amaranticolor*.

TRANSMISSION

ASSVd is transmitted by grafting to apple (Chen et al., 1988b; Kim et al., 2006) and pear seedlings (Chen et al., 1988b), by razor slashing of apple seedlings with viroid RNA (Koganezawa, 1985; Koganezawa et al., 1982), and by using (viroid-contaminated) pruning tools dipped in crude sap of infected stem to both lignified stems and green shoots of apple trees and seedlings (Kim et al., 2006). However, it was not transmitted by knife cutting to Japanese pear (Osaki et al., 1996). ASSVd may spread naturally from infected trees to uninfected neighboring trees (Desvignes et al., 1999; Ohtsuka, 1935; Ushirozawa et al., 1968). The mechanism of this transmission is unclear but it may be due to root grafting. The viroid is transmitted in Greece by grafting onto infected wild pear, traditional pear, and other pome fruit rootstocks and spreads through the use of infected propagation material (Kyriakopoulou et al., 2001).

ASSVd is seed borne in apple (Hadidi et al., 1991; Kim et al., 2006), but viroid transmission from apple seeds to seedlings was reported to be negative (Desvignes et al., 1999; Howell et al., 1995). However, Kim et al. (2006) showed that seedlings germinated from ASSVd-positive apple seeds demonstrated a 7.7% infection rate. ASSVd in oriental pear was either not seed transmitted or transmitted at a low rate (Hurtt and Podleckis, 1995).

ASSVd is transmitted by agroinfection of ASSVd recombinant constructs to apple and pear seedlings (Zhu et al., 1998) and to nine herbaceous plant species by mechanical inoculation of in vitro ASSVd dimeric transcripts and to a lesser degree by dimeric DNA plasmids or sap inoculation (Walia et al., 2014).

Recently it has been shown that ASSVd is transmitted by the greenhouse whitefly *Trialeurodes vaporariorum* to herbaceous plant species such as bean and cucumber (Walia et al., 2015). Both viroid RNA and DNA forms were identified in viruliferous insects. The viroid transfer efficiency was enhanced with the help of *Cucumis sativus* phloem protein 2 (CsPP2), which is known to translocate viroid RNAs. This PP2/ASSVd complex is stably present in the viroid infected cucumber plants (Walia et al., 2015).

GEOGRAPHICAL DISTRIBUTION AND EPIDEMIOLOGY

Fig. 21.3 shows the geographical distribution of ASSVd worldwide. Apple scar skin disease in apple is prevalent in China and Japan (Hadidi and Barba, 2011). The disease was also reported from South Korea (Kim et al., 2010), India (Behl et al., 1998; Thakur et al., 1995; Walia et al., 2009), and Iran (Yazarlou et al., 2012). It was also reported from the United States, Canada, and the United Kingdom, but its incidence is rare in these countries; it is extremely rare in Italy, France, and Germany (Hadidi and Barba, 2011). It was recently reported from Argentina (Nome et al., 2012).

Dapple apple disease was reported from the United States, Canada, the United Kingdom, China, and Japan (Hadidi and Barba, 2011), and South Korea (Kim et al., 2006, 2010; Lee et al., 2001). Many pear trees in China are latently infected with ASSVd (Liu et al., 1962; Zhu et al.,

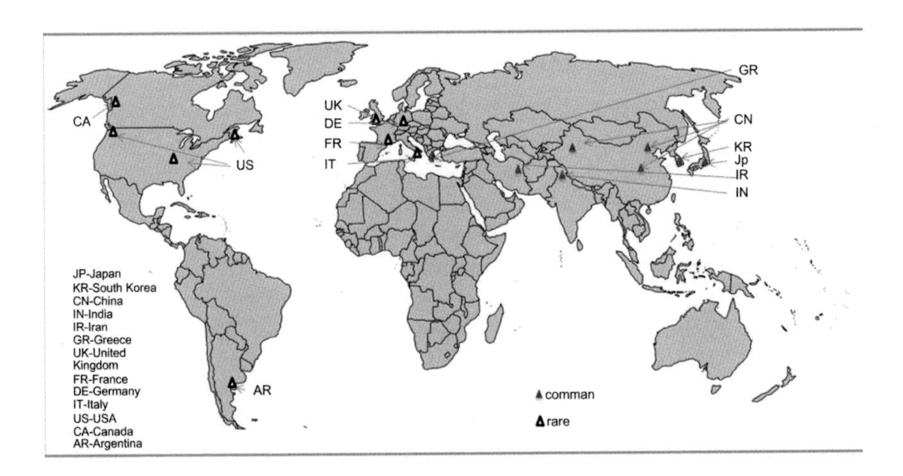

FIGURE 21.3 Geographical distribution of ASSVd worldwide.

1995). In the early 1960s nearly 90% of the pear trees were viroid-infected in China (Liu et al., 1962). ASSVd-infected pear fruits with rusty skin or fruit crinkle symptoms have been reported repeatedly from China during the last three decades (Chen et al., 1987; Shamloul et al., 2003, 2004; Zhu et al., 1995). About 14% of pear trees from several provinces in China were positive for ASSVd infection (Hong et al., 2012). The incidence of ASSVd-infected pear trees in Namyangju and Ulsan regions in South Korea were 1.2% and 1.4%, respectively (Kim et al., 2010).

Pear fruit dimple diseased trees infected with ASSVd were found in "Niitaka" and "Yoshino" cultivars in Chiba, Ibaraki, and Oita prefectures in Japan, and the viroid was thought to spread by grafting (Osaki et al., 1996). ASSVd in pear fruits affected with rusty skin-like symptoms was reported from central Italy (Kyriakopoulou et al., 2001). In Greece, several pear cultivars and wild pear trees were found infected with ASSVd (Kyriakopoulou et al., 2001).

ASSVd was reported from apricot and peach in China (Wang et al., 2012; Zhao and Niu, 2008), sweet cherry in Greece (Kaponi et al., 2010, 2013), and Himalayan wild cherry in India (Walia et al., 2012).

DETECTION

Field and greenhouse biological indexing of ASSVd from apple or pear trees has been described (Hadidi and Barba, 2011). These assays may need 2–3 years for developing fruit symptoms in the field or 2–3 months for apple indicators to develop leaf epinasty and curling in the greenhouse. Greenhouse indexing of ASSVd from pear trees can be done by biological indexing followed by molecular hybridization (Hurtt et al., 1992).

ASSVd may be detected in infected tissues by standard, return, two-dimensional, or bidirectional polyacrylamide gel electrophoretic analysis of the extracted RNAs (Hanold et al., 2011; Chapter 34: Gel Electrophoresis). Unlike biological indexing, these assays are rapid, and easy to perform. ASSVd variants may also be detected by different methods of molecular hybridization (Hadidi and Yang, 1990; Hadidi et al., 1990; Hurtt et al., 1992; Kyriakopoulou and Hadidi, 1998; Kyriakopoulou et al., 2001; Podleckis et al., 1993; Walia et al., 2014; Zhu et al., 1998; Chapter 35: Molecular Hybridization Techniques for Detecting and Studying Viroids). Different methods of RT-PCR (Hadidi and Candresse, 2003; Hadidi et al., 2011; Kim et al., 2010; Kumar et al., 2013; Chapter 36: Viroid Amplification Methods: RT-PCR, Real- Time RT-PCR, and RT-LAMP) are very sensitive for detecting ASSVd and other viroids from different plant species (Hadidi et al., 2011). ASSVd was also detected by nucleic acid sequence-based amplification with electrochemiluminescence (NASB-ECL) (Kim et al., 2006). Microarrays and next-generation sequencing have not yet been applied for detecting ASSVd, however, they have been used for detecting other viroids (Chapter 37: Detection and Identification of Viroids by Microarrays and Chapter 38: Application of Next-Generation Sequencing Technologies to Viroids).

CONTROL

Quarantine programs in many developed countries require ASSVd assays for imported pome fruit germplasm to prevent its introduction and/or spread to local susceptible plant species. Moreover, pome and stone fruit trees are among plants covered by certification schemes in many countries (Roy, 2011); thus, nursery stocks of apple, pear, and other host species must be propagated from mother trees indexed as free of ASSVd (Barba et al., 2003). ASSVd-infected trees in orchards should be removed to avoid possible spread of the viroid to neighboring trees by root grafting or other means.

The standard thermotherapy treatment (70 days exposure at 38°C) was successful in obtaining ASSVd-free apple plants (Howell et al., 1998). About 90% of plants were ASSVd-free when the thermotherapy treatment (48 days at 37°C) was preceded by keeping apple plants dormant for 3 months at 4°C in cool chambers (Desvignes et al., 1999). The viroid was successfully eliminated from infected pear by in vitro therapy and apical meristem culture (Postman and Hadidi, 1995). Dipping ASSVd-contaminated pruning tools in 2% sodium hypochlorite solution, (Clorox regular bleach production contains 6.0% solution of sodium hypochlorite), prevented viroid transmission (Kim et al., 2006).

References

Barba, M., Gumpf, D.J., Hadidi, A., 2003. Quarantine of imported germplasm. In: Hadidi, A., Flores, R., Randles, J.W., Semancik, J.S. (Eds.), Viroids. CSIRO Publishing, Collingwood, VIC, pp. 303–311.

Behl, M.K., Khurana, S.M.P., Parakh, D.B., 1998. Bumpy fruit and other viroid and viroid-like diseases of apple in HP, India. Acta Hortic. 472, 627–629.

Chen, W., Lin, L., Tien, P., Liu, F., Wang, G., Wang, H., 1988b. Grafting external healthy pear bud induces scar skin viroid in apple. Chin. J. Virol. 4, 367–370.

Chen, W., Lin, L., Yang, X., Tien, P., Liu, F., Wang, G., et al., 1988a. Infectivity of apple scar skin viroid. Acta Phytopathol. Sin. 18, 157–161.

Chen, W., Yang, X.C., Tien, P., 1987. Pear rusty skin, a viroid disease. Abstracts of the 7th International Virology Conference. Edmonton, Canada, p. 300.

Desvignes, J.C., Grasseau, N., Boyé, R., Cornaggia, D., Aparicio, F., Di Serio, F., et al., 1999. Biological properties of apple scar skin viroid: isolates, host range, different sensitivity of apple cultivars, elimination, and natural transmission. Plant Dis. 83, 768–772.

Hadidi, A., Barba, M., 2011. Apple scar skin viroid. In: Hadidi, A., Barba, M., Candresse, T., Jelkmann, W. (Eds.), Virus and Virus-Like Diseases of Pome and Stone Fruits. APS Press, St. Paul, MN, pp. 57–62.

Hadidi, A., Candresse, T., 2003. Polymerase chain reaction. In: Hadidi, A., Flores, R., Randles, J.W., Semancik, J.S. (Eds.), Viroids. CSIRO Publishing, Collingwood, VIC, pp. 115–122.

Hadidi, A., Hansen, A.J., Parish, C.L., Yang, X., 1991. Scar skin and dapple apple viroids are seed-borne and persistent in infected apple trees. Res. Virol. 142, 289–296.

Hadidi, A., Huang, C., Hammond, R.W., Hashimoto, J., 1990. Homology of the agent associated with dapple apple disease to apple scar skin viroid and molecular detection of these viroids. Phytopathology 80, 263–268.

Hadidi, A., Olmos, A., Pasquini, G., Barba, M., Martin, R.R., Shamloul, A.M., 2011. Polymerase chain reaction for detection of systemic plant pathogens. In: Hadidi, A., Barba, M., Candresse, T., Jelkmann, W. (Eds.), Virus and Virus-Like Diseases of Pome and Stone Fruits. APS Press, St. Paul, MN, pp. 341–359.

Hadidi, A., Yang, X., 1990. Detection of pome fruit viroids by enzymatic amplification. J. Virol. Methods 30, 261–270.

Hanold, D., Randles, J.W., Hadidi, A., 2011. Polyacrylamide gel electrophoresis for viroid detection. In: Hadidi, A., Barba, M., Candresse, T., Jelkmann, W. (Eds.), Virus and Virus-Like Diseases of Pome and Stone Fruits. APS Press, St. Paul, MN, pp. 327–332.

Hashimoto, J., Koganezawa, H., 1987. Nucleotide sequence and secondary structure of apple scar skin viroid. Nucleic Acids Res. 15, 7045–7051.

Hong, N., Ma, X.F., Hu, G.J., Zhu, H., Yao, B.Y., Song, Y.S., et al., 2012. Incidence of viral diseases on pear plants and the molecular characteristics of three pear viruses in China. Petria 22, 150–158.

Howell, W.E., Burgess, J., Mink, G.I., Zhang, Y.P., 1998. Elimination of apple fruit and bark deforming agents by heat therapy. Acta Hortic. 472, 641–646.

Howell, W.E., Skrzeczkowski, L.J., Wessels, T., Mink, G.I., Nunez, A., 1995. Non-transmission of apple scar skin viroid and peach latent mosaic viroid through seed. Acta Hortic. 472, 635–639.

Hurtt, S.S., Podleckis, E.V., 1995. Apple scar skin viroid is not seed transmitted or transmitted at a low rate in oriental pear. Acta Hortic. 386, 544–550.

Hurtt, S.S., Podleckis, E.V., Ibrahim, I.M., Hadidi, A., 1992. Early detection of apple scar skin group viroids from imported pear germplasm. Acta Hortic. 309, 311–318.

Ito, T., Yoshida, K., 1993. Factors for expression of leaf roll symptoms on young tree of apple cv. "Ralls Janet" caused by apple scar skin viroid. Bull. Fruit Tree Res. Stn. 25, 87–100.

Kaponi, M.S., Faggioli, F., Luigi, M., Barba, M., Sano, T., Kyriakopoulou, P.E., 2010. First report and molecular analysis of apple scar skin viroid in sweet cherry. Julius-Kühn Arch. 427, 361–365.

Kaponi, M.S., Sano, T., Kyriakopoulou, P.E., 2013. Natural infection of sweet cherry trees with apple scar skin viroid. J. Plant Pathol. 95, 429–433.

Kim, D.-H., Kim, H.-R., Heo, S., Kim, S.-H., Kim, M.-a, Shin, I.-S., et al., 2010. Occurrence of apple scar skin viroid and relative quantity analysis using real-time RT-PCR. Res. Plant Dis. 16, 247–253.

Kim, H.-R., Lee, S.-H., Lee, D.-H., Kim, J.-S., Park, J.-W., 2006. Transmission of apple scar skin viroid by grafting, using contaminated pruning equipment, and planting infected seeds. Plant Pathol. J. 22, 63–67.

Koganezawa, H., 1985. Transmission to apple seedlings of low molecular weight RNA from apple scar skin diseased trees. Ann. Phytopathol. Soc. Jpn. 51, 176–182.

Koganezawa, H., 1986. Further evidence for viroid etiology of apple scar skin and dapple diseases. Acta Hortic. 193, 29–33.

Koganezawa, H., Yanase, H., Sakuma, T., 1982. Viroid-like RNA associated with apple scar skin (or dapple apple) disease. Acta Hortic. 130, 193–197.

Koganezawa, H., Yang, X., Zhu, S.F., Hashimoto, J., Hadidi, A., 2003. Apple scar skin viroid in apple. In: Hadidi, A., Flores, R., Randles, J.W., Semancik, J.S. (Eds.), Viroids. CSIRO Publishing, Collingwood, VIC, Australia, pp. 137–141.

Kumar, S., Singh, L., Ram, R., Zaidi, A.A., Hallan, V., 2013. Simultaneous detection of major pome fruit viruses and a viroid. Indian J. Microbiol. 54, 431–440.

Kyriakopoulou, P.E., Giunchedi, L., Hadidi, A., 2001. Peach latent mosaic and pome fruit viroids in naturally infected cultivated pear *Pyrus communis* and wild pear *P. amygdaliformis*: implications on possible origin of these viroids in the Mediterranean region. J. Plant Pathol. 83, 51–62.

Kyriakopoulou, P.E., Hadidi, A., 1998. Natural infection of wild and cultivated pears with apple scar skin viroid in Greece. Acta Hortic. 472, 617–625.

Kyriakopoulou, P.E., Osaki, H., Zhu, S.F., Hadidi, A., 2003. Apple scar skin viroid in pear. In: Hadidi, A., Flores, R., Randles, J.W., Semancik, J.S. (Eds.), Viroids. CSIRO Publishing, Collingwood, VIC, Australia, pp. 142–145.

Lee, J.-H., Park, J.-K., Lee, D.-H., Uhm, J.-Y., Ghim, S.-Y., Lee, J.-Y., 2001. Occurrence of apple scar skin viroid-Korean strain (ASSVd-K) in apples cultivated in Korea. Plant Pathol. J. 17, 300–304.

Liu, F.-C., Chen, R.-F., Chen, Y.-X., 1957. Apple Scar Skin Disease. Science Press, Beijing, 43pp. (in Chinese).

Liu, F.-C., Wang, S.-Y., Chen, C., Chen, Y.-X., 1985. Research on the relationship between apple scar skin disease and pear trees. China Fruit. 1, 36–39 (in Chinese).

Liu, F.-C., Wang, S.-Y., Tian, C.-F., Chen, R.-F., 1962. Studies on apple scar skin disease (1960–1962). Ann. Rep. Fruit Inst. Chinese Acad. Agr. Sci. 1962, 71–79 (in Chinese).

Nome, C., Giagetto, A., Rossini, M., Di Feo, L., Nieto, A., 2012. First report of apple scar skin viroid (ASSVd) in apple trees in Argentina. New Dis. Repts. 25, 3.

Ohtsu, Y., Sakuma, T., Tanaka, Y., Takahashi, K., Isoda, T., Sekimoto, Y., et al., 1990. A few symptoms of "Kubomi" on fruits of Japanese pear. Ann. Phytopathol. Soc. Japan 56, 101 (Abstract, in Japanese).

Ohtsuka, Y., 1935. A new disease of apple, on the abnormality of fruit. J. Jap. Soc. Hort. Sci. 6, 44–53 (in Japanese).

Ohtsuka, Y., 1938. On Manshu-sabika-byo of apple, graft transmission and symptom variation in cultivars. J. Jap. Soc. Hort. Sci. 9, 282–286 (in Japanese).

Osaki, H., Kudo, A., Ohtsu, Y., 1996. Japanese pear fruit dimple disease caused by apple scar skin viroid (ASSVd). Ann. Phytopathol. Soc. Japan 62, 379–385.

Owens, R.A., Flores, R., Di Serio, F., Li, S.-F., Pallás, V., Randles, J.W., et al., 2012. Viroids. In: King, A.M.Q., Adams, M.J., Carstens, E.B., Lefkowitz, E.J. (Eds.), Virus Taxonomy – Ninth Report on the International Committee on Taxonomy of Viruses. Elsevier Academic Press, London, pp. 1221–1234.

Podleckis, E.V., Hammond, R.W., Hurtt, S.S., Hadidi, A., 1993. Chemiluminescent detection of potato and pome fruit viroids by digoxigenin-labeled dot blot and tissue blot hybridization. J. Virol. Methods 43, 147–158.

Postman, J.D., Hadidi, A., 1995. Elimination of apple scar skin viroid from pears by in vitro thermotherapy and apical meristem culture. Acta Hortic. 386, 536–543.

Puchta, H., Luckinger, R., Yang, X., Hadidi, A., Sänger, H.L., 1990. Nucleotide sequence and secondary structure of apple scar skin viroid (ASSVd) from China. Plant Mol. Biol. 14, 1065–1067.

Roy, A.-S., 2011. Control measures of pome and stone fruit viruses, viroids, and phytoplasmas: role of international organizations. In: Hadidi, A., Barba, M., Candresse, T.,

Jelkmann, W. (Eds.), Virus and Virus-Like Diseases of Pome and Stone Fruits. APS Press, St. Paul, MN, pp. 407–413.

Shamloul, A.M., Yang, X., Han, L., Hadidi, A., 2004. Characterization of a new variant of apple scar skin viroid associated with pear fruit crinkle disease. J. Plant. Pathol. 86, 249–256.

Shamloul, A.M., Yang, X., Han, L., Hadidi, X., 2003. Pear fruit crinkle. In: Hadidi, A., Flores, R., Randles, J.W., Semancik, J.S. (Eds.), Viroids. CSIRO Publishing, Collingwood, VIC, pp. 327–329.

Skrzeczkowski, L.J., Howell, W.E., Mink, G.I., 1993. Correlation between leaf epinasty symptoms on two apple cultivars and results of cRNA hybridization for detection of apple scar skin viroid. Plant Dis. 77, 919–921.

Thakur, P.D., Ito, T., Sharma, J.N., 1995. Natural occurrence of a viroid disease of apple in India. Indian J. Virol. 11, 73–75.

Ushirozawa, K., Tojo, Y., Takemae, S., Sekiguchi, A., 1968. Studies on apple scar skin disease (1) On transmission experiments. Bull. Nagano Hort. Res. Stn. Jpn. 9, 1–12 (in Japanese with English summary).

Walia, Y., Dhir, S., Bhadoria, S., Hallan, V., Zaidi, A.A., 2012. Molecular characterization of apple scar skin viroid from Himalayan wild cherry. Forest Pathol. 42, 84–87.

Walia, Y., Dhir, S., Ram, R., Zaidi, A.A., Hallan, V., 2014. Identification of the herbaceous host range of apple scar skin viroid and analysis of its progeny variants. Plant Pathol. 63, 684–690.

Walia, Y., Dhir, S., Zaidi, A.A., Hallan, V., 2015. Apple scar skin viroid naked RNA is actively transmitted by the whitefly *Trialeurodes vaporariorum*. RNA Biol. 12 (10), 1–8.

Walia, Y., Kumar, Y., Rana, T., Bhardwaj, P., Ram, R., Thakur, P.D., et al., 2009. Molecular characterization and variability analysis of apple scar skin viroid in India. J. Gen. Plant Pathol. 75, 307–311.

Wang, Y., Zhao, Y., Niu, J., 2012. Molecular identification and sequence analysis of the apple scar skin viroid (ASSVd) isolated from four kinds of fruit trees in Xinjiang province of China. Mol. Pathogens 3, 12–18.

Yamaguchi, A., Yanase, H., 1976. Possible relationship between the causal agent of dapple apple and scar skin. Acta Hortic. 67, 249–254.

Yang, X., Hadidi, A., Hammond, R.W., 1992. Nucleotide sequence of apple scar skin viroid reverse-transcribed in host extracts and amplified by the polymerase chain reaction. Acta Hortic. 309, 305–309.

Yazarlou, A., Jafarpour, B., Habili, N., Randles, J.W., 2012. First detection and molecular characterization of new apple scar skin viroid variants from apple and pear in Iran. Australasian Plant Dis. Notes. 7, 99–102.

Zhao, Y., Niu, J.X., 2008. Apricot is a new host of apple scar skin viroid. Australasian Plant Dis. Notes. 3, 98–100.

Zhu, S.F., Hadidi, A., Hammond, R.W., Yang, X., Hansen, A.J., 1995. Nucleotide sequence and secondary structure of pome fruit viroids from dapple diseased apples, pear rusty skin diseased pears and apple scar skin symptomless pears. Acta Hortic. 386, 554–559.

Zhu, S.F., Hammond, R.W., Hadidi, A., 1998. Agroinfection of pear and apple with dapple apple viroid results in systemic infection. Acta Hortic. 472, 613–616.

CHAPTER

22

Other Apscaviroids Infecting Pome Fruit Trees

Francesco Di Serio[1], *Enza M. Torchetti*[1], *Ricardo Flores*[2] *and Teruo Sano*[3]

[1]National Research Council, Bari, Italy
[2]Polytechnic University of Valencia-CSIC, Valencia, Spain
[3]Hirosaki University, Hirosaki, Japan

APPLE DIMPLE FRUIT VIROID (ADFVD)

INTRODUCTION

The first report of ADFVd dates back to over 20 years ago (Di Serio et al., 1996), when a small circular RNA was associated in Italy with an apple disease causing symptoms similar to those elicited in some apple cultivars by certain variants of apple scar skin viroid (ASSVd) (Hadidi et al., 1990). However, this circular RNA had relatively low sequence identity with ASSVd (63.5%) (Di Serio et al., 1996). Fulfillment of Koch's postulates conclusively identified ADFVd as the causal agent of apple dimple fruit disease (Di Serio et al., 2001) resulting in the new species *Apple dimple fruit viroid.*

TAXONOMIC POSITION AND NUCLEOTIDE SEQUENCE

Apple dimple fruit viroid is a species in the genus *Apscaviroid,* family *Pospiviroidae.* Most of the ADFVd sequence variants annotated in GenBank consist of 306–307 nt and have been isolated from apple trees in Italy (Di Serio et al., 1996, 1998) and China (Ye et al., 2013). However, variants composed of 303 and 310 nt have also been reported from

Viroids and Satellites.
DOI: http://dx.doi.org/10.1016/B978-0-12-801498-1.00022-X

apple trees in Japan (He et al., 2010) and fig in Italy (Chiumenti et al., 2014), respectively. ADFVd variants from apple trees cv. "Fuji" from China and Japan were clustered in distinct phylogenetic groups, while variants from apple and fig trees grown in Italy were clustered in the same group (Chiumenti et al., 2014; Ye et al., 2013), thus supporting the view that ADFVd sequence variability is more dependent on the geographic origin than on the host.

ECONOMIC SIGNIFICANCE

The incidence of the disease is limited. In orchards where trees harboring the viroid were found, only a few trees were actually infected. The limited progression of the infection in orchards (Malfitano et al., 2004a) contributes to lowering the risk of outbreaks, which so far have not been reported.

SYMPTOMS

ADFVd may alter the apple fruit skin by inducing yellow–green spots a few mm in diameter, sometimes slightly depressed (dimpled), and scattered prevalently around the calyx (Fig. 22.1A). Depressed spots may coalesce generating larger discolored and deformed areas. This symptomatology has been reported only in apple cultivars producing fruits with red skin. Fruit crinkle observed in two apple trees of cv. "Fuji" in China (Ye et al., 2013) may not be caused by ADFVd, but by other biotic and/or abiotic stresses, as previously shown for the internal necrotic regions underlying the spots initially observed in cv. "Starking Delicious" (Di Serio et al., 1996, 2001). No ADFVd symptoms have ever been reported on leaves, bark, or other organs of infected plants.

HOST RANGE

Natural hosts of ADFVd are apple and fig (*Ficus carica* L.). Pathogenicity in several apple cultivars has been shown (Di Serio et al., 2001), while the situation is less clear in the case of fig (Chiumenti et al., 2014). ADFVd has been experimentally transmitted to pear cv. "Fieud 37" (see below), which remained symptomless after becoming infected. This finding supports the contention that ADFVd may also latently infect pear trees in nature.

FIGURE 22.1 (A) Increasing severity of symptoms (from left to right) observed in fruit from apple trees (cv. "Starking Delicious") naturally infected with ADFVd; upper and lower panels are side and calyx-end views of the same apples. (B) Fruit crinkle disease on apple cv. "Ohrin" naturally infected with AFCVd. (C) Symptoms on pear cv. "A20" indicator 3 years after inoculation with three PBCVd isolates of different severity. Right; noninoculated cv. "A20". *Source: Panel (C) reproduced from Desvignes, J.C., Cornaggia, D., Grasseau, N., Ambrós, S., Flores, R., 1999. Pear blister canker viroid: studies on host range and improved bioassay with two new pear indicators, Fieud 37 and Fieud 110. Plant Dis. 83, 419–422.*

TRANSMISSION

ADFVd can be experimentally transmitted to apple and pear trees by grafting and budding, and by slashing the bark of young stems with a razor blade previously contaminated with a purified preparation of the viroid RNA. Seed transmission has been experimentally excluded for ADFVd (Malfitano et al., 2004a). In the field, the main transmission pathway is through infected propagation material.

However, in orchards in which the viroid has been found, the incidence was low, and no natural spread to adjacent or remote trees was recorded over a 3-year surveillance period (Malfitano et al., 2004a).

DETECTION

Visual inspection is considered an unreliable method for ADFVd detection because: (1) typical symptoms only appear on fruit; (2) these symptoms can be confused with those induced by ASSVd in certain apple cultivars; and (3) infection cannot be detected in tolerant apple cultivars. Similar limitations, besides the time for symptom expression (at least 2 years) and greenhouse costs, apply to bioassays with cvs. "Starkrimson" and "Braeburn," which are the best indicators for ADFVd (Di Serio et al., 2001). These bioassays, however, have been critical for fulfilling Koch's postulates for ADFVd (Di Serio et al., 2001).

Molecular diagnostic tools are used for routine detection of ADFVd. Polyacrylamide gel electrophoresis (PAGE) followed by hybridization with specific probes may discriminate ADFVd from closely related viroids on the basis of size and sequence (Di Serio et al., 1996). However, dot-blot and tissue printing hybridization with digoxigenin-labeled riboprobes have been used for large surveys (Di Serio et al., 2010; Lolić et al., 2007).

Several RT-PCR methods have been developed for large-scale detection of ADFVd alone (Di Serio et al., 1996) or concurrently with other viroids (Di Serio et al., 2002; Ragozzino et al., 2004). ADFVd was also one of the six viroids detected by a multiplex RT-PCR probe capture hybridization assay (RT-PCR-ELISA) (Shamloul et al., 2002). ADFVd was identified in a fig tree by next generation sequencing of small RNAs (Chiumenti et al., 2014).

GEOGRAPHICAL DISTRIBUTION AND EPIDEMIOLOGY

ADFVd has been reported in Italy (Chiumenti et al., 2014; Di Serio et al., 1996, 1998), Lebanon (Choueiri et al., 2007), China (Ye et al., 2013), and Japan (He et al., 2010). The viroid was not detected in surveys based on tissue printing hybridization of apple and pear trees in Bosnia-Herzegovina (Lolić et al., 2007), Albania (Navarro et al., 2011), Malta, Morocco, or Turkey (Di Serio et al., 2010).

APPLE FRUIT CRINKLE VIROID (AFCVD)

INTRODUCTION

AFCVd was first identified in 1976 in Japan from apples (*Malus pumila* Mill.) with a graft-transmissible fruit disorder called apple fruit crinkle disease (or *yuzu-ka byo*, in Japanese) (Koganezawa et al., 1989). A distinct isolate was detected in cultivated hops (*Humulus lupulus* L.) showing severe stunting and leaf curling similar to that incited by hop stunt viroid (Sano et al., 2004). A third isolate was detected in Japanese persimmon (*Diospyros kaki* Thunb.) (Nakaune and Nakano, 2008).

TAXONOMIC POSITION AND NUCLEOTIDE SEQUENCE

AFCVd is now listed as a tentative species of the genus *Apscaviroid*, family *Pospiviroidae*, because its overall sequence similarity (85%–87%) to Australian grapevine viroid is just below the border demarcating distinct viroid species (Di Serio et al., 2014). The isolates from apple and hop consist of 369–372 nt and 368–375 nt, respectively (Sano et al., 2004). Both isolates are genetically variable and cannot be clearly distinguished based on the nucleotide sequence, indicating that they comprise sequence variants of the same RNA species (Sano et al., 2008). The isolates from persimmon consist of 369–373 nt (Nakaune and Nakano, 2008). The overall nucleotide sequence similarity of the persimmon isolates (AB366019-AF366021) to the apple (E29032) and hop isolates (AB104531-AB104558) is 95.7%–96.0% and 91.4–96.3%, respectively (Nakaune and Nakano, 2008).

ECONOMIC SIGNIFICANCE

The apple isolate causes fruit crinkle disease on sensitive cultivars such as "Mutsu" or "Ohrin." The disease once prevailed in Nagano prefecture from the late 1980s to the early 1990s, mainly due to the unnoticed distribution of infected scions (Iijima, 1990), but its incidence is now under control and the economic losses to the apple industry in Japan are limited.

Affected hops were endemic during the late 1990s and the early 2000s in the hop growing areas of Akita prefecture in Japan, but the disease is now controlled by replanting with healthy cuttings (Sano et al., 2004).

SYMPTOMS

The major symptoms on apple consist of crinkling and dappling of the mature fruit surface (Fig. 22.1B), although the crinkling severity varies among the commercial apple cultivars. Apple cv. "Ohrin" exhibits the most serious fruit deformation accompanied by scattered brown necrotic areas in the flesh. Symptoms on cv. "Ohrin" are so similar to those from ASSVd infection that it is sometimes difficult to discriminate between them. Mixed infections with AFCVd and ASSVd are not common, but in one case with cv. "Fuji," more severe fruit crinkling symptoms were observed than in single infections with AFCVd (Ito et al., 1993). In contrast, cv. "Starking Delicious" shows only bark disorders resembling apple measles caused by manganese toxicity or boron deficiency, but no fruit symptoms (Ito et al., 1993). Similar bark disorders were also found in apple cv. "Nero 26" (Matsunaka and Machita, 1987), a hybrid with a sport of cv. "Delicious."

Hop cvs. "Kirin II" and "Shinshu-wase" infected with AFCVd show abnormal stunting and leaf curling similar to that caused by hop stunt viroid, in which the alpha acids content of the infected cones drops to nearly half that of healthy cones (Sano et al., 2004).

The persimmon isolate was detected in two asymptomatic cvs. "Tonewase" and "Fuyu," and one symptomatic cv. "Hiratanenashi" showing a fruit apex disorder. However, the relationship between AFCVd infection and occurrence of the disorder is not yet clear (Nakaune and Nakano, 2008).

HOST RANGE

Apple, hop, and Japanese persimmon are the natural hosts for AFCVd. The viroid infects tomato (*Solanum lycopersicum* L.) and cucumber (*Cucumis sativus* L.) experimentally (Fujibayashi et al., 2014).

TRANSMISSION

The apple isolate is easily transmitted by budding, grafting, and chip budding, and to apple seedlings by razor-slashing with purified RNA preparations (Ito et al., 1993). Inoculation of AFCVd to woody hosts other than *Malus* species has not been reported.

The hop isolate is spread via cuttings or rhizomes. Local transmission predominates along rows, and thus appears to be associated with farm operations that lead to mechanical injury, such as cutting, thinning, and mechanical leaf stripping. The persimmon isolate is graft-transmissible between persimmons.

DETECTION

AFCVd can be detected in crude or purified nucleic acid extracts from fruit pericarp, petiole, bark and leaves by return PAGE, sequential PAGE, northern-blot hybridization using a DIG-labeled AFCVd-specific riboprobe, and RT-PCR, the last either with primers specific for members of the genus *Apscaviroid*, or with AFCVd-specific primers (Sano et al., 2004). Nucleotide sequencing is essential for definitive diagnosis.

GEOGRAPHICAL DISTRIBUTION AND EPIDEMIOLOGY

The three AFCVd isolates have been reported only in Japan. The apple isolate spreads sporadically within the apple growing areas. The hop isolate was endemic until mid-2000s in a local hop growing area in Akita prefecture, but is not commonly observed today. Geographical distribution of the persimmon isolate is undetermined.

PEAR BLISTER CANKER VIROID (PBCVD)

INTRODUCTION

The viroid etiology of pear blister canker, a bark alteration observed specifically in the pear indicator cv. "A20" (Desvignes, 1970), was suspected from the close association between the disease and a small circular RNA displaying typical features of a viroid (Flores et al., 1991; Hernández et al., 1992). Infectivity was confirmed by mechanical inoculation of this RNA into A20 plants, expression of the characteristic symptoms, and the accumulation of an RNA with identical physical properties. This RNA was termed pear blister canker viroid (PBCVd) (Ambrós et al., 1995a).

TAXONOMIC POSITION AND NUCLEOTIDE SEQUENCE

Pear blister canker viroid belongs to the genus *Apscaviroid* (Hernández et al., 1992). The reference variant of PBCVd (S46812) comprises 315 nt and adopts a quasi-rod-like conformation of lowest free energy. Two bifurcations identified in silico (Hernández et al., 1992), are not present in a conformation determined in vitro by selective 2′-hydroxyl acylation analyzed by primer extension (Giguère et al., 2014). Variants from other isolates of different geographic origin (see below) have similar size (312–316 nt) and display some polymorphisms. PBCVd appears to be an example of a chimeric or mosaic RNA originating from recombination between several viroids coinfecting a common ancestor host (Hernández et al., 1992).

ECONOMIC SIGNIFICANCE

Most pear cultivars are tolerant to PBCVd, thus explaining its widespread and unnoticed distribution. About 10% of old French varieties are infected (Desvignes et al., 1999). The bark cankers incited in sensitive trees by PBCVd may be erroneously attributed to symptoms induced by fungi and bacteria, so the incidence and impact of the viroid needs to be estimated by specific diagnostic tests.

SYMPTOMS

Most commercial pear cultivars infected by PBCVd, including cvs. "Williams," "Comice," and "Conference," do not express bark symptoms (Desvignes et al., 1999); therefore, if expressed, they can be presumed to have a different etiology. Under field conditions, symptoms induced by PBCVd in the pear indicator cv. "A20" are restricted to bark pustules appearing 2 years after inoculation and evolving with time to bark scaling or bark splitting and ultimately to tree death. Differential symptom severity associated with several PBCVd isolates (Fig. 22.1C) may be related to minor divergences between their sequences (Desvignes et al., 1999), but this is yet to be tested.

HOST RANGE

In nature, PBCVd was initially reported in pear (*Pyrus communis* L.) and quince (*Cydonia oblonga* Mill.), then in wild pear (*P. amygdaliformis* Vill.) (Kyriakopoulou et al., 2001) and nashi (*P. serotina* Rehd.) (Joyce et al., 2006). The only herbaceous experimental host reported for PBCVd is cucumber (*Cucumis sativus* L.), to which it can be transmitted by leaf rubbing; however, this plant is not a good diagnostic host since infection results in just mild leaf rugosity (Flores et al., 1991). PBCVd can experimentally infect some species in the genera *Chaenomeles*, *Cydonia*, and *Sorbus*, five species of *Malus*, 15 species of *Pyrus* and 16 pear cultivars, without inciting symptoms (Desvignes et al., 1999). The apple cv. "Spy 277" is also a symptomless host (Lolić et al., 2007).

TRANSMISSION

PBCVd can be transmitted from pear to pear by grafting and also mechanically by making stem incisions with a razor blade immersed in nucleic acid preparations from infected tissues; these preparations also can be mechanically inoculated to carborundum-dusted leaves of cucumber (Flores et al., 1991). PBCVd is not seed-transmissible because when the indicator A20 was propagated onto 200 pear seedlings coming from a PBCVd-infected source it did not express symptoms during an observation period of 4 years (Desvignes, pers. comm.), with similar observations having been independently reproduced (Postman and Skrzeczkowski, pers. comm.). Transmission via infected propagative material appears quite feasible, particularly considering that the viroid does not incite any symptoms in many pear cultivars.

DETECTION

The long time required by the pear indicator cv. "A20" to develop the bark cankers under field conditions (2–3 years), promoted the search for an alternative indicator reacting in a shorter time. Inoculation of 250 pear seedlings cv. "Fieudière" resulted in two of them ("Fieud 37" and "Fieud 110") showing necrosis on petioles, leaves, and young shoots after 3–4 months. Young plants of these two selections, either grafted or propagated in vitro, can be used for PBCVd bioassay instead of A20 (Desvignes et al., 1999).

PBCVd can be detected in partially purified RNA preparations by a double PAGE technique specifically designed for small circular RNAs (Flores et al., 1991). Silver staining of the second denaturing gel usually provides sufficient sensitivity, because PBCVd accumulates at relatively high levels in bark and leaves. Dot-blot and tissue printing hybridization with radioactive and nonradioactive riboprobes, which are cheaper and quicker, as well as more sensitive and specific, were also developed (Ambrós et al., 1995b; Lolić et al., 2007).

Distinct RT-PCR protocols have been developed using specific primers for detecting PBCVd solely, or generic primers for the concurrent detection of PBCVd and other coinfecting pome fruit viroids (Faggioli and Ragozzino, 2002; Hassen et al., 2006). Other more sophisticated and sensitive formats consist of one-step multiplex RT-PCR (Lin et al., 2012; Ragozzino et al., 2004), RT-PCR-ELISA in single and multiplex variations (Shamloul et al., 2002), and one-step multiplex quantitative RT-PCR (Malandraki et al., 2015).

GEOGRAPHICAL DISTRIBUTION AND EPIDEMIOLOGY

PBCVd has been reported in the Mediterranean basin, including Europe (France, Spain, Italy, Bosnia-Herzegovina, Albania, and Greece) (Ambrós et al., 1995b; Desvignes, 1970; Di Serio et al., 2010; Lolić et al., 2007; Loreti et al., 1997; Malfitano et al., 2004b; Navarro et al., 2011), Malta (Attard et al., 2007), North-Africa (Tunisia) (Hassen et al., 2006), Near East (Turkey) (Yesilcollou et al., 2010), the Americas (Canada, the United States, and Argentina) (Hadidi and Postman, unpublished data; Nome et al., 2012; Torchetti et al., 2012), Australia (Joyce et al., 2006) and Japan (Sano et al., 1997). The use of quince clones as a pear rootstock in Southern Europe may have contributed to the spread of PBCVd because it is asymptomatic in this species.

CONTROL OF APSCAVIROIDS INFECTING POME FRUIT TREES

Propagation material and contaminated equipment are the major transmission pathways of apscaviroids. Since symptoms of apscaviroid infections are not always evident in pome fruit trees, primary infections in orchards can be avoided using viroid-free propagation materials certified by sensitive molecular techniques. Control of secondary spread of apscaviroids in orchards can be achieved by replacing infected plants with viroid-free tested plants and by decontaminating equipment between fields and/or plots. In particular, when AFCVd is detected in hop, it is encouraged to replace all plants in the yard because, due to the long latency, removal of just the symptomatic plants may leave sources of inoculum for further spread of the viroid. In addition, all operations in fields in which infected plants have been detected should be performed only after finishing operations in the viroid-free plots. For decontamination, formalin and caustic soda are effective, but these chemicals are corrosive and hazardous for workers. A 10% solution of household bleach or other disinfectants may also be used to treat equipment and hand tools (Eastwell and Sano, 2009; Sano, 2009).

References

Ambrós, S., Desvignes, J.C., Llácer, G., Flores, R., 1995a. Pear blister canker viroid: sequence variability and causal role in pear blister canker disease. J. Gen. Virol. 76, 2625–2629.

Ambrós, S., Desvignes, J.C., Llácer, G., Flores, R., 1995b. Peach latent mosaic and pear blister canker viroids: detection by molecular hybridization and relationships with specific maladies affecting peach and pear trees. Acta Hortic. 386, 515–521.

Attard, D., Afechtal, M., Agius, M., Matic, S., Gatt, M., Myrta, A., et al., 2007. First report of pear blister canker viroid in Malta. J. Plant Pathol. 89, S71.

Chiumenti, M., Torchetti, E.M., Di Serio, F., Minafra, A., 2014. Identification and characterization of a viroid resembling apple dimple fruit viroid in fig (*Ficus carica* L.) by next generation sequencing of small RNAs. Virus Res. 188, 54–59.

Choueiri, E., El Zammar, S., Jreijiri, F., Hobeika, C., Myrta, A., Di Serio, F., 2007. First report of apple dimple fruit viroid in Lebanon. J. Plant Pathol. 89, 304–404.

Desvignes, J.C., 1970. Les maladies à virus du poirier et leur détection. CTIFL Doc. 26, 1–12.

Desvignes, J.C., Cornaggia, D., Grasseau, N., Ambrós, S., Flores, R., 1999. Pear blister canker viroid: studies on host range and improved bioassay with two new pear indicators, Fieud 37 and Fieud 110. Plant Dis. 83, 419–422.

Di Serio, F., Afechtal, M., Attard, D., Choueiri, E., Gumus, M., Kaymak, S., et al., 2010. Detection by tissue printing hybridization of pome fruit viroids in the Mediterranean basin. Julius-Kühn Archiv. 427, 357–360.

Di Serio, F., Alioto, D., Ragozzino, A., Giunchedi, L., Flores, R., 1998. Identification of apple dimple fruit viroid in different commercial varieties of apple grown in Italy. Acta Hortic. 472, 595–601.

Di Serio, F., Aparicio, F., Alioto, D., Ragozzino, A., Flores, R., 1996. Identification and molecular properties of a 306 nucleotide viroid associated with apple dimple fruit disease. J. Gen. Virol. 77, 2833–2837.

Di Serio, F., Flores, R., Verhoeven, J.Th.J., Li, S.F., Pallás, V., et al., 2014. Current status of viroid taxonomy. Arch. Virol. 159, 3467–3478.

Di Serio, F., Malfitano, M., Alioto, D., Ragozzino, A., Desvignes, J.C., Flores, R., 2001. Apple dimple fruit viroid: fulfillment of Koch's postulates and symptom characteristics. Plant Dis. 85, 179–182.

Di Serio, F., Malfitano, M., Alioto, D., Ragozzino, A., Flores, R., 2002. Apple dimple fruit viroid: sequence variability and its specific detection by multiplex fluorescent RT-PCR in the presence of apple scar skin viroid. J. Plant Pathol. 84, 27–34.

Eastwell, K., Sano, T., 2009. Hop stunt disease (HSVd). In: Mahaffee, W.F., Pethybridge, S.J., Gent, D.H. (Eds.), The Compendium of Hop Diseases, Arthropod Pests and Other Disorders. APS Press, St. Paul, MN, pp. 48–51.

Faggioli, F., Ragozzino, E., 2002. Detection of pome fruit viroids by RT-PCR using a single primer pair. J. Plant Pathol. 84, 125–128.

Flores, R., Hernández, C., Llácer, G., Desvignes, J.C., 1991. Identification of a new viroid as the putative causal agent of pear blister canker disease. J. Gen. Virol. 72, 1199–1204.

Fujibayashi, M., He, Y.H., Hataya, T., Sano, T., 2014. Analysis on the pathogenicity of members of the genus *Apscaviroid* for herbaceous plants. Jpn. J. Phytopathol. 80, 297 (abstr. in Japanese).

Giguère, T., Raj Adkar-Purushothama, C., Perreault, J.P., 2014. Comprehensive secondary structure elucidation of four genera of the family *Pospiviroidae*. PLoS One 9, e98655.

Hadidi, A., Huang, C., Hammond, R.W., Hashimoto, J., 1990. Homology of the agent associated with dapple apple disease to apple scar skin viroid and molecular detection of these viroids. Phytopathology 80, 263–268.

Hassen, I.F., Roussel, S., Kummert, J., Fakhfakh, H., Marrakchi, M., Jijakli, M.H., 2006. Development of a rapid RT-PCR test for the detection of peach latent mosaic viroid, pear blister canker viroid, hop stunt viroid and apple scar skin viroid in fruit trees from Tunisia. J. Phytopathol. 154, 217–223.

He, Y., Isono, S., Kawaguchi-Ito, Y., Taneda, A., Kondo, K., Iijima, A., et al., 2010. Characterization of a new apple dimple fruit viroid variant that causes yellow dimple fruit formation in "Fuji" apple trees. J. Gen. Plant Pathol. 76, 324–330.

Hernández, C., Elena, S.F., Moya, A., Flores, R., 1992. Pear blister canker viroid is a member of the apple scar skin viroid subgroup (apscaviroids) and also has sequence homologies with viroids from other subgroups. J. Gen. Virol. 73, 2503–2507.

Iijima, A., 1990. Occurrence of a new viroid-like disease, apple fruit crinkle. Plant Prot. 44, 130–132.

Ito, T., Kanematsu, S., Koganezawa, H., Tsuchizaki, T., Yoshida, K., 1993. Detection of a viroid associated with apple fruit crinkle disease. Ann. Phytopath. Soc. Jpn. 59, 520–527.

Joyce, P.A., Constable, F.E., Crosslin, J., Eastwell, K., Howell, W.E., Rodoni, B.C., 2006. Characterisation of pear blister canker viroid isolates from Australian pome fruit orchards. Australas. Plant Path. 35, 465–471.

Koganezawa, H., Ohnuma, Y., Sakuma, T., Yanase, H., 1989. "Apple fruit crinkle," a new graft transmissible fruit disorder of apple. Bull. Fruit Tree Res. Stn. C16, 57–62 (in Japanese).

Kyriakopoulou, P.E., Giunchedi, L., Hadidi, A., 2001. Peach latent mosaic and pome fruit viroids in naturally infected cultivated pear *Pyrus communis* and wild pear *P. amygdaliformis:* implications on possible origin of these viroids in the Mediterranean region. J. Plant Pathol. 83, 51–62.

Lin, L., Li, R., Mock, R., Kinard, G., 2012. One-step multiplex RT-PCR for simultaneous detection of four pome tree viroids. Eur. J. Plant Pathol. 133, 765–772.

Lolić, B., Afechtal, M., Matic, S., Myrta, A., Di Serio, F., 2007. Detection by tissue-printing of pome fruit viroids and charaterization of pear blister canker viroid in Bosnia and Herzegovina. J. Plant Pathol. 89, 369–375.

Loreti, S., Faggioli, F., Barba, M., 1997. Identification and characterization of an Italian isolate of pear blister canker viroid. J. Phytopathol. 145, 541–544.

Malandraki, I., Varveri, C., Olmos, A., Vassilakos, N., 2015. One-step multiplex quantitative RT-PCR for the simultaneous detection of viroids and phytoplasmas of pome fruit trees. J. Virol. Methods 213, 12–77.

Malfitano, M., Alioto, D., Ragozzino, A., Flores, R., Di Serio, F., 2004a. Experimental evidence that apple dimple fruit viroid does not spread naturally. Acta Hortic. 657, 357–360.

Malfitano, M., Barone, M., Ragozzino, E., Flores, R., Alioto, D., 2004b. Identification and molecular characterization of pear blister canker viroid isolates in Campania (Southern Italy). Acta Hortic. 657, 367–371.

Matsunaka, K., Machita, I., 1987. A graft-transmissible blister bark occurring on apple cv. Nero 26. Ann. Rept. Plant Prot. North Jpn. 38, 186.

Nakaune, R., Nakano, M., 2008. Identification of a new apscaviroid from Japanese persimmon. Arch. Virol. 153, 969–972.

Navarro, B., Bacu, A., Torchetti, M., Kongjika, E., Susuri, L., Di Serio, F., et al., 2011. First record of pear blister canker viroid on pear in Albania. J. Plant Pathol. 93, S4–70.

Nome, C.F., Difeo, L.V., Giayetto, A., Rossini, M., Frayssinet, S., Nieto, A., 2012. First report of pear blister canker viroid in pear trees in Argentina. Plant Dis. 95, 882.

Ragozzino, E., Faggioli, F., Barba, M., 2004. Development of a one tube-one step RT-PCR protocol for the detection of seven viroids in four genera: *Apscaviroid, Hostuviroid, Pelamoviroid* and *Pospiviroid*. J. Virol. Methods 121, 25–29.

Sano, T., 2009. Apple fruit crinkle viroid (AFCVd). In: Mahaffee, W.F., Pethybridge, S.J., Gent, D.H. (Eds.), The Compendium of Hop Diseases, Arthropod Pests and Other Disorders. APS Press, St. Paul, MN, p. 39.

Sano, T., Isono, S., Matsuki, K., Kawaguchi-Ito, Y., Tanaka, K., Kond, K., et al., 2008. Vegetative propagation and its possible role as a genetic bottleneck in the shaping of the apple fruit crinkle viroid populations in apple and hop plants. Virus Genes 37, 298–303.

Sano, T., Li, S.F., Ogata, T., Ochiai, M., Suzuki, C., Ohnuma, S., et al., 1997. Pear blister canker viroid isolated from European pear in Japan. Ann. Phytopathol. Soc. Jpn. 63, 89–94.

Sano, T., Yoshida, H., Goshono, M., Monma, T., Kawasaki, H., Ishizaki, K., 2004. Characterization of a new viroid strain from hops: evidence for viroid speciation by isolation in different host species. J. Gen. Plant Pathol. 70, 181–187.

Shamloul, A.M., Faggioli, F., Keith, J.M., Hadidi, A., 2002. A novel multiplex RT-PCR probe capture hybridization (RT-PCR-ELISA) for simultaneous detection of six viroids in four genera: *Apscaviroid, Hostuviroid, Pelamoviroid,* and *Pospiviroid*. J. Virol. Methods 105, 115–121.

Torchetti, E.M., Birch, C., Cooper, M., Masters, C., Arocha, R.Y., Di Serio, F., et al., 2012. Detection and identification of pear blister canker viroid occurring in pear tress in Canada. Petria 22, 123–458.

Ye, T., Chen, S.Y., Wang, R., Hao, L., Chen, H., Wang, N., et al., 2013. Identification and molecular characterization of apple dimple fruit viroid in China. J. Plant Pathol. 95, 637–641.

Yesilcollou, S., Minoia, S., Torchetti, E.M., Kaymak, S., Gumus, M., Myrta, A., et al., 2010. Molecular characterization of Turkish isolates of pear blister canker viroid and assessment of the sequence variability of this viroid. J. Plant Pathol. 92, 813–819.

CHAPTER

23

Apscaviroids Infecting Citrus Trees

Matilde Tessitori

University of Catania, Catania, Italy

INTRODUCTION

To date seven viroids have been reported in *Citrus* spp., rarely in single infections but more often as mixtures of different viroids, with citrus exocortis viroid (CEVd), hop stunt viroid (HSVd), and citrus dwarfing viroid (CDVd) being the most widespread. CEVd and HSVd are the causal agents of exocortis and cachexia diseases, respectively, and are members of the genus *Pospiviroid* and *Hostuviroid*, respectively; citrus bark cracking viroid (CBCVd) belongs to the genus *Cocaviroid*; and citrus bent leaf viroid (CBLVd), CDVd, citrus viroid V (CVd-V), and citrus viroid VI (CVd-VI), are members of the genus *Apscaviroid*. In nature CDVd and CVd-V are restricted to citrus hosts, while CBCVd and CVd-VI are also present in natural infections of hop and persimmon, respectively (Jakse et al., 2015; Nakaune and Nakano, 2008). This chapter focuses on apscaviroids infecting citrus, whereas CEVd, HSVd, and CBCVd are addressed in separate chapters.

TAXONOMIC POSITION AND NUCLEOTIDE SEQUENCE

Originally categorized by electrophoretic mobility on sequential polyacrylamide gel electrophoresis in five different groups (CEVd, I, II, III, IV) (Duran-Vila et al., 1988), the citrus viroids were afterward renamed and classified according to the ICTV 9th report. The genus *Apscaviroid* includes four viroids infecting citrus: CBLVd (formerly citrus viroid I,

Viroids and Satellites.
DOI: http://dx.doi.org/10.1016/B978-0-12-801498-1.00023-1

318 nt, reference sequence M74065), CDVd (formerly citrus viroid IIIa, 327 nt, reference sequence S76452 and citrus viroid IIIb, 324 nt, reference sequence AF184147), CVd-V (294 nt, reference sequence EF617306), and CVd-VI (330 nt, reference sequence AB019508) (Owens et al., 2012).

Given the wide distribution of citrus viroids worldwide, a large number of complete genome sequences are available in sequence databases. Moreover, studies on the heterogeneous progeny derived from the inoculation of a single variant of citrus apscaviroids confirmed the quasispecies nature of these viroids, a characteristic that contributes to increase the number of sequences deposited (Gandía and Duran-Vila, 2004; Tessitori et al., 2013).

HOST RANGE

The apscaviroids infecting citrus have a natural host range restricted to woody plants, as with all the viroids of the genus. They naturally infect only citrus species apart from CVd-VI, which also has been detected in Japanese persimmon (*Diospyrus kaki* Thunb.) (Nakaune and Nakano, 2008).

Using chip-grafting Hadas et al. (1992) transmitted CBLVd (formerly CVd-Ib), a component of the Graft Transmissible Dwarfing (GTD) agent 225-T isolated in Israel, from alemow (*Citrus macrophylla*) to avocado (West Indian cultivar). Transmission trials of CVd-V on several herbaceous hosts gave negative results thus indicating its restricted host range (Barbosa et al., 2005).

SYMPTOMS AND ECONOMIC SIGNIFICANCE

Not all citrus apscaviroids cause disease or symptoms on susceptible species (e.g., *Poncirus trifoliata* (L.) Raf. and its hybrids, mandarins, *C. macrophylla* Wester, *Citrus jambhiri* Lush., and "Rangpur" lime (*Citrus limonia* Osbeck)). Given that most of the data on symptom induction by single viroids refer to experimental tests on the indicator Etrog citron, and that the effects on commercial citrus hosts is influenced by the species, variety, and rootstock, it is difficult to estimate their economic impact. However, they are often found in mixtures that may result in synergistic effects (exacerbated symptoms). A synergistic effect of citrus apscaviroids, with more severe symptoms than expected if the effects induced by each viroid were only additive, was demonstrated on Etrog citron as well as on field-grown trees coinfected with CBLVd and CDVd or with CBLVd and CVd-V (Serra et al., 2008a; Verniere et al., 2006).

In contrast, the dwarfing of trees by certain viroids can have a positive outcome in high-density citrus plantings, with increased production per unit of land area and reduced management costs. Dwarfing of citrus trees through the infection of a single viroid or a mixture of viroids, called GTD agents, in specific scion/rootstock combinations, has been described since the 1960s (Cohen, 1968; Fraser et al., 1961; Mendel, 1968). Although CDVd causes petiole necrosis and bending, resulting in general leaf drooping and mild stunting in Etrog citron, it induces dwarfing in commercial citrus plants grafted on *P. trifoliata* and its hybrids without any detrimental effects on the canopy, yield, or fruit quality (Hutton et al., 2000; Vidalakis et al., 2011). The lack of effects apart from tree size reduction has encouraged the use of CDVd as a dwarfing agent in high-density plantings in many citrus-growing countries (Hardy et al., 2007; Semancik, 2003; Tessitori et al., 2002; Vidalakis et al., 2011). Several types of GTD agents and their agronomic effects were later studied in field tests in Australia, California, Israel, and Italy (Hutton et al., 2000; Semancik, 2003). In 2000, two viroids were approved for commercial use by the California Department of Food and Agriculture and marketed under the names of TsnRNA-IIIb and TsnRNA-IIa (TsnRNA stands for transmissible small nuclear ribonucleic acid) for their use on trifoliate orange or citrange rootstocks (Vidalakis et al., 2011). In Australia, budwood containing GTD-IIIb is distributed by the New South Wales Department of Primary Industries for inducing a size reduction of about 50% in *P. trifoliata*, 25% in Troyer citrange, and 20% in Carrizo citrange (Hardy et al., 2007).

The genetic variability/stability of CDVd is important in determining its suitability as a dwarfing agent in high-density citrus plantings (Semancik, 2003; Vidalakis et al., 2011). The available data on CDVd sequences (more than 200 variants have been deposited in GenBank) support the high genetic stability of CDVd populations in natural conditions and in various geographical areas (Vidalakis et al., 2011). Recent studies on the selective pressure exerted by different citrus hosts and conditions on CDVd populations have shown that they are stable except in lemon and its hybrid relatives, thus excluding the use of CDVd as a dwarfing factor in the latter hosts (Tessitori et al., 2013).

The molecular mechanisms involved in the interaction between viroids and host plants have been investigated using the CDVd-Etrog citron system as a model. Eighteen up- and downregulated genes were identified, mainly related to cell wall structure, amino acid transport, signal transduction, and plant defence/stress response. Within the upregulated genes was a suppressor of RNA silencing, which may be involved in the pathogenicity response to viroid infection (Tessitori et al., 2007).

TRANSMISSION AND EPIDEMIOLOGY

The dissemination of citrus apscaviroids occurs mainly through infected propagation material such as infected budwood. Mechanical transmission in the field is generally low and can be avoided easily by disinfecting the cutting tools (Duran-Vila and Semancik, 2003). Natural vectors or seed transmission have never been demonstrated in citrus. The application of certification programs strongly contributed to reducing the spread of citrus viroids in the main citrus-growing countries.

DETECTION

As with other citrus viroids, apscaviroids can be diagnosed by biological indexing on "Arizona 861-S1" Etrog citron selection coupled with molecular analysis such as sequential polyacrylamide gel electrophoresis (Rivera-Bustamante et al., 1986), hybridization, RT-PCR, and real-time RT-PCR for the detection of single or multiple infections. Several protocols for RT-PCR or hybridization have been set up for diagnosing citrus viroids, including the apscaviroids CBLVd and CDVd (Bernard and Duran-Vila, 2006; Cohen et al., 2006; Ito et al., 2002; Murcia et al., 2008). Diagnosis of CVd-V and CVd-VI by RT-PCR uses specific primers (Serra et al., 2008a; Ito and Ohta, 2010; Ito et al., 2001). The first protocol for the real-time diagnosis of apscaviroids was developed for CDVd (Rizza et al., 2009), and more recently a real-time method for the multiple diagnosis of citrus viroids, including all apscaviroids, was reported (Vidalakis and Wang, 2013).

GEOGRAPHICAL DISTRIBUTION

CDVd, together with CEVd and HSVd, is present in all citrus-growing countries whereas CBLVd variants are less widespread than those of other citrus viroids, although they have been reported in almost all citrus-growing countries (Duran-Vila and Semancik, 2003). Natural infections of the variant CVd-I-LSS (Ito et al., 2000) have also been reported in Iran, Pakistan, and China (Alavi et al., 2005; Wu et al., 2014).

Since its discovery, there have been several reports of the spread of CVd-V worldwide. Numerous isolates have been characterized on commercial species of citrus in California (USA), Spain, Nepal, Oman (Serra et al., 2008b), Japan (Ito and Ohta, 2010), Iran (Bani Hashemian et al., 2010), China (Cao et al., 2010), Turkey (Önelge and Yurtmen, 2012), Pakistan (Cao et al., 2013), and Tunisia (Hamdi et al., 2015). Before these reports, CVd-V did not seem as widespread as other citrus viroids; however, more recent studies have revealed its presence in different species and varieties of citrus: 37 CVd-V positive samples out of 152 (24.3%) in Japan, 14 out of 38 (36.8%) in Tunisia, and 334 out of 359 (93%) in Punjab, Pakistan. The authors assumed that the high incidence in Punjab could have originated from the infection of the rootstock rough lemon (*C. jambhiri*) (Cao et al., 2013). Although almost all the studies have specifically focused on detecting CVd-V, it appears to be present in natural infections in mixtures with other viroids (Hamdi et al., 2015; Ito and Ohta, 2010).

CVd-VI was identified originally within a complex mixture of viroids in a tree of Shiranui (*C. reticulata* Blanco × *C. sinensis* (L.) Osb. × *C. reticulata*) in Nagasaki, Japan, and to date seems to be restricted to this country.

CONTROL

Viroid infections cannot be cured, and viroid control is by certification of budwood sources. Citrus certification programs and the use of sanitation in Australia, Spain, France, and Italy are applied, above all, to exclude CEVd and HSVd, and, as a consequence, all other citrus viroids. The viroid inoculum sources have therefore decreased over time. The citrus clones are efficiently freed of viroids by shoot-tip grafting in vitro (Duran-Vila and Semancik, 2003). In addition, the use of tolerant rootstocks can be considered as a control measure, although the trees may serve as a viroid reservoir.

Acknowledgments

I thank Moshe Bar-Joseph for critical reading of the CBLVd sections. Appreciation is expressed to Serena Rizza for reviewing and editing the manuscript.

References

Alavi, S.M., Sohi, H.H., Ahoonmanesh, A., Rahimian, H., Barzegar, A., 2005. Molecular characterization and identification of multiple viroids in Mazandaran province. Iran. J. Plant Pathol. 41, 303–304.

Bani Hashemian, S.M.B., Taheri, H., Duran-Vila, N., Serra, P., 2010. First report of citrus viroid V in Moro blood sweet orange in Iran. Plant Dis. 94, 129.

Barbosa, C.J., Pina, J.A., Perez-Panades, J., Bernad, L., Serra, P., Navarro, L., et al., 2005. Mechanical transmission of citrus viroids. Plant Dis. 89, 749–754.

Bernard, L., Duran-Vila, N., 2006. A novel RT-PCR approach for detection and characterization of citrus viroids. Mol. Cell. Probes 20, 105–113.

Cao, M.J., Liu, Y.Q., Wang, X.F., Yang, F.Y., Zhou, C.Y., 2010. First report of citrus bark cracking viroid and citrus viroid V infecting citrus in China. Plant Dis. 94, 922.

Cao, M.J., Su, H.N., Atta, S., Wang, X.F., Wu, Q., Li, Z.A., et al., 2013. Molecular characterization and phylogenetic analysis of citrus viroid V isolates from Pakistan. Eur. J. Plant Pathol. 135, 11–21.

Cohen, M., 1968. Exocortis virus as a possible factor in producing dwarf citrus trees. Proc. Florida State Horticultural Soc. 81, 115–119.

Cohen, O., Batuman, O., Stanbekova, G., Sano, T., Mawassi, M., Bar-Joseph, M., 2006. Construction of a multiprobe for the simultaneous detection of viroids infecting citrus trees. Virus Genes 33, 287–292.

Duran-Vila, N., Roistacher, C.N., Rivera-Bustamante, R., Semancik, J.S., 1988. A definition of citrus viroid groups and their relationship to the exocortis disease. J. Gen. Virol. 69, 3069–3080.

Duran-Vila, N., Semancik, J.S., 2003. Citrus viroids. In: Hadidi, A., Flores, R., Randles, J.W., Semancik, J.S. (Eds.), Viroids. CSIRO Publishing, Collingwood, VIC, pp. 178–194.

Fraser, L.R., Levitt, E.C., Cox, J.E., 1961. Relationship between exocortis and stunting of citrus varieties on *Poncirus trifoliata* rootstock. Proceedings of the 2nd Conference of International Organization of Citrus Virologists. University of Florida Press, Gainesville, pp. 34–39.

Gandía, M., Duran-Vila, N., 2004. Variability of the progeny of a sequence variant of citrus bent leaf viroid (CBLVd). Arch. Virol. 149, 407–416.

Hadas, R., Ashulin, L., Bar-Joseph, M., 1992. Transmission of a citrus viroid to avocado by heterologous grafting. Plant Dis. 76, 357–359.

Hamdi, I., Elleuch, A., Bessaies, N., Grubb, C.D., Fakhfakh, H., 2015. First report of citrus viroid V in North Africa. J. Gen. Plant Pathol. 81, 87–91.

Hardy, S., Sanderson, G., Barkley, P., Donovan, N., 2007. Dwarfing citrus trees using viroids. New South Wales Department of Primary Industries: Primefact 704 (NSW Agriculture).

Hutton, R.J., Broadbent, P., Bevington, K.B., 2000. Viroid dwarfing for high density citrus plantings. Hort. Rev. 24, 277–317.

Ito, T., Ieki, H., Ozaki, K., 2000. A population of variants of a viroid closely related to citrus viroid-I in citrus plants. Arch. Virol. 145, 2105–2114.

Ito, T., Ieki, H., Ozaki, K., 2001. Characterization of a new citrus viroid species tentatively termed citrus viroid OS. Arch. Virol. 146, 975–982.

Ito, T., Ieki, H., Ozaki, K., 2002. Simultaneous detection of six citrus viroids and apple stem grooving virus from citrus plants by multiplex reverse transcription polymerase chain reaction. J. Virol. Methods 106, 235–239.

Ito, T., Ohta, S., 2010. First report of citrus viroid V in Japan. J. Gen. Plant Pathol. 76, 348–350.

Jakse, J., Radisek, S., Pokorn, T., Matousek, J., Javornik, B., 2015. Deep-sequencing revealed citrus bark cracking viroid (CBCVd) as a highly aggressive pathogen on hop. Plant Pathol. 64, 831–842.

Mendel, K., 1968. Interrelations between tree performance and some virus diseases, in: Proceedings of the 4th Conference of the International Organization of Citrus Virologists, University of Florida Press, Gainesville, pp. 310–313.

Murcia, N., Serra, P., Olmos, A., Duran-Vila, N., 2008. A novel hybridization approach for detection of citrus viroids. Mol. Cell. Probes 23, 95–102.

Nakaune, R., Nakano, M., 2008. Identification of a new Apscaviroid from Japanese persimmon. Arch. Virol. 153, 969–972.

Önelge, N., Yurtmen, M., 2012. First report of citrus viroid V in Turkey. J. Plant Pathol. 94, S4.88.

Owens, R.A., Flores, R., Di Serio, F., Li, S.F., Pallas, V., Randles, J.W., et al., 2012. Viroids. In: King, A.M.Q., Adams, M.J., Carstens, E.B., Lefkowitz, E.J. (Eds.), Virus Taxonomy: Ninth Report of the International Committee on Taxonomy of Viruses. Elsevier/Academic Press, London, pp. 1221–1234.

Rivera-Bustamante, R.F., Gin, R., Semancik, J.S., 1986. Enhanced resolution of circular and linear molecular forms of viroid and viroid-like RNA by electrophoresis in a discontinuous-pH system. Anal. Biochem. 156, 91–95.

Rizza, S., Nobile, G., Tessitori, M., Catara, A., Conte, E., 2009. Real time RT-PCR assay for quantitative detection of citrus viroid III in plant tissues. Plant Pathol. 58, 181–185.

Semancik, J.S., 2003. Considerations for the introduction of viroids for economic advantage. In: Hadidi, A., Flores, R., Randles, J.W., Semancik, J.S. (Eds.), Viroids. CSIRO Publishing, Collingwood, VIC, pp. 357–362.

Serra, P., Barbosa, C.J., Daròs, J.A., Flores, R., Duran-Vila, N., 2008a. Citrus viroid V: molecular characterization and synergistic interactions with other members of the genus Apscaviroid. Virology 370, 102–112.

Serra, P., Eiras, M., Bani-Hashemian, S.M., Murcia, N., Kitajima, E.W., Daros, J.A., et al., 2008b. Citrus viroid V: occurrence, host range, diagnosis, and identification of new variants. Phytopathology 98, 1199–1204.

Tessitori, M., La Rosa, R., Di Serio, F., Albanese, G., Catara, A., 2002. Molecular characterization of a citrus viroid III (CVd-III) associated with citrus dwarfing in Italy. Proceedings of the 15th Conference of the International Organization of Citrus Virologists. Riverside, CA, pp. 387–389.

Tessitori, M., Maria, G., Capasso, C., Catara, G., Rizza, S., De Luca, V., et al., 2007. Identification of differentially expressed genes in Etrog citron in response to Citrus viroid III infection. Biochem. Biophys. Acta. 1769, 228–235.

Tessitori, M., Rizza, S., Reina, A., Causarano, G., Di Serio, F., 2013. The genetic diversity of citrus dwarfing viroid populations is mainly dependent on the infected host species. J. Gen. Virol. 94, 687–693.

Verniere, C., Perrier, X., Dubois, C., Dubois, A., Botella, L., Chabrier, C., et al., 2006. Interactions between citrus viroids affect symptom expression and field performance of clementine trees grafted on trifoliate orange. Phytopathology 96, 356–368.

Vidalakis, G., Pagliaccia, D., Bash, J.A., Afunian, M., Semancik, J.S., 2011. Citrus dwarfing viroid: Effects on tree size and scion performance specific to *Poncirus trifoliata* rootstock for high-density planting. Ann. Appl. Biol. 158, 204–217.

Vidalakis, G., Wang, J. Regents of the University of California. Molecular method for universal detection of citrus viroids. Patent US20130115591 A1. 2013 May 9.

Wu, Q., Cao, M., Su, H., Atta, S., Yang, F., Wang, X., et al., 2014. Molecular characterization and phylogenetic analysis of citrus viroid I-LSS variants from citrus in Pakistan and China reveals their possible geographic origin. Eur. J. Plant Pathol. 139, 13–17.

CHAPTER

24

Apscaviroids Infecting Grapevine

Nuredin Habili

The University of Adelaide, Waite Campus, Glen Osmond, SA, Australia

INTRODUCTION

Of the six viroids from the family *Pospiviroidae* detected in the grapevine, four belong to the genus *Apscaviroid* (Fig. 24.1; Table 24.1). These include grapevine yellow speckle viroids 1 and 2 (GYSVd-1 and GYSVd-2), Australian grapevine viroid (AGVd), and grapevine latent viroid (GLVd) (Zhang et al., 2014). GYSVd-1 occurs wherever grapevine is grown with the exception of wild grapevines in Japan, which contain no detectable viroids (Jiang et al., 2012).

Grapevine yellow speckle (GYS) disease, characterized by pinhead-size yellow spots along the main and lateral veins of the grapevine leaf (Fig. 24.2), was first described in Australia by Taylor and Woodham (1972). Later, Koltunow and Rezaian (1988) demonstrated the causal agent was a viroid by fulfilling Koch's postulates for GYSVd-1 and GYSVd-2, independently. This was achieved by inoculating monomeric RNA transcribed from cloned cDNA to viroid-free grapevine seedlings (Koltunow et al., 1989). These two apscaviroids are the only viroids that induce symptoms in the grapevine. GYSVd-2, which occurs less frequently, was first isolated from the grapevine cv. "Kyoho" on "Dogridge" rootstock in Australia (Little and Rezaian, 2003).

AGVd was first detected in grapevines in a mixture with other grapevine viroids (Rezaian et al., 1988). AGVd appears to be a mosaic of various segments from ASSVd, potato spindle tuber viroid, and GYSVd-1 and -2 (Little and Rezaian, 2003).

Viroids and Satellites.
DOI: http://dx.doi.org/10.1016/B978-0-12-801498-1.00024-3

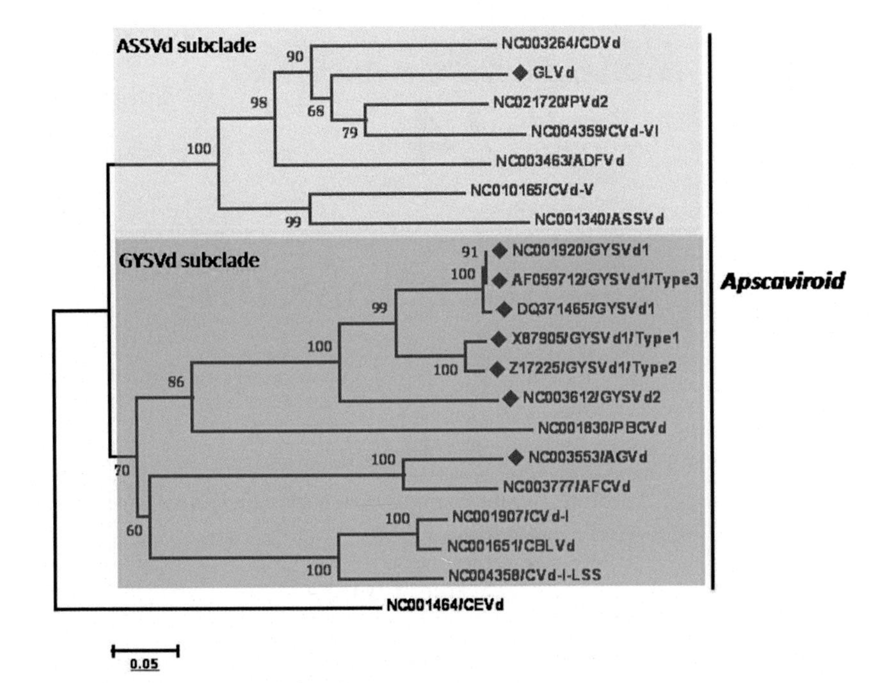

FIGURE 24.1 Phylogenetic tree of the apscaviroids using MEGA 6 (Tamura et al., 2013) with the neighbor-joining method. Bootstrap values (1000 replicates) are in the nodes of branches. The apscaviroids cluster into two subclades. Diamonds show apscaviroids infecting grapevine. CEVd was used as the outgroup. Viroids are identified by GenBank Accession No. and abbreviation.

TABLE 24.1 Apscaviroids of the Grapevine

Viroid	Size (nt)	Symptom in		Reference[a]
		Grapevine	Herbaceous host	
GYSVd-1	366	Yellow speckle	None known	Rigden and Rezaian (1993)
GYSVd-2	363	Yellow speckle	None known	Koltunow et al. (1989)
AGVd	369	None	Cucumber, tomato	Zaki-aghl and Izadpanah (2015)
GLVd	328	None	None known	Zhang et al. (2014)

[a] *The given reference is for the type isolate of each viroid.*

GLVd was detected in a 100-year old "Thompson Seedless" grapevine in China by next-generation sequencing (NGS) of overlapping viroid-derived small RNAs. It is a circular RNA that can infect and replicate independently in grapevine seedlings (Zhang et al., 2014).

FIGURE 24.2 Symptoms of yellow speckle viroid on grapevine cv. "Waltham Cross" (South Australia, April 2015).

TAXONOMIC POSITION AND NUCLEOTIDE SEQUENCE

GYSVd-1 comprises 366–368 nt arranged in a rod-like secondary structure (Koltunow and Rezaian, 1988) with 62% of its residues base-paired including 53% G + C. GYSVd-1 and apple scar skin viroid (ASSVd) share 37% sequence identity (Koltunow and Rezaian, 1988). A central conserved region (CCR) common to GYSVd-1 and ASSVd led to the establishment of the genus *Apscaviroid* (Little and Rezaian, 2003). The upper strand of the CCR can form a stem-loop structure that is conserved in all members of the family *Pospiviroidae* (Keese and Symons, 1985; for a review see Flores et al., 1997). Variations reported in the sequence of GYSVd-1 isolates occur mainly within the P domain (Jiang et al., 2012; Polivka et al., 1996; Rigden and Rezaian, 1993; Salman et al., 2014). Host passage does not alter the sequence of GYSVd-1 or AGVd (Elleuch et al., 2003). GYSVd-1 variants can be divided into three types, each adopting a distinct secondary structure within the P domain. Types 1 and 3 are associated with symptoms while type 2 is symptomless (Jiang et al., 2012; Rigden and Rezaian, 1993; Szychowski et al., 1998). GYSVd-1 displayed high sequence variation in a single vineyard and even within the same vine (Jo et al., 2015; Polivka et al., 1996; Salman et al., 2014). Jiang et al. (2009a) reported the occurrence of a variant of GYSVd-1 from China with sequence identities of 88% and 77% to the reference sequence variant of GYSVd-1 (NC_001920) and GYSVd-2 (NC_003612), respectively. Although they named it Chinese grapevine viroid or GYSVd-3, a pairwise comparison of one of its variants (DQ371465) with an Italian variant of GYSVd-1 (EU682452; Carra et al., 2009) showed 99% identity (N. Habili, unpublished). The full-length sequences of GYSVd-1 and AGVd have been detected in a sample of bottled wine (Habili et al., 2015). When compared with viroid sequence Accession No. KF916042, for an isolate from a native table grape in central Iran, the complete sequence of GYSVd-1 in the bottled wine revealed two nucleotide changes at positions 234 and 362 without affecting the secondary structure.

GYSVd-2 has 360–365 nt (67% base-paired), and 73% sequence identity with GYSVd-1 (Fig. 24.1). The CCR of this viroid is identical to that in other apscaviroids (Little and Rezaian, 2003). Since the T_L domain of GYSVd-2 has a high sequence similarity to that of some pospiviroids (Jiang et al., 2009b; Little and Rezaian, 2003), this suggests that GYSVd-2 originated from RNA recombination among other viroids. In comparison with the sequence of the type isolate, one point mutation ($A_{300}G$) is present in most Chinese, Iranian, and Italian variants suggesting that these may have a common origin (Hajizadeh et al., 2015). Unlike GYSVd-1, variants of GYSVd-2 do not show geographical diversity (Gambino et al., 2014; Jiang et al., 2009b).

AGVd reference variant NC_003553 has 369 nt with 54.7% G + C. This viroid has the longest stretch of uninterrupted base-pairing amongst grapevine apscaviroids (Little and Rezaian, 2003). The genome of AGVd, as originally described by Rezaian (1990), consists of a circular single-stranded RNA adopting a rod-like structure with a CCR identical to other apscaviroids. The reported size of the genome ranges from 361 (Elleuch et al., 2002) to 371 nt (Zaki-aghl et al., 2013), but most isolates have 369 nt. The two extra nucleotides in Iranian isolates with 371 nt form an extra loop in the P domain (Zaki-aghl et al., 2013). Eliminating this loop by substituting $U_{50}A_{51}$ for $A_{50}G_{51}$ did not alter the trafficking of the viroid in cucumber; however its infectivity was reduced by 24% (Zaki-aghl et al., 2013).

GLVd reference variant KR605505 has 328 nt of which 63.7% are base-paired with a 67.3% G + C content. The predicted secondary structure of GLVd is similar to most members of the family *Pospiviroidae* and contains five structural domains. The sequence of the upper and lower CCR is almost identical to that of ASSVd. The terminal conserved region and the poly-purine stretch in the P domain are present in GLVd as in other members of the family *Pospiviroidae*. Aspcaviroids are clustered into two subclades (Zhang et al., 2014), where GYSV-1, -2, and AGVd belong to the GYSVd subclade (Fig. 24.1; Table 24.2), while GLVd

TABLE 24.2 Assignment of Apscaviroids Into Two Subclades Using GLVd as a Reference Viroid

Viroid acronym	Viroid	Accession No.	Identity[a]
GLVd[b]	Grapevine latent viroid	KR_605505	100
CVdVI	Citrus viroid VI	NC_004359	74
PVd2	Persimmon viroid 2	NC_021720	72
CVdV	Citrus viroid V	NC_010165	66
CDVd	Citrus dwarfing viroid	NC_003264	64
ASSVd	Apple scar skin viroid	NC_001340	63
AGVd	Australian grapevine viroid	NC_003553	57
GYSVd-1	Grapevine yellow speckle viroid 1	NC_001920	54
GYSVd-2	Grapevine yellow speckle viroid 1	NC_003612	52
PBCVd	Pear blister canker viroid	NC_001830	43

[a] *Percentage identities are relative to grapevine latent viroid.*
[b] *Acronyms in bold indicate grapevine apscaviroids.*

belongs to the ASSVd subclade. This grouping is supported by a pairwise comparison of GLVd with selected apscaviroids (Table 24.2).

ECONOMIC SIGNIFICANCE

Since most grapevine viroids are latent, they are viewed as not being economically important. An exception is vein banding which is induced due to the synergism between grapevine fanleaf virus (GFLV) and GYSVd-1 or (possibly) GYSVd-2 (Krake and Woodham, 1983; Szychowski et al., 1995). Severe yellow speckle symptoms occur in certain years in Australia in the absence of GFLV, but no data on plant vigor and wine quality from affected vines are available.

GYSVd-2 causes GYS symptoms under certain conditions and this may affect vine health. However, no data on the economic significance of this viroid on vine performance are available.

AGVd and GLVd are latent in grapevine and hence appear to be economically insignificant.

SYMPTOMS AND HOST RANGE

GYSVd-1 requires heat and brilliant sunshine to cause symptoms (Habili and Randles, 2010). This situation is favored by the climate of the Australian region where GYS was first described (Taylor and Woodham, 1972). In a comparative study, a number of cultivars exhibited GYS symptoms in Australia, while vines of the same germplasm in California were symptom-free (Taylor and Woodham, 1972). The high variability in symptom expression of GYS can be related to seasonal variation in temperature, geographic location, and cultivar.

In 2010 an outbreak of GYS occurred in Southern Australia which coincided with a period of high temperature in spring. However, when in another year the mean maximum temperature in mid-spring was near normal, the incidence of GYS was low (Habili and Randles, 2010). When monitoring GYS symptoms along a row of 32 RT-PCR positive vines in cv. "Chardonnay" the GYS symptoms appeared only on 4%–15% of the mature leaves of the same shoot.

To study the GYS symptoms at the molecular level, an apscaviroid specific primer pair (Sano et al., 2000) covering the P domain associated with symptoms of GYSVd-1 (Rigden and Rezaian, 1993) was used. Three vines, which tested positive for GYSVd-1 by RT-PCR, were selected; the first never showed symptoms, the second always showed symptoms, and the third showed symptoms in the hot season but not in the cool season. A total of 120 clones were sequenced. As shown in Fig. 24.3, a three-base signature of $A_{309}CU_{311}$ in the P domain (Little and Rezaian, 2003), which was present in the symptomatic vines (Salman et al., 2014), was used as a reference.

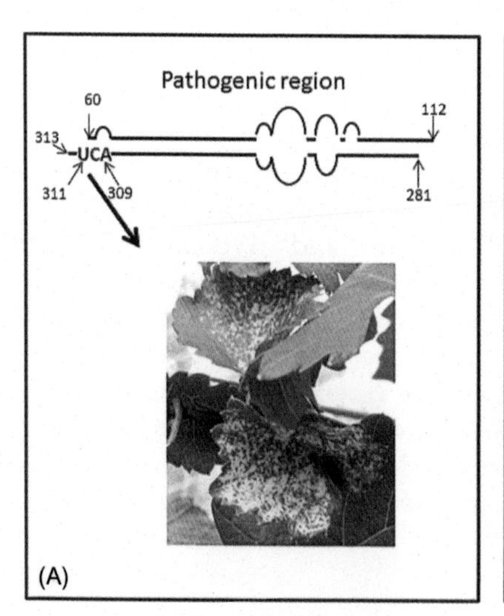

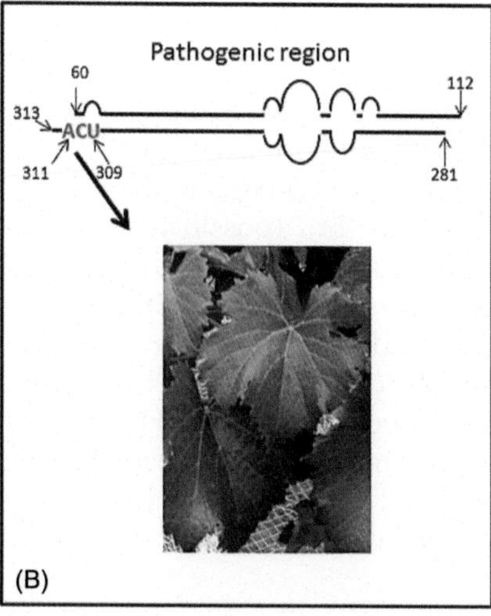

FIGURE 24.3 A natural base change from UCA to ACU (as indicated) in the P domain of GYSVd-1 was associated with the transition from symptomatic (A) in the hot summer of 2010 to asymptomatic (B) in the mild summer of 2012. Numbering of nucleotides in the sequence as described by Little and Rezaian (2003).

This signature was present in all clones of symptomatic vines in the hot season when they expressed symptoms. In nonsymptomatic vines during the cooler season, the sequence was reversed to $A_{311}CU_{309}$ in eight of 38 clones (Salman et al., 2014). The latter sequence, which was not found in GenBank, does not appear to be an error associated with reverse transcriptase or Taq DNA polymerase as it appears only in the P domain of viroids from certain vines (Fig. 24.3). Applying base changes, insertions, or deletions in targeted domains in infectious clones of GYSVd-1 followed by mechanical inoculation is needed to assign any specific sequence to the YS symptom (Zaki-aghl and Izadpanah, 2015).

The viroid variant in Chardonnay in Australia had the highest identity with a variant of GYSVd-1 in grapevine cv. "Nebbiolo" reported from Italy (Carra et al., 2009). It is interesting that an isolate of GYSVd-1 in *Vitis vinifera* cv. "Cabernet Sauvignon" from China, which was sequenced by de novo assembly of the grapevine transcriptome, was also highly similar to the Italian isolate (Jo et al., 2015), suggesting a recent introduction of the Chinese vine from a European source plant.

Apart from *V. vinifera*, GYSVd-1 has been detected in various *Vitis* spp. including *V. cinerea*, *V. coignetiae*, *V. aestivalis*, and hybrid rootstocks (Gambino et al., 2014). No herbaceous host for this viroid has been identified.

Infectious clones of GYSVd-2 induce yellow speckle symptoms in viroid-free grapevine seedlings (Koltunow et al., 1989). This viroid replicates only in the grapevine and, like GYSVd-1, it induces symptoms under strict environmental conditions. Similarly, a vein-banding syndrome occurs in coinfections of GYSVd-2 and GFLV. No herbaceous host is known for this viroid.

No symptoms have been attributed to AGVd in grapevine (Elleuch et al., 2003; Rezaian et al., 1988; Zaki-aghl et al., 2013). AGVd is transmissible to several herbaceous plants by artificial inoculation. Zaki-aghl et al. (2013) reported AGVd infection in tomato and cucumber by stem injection of nucleic acids extracted from grapevine. They also agroinfiltrated dimeric constructs into cucumber, tomato, *Cucurbita pepo* (squash), *Gynura aurantiaca*, *Calendula officinalis*, and *Nicotiana glutinosa*, in which symptoms were observed, whereas *Nicotiana tabacum* (cv. Turkish) was immune (Zaki-aghl et al., 2013). The viroid induced stunting in cucumber, stunting, leaflet deformation, and mottling in tomato, twisting and leaf edge sharpening in *G. aurantiaca*, and mottling and faint vein banding in *N. glutinosa*. Symptom severity in tomato was altered when the TL and P domains of AGVd were replaced by those of GYSVd-1. Exchange of the V domain also affected the symptom type whereas exchange of the TR domain did not have any effect on symptoms (Zaki-aghl and Izadpanah, 2015).

No symptoms or economic impact are associated with GLVd infection of grapevine, which is the only known host plant.

TRANSMISSION

Although no vector is known for GYSVd-1 and GYSVd-2, the natural spread of yellow speckle disease has been reported (Taylor and Woodham, 1972). The disease can spread via vegetative propagation. Spread by pruning tools in vineyards has been reported (Szychowski et al., 1988) but this could not be confirmed (Shanmuganathan and Fletcher, 1980; Staub et al., 1995). Transmission of GYSVd-1 via infected seeds in eight grapevine varieties has been shown (Wan Chow Wah and Symons, 1999; N. Habili, unpublished).

Natural transmission of AGVd has not been reported. The most common route of spread is by planting infected propagating material.

Healthy grapevine seedlings were infected following slash inoculation with in vitro transcripts of GLVd.

DETECTION

Gel electrophoresis, nucleic acid hybridization, RT-PCR, and NGS are used for the detection of GYSVd-1. Multiplex RT-PCR, as compared with single RT-PCR, was less sensitive for the detection of this viroid (Gambino et al., 2014). The most robust primers for the detection of GYSVd-1 and other apscaviroids are PBCVd94H and PBCVd100C (Sano et al., 2000). Using these primers, GYSVd-1 was detected in a "Cabernet Sauvignon" wine bottled for 10 years (Habili and Randles, 2013). Recently, NGS and de novo assembly of the grapevine transcriptome has been used to identify the full-length sequence of GYSVd-1 in grapevine (Jo et al., 2015). High throughput sequencing detected both GYSVd-1 and hop stunt viroid in Malbec, a variety of *V. vinifera* growing in Langhorne Creek, South Australia (Habili and Al Rwahnih, unpublished).

Gel electrophoresis, nucleic acid hybridization, and RT-PCR have been used for the detection of GYSVd-2 (Gambino et al., 2014; Hajizadeh et al., 2012; Little and Rezaian, 2003). Zhang et al. (2012) designed a polyprobe for the simultaneous detection of four grapevine viroids. GYSVd-1 and GYSVd-2 could be detected by northern-blot hybridization on a single membrane using a mixture of digoxygenin-labeled probes (Hajizadeh et al., 2012). The most commonly used RT-PCR for the detection of GYSVd-2 is that described by Jiang et al. (2009b).

AGVd is detected by RT-PCR (Gambino et al., 2014).

Northern-blot hybridization, RT-PCR, and NGS have been used for the detection of GLVd (Zhang et al., 2014).

GEOGRAPHICAL DISTRIBUTION AND EPIDEMIOLOGY

GYSVd-1 is present wherever *V. vinifera* is grown. Jiang et al. (2012) compared the sequences of GYSVd-1 in Japan and China and observed that each sequence variant was regionally distinct as was confirmed by Gambino et al. (2014).

GYSVd-2 has a high incidence in Iran with over 60% of tested samples positive and showing sequence variation (Hajizadeh et al., 2015; Zaki-aghl et al., 2013). GYSVd-2 isolates from China (Jiang et al., 2009b), Italy (Gambino et al., 2014), and Iran show high sequence similarity. No significant clustering of the GYSVd-2 sequence variants obtained from different geographical regions has been observed by phylogenetic analysis (Gambino et al., 2014; Hajizadeh et al., 2015). However, the highest incidence of the viroid within a wide range of local own-rooted grapevine cultivars has been reported from Iran, where grapevine was domesticated (see also Bar-Joseph, 2003).

AGVd has a worldwide distribution, but it seems to have a higher incidence in Tunisia, Iran, and India (Adkar-Purushothama et al., 2014; Elleuch et al., 2003; Hajizadeh et al., 2015; Zaki-aghl et al., 2013). The highest incidence (95%) in 137 samples of locally grown, own-rooted table grape cultivars has been reported from Iran (Hajizadeh et al., 2015). Geographically, AGVd is likely to have originated from Iran, where a very high viroid incidence has been recorded (Hajizadeh et al., 2015; Zaki-aghl et al., 2013). A phylogenetic tree shows that Iranian isolates are closer to the Chinese isolates, while Tunisian isolates clustered with Italian isolates (Gambino et al., 2014). There is insufficient evidence to correlate the differences in viroid size with the grapevine cultivar or geographical region. However, phylogenetic analysis showed that a close relationship exists among AGVd variants from Italy, Tunisia, and Australia (Gambino et al., 2014).

GLVd has only been reported from China.

CONTROL

Since elimination of GYSV-1 viroid by somatic embryogenesis is achievable (Gambino et al., 2011), it would be possible to use this technique to compare the yield and other characters in healthy and infected vines of the same vineyard. The plants must be present in an environment where the chance of symptom induction is high (e.g., Australia).

CONCLUDING REMARKS

Grapevine apscaviroids seem to have little economic impact on the productivity and quality of grapevines. This is probably the result of careful selection of symptomless vines for continued vegetative propagation by generations of traditional vine producers, long before the discovery of viroids (Bar-Joseph, 2003). With the recent advent of NGS and metagenomics, it seems possible that additional grapevine viroids will be discovered.

Acknowledgments

Thanks are due to Dr. Moshe Bar-Joseph for his valuable comments on the manuscript, to Ms. Qi Wu for technical assistance and to Dr. Mohammad Hajizadeh for his advice on bioinformatics.

References

Adkar-Purushothama, C.R., Kanchepalli, P.R., Yanjarappa, S.M., Zhang, Z., Sano, T., 2014. Detection, distribution, and genetic diversity of Australian grapevine viroid in grapevines in India. Virus Genes 49, 304–311.

Bar-Joseph, M., 2003. Natural history of viroids, horticultural aspects. In: Hadidi, A., Flores, R., Randles, J.W., Semancik, J.S. (Eds.), Viroids. CSIRO Publishing, Collingwood, VIC, pp. 246–251.

Carra, A., Mica, E., Gambino, G., Pindo, M., Moser, C., Schubert, A., 2009. Cloning and characterization of small non-coding RNAs from grape. Plant J. 59, 750–763.

Elleuch, A., Fakhfakh, H., Pelchat, M., Landry, P., Marrakchi, M., Perreault, J.P., 2002. Sequencing of Australian grapevine viroid and yellow speckle viroid isolated from a Tunisian grapevine without passage in an indicator plant. Eur. J. Plant Pathol. 108, 815–820.

Elleuch, A., Marrakchi, M., Perreault, J.P., Fakhfakh, H., 2003. First report of Australian grapevine viroid from the Mediterranean region. J. Plant Pathol. 85, 53–57.

Flores, R., Di Serio, F., Hernández, C., 1997. Viroids: the non-encoding genomes. Semin. Virol. 8, 65–73.

Gambino, G., Navarro, B., Torchetti, E.M., La Notte, P., Schneider, A., Mannini, F., et al., 2014. Survey on viroids infecting grapevine in Italy: identification and characterization of Australian grapevine viroid and grapevine yellow speckle viroid 2. Eur. J. Plant Pathol. 140, 199–205.

Gambino, G., Navarro, B., Vallania, R., Gribaudo, I., Di Serio, F., 2011. Somatic embryogenesis efficiently eliminates viroid infections from grapevines. Eur. J. Plant Pathol. 130, 511–519.

Habili, N., Randles, J.W., 2010. Sudden increase in the onset of grapevine yellow speckle disease in 2010. Aust. NZ Grapegrower and Winemaker 558, 23–26.

Habili, N., Randles, J.W., 2013. First report of the detection of grapevine yellow speckle viroid 1 in bottled wine. Abstracts of International Workshop on Viroids and Satellite RNAs (IWVdS). Beijing, China, p. 16.

Habili, N., Wu, Q. Hajizadeh, M., Randles, J.W., 2015. Identification and determination of full-length sequence of three grapevine viroids in a decade old bottled "Cabernet Sauvignon" wine. Proceedings of the 18th Congress of the International Council for the Study of Virus and Virus-Like Diseases of the Grapevine (ICVG), Ankara, Turkey, September 6–11, 2015, pp. 70–71.

Hajizadeh, M., Navarro, B., Sokhandan-Bashir, N., Torchetti, E.M., Di Serio, F., 2012. Development and validation of a multiplex RT-PCR method for the simultaneous detection of five grapevine viroids. J. Virol. Methods 179, 62–69.

Hajizadeh, M., Torchetti, E.M., Sokhandan-Bashir, N., Navarro, B., Doulati-Baneh, H., Martelli, G.P., et al., 2015. Grapevine viroids and grapevine fanleaf virus in north-west of Iran. J. Plant Pathol. 97, 363–369.

Jiang, D., Sano, T., Tsuji, M., Araki, H., Sagawa, K., Purushothama, C.R.A., et al., 2012. Comprehensive diversity analysis of viroids infecting grapevine in China and Japan. Virus Res. 169, 237–245.

Jiang, D., Zhang, Z., Wu, Z., Guo, R., Wang, H., Fan, P., et al., 2009b. Molecular characterization of grapevine yellow speckle viroid 2 (GYSVd-2). Virus Genes 38, 515–520.

Jiang, D.M., Guo, R., Wu, Z.J., Wang, H.Q., Li, S.F., 2009a. Molecular characterization of a member of a new species of grapevine viroid. Arch. Virol. 154, 1563–1566.

Jo, Y., Choi, H., Yoon, J.-Y., Choi, S.-K., Cho, W.K., 2015. *De novo* genome assembly of grapevine yellow speckle viroid 1 from a grapevine transcriptome. Genome Announc. 3, 1–2.

Keese, P., Symons, R.H., 1985. Domains in viroids: evidence of intermolecular RNA rearrangements and their contribution to viroid evolution. Proc. Natl. Acad. Sci. USA 82, 4582–4586.

Koltunow, A.M., Krake, L.R., Johnson, S.D., Rezaian, M.A., 1989. Two related viroids cause grapevine yellow speckle disease independently. J. Gen. Virol. 70, 3411–3419.

Koltunow, A.M., Rezaian, M.A., 1988. Grapevine yellow speckle viroid: structural features of a new viroid group. Nucleic Acids Res. 16, 849–864.

Krake, L.R., Woodham, R.C., 1983. Grapevine yellow speckle agent implicated in the aetiology of vein banding disease. Vitis 22, 40–50.

Little, A., Rezaian, M.A., 2003. Grapevine viroids. In: Hadidi, A., Flores, R., Randles, J.W., Semancik, J.S. (Eds.), Viroids. CSIRO Publishing, Collingwood, VIC, pp. 195–206.

Polivka, H., Staub, U., Gross, H.J., 1996. Variation of viroid profiles in individual grapevine plants: novel grapevine yellow speckle viroid 1 mutants show alterations in hairpin I. J. Gen. Virol. 77, 155–161.

Rezaian, M.A., 1990. Australian grapevine viroid – evidence for extensive recombination between viroids. Nucleic Acids Res. 18, 1813–1818.

Rezaian, M.A., Koltunow, A.M., Krake, L.R., 1988. Isolation of three viroids and a circular RNA from grapevine. J. Gen. Virol. 69, 413–422.

Rigden, J.E., Rezaian, M.A., 1993. Analysis of sequence variation in grapevine yellow speckle viroid 1 reveals two distinct alternative structures for the pathogenic domain. Virology 193, 474–477.

Salman, T.M., Habili, N., Shi, B.J., 2014. Effect of temperature on symptom expression and sequence polymorphism of grapevine yellow speckle viroid 1 in grapevine. Virus Res. 189, 243–247.

Sano, T., Kobayashi, T., Ishiguro, A., Motomura, Y., 2000. Two types of grapevine yellow speckle viroid 1 isolated from commercial grapevine had the nucleotide sequence of yellow speckle symptom-inducing type. J. Gen. Plant Pathol. 66, 68–70.

Shanmuganathan, N., Fletcher, G., 1980. The incidence of grapevine yellow speckle disease in Australian grapevines and the influence of inoculum on symptom expression. Aust. J. Agric. Res. 31, 327–333.

Staub, U., Polivka, H., Herrmann, J.V., Gross, H.J., 1995. Transmission of grapevine viroids is not likely to occur mechanically by regular pruning. Vitis 34, 119–123.

Szychowski, J.A., Credi, R., Reanwarakorn, K., Semancik, J.S., 1998. Population diversity in grapevine yellow speckle viroid 1 and the relationship to disease expression. Virology 248, 432–444.

Szychowski, J.A., Goheen, A.C., Semancik, J.S., 1988. Mechanical transmission and rootstock reservoirs as factors in the widespread distribution of viroids in grapevines. Amer. J. Enol. Viticult 39, 213–216.

Szychowski, J.A., McKenry, M.V., Walker, M.A., Wolpert, J.A., Credi, R., Semancik, J.S., 1995. The vein-banding disease syndrome: a synergistic reaction between grapevine viroids and fanleaf virus. Vitis 34, 229–232.

Tamura, K., Stecher, G., Peterson, D., Filipski, A., Kumar, S., 2013. MEGA6: molecular evolutionary genetics analysis version 6.0. Mol. Biol. Evol. 30, 2725–2729.

Taylor, R.H., Woodham, R.C., 1972. Grapevine yellow speckle – a newly recognized graft-transmissible disease of *Vitis*. Aust. J. Agric. Res. 23, 447–452.

Wan Chow Wah, Y.F., Symons, R.H., 1999. Transmission of viroids via grape seeds. J. Phytopathol. 147, 285–291.

Zaki-aghl, M., Izadpanah, K., 2015. Symptom alterations in Australian grapevine viroid chimera. Proceedings of the 18th Congress of the International Council for the Study of Virus and Virus-like Diseases of the Grapevine (ICVG), Ankara, Turkey, September 6–11, pp. 208–209.

Zaki-aghl, M., Izadpanah, K., Niazi, A., Behjatnia, S.A.A., Afsharifar, A.R., 2013. Molecular and biological characterization of the Iranian isolate of the Australian grapevine viroid. J. Agr. Sci. Tech. 15, 855–865.

Zhang, Z., Peng, S., Jiang, D., Pan, S., Wang, H., Li, S., 2012. Development of a polyprobe for the simultaneous detection of four grapevine viroids in grapevine plants. Eur. J. Plant Pathol. 132, 9–16.

Zhang, Z., Qi, S., Tang, N., Zhang, X., Chen, S., Zhu, P., et al., 2014. Discovery of replicating circular RNAs by RNA-seq and computational algorithms. PLoS Pathog. 10, e1004553.

CHAPTER

25

Coconut Cadang-Cadang Viroid and Coconut Tinangaja Viroid

Ganesan Vadamalai[1], *Sathis S. Thanarajoo*[1], *Hendry Joseph*[2], *Lih L. Kong*[1] *and John W. Randles*[3]

[1]Universiti Putra Malaysia, Selangor, Malaysia [2]Universiti Teknologi MARA (Sabah), Sabah, Malaysia [3]The University of Adelaide, Waite Campus, Glen Osmond, SA, Australia

INTRODUCTION

Coconut cadang-cadang viroid (CCCVd) causes cadang-cadang disease of coconut palms (*Cocos nucifera* L.) in the Philippines (Hanold and Randles, 1991). Chapter 4, Economic Significance of Palm Tree Viroids, describes the current disease epidemic. The name "cadang-cadang" derived from the word "gadan-gadan" of a Philippines' local dialect refers to a premature decline and gradual death of coconut palms (Randles and Rodriguez, 2003). Variants of CCCVd occur in commercial African oil palms (*Elaeis guineensis* Jacq.) with orange spotting (OS) of leaves. They show over 90% sequence identity to CCCVd from coconut in the Philippines but none are identical with CCCVd (Vadamalai et al., 2006; Wu et al., 2013).

Cadang-cadang is lethal and imposes considerable economic loss on farmers because diseased palms cease nut production years before they die (Randles, 2003). The effects of the newly recognized oil palm variants of CCCVd on oil palm production in South East Asia and the South Pacific region are yet to be determined (Hanold and Randles, 1991; Vadamalai, 2005).

Coconut tinangaja or "yellow mottle decline" disease in Guam (Reinking, 1961) is caused by coconut tinangaja viroid (CTiVd) (Wall

Viroids and Satellites.
DOI: http://dx.doi.org/10.1016/B978-0-12-801498-1.00025-5

and Wiecko, 2000). The molecular structure of CTiVd is similar to that of CCCVd but it has only about 64% sequence identity (Keese et al., 1988). Cadang-cadang and tinangaja diseases differ in symptomatology (Boccardo et al., 1981).

Here we describe the molecular structure and taxonomy of CCCVd and CiTVd, their associated diseases, identification, transmission, and epidemiology.

TAXONOMIC POSITION AND NUCLEOTIDE SEQUENCE

Coconut cadang-cadang viroid (Genbank accession number J02049) is the type species of the genus *Cocadviroid* in the family *Pospiviroidae* (Di Serio et al., 2014). The basic form of CCCVd is a 246 nt RNA (CCCVd 246), which is the smallest viroid size known (Fig. 25.1A). There are numerous variants of the basic 246 nt sequence (Haseloff et al., 1982; Imperial et al., 1981; Mohamed et al., 1982). Additional forms include a small 247 nt monomer, and larger molecules arising from the reiteration of 41, 50, or 55 nt in the terminal right domain to give 287, 296, 297, and 301 nt forms (Haseloff et al., 1982). Both small and large monomers are infectious (Randles et al., 1992). Dimeric forms accompany each monomer (Randles, 1985).

The oil palm variants (comprising 246, 270, 293, or 297 nt) are classified as CCCVd because they show greater than 90% sequence identity

(A)

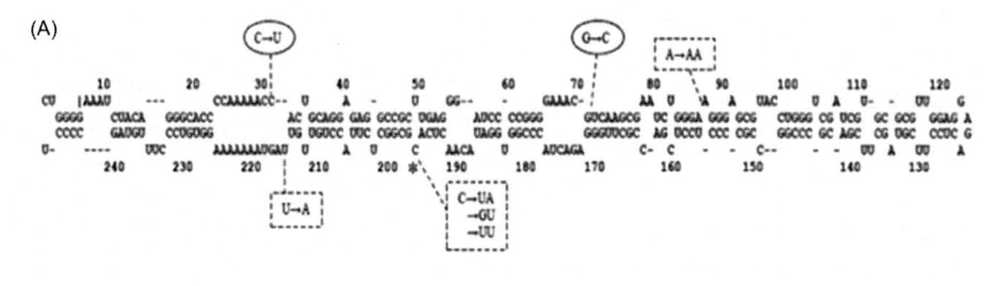

(B)

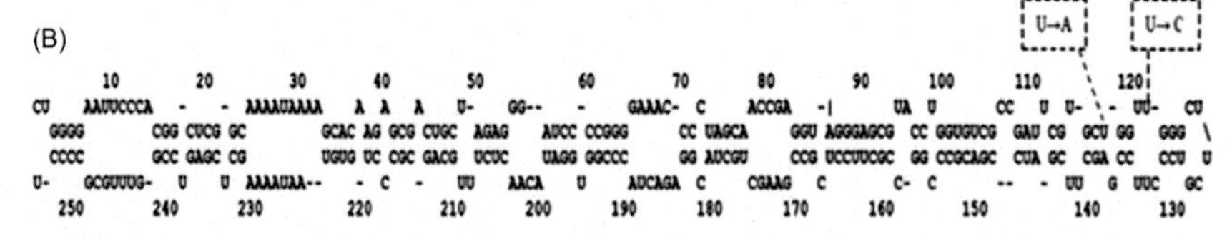

FIGURE 25.1 (A) Sequence and structure of CCCVd 246 (Genbank J02049). Substitutions in dashed text boxes are associated with the severe coconut leaflet "brooming" phenotype (Rodriguez and Randles, 1993), those in oval text boxes are for the CCCVd oil palm variants. (B) Sequence and structure of CTiVd (Genbank M20731). Text boxes show sequence variations between isolates.

with the coconut isolates of CCCVd. The oil palm variants have nucleotide substitutions at position 31 (C to U) and position 70 (G to C) compared to CCCVd from coconut (Vadamalai et al., 2006; Wu et al., 2013; Fig. 25.1A).

Coconut tinangaja viroid is a separate species in the genus *Cocadviroid*. It comprises 254 nt with 64% sequence identity to CCCVd (Keese et al., 1988) (Fig. 25.1B). It also has dimeric forms (Boccardo et al., 1981).

ECONOMIC SIGNIFICANCE

CCCVd is considered to be a serious economic threat to coconut and oil palm production in South East Asia and the South Pacific (see Chapter 4: Economic Significance of Palm Tree Viroids). Up to 2003, the total loss in the Philippines due to cadang-cadang was estimated to be 40 million coconut palms at a cost of $4 billion (Randles, 2003). Oil palms with both OS symptoms and infected with CCCVd-related RNA were reduced in height by 30% (Randles, 1998) and oil palms in Nigeria with OS symptoms reportedly showed a yield reduction of about 50% (Forde and Leyritz, 1968).

Tinangaja disease continues to damage coconut palm populations on Guam. Although Guam does not produce copra commercially, tinangaja still has an impact in reducing the supply of coconuts for the fresh fruit market (Wall and Randles, 2003).

SYMPTOMS

Cadang-Cadang

Coconut cadang-cadang disease progresses through early, medium, and late stages (Randles et al., 1992; Fig. 25.2). During the early stage, newly developing nuts become more rounded with equatorial scarifications, chlorotic leaf spots appear on expanded fronds, and inflorescences become stunted with necrotic tips. During the medium stage, spathe, inflorescence, and nut production decline, then cease, and leaf spots enlarge and become more numerous. During the late stage, fronds decline in size and number, leaflets become brittle, leaf spots coalesce to produce a general chlorosis, and trunk diameter and crown size are reduced. Then the palm dies. Cadang-cadang generally appears only after flowering commences. Palms developing the disease before flowering never produce nuts. "Winging" of petioles is a useful additional indicator of disease (Fig. 25.2C).

FIGURE 25.2 Cadang-cadang symptoms on coconut palm. (A) Affected plantation with range of disease stages. (B) Early stage with prolific production of rounded nuts. (C) "Winging" due to fiber retention on petiole. (D) Equatorial scarification of rounded nuts at early stage. (E) Chlorotic leaf spots on an older frond at mid to late stage. (F) Uncommon severe "brooming" symptom due to greatly reduced lamina.

Orange Spotting

OS is a bright orange nonnecrotic spotting on expanded leaflets of oil palm (Hanold and Randles, 1991). In young fronds, the spots are interveinal, vary in shape, and are about 2–3 mm long. In older fronds the spots have coalesced into larger patches, and distal necrosis of leaflets is seen in the oldest fronds (Forde and Leyritz, 1968; Hanold and Randles, 1991). OS palms are generally stunted, appear bronze-colored from a distance, and bear smaller or no bunches (Figs. 25.3 and 4.2).

Tinangaja

Tinangaja disease is distinguished by small elongated nuts without a kernel (Boccardo et al., 1981). It progresses through a reduction in the number of fronds and the size and number of fruit, shriveling and

FIGURE 25.3 Symptoms and effects of OS on oil palm. (A) Bright orange frond of an OS palm. (B) Young fronds (f3, left) are free of spots compared with older fronds (f10, center, f20, right). (C) OS translucent under the sunlight. (D) Oil soaked-like spotting underside of the OS leaflets. (E) A 4-year old healthy palm with fruit bunches. (F) A 4-year old OS palm with no fruit bunch production.

deformation of fruit, cessation of fruit production, failure to produce inflorescences, tapering of the distal end of the trunk, persistence of stipules, stippling of leaflets (fine chlorotic spots), and brittleness of leaflets and fronds (Fig. 25.4). Affected palms die within about 15 years (Wall and Randles, 2003).

HOST RANGE

Natural infection with CCCVd has been observed in coconut palm, oil palm, *Corypha elata* (buri palm), and *Livistona rotundifolia* in the

FIGURE 25.4 Tinangaja symptoms on coconut palm. (A) Affected nonbearing palms with reduction of crown. (B) Deformed nuts and sterile inflorescences. (C) Fine chlorotic leaf spots. (D) Shriveled nuts without endosperm, compared with a normal size nut (right).

Philippines (Randles et al., 1980), and with a CCCVd variant in oil palm in Malaysia (Wu et al., 2013). CCCVd has been successfully inoculated to the following species in the family *Arecaceae*: *Areca catechu* (betel nut palm), *C. nucifera* (82 coconut populations), *C. elata*, *Adonidia merrillii* (manila palm), *E. guineensis*, *Chrysalidocarpus lutescens* (palmera), *Oreodoxa regia* (royal palm), *Ptycosperma macarthuri* (Macarthur palm), and *Phoenix dactylifera* (date palm) (Imperial et al., 1985). Most became stunted, and all showed yellow leaf spotting. Related viroid-like RNA sequences have been found naturally in members of the families *Arecaceae* and *Pandanaceae*, as well as herbaceous monocotyledons in the families *Zingiberaceae*, *Marantaceae*, and *Commelinaceae* (Hanold and Randles, 1991, 1998).

No host range studies have been done with CTiVd. Both *A. catechu* and *Pandanus* sp. occur on Guam and are potential natural reservoirs of CTiVd (Wall and Randles, 2003).

TRANSMISSION

CCCVd and CTiVd are mechanically transmissible to coconut seedlings by high pressure injection and razor slashing with nucleic acid inoculum (Imperial et al., 1985; Wall and Wiecko, 2000). The mode of natural transmission of these viroids is not known. Experimental evidence has been obtained for a low rate of transmission to coconut palms of CCCVd through pollen and seed (Pacumbaba et al., 1994). Mechanical transmission in the field via harvesting scythes or machetes has not been demonstrated, although sanitation is nevertheless recommended (Randles and Rodriguez, 2003). Chewing and sap-sucking insects failed to transmit CCCVd to coconut (Zelazny et al., 1982). Pollen and seed transmission of CTiVd has not been demonstrated. Red scale insects and mealybugs feeding on shriveled nuts reacted positively when tested with digoxigenin-labeled and radiolabeled oligonucleotide probes (Wall and Randles, 2003) but their ability to transmit has not been reported.

Oil palm may be a useful model for studying CCCVd. A time-course survey of a site in an oil palm plantation in the Solomon Islands provided strong evidence for the natural spread of OS and the associated CCCVd-related viroid from a primary focus (Randles, 1998). Moreover, a cDNA clone of the oil palm variant CCCVd 246_{OP} has been mechanically transmitted to oil palm seedlings. OS was observed within 6 months of inoculation, confirming the pathogenicity of CCCVd 246_{OP} (Thanarajoo, 2014). Other variants remain to be tested.

DISEASE IDENTIFICATION

Diagnosis of cadang-cadang requires detection of CCCVd for confirmation. Steps include extraction of total RNA from fresh or preserved leaf samples and fractionation of the viroid molecules using polyacrylamide gel electrophoresis followed by silver staining (Hanold and Randles, 1991; Imperial et al., 1985). Gels of 5%–20% can be used in one- or two-dimensional assays (Hanold et al., 2003; and Chapter 34: Gel Electrophoresis). Hybridization with labeled complementary RNA or DNA probes is more sensitive and specific, and is used in either dot-blot or northern blot assays at varying stringencies (Hanold and Randles, 1991; Imperial et al., 1985).

Reverse transcription PCR (RT-PCR) specific for CCCVd (Hodgson et al., 1998) is used for the detection of CCCVd variants in oil palm with target specific primers. PCR products are cloned and sequenced for identification of variants (Vadamalai et al., 2006; Wu et al., 2013). A ribonuclease protection assay (RPA) has been developed for detecting the 246 nt sequence variant of CCCVd (Vadamalai et al., 2009). Reverse transcription loop-mediated isothermal amplification, a highly specific method using four inner and outer primers which recognize six distinct regions of the sequence, has been developed for the rapid detection of CCCVd variants in the oil palm (Thanarajoo et al., 2014). Thus, a range of techniques is now available for identifying CCCVd and its variants.

For the specific detection of CTiVd, Hodgson et al. (1998) developed an extraction and analytical agarose gel electrophoresis method combined with either diagnostic oligonucleotide probe hybridization or RT-PCR using the specific oligonucleotide as one of the two primers.

GEOGRAPHICAL DISTRIBUTION AND EPIDEMIOLOGY

Coconut cadang-cadang disease and CCCVd are found in the central eastern Philippines, (Hanold and Randles, 1991; Randles and Rodriguez, 2003; and Chapter 4: Economic Significance of Palm Tree Viroids). The highest incidence is in the Bicol region, Masbate, Catanduanes, Samar, and other smaller islands in the central Philippines zone. No cadang-cadang has been reported in the whole of Mindanao. Surveys of incidence indicate that cadang-cadang spreads at about 0.5 km p.a., and thus the distribution is relatively stable.

Sequence variants of CCCVd have been identified in African oil palm in Malaysia in association with OS (Vadamalai, 2005; Wu et al., 2013). OS was first recognized in West Africa (Forde and Leyritz, 1968) and since then has been observed in commercial oil palm plantations in Indonesia, Malaysia, Thailand, Papua New Guinea, the Philippines, the Solomon Islands, and Central and South America at incidences of 0.1%–10% (Hanold and Randles, 1991; Randles, 1998). Incidences can be much higher, and Selvaraja et al. (2012) studied a site with an incidence of 74.3% and reported that severity of OS symptoms ranged from low to high, with evidence for spatial clustering. The global distribution of OS could be explained by vertical transmission of the causal agents and their distribution with oil palm seed during the international expansion of commercial plantations which have a narrow genetic base (Randles, 1998). Although OS is not lethal and considered to be a relatively minor disease, the removal of OS and the viroid from plantations may lead to a significant economic gain in yield and quality (Randles, 2003).

In addition, CCCVd-related RNA has been identified by hybridization analysis in a number of other monocotyledons in Asia and Oceania (Hanold and Randles, 1991, 1998). Sequence information has not been obtained. Also, a number of coconut palm diseases of unknown etiology occur in Sri Lanka and CCCVd-related RNA was detected by hybridization and RPA (Hanold and Randles, 1991, 1998; Vadamalai et al., 2009). However, much more research is needed to understand the biological significance of these results especially given the widespread distribution of CCCVd-related RNA sequences in a range of plant species across Asia and Oceania.

Tinangaja is found only on Guam, and possibly also Anatahan, both in the Marianas archipelago (Boccardo et al., 1981). The distribution of the disease on Guam appears to be clustered, not random (Wall and Randles, 2003).

CONTROL

CCCVd is not controlled. Specific control recommendations cannot be developed until the epidemiology of CCCVd is more clearly understood. Exclusion is the only method considered to be effective in controlling the spread of CCCVd. No genetically resistant coconut cultivars have been identified experimentally, but some individuals exhibit resistance to multiple inoculation (Bonaobra et al., 1998) and field resistance may be available (Randles and Rodriguez, 2003). The present approach is to remove symptomatic palms within the disease zone, and to supply improved hybrid seedlings as replacements. This is expected to reduce disease incidence with time (Philippine Coconut Authority, 2014). Movement of living coconut tissue, such as seedlings, nuts for germination, and pollen, out of the general cadang-cadang region in the central Philippines is regulated. Molecular indexing is required for quarantine and for identifying areas free of disease for international movement of coconut materials (Carpio, 2011).

The lack of information on the epidemiology of CTiVd precludes the development of specific and targeted control measures. In summary, recommendations for control are (1) elimination of infected trees and their replacement with healthy ones; (2) disinfection of cutting and pruning tools between trees with 10% commercial bleach solution plus 1% mineral or vegetable oil; and (3) collection of seed nuts from areas which appear to be free, or with a very low incidence of tinangaja disease. Long distance exchange of coconut germplasm in the form of embryo cultures is not recommended, until further tests can determine whether or not embryos can be infected (Wall and Randles, 2003).

References

Boccardo, G., Beaver, R.G., Randles, J.W., Imperial, J.S., 1981. Tinangaja and bristle top, coconut diseases of uncertain etiology in Guam, and their relationship to cadang-cadang disease of coconut in the Philippines. Phytopathology 71, 1104–1107.

Bonaobra, Z.S., Rodriguez, M.J.B., Estioko, L.P., Baylon, G.B., Cueto, C.A., Namia, M.T.I., 1998. Screening of coconut populations for resistance to CCCVd using coconut seedlings. In: Hanold, D., Randles, J.W. (Eds). Report on ACIAR-funded research on viroids and viruses of coconut palm and other tropical monocotyledons 1985–1993. ACIAR Working Paper No. 51, Canberra, Australia, pp. 69–75.

Carpio, C.B., 2011. Practical strategies and regulatory measures adopted in the control, management and containment of the cadang-cadang (CCRNA) disease in the Philippines. Report, APCC/MCD & JED/CRI Consultative Meeting on Phytoplasma/ Wilt Diseases in Coconut. Lunuwila, Sri Lanka. 15–17 June 2011, pp. 160–170.

Di Serio, F., Flores, R., Verhoeven, J.T., Li, S.F., Pallás, V., Randles, J.W., et al., 2014. Current status of viroid taxonomy. Arch. Virol. 159, 3467–3478.

Forde, S.C.M., Leyritz, M.J.P., 1968. A study of confluent orange spotting of the oil palm in Nigeria. J. Nigerian Inst. Oil Palm Res. 4, 371–380.

Hanold, D., Randles, J.W., 1991. Detection of coconut cadang-cadang viroid-like sequences in oil and coconut palm and other monocotyledons in the south-west Pacific. Ann. Appl. Biol. 118, 139–151.

Hanold, D., Randles, J.W. (Eds.), 1998. Report on ACIAR-Funded Research on Viroids and Viruses of Coconut Palm and Other Tropical Monocotyledons 1985–1993. ACIAR Working Paper No. 51, Canberra, Australia.

Hanold., D., Semancik, J.S., Owens, R.A., 2003. Polyacrylamide gel electrophoresis. In: Hadidi, A., Flores, R., Randles, J.W., Semancik, J.S. (Eds.), Viroids. CSIRO Publishing, Collingwood, VIC, pp. 95–102.

Haseloff, J., Mohamed, N.A., Symons, R.H., 1982. Viroid RNAs of cadang-cadang disease of coconuts. Nature 299, 316–321.

Hodgson, R.A.J., Wall, G.C., Randles, J.W., 1998. Specific identification of coconut tinangaja viroid for differential field diagnosis of viroids in coconut palms. Phytopathology 88, 774–781.

Imperial, J.S., Rodriguez, M.J.B., Randles, J.W., 1981. Variation in the viroid-like RNA associated with cadang-cadang disease: evidence for an increase in molecular weight with disease progress. J. Gen. Virol. 56, 77–85.

Imperial, J.S., Bautista, R.M., Randles, J.W., 1985. Transmission of the coconut cadang-cadang viroid to six species of palm by inoculation with nucleic acid extracts. Plant Pathol. 34, 391–401.

Keese, P., Osorio-Keese, M.E., Symons, R.H., 1988. Coconut tinangaja viroid: sequence homology with coconut cadang-cadang viroid and other potato spindle tuber viroid related RNAs. Virology 162, 508–510.

Mohamed, N.A., Haseloff, J., Imperial, J.S., Symons, R.H., 1982. Characterisation of the different electrophoretic forms of the cadang-cadang viroid. J. Gen. Virol. 63, 181–188.

Pacumbaba, E.B., Zelazny, B., Orense, J.C., Rillo, E.P., 1994. Evidence for pollen and seed transmission of the coconut cadang-cadang viroid in *Cocos nucifera*. J. Phytopathol. 142, 37–42.

Philippine Coconut Authority, 2014. Terminal report. Coconut *cadang-cadang* Coconut cadang-cadang disease surveillance survey 2012-2013. Philippine Coconut Authority, Quezon City, p. 70.

Randles, J.W., 1985. Coconut cadang-cadang viroid. In: Maramorosch, K., McKelvey, J.J. (Eds.), Subviral Pathogens of Plants and Animals: Viroids and Prions. Academic Press, Orlando, FL, pp. 39–74.

Randles, J.W., 1998. CCCVd-related sequences in species other than coconut. In: Hanold, D., Randles, J.W. (Eds.), Report on ACIAR-Funded Research on Viroids and Viruses of Coconut Palm and Other Tropical Monocotyledons 1985-1993. ACIAR Working Paper No. 51, ACIAR, Canberra, Australia, pp. 144–152.

Randles, J.W., 2003. Economic impact of viroid diseases. In: Hadidi, A., Flores, R., Randles., J.W., Semancik, J.S. (Eds.), Viroids. CSIRO Publishing, Collingwood, VIC, pp. 3–11.

Randles, J.W., Rodriguez, M.J.B., 2003. Coconut cadang-cadang viroid. In: Hadidi, A., Flores, R., Randles., J.W., Semancik, J.S. (Eds.), Viroids. CSIRO Publishing, Collingwood, VIC, pp. 233–241.

Randles, J.W., Boccardo, G., Imperial, J.S., 1980. Detection of the cadang-cadang RNA in African oil palm and buri palm. Phytopathology 70, 185–189.

Randles, J.W., Hanold, D., Pacumbaba, E.P., Rodriguez, M.J.B., 1992. Cadang-cadang disease of coconut palm. In: Mukhopadhyay, A.N., Kumar, J., Chaube, H.S., Singh, U.S. (Eds.), Plant Diseases of International Importance. Diseases of Sugar, Forest and Plantation Crops, vol. 4. Prentice Hall Inc, New Jersey, pp. 277–295.

Reinking, O.A., 1961. Yellow mottle decline in the territory of Guam. Plant Dis. Rep. 45, 599–604.

Rodriguez, M.J.B., Randles, J.W., 1993. Coconut cadang-cadang viroid (CCCVd) mutants associated with severe disease vary in both the pathogenicity domain and the central conserved region. Nucleic Acids Res. 21, 2771.

Selvaraja, S., Balasundram, S.K., Vadamalai, G., Husni, M.H.A., 2012. Spatial variability of orange spotting disease in oil palm. J. Biol. Sci. 12, 232–238.

Thanarajoo, S.S., 2014. Rapid detection, accumulation and translocation of coconut cadang-cadang viroid (CCCVd) variants in oil palm. PhD Thesis, Universiti Putra Malaysia, 122 pp.

Thanarajoo, S.S., Kong, L.L., Kadir, J., Lau, W.H., Vadamalai, G., 2014. Detection of coconut cadang-cadang viroid (CCCVd) in oil palm by reverse transcription loop-mediated isothermal amplification (RT-LAMP). J. Virol. Methods 202, 19–23.

Vadamalai, G., 2005. An investigation of oil palm orange spotting disorder. PhD Thesis, University of Adelaide. 155 pp.

Vadamalai, G., Hanold, D., Rezaian, M.A., Randles, J.W., 2006. Variants of the coconut cadang-cadang viroid isolated from an African oil palm (*Elaeis guineensis* Jacq.) in Malaysia. Arch. Virol. 151, 1447–1456.

Vadamalai, G., Perera, A.A.F.L.K., Hanold, D., Rezaian, M.A., Randles, J.W., 2009. Detection of coconut cadang-cadang viroid sequences in oil and coconut palm by ribonuclease protection assay. Ann. Appl. Biol. 154, 117–125.

Wall, G.C., Randles, J.W., 2003. Coconut tinangaja viroid. In: Hadidi, A., Flores, R., Randles, J.W., Semancik, J.S. (Eds.), Viroids. CSIRO Publishing, Collingwood, VIC, pp. 242–245.

Wall, G.C., Wiecko, A.T., 2000. Molecular properties of the coconut tinangaja viroid (CTiVd) and its pathogenicity. Proceedings of the CAS Research Conference, April 26–27, 2000. University of Guam.

Wu, Y.H., Cheong, L.C., Meon, S., Lau, W.H., Kong, L.L., Joseph, H., et al., 2013. Characterization of coconut cadang-cadang viroid variants from oil palm affected by orange spotting disease in Malaysia. Arch. Virol. 158, 1407–1410.

Zelazny, B., Randles, J.W., Boccardo., G., Imperial,, J.S., 1982. The viroid nature of the cadang-cadang disease of coconut palm. Scientia Filipinas. 2, 45–63.

CHAPTER

26

Other Cocadviroids

Irene Lavagi[1], *Jaroslav Matoušek*[2] *and Georgios Vidalakis*[1]

[1]University of California, Riverside, CA, United States [2]Academy of Sciences of the Czech Republic (ASCR), České Budějovice, Czech Republic

INTRODUCTION

In addition to the coconut cadang-cadang viroid and the coconut tinangaja viroid described in Chapter 4, Economic Significance of Palm Tree Viroids and Chapter 25, Coconut Cadang-Cadang Viroid and Coconut Tinangaja Viroid, two additional viroids belong to the genus *Cocadviroid*, namely the citrus bark cracking viroid (CBCVd) and the hop latent viroid (HLVd) (Owens et al., 2012; Di Serio et al., 2014).

CBCVd (formerly CVd-IV) was originally identified after sequential polyacrylamide gel electrophoresis analysis of symptomatic tissues from the bioamplification and bioindicator host "Etrog" citron (*Citrus medica* L.) Arizona 861-S-1 (Duran-Vila and Semancik, 2003). CBCVd induces bark cracking symptoms on trifoliate orange (*Poncirus trifoliata* (L.) Raf.) rootstock and the hybrid rootstock citrange (*P. trifoliata* × *C. sinensis*) (Murcia et al., 2015; Vernière et al., 2004, 2006). CBCVd has also been reported in hops (*Humulus lupulus* L.), where it induces severe symptoms affecting vegetative growth and cone production (Jakse et al., 2015).

Pallás et al. (1987) reported a viroid-like RNA from hops, and Puchta et al. (1988) described this in detail and named it HLVd. Hops are propagated vegetatively and the efficient mechanical transmission of HLVd, in combination with the absence of acute symptoms, have contributed to its worldwide distribution (Barbara and Adams, 2003; Eastwell and Nelson, 2007; Mahaffee et al., 2009; Pethybridge et al., 2008). The high economic value of the hop cones can be jeopardized by HLVd, which

Viroids and Satellites.
DOI: http://dx.doi.org/10.1016/B978-0-12-801498-1.00026-7

affects acids, resins, and essential oils critical for beer brewing (Barbara and Adams, 2003; Pethybridge et al., 2008).

TAXONOMIC POSITION AND NUCLEOTIDE SEQUENCE

CBCVd and HLVd are members of the genus *Cocadviroid*, in the family *Pospiviroidae*. The taxonomic classification of CBCVd and HLVd is based on the lack of hammerhead ribozymes, phylogenetic relationships, and the presence of specific structural motifs in their sequences, particularly a central conserved region (see Chapter 13: Viroid Taxonomy; Di Serio et al., 2014).

CBCVd (Fig. 26.1) is considered a chimeric-recombinant of the citrus exocortis viroid (CEVd) and hop stunt viroid (Duran-Vila and Semancik, 2003), which may explain CBCVd infectivity in citrus and hops (Jakse et al., 2015). This potential evolutionary relationship between CBCVd and CEVd, and some common structural and sequence features, has raised the question of including CBCVd in the genus *Pospiviroid* along with CEVd (Semancik and Vidalakis, 2005; Vernière et al., 2006). However, the recent finding of CBCVd naturally infecting hops supports its present classification along with HLVd. The HLVd sequence is relatively conserved (Fig. 26.1), with a small number of sequence variants (Eastwell and Nelson, 2007; Puchta et al., 1988).

ECONOMIC SIGNIFICANCE

CBCVd infects a large number of citrus species and citrus relatives with no apparent effect and, therefore, there is no measurable economic impact (see Chapter 2: Economic Significance of Fruit Tree and Grapevine Viroids; Vernière et al., 2006). However, when sensitive citrus species and hop cultivars were infected with CBCVd, effects on plant growth and production of significant economic impact have been reported. For example, CBCVd has been associated with an aggressive hop disease that kills plants in 3–5 years (EPPO, 2015; Jakse et al., 2015).

HLVd affects the yield of hop cones (e.g., 37% reduction for cv. "Omega") and the weight of individual cones as well as the concentration of resins such as alpha acids (e.g., reduction on average 40%, Osvald's clone 31, 72, 114, and cv. "Premiant"). Alpha acids affect the value of the hop crop and the beer bitterness and flavor. HLVd effects on hops with economic implications also include changes in essential

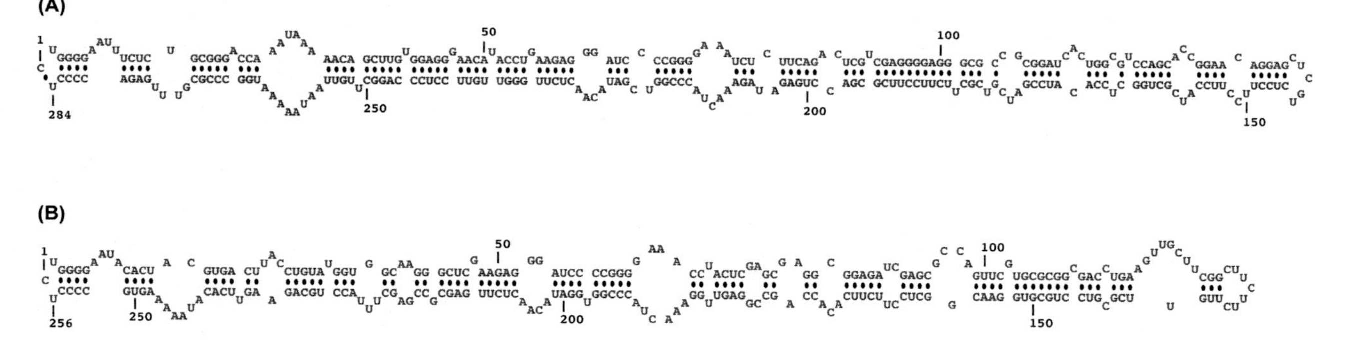

FIGURE 26.1 (A) Nucleotide sequence of citrus bark cracking viroid (CBCVd) arranged in a thermodynamically optimized secondary structure (GenBank NC_003539). (B) Nucleotide sequence of hop latent viroid (HLVd) arranged in a thermodynamically optimized secondary structure (GenBank NC_003611).

oils, where the slightest difference in content or composition can change the beer aroma. HLVd also affects nursery production and breeding programs since it significantly reduces rooting and establishment of softwood cuttings (Pethybridge et al., 2008).

SYMPTOMS

CBCVd induces moderate stunting and random leaf epinasty on "Etrog" citron Arizona 861-S-1, typically associated with mid-vein and petiole necrosis (Fig. 26.2) (Timmer et al., 2000). CBCVd causes bark cracking, with corresponding green streaks, and reduces tree size on the sensitive rootstocks trifoliate orange and Carrizo citrange (Fig. 26.2) (Murcia et al., 2015; Vernière et al., 2004, 2006). In hops, CBCVd has been associated with stunting, internode shortening, leaf yellowing and downward curling, small cone formation, and dry root rot (Fig. 26.3) (EPPO, 2015; Jakse et al., 2015).

HLVd symptoms are evident only by comparison of infected and healthy hops and include chlorosis, slow growth, and fewer smaller cones. Despite the absence of any characteristic symptoms in somatic tissues, HLVd induces physiological changes that affect quantitatively and qualitatively the metabolites of the lupulin secretory glands and essential oils, thus indicating that HLVd is not truly latent (Adams et al., 1991; Barbara et al., 1990; Matoušek et al., 2001; Patzak et al., 2001; Pethybridge et al., 2008).

FIGURE 26.2 Symptoms of citrus bark cracking viroid (CBCVd) on citrus. (A) "Etrog" citron Arizona 861-S-1, non-inoculated (left) and CBCVd-inoculated (right) expressing moderate stunting and random leaf epinasty. (B) Trifoliate orange rootstock expressing CBCVd-induced bark cracking. (C) Green streaks associated with CBCVd bark cracks on the cambial face of the bark of trifoliate orange rootstock.
Source: Photos courtesy of Nuria Duran-Vila.

FIGURE 26.3 Symptoms of citrus bark cracking viroid (CBCVd) on hops. (A) Infected hop plants develop stunted and reduced growth (three plants front row). (B) Cracking of primary bines. (C) Stunted and reduced development of leaves, lateral branches, and cones. (D) Dry root rot of infected hop (left) and roots from healthy hop (right).
Source: Photos courtesy of Sebastjan Radisek.

HOST RANGE

CBCVd has a broad natural host range, including over 30 citrus species and 12 citrus relatives and citrus hybrids, while several herbaceous plants have been reported as experimental hosts (Barbosa et al., 2002).

HLVd has only three natural hosts: the commercial hop, the Oriental *Humulus japonicus* Sieb. and Zucc., and the stinging nettle (*Urtica dioica* L.) (Pethybridge et al., 2008). Heat-generated HLVd variants (i.e., thermomutants) can infect what otherwise are considered non-susceptible species such as *Solanum lycopersicum* and *Nicotiana benthamiana* (Matoušek, 2003).

TRANSMISSION

CBCVd and HLVd are transmitted over long distances and introduced into citrus orchards and hop gardens with infected propagative materials. Subsequently they are transmitted mechanically, by grafting and vegetative propagation, and by contaminated tools or machinery (Barbosa et al., 2005).

There are no reports of pollen, seed, or vector transmission of CBCVd in citrus or hops (Duran-Vila and Semancik, 2003; EPPO, 2015). The reports for pollen, seed, and vector transmission of HLVd vary. There is no evidence for insect transmission (Adams et al., 1996; Mahaffee et al., 2009). However, new infections away from neighboring infected hops, led to the hypothesis that a vector may be involved in HLVd spread in hop yards (Adams et al., 1992). HLVd transmission by pollen or seed has been reported as low (Mahaffee et al., 2009; Pethybridge et al., 2008) or none (Matoušek and Patzak, 2000; Matoušek et al., 2008).

DETECTION

CBCVd in citrus is detected biologically with "Etrog" citron Arizona 861-S-1 followed by sequential polyacrylamide gel electrophoresis, northern-blot hybridization, or reverse transcription polymerase chain reaction (RT-PCR) for a definite viroid identification (Duran-Vila and Semancik, 2003; Malfitano et al., 2005; Murcia et al., 2009; Timmer et al., 2000; Wang et al., 2013). This approach is time-consuming and not high throughput. The need for fast, reliable, high throughput CBCVd detection in field samples of various citrus species and environmental conditions has led to the development of different RT-PCR assays with various applications (Bernad and Duran-Vila, 2006; Garnsey et al., 2002; Ito et al., 2002, 2003; Wang et al., 2009, 2013). Vidalakis and Wang (2014) developed a quantitative PCR method for the universal detection of citrus viroids, including CBCVd, used since 2010 by the California Department of Food and Agriculture for the "Citrus Nursery Stock Pest Cleanliness Program," while several citrus programs around the world are experimenting with the method (Bostock et al., 2014). CBCVd was detected in hops using next-generation sequencing (Jakse et al., 2015) and most likely this technology will be used soon for routine viroid detection (Al Rwahnih et al., 2015; Barba et al., 2014; Hadidi et al., 2016; Su et al., 2015; Wu et al., 2015).

Biological indexing of HLVd is not viable because of the limited host range and lack of clear symptomatology. HLVd is detected by bidirectional gel electrophoresis, however, dot-blot hybridization and RT-PCR are preferred for routine detection (Eastwell and Nelson, 2007; Guo et al., 2012; Sue et al., 2014).

GEOGRAPHICAL DISTRIBUTION AND EPIDEMIOLOGY

CBCVd and HLVd have worldwide distribution (Table 26.1). The increasing number of recent CBCVd and HLVd reports demonstrates the spread of these two viroids in new areas and is also indicative of the availability of sensitive and high throughput viroid detection methods (Duran-Vila and Semancik, 2003; Jakse et al., 2015; Pethybridge et al., 2008). Spread and transmission efficiency of CBCVd has not been studied in citrus orchards, but in hop gardens CBCVd progresses rapidly (up to 20% every year) (Barbosa et al., 2005; EPPO, 2015; Jakse et al., 2015).

CONTROL

Production of viroid-free citrus and hop propagative materials is the basis of CBCVd and HLVd control. CBCVd can be eliminated from citrus with shoot-tip-grafting but not thermotherapy (Calavan et al., 1972; Navarro et al., 1975; Roistacher et al., 1976; Stubbs, 1968). HLVd titer can be reduced by both heat and cold treatment in hops; however, subsequent culturing of meristematic tissues is required for complete viroid elimination (Pethybridge et al., 2008).

Removal of viroid-infected plants and replanting with viroid-free ones, as well as equipment sanitation, are necessary to minimize viroid spread within and among plantings. Among the different heat and chemical treatments tested for tool sanitation against citrus and hops viroids, the most reliable and widely used one is an aqueous solution of at least 5% household bleach (sodium hypochlorite, minimum 1% available chlorine) (Barbara and Adams, 2003; Duran-Vila and Semancik, 2003; Roistacher et al., 1969).

Control measures for CBCVd and HLVd are typically included within comprehensive germplasm programs. Citrus and hops have a long tradition in such programs (e.g., 1937-Citrus Psorosis Freedom Program, California, United States and 1947-Progressive Verticillium Wilt of Hops Order, United Kingdom) (Calavan et al., 1978; Pethybridge et al., 2008). Today, there is a variety of certification, nuclear, or registration programs in citrus and hop producing countries, including consortia of programs such as the "U.S.A. National Clean Plant Network (NCPN)" (Bostock et al., 2014; EPPO, 2009; Gergerich et al., 2015; Navarro, 2015; Vidalakis et al., 2010a,b).

TABLE 26.1 Reports of Geographic Distribution for Citrus Bark Cracking Viroid and Hop Latent Viroid

Citrus bark cracking viroid	
Country	**Reference**
Australia	Nerida Donovan (personal communication)
Cuba	Velázquez et al. (2002)
Egypt	Al-Harthi et al. (2013)
Greece	Wang et al. (2013)
Iran	Hashemian et al. (2013
Israel	Puchta et al. (1991)
Italy	Malfitano et al. (2005)
Jamaica	Fisher et al. (2010)
Japan	Ito et al. (2002, 2003)
Jordan	Al-Harthi et al. (2013)
Lebanon	Al-Harthi et al. (2013)
Oman	Al-Harthi et al. (2013)
PR China (Sichuan and Zhejiang)	Cao et al. (2010)
Slovenia	EPPO (2015), Jakse et al. (2015)
South Africa	Cook et al. (2012)
Spain	Duran-Vila et al. (1988), Navarro (2015)
Sudan	Mohamed et al. (2009)
Syria	Al-Harthi et al. (2013)
Tunisia	Najar and Duran-Vila (2004)
Turkey	Onelge et al. (2000)
United States of America (California, Florida, and Texas)	Duran-Vila et al. (1988), Garnsey et al. (2002), Kunta et al. (2007)
Hop latent viroid	
Country	**Reference**
Australia	Pethybridge et al. (2008)
Belgium	Puchta et al. (1988)
Brazil	Fonseca et al. (1993)
China	Puchta et al. (1988)
Czech Republic	Puchta et al. (1988)

(*Continued*)

TABLE 26.1 (Continued)

Hop latent viroid	
Country	**Reference**
Ex. Yugoslavia	Puchta et al. (1988)
France	Puchta et al. (1988)
Germany	Puchta et al. (1988)
Hungary	Puchta et al. (1988)
Japan	Puchta et al. (1988)
New Zealand	Hay (1989)
Poland	Puchta et al. (1988)
Portugal	Puchta et al. (1988)
Russia	Puchta et al. (1988)
Slovenia	Puchta et al. (1988), Knapic and Javornik (1998)
South Africa	Puchta et al. (1988)
South Korea	Puchta et al. (1988), Lee et al. (1990)
Spain	Pallás et al. (1987), Puchta et al. (1988)
United Kingdom	Puchta et al. (1988), Barbara et al. (1990); Adams et al. (1991), Adams et al. (1992)
United States of America (Washington, Oregon, and Wisconsin)	Puchta et al. (1988), Singh et al. (2003), Eastwell and Nelson (2007)

Acknowledgments

The authors would like to acknowledge Noora Siddiqui, Shih-hua Tan, and Brandon Stuardo Ramirez for their assistance. We are grateful to Joseph Semancik and Ken Eastwell for providing information and editorial comments.

References

Adams, A.N., Barbara, D.J., Morton, A., 1991. Effects of hop latent viroid on weight and quality of the cones of the hop cultivar Wye Challenger. Tests of agrochemicals and cultivars. Ann. Appl. Biol. 118, 126–127.

Adams, A.N., Barbara, D.J., Morton, A., Darby, P., 1996. The experimental transmission of hop latent viroid and its elimination by low temperature treatment and meristem culture. Ann. Appl. Biol. 128, 37–44.

Adams, A.N., Morton, A., Barbara, D.J., Ridout, M.S., 1992. The distribution and spread of hop latent viroid within two commercial plantings of hop (*Humulus lupulus*). Ann. Appl. Biol. 121, 585–592.

Al Rwahnih, M., Daubert, S., Golino, D., Islas, C., Rowhani, A., 2015. Comparison of next-generation sequencing versus biological indexing for the optimal detection of viral pathogens in grapevine. Phytopathology 105, 758–763.

Al-Harthi, S.A., Al-Sadi, A.M., Al-Saady, A.A., 2013. Potential of citrus budlings originating in the Middle East as sources of citrus viroids. Crop Protec. 48, 13–15.

Barba, M., Czosnek, H., Hadidi, A., 2014. Historical perspective, development and applications of next-generation sequencing in plant virology. Viruses 6, 106–136.

Barbara, D., Adams, A., 2003. Hop latent viroid. In: Hadidi, A., Flores, R., Randles, J.W., Semancik, J.S. (Eds.), Viroids. CSIRO Publishing, Collingwood, VIC, pp. 213–217.

Barbara, D.J., Morton, A., Adams, A.N., Green, C.P., 1990. Some effects of hop latent viroid on two cultivars of hop (*Humulus lupulus*) in the UK. Ann. Appl. Biol. 117, 359–366.

Barbosa, C., Pina, J., Navarro, L., Duran-Vila, N., 2002. Replication/accumulation and symptom expression of citrus viroids on some species of citrus and related genera. In: Duran-Vila, N., Milne, R.G., da Graça, J.V., (Eds), Proceedings of the 15th IOCV, pp. 264–271.

Barbosa, C.J., Pina, J.A., Pérez-Panadés, J., Bernad, L., Serra, P., Navarro, L., et al., 2005. Mechanical transmission of citrus viroids. Plant Dis. 89, 749–754.

Bernad, L., Duran-Vila, N., 2006. A novel RT-PCR approach for detection and characterization of citrus viroids. Mol. Cell. Probes 20, 105–113.

Bostock, R.M., Thomas, C., Hoenisch, R., Golino, D.A., Vidalakis, G., 2014. Plant health: how diagnostic networks and interagency partnerships protect plant systems from pests and pathogens. California Agricult. 68, 117–124.

Calavan, C., Mather, S.M., McEachern, E., Reuther, W., Calavan, E., Carman, G., 1978. Registration, certification, and indexing of citrus trees. In: Reuther, W., Calavan, E.C., Carman, G.E. (Eds.), The Citrus Industry, vol. 4. University of California, Division of Agricultural Sciences, Berkeley, CA, pp. 185–222.

Calavan, E., Roistacher, C., Nauer, E., 1972. Thermotherapy of citrus for inactivation of certain viruses. Plant Dis. Rep. 56, 976–980.

Cao, M.J., Liu, Y.Q., Wang, X.F., Yang, F.Y., Zhou, C.Y., 2010. First report of citrus bark cracking viroid and citrus viroid V infecting citrus in China. Plant Dis. 94, 922–1122.

Cook, G., van Vuuren, S.P., Breytenbach, J.H.J., Manicom, B.Q., 2012. Citrus viroid IV detected in *Citrus sinensis* and *C. reticulata* in South Africa. Plant Dis. 96, 772–872.

Di Serio, F., Flores, R., Verhoeven, J.T.J., Li, S.F., Pallás, V., Randles, J.W., et al., 2014. Current status of viroid taxonomy. Arch. Virol. 159, 3467–3478.

Duran-Vila, N., Semancik, J., 2003. Citrus viroids. In: Hadidi, A., Flores, R., Randles, J.W., Semancik, J.S. (Eds.), Viroids. CSIRO Publishing, Collingwood, VIC, pp. 178–194.

Duran-Vila, N., Roistacher, C.N., Rivera-Bustamante, R., Semancik, J.S., 1988. A definition of citrus viroid groups and their relationship to the exocortis disease. J. Gen. Virol. 69, 3069–3080.

Eastwell, K., Nelson, M., 2007. Occurrence of viroids in commercial hop (*Humulus lupulus L.*) production areas of Washington State. Plant Health Progress. Available from: http://dx.doi.org/10.1094/PHP-2007-1127-01-RS.

EPPO, 2009. Certification scheme for hop. EPPO Bull. 39, 278–283.

EPPO, 2015. Citrus bark cracking viroid. A viroid causing a severe disease on a new host, hop *(Humulus lupulus)*. In EPPO Alert List.

Fisher, L., Bennett, S., Tennant, P., Mc Laughlin, W., 2010. Detection of citrus tristeza virus and citrus viroid species in Jamaica. In: Benkeblia, N. (Ed.), International Symposium on Tropical Horticulture, pp. 117–122.

Fonseca, M.E., Marhino, V.L.A., Nagata, T., 1993. Hop latent viroid in hop germplasm introduced into Brazil from the United States. Plant Dis. 77, 952.

Garnsey, S., Zies, D., Irey, M., Sieburth, P., Semancik, J., Levy, L., et al., 2002. Practical field detection of citrus viroids in Florida by RT-PCR. In: Duran-Vila, N., Milne, R.G., da Graça, J.V., (Eds.), Proceedings of the 15th IOCV, pp. 219–229.

Gergerich, R.C., Welliver, R.A., Gettys, S., Osterbauer, N.K., Kamenidou, S., Martin, R.R., et al., 2015. Safeguarding fruit crops in the age of agricultural globalization. Plant Dis. 99, 176–187.

Guo, L.-X., Duan, W.J., Zhang, X.L., Deng, C.L., Wen, W.G., Li, S.F., 2012. Detection of hop latent viroid by real-time fluorescent RT-PCR assay. Acta Phytopathol. Sinica. 42, 466–473.

Hadidi, A., Flores, R., Candresse, T., Barba, M., 2016. Next-generation sequencing and genome editing in plant virology. Front. Microbiol. 7, 1325.

Hashemian, S.M.B., Taheri, H., Alian, Y.M., Bové, J.M., Duran-Vila, N., 2013. Complex mixtures of viroids identified in the two main citrus growing areas of Iran. J. Plant Pathol. 95, 647–654.

Hay, F., 1989. Studies on the viruses of hop *(Humulus lupulus L.)* in New Zealand. Lincoln College, University of Canterbury.

Ito, T., Ieki, H., Ozaki, K., Iwanami, T., Nakahara, K., Hataya, T., et al., 2002. Multiple citrus viroids in citrus from Japan and their ability to produce exocortis-like symptoms in citron. Phytopathology 92, 542–547.

Ito, T., Namba, N., Ito, T., 2003. Distribution of citrus viroids and apple stem grooving virus on citrus trees in Japan using multiplex reverse transcription polymerase chain reaction. J. Gen. Plant Pathol. 69, 205–207.

Jakse, J., Radisek, S., Pokorn, T., Matousek, J., Javornik, B., 2015. Deep-sequencing revealed citrus bark cracking viroid (CBCVd) as a highly aggressive pathogen on hop. Plant Pathol. 64, 831–842.

Knapic, V., Javornik, B., 1998. Viroidi-Povročitelji Rastlinskih Bolezni. Sodobno Kmetijstvo. 31, 462–465.

Kunta, M., Da Graca, J., Skaria, M., 2007. Molecular detection and prevalence of citrus viroids in Texas. Hortic. Sci. 42, 600–604.

Lee, J.Y., Lee, S., Sänger, H.L., 1990. Viroid diseases occurring on Korean hop plants. Kor. J. Plant Pathol. 6, 256–260.

Mahaffee, W.F., Pethybridge, S.J., Gent, D.H. Compendium of hop diseases and pests. Amer. Phytopathol. Soc. (APS Press), St Paul, MN. 2009, 93 pp.

Malfitano, M., Barone, M., Duran-Vila, N., Alioto, D., 2005. Indexing of viroids in citrus orchards of Campania, southern Italy. J. Plant Pathol. 87, 115–121.

Matoušek, J., 2003. Hop latent viroid (HLVd) microevolution: an experimental transmission of HLVd "thermomutants" to solanaceous species. Biol. Plant. 46, 607–610.

Matoušek, J., Orctová, L., Škopek, J., Pešina, K., Steger, G., 2008. Elimination of hop latent viroid upon developmental activation of pollen nucleases. Biol. Chem. 389, 915–918.

Matoušek, J., Patzak, J., 2000. A low transmissibility of hop latent viroid through a generative phase of *Humulus lupulus L.* Biol. Plant. 43, 145–148.

Matoušek, J., Patzak, J., Orctová, L., Schubert, J., Vrba, L., Steger, G., et al., 2001. The variability of hop latent viroid as induced upon heat treatment. Virology 287, 349–358.

Mohamed, M.E., Hashemian, S.M.B., Dafalla, G., Bové, J.M., Duran-Vila, N., 2009. Occurrence and identification of citrus viroids from Sudan. J. Plant Pathol. 91, 185–190.

Murcia, N., Hashemian, S.M.B., Serra, P., Pina, J.A., Duran-Vila, N., 2015. Citrus viroids: Symptom expression and performance of Washington navel sweet orange trees grafted on carrizo citrange. Plant Dis. 99, 125–136.

Murcia, N., Serra, P., Olmos, A., Duran-Vila, N., 2009. A novel hybridization approach for detection of citrus viroids. Mol. Cell. Probes 23, 95–102.

Najar, A., Duran-Vila, N., 2004. Viroid prevalence in Tunisian citrus. Plant Dis. 88, 1286–2286.

Navarro, L., 2015. The Spanish citrus industry. In: Sabater-Muñoz, B., Moreno, P., Peña, L., Navarro, L. (Eds.), XIIth Intl Citrus Congress. International Society for Horticultural Science (ISHS). Leuven, Belgium, pp. 41–48.

Navarro, L., Roistacher, C.N., Murashige, T., 1975. Improvement of shoot tip grafting in vitro for virus-free citrus. J. Amer. Soc. Hortic. Sci. 100, 471–479.

Onelge, N., Kersting, U., Guang, Y., Bar-Joseph, M., Bozan, O., 2000. Nucleotide sequence of citrus viroids CVd-IIIa and CVd-IV obtained from dwarfed Meyer lemon trees grafted on sour orange. J. Plant Dis. Protect. 107, 387–391.

Owens, R.A., Flores, R., Di Serio, F., Li, S.F., Palla's, V., Randles, J.W., et al., 2012. Viroids. In: King, A.M.Q., Adams, M.J., Carstens, E.B., Lefkowitz, E.J. (Eds.), Virus Taxonomy, Ninth Report of the International Committee on Taxonomy of Viruses. Elsevier/Academic Press, London UK, pp. 1221–1234.

Pallás, V., Navarro, L., Flores, R., 1987. Isolation of a viroid-like RNA from hop different from hop stunt viroid. J. Gen. Virol. 68, 3201–3205.

Patzak, J., Matoušek, J., Krofta, K., Svoboda, P., 2001. Hop latent viroid (HLVd)-caused pathogenesis: effects of HLVd infection on lupulin composition of meristem culture-derived *Humulus lupulus*. Biol. Plant. 44, 579–585.

Pethybridge, S.J., Hay, F.S., Barbara, D.J., Eastwell, K.C., Wilson, C.R., 2008. Viruses and viroids infecting hop: significance, epidemiology, and management. Plant Dis. 92, 324–338.

Puchta, H., Ramm, K., Luckinger, R., Hadas, R., Bar-Joseph, M., Sänger, H.L., 1991. Primary and secondary structure of citrus viroid IV (CVd IV), a new chimeric viroid present in dwarfed grapefruit in Israel. Nucleic Acids Res. 19, 6640.

Puchta, H., Ramm, K., Sänger, H.L., 1988. The molecular structure of hop latent viroid (HLV), a new viroid occurring worldwide in hops. Nucleic Acids Res. 16, 4197–4216.

Roistacher, C., Calavan, E., Blue, R., 1969. Citrus exocortis virus-chemical inactivation on tools, tolerance to heat and separation of isolates. Plant Dis. Rep. 53, 333–336.

Roistacher, C., Navarro, L., Murashige, T., 1976. Recovery of citrus selections free of several viruses, exocortis viroid and *Spiroplasma citri* by shoot tip grafting in vitro. In: Calavan, E.C (Ed). Proceedings of the 7th IOCV, Riverside, CA. pp. 186–193.

Semancik, J.S., Vidalakis, G., 2005. The question of citrus viroid IV as a cocadviroid. Arch. Virol. 150, 1059–1067.

Singh, R., Ready, K., Hadidi, A., 2003. Viroids in North America and global distribution of viroid diseases. In: Hadidi, A., Flores, R., Randles, J.W., Semancik, J.S. (Eds.), Viroids. CSIRO Publishing, Collingwood, VIC, pp. 255–264.

Stubbs, L., 1968. Apparent elimination of exocortis and yellowing viruses in lemon by heat therapy and shoot-tip propagation. In: Cilds, J.F. (Ed.), Proceedings of the 4th IOCV. Univerity of Florida Press, Gainesville, pp. 96–99.

Su, X., Fu, S., Qian, Y., Xu, Y., Zhou, X., 2015. Identification of hop stunt viroid infecting *Citrus limon* in China using small RNAs deep sequencing approach. Virol. J. 12, 103.

Sue, M.J., Yeap, S.K., Omar, A.R., Tan, S.W., 2014. Application of PCR-ELISA in Molecular Diagnosis. BioMed Res. Int. 2014, 6.

Timmer, L.W., Garnsey, S.M., Graham, J.H., 2000. Compendium of citrus diseases. APS Press, St. Paul, MN.

Velázquez, K., Soto, M., Pérez, R., Pérez, J., Rodríguez, D., Duran-Vila, N., 2002. Biological and molecular characterization of two isolates of citrus viroids recovered from Cuban plantations. In: Duran-Vila, N., Milne, R.G., da Graça, J.V., (Eds.), Proceedings of the 15th IOCV. pp. 11–19.

Vernière, C., Perrier, X., Dubois, C., Dubois, A., Botella, L., Chabrier, C., et al., 2004. Citrus viroids: symptom expression and effect on vegetative growth and yield of clementine trees grafted on trifoliate orange. Plant Dis. 88, 1189–1197.

Vernière, C., Perrier, X., Dubois, C., Dubois, A., Botella, L., Chabrier, C., et al., 2006. Interactions between citrus viroids affect symptom expression and field performance of clementine trees grafted on trifoliate orange. Phytopathology 96, 356–368.

Vidalakis, G., da Graça, J., Dixon, W., Ferrin, D., Kesinger, M., Krueger, R., et al., 2010a. Citrus quarantine, sanitary and certification programs in the USA. Prevention of introduction and distribution of citrus pests. Part 1. Citrograph 1, 26–35.

Vidalakis, G., da Graça, J., Dixon, W., Ferrin, D., Kesinger, M., Krueger, R., et al., 2010b. Citrus quarantine, sanitary and certification programs in the USA. Prevention of introduction and distribution of citrus pests. Part 2. Citrograph 1, 27–39.

Vidalakis, G., Wang, J., 2014. Molecular method for universal detection of citrus viroids. United States Patent US 8815547 B2, USA, p. 19.

Wang, J., Boubourakas, I.N., Voloudakis, A.E., Agorastou, T., Magripis, G., Rucker, T.L., et al., 2013. Identification and characterization of known and novel viroid variants in the Greek national citrus germplasm collection: threats to the industry. Eur. J. Plant Pathol. 137, 17–27.

Wang, X., Zhou, C., Tang, K., Zhou, Y., Li, Z., 2009. A rapid one-step multiplex RT-PCR assay for the simultaneous detection of five citrus viroids in China. Eur. J. Plant Pathol. 124, 175–180.

Wu, Q., Ding, S., Zhang, Y., Zhu, S., 2015. Identification of viruses and viroids by next-generation sequencing and homology dependent and homology independent algorithms. Ann. Rev. Phytopathol. 53, 1–20.

CHAPTER

27

Coleus Blumei Viroids

Xianzhou Nie and Rudra P. Singh

Agriculture and Agri-Food Canada, Fredericton, NB, Canada

INTRODUCTION

Coleus blumei is an ornamental plant that originated in Indonesia. Six viroids, namely Coleus blumei viroid 1, 2, 3, 4, 5, and 6 (CbVd-1, -2, -3, -4, -5, and -6), have been reported in *Coleus* spp. to date. CbVd-1 was first reported in a commercially grown coleus in Brazil in 1989 (Fonseca et al., 1989), and subsequently in many other countries including Germany, Canada, Japan, China, Korea, and India (Adkar-Purushothama et al., 2013; Chung and Choi, 2008; Fu et al., 2011; Hou et al., 2009a,b; Ishiguro et al., 1996; Li et al., 2006; Singh and Boucher, 1991; Spieker et al., 1990). Although coleviroids do not appear to cause serious diseases in their host plants, they display interesting features that assist our understanding in viroid diversification and evolution (Jiang et al., 2014).

TAXONOMIC POSITION AND NUCLEOTIDE SEQUENCE

CbVd-1 is the type member of the genus *Coleviroid*, family *Pospiviroidae* (Flores et al., 1998; Mayo, 1999). Of the six viroids described from *Coleus* (CbVd-1 to 6), only the first three (i.e., CbVd-1, CbVd-2, and CbVd-3) have been assigned to the genus *Coleviroid* to date (Di Serio et al., 2014). The genome sizes for CbVd-1 to -6 are: 248–251; 301; 361–364; 295; 274; and 340–343 nucleotides, respectively (Table 27.1). Like other members of the family *Pospiviroidae*, coleviroids adopt a rod-shaped secondary structure of minimum free energy (Fig. 27.1). Phylogenetic analysis of the available coleviroid sequence variants identified six coleviroid clusters, each representing the six

Viroids and Satellites.
DOI: http://dx.doi.org/10.1016/B978-0-12-801498-1.00027-9

TABLE 27.1 Coleviroids in *Coleus blumei*

Viroid	Genome size (nt)	Putative parental viroid	Country of incidence	Coleus cultivar	First report/ reference
CbVd-1	248–251	–	Many countries	Bienvenue, Golden Bedder, Ruhm von Luxemburg, Rainbow Gold, Wizard	Fonseca et al. (1989)
CbVd-2	301	CbVd-1 (right half) and CbVd-3 (left half)	Germany, China	Ruhm von Luxemburg	Spieker (1996)
CbVd-3	361–364	–	Germany, China	Bienvenue, Ruhm von Luxemburg; Fairway Ruby	Spieker et al. (1996)
CbVd-4	295	CbVd-1 (left half) and CbVd-3 (right half)	Germany	Ruhm von Luxemburg	Spieker (1996)
CbVd-5	274	–	China	Unknown	Hou et al. (2009a)
CbVd-6	340–343	CbVd3 (left half) and CbVd-5 (right half)	China	Unknown	Hou et al. (2009b)

species placed in the genus (Fig. 27.2). It is noteworthy that all the coleviroids share a common central conserved region (CCR) (Hou et al., 2009a,b; Jiang et al., 2011), which enabled the development of a universal probe for their rapid detection and identification (Jiang et al., 2013). Moreover, CbVd-2, CbVd-4, and CbVd-6 are recombinants with genome segments/elements contributed by CbVd-1, CbVd-3, and CbVd-5 (Hou et al., 2009a,b; Spieker, 1996). The left and the right halves flanking the CCR of CbVd-2 are likely contributed by CbVd-3 and CbVd-1, respectively. CbVd-4 includes the left half of CbVd-1 and the right half of CbVd-3. For CbVd-6, the left and right halves are from CbVd-3 and CbVd-5, respectively.

Sequence variation also exists at the species level. For instance, in CbVd-1, the isolate from the coleus cultivar "Bienvenue" (accession number X95960), shares a sequence identity of 88–99% with other CbVd-1 isolates: 99% with Brazilian (X69293), clone B (DQ178395), and clone F (DQ178396); 98% with clone I2 (DQ178397), clone J (DQ178398),

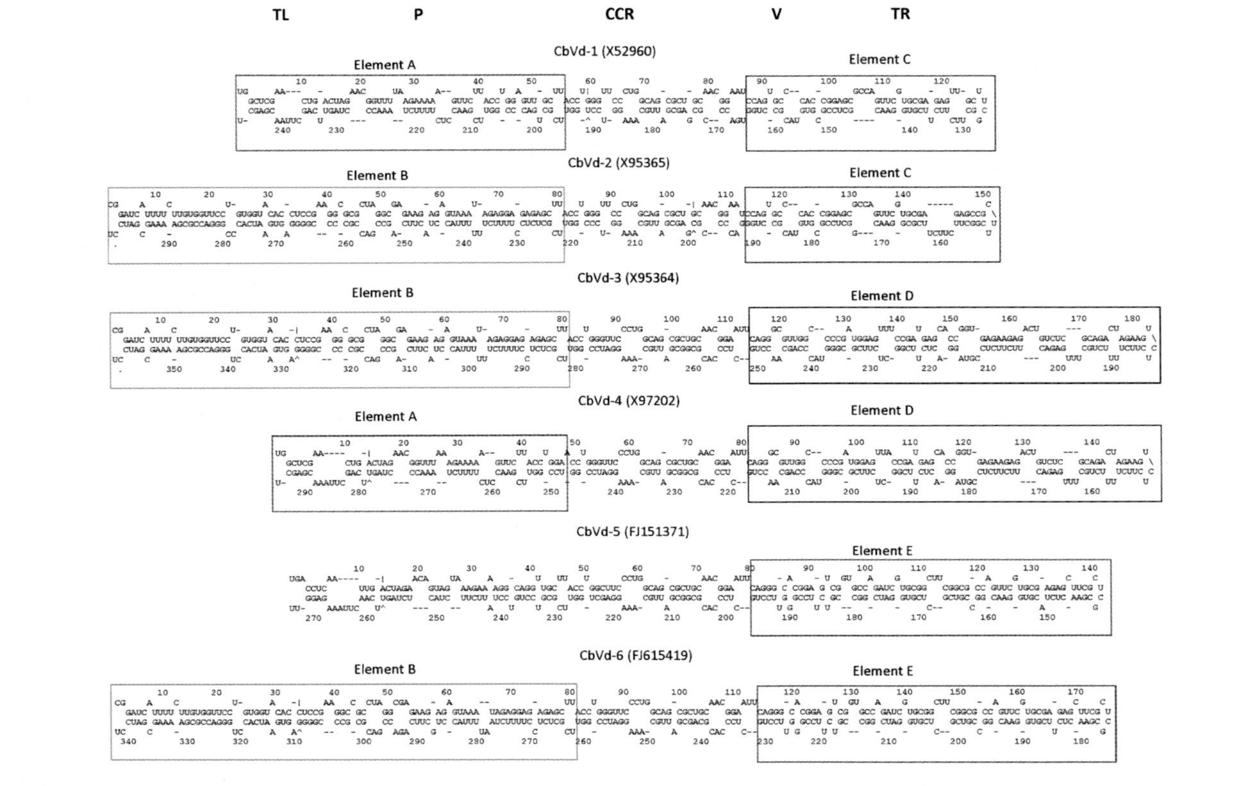

FIGURE 27.1 Nucleotide sequence and the predicted secondary structure of representative isolates of coleviroids. Elements that are shared by different viroids are marked by same-colored boxes.

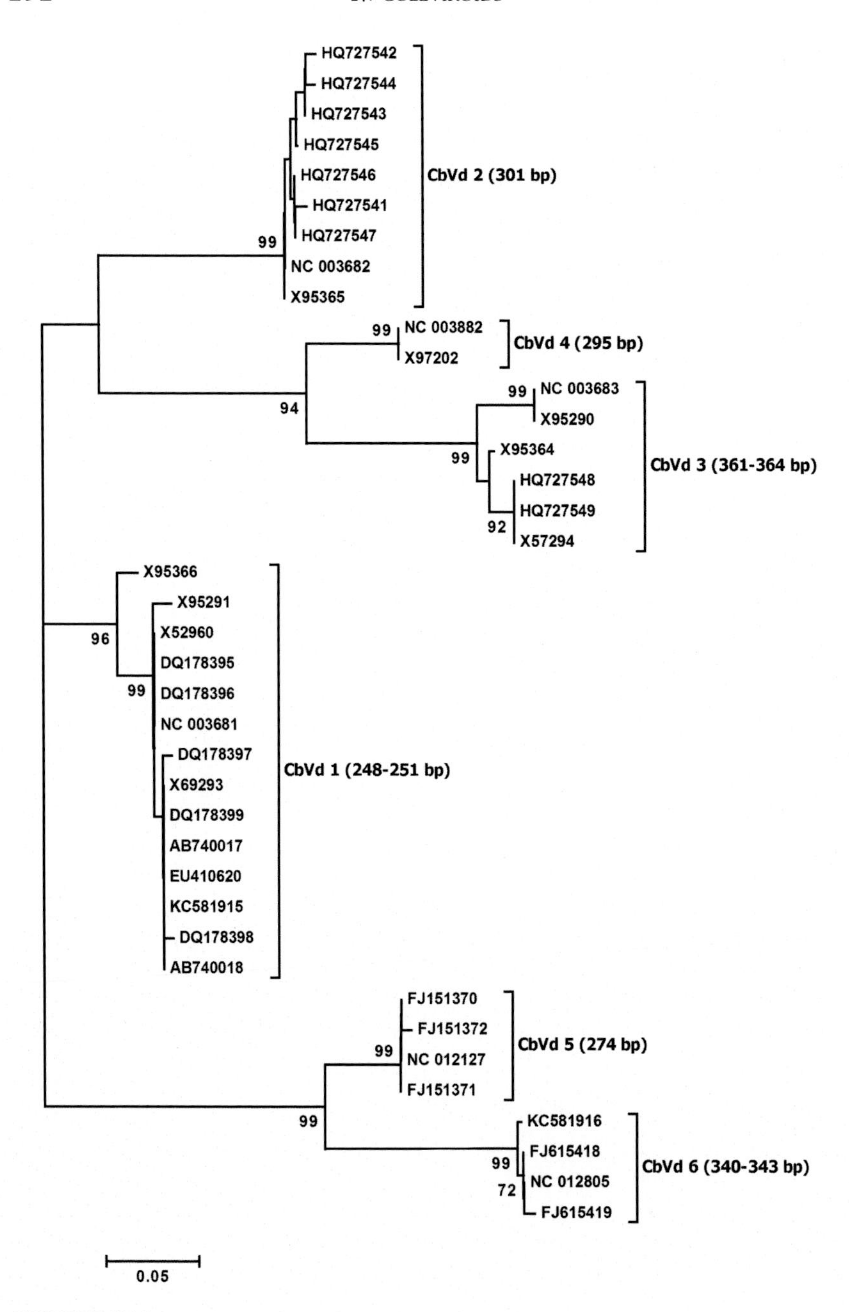

FIGURE 27.2 Phylogenetic relationship of coleviroids.

clone I1 (DQ178399), Ind-1 (AB740017), K (EU410620), Ind-2 (AB740018), and BJ-1–1 (KC581915); 97% with RJ (X95366); and 88% with RG (X95291).

ECONOMIC SIGNIFICANCE

No direct economic losses due to CbVd infections have been reported to date on *Coleus* or other plant species.

SYMPTOMS

The symptoms range from absence of a syndrome to a change in leaf color and a reduction in leaf size. For instance, CbVd-1-K infection in cultivar Highway Rose leads to plant dwarfing, leaf color fading and leaf size reduction (Chung and Choi, 2008). Nevertheless, most cultivars are symptomless when infected.

HOST RANGE

Several plant species including *Mentha spicata*, *M. arvensis* var. *piperascens*, *Ocimum basilicum*, and *Melissa officinalis* have been reported to be infected naturally with CbVd-1 (Ishiguro et al., 1996). CbVd-1 can be transmitted mechanically to *Ocimum sanctum* but not to potato, tomato, and *Scopolia sinensis* (Singh and Boucher, 1991).

TRANSMISSION

CbVd-1 can be transmitted to *Coleus* by mechanical wounding and grafting (Chung and Choi, 2008; Singh and Boucher, 1991; Singh et al., 2003). CbVd-1 is also seed transmissible, with the infection rate ranging from 0% to 100% depending on the cultivar (Chung and Choi, 2008).

DETECTION

Due to the lack of characteristic symptoms, *C. blumei* viroids can only be detected unambiguously by molecular methodologies including RT-PCR (Jiang et al., 2011), dot-blot hybridization (Li et al., 2006), microarray (Zhang et al., 2013), and next-generation sequencing

(Zhang et al., 2014). Return-polyacrylamide gel electrophoresis (R-PAGE) has been successfully used to detect coleviroids in *Coleus* plants (Ishiguro et al., 1996; Singh and Boucher, 1991). However, R-PAGE cannot differentiate the different viroids, with sequencing being the most reliable method for their ultimate determination. Recently, a cRNA probe containing an octamer of a 32-nt sequence derived from the CCR, named as the 8-central-conserved-region probe (8CCR probe), has been developed for the detection of coleviroids by dot-blot hybridization and the identification of specific coleviroids by 2D-PAGE/denaturing PAGE-northern hybridization (Jiang et al., 2013).

GEOGRAPHICAL DISTRIBUTION AND EPIDEMIOLOGY

CbVds have been reported in Germany, Brazil, Canada, Japan, China, Korea, and India. Interestingly, isolates of CbVd-1, such as CbVd-1-K (EU410620) from Korea, clone I1 (DQ178399) from China, and isolate Ind-1 (AB740017) from India are identical. It is possible that the variant/isolate shared the same source or origin, and spread to these countries via the movement of infected seeds.

CONTROL

No known control measures have been implemented for managing these viroids.

References

Adkar-Purushothama, C.R., Nagaraja, H., Sreenivasa, M.Y., Sano, T., 2013. First report of *Coleus blumei* viroid infecting coleus in India. Plant Dis. 97, 149.

Chung, B.N., Choi, G.S., 2008. Incidence of *Coleus blumei* viroid 1 in seeds of commercial coleus in Korea. Plant Pathol. J. 24, 305–308.

Di Serio, F., Flores, R., Verhoeven, J.Th.J., Li, S.-F., Pallás, V., et al., 2014. Current status of viroid taxonomy. Arch. Virol. 159, 3467–3478.

Flores, R., Randles, J.W., Bar-Joseph, M., Diener, T.O., 1998. A proposed scheme for viroid classification and nomenclature. Arch. Virol. 143, 623–629.

Fonseca, M.E.N., Boiteux, L.S., Singh, R.P., Kitajima, E.W., 1989. A small viroid in coleus species from Brazil. Fitopatol. Bras. 14, 94–96.

Fu, F.H., Li, S.F., Jiang, D.M., Wang, H.Q., Liu, A.Q., Sang, L.W., 2011. First report of *Coleus blumei* viroid 2 from commercial coleus in China. Plant Dis. 95, 494–594.

Hou, W.Y., Li, S.F., Wu, Z.J., Jiang, D.M., Sano, T., 2009b. Coleus blumei viroid 6: a new tentative member of the genus *Coleviroid* derived from natural genome shuffling. Arch. Virol. 154, 993–997.

Hou, W.Y., Sano, T., Li, S.F., Li, F., Li, L., Wu, Z.J., 2009a. Identification and characterization of a new *Coleviroid* (CbVd-5). Arch. Virol. 154, 315–320.

Ishiguro, A., Sano, T., Harada, Y., 1996. Nucleotide sequence and host range of a coleus viroid isolated from coleus (*Coleus blumei* Benth) in Japan. Ann. Phytopathol. Soc. Jpn. 62, 84–86.

Jiang, D., Gao, R., Qin, L., Wu, Z., Xie, L., Hou, W., et al., 2014. Infectious cDNA clones of four viroids in *Coleus blumei* and molecular characterization of their progeny. Virus Res. 180, 97–101.

Jiang, D., Hou, W., Sano, T., Kang, N., Lü, Q., Wu, Z., et al., 2013. Rapid detection and identification of viroids in the genus *Coleviroid* using a universal probe. J. Virol. Methods 187, 321–326.

Jiang, D.M., Wu, Z.J., Xie, L.H., Sano, T., Li, S., 2011. Sap-direct RT-PCR for the rapid detection of *Coleus blumei* viroids of the genus *Coleviroid* from natural host plants. J. Virol. Methods 174, 123–127.

Li, S.F., Su, Q.F., Guo, R., Tsuji, M., Sano, T., 2006. First report of *Coleus blumei* viroid from coleus in China. Plant Pathol. 55, 565.

Mayo, M.A., 1999. Developments in plant virus taxonomy since the publication of the 6th ICTV report. Arch. Virol. 144, 1659–1666.

Singh, R.P., Bioteus, M.E.N.F., Ready, K.F.M., Nie, X., 2003. *Coleus blumei* viroid. In: Hadidi, A., Flores, R., Randles, J.W., Semancik, J.S. (Eds.), Viroids. CSIRO Publishing, Collingwood, VIC, pp. 228–230.

Singh, R.P., Boucher, A., 1991. High incidence of transmission and occurrence of a viroid in commercial seeds of Coleus in Canada. Plant Dis. 75, 184–187.

Spieker, R.L., 1996. In vitro-generated "inverse" chimeric *Coleus blumei* viroids evolve in vivo into infectious RNA replicons. J. Gen. Virol. 77, 2839–2846.

Spieker, R.L., Haas, B., Charng, Y.C., Freimuller, K., Sänger, H.L., 1990. Primary and secondary structure of a new viroid "species" (CbVd1) present in the *Coleus blumei* cultivar "Bienvenue." Nucleic Acids Res. 18, 3998.

Spieker, R.L., Marinkovic, S., Sänger, H.L., 1996. A new sequence variant of *Coleus blumei* viroid 3 from the *Coleus blumei* cultivar "Fairway Ruby." Arch. Virol. 141, 1377–1386.

Zhang, Y., Yin, J., Jiang, D., Xin, Y., Ding, F., Deng, Z., et al., 2013. A universal oligonucleotide array with a minimal number of probes for the detection and identification of viroids at the genus level. PLoS One 8, e64474.

Zhang, Z., Qi, S., Tang, N., Zhang, X., Chen, S., Zhu, P., et al., 2014. Discovery of replicating circular RNAs by RNA-seq and computational algorithms. PLoS Pathog. 10, e1004553.

CHAPTER

28

Avocado Sunblotch Viroid

David N. Kuhn[1], *Andrew D.W. Geering*[2] *and Jonathan Dixon*[3]

[1]U.S. Department of Agriculture, Miami, FL, United States
[2]The University of Queensland, St Lucia, QLD, Australia
[3]Seeka Kiwifruit Industries Limited, Te Puke, New Zealand

INTRODUCTION

Avocado sunblotch disease was observed as early as 1914 in the West Indian Gardens Nursery at Altadena (Whitsell, 1952), which was owned by Frederick Popenoe, a pioneer of the Californian avocado industry (Shepherd and Bender, 2004). However, it took until 1928 before an official description of the disease was published, and the symptoms were initially attributed to sunburn (Coit, 1928). In 1932, graft-transmissibility of the disease was demonstrated, suggesting a viral etiology, and this hypothesis persisted for nearly 50 years until avocado sunblotch viroid (ASBVd) was discovered (Allen et al., 1981; Dale and Allen, 1979; Desjardins et al., 1980; Palukaitis et al., 1979; Semancik and Desjardins, 1980; Thomas and Mohamed, 1979).

As avocado is the only known natural host of ASBVd, it is likely that the viroid originated in the same region as its host, the eastern and central highlands of Mexico, through Guatemala to the Pacific coast of Central America (Knight, 2002). In 1911, Popenoe organized a collecting mission to Atlixco, Mexico, in search of superior varieties, including the original "Fuerte" plant, which was to become the mainstay of the Californian industry during its first decades. "Fuerte" was among the worst affected varieties and 30-year old "Fuerte-like" trees with sunblotch symptoms were observed near Atlixco in 1948 (Trask, 1948). During the 1930s and 1940s, ASBVd became prevalent in San Diego County, where a little over one-half of the total acreage of avocado trees in California was to be found, and ranked second in importance to

Viroids and Satellites.
DOI: http://dx.doi.org/10.1016/B978-0-12-801498-1.00028-0

Phytophthora cinnamomi as a pathogen (Whitsell, 1952). From California, ASBVd was likely transported to many countries in the world like Australia (see Chapter 46: Geographical Distribution of Viroids in Oceania), Israel (Whitsell, 1952), and Peru (Zentmyer, 1957), as new varieties developed there were in strong demand elsewhere.

TAXONOMIC POSITION AND NUCLEOTIDE SEQUENCE

Family: *Avsunviroidae*
Genus: *Avsunviroid*
Species: *Avocado sunblotch viroid* (reference sequence variant GenBank J02020.1) is a 247 nucleotide, single-stranded, circular RNA. Members in the *Avsunviroidae* are distinguished from those in the *Pospiviroidae* by the site of replication in the chloroplast, the absence of a central conserved region and the presence of distinctive hammerhead ribozyme structures that autocatalyze RNA cleavage (Di Serio et al., 2014). Genera in the *Avsunviroidae* are distinguished by the morphology of the hammerhead structure, G + C content, solubility in 2M LiCl and phylogenetic position using whole viroid genome sequences (Di Serio et al., 2014). ASBVd has the highest A + U content (62%) of all viroids, and its circular, single-stranded, (+) polarity RNA folds in vitro into a rod-like structure that includes a large bulge located in the left domain, a central domain that includes the conserved hammerhead nucleotides and a relatively large right terminal loop (Giguère et al., 2014). Alternative (+) and (−) RNA conformations in vitro have been proposed for a different sequence variant of ASBVd, although the conformations are still quasi rod-shaped (Delan-Forino et al., 2014). In common with members of the *Pospiviroidae* but not other members of the *Avsunviroidae*, ASBVd is soluble in 2M LiCl, a property that likely reflects its rod-like structure in vitro (Flores et al., 2000).

Sequence variation of ASBVd is common, observed within a single tree (Rakowski and Symons, 1989), between trees with different symptoms (Semancik and Szychowski, 1994), and from a single tree sampled over a period of years (Schnell et al., 2001a). Three types of sequence variants of ASBVd have been described: B (from trees with bleached leaves), V (from trees with variegated leaves), and SC (from symptomless carrier trees) (Semancik and Szychowski, 1994). ASBVd variants causing different symptoms have similar nucleotide substitutions in specific regions of the molecule, most frequently within the terminal loop structures outside the double stranded complementary regions. When multiple viroid sequences were determined for individual trees over time, it was observed that the most frequently identified sequence variant for a tree in one year was not necessarily the most frequently identified variant in a subsequent year (Schnell et al., 2001a). As there was no discernible pattern to sequence inheritance, it was best described as quasispecies in each tree, because the generation of a B type genome from a V or SC genome was not observed. Although there was a great deal of sequence variation, it did not produce enough mutation or sequence drift to generate or interconvert any of the three types of variants (Schnell et al., 2001a,b).

ECONOMIC SIGNIFICANCE

If left unmanaged, sunblotch disease can have a very large economic impact. Apparently asymptomatic "Hass" trees can have large yield reductions of 15%–30%, and yield reductions are even greater in symptomatic trees, at 50%–80% (Saucedo-Carabez et al., 2014). About half of the fruit produced by a symptomatic "Hass" tree may also be scarred and therefore significantly downgraded in quality (Saucedo-Carabez et al., 2014). However, ASBVd is easy to control through clean planting material schemes, is now very rare in many countries, and therefore is relatively unimportant.

An indirect economic impact of ASBVd is from the quarantine conditions imposed by countries on imports of fresh fruit (not processed pulp) from countries where the pathogen is recorded. This issue is exemplified in 2015 by actions from the Costa Rican Government to close off all imports of avocados from nine countries due to the perceived risk of introducing ASBVd (Dyer, 2015).

SYMPTOMS

The most recognizable symptoms of sunblotch disease (Fig. 28.1) are the sunken, yellow to purplish longitudinal scars or broad spots that appear on the surface of the fruit, which are most pronounced from the stem end to the middle and often cause distortion. Yellow or lighter-colored streaks may also occur on the younger limbs and shoots of the tree, and are usually most prominent in the new flush growth. The leaves of infected trees are normally asymptomatic, but occasionally white, yellow, or gray–green variegation is observed, which may be more prominent on one side of the leaf, leading to a ruffled or distorted appearance. Bark symptoms are more common on mature trees, and rectangular checking or cracking (crocodile bark) may be observed, particularly on

FIGURE 28.1 ASBVd symptoms: variegation on leaves (A, C, D), yellow depressions and distortions on fruit (B, C). Trees are from USDA-ARS Subtropical Horticulture Research Station, Miami, FL.

the trunk. Diseased trees may also be dwarfed with a diminished canopy and a distinctively decumbent or sprawling growth habit. Many infected trees may remain asymptomatic for their whole life, while diseased trees can become asymptomatic, and vice versa, symptoms can reemerge following significant stresses such as hard pruning.

Viroids do not encode proteins and it has long been wondered by what mechanism(s) they can cause disease in plants. A review of disease mechanisms of viroids proposes a wide variety of possibilities including interaction with host proteins to change their activity, activation, or inhibition of protein kinases, hormone-mediated responses, and perhaps most reasonably RNA-silencing through generation of small RNA fragments that can regulate host gene expression (Navarro et al., 2012). Interaction with host proteins is required for viroid replication and so can never be ruled out as the cause of symptoms. There is evidence for generation of small RNAs by viroids in the *Avsunviroidae* (Martínez de Alba et al., 2002), however no small RNAs were found to be generated from ASBVd. Thus, the disease mechanism generating the symptoms of sunblotch disease in avocado is yet unknown.

HOST RANGE

Avocado (*Persea americanum*) is the only known natural host of ASBVd, although its experimental host range includes other species in the family *Lauraceae*, including *Cinnamomum camphora*, *Cinnamomum zeylanicum*, *Ocotea bullata*, and *Persea schiedeana* (da Graça and Van Vuuren, 1980, 1981).

ASBVd is able to replicate in the yeast *Saccharomyces cerevisiae*, and while having no epidemiological significance, this does suggest that the cellular factors necessary to support replication are conserved across kingdoms (Delan-Forino et al., 2011).

TRANSMISSION

ASBVd is transmitted in seed, by grafting, and by mechanical means. Transmission rates of 86%–100% have been observed in seed from asymptomatic carrier trees but rates are much lower in seed from symptomatic trees (0%–5.5%) (Wallace and Drake, 1962). Seed transmission may result from infection of either the ovule or pollen, although the pollen-recipient tree does not become infected as a result of being cross-pollinated by pollen from an infected tree (Desjardins et al., 1984). Plants with seed-borne infection are often asymptomatic and infection may not become apparent until grafted with a scion of a susceptible variety (Desjardins et al., 1984).

ASBVd can remain infective on sap-contaminated cutting equipment such as grafting tools, and on surfaces where avocado plants are propagated (Desjardins et al., 1980, 1987). The unintentional infection of grafted trees using contaminated tools and working surfaces, as well as the use of infected seedlings and budwood in avocado nurseries is likely to have spread asymptomatic ASBVd variants widely throughout some avocado industries.

A slow increase in the incidence of ASBVd infection has been observed in the USDA Subtropical Horticultural Research Station Avocado Germplasm Collection in Florida, despite strict sanitization procedures in field operations. Significant clustering of infected trees was also observed, suggesting that root grafting between avocado trees first noted in California groves (Whitsell, 1952) may be another means of transmission (Schnell et al., 2011).

DETECTION

At first, sunblotch disease indexing was done by grafting bark pieces onto seedling indicator plants of susceptible varieties such as "Hass" and observing symptom development over a period of at least 2 years (Allen and Firth, 1980; Whitsell, 1952). Since the characterization of

ASBVd, detection methods have progressively improved, from polyacrylamide gel electrophoresis (da Graça and Mason, 1983; López-Herrera et al., 1987; Spiegel et al., 1984), dot-blot hybridization with labeled complementary cDNA (Bar-Joseph et al., 1985; Lima et al., 1994; Palukaitis and Symons, 1980; Rosner et al., 1983), to reverse transcription PCR amplification (RT-PCR) (Schnell et al., 1997; Schnell et al., 2001b; Semancik and Szychowski, 1994). Flower buds are preferable over leaves for diagnosis, as the viroid titer is higher and carbohydrate contamination of the RNA extract less of a problem (da Graça and Mason, 1983). Commercial detection kits for ASBVd are available through Agdia Inc. (nucleic acid hybridization assay) and Norgen Biotek Corp. (RT-PCR).

GEOGRAPHICAL DISTRIBUTION AND EPIDEMIOLOGY

ASBVd is recorded from Australia (Palukaitis et al., 1981), Israel (Spiegel et al., 1984), Ghana (Acheampong et al., 2008), Mexico (De La Torre-A et al., 2009; Saucedo-Carabez et al., 2014), Peru (Querci et al., 1995; Zentmyer, 1959), South Africa (Korsten et al., 1986), Spain (López-Herrera et al., 1987), Venezuela (Rondon and Figueroa, 1976), and the United States (Coit, 1928; Olano et al., 2002). Countries that have declared area-freedom from ASBVd include Costa Rica (World Trade Organization, Notification of Emergency Measures G/SPS/N/CRI/160, 5 May 2015) and New Zealand (Pugh and Thomson, 2009). It is likely that the geographic distribution of ASBVd is wider than stated, and the most likely areas to find the viroid will be those that imported untested avocado germplasm before highly sensitive testing methods became available.

The most critical point in the epidemiology of ASBVd is the avocado nursery. If infection can be prevented during propagation, then it is highly probable that the trees will remain viroid-free for the remainder of their lives, as long as they are not planted into existing diseased blocks, and basic hygiene is practiced with pruning tools. In countries such as Australia, Israel, and South Africa, where clean planting material schemes have been operational for many decades, ASBVd is now found only rarely.

CONTROL

There is no cure for avocado trees infected with ASBVd, and prevention or eradication is the only control option. Viroid elimination by micrografting and somatic embryo generation has been unsuccessful (Suarez et al., 2005, 2006). The avocado industries of Australia, New Zealand, and South Africa have implemented accredited nursery schemes, which require the registration and segregation of nuclear and multiplication trees used to source budwood and seed, and then the regular testing of these trees for ASBVd (every 3–5 years).

The most effective cleaning solution for ASBVd is a 10% solution of commercial bleach, which will readily break down RNA molecules. Implements cleaned with the 10% bleach solution also need to be rinsed with clean water to prevent high rates of corrosion occurring on metal and degradation of other surfaces. Disposable gloves must also be worn by propagators for plant material from each tree and these are then discarded when using plant material from a different tree.

Acknowledgments

D.K. would like to acknowledge Dr. Raymond J. Schnell (former USDA-ARS, currently Mars, Inc.) for his extensive and excellent work in the development of ASBVd detection methods and their application to disease management.

References

Acheampong, A.K., Akromah, R., Ofori, F.A., Takrama, J.F., Zeidan, M., 2008. Is there avocado sunblotch viroid in Ghana? Afr. J. Biotechnol. 7, 3540–3545.

Allen, R., Palukaitis, P., Symons, R., 1981. Purified avocado sunblotch viroid causes disease in avocado seedlings. Australas. Plant Path. 10, 31–32.

Allen, R.N., Firth, D.J., 1980. Sensitivity of transmission tests for avocado sunblotch viroid and other pathogens. Australas. Plant Path. 9, 2–3.

Bar-Joseph, M., Segev, D., Twizer, S., Rosner, A., 1985. Detection of avocado sunblotch viroid by hybridization with synthetic oligonucleotide probes. J. Virol. Methods 10, 69–73.

Coit, J.E., 1928. Sun-blotch of the avocado. California Avocado Association Yearbook. 12, 26–29.

da Graça, J.V., Mason, T.E., 1983. Detection of avocado sunblotch viroid in flower buds by polyacrylamide gel electrophoresis. J. Phytopathol. 108, 262–266.

da Graça, J.V., Van Vuuren, S.P., 1980. Transmission of avocado sunblotch disease to cinnamon. Plant Dis. 64, 475.

da Graça, J.V., Van Vuuren, S.P., 1981. Host range studies on avocado sunblotch. South African Avocado Growers' Association Yearbook. 1981 (4), 81–82.

Dale, J.L., Allen, R.N., 1979. Avocado afffected by sunblotch disease contains low molecular weight ribonucleic acid. Australas. Plant Path. 8, 3–4.

De La Torre-A, R., Téliz-Ortiz, D., Pallás, V., Sánchez-Navarro, J.A., 2009. First report of avocado sunblotch viroid in avocado from Michoacán, México. Plant Dis. 93, 202.

Delan-Forino, C., Deforges, J., Benard, L., Sargueil, B., Maurel, M.-C., Torchet, C., 2014. Structural analyses of avocado sunblotch viroid reveal differences in the folding of plus and minus RNA strands. Viruses 6, 489.

Delan-Forino, C., Maurel, M.-C., Torchet, C., 2011. Replication of avocado sunblotch viroid in the yeast *Saccharomyces cerevisiae*. J. Virol. 85, 3229–3238.

Desjardins, P.R., Drake, R.J., Sasaki, P.J., Atkins, E.L., Bergh, B.O., 1984. Pollen transmission of avocado sunblotch viroid and the fate of the pollen recipient tree. Phytopathology 74, 845.

Desjardins, P.R., Drake, R.J., Swiecki, S.A., 1980. Infectivity studies of avocado sunblotch disease causal agent, possibly a viroid rather than a virus. Plant Dis. 64, 313–315.

Desjardins, P.R., Saski, P.J., Drake, R.J., 1987. Chemical inactivation of avocado sunblotch viroid on pruning and propagation tools. California Avocado Association Yearbook. 71, 259–262.

Di Serio, F., Flores, R., Verhoeven, J.T.J., Li, S.-F., Pallás, V., Randles, J., et al., 2014. Current status of viroid taxonomy. Arch. Virol. 159, 3467–3478.

Dyer, Z., 2015. Costa Rica bans avocado imports from 9 countries. THE TICO TIMES *News* May 12.

Flores, R., Daròs, J.-A., Hernández, C., 2000. Avsunviroidae family: viroids with hammerhead ribozymes. Adv. Virus Res. 55, 271–323.

Giguère, T., Adkar-Purushothama, C.R., Bolduc, F., Perreault, J.-P., 2014. Elucidation of the structures of all members of the *Avsunviroidae* family. Mol. Plant Pathol. 15, 767–779.

Knight, R.J., 2002. History, distribution and uses. In: Whiley, A.W., Schaffer, B., Wolstenholme, B.N. (Eds.), The Avocado: Botany, Production and Uses. CABI Publishing, Wallingford, pp. 1–14.

Korsten, L., Bar-Joseph, M., Botha, A.D., Haycock, L.S., Kotzé, J.M., 1986. Commercial monitoring of avocado sunblotch viroid. South African Avocado Growers' Association Yearbook. 1986 (9), 63.

Lima, M.I., Fonseca, M.E.N., Flores, R., Kitajima, E.W., 1994. Detection of avocado sunblotch viroid in chloroplasts of avocado leaves by *in situ* hybridization. Arch. Virol. 138, 385–390.

López-Herrera, C., Pliego, F., Flores, R., 1987. Detection of avocado sunblotch viroid in Spain by double polyacrylamide-gel electrophoresis. J. Phytopathol. 119, 184–189.

Martínez de Alba, A.E., Flores, R., Hernández, C., 2002. Two chloroplastic viroids induce the accumulation of small RNAs associated with posttranscriptional gene silencing. J. Virol. 76, 13094–13096.

Navarro, B., Gisel, A., Rodio, M.-E., Delgado, S., Flores, R., Di Serio, F., 2012. Viroids: how to infect a host and cause disease without encoding proteins. Biochimie 94, 1474–1480.

Olano, C.T., Schnell, R.J., Kuhn, D.N., 2002. Current status of ASBVd infection among avocado accessions in the National Germplasm Collection. P. Fl. State Hortic. Soc. 115, 280–282.

Palukaitis, P., Hatta, T., Alexander, D.M., Symons, R.H., 1979. Characterization of a viroid associated with avocado sunblotch disease. Virology 99, 145–151.

Palukaitis, P., Rakowski, A.G., Alexander, D.M., Symons, R.H., 1981. Rapid indexing of the sunblotch disease of avocados using a complementary DNA probe to avocado sunblotch viroid. Ann. Appl. Biol. 98, 439–449.

Palukaitis, P., Symons, R.H., 1980. Avocado sunblotch viroid - detection and quantitation in plant-extracts by hybridization analysis with labeled complementary-DNA. P. Aust. Biochem. Soc. 13, 88–98.

Pugh, K., Thomson, V., 2009. Protecting our avocados. Biosecurity Magazine 93, 16–17.

Querci, M., Owens, R.A., Vargas, C., Salazar, L.F., 1995. Detection of potato spindle tuber viroid in avocado growing in Peru. Plant Dis. 79, 196–202.

Rakowski, A.G., Symons, R.H., 1989. Comparative sequence studies of variants of avocado sunblotch viroid. Virology 173, 352–356.

Rondon, A., Figueroa, M., 1976. Sunblotch of avocado in Venezuela. Agron. Trop. 26, 463–466.

Rosner, A., Spiegel, S., Alper, M., Bar-Joseph, M., 1983. Detection of avocado sunblotch viroid (ASBV) by dot-spot self-hybridization with a [32P]-labelled ASBV-RNA. Plant Mol. Biol. 2, 15–18.

Saucedo-Carabez, J.R., Teliz-Ortiz, D., Ochoa-Ascencio, S., Ochoa-Martinez, D., Vallejo-Perez, M.R., Beltran-Pena, H., 2014. Effect of avocado sunblotch viroid (ASBVd) on avocado yield in Michoacan, Mexico. Eur. J. Plant Pathol. 138, 799–805.

Schnell, R.J., Kuhn, D.N., Olano, C.T., Quintanilla, W.E., 2001a. Sequence diversity among avocado sunblotch viroids isolated from single avocado trees. Phytoparasitica 29, 451–460.

Schnell, R.J., Kuhn, D.N., Ronning, C.M., Harkins, D., 1997. Application of RT-PCR for indexing avocado sunblotch viroid. Plant Dis. 81, 1023–1026.

Schnell, R.J., Olano, C.T., Kuhn, D.N., 2001b. Detection of avocado sunblotch viroid variants using fluorescent single-strand conformation polymorphism analysis. Electrophoresis. 22, 427–432.

Schnell, R.J., Tondo, C.L., Kuhn, D.N., Winterstein, M.C., Ayala-Silva, T., Moore, J.M., 2011. Spatial analysis of avocado sunblotch disease in an avocado germplasm collection. J. Phytopathol. 159, 773–781.

Semancik, J.S., Desjardins, P.R., 1980. Multiple small RNA species and the viroid hypothesis for the sunblotch avocado disease. Virology 104, 117–121.

Semancik, J.S., Szychowski, J.A., 1994. Avocado sunblotch disease – a persistent viroid infection in which variants are associated with differential symptoms. J. Gen. Virol. 75, 1543–1549.

Shepherd, J.S., Bender, G.S., 2004. History of the avocado industry in California, second ed. The University of California Cooperative Extension, San Diego County and The California Avocado Society, San Diego, CA.

Spiegel, S., Alper, M., Allen, R.N., 1984. Evaluation of biochemical methods for the diagnosis of the avocado sunblotch viroid in Israel. Phytoparasitica 12, 37–43.

Suarez, I.E., Schnell, R.A., Kuhn, D.N., Litz, R.E., 2005. Micrografting of ASBVd-infected avocado (Persea americana) plants. Plant Cell Tiss. Org. Cult. 80, 179–185.

Suarez, I.E., Schnell, R.A., Kuhn, D.N., Litz, R.E., 2006. Recovery and indexing of avocado plants (Persea americana) from embryogenic nucellar cultures of an avocado sunblotch viroid-infected tree. Plant Cell Tiss. Org. Cult. 84, 27–37.

Thomas, W., Mohamed, N., 1979. Avocado sunblotch—a viroid disease? Australas. Plant Path. 8, 1–3.

Trask, E.E., 1948. Observations on the avocado industry in Mexico. California Avocado Society 1948 Yearbook. 33, 50–53.

Wallace, J.M., Drake, R.J., 1962. The high rate of seed transmission of avocado sun-blotch virus from symptomless trees and the origin of such trees. Phytopathology 52, 237–241.

Whitsell, R., 1952. Sun-blotch disease of avocados. California Avocado Society 1952 Yearbook. 37, 215–240.

Zentmyer, G.A., 1957. The search for resistant rootstocks in Latin America. California Avocado Society Yearbook. 41, 101–106.

Zentmyer, G.A., 1959. Avocado diseases in Latin America. Plant Dis. Rep. 43, 1229.

CHAPTER

29

Peach Latent Mosaic Viroid in Infected Peach

Ricardo Flores[1], *Beatriz Navarro*[2], *Sonia Delgado*[1], *Carmen Hernández*[1], *Wen-Xing Xu*[3], *Marina Barba*[4], *Ahmed Hadidi*[5] *and Francesco Di Serio*[2]

[1]Polytechnic University of Valencia-CSIC, Valencia, Spain [2]National Research Council, Bari, Italy [3]Huazhong Agricultural University, Wuhan, China [4]CREA-Research Centre for Plant Protection and Certification, Rome, Italy [5]U.S. Department of Agriculture, Beltsville, MD, United States

INTRODUCTION

Although the related diseases peach calico (PC) (Blodgett, 1944) and peach blotch (PB) (Willison, 1946) in the United States, and peach yellow mosaic (PYM) (Kishi et al., 1973) in Japan, were described previously, peach latent mosaic (PLM) disease was first reported in France on material imported from these two countries (Desvignes, 1976). The term latent refers to the lack in most natural infections of leaf symptoms and to their instability, as well as to the prolonged time that these and other phenotypic alterations take to appear. In the greenhouse, isolates of PLM are graded as severe or latent depending on whether or not they incite leaf symptoms on seedlings of the peach indicator GF-305 (Desvignes, 1976, 1986). Cross-protection assays between isolates of PC, PB, PYM, and PLM suggested a common causal agent (Desvignes, 1986), but attempts to identify virions failed. This failure, and the unusual heat resistance of the agent, suggested a viroid etiology. This hypothesis gained credibility when a viroid-like RNA was found in samples infected with severe and latent isolates of the PLM agent but not in healthy controls (Flores and Llácer, 1988), and it was finally substantiated by inoculating the purified viroid-like RNA to GF-305 peach

Viroids and Satellites.
DOI: http://dx.doi.org/10.1016/B978-0-12-801498-1.00029-2

seedlings: they expressed the characteristic symptoms and an RNA with identical physical properties was retrieved and named peach latent mosaic viroid (PLMVd) (Flores et al., 1990).

TAXONOMIC POSITION AND NUCLEOTIDE SEQUENCE

PLMVd (Hernández and Flores, 1992), together with avocado sunblotch viroid (Hutchins et al., 1986; Symons, 1981), chrysanthemum chlorotic mottle viroid (Navarro and Flores, 1997), and eggplant latent viroid (Fadda et al., 2003), form the family *Avsunviroidae* (see Chapter 13: Viroid Taxonomy). Both polarity strands of these four viroids can adopt hammerhead structures that mediate their self-cleavage in vitro and in vivo during replication in plastids through a symmetric rolling circle (see Chapter 7: Viroid Replication). PLMVd and chrysanthemum chlorotic mottle viroid form the genus *Pelamoviroid* because of some common characteristics: a multibranched secondary structure (Ambrós et al., 1998; Hernández and Flores, 1992), stabilized in the plus strands by a kissing-loop interaction (Bussière et al., 2000; Dubé et al., 2011; Gago et al., 2005) (Fig. 29.1), and insolubility in 2M LiCl (Navarro and Flores, 1997).

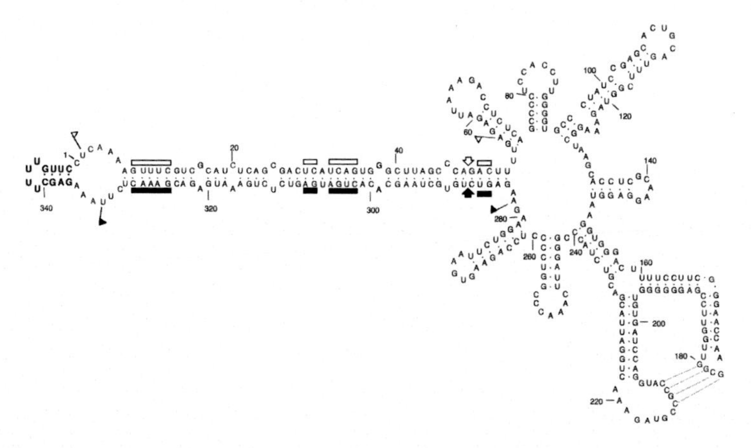

FIGURE 29.1 Sequence and predicted most-stable secondary structure of the PLMVd variant PC-C40 inducing peach calico (PC). The nucleotides of the insertion strictly associated with PC are denoted in boldface. Variants without this insertion are latent or induce a mosaic/blotch phenotype. The (+) and (−) self-cleaving domains are delimited by flags, residues conserved in most natural hammerhead structures are indicated by bars, and the self-cleavage sites are marked by arrows. Closed and open symbols refer to (+) and (−) polarities, respectively. Nucleotides involved in a kissing-loop interaction between positions 177–180 and 210–213 are indicated by broken lines. *Source: Reproduced with modifications from Rodio, M.E., Delgado, S., De Stradis, A.E., Gómez, M.D., Flores, R., Di Serio, F., 2007. A viroid RNA with a specific structural motif inhibits chloroplast development. Plant Cell 19, 3610–3626.*

The reference PLMVd variant (GenBank M83545.1), has 338 nt with 90 G (26.7%), 87 C (25.8%), 80 A (23.7%), and 80 U (23.7%) (Ambrós et al., 1998; Hernández and Flores, 1992). Sequencing of other variants has expanded the size from 335 to 351 nt (Ambrós et al., 1998; Fekih Hassen et al., 2007; Malfitano et al., 2003; Pelchat et al., 2000; Rodio et al., 2006; Shamloul et al., 1995; Wang et al., 2013; Yazarlou et al., 2012). However, only a few have been biologically tested. Despite the high sequence variability of PLMVd, observed in natural isolates and in the progeny from single variants (Ambrós et al., 1998, 1999; Di Serio et al., 2009; Glouzon et al., 2014; Malfitano et al., 2003; Navarro et al., 2012; Rodio et al., 2006; Wang et al., 2013; Yazarlou et al., 2012), the multibranched conformation and the hammerhead structures are preserved, thus supporting their significance in vivo. Moreover, part of the (+) and (−) linear forms from infected tissues are opened at the hammerhead-predicted sites (Delgado et al., 2005). UV irradiation has revealed additional elements of tertiary structure in the PLMVd (+) strand (Hernández et al., 2006), and bioinformatic analyses have clustered PLMVd variants into several groups (Ambrós et al., 1998; Fekih Hassen et al., 2007).

ECONOMIC SIGNIFICANCE

PLM disease diminishes the fruit quality (due to skin discolorations and suture cracks), shortens longevity of trees, and makes them more susceptible to other stresses. Ironically, some PLMVd-incited symptoms, particularly flowering delay, protect against late frosts in specific geographic areas (Gibson et al., 2001), although the delayed fruit ripening may reduce the economic value.

SYMPTOMS

Only reported in peach cultivars, including nectarines. Even if planting with infected material, symptoms do not usually appear before 2 years. They include: leaf chlorotic blotches or mosaics, yellow-creamy mosaics, and calico (albinism), flower streaking, delays in foliar sprouting, flowering, and ripening, malformation and discoloration of fruit, which frequently display cracked sutures and enlarged roundish stones, bud necrosis, stem pitting, and early aging of trees, which adopt a characteristic developmental pattern (open habit) (Fig. 29.2) (Desvignes, 1986; Flores et al., 2006).

Different PLMVd variants incite different symptoms, as shown by inoculating GF-305 with in vitro transcripts of the corresponding cDNA

FIGURE 29.2 Symptoms in different peach organs induced by PLMVd. (A) Typical leaf mosaic/blotch. (B) Calico. (C) Flowers with pink broken-line pattern (upper rows) and healthy controls (lower row). (D) Fruits displaying suture cracking. Photographs courtesy of J.C. Desvignes. *Source: Reproduced with modifications from Flores, R., Delgado, S., Rodio, M.E., Ambrós, S., Hernández, C., Di Serio, F., 2006. Peach latent mosaic viroid: not so latent. Mol. Plant Pathol. 7, 209–221.*

clones (Ambrós et al., 1998, 1999; Wang et al., 2013). PC is strictly associated with variants containing a 12–14 nt hairpin insertion with a U-rich capping loop (Malfitano et al., 2003) (Fig. 29.1). Removing the insertion from a PC-inducing variant results in replication without symptoms, revealing that the insertion contains the PC pathogenic determinant (Malfitano et al., 2003). Studies with other PC-inducing variants support this notion (Rodio et al., 2006). Moreover, insertions can be acquired and lost spontaneously, explaining the variable phenotype of some PLMVd infections.

Symptoms induced by PC variants have been further dissected (Rodio et al., 2007). Albino leaf sectors display malformed plastids resembling proplastids, wherein the plastid translation machinery is impaired due to deficient plastid rRNA processing but, nevertheless, they can import nuclear-encoded proteins. PLMVd replicates in the albino leaf sectors, invades the shoot apical meristem and induces proplastid alterations, thus interfering with chloroplast development that ultimately results in a phenotype resembling that of variegated mutants in which plastid rRNA maturation is also impaired (Rodio et al., 2007). Moreover, two small RNAs containing the pathogenic determinant for PC direct cleavage (as predicted by RNA silencing) of the mRNA encoding a chloroplastic heat-shock protein; this protein participates in chloroplast biogenesis in Arabidopsis and most likely in peach (Navarro et al., 2012). (see Chapter 9: Viroid Pathogenesis). Identification of the pathogenic determinant(s) in mosaic-inducing variants is more difficult because it is not associated with an insertion and because PLMVd populations evolve rapidly. A single nucleotide change has been proposed to induce yellowing or chlorotic-edge symptoms (Wang et al., 2013), although the peach mRNA(s) potentially targeted by small RNAs containing this polymorphism remain unknown.

HOST RANGE

This chapter focuses on PLMVd infecting peach (*Prunus persica* Batsch). The viroid has been graft-transmitted to peach hybrids (e.g., almond × peach and plum × peach), (Desvignes, 1986). For other PLMVd hosts, see Chapter 30, Peach Latent Mosaic Viroid in Temperate Fruit Trees Other Than Peach.

TRANSMISSION

PLMVd is horizontally transmitted by grafting and budding with infected material (Desvignes, 1986). *Myzus persicae* can experimentally transmit PLMVd (Desvignes, 1986), but at a low rate suggesting minor relevance under natural conditions (Flores et al., 1992). Mechanically, the viroid has also been transmitted by stem slashing or high pressure injection with PLMVd either purified from infected tissue or PLMVd cDNA clones/in vitro transcripts (Ambrós et al., 1998; Flores et al., 1990; Yazarlou et al., 2012), and by slashing infected plants and then the healthy ones (Hadidi et al., 1997), which may facilitate local spread.

PLMVd is not vertically transmitted through seed (Barba et al., 2007; Howell et al., 1998), but it is pollen-borne and -transmitted (Barba et al., 2007). Specifically, PLMVd was detected in supernatants and washed pollen from nine infected peach cultivars (Barba et al., 2007). Pollination of healthy plants with PLMVd-infected pollen transmitted the viroid at a rate dependent on the peach cultivar: after 6 years, PLMVd was detected in five of the cultivars pollinated experimentally (Barba et al., 2007).

DETECTION

PLM disease was initially detected and controlled by a cross-protection assay (Desvignes, 1976, 1986). Greenhouse-grown GF-305 seedlings were graft-inoculated with material from suspected trees, and 1.5–2 months later regrafted with material from seedlings infected with a symptomatic (marking) PLMVd strain. One month after the challenge inoculation, seedlings preinfected with a PLMVd latent strain did not display the leaf symptoms incited by the marking strain, while the non-infected seedlings did.

In the laboratory, PLMVd was first identified in RNA preparations by a polyacrylamide gel electrophoresis technique for detecting small circular RNAs (Albanese et al., 1992; Flores and Llácer, 1988; Flores et al., 1990). PLMVd cloning and sequencing (Hernández and Flores, 1992) led to more sensitive methods based on molecular hybridization with cDNA or cRNA probes (Ambrós et al., 1995; Hadidi et al., 1997; Loreti et al., 1995; Skrzeczkowski et al., 1996; Xu et al., 2009) and RT-PCR with specific primers (Shamloul et al., 1995). Afterward, similar or alternative formats of RT-PCR (Hassan et al., 2004, 2006; Ragozzino et al., 2004; Shamloul and Hadidi, 1999; Shamloul et al., 2002; Zhang et al., 1998), reverse transcription loop-mediated isothermal amplification (Boubourakas et al., 2009), liquid phase in situ RT-PCR (Boubourakas et al., 2011), and quantitative RT-PCR (Lin et al., 2013; Luigi and Faggioli, 2011; Parisi et al., 2011) were developed. PLMVd has been also detected by microarrays (Zhang et al., 2013) and next-generation sequencing (Di Serio et al., 2009; Glouzon et al., 2014).

GEOGRAPHICAL DISTRIBUTION AND EPIDEMIOLOGY

Reported in all peach-growing areas: Central Europe (Austria, former Yugoslavia, and Romania), the Mediterranean basin (France, Spain, Italy, Greece, Turkey, Syria, Lebanon, Jordan, Egypt, Tunisia, Algeria, and Morocco), North America (United States, Canada, and Mexico), South-America (Brazil, Argentina, Chile), Asia (Iran, Pakistan, Nepal, Japan, China), and Australia (Albanese et al., 1992; Boubourakas et al., 2009; Choueiri et al., 2001; De la Torre-Almaráz et al., 2015; Desvignes, 1986; Fekih Hassen et al., 2007; Hassen et al., 2006; Herranz et al., 2013; Ismaeil et al., 2001; Nieto et al., 2008; Pelchat et al., 2000; Shamloul et al., 1995; Skrzeczkowski et al., 1996; Wang et al., 2013; Yazarlou et al., 2012). The high incidence of PLMVd in some surveys (Badenes and Llácer, 1998; Hadidi et al., 1997; Skrzeczkowski et al., 1996) presumably results from horticultural practices.

CONTROL

Use of viroid-free propagation material and recurrent disinfection of pruning implements. Thermotherapy (maintaining the plants at 37°C for 35–45 days before sampling the tips) (Desvignes, 1986), or in vitro micrografting (Barba et al., 1995), may lead to PLMVd elimination. Reducing the tip size increases the proportion of viroid-free plants (Barba et al., 1995), which most likely is the consequence of PLMVd ability to invade regions proximal to the shoot apical meristem (Rodio et al., 2007).

Acknowledgments

Research in R.F. laboratory is currently funded by grant BFU2014–56812-P from the Spanish Ministerio de Economía y Competitividad (MINECO).

References

Albanese, G., Giunchedi, L., La Rosa, R., Poggi-Pollini, C., 1992. Peach latent mosaic viroid in Italy. Acta Hortic. 309, 331–338.

Ambrós, S., Desvignes, J.C., Llácer, G., Flores, R., 1995. Peach latent mosaic and pear blister canker viroids: detection by molecular hybridization and relationships with specific maladies affecting peach and pear trees. Acta Hortic. 386, 515–521.

Ambrós, S., Hernández, C., Desvignes, J.C., Flores, R., 1998. Genomic structure of three phenotypically different isolates of peach latent mosaic viroid: implications of the existence of constraints limiting the heterogeneity of viroid quasi-species. J. Virol. 72, 7397–7406.

Ambrós, S., Hernández, C., Flores, R., 1999. Rapid generation of genetic heterogeneity in progenies from individual cDNA clones of peach latent mosaic viroid in its natural host. J. Gen. Virol. 80, 2239–2252.

Badenes, M.L., Llácer, G., 1998. Occurrence of peach latent mosaic viroid in American peach and nectarine cultivars in Valencia, Spain. Acta Hortic. 472, 565–570.

Barba, M., Cupidi, A., Loreti, S., Faggioli, F., Martino, L., 1995. *In vitro* micrografting: a technique to eliminate peach latent mosaic viroid from peach. Acta Hortic. 386, 531–535.

Barba, M., Ragozzino, E., Faggioli, F., 2007. Pollen transmission of peach latent mosaic viroid. J. Plant Pathol. 89, 287–289.

Blodgett, E.C., 1944. Peach calico. Phytopathology 34, 650–657.

Boubourakas, I.N., Fukuta, S., Kyriakopoulou, P.E., 2009. Sensitive and rapid detection of peach latent mosaic viroid by the reverse transcription loop-mediated isothermal amplification. J. Virol. Methods 160, 63–68.

Boubourakas, I.N., Voloudakis, A.E., Fasseas, K., Resnick, N., Koltai, H., Kyriakopoulou, P. E., 2011. Cellular localization of peach latent mosaic viroid in peach sections by liquid phase in situ RT-PCR. Plant Pathol. 60, 468–473.

Bussière, F., Ouellet, J., Côté, F., Lévesque, D., Perreault, J.P., 2000. Mapping in solution shows the peach latent mosaic viroid to possess a new pseudoknot in a complex, branched secondary structure. J. Virol. 74, 2647–2654.

Choueiri, E., Abo Ghanem-Sabanadzovic, N., Khazzaka, K., Sabanadzovic, S., Di Terlizzi, B., Jreijiri, F., et al., 2001. Identification of peach latent mosaic viroid in Lebanon. J. Plant Pathol. 83, 225.

De la Torre-Almaráz, R., Pallás, V., Sánchez-Navarro, J., 2015. First report of peach latent mosaic viroid in peach trees from Mexico. Plant Dis. 99, 899.

Delgado, S., Martínez de Alba, A.E., Hernández, C., Flores, R., 2005. A short double-stranded RNA motif of peach latent mosaic viroid contains the initiation and the self-cleavage sites of both polarity strands. J. Virol. 79, 12934–12943.

Desvignes, J.C., 1976. The virus diseases detected in greenhouse and in the field by the peach seedling GF 305 indicator. Acta Hortic. 67, 315–323.

Desvignes, J.C., 1986. Peach latent mosaic and its relation to peach mosaic and peach yellow mosaic virus diseases. Acta Hortic. 193, 51–57.

Di Serio, F., Gisel, A., Navarro, B., Delgado, S., Martínez de Alba, A.E., Donvito, G., et al., 2009. Deep sequencing of the small RNAs derived from two symptomatic variants of a chloroplastic viroid: implications for their genesis and for pathogenesis. PLoS One 4, e7539.

Dubé, A., Bolduc, F., Bisaillon, M., Perreault, J.P., 2011. Mapping studies of the peach latent mosaic viroid reveal novel structural features. Mol. Plant Pathol. 12, 688–701.

Fadda, Z., Daròs, J.A., Fagoaga, C., Flores, R., Duran-Vila, N., 2003. Eggplant latent viroid, the candidate type species for a new genus within the family Avsunviroidae (hammerhead viroids). J. Virol. 77, 6528–6532.

Fekih Hassen, I., Massart, S., Motard, J., Roussel, S., Parisi, O., Kummert, J., et al., 2007. Molecular features of new peach latent mosaic viroid variants suggest that recombination may have contributed to the evolution of this infectious RNA. Virology 360, 50–57.

Flores, R., Delgado, S., Rodio, M.E., Ambrós, S., Hernández, C., Di Serio, F., 2006. Peach latent mosaic viroid: not so latent. Mol. Plant Pathol. 7, 209–221.

Flores, R., Hernández, C., Avinent, L., Hermoso, A., Llácer, G., Juárez, J., et al., 1992. Studies on the detection, transmission and distribution of peach latent mosaic viroid in peach trees. Acta Hortic. 309, 325–330.

Flores, R., Hernández, C., Desvignes, J.C., Llácer, G., 1990. Some properties of the viroid inducing the peach latent mosaic disease. Res. Virol. 141, 109–118.

Flores, R., Llácer, G., 1988. Isolation of a viroid-like RNA associated with peach latent mosaic disease. Acta Hortic. 235, 325–332.

Gago, S., De la Peña, M., Flores, R., 2005. A kissing-loop interaction in a hammerhead viroid RNA critical for its in vitro folding and in vivo viability. RNA. 11, 1073–1083.

Gibson, P.G., Reighard, G.L., Scott, S.W., Zimmerman, M.T., 2001. Identification of graft-transmissible agents from "Ta-Tao 5" peach and their effect on "Coronet" peach. Acta Hortic. 550, 309–314.

Glouzon, J.P.S., Bolduc, F., Wang, S., Najmanovich, R.J., Perreault, J.P., 2014. Deep-sequencing of the peach latent mosaic viroid reveals new aspects of population heterogeneity. PLoS One 9, e87297.

Hadidi, A., Giunchedi, L., Shamloul, A.M., Poggi-Pollini, C., Amer, M.A., 1997. Occurrence of peach latent mosaic viroid in stone fruits and its transmission with contaminated blades. Plant Dis. 81, 154–158.

Hassan, M., Zouhar, M., Rysanek, P., 2004. Development of a PCR method of peach latent mosaic viroid and hop stunt viroid detection for certification of planting material. Acta Hortic. 657, 391–395.

Hassen, I.F., Roussel, S., Kummert, J., Fakhfakh, H., Marrakchi, M., Jijakli, M.H., 2006. Development of a rapid RT-PCR test for the detection of peach latent mosaic viroid, pear blister canker viroid, hop stunt viroid and apple scar skin viroid in fruit trees from Tunisia. J. Phytopathol. 154, 217–223.

Hernández, C., Di Serio, F., Darès, J.A., Flores, R., 2006. An element of tertiary structure of peach latent mosaic viroid RNA revealed by UV irradiation. J. Virol. 80, 9336–9340.

Hernández, C., Flores, R., 1992. Plus and minus RNAs of peach latent mosaic viroid self-cleave *in vitro* via hammerhead structures. Proc. Natl. Acad. Sci. USA 89, 3711–3715.

Herranz, M.C., Niehl, A., Rosales, M., Fiore, N., Zamorano, A., Granell, A., et al., 2013. A remarkable synergistic effect at the transcriptomic level in peach fruits doubly infected by prunus necrotic ringspot virus and peach latent mosaic viroid. Virol J. 10, 164.

Howell, W.E., Skrzeczkowski, L.J., Mink, G.I., Nunez, A., Wessels, T., 1998. Non-transmission of apple scar skin viroid and peach latent mosaic viroid through seed. Acta Hortic. 472, 635–639.

Hutchins, C., Rathjen, P.D., Forster, A.C., Symons, R.H., 1986. Self-cleavage of plus and minus RNA transcripts of avocado sunblotch viroid. Nucleic Acids Res. 14, 3627–3640.

Ismaeil, F., Abou-Ghanem-Sabanadzovic, N., Myrta, A., Di Terlizzi, B., Savino, V., 2001. First record of peach latent mosaic viroid and hop stunt viroid in Syria. J. Plant Pathol. 83, 227.

Kishi, K., Takanashi, K., Abiko, K., 1973. New virus diseases of peach, yellow mosaic, oil blotch and star mosaic. Bull. Hort. Res. Sta. Japan, Ser. A. 12, 197–208.

Lin, L., Li, R., Bateman, M., Mock, R., Kinard, G., 2013. Development of a multiplex TaqMan real-time RT-PCR assay for simultaneous detection of Asian prunus viruses, plum bark necrosis stem pitting associated virus, and peach latent mosaic viroid. Eur. J. Plant Pathol. 137, 797–804.

Loreti, S., Faggioli, F., Barba, M., 1995. A rapid extraction method to detect peach latent mosaic viroid by molecular hybridization. Acta Hortic. 386, 560–564.

Luigi, M., Faggioli, F., 2011. Development of quantitative real-time RT-PCR for the detection and quantification of peach latent mosaic viroid. Eur. J. Plant Pathol. 130, 109–116.

Malfitano, M., Di Serio, F., Covelli, L., Ragozzino, A., Hernández, C., Flores, R., 2003. Peach latent mosaic viroid variants inducing peach calico contain a characteristic insertion that is responsible for this symptomatology. Virology 313, 492–501.

Navarro, B., Flores, R., 1997. Chrysanthemum chlorotic mottle viroid: unusual structural properties of a subgroup of self-cleaving viroids with hammerhead ribozymes. Proc. Natl. Acad. Sci. USA 94, 11262–11267.

Navarro, B., Gisel, A., Rodio, M.E., Delgado, S., Flores, R., Di Serio, F., 2012. Small RNAs containing the pathogenic determinant of a chloroplast-replicating viroid guide the degradation of a host mRNA as predicted by RNA silencing. Plant J. 70, 991–1003.

Nieto, A.M., Di Feo, L., Nome, C.F., 2008. First report of peach latent mosaic viroid in peach trees in Argentina. Plant Dis. 92, 1137.

Parisi, O., Lepoivre, P., Jijakli, M.H., 2011. Development of a quick quantitative real-time PCR for the in vivo detection and quantification of peach latent mosaic viroid. Plant Dis. 95, 137–142.

Pelchat, M., Levesque, D., Ouellet, J., Laurendeau, S., Levesque, S., Lehoux, J., et al., 2000. Sequencing of peach latent mosaic viroid variants from nine North American peach cultivars shows that this RNA folds into a complex secondary structure. Virology 271, 37–45.

Ragozzino, E., Faggioli, F., Barba, M., 2004. Development of a one tube-one step RT-PCR protocol for the detection of seven viroids in four genera: *Apscaviroid, Hostuviroid, Pelamoviroid* and *Pospiviroid*. J. Virol. Methods 121, 25–29.

Rodio, M.E., Delgado, S., De Stradis, A.E., Gómez, M.D., Flores, R., Di Serio, F., 2007. A viroid RNA with a specific structural motif inhibits chloroplast development. Plant Cell 19, 3610–3626.

Rodio, M.E., Delgado, S., Flores, R., Di Serio, F., 2006. Variants of peach latent mosaic viroid inducing peach calico: uneven distribution in infected plants and requirements of the insertion containing the pathogenicity determinant. J. Gen. Virol. 87, 231–240.

Shamloul, A.M., Faggioli, F., Keith, J.M., Hadidi, A., 2002. A novel multiplex RT-PCR probe capture hybridization (RT-PCR-ELISA) for simultaneous detection of six viroids in four genera: *Apscaviroid, Hostuviroid, Pelamoviroid,* and *Pospiviroid.* J. Virol. Methods 105, 115–121.

Shamloul, A.M., Hadidi, A., 1999. Sensitive detection of potato spindle tuber and temperate fruit tree viroids by reverse transcription-polymerase chain reaction-probe capture hybridization. J. Virol. Methods 80, 145–155.

Shamloul, A.M., Minafra, A., Hadidi, A., Giunchedi, L., Waterworth, H.E., Allam, E.K., 1995. Peach latent mosaic viroid: nucleotide sequence of an Italian isolate, sensitive detection using RT-PCR and geographic distribution. Acta Hortic. 386, 522–530.

Skrzeczkowski, L.J., Howell, W.E., Mink, G.I., 1996. Occurrence of peach latent mosaic viroid in commercial peach and nectarine cultivars in the U.S. Plant Dis. 80, 823.

Symons, R.H., 1981. Avocado sunblotch viroid: primary sequence and proposed secondary structure. Nucleic Acids Res. 9, 6527–6537.

Wang, L.P., He, Y., Kang, Y.P., Hong, N., Farooq, A.B.U., Wang, G.P., et al., 2013. Virulence determination and molecular features of peach latent mosaic viroid isolates derived from phenotypically different peach leaves: a nucleotide polymorphism in L11 contributes to symptom alteration. Virus Res. 177, 171–178.

Willison, R.S., 1946. Peach blotch. Phytopathology 36, 273–276.

Xu, W.X., Hong, N., Jin, Q.T., Farooq, A.B., Wang, Z.Q., Song, Y., et al., 2009. Probe binding to host proteins: a cause for false positive signals in viroid detection by tissue hybridization. Virus Res. 145, 26–30.

Yazarlou, A., Jafarpour, B., Tarighi, S., Habili, N., Randles, J.W., 2012. New Iranian and Australian peach latent mosaic viroid variants and evidence for rapid sequence evolution. Arch. Virol. 157, 343–347.

Zhang, Y.J., Yin, J., Jiang, D.M., Xin, Y.Y., Ding, F., Deng, Z.N., et al., 2013. A universal oligonucleotide microarray with a minimal number of probes for the detection and identification of viroids at the genus level. PLoS One 8, e64474.

Zhang, Y.P., Uyemoto, J.K., Kirkpatrick, B.C., 1998. A small-scale procedure for extracting nucleic acids from woody plants infected with various phytopathogens for PCR assay. J. Virol. Methods 71, 45–50.

CHAPTER

30

Peach Latent Mosaic Viroid in Temperate Fruit Trees Other Than Peach

Panayota E. Kyriakopoulou[1], Luciano Giunchedi[2], Marina Barba[3], Iraklis N. Boubourakas[4], Maria S. Kaponi[5] and Ahmed Hadidi[6]

[1]Agricultural University of Athens, Athens, Greece [2]University of Bologna, Bologna, Italy [3]CREA-Research Centre for Plant Protection and Certification, Rome, Italy [4]Directorate of Rural Economy and Veterinary of Piraeus, Attica Prefecture, Attica, Greece [5]Formerly, Hirosaki University, Hirosaki, Japan [6]U.S. Department of Agriculture, Beltsville, MD, United States

INTRODUCTION

Desvignes (1976) described "peach latent mosaic disease", then Flores and Llácer (1988) and Flores et al. (1990) proved that it was due to a viroid, peach latent mosaic viroid (PLMVd). The first report extending the natural host range of PLMVd to species other than peach (*Prunus persica*) was published in 1997 (Hadidi et al., 1997). Since then, PLMVd non-peach has been found to naturally infect additional species of *Rosaceae*, both stone and pome fruits, cultivated and wild, as well as in a few non-*Rosaceae* species in all continents except Australia/Oceania. PLMVd in peach, however, was reported in Australia (Di Serio et al., 1999).

Viroids and Satellites.
DOI: http://dx.doi.org/10.1016/B978-0-12-801498-1.00030-9

TAXONOMIC POSITION, NUCLEOTIDE SEQUENCE, AND PHYLOGENY

Peach latent mosaic viroid is the type species of genus *Pelamoviroid* in the family *Avsunviroidae* (Di Serio et al., 2014; Chapter 13: Viroid Taxonomy). Analyses of progeny of individual PLMVd cDNA clones have revealed its extreme heterogeneity, probably due to a high mutation rate (Ambrós et al., 1999; Glouzon et al., 2014; Malifano et al., 2003; Yazarlou et al., 2012).

PLMVd sequences from trees other than peach are 337–340 nt long (Table 30.1), and share 90%–95% sequence identity with the 338 nt peach reference variant M83545.1 (Hernández and Flores, 1992). In particular, 24 stone fruit sequences from seven Tunisian almond, four Greek apricot, seven sweet cherry (one from Canada and six from Greece), one prune and five plum trees from Greece, Iran, Italy, and Romania, share sequence identities of 92%–95%, 93%–94%, 92%–95%, and 90%–94%, respectively, with the reference variant. Twelve pome fruit variant sequences from one Indian apple tree, five from pear trees (three from Greece, one from Italy, and one from Tunisia), two Greek quince trees, and four Greek wild pear trees share sequence identities of 91%, 92%–95%, 95%–96%, and 92%–94%, respectively, with the reference viroid variant (M.S. Kaponi, unpublished). The sequence and secondary structure of PLMVd-plum and PLMVd-wild pear (not shown) are similar to those of PLMVd-peach (see Chapter 29: Peach Latent Mosaic Viroid in Infected Peach).

Boubourakas (2010) found no correlation between the percentage of sequence identity and plant species, variety, or geographic origin of 21 non-peach PLMVd clones from Greece (Table 30.1). In phylogenetic analysis comprising previously reported peach PLMVd sequences (Ambrós et al., 1998), these clones were grouped separately from the peach ones, except for all plum isolates and one pear isolate, and all Greek peach and non-peach isolates were clustered in group III. According to these findings, in two pear variants, three informative nucleotide substitutions within the highly conserved regions of the plus-strand hammerhead structure did not change the overall hammerhead conformation.

Yazarlou et al. (2012) noticed a tendency for geographic clustering in PLMVd sequences from peach and plum in Iran, as well as significant variation within local Iranian sequences. This variation was attributed to the evolutionary potential of PLMVd and to its long coexistence with endemic host species, such as peach.

Alignment and phylogenetic analysis of the 36 worldwide non-peach sequences presented in Table 30.1 shows that they are clustered in almost all phylogenetic groups, among 303 complete PLMVd sequences from peach (data not shown), indicating that non-peach PLMVd variants are not host- or geographic origin-specific (M.S. Kaponi, unpublished).

TABLE 30.1 Hosts, Geographical Distribution, and Detection Methods for PLMVd in Plant Species Other Than Peach

Host	Country	Detection method	GenBank accession number	Sequence length (nt)	Reference
Apricot	Albania	TPH[a]	—	—	Torres et al. (2004), Musa et al. (2010)
European plum	Albania	TPH	—	—	Torres et al. (2004), Musa et al. (2010)
Apricot	Algeria	TPH, RT-PCR[b]	—	—	Rouag et al. (2008), Meziani et al. (2010)
European plum	Algeria	TPH, RT-PCR	—	—	Rouag et al. (2008), Meziani et al. (2010)
Sweet cherry	Algeria	TPH, RT-PCR	—	—	Rouag et al. (2008), Meziani et al. (2010)
Sweet cherry	Canada	RT-PCR, DBH,[c] SEQ[d]	AF339741	337	Hadidi et al. (1997)
European plum	Chile	DBH	—	—	Fiore et al. (2003)
Apricot	Egypt	TPH	—	—	El-Dougdoug et al. (2012)
European plum	Egypt	TPH	—	—	El-Dougdoug et al. (2012)
Apple	Egypt	TPH	—	—	El-Dougdoug (1998), El-Dougdoug et al. (2012)
Pear	Egypt	TPH	—	—	El-Dougdoug et al. (2012)
Grapevine	Egypt	TPH	—	—	El-Dougdoug et al. (2012)
Mango	Egypt	TPH	—	—	El-Dougdoug et al. (2012)
Apricot	Greece	DBH, RT-PCR, EQ, RT-LAMP[e]	KU048779	338	Boubourakas (2010)
			KU048778	337	
			KU048780	339	
			KU048781	339	
Sweet cherry	Greece	DBH, RT-PCR, SEQ	KU048785	338	Boubourakas (2010)
			KU048786	338	
			KU048787	338	
			KU048782	339	
			KU048873	339	
			KU048784	339	

(*Continued*)

TABLE 30.1 (Continued)

Host	Country	Detection method	GenBank accession number	Sequence length (nt)	Reference
European plum	Greece	DBH, RT-PCR, SEQ, RT-LAMP	KU048791	339	Boubourakas (2010)
			KU048792	339	
			KU048793	339	
Quince	Greece	RT-PCR, SEQ, RT-LAMP	KU048794	339	Boubourakas (2010)
			KU048795	339	
Pear	Greece	DBH, SH,[f] RT-PCR, SEQ	—		Kyriakopoulou et al. (1998, 2001), Boubourakas (2010)
			KU048788	339	
			KU048789	339	
			KU048790	339	
Wild pear	Greece	DBH, SH, RT-PCR, SEQ, RT-LAMP	AF339740	338	Kyriakopoulou et al. (1998, 2001), Boubourakas (2010)
			KU048976	339	
			KU048797	339	
			KU048798	339	
Apple	India	SEQ	FM955277	337	Walia et al. (2008)
European plum	Iran	RT-PCR, SEQ, transmission to peach cv. Nemaguard (BI)[g]	HM185115	338	Yazarlou et al. (2012)
Japanese plum	Italy	DBH, SH, RT-PCR, SEQ	AF339742	338	Kyriakopoulou et al. (2001), Giunchedi et al. (1998), Hadidi et al. (1997), Faggioli et al. (1997)
Sweet cherry	Italy	DBH, dPAGE,[h] GF305 (BI)	—	—	Crescenzi et al. (2002)
Japanese plum	Italy	MH, RT-PCR, BI	—	338	Giunchedi et al. (1998)
Apricot	Italy	TPH. RT-PCR	—	—	Faggioli and Barba (2008)
Pear	Italy	DBH, SH, RT-PCR, SEQ	—	338	Kyriakopoulou et al. (2001), Kyriakopoulou et al. (1998)

(Continued)

TABLE 30.1 (Continued)

Host	Country	Detection method	GenBank accession number	Sequence length (nt)	Reference
Apricot	Japan	RT-PCR, DBH, NBH[i] RT-PCR	—	—	Hadidi et al. (1997), Osaki et al. (1999)
Sweet cherry	Japan	RT-PCR	—	—	Osaki et al. (1999)
Japanese apricot	Japan	RT-PCR	—	—	Osaki et al. (1999)
Japanese plum	Japan	RT-PCR	—	—	Osaki et al. (1999)
Apricot	Jordan	DBH	—	—	Al Rwahnih et al. (2001)
Apricot	Morocco	DBH, TPH	—	—	Bouani et al. (2004)
Sweet cherry	Morocco	DBH, TPH	—	—	Bouani et al. (2004)
Apricot	Nepal	RT-PCR, DBH, NBH	—	—	Hadidi et al. (1997)
Apricot	Poland	TPH	—	—	Paduch-Cichal and Skrzeczkowski (2001)
Sweet cherry	Romania	RT-PCR, DBH, NBH	—	—	Hadidi et al. (1997)
Nanking cherry	Romania	RT-PCR, SEQ	—	338	Giunchedi et al. (1998)
European plum	Serbia	RT-PCR, DBH, NBH	—	—	Hadidi et al. (1997)
European plum	Slovenia	TPH		—	Virscek-Marn (2009)
Almond	Tunisia	RT-PCR, SEQ	DQ680724	339	Fekih Hassen et al. (2005, 2007)
			DQ680725	339	
			DQ680731	338	
			DQ680748	340	
			DQ680777	340	
			DQ680778	339	
			DQ680779	340	

(Continued)

TABLE 30.1 (Continued)

Host	Country	Detection method	GenBank accession number	Sequence length (nt)	Reference
Pear	Tunisia	RT-PCR, SEQ	DQ680768	339	Fekih Hassen et al. (2006, 2007)
Apricot	Turkey	TPH	—	—	Gümüs et al. (2007)
Sweet cherry	USA-California	RT-PCR	—	—	Osman et al. (2012)

[a] *TPH, Tissue print hybridization = Tissue blot hybridization.*
[b] *RT-PCR, Reverse transcription-polymerase chain reaction.*
[c] *DBH, Dot-blot hybridization.*
[d] *SEQ, Sequencing of cDNA clones.*
[e] *RT-LAMP, Reverse transcription—loop mediated isothermal amplification.*
[f] *SH, Southern hybridization using PLMVd cDNA.*
[g] *BI, Biological Indexing.*
[h] *dPAGE, Denaturing polyacrylamide gel electrophoresis.*
[i] *NBH, Northern-blot hybridization.*

ECONOMIC SIGNIFICANCE

Generally, evaluation of PLMVd economic significance in non-peach species is currently not possible, considering the lack both of information on symptomatology and of extensive national and international surveys to indicate distribution. In surveys not exclusive for PLMVd, samples were collected from unspecified symptomatic or healthy-looking plants, plants with symptoms attributed to other pathogens (see Section "Symptoms") or with symptoms not recorded, and sampling was limited or very limited. Even in cases of extended or relatively extended surveys in Algeria (Meziani et al., 2010) and Jordan (Al Rwahnih et al., 2001), no economic data are recorded. However, fruit quality of the Japanese plum "Angeleno" is reduced due to the spotted fruit symptom (Fig. 30.1), which is possibly associated with PLMVd etiology in Italy (Giunchedi et al., 1998) and Greece (Boubourakas, 2010).

SYMPTOMS

PLMVd is known to be a damaging and economically important pathogen of peach and peach hybrids, with well studied symptomatology ranging from latent to severe (see Chapter 29: Peach Latent Mosaic Viroid in Infected Peach). Its symptomatology on other species, such as almond (*Prunus dulcis*, syn. *P. amygdalus, Amygdalus communis*), apricot (*Prunus armeniaca*), sweet cherry (*Prunus avium*), Japanese apricot

FIGURE 30.1 Spotted fruit symptom of Angeleno plum associated with PLMVd infection (Boubourakas, 2010).

(mume, *Prunus mume*), Japanese plum (*Prunus salicina*), European plum (prune, *Prunus domestica*), apple (*Malus domestica*), pear (*Pyrus communis*), quince (*Cydonia oblonga*), wild pear (*Pyrus amygdaliformis*), grapevine (*Vitis vinifera*), and mango (*Mangifera indica*) has not been studied and the data so far comprise only field observations. Reports on the detection of PLMVd in these species refer to samples collected from trees showing various virus-like or general symptoms, such as for trees with mixed infections with other viroids, viruses, phytoplasmas and/or other pathogens and stresses, as well as where no symptoms were recorded. Thus, there are no conclusive data on the symptomatology of PLMVd in non-peach species. Experimental work is needed, especially fulfillment of Koch's postulates, to reach symptomatological conclusions on the species under discussion. It is suggested to inoculate viroid clones to the selected species and test symptom expression under permissive high temperature conditions as with peach. The only known suggestive case of PLMVd symptomatology in non-peach species is "spotted fruit disease" in Japanese plum cv. Angeleno (Fig. 30.1) described both in Italy (Giunchedi et al., 1997, 1998) and Greece (Boubourakas, 2010). It is characterized by the presence of numerous discolored areas with isodiametric shape (1–2 mm in diameter), irregular border, randomly scattered on the fruit surface but mainly in the petiole area, which may merge to form spots 6–8 mm in diameter.

HOST RANGE

Hosts of PLMVd other than peach and peach hybrids discovered so far belong mainly to the family *Rosaceae*. Stone fruit hosts include almond, apricot, sweet cherry, Japanese apricot, Japanese plum, European plum, and pome fruit hosts include apple, pear, quince, and wild pear; PLMVd has been transmitted to Nanking cherry (*Prunus tomentosa*) in Italy (Table 30.1). Wild pear is widespread among natural plant communities of mountainous areas of Greece. In Egypt, besides apricot, prune, apple, and pear (El-Dougdoug, 1998; El-Dougdoug et al., 2012), plant species outside the *Rosaceae*, such as grapevine and mango, have been reported as hosts of PLMVd, as revealed by tissue print hybridization, which may require confirmation by amplification using reverse-transcription polymerase chain reaction (RT-PCR) and sequencing.

TRANSMISSION

In non-peach plant species, PLMVd has been transmitted by grafting (chip budding) from infected plum and sweet cherry to the indicators *Prunus tomentosa* and peach GF-305, respectively, without showing symptoms, whereas the viroid was detected by molecular techniques 5–6 months after inoculation (Crescenzi et al., 2002; Giunchedi et al., 1998, 2011). The presence of PLMVd in almond and pear trees neighboring infected peach in Tunisia has been attributed to mechanical transmission with contaminated tools (Fekih Hassen et al., 2005). Similarly, in Greece, most PLMVd-positive non-peach samples were collected from trees distributed within PLMVd-infected peach orchards (where PLMVd infection frequency in peach was found to be very high, >75%, in some early varieties such as May Crest and Spring Crest), or close to PLMVd-infected trees (Boubourakas, 2010). This may be explained by mechanical transmission associated with some horticultural practices like pruning (Flores et al., 1990; Hadidi et al., 1997), and with pollen from peach to peach trees (Barba et al., 2007) or to other stone fruit trees where peach pollen germinates on stigmas, irrespective of whether or not fertilization occurs. Desvignes (1999) reported that PLMVd is a damaging viroid and its spreading in peach orchards may occur in a radius of 20–30 m from an infected peach tree.

In Greece, PLMVd was found distributed in mountainous areas in wild pear. It was detected in this species by dot-blot hybridization, northern-blot hybridization, RT-PCR, RT-loop mediated isothermal amplification, real-time RT-PCR, cloning and sequencing in Peloponnesus, namely in the prefectures of Korinthia, Argolis-Epidaurus and Troizinia, Arcadia, and Eleia in Olympia province (Boubourakas, 2010; Boubourakas et al., 2006; Kyriakopoulou et al., 1998, 2001), but no systematic survey was done throughout the country. This thorny bush-tree grows away from cultivated areas, as well as within fields on mountainsides as a weed tree or in their margins. In Greece, wild pear was the traditional rootstock of pear, and to a limited extent of apple, for centuries and possibly for millennia, until a few decades ago when modern agriculture arrived; this seems to have contributed to PLMVd spread, at least in pear. Finding PLMVd geographically distributed in wild pear on the mountains, away from cultivated areas and in pear trees grafted on this rootstock, the last both in orchards and in isolated trees, indicates that this species is a primary host of PLMVd, and leads to the hypothesis that Greece is a geographical origin of PLMVd. Viroid transmission by goats (through their horns) (Gohen et al., 2005) might have contributed to the spread of PLMVd in wild pear, pear, and other species, considering the extensive goat husbandry practiced by farmers throughout the centuries in Greece, especially in the mountainous areas.

DETECTION

PLMVd detection methods used in naturally infected non-peach species are included in Table 30.1. Here, it should be pointed out that in quantitative real-time PCR tests, the titer of PLMVd in non-peach hosts was about 0.4% of that in cultivated peach (Boubourakas, 2010), indicating that more sensitive detection methods should be applied in surveys to estimate the actual frequency of PLMVd in non-peach species.

GEOGRAPHICAL DISTRIBUTION AND EPIDEMIOLOGY

As shown in Fig. 30.2, PLMVd in non-peach species has been detected in 20 countries in Europe, Africa, Asia, North and South America. Pome fruit trees include apple in Egypt and India, pear in Egypt, Greece, Italy, and Tunisia, quince in Greece, and wild pear in Greece; grapevine and mango in Egypt (Table 30.1; Fig. 30.2).

In general, there are no systematic epidemiological surveys for PLMVd in its non-peach hosts. Data so far show a much lower frequency of PLMVd in these species compared to peach. In Algeria (Meziani et al., 2010), large scale surveys for virus and viroid infection in the main stone fruit growing areas, using samples from 553 trees in mother blocks, 354 from variety collections, 227 from nurseries, and 574 from commercial orchards (in total 1715), showed PLMVd infection in 26/357 tested apricot trees, 8/341 plums and 13/247 cherries, besides peach (108 infected/418 tested). The virus and viroid infections in Algeria were found mainly in mother blocks. In Egypt, El-Dougdoug et al. (2012), surveying 50 apple, 72 apricot, 75 grapevine, 50 mango,

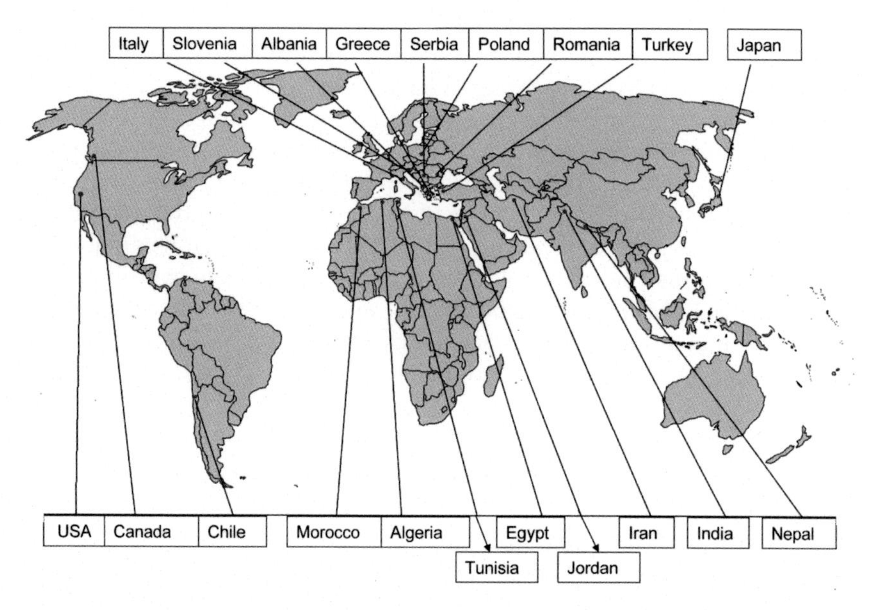

FIGURE 30.2 Worldwide distribution of PLMVd in naturally infected hosts other than peach.

25 pear, and 25 plum leaf samples, reported 5.4%, 2.5%, 0.5%, 13.5%, 23.4%, and 3.5% PLMVd occurrence, respectively, compared to 45% in peach (100 trees examined). In Greece, besides peach and peach hybrids, where PLMVd was detected at frequencies above 75%, both in orchards and in mother blocks (Boubourakas, 2010), no systematic surveys have been done with respect to its frequency in nonpeach species. Available data for Greece (Boubourakas, 2010; Boubourakas et al., 2006; Kyriakopoulou et al., 1998, 2001) show its occurrence in every area surveyed: widespread in Peloponnesus (wild pear, apricot, pear, plum/prune, quince), Thessaly (apricot, cherry, pear, plum/prune, quince) and Macedonia (sweet cherry, plum/prune). Details on its transmission and spread are described in Section "Transmission".

In general, data so far show much lower frequency of PLMVd in non-peach hosts compared to peach. However, improving detection methods in these species may change this picture.

CONTROL

Viroid-tested propagating material is highly recommended to control PLMVd in non-peach plant species.

Acknowledgments

The authors would like to thank growers and nurserymen of Western Macedonia, Central Greece, and Peloponnesus for providing numerous plant samples infected with viroids which were used for research investigations reported in this chapter.

References

Al Rwahnih, M., Myrta, A., Abou-Ghanem, N., Di Terlizzi, B., Savino, V., 2001. Viruses and viroids of stone fruits in Jordan. Bull. OEPP-EPPO. 31, 95–98.

Ambrós, S., Hernández, C., Desvignes, J.C., Flores, R., 1998. Genomic structure of three phenotypically different isolates of peach latent mosaic viroid: implications of the existence of constraints limiting the heterogeneity of viroid quasispecies. J. Virol. 72, 7397–7406.

Ambrós, S., Hernández, C., Flores, R., 1999. Rapid generation of genetic heterogeneity in progenies from individual cDNA clones of peach latent mosaic viroid in its natural host. J. Gen. Virol. 80, 2239–2252.

Barba, M., Ragozzino, E., Faggioli, F., 2007. Pollen transmission of peach latent mosaic viroid. J. Plant Pathol. 89, 287–289.

Bouani, A., Al-Rwahnih, M., Ghanem-Sabanadzovic, N.A., Alami, I., Zemzami, M., Myrta, A., et al., 2004. A preliminary account of the sanitary status of stone fruit trees in Morocco. Bull. EPPO-OEPP. 34, 399–402.

Boubourakas, I.N., 2010. Peach latent mosaic viroid (PLMVd) and pome fruit viroids in Greece. PhD Thesis, Agricultural University of Athens, Athens, Greece (English summary), 217 pp.

Boubourakas, I.N., Hadidi, A., Kyriakopoulou, P.E., 2006. The presence of ASSVd, PBCVd and PLMVd viroids in cultivated and wild pome and stone fruits in Greece. Phytopathol. Mediterr. 45, 173–174.

Crescenzi, A., Piazzolla, P., Hadidi, A., 2002. First report of peach latent mosaic viroid in sweet cherry in Italy. J. Plant Pathol. 84, 6.

Desvignes, J.C., 1976. The virus diseases detected in greenhouse and in the field by the peach seedling GF-305 indicator. Acta Hortic. 67, 315–323.

Desvignes, J.C., 1999. Virus diseases of fruit trees. Centre Technique Interprofessionnel des Fruits et Légumes. CTIFL, Paris. 386, 115–118.

Di Serio, F., Flores, R., Verhoeven, J.Th.J., Li, S.-F., Pallás, V., et al., 2014. Current status of viroid taxonomy. Arch. Virol. 159, 3467–3478.

Di Serio, F., Malfitano, M., Flores, R., Randles, J.W., 1999. Detection of peach latent mosaic viroid in Australia. Australas. Plant Pathol. 28, 80–81.

El-Dougdoug, K.A., 1998. Occurrence of peach latent mosaic viroid (PLMVd) in apple fruit (*Malus domestica*). Ann. Agric. Sci., Ain Shams Univ. Cairo, Egypt 43, 21–30.

El-Dougdoug, K.A., Dawoud, R.A., Rezk, A.A., Sofy, A.R., 2012. Incidence of fruit trees viroid diseases by tissue print hybridization in Egypt. Int. J. Virol. 8, 114–120.

Faggioli, F., Barba, M., Ghanem-Sabanadzovic, N.A., Alami, I., Zemzami, M., Myrta, A., et al., 2008. Peach latent mosaic viroid: major findings of our studies over a period of fifteen years in Italy. Acta Hortic. 781, 529–534.

Faggioli, F., Loretti, S., Barba, M., 1997. Occurrence of peach latent mosaic viroid (PLMVd) on plum in Italy. Plant Dis. 81, 423.

Fekih Hassen, I., Massart, S., Motard, J., Roussel, S., Parisi, O., Kummert, J., et al., 2007. Molecular features of new peach latent mosaic viroid variants suggest that recombination may have contributed to the evolution of this infectious RNA. Virology 360, 50–57.

Fekih Hassen, I., Roussel, S., Kummert, J., Fakhfakh, H., Marrakchi, M., Jijakli, M.H., 2005. Peach latent mosaic viroid detected for the first time on almond trees in Tunisia. Plant Dis. 89, 1244.

Fekih Hassen, I., Roussel., S., Kummert, J., Fakhfakh, H., Marrakchi, M., Jijakli, M.H., 2006. Development of a rapid RT-PCR test for the detection of peach latent mosaic viroid, pear blister canker viroid, hop stunt viroid and apple scar skin viroid in fruit trees from Tunisia. J. Phytopathol. 154, 217–223.

Fiore, N., Abou Ghanem-Sabanadzovic, N., Infante, R., Myrta, A., Pallás, V., 2003. Detection of peach latent mosaic viroid in stone fruits from Chile. CIHEAM-Options Méditerranéennes Serie B No 45. Bari, Italy, pp. 143–145.

Flores, R., Hernández, C., Desvignes, J.C., Llácer, G., 1990. Some properties of the viroid inducing peach latent mosaic disease. Res. Virol. 141, 109–118.

Flores, R., Llácer, G., 1988. Isolation of a viroid-like RNA associated with peach latent mosaic disease. Acta Hortic. 235, 325–331.

Giunchedi, L., Gentit, P., Nemchinov, L., Poggi-Pollini, C., Hadidi, A., 1998. Plum spotted fruit: a disease associated with peach latent mosaic viroid. Acta Hortic. 472, 571–579.

Giunchedi, L., Kyriakopoulou, P.E., Barba, M., Hadidi, A., 2011. Peach latent mosaic viroid in naturally infected temperate fruit trees. In: Hadidi, A., Barba, M., Candresse, T., Jelkmann, W. (Eds.), Virus and Virus-Like Diseases of Pome and Stone Fruits. APS Press, St. Paul, MN, pp. 225–227.

Glouzon, J.-P.S., Bolduc, F., Wang, S., Najmanovich, R.J., Perreault, J.-P., 2014. Deep sequencing of the peach latent mosaic viroid reveals new aspects of population heterogeneity. PLoS One 9, e87297.

Gohen, O., Batuman, O., Moskowits, Y., Rozov., A., Gootwine, E., Mawassi, M., et al., 2005. Goat horns: platforms for viroid transmission to fruit trees? Phytoparasitica 33, 141–148.

Gümüs, M., Paylan, I.C., Matic, S., Myrta, A., Sipahioglu, H.M., Erkan, S., 2007. Occurrence and distribution of stone fruit viruses and viroids in commercial plantings of Prunus species in western Anatolia. Turkey J. Plant Pathol. 89, 265–268.

Hadidi, A., Giunchedi, L., Shamloul, A.M., Poggi-Pollini, C., Amer, M.A., 1997. Occurrence of peach latent mosaic viroid in stone fruits and its transmission with contaminated blades. Plant Dis. 81, 154–158.

Hernández, C., Flores, R., 1992. Plus and minus RNAs of peach latent mosaic viroid self cleave in vitro via hammerhead structures. Proc. Natl. Acad. Sci. USA 89, 3711–3715.

Kyriakopoulou, P.E., Giunchedi, L., Hadidi, A., 2001. Peach latent mosaic viroid and pome fruit viroids in naturally infected cultivated *Pyrus communis* and wild pear *P. amygdaliformis*: implications on possible origin of these viroids in the Mediterranean region. J. Plant Pathol. 83, 51–62.

Kyriakopoulou, P.E., Hadidi, A., Dougdoug, K., Giunchedi, L., 1998. Natural infection of wild and cultivated pears in Greece with peach latent mosaic viroid and pome fruit viroids. In: Abstracts of 7th Intl. Congress of Plant Pathol. (ICPP98), Edinburgh, UK, paper number 3.7.49.

Malifano, M., Di Serio, F., Covelli, L., Raggozino, A., Hernández, C., Flores, R., 2003. Peach latent mosaic viroid variants inducing peach calico (extreme chlorosis) contain a characteristic insertion that is responsible for this symptomatology. Virology 313, 492–501.

Meziani, S., Rouag, N., Milano, R., Kheddam, M., Djelouah, K., 2010. Assessment of the main stone fruit viruses and viroids in Algeria. In: Jelkmann, W., Krczal, G., Feldmann, F. (Eds.), Proc. 21st Intl. Conference on Virus and Other Graft-Transmissible Diseases of Fruit Crops. Neüstadt, Germany, pp. 289–292.

Musa, A., Merkuri, J., Milano, R., Djelouah, K., 2010. Investigation on the phytosanitary status of the main stone fruit nurseries and mother plots in Albania. In: Jelkmann, W.,

Krczal, G., Feldmann, F. (Eds.), Proc. 21st Intl. Conference on Virus and Other Graft Transmissible Diseases of Fruit Crops. Neüstadt, Germany, pp. 304–308.

Osaki, H., Yamamuchi, Y., Sato, Y., Tomita, Y., Kawai, Y., Miyamoto, Y., et al., 1999. Peach latent mosaic isolated from stone fruits in Japan. Ann. Phytopathol. Soc. Jpn. 65, 3–8.

Osman, F., Al Rwahnih, M., Golino, D., Pitman, T., Cordero, F., Preece, J.E., et al., 2012. Evaluation of the phytosanitary status of the *Prunus* species in the National clonal germplasm repository in California: survey of viruses and viroids. J. Plant Pathol. 94, 249–253.

Paduch-Cichal, E., Skrzeczkowski, J.L., 2001. Occurrence of peach latent mosaic viroid in Polish orchards. Phytopathol. Polonica. 22, 137–143.

Rouag, N., Guechi, A., Matic, S., Myrta, A., 2008. Viruses and viroids of stone fruits in Algeria. J. Plant Pathol. 90, 393–395.

Torres, H., Gómez, G., Pallás, V., Stamo, B., Shalaby, A., Aouane, B., et al., 2004. Detection by tissue printing of stone fruit viroids, from Europe, the Mediterranean and North and South America. Acta Hortic. 657, 379–383.

Virscek-Marn, M., 2009. Viroidi. Presentation published online on 12/14/2009. http://www.dvrs.bf.uni-lj.si/Virscek_Marn_Dvrs_09.pdf.

Walia,,Y., Kumar,Y., Bhardwaj, P., Negi, A., Hallan,V., Zaidi, A.A., 2008. Molecular characterization of peach latent mosaic viroid from apple. NCBI-GenBank http://www.ncbi.nlm.nih.gov/nuccore/FM955277.

Yazarlou, A., Jafarpour, B., Tarighi, S., Habili, N., Randles, J.W., 2012. New Iranian and Australian peach latent mosaic viroid variants and evidence for rapid sequence evolution. Arch. Virol. 157, 343–347.

CHAPTER

31

Chrysanthemum Chlorotic Mottle Viroid

Ricardo Flores[1], *Selma Gago-Zachert*[2], *Pedro Serra*[1], *Marcos De la Peña*[1] *and Beatriz Navarro*[3]

[1]**Polytechnic University of Valencia-CSIC, Valencia, Spain**
[2]**Leibniz Institute of Plant Biochemistry, Halle, Germany**
[3]**National Research Council, Bari, Italy**

INTRODUCTION

Chrysanthemum chlorotic mottle (CChM) disease was first reported 45 years ago and found to be caused by a graft-transmissible agent (Dimock et al., 1971). Because bioassays on indicator plants and examination by electron microscopy failed to reveal the presence of a virus associated with the affected plants, in which previous attempts to isolate bacteria or fungi had also failed (Dimock et al., 1971), a viroid etiology was suspected. In support of this view the infectious agent: (1) did not sediment after high speed centrifugation; (2) displayed in sucrose gradient centrifugation the sedimentation coefficient (6–14S) predicted for a small nucleic acid; and (3) was sensitive to RNase but not to DNase (Romaine and Horst, 1975). However, in contrast with previous results obtained with other viroids, the agent of CChM disease could not be identified as a differential band in polyacrylamide gel electrophoresis (PAGE) and, following fractionation with 2 M LiCl, it was preferentially found in the precipitate (Kawamoto et al., 1985). Moreover, while interference among chrysanthemum stunt viroid, citrus exocortis viroid, and a mild and a severe strain of potato spindle tuber viroid was observed in chrysanthemum plants coinoculated in a variety of combinations, this effect was not observed with the agent of

Viroids and Satellites.
DOI: http://dx.doi.org/10.1016/B978-0-12-801498-1.00031-0

CChM disease, further supporting the existence of key differences between the latter and the other three known viroids (Niblett et al., 1978).

To explore the possibility that the expected viroid-like RNA inciting the CChM disease could have been masked by comigrating host RNAs in the short gels initially used, preparations enriched in the pathogenic RNA were reexamined in long sequencing gels. With this modification, a distinct RNA was identified in the affected plants, which when eluted and inoculated mechanically, replicated and caused the typical CChM symptoms, thus fulfilling Koch's postulates (Navarro and Flores, 1997). The extremely low accumulation in vivo of the linear, and especially of the circular forms of chrysanthemum chlorotic mottle viroid (CChMVd), accounted in part for the negative results of the first tests.

TAXONOMIC POSITION AND NUCLEOTIDE SEQUENCE

The reference sequence variant of CChMVd (CM5, GenBank accession number Y14700), has 399 nt (in agreement with the size inferred from the mobility in denaturing PAGE) and consists of: 87 A (21.8%), 109 C (27.3%), 112 G (28.1%), and 91 U (22.8%). Other sequence variants range from 398 to 401 nt, thus making CChMVd the largest known viroid other than those with sequence repeats. The CChMVd (+) strand, by convention the most abundant in vivo, adopts a predicted most stable secondary structure that is highly branched (Fig. 31.1) (Navarro and Flores, 1997). This structure is most likely biologically significant because the sequence polymorphisms observed initially (Navarro and Flores, 1997), and subsequently in many variants (De la Peña and Flores, 2002; De la Peña et al., 1999; Gago et al., 2005), do not alter its stability: the changes map at loops or, when affecting a base-pair, the substitutions are covariations or compensatory mutations.

CChMVd, along with avocado sunblotch viroid (Hutchins et al., 1986; Symons, 1981), peach latent mosaic viroid (PLMVd) (Hernández and Flores, 1992), and eggplant latent viroid (Fadda et al., 2003), are grouped in the family *Avsunviroidae* (see Chapter 13: Viroid Taxonomy). The most distinctive feature of these four viroids is that the strands of both polarities can fold into hammerhead structures and self-cleave in vitro—and in vivo—as predicted by these ribozymes, which play a key role during replication in plastids through a symmetric rolling circle (see Chapter 7: Viroid Replication). Within the family *Avsunviroidae*, CChMVd has been classified together with PLMVd in the genus *Pelamoviroid* because the two viroids present some common characteristics (Navarro and Flores, 1997), prominent among which are their insolubility in 2 M LiCl and the adoption of a multibranched secondary structure stabilized by a kissing-loop interaction between two hairpin

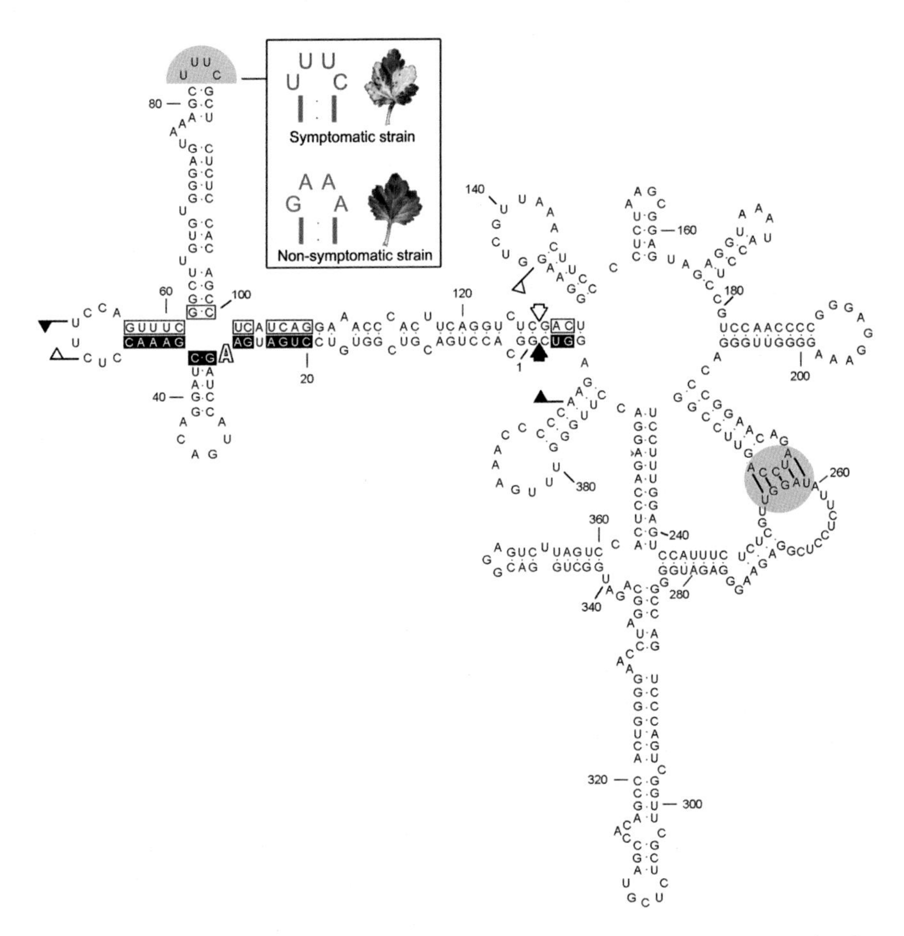

FIGURE 31.1 Primary and predicted secondary structure of lowest free energy for the CChMVd symptomatic variant CM5 (De la Peña et al., 1999; Navarro and Flores, 1997). Sequences forming plus and minus hammerhead structures are delimited by flags, nucleotides conserved in most natural hammerhead structures by bars, and self-cleavage sites by arrows. Solid and open symbols refer to plus and minus polarities, respectively. The outlined font indicates the position in the CChMVd secondary structure of the extra A (A10) that forms part of the plus hammerhead structure (De la Peña and Flores, 2001). The upper gray semicircle denotes the tetraloop at which the pathogenicity determinant has been mapped, with the two natural variants and associated symptoms being showed in the adjacent inset. The lower gray oval denotes a kissing-loop interaction critical for in vitro folding and in vivo viability (Gago et al., 2005). *Source: Reproduced with modifications from Flores, R., Navarro, B., Gago, S., De la Peña, M., 2007. Chrysanthemum chlorotic mottle viroid: a system for reverse genetics in the family Avsunviroidae (hammerhead viroids). Plant Viruses 1, 27–32.*

loops. This interaction, observed in PLMVd in vitro and proposed also in CChMVd (Bussière et al., 2000), has been detected experimentally in the latter and found to be critical for in vitro folding and in vivo viability (Gago et al., 2005); moreover, such interaction apparently exists only in the plus polarity strand (Gago et al., 2005; Giguère et al., 2014).

Interestingly, CChMVd has the highest mutation rate described for any biological entity (Gago et al., 2009).

ECONOMIC SIGNIFICANCE

Because CChMVd induces alterations in the vegetative and floral parts of a very important ornamental crop cultivated worldwide, this viroid is economically relevant and must be controlled. Moreover, the presence of severe strains can be masked to a certain extent by preexisting nonsymptomatic strains (see below).

SYMPTOMS

CChM disease, first reported in the cultivar "Yellow Delaware" grown in greenhouses in the United States, was characterized by leaf mottling and chlorosis, delay in flowering, and dwarfing (Dimock et al., 1971). Symptoms in cultivars "Deep Ridge" and "Bonnie Jean" inoculated with grafts of infected tissue were expressed earlier (12–15 days) and more reproducibly at relatively high temperature (24–27°C) and light intensity, and displayed a defined temporal pattern: mottling, light- and dark-green sectors sharply defined in the next leaves (Fig. 31.1), and complete chlorosis in leaves that developed subsequently (Dimock et al., 1971; Horst, 1987). Cuttings of cultivar "Bonnie Jean" reacted similarly to infection, but within a shorter time interval (8–10 days) when grown in controlled chambers at 28°C and under constant fluorescent illumination (Navarro and Flores, 1997). Moreover, the same severe symptoms developed following mechanical inoculation either with a dimeric tandem cDNA or an in vitro transcript of the CChMVd reference sequence variant (Navarro and Flores, 1997) cloned from the original severe isolate (Dimock et al., 1971).

Unexpectedly, when inoculated with material from plants infected with the severe strain (CChMVd-S), certain sources of chrysanthemum cultivars sensitive to the CChM disease did not express the typical symptoms (Horst, 1975). This finding led to prediction of the existence of a CChMVd nonsymptomatic strain (CChMVd-NS) that would exert a superinfection exclusion effect on the CChMVd-S strain. In agreement with this prediction, experiments performed in a cross-protection format proved the existence of the postulated transmissible agent (Horst, 1975), which was physically identified by northern-blot hybridization shortly after the CChMVd-S was cloned (Navarro and Flores, 1997) and a probe became available (De la Peña et al., 1999). Reverse-transcription RNA polymerase reaction (RT-PCR) amplification, cloning, and sequencing showed that CChMVd-S and -NS RNAs had very similar size (as anticipated from northern-blot hybridization) and sequence, with CChMVd-NS variants presenting idiosyncratic mutations not previously observed in those from the CChMVd-S strain. The cDNA clones containing the CChMVd-NS specific mutations were infectious, but nonsymptomatic, in chrysanthemum. Moreover, site-directed mutagenesis revealed that one of the CChMVd-NS characteristic mutations, a UUUC to GAAA substitution mapping at a tetraloop capping a hairpin in the predicted branched conformation (Fig. 31.1), reverted the symptomatic phenotype into nonsymptomatic without affecting the viroid RNA titer (De la Peña et al., 1999). Experimental coinoculations with typical CChMVd-S and -NS symptomatic variants result in symptomless phenotypes only when the latter was in more than 100-fold excess, thus indicating the higher fitness of the S variant (De la Peña and Flores, 2002).

HOST RANGE

In an initial test of 51 species and cultivars of potential hosts belonging to 35 different genera, including nine species of chrysanthemum, CChMVd displayed a very narrow host range. Only some cultivars, i.e., "Bonnie Jean," "Deep Ridge," and "Yellow Delaware" of the natural host, chrysanthemum (*Chrysanthemum morifolium*, Ramat., syn. *Dendranthema grandiflora* Ramat.) and one cultivar, "Clara Curtiss," of *Chrysanthemum zawadskii* subsp. *latilobum* (Maxim.), were susceptible, while CChMVd did not infect 11 plant species supporting replication of other viroids (Horst, 1987). No other systematic analysis has been performed subsequently, but additional chrysanthemum cultivars have been reported to be susceptible to CChMVd: i.e., "Sharotte" (Chung et al., 2006), "Piato," and "Sttetsuman" (Hosokawa et al., 2005) and 21 unnamed cultivars in Japan (Yamamoto and Sano, 2006).

TRANSMISSION

In an experimental context CChMVd was first transmitted by tissue grafting from infected plants into healthy ones (Dimock et al., 1971). Subsequent mechanical transmission was achieved by applying to carborundum-dusted leaves crude extracts from infected tissue made in an alkaline buffer or total nucleic acid preparations resulting from phenol-mediated extraction (Romaine and Horst, 1975), as well as with plasmids containing dimeric head-to-tail cDNA inserts or their in vitro transcripts (Navarro and Flores, 1997). In agricultural contexts CChMVd is spread short distances as a consequence of repeated plant manipulation, and long distances by propagation of infected stocks (Horst, 1987), particularly in the case of nonsymptomatic strains that may remain unnoticed. There is no reported evidence of CChMVd transmission by either insects or pollen, the latter of which is produced in large amounts by this ornamental.

DETECTION

CChMVd was initially detected by bioassay on susceptible chrysanthemum cultivars such as "Bonnie Jean," "Yellow Delaware," and "Deep Ridge." Molecular characterization and cloning (Navarro and Flores, 1997) opened the possibility of detecting CChMVd by dot- and northern-blot hybridization, as well as RT-PCR (De la Peña et al., 1999). Subsequently, many RT-PCR variations (including single and multiplex formats) have been proposed (Hosokawa et al., 2006; Hosokawa et al., 2007; Park et al., 2013; Song et al., 2013). In our experience, two pairs of RT-PCR primers that can be used for detecting and cloning the full-length CChMVd-cDNA are: (5′-TCCAGTCGAGACCTGAAGTGGGTTT CC-3′) and (5′-AGGTCGTAAAACTTCCCCTCTAAGCGG-3′), complementary and homologous to positions 133–107 and 134–160, respectively, of the CM20 variant (Gago et al., 2005), and (5′-CTACTCCGCTT AGAGGGGAAGTTTTACGA-3′) and (5′-AGGTAAATACCTCCGTCCA ACCCC-3′), complementary and homologous, respectively, to positions 165–137 and 166–189 of the CM20 reference variant (Dufour et al., 2009).

GEOGRAPHICAL DISTRIBUTION AND EPIDEMIOLOGY

Following the first report of CChM disease in the United States and the subsequent characterization of its causal agent, similar disorders affecting chrysanthemum were described in Europe (Denmark and France) and Asia (India) (Horst, 1987). More recently, CChMVd has been reported in Korea (Chung et al., 2006) and China (Zhang et al., 2011). Nonsymptomatic strains of this agent may even be more prevalent worldwide than anticipated, as revealed by their detection in material shipped from various areas in Japan and The Netherlands (Hosokawa et al., 2005; Yamamoto and Sano, 2006; our unpublished data).

CONTROL

The most important preventive measures include using CChMVd-free propagation material, careful handling during cultivation to avoid plant-to-plant contact, frequent disinfection of pruning implements with diluted commercial bleach, and regular cleaning of greenhouses. Special attention should be paid to the early detection (and exclusion) of non-symptomatic strains in mother stock plants, which should be tested recurrently by RT-PCR. CChMVd can be removed from infected cultivars, although at a low rate, by leaf primordia-free shoot apical meristem culture (Hosokawa et al., 2005).

Acknowledgments

Research in R.F. laboratory is currently funded by grant BFU2014–56812-P from the Spanish Ministerio de Economía y Competitividad (MINECO).

References

Bussière, F., Ouellet, J., Côté, F., Lévesque, D., Perreault, J.P., 2000. Mapping in solution shows the peach latent mosaic viroid to possess a new pseudoknot in a complex, branched secondary structure. J. Virol. 74, 2647–2654.

Chung, B.N., Kim, D.C., Kim, J.S., Cho, J.D., 2006. Occurrence of chrysanthemum chlorotic mottle viroid in chrysanthemum *(Dendranthema grandiflorum)* in Korea. Plant Pathol. J. 22, 334–338.

De la Peña, M., Flores, R., 2001. An extra nucleotide in the consensus catalytic core of a viroid hammerhead ribozyme: implications for the design of more efficient ribozymes. J. Biol. Chem. 276, 34586–34593.

De la Peña, M., Flores, R., 2002. Chrysanthemum chlorotic mottle viroid RNA: dissection of the pathogenicity determinant and comparative fitness of symptomatic and non-symptomatic variants. J. Mol. Biol. 321, 411–421.

De la Peña, M., Navarro, B., Flores, R., 1999. Mapping the molecular determinant of pathogenicity in a hammerhead viroid: a tetraloop within the *in vivo* branched RNA conformation. Proc. Natl. Acad. Sci. USA 96, 9960–9965.

Dimock, A.W., Geissinger, C.M., Horst, R.K., 1971. Chlorotic mottle: a newly recognized disease of chrysanthemum. Phytopathology 61, 415–419.

Dufour, D., la Peña, De, Gago, M., Flores, S., Gallego, J., R., 2009. Structure-function analysis of the ribozymes of chrysanthemum chlorotic mottle viroid: a loop-loop interaction motif conserved in most natural hammerheads. Nucleic Acids Res. 37, 368–381.

Fadda, Z., Daròs, J.A., Fagoaga, C., Flores, R., Duran-Vila, N., 2003. Eggplant latent viroid, the candidate type species for a new genus within the family *Avsunviroidae* (hammerhead viroids). J. Virol. 77, 6528–6532.

Flores, R., Navarro, B., Gago, S., De la Peña, M., 2007. Chrysanthemum chlorotic mottle viroid: a system for reverse genetics in the family *Avsunviroidae* (hammerhead viroids). Plant Viruses 1, 27–32.

Gago, S., De la Peña, M., Flores, R., 2005. A kissing-loop interaction in a hammerhead viroid RNA critical for its in vitro folding and in vivo viability. RNA. 11, 1073–1083.

Gago, S., Elena, S.F., Flores, R., Sanjuán, R., 2009. Extremely high variability of a hammerhead viroid. Science 323, 1308.

Giguère, T., Adkar-Purushothama, C.R., Bolduc, F., Perreault, J.P., 2014. Elucidation of the structures of all members of the *Avsunviroidae* family. Molecular Plant Pathol. 15, 767–779.

Hernández, C., Flores, R., 1992. Plus and minus RNAs of peach latent mosaic viroid self-cleave *in vitro* via hammerhead structures. Proc. Natl. Acad. Sci. USA 89, 3711–3715.

Horst, R.K., 1975. Detection of a latent infectious agent that protects against infection by chrysanthemum chlorotic mottle viroid. Phytopathology 65, 1000–1003.

Horst, R.K., 1987. Chrysanthemum chlorotic mottle. In: Diener, T.O. (Ed.), The Viroids. Plenum, New York, pp. 291–295.

Hosokawa, M., Matsushita, Y., Ohishi, K., Yazawa, S., 2005. Elimination of chrysanthemum chlorotic mottle viroid (CChMVd) recently detected in Japan by leaf-primordia free shoot apical meristem culture from infected cultivars. J. Jap. Soc. Hortic. Sci. 74, 386–391.

Hosokawa, M., Matsushita, Y., Uchida, H., Yazawa, S., 2006. Direct RT-PCR method for detecting two chrysanthemum viroids using minimal amounts of plant tissue. J. Virol. Meth. 131, 28–33.

Hosokawa, M., Shiba, H., Kawabe, T., Nakashima, A., Yazawa, S., 2007. A simple and simultaneous detection method for two different viroids infecting chrysanthemum by multiplex direct RT-PCR. J. Jap. Soc. Hortic. Sci. 76, 60–65.

Hutchins, C., Rathjen, P.D., Forster, A.C., Symons, R.H., 1986. Self-cleavage of plus and minus RNA transcripts of avocado sunblotch viroid. Nucleic Acids Res. 14, 3627–3640.

Kawamoto, S.O., Horst, R.K., Wong, S.M., 1985. Solubility of chrysanthemum chlorotic mottle viroid in LiCl solutions. Acta Hortic. 164, 333–340.

Navarro, B., Flores, R., 1997. Chrysanthemum chlorotic mottle viroid: unusual structural properties of a subgroup of self-cleaving viroids with hammerhead ribozymes. Proc. Natl. Acad. Sci. USA 94, 11262–11267.

Niblett, C.L., Dickson, E., Fernow, K.H., Horst, R.K., Zaitlin, M., 1978. Cross-protection among four viroids. Virology 91, 198–203.

Park, J., Jung, Y., Kil, E.J., Kim, J., Tran, D.T., Choi, S.K., et al., 2013. Loop-mediated isothermal amplification for the rapid detection of chrysanthemum chlorotic mottle viroid (CChMVd). J. Virol. Methods 193, 232–237.

Romaine, C.P., Horst, R.K., 1975. Suggested viroid etiology for chrysanthemum chlorotic mottle disease. Virology 64, 86–95.

Song, A., You, Y., Chen, F., Li, P., Jiang, J., Chen, S., 2013. A multiplex RT-PCR for rapid and simultaneous detection of viruses and viroids in chrysanthemum. Lett. Appl. Microbiol. 56, 8–13.

Symons, R.H., 1981. Avocado sunblotch viroid: primary sequence and proposed secondary structure. Nucleic Acids Res. 9, 6527–6537.

Yamamoto, H., Sano, T., 2006. An epidemiological survey of chrysanthemum chlorotic mottle viroid in Akita Prefecture as a model region in Japan. J. Gen. Plant Pathol. 72, 387–390.

Zhang, Z.Z., Pan, S., Li, S.F., 2011. First report of chrysanthemum chlorotic mottle viroid in chrysanthemum in China. Plant Dis. 95, 1320.

CHAPTER

32

Eggplant Latent Viroid

José-Antonio Daròs

Polytechnic University of Valencia-CSIC, Valencia, Spain

INTRODUCTION

Eggplant latent viroid (ELVd) was discovered during a survey of viroid and viroid-like agents in vegetable crops that grow typically in eastern and southern Spain. Two of the analyzed samples from eggplant (*Solanum melongena* L.) cv. "Sonja," were found to contain a viroid-like RNA (Fagoaga et al., 1994). Analysis of the viroid-like species, purified from these two samples, indicated a likely circular RNA molecule whose size did not exactly match those of other well-known viroids. Interestingly, it was possible to transmit the viroid-like RNA to new eggplants by mechanical inoculation, and the infected plants remained symptomless. Therefore, the viroid-like RNA was considered to be a new viroid (Fagoaga et al., 1994).

NUCLEOTIDE SEQUENCE AND SECONDARY STRUCTURE

ELVd cDNAs were amplified by reverse transcription-polymerase chain reaction (RT-PCR) from RNA preparations purified from the infected eggplant cv. "Redonda Morada" (Fadda et al., 2003). These cDNAs were cloned and sequenced. Analysis of the sequences of 10 full-length cDNA clones supported the existence of a new viroid with high sequence variability since only two of the 10 clones were identical. The ELVd sequence variants showed a guanine plus cytosine (G + C) content of 53%–54% and, remarkably, they contained the conserved nucleotide residues and flanking stems characteristic of hammerhead structures in the strands of both polarities (Fadda et al., 2003).

Viroids and Satellites.
DOI: http://dx.doi.org/10.1016/B978-0-12-801498-1.00032-2

The ELVd sequence variants were classified into four groups: I, II, III, and IV. Group I consisted of three variants of 335 nt, which shared more than 98% sequence similarity. Group II comprised three variants of 332, 333, and 334 nt, which shared more than 98% similarity. Group III contained two variants of 335 nt, which were 98% similar. Group IV included a single sequence of 334 nt. Representative sequences in each group were selected, namely ELVd-1, 2, 3, and 4, and were deposited in GenBank with the respective accession numbers AJ536612, AJ536613, AJ536614, and AJ536615. The representative variants displayed rather low intergroup similarity, despite them forming part of the same species: ELVd-1 was 88% similar to ELVd-2, 3, or 4; ELVd-2 was 91% similar to ELVd-3 or 4; and ELVd-3 was 92% similar to ELVd-4.

ELVd-2 (AJ536613) is probably the most representative sequence variant of the species given its somewhat intermediate properties, and has been used in most subsequent research (Carbonell et al., 2006; Gómez and Pallás, 2010b, 2012b; Martínez et al., 2009; Molina-Serrano et al., 2007; Nohales et al., 2012). However, ELVd-1 (AJ536612) has been selected as the reference variant of the species (Owens et al., 2012). Fig. 32.1 represents the minimum free energy conformation of ELVd-2 (AJ536613) obtained by the Mfold algorithm (Zuker, 2003). It is a quasirod-like structure with two bifurcations at both ends: upper left and upper right hairpins. Most of the variability found in the other ELVd sequence variants maps at single-stranded regions or, when affecting double-stranded segments, occurs as compensatory mutations

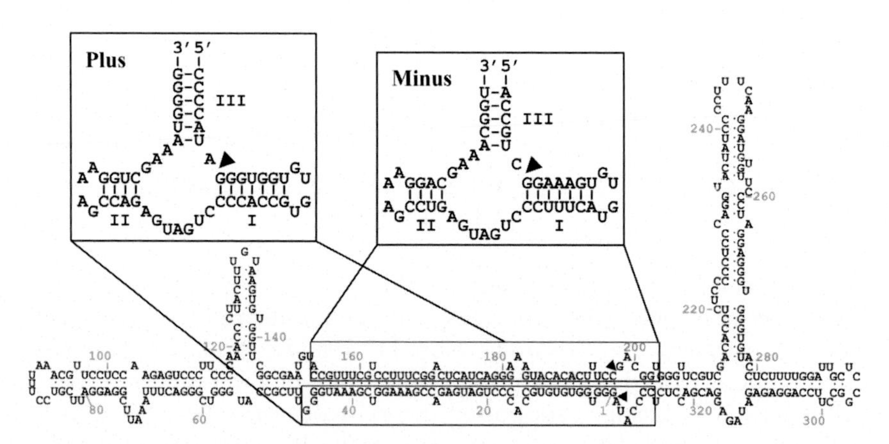

FIGURE 32.1 Sequence of ELVd-2 (AJ536613) folded in the secondary structure of minimum free energy predicted by Mfold. The + and − hammerhead ribozymes (boxed) and the self-cleavage sites (arrowheads) are indicated. The structures of both ribozymes, folded in the classic hammerhead representation, are also shown. Helices I, II, and III of these structures are indicated.

or covariations, which suggests that the proposed conformation, or another similar structure, most likely exists in vivo (Fadda et al., 2003). The experimental determination in vitro of the secondary structures of both strands of ELVd-1 (NC_004728, AJ536612), by selective 2′-hydroxyl acylation analyzed by primer extension (SHAPE), indicated that they fold in similar structures composed of a relatively long rod-like central domain, which ends with a three-way junction on the left and three- (+ strand) or five-way (– strand) junctions on the right (Giguère et al., 2014). No evidence of the kissing-loop interaction present in other hammerhead-containing viroids, like peach latent mosaic viroid (PLMVd) and chrysanthemum chlorotic mottle viroid (CChMVd), was found in this analysis (Giguère et al., 2014). A subsequent in vitro SHAPE analysis of both strands of ELVd-2 (AJ536613) corroborated the conformations derived from thermodynamics-based predictions in silico for this sequence variant (López-Carrasco et al., 2016).

ELVd hammerhead ribozymes were shown to be functional in vitro (Fadda et al., 2003). Monomeric ELVd-2 (AJ536613) RNAs of both polarities self-cleaved spontaneously during transcription and produced fragments with sizes according to the predicted cleavage sites of both hammerhead ribozymes. In addition, when the uncleaved monomeric transcripts were purified and subjected to in vitro self-cleavage without any proteins, these RNAs self-cleaved again, but only in the presence of 5 mM $MgCl_2$. A primer extension analysis of the cleavage products confirmed the predicted ribozyme self-cleavage sites (Fadda et al., 2003). The ELVd-2 (AJ536613) hammerhead ribozymes are schematically represented in Fig. 32.1. They contain long helices I and stable helices III, as in PLMVd and CChMVd, but unlike avocado sunblotch viroid (ASBVd) ribozymes, which self-cleave in dimeric molecules through the formation of double hammerhead structures (Davies et al., 1991; Forster et al., 1988). Both ELVd hammerhead structures more closely resemble each other than those from other RNAs, which is a quite common situation in ribozyme-containing pathogenic RNAs.

TAXONOMIC POSITION

ELVd is a member of the family *Avsunviroidae*. This allocation is based on not only the presence of hammerhead ribozymes in the strands of both polarities, like the other members of the family, but also on the absence of the central conserved region present in all members of the family *Pospiviroidae*. The restricted host range is also a common property of viroids in the family *Avsunviroidae* (Flores et al., 2000). The ability of ELVd RNA to traffic into chloroplasts (Gómez and Pallás, 2010a,b, 2012a,b) may also indicate that the replication and accumulation site of this viroid is in this organelle, a property that is also considered a hallmark of the family *Avsunviroidae*.

Within the family *Avsunviroidae*, ELVd structural properties are intermediate between those of ASBVd (the only member of the genus *Avsunviroid*), and those of PLMVd and CChMVd, which form the genus *Pelamoviroid*. The ELVd secondary structure is quasirod-like, similarly to ASBVd. PLMVd and CChMVd have branched conformations and include a kissing-loop interaction (Bussière et al., 2000; Gago et al., 2005; Giguère et al., 2014). However, the ELVd hammerhead structures have stable helices III, like those of the two last viroids, but unlike the ASBVd hammerheads. In addition, ELVd G + C content is high, as in PLMVd and CChMVd, but unlike ASBVd. These intermediate properties led to ELVd being assigned to a new genus, *Elaviroid*, in the family *Avsunviroidae* (Fadda et al., 2003). This genus is currently accepted by the International Committee on Virus Taxonomy, and ELVd remains its only species to date (Owens et al., 2012).

SYMPTOMS

ELVd infections of eggplant are apparently symptomless. During the initial characterization of the viroid, weekly observations of the growing pattern, flower number and weight, and number and aspect of fruits over three months, all failed to reveal any symptom in infected eggplants (Fadda et al., 2003). Fig. 32.2 compares mock-inoculated and ELVd-infected eggplants of cv. "Black Beauty." No noticeable symptoms

FIGURE 32.2 Mock-inoculated (left) and ELVd-infected (right) eggplants, cv. "Black Beauty." No symptoms of infection were detected for 2 months postinoculation as shown.

were detected in the infected plants. Therefore, there is no economic significance associated so far with infection by this viroid.

HOST RANGE

ELVd RNA was mechanically inoculated into some typical hosts of known viroids (tomato, cucumber, chrysanthemum, and citron). No infection was detected (Fagoaga et al., 1994). Nor did *Agrobacterium tumefaciens*-mediated inoculation (agroinoculation) of *Nicotiana benthamiana* Domin plants produce infection. However, ELVd infectivity has been demonstrated in a series of tested eggplant cultivars, all of which remained symptomless (Fagoaga and Duran-Vila, 2003).

TRANSMISSION

ELVd has been transmitted mechanically by cutting tools with 55% efficiency, and also is seed-transmitted with an efficiency of approximately 20% (Fadda et al., 2003; Fagoaga and Duran-Vila, 2003). Seed treatment with 1% sodium hypochlorite did not affect the seed transmission ratio, which indicates that transmission does not result from accidental seedling inoculation with the remains of contaminated tissues during the germination process.

CONTROL

Despite ELVd only inducing apparently symptomless infections, its control is needed to avoid eventual emergence of symptomatic variants or synergistic interactions with other pathogens. The relatively high rate of ELVd seed transmission makes it necessary to test parental plants used to obtain seeds for commercial purposes.

Acknowledgments

This work has been supported by Grants AGL2013–49919-EXP and BIO2014-54269-R from the Spanish Ministerio de Economía y Competitividad (MINECO).

References

Bussière, F., Ouellet, J., Côté, F., Lévesque, D., Perreault, J.P., 2000. Mapping in solution shows the peach latent mosaic viroid to possess a new pseudoknot in a complex, branched secondary structure. J. Virol. 74, 2647–2654.

Carbonell, A., De la Peña, M., Flores, R., Gago, S., 2006. Effects of the trinucleotide preceding the self-cleavage site on eggplant latent viroid hammerheads: differences in co- and post-transcriptional self-cleavage may explain the lack of trinucleotide AUC in most natural hammerheads. Nucleic Acids Res. 34, 5613–5622.

Davies, C., Sheldon, C.C., Symons, R.H., 1991. Alternative hammerhead structures in the self-cleavage of avocado sunblotch viroid RNAs. Nucleic Acids Res. 19, 1893–1898.

Fadda, Z., Daròs, J.A., Fagoaga, C., Flores, R., Duran-Vila, N., 2003. *Eggplant latent viroid*, the candidate type species for a new genus within the family *Avsunviroidae* (hammerhead viroids). J. Virol. 77, 6528–6532.

Fagoaga, C., Duran-Vila, N., 2003. Eggplant latent. In: Hadidi, A., Flores, R., Randles, J.W., Semancik, J.S. (Eds.), Viroids. CSIRO Publishing, Collingwood, VIC, p. 333.

Fagoaga, C., Pina, J.A., Duran-Vila, N., 1994. Occurrence of small RNAs in severely diseased vegetable crops. Plant Dis. 78, 749–753.

Flores, R., Daròs, J.A., Hernández, C., 2000. *Avsunviroidae* family: viroids containing hammerhead ribozymes. Adv. Virus Res. 55, 271–323.

Forster, A.C., Davies, C., Sheldon, C.C., Jeffries, A.C., Symons, R.H., 1988. Self-cleaving viroid and newt RNAs may only be active as dimers. Nature 334, 265–267.

Gago, S., De la Peña, M., Flores, R., 2005. A kissing-loop interaction in a hammerhead viroid RNA critical for its in vitro folding and in vivo viability. RNA. 11, 1073–1083.

Giguère, T., Adkar-Purushothama, C.R., Bolduc, F., Perreault, J.P., 2014. Elucidation of the structures of all members of the *Avsunviroidae* family. Mol. Plant Pathol. 15, 767–779.

Gómez, G., Pallás, V., 2010a. Can the import of mRNA into chloroplasts be mediated by a secondary structure of a small non-coding RNA? Plant Signal. Behav. 5, 1517–1519.

Gómez, G., Pallás, V., 2010b. Noncoding RNA mediated traffic of foreign mRNA into chloroplasts reveals a novel signaling mechanism in plants. PLoS One 5, e12269.

Gómez, G., Pallás, V., 2012a. A pathogenic non coding RNA that replicates and accumulates in chloroplasts traffics to this organelle through a nuclear-dependent step. Plant Signal. Behav. 7, 882–884.

Gómez, G., Pallás, V., 2012b. Studies on subcellular compartmentalization of plant pathogenic noncoding RNAs give new insights into the intracellular RNA-traffic mechanisms. Plant Physiol. 159, 558–564.

López-Carrasco, A., Gago-Zachert, S., Mileti, G., Minoia, S., Flores, R., Delgado, S., 2016. The transcription initiation sites of eggplant latent viroid strands map within distinct motifs in their *in vivo* RNA conformations. RNA Biol. 13, 83–97.

Martínez, F., Marqués, J., Salvador, M.L., Daròs, J.A., 2009. Mutational analysis of eggplant latent viroid RNA processing in *Chlamydomonas reinhardtii* chloroplast. J. Gen. Virol. 90, 3057–3065.

Molina-Serrano, D., Suay, L., Salvador, M.L., Flores, R., Daròs, J.A., 2007. Processing of RNAs of the family *Avsunviroidae* in *Chlamydomonas reinhardtii* chloroplasts. J. Virol. 81, 4363–4366.

Nohales, M.A., Molina-Serrano, D., Flores, R., Daròs, J.A., 2012. Involvement of the chloroplastic isoform of tRNA ligase in the replication of viroids belonging to the family *Avsunviroidae*. J. Virol. 86, 8269–8276.

Owens, R.A., Flores, R., Di Serio, F., Li, S.F., Pallás, V., Randles, J.W., et al., 2012. Viroids. In: King, A.M.Q., Adams, M.J., Carstens, E.B., Lefkowitz, E.J. (Eds.), Virus Taxonomy: Ninth Report of the International Committee on Taxonomy of Viruses. Elsevier, Academic Press, London, pp. 1221–1234.

Zuker, M., 2003. Mfold web server for nucleic acid folding and hybridization prediction. Nucleic Acids Res. 31, 3406–3415.

PART IV

DETECTION AND IDENTIFICATION METHODS

CHAPTER

33

Viroid Detection and Identification by Bioassay

Xianzhou Nie and Rudra P. Singh

Agriculture and Agri-Food Canada, Fredericton, NB, Canada

INTRODUCTION

Biological assays have played a pivotal role in viroid disease recognition and indexing especially before molecular tools became available (Hammond and Owens, 2006; Singh and Ready, 2003). Although bioassays have gradually become complementary to molecular assays for viroid detection and identification, they are still widely used for viroid characterization and pathogenicity analyses, including fulfilment of Koch's postulates, as well as for viroid species clarification (Semancik and Vidalakis, 2005; Verhoeven et al., 2011). To date, bioassays are still used in some certification programs (Hammond and Owens, 2006).

VIROID SPECIES, PRIMARY HOSTS, AND INDICATOR HOSTS

A successful biological assay of plant viroid pathogens relies on the selection of suitable indicator hosts and the recognition of unique symptoms induced by the viroid in the hosts (Legrand, 2015). For successful biological indexing, both the host range as well as the symptoms produced on that host by specific viroids must be known. Table 33.1 summarizes the primary and indicator hosts for viroids where a bioassay is available.

The primary and indicator hosts of viroids may differ, depending on the viroid and its host range. Avocado sunblotch viroid and chrysanthemum chlorotic mottle viroid only infect their respective primary hosts (i.e., avocado and chrysanthemum) (Owens, 2015), and therefore, the

Viroids and Satellites.
DOI: http://dx.doi.org/10.1016/B978-0-12-801498-1.00033-4

TABLE 33.1 Viroids and Their Detection by Bioassay

Viroid	Primary hosts	Indicator hosts	References
Apple dimple fruit viroid	Apple	*Malus pumila* cvs. "Starkrimson", "Braeburn"	Di Serio et al. (2001)
Apple fruit crinkle viroid	Apple	*Malus pumila* cvs. "Ohrin", "NY58–22"	Ito and Yoshida (1998)
Apple scar skin viroid	Apple	*Malus pumila* cvs. "Stark's Earliest", "Sugar Crab"	Howell and Mink (1992)
Avocado sunblotch viroid	Avocado	*Persea americana* cv. "Hass"	Allen et al. (1981)
Chrysanthemum chlorotic mottle viroid	Chrysanthemum	Chrysanthemum cvs. "Deep Ridge", "Bonnie Jean", "Yellow Delaware"	Horst (1975)
Chrysanthemum stunt viroid	Chrysanthemum	Chrysanthemum cv. "Mistletoe"	Bachelier et al. (1976), Hollings and Stone (1973)
Citrus bark cracking viroid	Citrus, hop	Etrog citron Arizona 861-S1	Duran-Vila and Semancik (2003), Wang et al. (2013)
Citrus dwarfing viroid	Citrus	Etrog citron Arizona 861-S1	Serra et al. (2008a,b)
Citrus bent leaf viroid	Citrus	Etrog citron Arizona 861-S1	Serra et al. (2008a,b)
Citrus exocortis viroid	Citrus, tomato	Etrog citron Arizona 861-S1; *Gynura aurantiaca*	Chaffai et al. (2007), Duran-Vila et al. (1988)
Citrus viroid V	Citrus	Etrog citron Arizona 861-S1	Serra et al. (2008a,b)
Coconut cadang-cadang viroid	Coconut	*Cocos nucifera; Corypha elata; Adonidia merrillii; Areca catechu; Elaeis guineensis; Chrysalidocarpus lutescens; Oreodoxa regia*	Imperial et al. (1985)
Columnea latent viroid	Columnea	*Calendula officinalis*	Matsushita and Tsuda (2015)
Hop stunt viroid	Hop, grape, citrus, cucumber, plum, peach	*Cucumis sativus* L. "Suuyou"; *Luffa aegyptiaca* Mill; Parson's special mandarin	Palacio-Bielsa et al. (2004), Roistacher et al. (1973), Yaguchi and Takahashi (1984)

(Continued)

TABLE 33.1 (Continued)

Viroid	Primary hosts	Indicator hosts	References
Mexican papita viroid	*Solanum cardiophyllum*	*Solanum lycopersicum* cv. "Rutgers"; *Nicotiana glutinosa*	Martínez-Soriano et al. (1996), Singh and Ready (2003)
Peach latent mosaic viroid	Peach	*Prunus persica* cv. "GF-305"	Desvignes (1976)
Pear blister canker viroid	Pear	*Pyrus communis* cv. "A20", "Fieud 37", "Fieud 110"	Desvignes et al. (1999)
Pepper chat fruit viroid	Sweet pepper	*Solanum lycopersicum* cv. "Money-maker"; *S. tuberosum* cv. "Nicola"	Verhoeven et al. (2009)
Potato spindle tuber viroid	Potato, tomato	*Solanum lycopersicum* cvs. "Sheyenne", "Rutgers"; *Scopolia sinesis; S. berthaultii; Datura stramonium*	Matsushita and Tsuda (2015), Singh (1973)
Tomato apical stunt viroid	Tomato	*Solanum lycopersicum; Nicotiana glutinosa; Nicotiana tabacum* cv. "White Burley"; *Solanum melongena; Calendula officinalis*	Matsushita and Tsuda (2015), Verhoeven et al. (2011), Walter (1987)
Tomato chlorotic dwarf viroid	Tomato	*Solanum lycopersicum* cvs. "Sheyenne", "Rutgers"; *Nicotiana glutinosa; Solanum melongena; Calendula officinalis*	Matsushita and Tsuda (2015), Singh et al. (1999)
Tomato planta macho viroid	Tomato	*Solanum lycopersicum* cv. "Rutgers"	Galindo et al. (1982), Verhoeven et al. (2011)

primary hosts can be used as indicator hosts for their bioassay (Allen et al., 1981; Horst, 1975). On the other hand, for viroids that do not cause visible symptoms (latent) on either primary or experimental hosts, bioassay-based detection and diagnosis are not possible. These viroids include Iresine viroid 1, hop latent viroid, coleus blumei viroids, Australian grapevine viroid, eggplant latent viroid, and dahlia latent viroid. In the case of *Solanaceae*-infecting pospiviroids, including potato spindle tuber viroid (PSTVd), tomato chlorotic dwarf viroid, citrus exocortis viroid (CEVd), tomato planta macho viroid, tomato apical stunt viroid, Columnea latent viroid, and pepper chat fruit viroid, some indicator hosts are shared, and moreover, they may elicit similar types of symptoms in these plants (Matsushita and Tsuda, 2015; Singh et al., 1999). Therefore, a panel of indicator hosts may be needed to provide better detection and identification of a specific viroid.

Citrus is a natural host to at least seven viroids belonging to four genera. These viroids are CEVd (genus *Pospiviroid*), citrus bent leaf viroid (CBLVd, genus *Apscaviroid*), hop stunt viroid (HSVd, genus *Hostuviroid*), citrus dwarfing viroid (CDVd, genus *Apscaviroid*), citrus bark cracking viroid (genus *Cocadviroid*), citrus viroid V (CVd-V, genus *Apscaviroid*), and citrus viroid VI (CVd-VI, genus *Apscaviroid*). CEVd and certain variants of HSVd are the respective causal agents of the citrus diseases exocortis and cachexia (Vernière et al., 2004). CEVd can be diagnosed based on severe stunting, epinasty, and necrosis on Etrog citron (Arizona 861-S1) and severe stunting, necrosis, epinasty, and leaf rugosity on *Gynura aurantiaca* (Chaffai et al., 2007); whereas HSVd can be diagnosed based on stunting symptoms in cucurbit species such as *Cucumis sativus* (cv. Suyo) and *Luffa aegyptiaca* Mill (Duran-Vila et al., 2000; Palacio-Bielsa et al., 2004; Reanwarakorn and Semancik, 1998, 1999a,b). The biological assay for other citrus viroids is mainly carried out on Etrog citron (Murcia et al., 2009; Serra et al., 2008a,b). Upon infection with CBLVd, citron plants develop "variable syndrome" characterized by flushes of tissue showing mild leaf bending/epinasty resulting from local midvein necrosis on the underside of the leaf, severe necrotic lesions and cracks in the stems. As the disease progresses, the plants become excessively branched. CDVd induces mild epinasty due to petiole and midvein necrosis in the leaves. CVd-V on the other hand incites very small necrotic lesions and cracks in the stems. CVd-VI induces mild petiole necrosis and very mild leaf bending, which differs from the clear leaf bending/epinasty elicited by other citrus viroids (CEVd, CBLVd, CDVd, and CVd-V) (Ito et al., 2001).

FACTORS AFFECTING DETECTION AND IDENTIFICATION OF VIROID BY BIOASSAY

Viroid Strains

Viroids exist in host plants as a group of sequence variants (Góra et al., 1994) and can be described as quasispecies. Symptom severity in certain indicator plants may depend on the dominant sequence variant/haplotype/strain in the population (Singh and Ready, 2003). Different viroid strains exhibiting different pathogenic properties in host species also exist. For instance, mild and severe PSTVd strains have been recognized in potatoes (Singh et al., 1970). The most noteworthy species in terms of strain diversity are HSVd and CEVd, which can infect a wide range of hosts ranging from woody plants to herbaceous species and cause diverse symptoms on their hosts (Chaffai et al., 2007; Palacio-Bielsa et al., 2004). Bioassays for their detection and identification may thus require a panel of indicator hosts. For example, with citrus-infecting HSVd, two distinct strain groups, namely the cachexia-inducing (formerly CVd-IIb) and the noncachexia inducing (formerly CVd-IIa) strains have been recognized. Cucurbit-based bioassays can detect this viroid at the species level, but cannot determine the strain status. To determine the viroid strain type, indicator citrus varieties such as Parson's special mandarin (PSM) have to be used (Duran-Vila et al., 2000; Reanwarakorn and Semancik, 1998, 1999a,b).

Host Species and Cultivar/Genotype

A specific symptom incited by a viroid is the result of the viroid–host interaction, and is therefore determined largely by properties of the viroid and the host. Clearly, host properties including genotype are as important as the viroid pathotype. Therefore, selecting a compatible cultivar/genotype for a specific viroid may be critical for successful biological detection and identification of a viroid. A recent study by Matsushita and Tsuda (2015) revealed that the dihaploid *Petunia* × *hybrida* Mitchell developed noticeable symptoms ranging from leaf necrosis to severe stunting 1–2 month after infection with PSTVd, tomato chlorotic dwarf, and tomato apical stunt viroids. However, no other commercial cultivars of petunia developed noticeable symptoms of infection with these viroids. Another example comes from chrysanthemum stunt viroid, whose biological assay is normally performed using several *Chrysanthemum morifolium* cultivars such as "Mistletoe," "Blanche," "Sunfire," and "Bonnie Jean" (Bachelier et al., 1976; Hollings and Stone, 1973), even though the viroid can infect many species in the family *Compositae* (Singh and Ready, 2003).

Mixed Infections

Coinfections with more than one different species of viroid or virus are common in horticultural and ornamental plants. Mixed infections can often produce more complex symptoms, frequently more severe than those expected for purely additive combined effects (Serra et al., 2008a,b). In potato, mixed infections with PSTVd and potato virus Y result in severe necrosis not displayed with either pathogen alone (Singh and Somerville, 1987). In addition to the enhanced visual symptoms, the titer of the pathogens may be significantly elevated (Valkonen, 1992). Moreover, remarkable synergisms at the transcriptomic level also occur in mixed infections, as evidenced in peach fruits doubly-infected with peach latent mosaic viroid and prunus necrotic ringspot virus (Herranz et al., 2013).

The synergism between citrus viroids in citrus is probably the most notable among all viroids known to date. Simultaneous infections with up to five viroids have been reported in citrus (Duran-Vila et al., 1988; Gillings et al., 1991; Semancik et al., 1988; Vernière et al., 2006). Interestingly, coinfection with CVd-V and either CBLVd or CDVd in Etrog citron produces significantly more severe symptoms than either infection alone (Serra et al., 2008a,b). Similar phenomena have also been observed in citron coinfected with CBLVd and CDVd (Duran-Vila et al., 1988).

Environmental Conditions

The effects of environmental conditions, including temperature, light intensity, day length, and nutritional status, on viroid symptom development and successful bioassay have been reviewed and summarized by Singh and Ready (2003). Generally, higher temperature (≥28°C) and high light intensity favor viroid accumulation and symptom development. In the bioassay for cachexia-inducing HSVd in PSM, a lower temperature was employed when other citrus viroids were present (Semancik et al., 1988). Nevertheless, elevated temperatures at 28–38°C were used for biological indexing of HSVd sequence variants in PSM (Ito et al., 2006; Reanwarakorn and Semancik, 1999a,b).

Inoculation Methods

Inoculation method can affect symptom development. Typically, inoculation is performed via mechanical wounding or grafting. Mechanical inoculation includes cutting, slashing, and rubbing, and is the only procedure for fulfilling Koch's postulates. Adams et al. (1995) reported that cutting was more effective than rub inoculation for infection of hop plants with hop latent viroid. Generally, graft-inoculation offers the most reliable inoculation means for viroid infection and biological indexing (Singh and Ready, 2003). In addition to mechanical and graft-inoculation, inoculum injection has been used for bioassays (Imperial et al., 1985; Kapari-Isaia et al., 2008).

CROSS-PROTECTION AS A MEANS OF BIOASSAY

Cross-protection, in which the infection of a plant with a strain of a viroid provides protection of the plant from infection with another strain of the same viroid, can be used for indirect biological indexing of a viroid. This technique has been applied in several viroids including PSTVd, peach latent mosaic viroid, CEVd, HSVd, and chrysanthemum chlorotic mottle viroid (Desvignes, 1976; Duran-Vila and Semancik, 1990, 2003; Fernow, 1967; Horst, 1975; Khoury et al., 1988; Semancik et al., 1992). Typically, a plant that has been infected with a mild strain of a viroid is challenge-inoculated with an inoculum from a plant that is suspected of being infected with a severe strain of the same viroid. Absence of symptoms indicates a positive indexing of the viroid. However, since cross-protection can occur between different but closely related viroids (Niblett et al., 1978; Pallás and Flores, 1989; Tomassoli et al., 2015), precautions should be taken when this technique is used for biological indexing of a viroid.

COMBINING MOLECULAR AND BIOLOGICAL ASSAYS FOR VIROID DETECTION AND IDENTIFICATION

Biological assay has several disadvantages. It is labor-intensive and time-consuming, taking several weeks or even years depending on viroid and indicator host. It also requires significant greenhouse/growth room space. Moreover, its efficacy and accuracy are affected by numerous factors that affect the viroid-associated symptom expression. Symptom expression is also influenced by the titer of the viroid in the inoculum and by the developmental stage at which the host plant is inoculated. Molecular assays, including polyacrylamide gel electrophoresis (PAGE), hybridization (e.g., dot- and northern-blots, micro-/macroarrays), amplification (e.g., reverse transcription-polymerase chain reaction (RT-PCR) and reverse transcription loop-mediated isothermal amplification (RT-LAMP)) and sequencing (e.g., next-generation sequencing and Sanger sequencing), offer rapid, cost-effective, and reliable diagnosis of viroids. These tools are particularly effective for detection of viroids that do not incite any recognizable symptoms on plants as well as for discovery of new viroids (Ito et al., 2013), new viroid strains, or new hosts (Jakse et al., 2015).

Some of the assays, such as PAGE, RT-PCR, and RT-LAMP, are suitable for large-scale testing of samples, and thus have been widely used for viroid indexing (Chapters 34–36). However, molecular assays cannot ascertain the biological properties such as infectivity and pathogenicity of the viroid. Clearly, biological assays are needed to fill the gap. Moreover, plant pathology requires the application of Koch's postulates in which an RNA with the physicochemical properties of a viroid would be cloned and then has its biological activity determined by inoculation and demonstration of infectivity. This may also apply where multiple variants (quasispecies) occur, to assess the properties of individual variants.

Thus, bioassays will always have a role in viroid research. A combination of molecular and biological assays should provide the most effective means for viroid identification and characterization.

References

Adams, A.N., Barbara, D.J., Morton, A., Darby, P., Green, C.P., 1995. The control of hop latent viroid in UK hops. Acta Hortic. 385, 91–97.

Allen, R.N., Palukaitis, P., Symons, R.H., 1981. Purified avocado sunblotch viroid causes disease in avocado seedlings. Australas. Plant Pathol. 10, 31–32.

Bachelier, J.C., Monsion, M., Dunez, J., 1976. Possibilities of improving detection of chrysanthemum stunt and obtention of viroid free plants by meristem tip culture. Acta Hortic. 59, 63–70.

Chaffai, M., Serra, P., Gandía, M., Hernández, C., Duran-Vila, N., 2007. Molecular characterization of CEVd strains that induce different phenotypes in *Gynura aurantiaca*: structure-pathogenicity relationships. Arch. Virol. 152, 1283–1294.

Desvignes, J.C., 1976. The virus diseases detected in greenhouse and in the field by the peach seedling GF 305 indicator. Acta Hortic. 67, 315–323.

Desvignes, J.C., Cornaggia, D., Grasseau, N., Ambrós, S., Flores, R., 1999. Pear blister canker viroid: host range and improved bioassay with two new pear indicators, Fieud 37 and Fieud 110. Plant Dis. 83, 419–422.

Di Serio, F., Malfitano, M., Alioto, D., Ragozzino, A., Desvignes, J.C., Flores, R., 2001. Apple dimple fruit viroid: fulfillment of Koch's postulates and symptom characteristics. Plant Dis. 85, 179–182.

Duran-Vila, N., Roistacher, C.N., Rivera-Bustamante, R., Semancik, J.S., 1988. A definition of citrus viroid groups and their relationship to the exocortis disease. J. Gen. Virol. 69, 3069–3080.

Duran-Vila, N., Semancik, J.S., 1990. Variations on the "cross protection" effect between two strains of citrus exocortis viroid. Ann. Appl. Biol. 17, 367–377.

Duran-Vila, N., Semancik, J.S., 2003. Citrus viroids. In: Hadidi, A., Flores, R., Randles, J.W., Semancik, J.S. (Eds.), Viroids. CSIRO Publishing, Collingwood, VIC, pp. 178–194.

Duran-Vila, N., Semancik, J.S., Broadbent, P., 2000. Viroid diseases, cachexia, and exocortis. In: Timmer, L.W., Garnsey, S.M., Graham, J.H. (Eds.), Compendium of Citrus Diseases, second ed. American Phytopathology Society, St. Paul, MN, pp. 51–54.

Fernow, K.H., 1967. Tomato as a test plant for detecting mild strains of potato spindle tuber virus. Phytopathology 57, 1347–1352.

Galindo, J., Smith, D.R., Diener, T.O., 1982. Etiology of planta macho, a viroid disease of tomato. Phytopathology 72, 49–54.

Gillings, M.R., Broadbent, P., Gollnow, B.L., 1991. Viroids in Australian citrus: relationship to exocortis, cachexia and citrus dwarfing. Aust. J. Plant Physiol. 18, 559–570.

Góra, A., Candresse, T., Zagórski, W., 1994. Analysis of the population structure of three phenotypically different PSTVd isolates. Arch. Virol. 138, 233–245.

Hammond, R.W., Owens, R.A., 2006. Viroids: new and continuing risks for horticultural and agricultural crops. APSnet Features.

Herranz, M.C., Niehl, A., Rosales, M., Fiore, N., Zamorano, A., Granell, A., et al., 2013. A remarkable synergistic effect at the transcriptomic level in peach fruits doubly infected by prunus necrotic ringspot virus and peach latent mosaic viroid. Virol. J. 10, 164.

Hollings, M., Stone, O.M., 1973. Some properties of chrysanthemum stunt, a virus with the characteristics of an uncoated ribonucleic acid. Ann. Appl. Biol. 74, 333–348.

Horst, R.K., 1975. Detection of a latent infectious agent that protects against infection by chrysanthemum chlorotic mottle viroid. Phytopathology 65, 1000–1003.

Howell, W.E., Mink, G.I., 1992. Rapid biological detection of apple scar skin viroid. Acta Hortic. 309, 291–296.

Imperial, J.S., Bautista, R.M., Randles, J.W., 1985. Transmission of the coconut cadang-cadang viroid to six species of palm by inoculation with nucleic acid extracts. Plant Pathol. 34, 391–401.

Ito, T., Furuta, T., Ito, T., Isaka, M., Ide, Y., Kaneyoshi, J., 2006. Identification of cachexia-inducible hop stunt viroid variants in citrus orchards in Japan using biological indexing and improved reverse transcription polymerase chain reaction. J. Gen. Plant Pathol. 72, 378–382.

Ito, T., Ieki, H., Ozaki, K., Ito, T., 2001. Characterization of a new citrus viroid species tentatively termed citrus viroid OS. Arch. Virol. 146, 975–982.

Ito, T., Suzaki, K., Nakano, N., Sato, A., 2013. Characterization of a new apscaviroid from American persimmon. Arch. Virol. 158, 2629–2631.

Ito, T., Yoshida, K., 1998. Reproduction of apple fruit crinkle disease symptoms by apple fruit crinkle viroid. Acta Hortic. 472, 587–594.

Jakse, J., Radisek, S., Pokorn, T., Matousek, J., Javornik, B., 2015. Deep-sequencing revealed citrus bark cracking viroid (CBCVd) as a highly aggressive pathogen on hop. Plant Pathol. 64, 831–842.

Kapari-Isaia, T., Kyriakou, A., Papayiannis, L., Tsaltas, D., Gregoriou, S., Psaltis, I., 2008. Rapid in vitro microindexing of viroids in citrus. Plant Pathol. 57, 348–353.

Khoury, J., Singh, R.P., Boucher, A., Coombs, D.H., 1988. Concentration and distribution of mild and severe strains of potato spindle tuber viroid in cross-protected tomato plants. Phytopathology 78, 1331–1336.

Legrand, P., 2015. Biological assays for plant viruses and other graft-transmissible pathogens diagnoses: a review. EPPO Bull. 45, 240–251.

Martínez-Soriano, J.P., Galindo-Alonso, J., Maroon, C.J., Yucel, I., Smith, D.R., Diener, T.O., 1996. Mexican papita viroid: putative ancestor of crop viroids. Proc. Natl. Acad. Sci. USA 93, 9397–9401.

Matsushita, Y., Tsuda, S., 2015. Host ranges of potato spindle tuber viroid, tomato chlorotic dwarf viroid, tomato apical stunt viroid, and columnea latent viroid in horticultural plants. Eur. J. Plant Pathol. 141, 193–197.

Murcia, N., Bernad, L., Serra, P., Hashemian, S.B., Duran-Vila, N., 2009. Molecular and biological characterization of natural variants of citrus dwarfing viroid. Arch. Virol. 154, 1329–1334.

Niblett, C.L., Dickson, E., Fernow, K.H., Horst, R.K., Zaitlin, M., 1978. Cross-protection among four viroids. Virology 91, 198–203.

Owens, R.A., 2015. Viroid discovery – past accomplishments, future challenges. Acta Hortic. 1072, 15–28.

Palacio-Bielsa, A., Romero-Durban, J., Duran-Vila, N., 2004. Characterization of citrus HSVd isolates. Arch. Virol. 149, 537–552.

Pallás, V., Flores, R., 1989. Interactions between citrus exocortis and potato spindle tuber viroids in plants of *Gynura aurantiaca* and *Lycopersicon esculentum*. Intervirology 30, 10–17.

Reanwarakorn, K., Semancik, J.S., 1998. Regulation of pathogenicity in hop stunt viroid-related group II citrus viroids. J. Gen. Virol. 79, 3163–3171.

Reanwarakorn, K., Semancik, J.S., 1999a. Correlation of hop stunt viroid variants to cachexia and xyloporosis diseases of citrus. Phytopathology 897, 568–574.

Reanwarakorn, K., Semancik, J.S., 1999b. Discrimination of cachexia disease agents among citrus variants of hop stunt viroid. Ann. Appl. Biol. 135, 481–487.

Roistacher, C.N., Blue, R.L., Calavan, E.C., 1973. A new test for citrus cachexia. Citrograph 58, 261–262.

Semancik, J.S., Gumpf, D.J., Bash, J.A., 1992. Interference between viroids inducing exocortis and cachexia diseases of citrus. Ann. Appl. Biol. 121, 577–583.

Semancik, J.S., Roistacher, C.N., Rivera-Bustamante, R., Duran-Vila, N., 1988. Citrus cachexia viroid, a new viroid of citrus: relationship to viroids of the exocortis disease complex. J. Gen. Virol. 69, 3059–3068.

Semancik, J.S., Vidalakis, G., 2005. The question of citrus viroid IV as a *Cocadviroid*. Arch. Virol. 150, 1059–1067.

Serra, P., Barbosa, C.J., Daròs, J.A., Flores, R., Duran-Vila, N., 2008a. Citrus viroid V: molecular characterization and synergistic interactions with other members of the genus *Apscaviroid*. Virology 370, 102–112.

Serra, P., Eiras, M., Bani-Hashemian, S.M., Murcia, N., Kitajima, E.W., Daròs, J.A., et al., 2008b. Citrus viroid V: Occurrence, host range, diagnosis, and identification of new variants. Phytopathology 98, 1199–1204.

Singh, R.P., 1973. Experimental host range of the potato spindle tuber "virus." Amer. Potato J. 50, 111–123.

Singh, R.P., Finnie, R.E., Bagnall, R.H., 1970. Relative prevalence of mild and severe strains of potato spindle tuber virus in eastern Canada. Amer. Potato J. 47, 289–293.

Singh, R.P., Nie, X., Singh, M., 1999. Tomato chlorotic dwarf viroid: an evolutionary link in the origin of pospiviroids. J. Gen.Virol. 80, 2823–2828.

Singh, R.P., Ready, K.F.M., 2003. Biological indexing. In: Hadidi, A., Flores, R., Randles, J.W., Semancik, J.S. (Eds.), Viroids. CSIRO Publishing, Collingwood, VIC, pp. 89–94.

Singh, R.P., Somerville, T.H., 1987. New disease symptoms observed on field-grown potato plants with potato spindle tuber viroid and potato virus Y infections. Potato Res. 30, 127–133.

Tomassoli, L., Luison, D., Luigi, M., Costantini, E., Mangiaracina, P., Faggioli, F., 2015. Supportive and antagonistic interactions among pospiviroids infecting solanaceous ornamentals. Acta Hortic. 1072, 71–77.

Valkonen, J., 1992. Accumulation of potato virus Y is enhanced in *Solatium brevidens* also infected with tobacco mosaic virus or potato spindle tuber viroid. Ann. Appl. Biol. 121, 321–327.

Verhoeven, J.T.J., Jansen, C.C.C., Roenhorst, J.W., Flores, R., De la Peña, M., 2009. Pepper chat fruit viroid: biological and molecular properties of a proposed new species of the genus *Pospiviroid*. Virus Res. 144, 209–214.

Verhoeven, J.T.J., Roenhorst, J.W., Owens, R.A., 2011. Mexican papita viroid and tomato planta macho viroid belong to a single species in the genus *Pospiviroid*. Arch. Virol. 156, 1433–1437.

Vernière, C., Perrier, X., Dubois, C., Dubois, A., Botella, L., Chabrier, C., et al., 2004. Citrus viroids: symptom expression and effect on vegetative growth and yield of clementine trees grafted on trifoliate orange. Plant Dis. 88, 1189–1197.

Vernière, C., Perrier, X., Dubois, C., Dubois, A., Botella, L., Chabrier, C., et al., 2006. Interactions between citrus viroids affect symptom expression and field performance of clementine trees grafted on trifoliate orange. Phytopathology 96, 356–368.

Walter, B., 1987. Tomato apical stunt. In: Diener, T.O. (Ed.), The Viroids. Plenum Press, New York, NY, pp. 321–327.

Wang, J., Boubourakas, I.N., Voloudakis, A.E., Agorastou, T., Magripis, G., Rucker, T.L., et al., 2013. Identification and characterization of known and novel viroid variants in the Greek national citrus germplasm collection: threats to the industry. Eur. J. Plant Pathol. 137, 17–27.

Yaguchi, S., Takahashi, T., 1984. Response of cucumber cultivars and other cucurbitaceous species to infection by hop stunt viroid. Phytopath. Z. 109, 21–32.

CHAPTER

34

Gel Electrophoresis

Dagmar Hanold[1] *and Ganesan Vadamalai*[2]

[1]The University of Adelaide, Waite Campus, Glen Osmond, SA, Australia

[2]Universiti Putra Malaysia, Selangor, Malaysia

INTRODUCTION

Gel electrophoresis (GE) has been an invaluable tool for viroid research ever since it led to the discovery of this group of plant pathogens (Diener, 1971). Different GE systems are applied both for molecular characterization and in routine diagnosis of viroids (Randles, 1993). Since GE fractionation is based on physical properties and sequence information is not required, it is suitable for recognizing circular molecules that may represent a novel viroid. Furthermore, it is useful for identifying a viroid that may be associated with a disease of unknown etiology (Hadidi et al., 1990; Randles, 1993; Shamloul et al., 2004). GE can also be used to purify viroids, e.g., for inoculation experiments to fulfill Koch's postulates, or for sequencing (Hanold and Randles, 1998a).

Preliminary identification of viroids or their characterization by GE methods relies on the specific molecular properties of these pathogens as small, circular, single-stranded RNAs with strong secondary structure, resulting in specific separation patterns on gels. Even single nucleotide differences in a viroid molecule can lead to changes in molecular structure and thus electrophoretic patterns, as for the 246 and 247 nt forms of coconut cadang-cadang viroid (CCCVd) (Imperial and Rodriguez, 1983; Randles et al., 1998). GE can also be used to detect mutants such as those associated with the severe "brooming" symptoms of CCCVd in coconut palm (Rodriguez and Randles, 1993) or the CCCVd variant found in oil palm (Vadamalai, 2005; Vadamalai et al., 2006).

GE methods are currently the most reliable and simplest strategy for identifying and isolating viroids, e.g., for cloning and bioassays.

Viroids and Satellites.
DOI: http://dx.doi.org/10.1016/B978-0-12-801498-1.00034-6

GE SYSTEMS

GE is a flexible methodology because a range of parameters can be varied, including gel matrix, running conditions, and optional addition of denaturants (Sambrook and Russell, 2001). A wide range of GE systems can be devised to suit different applications and to achieve the best separation relevant to the size and type of molecules under investigation. For systems involving more than one separation run, variables can be altered between runs. GE thus offers the opportunity to influence molecular structure of nucleic acids during electrophoresis and can be fine-tuned to optimize conditions for specific research goals.

Nondenatured viroid RNAs migrate as partially base-paired structures in native gels but as open circles under denaturing conditions, resulting in markedly decreased mobility relative to denatured linear molecules, which may show the same apparent size under native conditions. This variation in electrophoretic mobility has been used for the efficient detection of circular molecules such as viroids, circular satellite RNAs, and circular single-stranded virus DNA, even in complex mixtures of nucleic acids (Randles et al., 1987; Shamloul et al., 2004).

Due to its superior power of resolution for small molecules and higher sensitivity of detection because silver stains can be used, polyacrylamide gel electrophoresis (PAGE) is more widely used in viroid research than other GE systems. Tools used in viroid research include native and denaturing PAGE as single dimension runs or in consecutive (Flores et al., 1985; Rivera-Bustamante et al., 1986; Schumacher et al., 1986), bidirectional (Hadidi et al., 1990; Singh and Boucher, 1987), or two-dimensional (Feldstein et al., 1997; Schumacher et al., 1983) combinations. PAGE with gradients in temperature, pH, or urea concentration can be used to visualize the specific melting patterns characteristic of viroids (Riesner, 1987; Riesner et al., 1989). PAGE can be followed by RNA transfer to membranes, with most currently used systems being modified from the original northern-blotting (Alwine et al., 1977; Arrand, 1985). This approach eliminates problems due to interfering substances in the samples that may produce false results, e.g., in dot-blot hybridization assays. Table 34.1 provides an overview of GE systems generally used in research on viroids and other small circular nucleic acids.

It is crucial for PAGE analysis of viroid RNAs to remove host compounds which interfere with, distort, or conceal stained viroid bands. To obtain extracts suitable for PAGE analysis, preparation methods must be adjusted to suit each specific host/viroid system. Furthermore, since the concentration of some viroids in their hosts may be low, fractions enriched in low molecular weight RNA may be needed to enable detection by PAGE (Diener, 1987). For well-characterized viroid/host systems, short extraction

TABLE 34.1 Electrophoretic Methods Used in Viroid Research

Type of electrophoresis	1D PAGE[a]		Consecutive PAGE		Bidirectional PAGE	2D PAGE		Gradient PAGE	AGE[a]
	Native	Denaturing	Native/native	Native/denaturing		Native/native	Native/denaturing		
VARIABLES									
First run									
Gel matrix	✓	✓	✓	✓	✓	✓	✓	✓	✓
Run parameters	✓	✓	✓	✓	✓	✓	✓	✓	✓
Type of denaturant	na	✓	na	na	na	na	na	✓	na
Second run									
Gel matrix	na	na	✓	✓	na	✓	✓	na	na
Run parameters	na	na	✓	✓	✓	✓	✓	na	na
Type of denaturant	na	na	na	✓	✓	na	✓	na	na
Vary from 1st run	na	na	✓	✓	✓	✓	✓	na	na
APPLICATIONS									
Apparent size	✓	✓	na	na	✓	✓	✓	na	na
Molecular forms	✓	na	✓	✓	na	✓	✓	na	na
Sec/tert structure	na	na	✓	✓	✓	✓	✓	✓	na
ds/ss	na	na	na	✓	na	na	✓	na	na

(Continued)

TABLE 34.1 (Continued)

Type of electrophoresis	1D PAGE[a]		Consecutive PAGE		Bidirectional PAGE	2D PAGE		Gradient PAGE	AGE[a]
	Native	Denaturing	Native/native	Native/denaturing		Native/native	Native/denaturing		
Circular/linear	na	na	na	✓	na	na	✓	na	na
Variants/strains	✓	na	na	✓	na	✓	✓	na	na
Mutants	na	na	na	✓	na	✓	✓	na	na
Diagnostic	✓	✓	na	✓	✓	✓	✓	na	na
Preparative	na	na	✓	✓	na	✓	✓	na	na
Analysis of clones	na	na	na	na	na	na	na	na	✓
RT-PCR analysis	na	na	na	na	na	na	na	na	✓
RPA	na	✓	na	na	na	na	na	na	na

[a] *PAGE, polyacrylamide gel electrophoresis; AGE, agarose gel electrophoresis.*

✓, indicates available option; na, not applicable.

Hanold, D., 1993. Methods applicable to viroids. In: Matthews, R.E.F. (Ed.), Diagnosis of Plant Virus Diseases. CRC Press, Boca Raton, FL, pp. 295–314; Hanold, D., 1998. Diagnostic methods applicable to viroids. In: Hanold, D., Randles, J.W. (Eds.), Report on ACIAR-funded research on viroids and viruses of coconut palm and other tropical monocotyledons 1985-1993. Australian Centre for International Agricultural Research, pp. 27–39; Hanold, D., Randles, J.W. (Eds.), 1998a. Report on ACIAR - funded research on viroids and viruses of coconut palm and other tropical monocotyledons 1985-1993. Australian Centre for International Agricultural Research, Canberra. http://aciar.gov.au/files/node/2259/report_on_aciar_funded_research_on_viroids_and_vir_88070.pdf; Hanold, D., Semancik, J.S., Owens, R.A., 2003. Polyacrylamide gel electrophoresis. In: Hadidi, A., Flores, R., Randles, J.W., Semancik, J. (Eds.), Viroids. CSIRO Publishing, Collingwood, VIC, Australia, pp. 95–102; Hanold, D., Randles, J.W., Hadidi, A., 2011. Polyacrylamide gel electrophoresis for viroid detection. In: Hadidi, A., Barba, M., Candresse, T., Jelkmann, W. (Eds.), Virus and Virus-Like Diseases of Pome and Stone Fruits. APS Press, St. Paul, MN, pp. 327–332.

methods can sometimes be devised for routine diagnosis (Hanold, 1993, 1998; Hanold and Randles, 1998c; Hanold et al., 2003, 2011).

Detection thresholds for GE depend on the gel system and method of visualization. The most sensitive staining method for PAGE is silver staining (Imperial et al., 1985; Sammons et al., 1981) which can detect as little as 50 pg of RNA per band (Hanold, 1993, 1998). However, if specific probes are available, a few pg of target RNA may be detected on gel-blots (Hanold, 1993, 1998; Hanold and Randles, 1991, 1998b).

For use as controls, estimating detection limits, or equilibrating a GE system, small circular viroid-like satellite RNAs may be suitable. For example, satellite RNAs (virusoids) of velvet tobacco mottle virus and solanum nodiflorum mottle virus (Gould, 1981; Gould and Hatta, 1981; Randles et al., 1981) can be isolated in relatively large amounts from their herbaceous experimental hosts without host compounds likely to distort banding patterns (Francki et al., 1986).

EXAMPLES OF APPLICATIONS

Viroid Identification, Purification, and Molecular Characterization

Viroid RNAs purified through several cycles of consecutive or two-dimensional PAGE (Table 34.1) are suitable for infectivity tests (e.g., CCCVd; Hanold, 1998; Imperial et al., 1985), and for molecular, characterization, including cloning and sequencing (Diener, 1987; Feldstein et al., 1997; Hadidi et al., 1990; Hanold, 1993; Hanold and Randles, 1998a; Hanold et al., 2003; Mohamed et al., 1985; Riesner, 1987; Riesner et al., 1989; Rivera-Bustamante et al., 1986; Schumacher et al., 1983, 1986; Singh and Boucher, 1987).

Viroid Diagnosis and Surveys

Even in a poorly resourced laboratory, PAGE is a reliable tool for viroid diagnosis since it does not require radioactive substances or elaborate equipment. For example, it is applied routinely for field diagnosis of CCCVd in the Philippines using a "mobile laboratory" that is driven to survey areas for on-site analysis of palms (Hanold et al., 2003; Randles et al., 1992). Since fresh samples are available, and spoilage of samples during transport from remote areas is eliminated, this method ensures good reproducibility of results (Hanold et al., 2011; Namia et al., 1998; Rodriguez et al., 1998).

Various RNA or DNA probes, different hybridization conditions and differential stringency of posthybridization washes can be used to detect low or high sequence similarity targets as needed. For example, this method has identified viroid-like molecules with varying levels of similarity to CCCVd in palms and other monocotyledons during a survey in the Pacific region (Hanold and Randles, 1991, 1998a).

Ribonuclease Protection Assay (RPA)

RPA uses liquid hybridization of target nucleic acids with a labeled complementary RNA probe, digestion of unpaired regions with single-strand-specific ribonucleases, fractionation by PAGE, and autoradiography to detect the hybrids. Nucleotide sequence mismatches in viroid isolates due to mutations are identified because they lead to cleavage of hybrids at single-stranded loci and additional bands, compared to the results when using perfectly complementary target and probe. Unlike RT-PCR (Hodgson et al., 1998), RPA is robust and not affected by inhibitors of transcriptases and polymerases in extracts.

RPA has been used to detect and quantify specific RNAs in complex nucleic acid mixtures (Ahmad et al., 1993; Aranda et al., 1993; Kurath and Palukaitis, 1989b; Rosenau et al., 2002; Winter et al., 1985), to identify sequence variants in populations of RNAs (Cabrera et al., 2000; Kurath and Palukaitis, 1989a; Lakshman and Tavantzis, 1992; Lopez-Galindez et al., 1988), to assess genetic heterogeneity in populations of RNA viruses (Ali and Randles, 2001; Aranda et al., 1995), and to study plant virus evolution and epidemiology (Palukaitis et al., 1994). Targets partially protecting a full-length probe, e.g., for $CCCVd_{246}$, have been described as quasispecies (Flores et al., 2003).

RPA was used to detect CCCVd variants in oil palm and in coconut samples from Malaysia and Sri Lanka (Vadamalai, 2005; Vadamalai et al., 2009). It was considered suitable for detecting closely related variants in which mismatched sites were unlikely to exceed two contiguous nucleotides. A survey of oil palms (Vadamalai et al., 2009) confirmed direct sequencing results that showed that multiple variants of CCCVd can coexist in a single palm, and that both asymptomatic palms and palms with orange spotting symptoms contain CCCVd (Vadamalai et al., 2006; Wu et al., 2013). For Sri Lanka, where CCCVd-related sequences had been previously reported in coconut palm by Hanold and Randles (1998b), the presence of a viroid with high similarity to $CCCVd_{246}$ was confirmed (Vadamalai et al., 2009).

Detection of Small Interfering RNAs in CCCVd-Infected Coconut Palm

Denaturing PAGE in combination with gel-blot hybridization assay was used for detection of small interfering RNAs (siRNAs) in CCCVd-infected

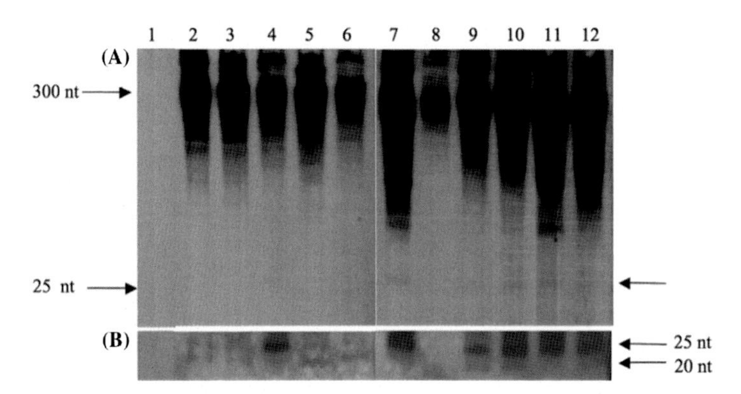

FIGURE 34.1 Northern-blot of total leaf nucleic acid extracts from coconut palm fractionated by 15% denaturing PAGE and hybridized with labeled full-length CCCVd cDNA at high stringency. Healthy palm (lane 1); CCCVd-infected palms at early (lanes 2–4), medium (lanes 5,6), and late (lanes 7–9) disease stages; palms infected with a severe CCCVd variant and showing brooming symptoms (lanes 10–12). The autoradiograph in panel A was overexposed to illustrate the low amounts of siRNA relative to the amount of viroid sequences present, whereas panel B shows a sixfold longer exposure of the small RNA zone. Arrows indicate positions of the 300 nt full viroid sequence and the 20–25 nt viroid small RNAs.

coconut palms (Vadamalai, 2005). Posttranscriptional gene silencing, indicated by the presence of siRNAs, had previously been shown to occur in plants infected with different viroids (Itaya et al., 2001; Markarian et al., 2004; Martinez de Alba et al., 2002; Papaefthimiou et al., 2001; Sano et al., 2010). Results from CCCVd-infected coconut palms at different stages of disease from the Philippines suggest that CCCVd also induces posttranscriptional gene silencing (Fig. 34.1).

References

Ahmad, N., Kuramoto, I.K., Baroudy, B.M., 1993. A ribonuclease protection assay for the direct detection and quantitation of hepatitis C virus RNA. Clin. Diagn. Virol. 1, 233–244.

Ali, A., Randles, J.W., 2001. Genomic heterogeneity in pea seed-borne mosaic virus isolates from Pakistan, the centre of diversity of the host species, Pisum sativum. Arch. Virol. 146, 1855–1870.

Alwine, J.C., Kemp, D.J., Stark, G.R., 1977. Method for detection of specific RNAs in agarose gels by transfer to diazobenzyloxymethyl paper and hybridisation with DNA probes. Proc. Natl. Acad. Sci. USA 74, 5350–5354.

Aranda, M.A., Fraile, A., Garcia-Arenal, F., 1993. Genetic variability and evolution of the satellite RNA of cucumber mosaic virus during natural epidemics. J. Virol. 67, 5896–5901.

Aranda, M.A., Fraile, A., Garcia-Arenal, F., Malpica, J.M., 1995. Experimental evaluation of the ribonuclease protection assay method for the assessment of genetic heterogeneity in populations of RNA viruses. Arch. Virol. 140, 1373–1383.

Arrand, J., 1985. Preparation of nucleic acid probes. In: Hames, B., Higgins, S. (Eds.), Nucleic Acid Hybridisation: A Practical Approach. IRL Press, Oxford, pp. 17–45.

Cabrera, O., Roossinck, M.J., Scholthof, K.-B.G., 2000. Genetic diversity of panicum mosaic virus satellite RNAs in St. Augustine grass. Phytopathology 90, 977–980.

Diener, T.O., 1971. Potato spindle tuber "virus." IV. A replicating, low molecular weight RNA. Virology 45, 411–428.

Diener, T.O., 1987. Biological properties. In: Diener, T.O. (Ed.), The Viroids. Plenum Press, New York, NY, pp. 9–35.

Feldstein, P.A., Levy, L., Randles, J.W., Owens, R.A., 1997. Synthesis and two-dimensional electrophoretic analysis of mixed populations of circular and linear RNAs. Nucleic Acids Res. 25, 4850–4854.

Flores, R., Duran-Vila, N., Pallas, V., Semancik, J.S., 1985. Detection of viroid and viroid-like RNAs from grapevine. J. Gen. Virol. 66, 2095–2102.

Flores, R., Randles, J.W., Owens, R.A., 2003. Classification. In: Hadidi, A., Flores, R., Randles, J.W., Semancik, J. (Eds.), Viroids. CSIRO Publishing, Collingwood, VIC, pp. 71–75.

Francki, R.I.B., Grivell, C.J., Gibb, K.S., 1986. Isolation of velvet tobacco mottle virus capable of replication with and without a viroid-like RNA. Virology 148, 381–384.

Gould, A.R., 1981. Studies on encapsidated viroid-like RNA II. Purification and characterization of a viroid-like RNA associated with velvet tobacco mottle virus (VTMoV). Virology 108, 123–133.

Gould, A.R., Hatta, T., 1981. Studies on encapsidated viroid-like RNA III. Comparative studies on RNAs isolated from velvet tobacco mottle virus and solanum nodiflorum mottle virus. Virology 109, 137–147.

Hadidi, A., Huang, C., Hammond, R.W., Hashimoto, J., 1990. Homology of the agent associated with dapple apple disease to apple scar skin viroid and molecular detection of these viroids. Phytopathology 80, 263–268.

Hanold, D., 1993. Methods applicable to viroids. In: Matthews, R.E.F. (Ed.), Diagnosis of Plant Virus Diseases. CRC Press, Boca Raton, FL, pp. 295–314.

Hanold, D., 1998. Diagnostic methods applicable to viroids. In: Hanold, D., Randles, J.W. (Eds.), Report on ACIAR-Funded Research on Viroids and Viruses of Coconut Palm and Other Tropical Monocotyledons 1985–1993. Australian Centre for International Agricultural Research, pp. 27–39.

Hanold, D., Randles, J.W., 1991. Detection of coconut cadang-cadang viroid-like sequences in oil and coconut palm and other monocotyledons in the south-west Pacific. Ann. Appl. Biol. 118, 139–151.

Hanold, D., Randles, J.W. (Eds.), 1998a. Report on ACIAR-Funded Research on Viroids and Viruses of Coconut Palm and Other Tropical Monocotyledons 1985-1993. Australian Centre for International Agricultural Research, Canberra. <http://aciar.gov.au/files/node/2259/report_on_aciar_funded_research_on_viroids_and_vir_88070.pdf>.

Hanold, D., Randles, J.W., 1998b. Results of the survey for coconut palm. In: Hanold, D., Randles, J.W. (Eds.), Report on ACIAR-Funded Research on Viroids and Viruses of Coconut Palm and Other Tropical Monocotyledons 1985–1993. Australian Centre for International Agricultural Research, pp. 134–143.

Hanold, D., Randles, J.W., 1998c. Methods used for assay of the samples collected during the survey. In: Hanold, D., Randles, J.W. (Eds.), Report on ACIAR-Funded Research on Viroids and Viruses of Coconut Palm and Other Tropical Monocotyledons 1985–1993. Australian Centre for International Agricultural Research, pp. 131–133.

Hanold, D., Randles, J.W., Hadidi, A., 2011. Polyacrylamide gel electrophoresis for viroid detection. In: Hadidi, A., Barba, M., Candresse, T., Jelkmann, W. (Eds.), Virus and Virus-Like Diseases of Pome and Stone Fruits. APS Press, St. Paul, MN, pp. 327–332.

Hanold, D., Semancik, J.S., Owens, R.A., 2003. Polyacrylamide gel electrophoresis. In: Hadidi, A., Flores, R., Randles, J.W., Semancik, J. (Eds.), Viroids. CSIRO Publishing, Collingwood, VIC, pp. 95–102.

Hodgson, R.A.J., Wall, G.C., Randles, J.W., 1998. Specific identification of coconut tinangaja viroid (CTiVd) for differential field diagnosis of viroids in coconut palm. Phytopathology 88, 774–781.

Imperial, J.S., Bautista, R.M., Randles, J.W., 1985. Transmission of the coconut cadang cadang viroid to six species of palm by inoculation with nucleic acid extracts. Plant Pathol. 34, 391–401.

Imperial, J.S., Rodriguez, M.J.B., 1983. Variation in the coconut cadang-cadang viroid: evidence for single base addition with disease progress. Phil. J. Crop Sci. 8, 87–91.

Itaya, A., Folimonov, A., Matsuda, Y., Nelson, R.S., Ding, B., 2001. Potato spindle tuber viroid as inducer of RNA silencing in infected tomato. Mol. Plant Microbe Interact. 14, 1332–1334.

Kurath, G., Palukaitis, P., 1989a. RNA sequence heterogeneity in natural populations of three satellite RNAs of cucumber mosaic virus. Virology 173, 231–240.

Kurath, G., Palukaitis, P., 1989b. Satellite RNAs of cucumber mosaic virus: recombinants constructed *in vitro* reveal independent functional domains for chlorosis and necrosis in tomato. Mol. Plant Microbe Interact. 2, 91–96.

Lakshman, D.K., Tavantzis, S.M., 1992. RNA progeny of an infectious two-base deletion cDNA mutant of potato spindle tuber viroid (PSTV) acquire two nucleotides *in planta*. Virology 187, 565–572.

Lopez-Galindez, C., Lopez, J.A., Melero, J.A., De La Fuente, L., Martinez, C., Ortin, J., et al., 1988. Analysis of genetic variability and mapping of point mutations in influenza virus by the RNAse A mismatch cleavage method. Proc. Natl. Acad. Sci. USA 85, 3522–3526.

Markarian, N., Li, H.W., Ding, S.W., Semancik, J.S., 2004. RNA silencing as related to viroid induced symptom expression. Arch. Virol. 149, 397–406.

Martinez de Alba, A.E., Flores, R., Hernandez, C., 2002. Two chloroplastic viroids induce the accumulation of the small RNAs associated with post-transcriptional gene silencing. J. Virol. 76, 13094–13096.

Mohamed, N.A., Bautista, R., Buenaflor, G., Imperial, J.S., 1985. Purification and infectivity of the coconut cadang-cadang viroid. Phytopathology 75, 79–83.

Namia, M.T.I., Rodriguez, M.J.B., Randles, J.W., 1998. Diagnosis of cadang-cadang by rapid polyacrylamide gel electrophoresis. In: Hanold, D., Randles, J.W. (Eds.), Report on ACIAR-Funded Research on Viroids and Viruses of Coconut Palm and Other Tropical Monocotyledons 1985-1993. Australian Centre for International Agricultural Research, Canberra, pp. 40–43.

Palukaitis, P., Roossinck, M.J., Garcia-Arenal, F., 1994. Applications of ribonuclease protection assay in plant virology. In: Adolph, K.W. (Ed.), Molecular Virology Techniques Part A. Academic Press Inc, San Diego, CA, pp. 237–250.

Papaefthimiou, I., Hamilton, A.J., Denti, M.A., Baulcombe, D.C., Tsagris, M., Tabler, M., 2001. Replicating potato spindle tuber viroid RNA is accompanied by short RNA fragments that are characteristic of post-transcriptional gene silencing. Nucleic Acids Res. 29, 2395–2400.

Randles, J.W., 1993. Strategies for implicating virus-like pathogens as the cause of diseases of unknown etiology. In: Matthews, R.E.F. (Ed.), Diagnosis of Plant Virus Diseases. CRC Press, Boca Raton, FL, pp. 315–332.

Randles, J.W., Davies, C., Hatta, T., Gould, A.R., Francki, R.I.B., 1981. Studies on encapsidated viroid-like RNA I. Characterization of velvet tobacco mottle virus. Virology 108, 111–122.

Randles, J.W., Hanold, D., Julia, J.F., 1987. Small circular single-stranded DNA associated with foliar decay disease of coconut palm in Vanuatu. J. Gen. Virol. 68, 273–280.

Randles, J.W., Hanold, D., Pacumbaba, E.P., Rodriguez, M.J.B., 1992. Cadang-cadang disease of coconut palm. In: Mukhopadhyay, A.N., Kumar, J., Chaube, H.S., Singh, U.S. (Eds.), Plant Diseases of International Importance, vol. IV. Prentice Hall, Englewood Cliffs, NJ, pp. 277–295.

Randles, J.W., Hanold, D., Pacumbaba, E.P., Rodriguez, M.J.B., 1998. Cadang-cadang disease of coconut palm – an overview. In: Hanold, D., Randles, J.W. (Eds.), Report on ACIAR-Funded Research on Viroids and Viruses of Coconut Palm and Other Tropical Monocotyledons 1985-1993. Australian Centre for International Agricultural Research, Canberra, pp. 11–26.

Riesner, D., 1987. Physical-chemical properties. In: Diener, T.O. (Ed.), The Viroids. Plenum Press, New York, NY, pp. 63–98.

Riesner, D., Steger, G., Zimmat, R., Owens, R.A., Wagenhöfer, M., Hillen, W., et al., 1989. Temperature-gradient gel electrophoresis of nucleic acids: analysis of conformational transitions, sequence variations, and protein-nucleic acid interactions. Electrophoresis 10, 377–389.

Rivera-Bustamante, R., Gin, R., Semancik, J.S., 1986. Enhanced resolution of circular and linear molecular forms of viroid and viroid-like RNA by electrophoresis in a discontinuous-pH system. Anal. Biochem. 156, 91–95.

Rodriguez, M.J.B., Namia, M.T.I., Estioko, L.P., 1998. The mobile diagnostic laboratory. In: Hanold, D., Randles, J.W. (Eds.), Report on ACIAR-Funded Research on Viroids and Viruses of Coconut Palm and Other Tropical Monocotyledons 1985–1993. Australian Centre for International Agricultural Research, Canberra, pp. 44–47.

Rodriguez, M.J.B., Randles, J.W., 1993. Coconut cadang-cadang viroid (CCCVd) mutants associated with severe disease vary in both the pathogenicity domain and the central conserved region. Nucleic Acids Res. 21, 2771.

Rosenau, C., Kaboord, B., Qoronfleh, M.W., 2002. Development of a chemiluminescence-based ribonuclease protection assay. BioTechniques 33, 1354–1358.

Sambrook, J., Russell, D.W., 2001. Molecular Cloning: A Laboratory Manual, third ed. Cold Spring Harbor Laboratory Press, New York, NY.

Sammons, D.W., Adams, L.P., Nishazawa, E.E., 1981. Ultrasensitive silver-based colour staining of polypeptides in polyacrylamide gels. Electrophoresis 2, 135–141.

Sano, T., Barba, M., Li, S.-F., Hadidi, A., 2010. Viroids and RNA silencing: mechanism, role in viroid pathogenicity and development of viroid-resistant plants. GM Crops. 1, 1–7.

Schumacher, J., Meyer, N., Riesner, D., Weideman, H.L., 1986. Diagnostic procedure for detection of viroids and viruses with circular RNAs by "return"-gel electrophoresis. J. Phytopathol. 115, 332–343.

Schumacher, J., Randles, J.W., Riesner, D., 1983. A two-dimensional electrophoretic technique for the detection of circular viroids and virusoids. Anal. Biochem. 135, 228–295.

Shamloul, A.M., Yang, X., Han, L., Hadidi, A., 2004. Characterization of a new variant of apple scar skin viroid associated with pear fruit crinkle disease. J. Plant Pathol. 86, 249–256.

Singh, R.P., Boucher, A., 1987. Electrophoretic separation of a severe from mild strain of potato spindle tuber viroid. Phytopathology 77, 1588–1591.

Vadamalai, G., 2005. An investigation of oil palm orange spotting disorder. Ph.D Thesis. The University of Adelaide, South Australia. <https://digital.library.adelaide.edu.au/dspace/handle/2440/37756>.

Vadamalai, G., Hanold, D., Rezaian, M.A., Randles, J.W., 2006. Variants of coconut cadang-cadang viroid isolated from an African oil palm (*Elaies guineensis* Jacq.) in Malaysia. Arch. Virol. 151, 1447–1456.

Vadamalai, G., Perera, A.A.F.L.K., Hanold, D., Rezaian, M.A., Randles, J.W., 2009. Detection of coconut cadang-cadang viroid sequences in oil and coconut palm by ribonuclease protection assay. Ann. Appl. Biol. 154, 117–125.

Winter, E., Yamamoto, F., Almoguera, C., Perucho, M., 1985. A method to detect and characterize point mutations in transcribed genes: amplification and overexpression of the mutant c-Ki-*ras* allele in human tumor cells. Proc. Natl. Acad. Sci. USA 82, 7575–7579.

Wu, Y.H., Cheong, L.C., Meon, S., Lau, W.H., Kong, L.L., Joseph, H., et al., 2013. Characterization of coconut cadang-cadang viroid variants from oil palm affected by orange spotting disease in Malaysia. Arch. Virol. 158, 1407–1410.

CHAPTER

35

Molecular Hybridization Techniques for Detecting and Studying Viroids

Vicente Pallás[1], Jesus A. Sánchez-Navarro[1], Gary R. Kinard[2] and Francesco Di Serio[3]

[1]Polytechnic University of Valencia-CSIC, Valencia, Spain
[2]U.S. Department of Agriculture, Beltsville, MD, United States
[3]National Research Council, Bari, Italy

INTRODUCTION

Serological methods cannot detect viroids because they do not code for any proteins. It is not surprising, therefore, that viroids were the first obligate plant pathogen for which molecular hybridization was used as a detection tool (Owens and Diener, 1981). Molecular hybridization has also facilitated considerable progress in identifying viroid replication intermediates, ribozyme activities, and RNA structure. Many laboratories have replaced molecular hybridization as a routine diagnostic tool with reverse transcription polymerase chain reaction (RT-PCR) (see Hadidi et al., 2011; James et al., 2006 for review). However, hybridization remains a useful technique because it balances sensitivity with ease of use, time, and cost.

The basic principles of nucleic acid hybridization on solid supports are described in previous reviews (Hull, 1993; Mühlbach et al., 2003; Pallás et al., 1998, 2011) and will not be repeated here. In this chapter we update the last two decades of progress using molecular hybridization methodologies (dot-blot, gel-blot, multiplex, tissue-printing, and in situ hybridization) to detect and study viroids.

Viroids and Satellites.
DOI: http://dx.doi.org/10.1016/B978-0-12-801498-1.00035-8

DOT-BLOT HYBRIDIZATION

This technique involves the direct application of a nucleic acid onto nitrocellulose or nylon membranes, and its subsequent detection with viroid-specific probes (Fig. 35.1A). The development of nonradioactive precursors to label nucleic acid probes made molecular hybridization a more widely used technique. The hybrids formed between viroid RNA and the RNA probe are more stable than RNA–DNA hybrids. Therefore, more stringent hybridization conditions can be applied when using riboprobes facilitating reduced nonspecific background (see Mühlbach et al., 2003; Pallás et al., 2011 for reviews). Nonradioactive riboprobes are synthesized by incorporating digoxigenin (DIG) into cRNA by in vitro transcription of cloned viroid cDNA. Nucleic acid targets are fixed to the membrane by baking or by UV cross-linking, with the latter resulting in a 5–10-fold sensitivity increase (Pallás et al., 1998). Hybridization efficiency depends on the complexity (length and composition) and concentration of the probe, temperature, salt concentration, base mismatches, and inclusion of hybridization accelerators. In general, higher temperatures and lower salt concentrations increase stringency. Including formamide in the hybridization solution also increases stringency by maximizing correct base-pairing, decreasing the background, and minimizing riboprobe degradation. Good signal/background ratio is usually achieved at 70–72°C in 50% formamide. Sample preparation can use organic or nonorganic solvents, special matrices and/or other reagents, or procedures that separate nucleic acids from other plant cell components. Depending on the extraction and fractionation method, targets are either total nucleic acids, total RNAs, or purified viroid RNA. The use of phenol and other hazardous organic solvents for extractions is decreasing. Small-scale procedures minimize the amount used (Li et al., 2008) and alternative protocols that avoid organic solvents altogether have been developed (Astruc et al., 1996; Cañizares et al., 1998).

GEL-BLOT HYBRIDIZATION

This method, also called northern-blot hybridization (NBH), detects RNAs that are separated by gel electrophoresis and transferred to a solid support. Probes are then applied whose subsequent detection generates radioactive, colorimetric or chemiluminescent signals (Fig. 35.1B). NBH is generally more informative than either electrophoresis or dot-blot hybridization alone. Data on the molecular size of the target and, to some extent, its sequence composition can be obtained. Denaturing gels are preferred for NBH, although viroid RNAs separated in nondenaturing

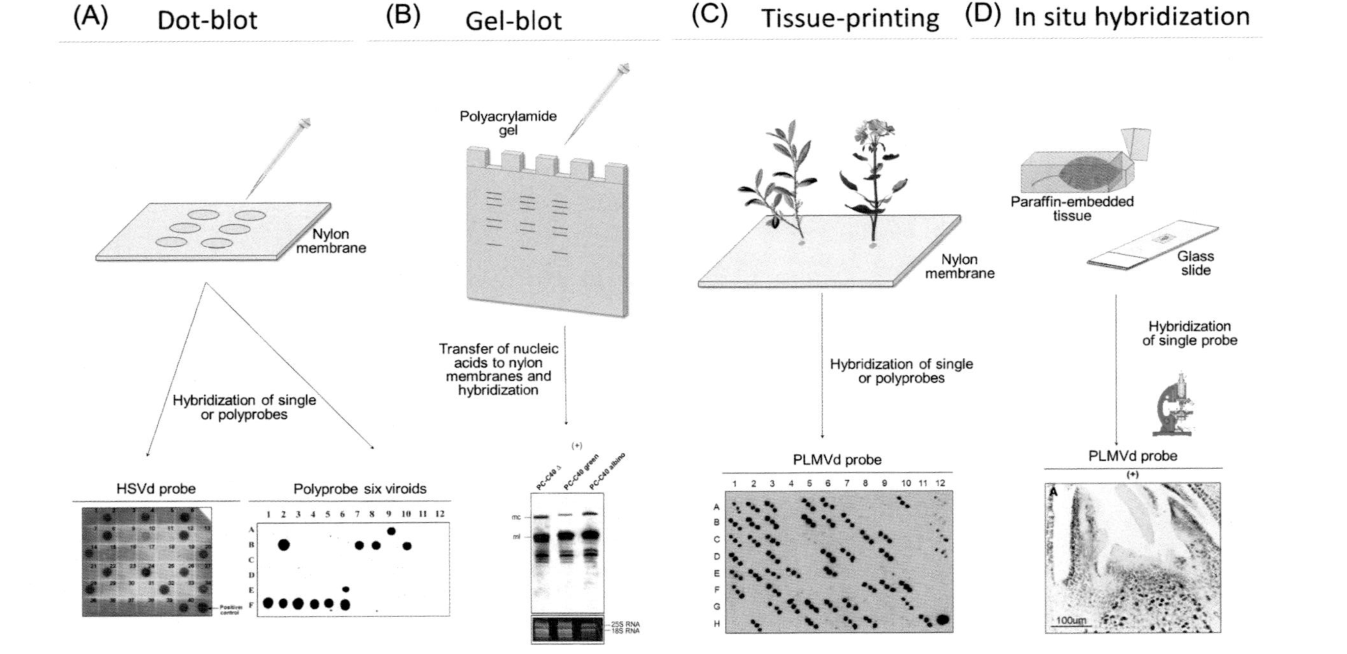

FIGURE 35.1 Nucleic acid hybridization based methods for viroid detection: (A) Dot-blot detection of HSVd infected apricot trees using a single probe (*Source: Amari, K., Ruiz, G., Gomez, G., Sanchez-Pina, M.A., Palla's, V., Egea, J., 2007. An important new apricot disease in Spain is associated with hop stunt viroid infection. Eur. J. Plant Pathol. 118, 173–181*, with permission of Springer) and polyprobe detection of ADFVd, AFCVd, ASSVd, HSVd, PBCVd, and PLMVd (*Source: Lin, L., Li, R., Mock, R., Kinard, G., 2011. Development of a polyprobe to detect six viroids of pome and stone fruit trees. J. Virol. Methods 171, 91–97*, with permission of Elsevier). Polyprobe samples F1-F6 are pome or stone fruit trees individually infected with the six viroids. The other positive samples are commercial peach trees infected with one or more of the six viroids. (B) Gel-blot hybridization to detect PLMVd following denaturing polyacrylamide gel electrophoresis (*Source: Rodio, M.E., Delgado, S., De Stradis, A., Gómez, M.D., Flores, R., Di Serio, F., 2007. A viroid RNA with a specific structural motif inhibits chloroplast development. Plant Cell 19, 3610–3626*, with permission of American Society of Plant Biologists (ASPB). The monomeric circular (mc) and linear (ml) forms of the PLMVd RNAs are indicated. (C) Tissue print hybridization to detect PLMVd infected stone fruit trees (*Source: Mandic, B., Al Rwahnih, M., Myrta, A., Gómez, G., Pallás, V., 2008. Incidence and genetic diversity of peach latent mosaic viroid and hop stunt viroid in stone fruits in Serbia. Eur. J. Plant Pathol. 120, 167–176*; with permission from Springer). Columns 1–11 are field samples; column 12 contains positive and negative controls. (D) In situ hybridization to detect PLMVd in longitudinal sections of peach apical meristems (*Source: Rodio, M.E., Delgado, S., De Stradis, A., Gómez, M.D., Flores, R., Di Serio, F., 2007. A viroid RNA with a specific structural motif inhibits chloroplast development. Plant Cell 19, 3610–3626*, with permission of ASPB).

gels are also efficiently detected (Murcia et al., 2009). Denaturing gels provides the additional advantage of separating circular and linear viroid forms (see Chapter 34: Gel Electrophoresis).

Many aspects of viroid biology have been investigated by NBH, including the seminal studies on viroid replication and elucidation of the rolling-circle mechanisms described in Chapter 7, Viroid Replication (see Mühlbach et al., 2003 for review). NBH-detection of PSTVd subgenomic RNAs helped to identify a novel degradation pathway targeting viroids in the nucleus (Minoia et al., 2015). Viroid subgenomic RNAs, generated by other mechanisms, were previously identified in tissues infected by avocado sunblotch viroid and peach latent mosaic viroid (PLMVd), two chloroplast-replicating viroids (Daròs et al., 1994; Delgado et al., 2005). NBH detected viroid-derived small RNAs of 21–24 nt in infected tissue (Itaya et al., 2001; Martínez de Alba et al., 2002), thus supplying the first solid evidence of RNA silencing in plant-viroid interactions (see Chapter 11: Viroids and RNA Silencing).

NBH is less sensitive than RT-PCR; however, it is less prone to generating false positives and is therefore especially useful for associating viroids with diseases. Viroid infection of new hosts and fulfilling Koch's postulates for viroid-induced diseases usually includes detecting the circular and linear viroid RNAs by NBH. This technique also allows efficient discrimination of viroid variants characterized by insertion/deletions (Malfitano et al., 2003; Semancik et al., 1994). Mixed infections by viroids differing in size can also be identified by NBH using: (1) a probe specific for only one viroid (Di Serio et al., 2001); (2) a probe specific for a region highly conserved in the targeted viroids (Jiang et al., 2013); (3) a mixture of individual probes specific for each viroid (Hajizadeh et al., 2012); or, (4) a polyprobe (see below).

NBH can be applied to a limited number of samples, is time-consuming, and requires technical expertise. For large-scale viroid surveys, RT-PCR amplification (see Chapter 36: Viroid Amplification Methods: RT-PCR, Real-Time RT-PCR, and RT-LAMP) or tissue-print and polyprobe hybridization (see below) may be more appropriate. However, NBH is still an important tool, especially in conjunction with other techniques. For example, RT-PCR products can be transferred to a membrane after gel electrophoresis for hybridization with specific probes (Southern-blot) to confirm and/or increase specificity of detection (Hadidi and Yang, 1990).

POLYVALENT DETECTION USING POLYPROBES

An extension of nonisotopic nucleic acid hybridization is the development of riboprobes that detect more than one viroid simultaneously (Fig. 35.1A). The term polyprobe was first used to describe a probe to detect six stone fruit tree RNA viruses (Herranz et al., 2005). Polyprobes have been subsequently developed to detect viroids of citrus (Cohen et al., 2006), pome and stone fruit trees (Lin et al., 2011), grapevines (Zhang et al., 2012), and ornamentals and vegetables (Jiang et al., 2013; Torchetti et al., 2012). Polyprobes have also been developed to detect eight RNA viruses and two viroids of stone fruit trees simultaneously (Peiró et al., 2012). Polyprobes are constructed by cloning in tandem partial or full-length sequences of viroid cDNA, usually generated by RT-PCR, into an expression vector from which DIG-labeled probes are transcribed. Alternatively, for constructing a universal probe for members of the genus *Coleviroid*, the DNA insert was synthesized artificially based on the highly conserved 32 nt of the central conserved region. Eight copies were then inserted into the vector, the transcription of which yielded an octameric probe (Jiang et al., 2013).

Sample preparation techniques with polyprobes are generally the same as with single probes. The sensitivity of polyprobes is generally comparable to that of individual probes, although assay conditions may require optimization with increasing probe length and/or number of different targets. Reducing the hybridization temperature from 68 to 60°C yielded a comparable dilution end point of detection for virus–viroid polyprobe as compared to individual probes (Peiró et al., 2012).

Polyprobes offer numerous advantages for viroid detection, especially the efficient and cost-effective use of reagents, sample materials, time, and labor. The small target size of viroid RNAs makes them particularly amenable to polyprobes, which can be designed based on the crop, host range, or viroid diversity. Any sample that generates a positive signal using the polyprobe can subsequently be examined by individual probes (including NBH) or a nucleic acid amplification procedure. Polyprobes may also facilitate detection of related strains or the discovery of new viroids if the sequence similarity and the reaction conditions allow for adequate hybridization. A polyprobe constructed from hop stunt viroid, Australian grapevine viroid, and grapevine yellow speckle viroid 1 also detected grapevine yellow speckle viroid 2 (Zhang et al., 2012), and a polyprobe designed to detect eight pospiviroids (Torchetti et al., 2012) also detected an additional species (Sorrentino et al., 2015) and theoretically should detect all known pospiviroids.

Polyprobes are especially useful for large-scale screening and survey research. The protocol is simple and less susceptible to contamination than nucleic acid amplification or sequencing methods. Furthermore, only one synthesis reaction is needed to detect up to 6–8 viroids, which saves resources and avoids potential variability associated with repeated individual riboprobe production.

TISSUE-PRINT HYBRIDIZATION

Minimizing sample preparation is advantageous for any detection technique for obligate plant pathogens. For viroid detection, this can be achieved using tissue-printing that only requires the transfer of plant sap (from cutting the stem, leaf, or fruit) directly onto a membrane (Fig. 35.1C). After the samples are fixed, the membrane is processed as with dot-blot hybridization. This technique has been reported to detect numerous viroids (Astruc et al., 1996; Hooftman et al., 1996; Hurtt et al., 1996; Mandic et al., 2008; Palacio et al., 2000; Podleckis et al., 1993; Romero-Durban et al., 1995; Torres et al., 2004).

Tissue-printing hybridization has also been applied to field conditions for large-scale indexing of stone fruit viroids (Amari et al., 2001; Astruc et al., 1996; Bouani et al., 2004; Cañizares et al., 2001; El-Dougdoug et al., 2012; Gümüs et al., 2007; Michelutti et al., 2005; Torres et al., 2004), pome fruit viroids (Hurtt et al., 1996; Lolic et al., 2007), and CSVd (Tomassoli et al., 2004). The technique can be applied not only for diagnostic purposes, but also to study viroid distribution within infected plants (Amari et al., 2007; Stark-Lorenzen et al., 1997).

Tissue-printing hybridization has some limitations, the most prominent being the false negatives caused by low viroid titer in plant sap. Detection of citrus viroids by tissue printing is hampered by seasonal fluctuations in viroid titers, a limitation overcome using citron (*Citrus medica*) as an amplification host in which viroids accumulate to higher levels (Palacio et al., 2000). Subsequent analysis of the inoculated citrons was simplified by using tissue printing, which currently is the method for routine citrus indexing in Spain (Palacio et al., 2000). When detecting PLMVd by tissue-printing hybridization, WenXing et al. (2009) observed strong signals that could arise by interactions between the viroid probe and plant proteins, thus generating false positives. This problem was mitigated by including an additional step to remove host proteins.

IN SITU HYBRIDIZATION

Molecular hybridization techniques coupled with electron, light or confocal laser scanning microscopy are called in situ hybridization assays (Fig. 35.1D). They have been mostly applied to study viroid subcellular localization and trafficking. In situ hybridization revealed the nuclear and/or nucleolar localization of several members of the family *Pospiviroidae* (Bonfiglioli et al., 1996; Gambino et al., 2011; Matsushita et al., 2011; Zhang et al., 2014). Additionally, two representative members of the family *Avsunviroidae* (avocado sunblotch viroid and PLMVd) have been localized to host chloroplasts by in situ hybridization coupled to electron microscopy (Bonfiglioli et al., 1994; Bussière et al., 1999; Lima et al., 1994).

Viroid trafficking through plasmodesmata and vascular tissues has also been studied by in situ hybridization (for a review see Wang and Ding, 2010). The technique was used to show that PSTVd is generally excluded from shoot apical meristems (SAM) and other flower organs (excluding sepals) in tomato and *Nicotiana benthamiana* because of an antiviroid barrier likely based on RNA silencing (Di Serio et al., 2010; Zhu et al., 2001). However, under certain conditions the viroid is able to overcome this barrier (Matsushita et al., 2011). In situ hybridization has recently shown that PSTVd in petunia invades the embryo through the ovule or pollen before embryogenesis (Matsushita and Tsuda, 2014). PSTVd and tomato chlorotic dwarf viroid, sharing >85% nucleotide identity, exhibit different invasion patterns of the SAM in tomato plants. Only the former was detected by in situ hybridization in the ovules (Matsushita et al., 2011).

In situ hybridization also revealed that the extent of SAM invasion by CSVd in *Argyranthemum* plants depends on the infected cultivar (Zhang et al., 2014). In this same context, PLMVd, a chloroplast-replicating viroid, was shown to deeply invade the SAM of its natural host peach, leaving only a few uppermost cell layers free of viroid (Rodio et al., 2007). Therefore, in situ hybridization highlighted that several factors affect SAM invasion by viroids, prevalent among which is the viroid–host combination. This detection method also revealed that the high efficiency of somatic embryogenesis in eliminating viroids from several infected grapevine cultivars is due to embryos formed in viroid-infected calli remaining viroid-free (Gambino et al., 2011).

Acknowledgments

The work done in V.P. and J.A.S-N laboratories has been funded by the National Program of the Spanish Ministry of Innovation and Science (project BIO2014–54862-R) and in F.D.S. laboratory was supported by the Ministero dell'Economia e Finanze Italiano to the CNR (CISIA, Legge 191/ 2009).

References

Amari, K., Cañizares, M.C., Myrta, A., Sabanadzovic, S., Di Terlizzi, B., Pallás, V., 2001. Tracking hop stunt viroid (HSVd) infection in apricot trees during a whole year by non-isotopic tissue printing hybridization. Acta Hortic. 550, 315–320.

Amari, K., Ruiz, G., Gomez, G., Sanchez-Pina, M.A., Pallás, V., Egea, J., 2007. An important new apricot disease in Spain is associated with hop stunt viroid infection. Eur. J. Plant Pathol. 118, 173–181.

Astruc, N., Marcos, J.F., Macquaire, G., Candresse, T., Pallás, V., 1996. Studies on the diagnosis of hop stunt viroid in fruit trees: identification of new hosts and application of a nucleic acid extraction procedure based on non-organic solvents. Eur. J. Plant Pathol. 102, 837–846.

Bonfiglioli, R.G., Webb, D.R., Symons, R.H., 1996. Tissue and intracellular distribution of coconut cadang cadang viroid and citrus exocortis viroid determined by *in situ* hybridization and confocal laser scanning and transmission electron microscopy. Plant J. 9, 457–465.

Bonfiglioli, R.G., McFadden, G.I., Symons, R.H., 1994. In situ hybridization localizes avocado sunblotch viroid on chloroplast thylakoid membranes and coconut cadang cadang viroid in the nucleus. Plant J. 6, 99–103.

Bouani, A., Al Rwahnih, M., Abou Ghanem-Sabanadzovic, N., Alami, I., Zemzami, M., Myrta, A., et al., 2004. A preliminary account of the sanitary status of stone-fruit trees in Morocco. EPPO Bull. 34, 399–402.

Bussière, F., Lehoux, J., Thompson, D.A., Skrzeczkowski, L.J., Perreault, J.P., 1999. Subcellular localization and rolling circle replication of peach latent mosaic viroid: hallmarks of group A viroids. J. Virol. 73, 6357–6360.

Cañizares, M.C., Marcos, J.F., Pallás, V., 1998. Studies on the incidence of hop stunt viroid in apricot trees by using an easy and short extraction method to analyze a large number of plants. Acta Hortic. 472, 581–585.

Cañizares, M.C., Aparicio, F., Amari, K., Pallás, V., 2001. Studies on the aetiology of apricot "viruela" disease. Acta Hortic. 550, 249–258.

Cohen, O., Batuman, O., Stanbekova, G., Sano, T., Mawassi, M., Bar-Joseph, M., 2006. Construction of a multiprobe for the simultaneous detection of viroids infecting citrus trees. Virus Genes 33, 287–292.

Daròs, J.A., Marcos, J.F., Hernández, C., Flores, R., 1994. Replication of avocado sunblotch viroid: evidence for a symmetric pathway with two rolling circles and hammerhead ribozyme processing. Proc. Natl. Acad. Sci. USA 91, 12813–12817.

Delgado, S., Martínez de Alba, A.E., Hernández, C., Flores, R., 2005. A short double stranded RNA motif of peach latent mosaic viroid contains the initiation and the self cleavage sites of both polarity strands. J. Virol. 79, 12934–12943.

Di Serio, F., Malfitano, M., Alioto, D., Ragozzino, A., Desvignes, J.C., Flores, R., 2001. Apple dimple fruit viroid: fulfillment of Koch's postulates and symptom characteristics. Plant Dis. 85, 179–182.

Di Serio, F., Martínez de Alba, A.E., Navarro, B., Gisel, A., Flores, R., 2010. RNA-dependent RNA polymerase 6 delays accumulation and precludes meristem invasion of a viroid that replicates in the nucleus. J. Virol. 84, 2477–2489.

El-Dougdoug, K.A., Dawoud, R.A., Rezk, A.A., Sofy, A.R., 2012. Incidence of fruit tree viroid diseases by tissue print hybridization in Egypt. Internatl. J. Virol. 8, 114–120.

Gambino, G., Navarro, B., Vallania, R., Gribaudo, I., Di Serio, F., 2011. Somatic embryogenesis efficiently eliminates viroid infections from grapevines. Eur. J. Plant Pathol. 130, 511–519.

Gümüs, M., Paylan, I.C., Matic, S., Myrta, A., Sipahioglu, H.M., Erkan, S., 2007. Occurrence and distribution of stone fruit viruses and viroids in commercial plantings of Prunus species in western Anatolia, Turkey. J. Plant Path. 89, 265–268.

Hadidi, A., Olmos, A., Pasquini, G., Barba, M., Martin, R.R., Shamloul, A.M., 2011. Polymerase chain reaction for detection of systemic plant pathogens. In: Hadidi, A., Barba, M., Candresse, T., Jelkmann, W. (Eds.), Virus and Virus-Like Diseases of Pome and Stone Fruits. APS press, St. Paul, MN, pp. 343–362.

Hadidi, A., Yang, X., 1990. Detection of pome fruit viroids by enzymatic cDNA amplification. J. Virol. Methods 30, 261–269.

Hajizadeh, M., Navarro, B., Bashir, N.S., Torchetti, E.M., Di Serio, F., 2012. Development and validation of a multiplex RT-PCR method for the simultaneous detection of five grapevine viroids. J. Virol. Methods 179, 62–69.

Herranz, M.C., Sánchez-Navarro, J.A., Aparicio, F., Pallás, V., 2005. Simultaneous detection of six stone fruit viruses by non-isotopic molecular hybridization using a unique riboprobe or "polyprobe." J. Virol. Methods 124, 49–55.

Hooftman, R., Arts, M.J., Shamloul, A.M., Van Zaayen, A., Hadidi, A., 1996. Detection of chrysanthemum stunt viroid by reverse transcription polymerase chain reaction and by tissue blot hybridization. Acta Hortic. 432, 120–128.

Hull, R., 1993. Nucleic acid hybridization procedures. In: Matthews, R.E.F. (Ed.), Diagnosis of Plant Virus Diseases. CRC Press, Boca Raton, FL, pp. 295–314.

Hurtt, S.S., Podlekis, E.V., Howell, W.E., 1996. Integrated molecular and biological assays for rapid detection of apple scar skin viroid in pear. Plant Dis. 80, 458–462.

Itaya, A., Folimonov, A., Matsuda, Y., Nelson, R.S., Ding, B., 2001. Potato spindle tuber viroid as inducer of RNA silencing in infected tomato. Mol. Plant-Microbe Interact. 14, 1332–1334.

James, D., Varga, A., Pallás, V., Candresse, T., 2006. Strategies for simultaneous detection of multiple plant viruses. Can. J. Plant Pathol. 28, 1–15.

Jiang, D., Hou, W., Sano, T., Kang, N., Qin, L., Wu, Z., et al., 2013. Rapid detection and identification of viroids in the genus *Coleviroid* using a universal probe. J. Virol. Methods 187, 321–326.

Li, R., Mock, R., Huang, Q., Abad, J., Hartung, J., Kinard, G., 2008. A reliable and inexpensive method of nucleic acid extraction for the PCR-based detection of diverse plant pathogens. J. Virol. Methods 154, 48–55.

Lima, M.I., Fonseca, M.E., Flores, R., Kitajima, E.W., 1994. Detection of avocado sunblotch viroid in chloroplasts of avocado leaves by in situ hybridization. Arch. Virol. 138, 385–390.

Lin, L., Li, R., Mock, R., Kinard, G., 2011. Development of a polyprobe to detect six viroids of pome and stone fruit trees. J. Virol. Methods 171, 91–97.

Lolic, B., Afechtal, M., Matic, S., Myrta, A., Di Serio, F., 2007. Detection by tissue printing of pome fruit viroids and characterization of pear blister canker viroid in Bosnia and Herzegovina. J. Plant Pathol. 89, 369–375.

Malfitano, M., Di Serio, F., Covelli, L., Ragozzino, A., Hernández, C., Flores, R., 2003. Peach latent mosaic viroid variants inducing peach calico (extreme chlorosis) contain a characteristic insertion that is responsible for this symptomatology. Virology 313, 492–501.

Mandic, B., Al Rwahnih, M., Myrta, A., Gómez, G., Pallás, V., 2008. Incidence and genetic diversity of peach latent mosaic viroid and hop stunt viroid in stone fruits in Serbia. Eur. J. Plant Pathol. 120, 167–176.

Martínez de Alba, A.E., Flores, R., Hernández, C., 2002. Two chloroplastic viroids induce the accumulation of small RNAs associated with post transcriptional gene silencing. J. Virol. 76, 13094–13096.

Matsushita, Y., Tsuda, S., 2014. Distribution of potato spindle tuber viroid in reproductive organs of petunia during its developmental stages. Phytopathology 104, 964–969.

Matsushita, Y., Usugi, T., Tsuda, S., 2011. Distribution of tomato chlorotic dwarf viroid in floral organs of tomato. Eur. J. Plant Pathol. 130, 441–447.

Michelutti, R., Myrta, A., Pallás, V., 2005. A preliminary account on the sanitary status of stone fruits at the clonal genebank in Harrow, Canada. Phytopathol. Mediterr. 44, 71–74.

Minoia, S., Navarro, B., Delgado, S., Di Serio, F., Flores, R., 2015. Viroid RNA turnover: characterization of the subgenomic RNAs of potato spindle tuber viroid accumulating in infected tissues provides insights into decay pathways operating in vivo. Nucleic Acids Res. 43, 2313–2325.

Mühlbach, H.P., Weber, U., Gómez, G., Pallás, V., Duran-Vila, N., Hadidi, A., 2003. Molecular hybridization. In: Hadidi, A., Flores, R., Randles, J.W., Semancik, J.S. (Eds.), Viroids. CSIRO, Collingwood, VIC, pp. 103–114.

Murcia, N., Serra, P., Olmos, A., Duran-Vila, N., 2009. A novel hybridization approach for detection of citrus viroids. Mol. Cell Probes 23, 95–102.

Owens, R.A., Diener, T.O., 1981. Sensitive and rapid diagnosis of potato spindle tuber viroid disease by nucleic acid hybridization. Science 213, 670–672.

Palacio, A., Foissac, X., Duran-Vila, N., 2000. Indexing of citrus viroids by imprint hybridization. Eur. J. Plant Pathol. 105, 897–903.

Pallás, V., Faggioli, F., Aparico, F., Sánchez-Navarro, J.A., 2011. Molecular hybridization techniques for detecting and studying fruit tree viruses and viroids. In: Hadidi, A., Barba, M., Candresse, T., Jelkmann, W. (Eds.), Virus and Virus-Like Diseases of Pome and Stone Fruits. APS Press, St. Paul, MN, pp. 335–342.

Pallás, V., Más, P., Sánchez-Navarro, J.A., 1998. Detection of plant RNA viruses by non-isotopic dot-blot hybridization. In: Foster, G., Taylor, S. (Eds.), Plant Virus Protocols: From Virus Isolation to Transgenic Resistance. Humana Press, Totowa, NJ, pp. 461–468.

Peiró, A., Pallás, V., Sánchez-Navarro, J., 2012. Simultaneous detection of eight viruses and two viroids affecting stone fruit tree by using a unique polyprobe. Eur. J. Plant Pathol. 132, 469–475.

Podleckis, E.V., Hammond, R.W., Hurtt, S.S., Hadidi, A., 1993. Chemiluminescent detection of potato and pome fruit viroids by digoxigenin-labeled dot blot tissue blot hybridization. J. Virol. Methods 43, 147–158.

Rodio, M.E., Delgado, S., De Stradis, A., Gómez, M.D., Flores, R., Di Serio, F., 2007. A viroid RNA with a specific structural motif inhibits chloroplast development. Plant Cell 19, 3610–3626.

Romero-Durban, J., Cambra, M., Duran-Vila, N., 1995. A simple imprint hybridization method for detection of viroids. J. Virol. Methods 55, 37–47.

Semancik, J.S., Szychowski, J.A., Rakowski, A.G., Symons, R.H., 1994. A stable 463 nucleotide variant of citrus exocortis viroid produced by terminal repeats. J. Gen. Virol. 75, 727–732.

Sorrentino, R., Torchetti, E.M., Minutolo, M., Di Serio, F., Alioto, D., 2015. First report of iresine viroid 1 in ornamental plants in Italy and of Celosia cristata as a novel natural host. Plant Dis. 99, 1655.

Stark-Lorenzen, P., Guitton, M.-C., Werner, R., Mühlbach, H.-P., 1997. Detection and tissue distribution of potato spindle tuber viroid in infected tomato plants by tissue print hybridization. Arch. Virol. 142, 1289–1296.

Tomassoli, L., Faggioli, F., Zaccaria, A., Caccia, R., Albani, M., Barba, M., 2004. Molecular diagnosis of chrysanthemum stunt viroid for routine indexing. Phytopath. Mediterr. 43, 285–288.

Torchetti, E., Navarro, B., Di Serio, F., 2012. A single polyprobe for detecting simultaneously eight pospiviroids infecting ornamentals and vegetables. J. Virol. Methods 186, 141–146.

Torres, H., Gómez, G., Stamo, B., Shalaby, A., Aoune, B., Gavriel, I., et al., 2004. Detection by tissue printing of stone fruit viroids from Europe, the Mediterranean and North and South America. Acta Hortic. 657, 379–384.

Wang, Y., Ding, B., 2010. Viroids: small probes for exploring the vast universe of RNA trafficking in plants. J. Integr. Plant Biol. 52, 28–39.

WenXing, X., Ni, H., QiuTing, J., Farooq, A.B.U., ZeQiong, W., YanSu, S., et al., 2009. Probe binding to host proteins: a cause for false positive signals in viroid detection by tissue hybridization. Virus Res. 145, 26–30.

Zhang, Z., Lee, Y., Spetz, C., Clarke, J.L., Wang, Q., Blystad, D.R., 2014. Invasion of shoot apical meristems by chrysanthemum stunt viroid differs among Argyranthemum cultivars. Front Plant Sci. 6, 53.

Zhang, Z., Peng, S., Jiang, D., Pan, S., Wang, H., Li, S., 2012. Development of a polyprobe for the simultaneous detection of four grapevine viroids in grapevine plants. Eur. J. Plant Pathol. 132, 9–16.

Zhu, Y., Green, L., Woo, Y.M., Owens, R., Ding, B., 2001. Cellular basis of potato spindle tuber viroid systemic movement. Virology 279, 69–77.

CHAPTER

36

Viroid Amplification Methods: RT-PCR, Real-Time RT-PCR, and RT-LAMP

Francesco Faggioli[1], *Marta Luigi*[1] *and Iraklis N. Boubourakas*[2]

[1]CREA-Research Centre for Plant Protection and Certification, Rome, Italy
[2]Directorate of Rural Economy and Veterinary of Piraeus, Attica Prefecture, Attica, Greece

INTRODUCTION

Viroids are the first plant pathogens for which the reverse transcription-polymerase chain reaction (RT-PCR) was utilized (Hadidi and Yang, 1990). Their amplification methods have increased in number with the development and application of real-time RT-PCR (Boonham et al., 2004), RT-Loop mediated isothermal AMPlification (RT-LAMP) (Fukuta et al., 2005), and the combination of different techniques such as RT-PCR-Enzyme-Linked Immuno Sorbent Assay (ELISA) (Shamloul et al., 2002), multiplex RT-PCR (Ito et al., 2002; Levy et al., 1992), multiplex bead-based array (van Brunschot et al., 2014), and in situ RT-PCR (Boubourakas et al., 2011).

VIROID AMPLIFICATION PROCEDURES

The first common critical step of the viroid amplification procedures is the efficient recovery of target viroid RNA with minimal amounts of plant inhibitors such as polyphenols and polysaccharides (Li et al., 2008).

The next step is the production of a single-stranded cDNA of the RNA target using reverse transcriptase and an oligonucleotide primer. The primer anneals to the RNA, and the cDNA is extended toward the 5′ end of the RNA through the RNA-dependent DNA polymerase activity of

Viroids and Satellites.
DOI: http://dx.doi.org/10.1016/B978-0-12-801498-1.00036-X

reverse transcriptase. Primer design is a key step for developing an amplification protocol. Designing the optimal primers entails a trade-off of a variety of parameters, including melting temperature, string-based alignment scores for complementarity, primer length, and GC content (Burpo, 2001). Many bioinformatic tools exist for primer design, but most viroid genomes have some specific characteristics: they are short, circular, rich in GC, have a stable secondary structure, and contain short palindromic sequences (Flores et al., 2005).

The PCR mix consists of the cDNA, DNA polymerase buffer, deoxyribonucleotide triphosphates (dNTPs), primers, and Taq DNA polymerase, but the dynamics and physical chemistry of the amplification reaction itself are complex and many factors affect the specificity and efficiency of DNA amplification.

VIROID RNA PREPARATION

Many methods have been developed for nucleic acids extraction, most of them are specific for a pathogen/host combination but some characteristics are universal.

The first challenge for RNA isolation is the disruption of the host cell wall and the inactivation of ribonuclease. The second is the purification of nucleic acid from other plant metabolites. Two different approaches can be used: a separation between two liquid phases or on solid-phase. One of the most suitable liquid mixtures used is water–saturated phenol (or derivations) at neutral pH. A detergent like cetyltrimethylammonium bromide can be used especially in the case of tissues that produce large quantities of polysaccharides (Sambrook and Russell, 2001).

The separation in solid-phase systems takes advantage of the adsorption properties of silica matrices for selective binding with high affinity of the negatively charged DNA or RNA backbone. Silica capture-based methods can be either homemade (Foissac et al., 2001) or obtained using several commercial extraction kits.

Nitrocellulose and polyamide membranes (especially if positively charged) also bind nucleic acids, but with less specificity (Olmos et al., 2005). This method is fast and cheap, but susceptible to false negative results due to low amounts of RNA released or high amounts of amplification inhibitors. Nevertheless, there are cases where plant sap or even traces of plant tissue taken with a sterilized toothpick were also used as template source for the detection of viroids by RT-PCR and RT-LAMP, respectively (Boubourakas et al., 2010; Fukuta et al., 2005; Jiang et al., 2011; Tsutsumi et al., 2010).

END POINT RT-PCR

After reverse transcription, the amplification reaction is composed of three essential steps: (1) melting of the target; (2) annealing of two oligonucleotide primers to the denatured DNA strands; and (3) primer extension by a thermostable DNA polymerase (Saiki et al., 1988).

Generally the primers are designed to be exactly complementary or homologous to the target sequence and able to produce an amplicon with a known length only when the specific targets are present. Sometimes, to have a more generic detection (genus or family identification), the primers can be designed for conserved regions and some degenerate nucleotides can be utilized (Faggioli et al., 2001; Luigi et al., 2014; Olivier et al., 2014; Verhoeven et al., 2004). Another interesting characteristic of viroid cDNA amplification is linked to the small size and circularity of viroid genomes, which permits amplification of their full-length cDNAs by using adjacent primers.

Detection of amplification products is performed at the end of the reaction, hence the name "end point RT-PCR," generally using gel electrophoresis followed by staining with an intercalating dye.

In viroid diagnosis, end point RT-PCR is used in two- (Hadidi and Yang, 1990) or one-step reactions (Mumford et al., 2000; Ragozzino et al., 2004; Shamloul et al., 1997), or in the polyvalent detection by a mixture of specific primer sets in multiplex amplification (Hajizadeh et al., 2012; Ito et al., 2002; Levy et al., 1992; Wang et al., 2009). Positive results can be visualized using classical gel electrophoresis or with fluorescent (Di Serio et al., 2002) and standard or multiplex RT-PCR-ELISA (Shamloul and Hadidi, 1999; Shamloul et al., 2002). Recently, another modified RT-PCR protocol based on the utilization of SYBR Green and pepsin, DNase I pretreated, and formalin-fixed peach leaf sections was developed for the *in situ* localization of peach latent mosaic viroid (Boubourakas et al., 2011).

REAL-TIME RT-PCR

Real-time RT-PCR has become one of the most widely used tools for nucleic acid detection and quantification (when it is known as qRT-PCR; Luigi and Faggioli, 2011, 2013). Two types of chemicals are preferentially used to detect PCR products. In the first case, nonspecific dyes (generally SYBR green I; Ramos-Payan et al., 2003) are used, which are able to bind to generic dsDNA and emit fluorescence in proportion to the dsDNA concentration (Wittwer et al., 1997). In the second case, a third oligonucleotide probe, dually labeled with a reporter fluorescent dye at the 5′ end and a quencher to the 3′ end, is added to the PCR reaction mixture. During the reaction, the 5′ to 3′ exonuclease activity of the Taq DNA polymerase degrades the probe and releases the reporter dye, which is no longer silenced by the quencher. The resulting fluorescent signal emitted is detected in "real time" during each amplification cycles, and increases proportionally to the amount of amplicon produced (Fig. 36.1) (Gibson et al., 1996; Heid et al., 1996).

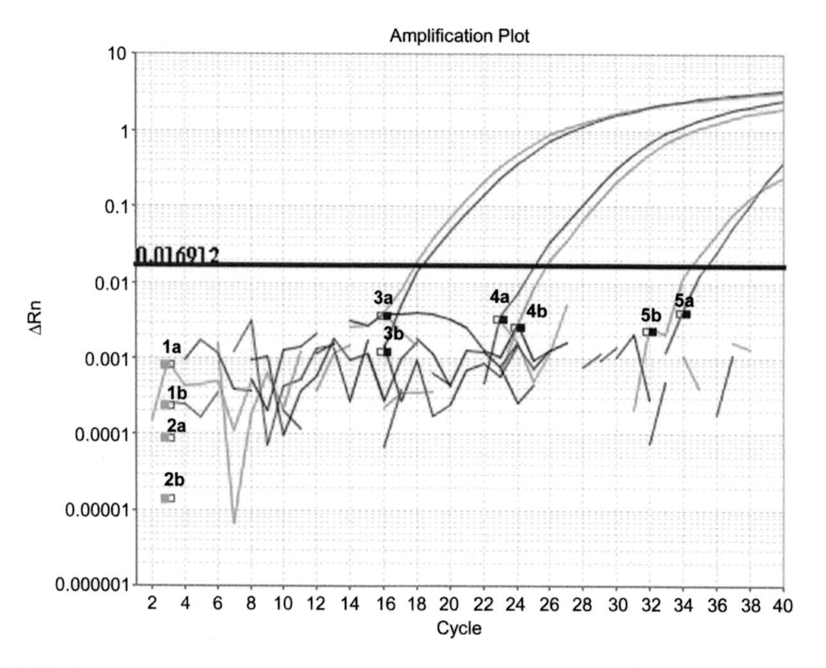

FIGURE 36.1 Results of a real-time RT-PCR experiment for the detection of peach latent mosaic viroid in unknown peach samples. Samples 1, 2 are negative; samples 3, 4, and 5 are positive. The samples were loaded in duplicate (a and b). *Source: Photo by F. Faggioli—CREA.*

Real-time RT-PCR assays have been applied to viroid research for detection, evaluation of plant resistance, transmission, and genotyping (Boonham et al., 2004; Botermans et al., 2013; Loconsole et al., 2013; Luigi and Faggioli, 2011, 2013; Nielsen et al., 2012; Tessitori et al., 2005). Since this technique allows quantification of nucleic acid targets, there are some factors that are required for the success of the analysis: (1) high RNA quality, which is a prerequisite to generate valid quantification data; (2) primer and probe design, since the optimal melting temperature influences the specificity and can affect the quantification results; and (3) instrument calibration, which determines the threshold settings (Bustin and Mueller, 2005).

RT-LAMP

Loop-mediated isothermal AMPlification (LAMP), (Notomi et al., 2000) is a nucleic acid amplification method performed in the presence of at least four different primers (Fig. 36.2), specifically designed to recognize six distinct regions on the target under isothermal conditions (60–65°C) (Nagamine et al., 2002; Notomi et al., 2000). The amplification

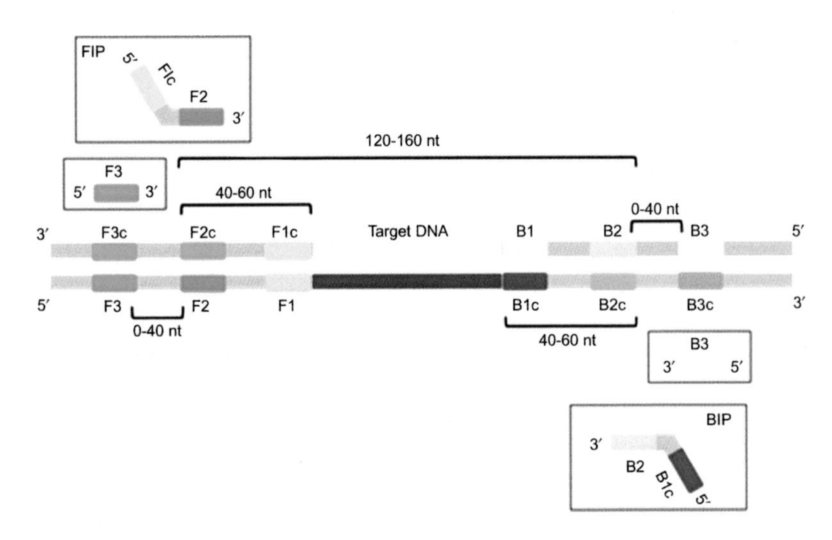

FIGURE 36.2 Target region and distances of RT-LAMP primers. *Source: The figure is a modification of that presented on the EIKEN GENOME SITE: http://loopamp.eiken.co.jp/e/lamp/primer.html.*

can be preceded by reverse transcription (RT-LAMP) either in a separate tube (Park et al., 2013; Thanarajoo et al., 2014) or simultaneously (Boubourakas et al., 2009; Tsutsumi et al., 2010) using reverse transcriptases remaining active at 60–65°C.

Because of its speed, robustness, and simplicity of use, RT-LAMP is gaining popularity, first in diagnostics in human medicine (Parida et al., 2008) and then in plant health (Boubourakas et al., 2009; Fukuta et al., 2005; Lenarcic et al., 2012; Liu et al., 2014; Park et al., 2013; Thanarajoo et al., 2014; Tsutsumi et al., 2010).

F3 and B3 primers have a role only during the initial noncycling step of the LAMP, when DNA with stem loops at each end serve as the starting structure for the amplification. During the cycling FIP and BIP, composed of two distinct sequences corresponding to the sense and antisense sequences of the target, act in loop formation (Fig. 36.2) (Notomi et al., 2000). To accelerate the reaction and increase the specificity an optional couple of primers designed between the regions B1 and B2, or the regions F1 and F2, can be added to the reaction (Nagamine et al., 2002).

The key factors for the LAMP primer design are the following: (1) the Tm for F3, B3, F2, B2, and loop primers should be approximately 60°C, while for F1 and B1 the Tm should be around 65°C; (2) sequence regions with substantial secondary structure should be avoided; (3) GC content of the primers should be about 50%–60% in the case of GC-rich and about 40%–50% for AT-rich targets; (4) special care also must be taken so that the distance between primer ends of F2 and B2 is between 120 and 160 bases and that of F2 and F1 is between 40 and 60 bases

(Fig. 36.2). Primer design may be performed manually, or using software design programs: Primer Explorer V4 (Eiken Chemical Company Ltd, Japan), LAMP Designer (Premier Biosoft, USA), and LAMP Designer (OptiGene, UK).

The role of the DNA polymerase is crucial. The *Bst* polymerase isolated from *Bacillus stearothermophilus*, exhibits DNA polymerizing capacity along with the ability to displace the DNA strand. The recently released hot-start *Bst* 2.0 WarmStart is considered appropriate for templates with high GC content like viroids (Tanner et al., 2012). Other components required for RT-LAMP are: a high amount of dNTPs, magnesium sulfate and betaine used to reduce base stacking and to also increase target selectivity (Soliman and El-Matbouli, 2005). The final products of the LAMP are double-stranded stem-loop DNAs with several inverted repeats of the target and cauliflower-like structures with multiple loops (Nagamine et al., 2002; Notomi et al., 2000).

RT-LAMP reaction produces a high DNA yield (>10 μg) and the large quantity of magnesium pyrophosphate generated as a by-product allows direct visual observation as a white precipitate in positive reactions (Tsutsumi et al., 2010). End point or even real-time turbidity measurements at 400 nm are applicable (Boubourakas et al., 2009; Thanarajoo et al., 2014). LAMP products also can be analyzed by gel electrophoresis; staining reveals a ladder-like pattern due to the cauliflower-like structures obtained (Parida et al., 2008). Color change methods using SYBR Green or other intercalating dyes can be also used (Park et al., 2013). Unlike SYBR Green, metal ion indicators like calcein (Loopamp Fluorescent Detection Reagent) and hydroxynaphthol blue are added directly to the initial reaction mixture, so that the system can be enclosed in a single tube, and carry-over contamination is minimized (Boubourakas et al., 2010; Thanarajoo et al., 2014).

More protocols for the RT-LAMP diagnosis of viroids and other plant pathogens are likely to become available. RT-LAMP can contribute greatly to the transition of the diagnosis procedure from the laboratory bench to the field. Its simplicity, ease of operation, and robustness can be exploited for the enhancement of viroid diagnosis in developing countries.

FINAL CONSIDERATIONS

It is not easy to make an evaluation of different viroid amplification methods, because each has its own characteristics and should be chosen depending on the purpose to be achieved. The main advantages and disadvantages of each technique are presented in Table 36.1, so that it is possible to make an evaluation and a comparison of methods described in this chapter.

TABLE 36.1 Comparison of the Main Advantages and Disadvantages of Viroid Amplification Methods

	End point RT-PCR	**Real-time RT-PCR**	**RT-LAMP**
Advantages	Widely used and well established molecular diagnosis format	Simultaneous amplification and detection	Isothermal amplification, no requirement for expensive thermal cyclers
	Variety of PCR equipment consumables and primer design software	Closed tube system, fewer contaminations	High sensitivity
	Easy design of primers	Quantification of viroid load	High specificity
	Multiplex protocols can be easily developed	Higher sensitivity	Fast
	Possibility to sequence the entire genome	High throughput analysis with specific software	Cheap equipment and reagents
		High sequence specificity	Closed tube system and naked eye observation
		No post-PCR processing	Field applied diagnosis
			Bst polymerase is not easily affected by plant polyphenols and polysaccharides
			Qualitative and real time quantification
Disadvantages	Time-consuming, post-PCR processing	Expensive detection equipment and consumables	Complicated primer design
	Qualitative, results are not expressed as numbers and are based on size discrimination	Equipment not portable	Restricted availability of equipment and reagents
	Requirement of thermal cycler and gel documentation system	Need for high template quality	Multiplex protocols are difficult to be developed

(*Continued*)

TABLE 36.1 (Continued)

	End point RT-PCR	Real-time RT-PCR	RT-LAMP
	Negatively affected by plant polyphenols and polysaccharides Open tube system Less sensitive Low resolution		Susceptible to carry-over and cross-contamination

Acknowledgments

The authors would like to thank Dr. Ahmed Hadidi for his suggestions and critical reading of the chapter.

References

Boonham, N., González Pérez, L., Mendez, M.S., Lilia Peralta, E., Blockley, A., Walsh, K., et al., 2004. Development of a real-time RT-PCR assay for the detection of potato spindle tuber viroid. J. Virol. Methods 116, 139–146.

Botermans, M., van de Vossenberg, B.T.L.H., Verhoeven, J.Th.J., Roenhorst, J.W., Hooftman, M., et al., 2013. Development and validation of a real-time RT-PCR assay for generic detection of Pospiviroids. J. Virol. Methods 187, 43–50.

Boubourakas, I.N., Fukuta, S., Kyriakopoulou, P., 2009. Sensitive and rapid detection of peach latent mosaic viroid by the reverse transcription loop-mediated isothermal amplification. J. Virol. Methods 160, 63–68.

Boubourakas, I.N., Fukuta, S., Luigi, M., Faggioli, F., Barba, M., Kyriakopoulou, P.E., 2010. Improvement of the reverse transcription loop mediated isothermal amplification (RT-LAMP) method for the detection of peach latent mosaic viroid (PLMVd). Julius-Kuhn-Archiv. 427, 65–69.

Boubourakas, I.N., Voloudakis, A.E., Fasseas, K., Resnick, N., Koltai, H., Kyriakopoulou, P.E., 2011. Cellular localization of peach latent mosaic viroid in peach sections by liquid phase *in situ* RT-PCR. Plant Pathol. 60, 468–473.

Burpo, F.J., 2001. A critical review of PCR primer design algorithms and cross hybridization case study. Biochemistry 218, 1–10.

Bustin, S.A., Mueller, R., 2005. Real-time reverse transcription PCR (qRT-PCR) and its potential use in clinical diagnosis. Clin. Sci. 109, 365–379.

Di Serio, F., Malfitano, M., Alioto, D., Ragozzino, A., Flores, R., 2002. Apple dimple fruit viroid: sequence variability and its specific detection by multiplex fluorescent RT-PCR in the presence of apple scar skin viroid. J. Plant Pathol. 84, 27–34.

Faggioli, F., Ragozzino, E., Barba, M., 2001. Simultaneous detection of pome fruit viroids by single tube RT-PCR. Acta Hortic. 550, 59–63.

Flores, R., Hernández, C., Martínez de Alba, A.E., Daròs, J.A., Di Serio, F., 2005. Viroids and viroid–host interactions. Annu. Rev. Phytopathol. 43, 117–139.

Foissac, X., Svanella-Dumas, L., Dulucq, M.J., Gentit, P., Candresse, T., 2001. Polyvalent detection of fruit tree Tricho, Capillo and Foveaviruses by nested RT-PCR using degenerated and inosine containing primer (PDO RT-PCR). Acta Hortic. 550, 37–43.

Fukuta, S., Niimi, Y., Oishi, Y., Yoshimura, Y., Anai, N., Hotta, M., et al., 2005. Development of reverse transcription loop-mediated isothermal amplification (RT-LAMP) method for detection of two viruses and chrysanthemum stunt viroid. Annu. Rep. Kansai Plant Protect. Soc. 47, 31–36.

Gibson, U.E., Heid, C.A., Williams, P.M., 1996. A novel method for real time quantitative RT-PCR. Genome Res. 6, 995–1001.

Hadidi, A., Yang, X., 1990. Detection of pome fruit viroids by enzymatic cDNA amplification. J. Virol. Methods 30, 261–270.

Hajizadeh, M., Navarro, B., Bashir, N.S., Torchetti, E.M., Di Serio, F., 2012. Development and validation of a multiplex RT-PCR method for the simultaneous detection of five grapevine viroids. J. Virol. Methods 179, 62–69.

Heid, C.A., Stevens, J., Livak, K.J., Williams, P.M., 1996. Real time quantitative PCR. Genome Res. 6, 986–994.

Ito, T., Ieki, H., Ozaki, K., 2002. Simultaneous detection of six citrus viroids and apple stem grooving virus from citrus plants by multiplex reverse transcription polymerase chain reaction. J. Virol. Methods 106, 235–239.

Jiang, D., Wu, Z., Sano, T., Li, S., 2011. Sap-direct RT-PCR for the rapid detection of coleus blumei viroids of the genus *Coleviroid* from natural host plants. J. Virol. Methods 174, 123–127.

Lenarcic, R., Morisset, D., Mehle, N., Ravnikar, M., 2012. Fast real-time detection of potato spindle tuber viroid by RT-LAMP. Plant Pathol. 62, 1147–1156.

Levy, L., Hadidi, A., Garnsey, S.M., 1992. Reverse-transcription polymerase chain reaction assays for the rapid detection of citrus viroids using multiplex primer sets. Proc. Int. Soc. Citriculture 2, 800–803.

Li, R., Mock, R., Huang, Q., Abad, J., Hartung, J., Kinard, G., 2008. A reliable and inexpensive method of nucleic acid extraction for the PCR-based detection of diverse plant pathogens. J. Virol. Methods 154, 48–55.

Liu, X.L., Zhao, X.T., Muhammad, I., Ge, B.B., Hong, B., 2014. Multiplex reverse transcription loop-mediated isothermal amplification for the simultaneous detection of CBV and CSVd in chrysanthemum. J. Virol. Methods 210, 26–31.

Loconsole, G., Onelge, N., Yokomi, R.K., Kubaa, R.A., Savino, V., Saponari, M., 2013. Rapid differentiation of citrus hop stunt viroid variants by real-time RT-PCR and high resolution melting analysis. Mol. Cell. Probes 27, 221–229.

Luigi, M., Costantini, E., Luison, D., Mangiaracina, P., Tomassoli, L., Faggioli, F., 2014. A diagnostic method for the simultaneous detection and identification of pospiviroids. J. Plant Pathol. 96, 151–158.

Luigi, M., Faggioli, F., 2011. Development of quantitative real-time RT-PCR for the detection and quantification of peach latent mosaic viroid. Eur. J. Plant Pathol. 130, 109–116.

Luigi, M., Faggioli, F., 2013. Development of a quantitative Real-Time RT-PCR (qRT-PCR) for the detection of hop stunt viroid. Eur. J. Plant Pathol. 137, 231–235.

Mumford, R.A., Walsh, K., Boonham, N., 2000. A comparison of molecular methods for the routine detection of viroids. Bull. OEPP/EPPO Bull. 30, 431–435.

Nagamine, K., Kuzihara, Y., Notomi, T., 2002. Isolation of single-stranded DNA from loop-mediated isothermal amplification products. Biochem. Biophys. Res. Commun. 290, 1195–1198.

Nielsen, S.L., Enkegaard, A., Nicolasen, M., Kryger, P., Virscek Marn, M., Plesko, I.M., et al., 2012. No transmission of potato spindle tuber viroid shown in experiments with thrips (*Frankliniella occidentalis, Thrips tabaci*), honey bees (*Apis mellifera*) and bumblebees (*Bombus terrestris*). Eur. J. Plant Pathol. 133, 505–509.

Notomi, T., Okayama, H., Masubuchi, H., Yonekawa, T., Watanabe, K., Amino, N., et al., 2000. Loop-mediated isothermal amplification of DNA. Nucleic Acids Res. 28, 63–73.

Olivier, T., Demonty, E., Fauche, F., Steyer, S., 2014. Generic detection and identification of pospiviroids. Arch. Virol. 159, 2097–2102.

Olmos, A., Bertolini, E., Gil, M., Cambra, M., 2005. Real time assay for quantitative detection of non-persistently transmitted plum pox virus RNA targets in single aphids. J. Virol. Methods 128, 151–155.

Parida, M., Sannarangaiah, S., Dash, P.K., Rao, P.V.L., Morita, K., 2008. Loop mediated isothermal amplification (LAMP): a new generation of innovative gene amplification technique; perspectives in clinical diagnosis of infectious diseases. Rev. Med. Virol. 18, 407–421.

Park, J., Jung, Y., Kil, E.-J., Kim, J., Thi Tran, D., Choi, S.-K., et al., 2013. Loop-mediated isothermal amplification for the rapid detection of chrysanthemum chlorotic mottle viroid (CChMVd). J. Virol. Methods 193, 232–237.

Ragozzino, E., Faggioli, F., Barba, M., 2004. Development of a one tube-one step RT-PCR protocol for the detection of seven viroids in four genera: *Apscaviroid*, *Hostuviroid*, *Pelamoviroid* and *Pospiviroid*. J. Virol. Methods 121, 25–29.

Ramos-Payan, R., Aguilar-Medina, M., Estrada-Parra, S., Gonzalez, Y.M.J.A., Favila-Castillo, L., Monroy-Ostria, A., et al., 2003. Quantification of cytokine gene expression using an economical real-time polymerase chain reaction method based on SYBR Green I. Scandin. J. Immunol. 57, 439–445.

Saiki, R.K., Gelfand, G.H., Stoffel, S., Scharf, S.J., Higuchi, R., Horn, G., 1988. Primer directed enzymatic amplification of DNA with a thermostable DNA polymerase. Science 239, 487–491.

Sambrook, J., Russell, D., 2001. Molecular Cloning: A Laboratory Manual, vol. 1, third ed. Cold Spring Harbor Laboratory Press, New York, NY.

Shamloul, A.M., Faggioli, F., Keith, J.M., Hadidi, A., 2002. A novel multiplex RT-PCR probe capture hybridization (RT-PCR-ELISA) for simultaneous detection of six viroids in four genera: *Apscaviroid*, *Hostuviroid*, *Pelamoviroid* and *Pospiviroid*. J. Virol. Methods 105, 115–121.

Shamloul, A.M., Hadidi, A., 1999. Sensitive detection of potato spindle tuber and temperature fruit tree viroids by reverse transcription-polymerase chain reaction-probe capture hybridization. J. Virol. Methods 80, 145–155.

Shamloul, A.M., Hadidi, A., Zhu, S.F., Singh, R.P., Sagredo, B., 1997. Sensitive detection of potato spindle tuber viroid using RT-PCR and identification of a viroid variant naturally infecting pepino plants. Can. J. Plant Pathol. 19, 89–96.

Soliman, H., El-Matbouli, M., 2005. An inexpensive and rapid diagnostic method of Koi Herpesvirus (KHV) infection by loop-mediated isothermal amplification. Virol. J. 2, 83.

Tanner, N.A., Zhang, Y., Evans Jr., T.C., 2012. Simultaneous multiple target detection in real-time loop-mediated isothermal amplification. BioTechniques 53, 81–89.

Tessitori, M., Rizza, S., Reina, A., Catara, A., 2005. Real-Time RT-PCR based on SYBR-Green I for the detection of citrus exocortis and citrus cachexia diseases. In: Hilf, M.E., Duran Vila, N., Rocha, P.M.A. (Eds.). Proc. 16th IOCV Conf. Riverside, CA, pp. 456–459.

Thanarajoo, S.S., Kong, L.L., Kadir, J., Lau, W.H., Vadamalai, G., 2014. Detection of coconut cadang-cadang viroid (CCCVd) in oil palm by reverse transcription loop-mediated isothermal amplification (RT-LAMP). J. Virol. Methods 202, 19–23.

Tsutsumi, N., Yanagisawa, H., Fujiwara, Y., Ohara, T., 2010. Detection of potato spindle tuber viroid by reverse transcription loop mediated isothermal amplification. Res. Bull. Plant Prot. Serv. Jpn. 46, 61–67.

van Brunschot, S.L., Bergervoet, J.H.W., Pagendam, D.E., de Weerdt, M., Geering, A.D.W., Drenth, A., et al., 2014. Development of a multiplexed bead-based suspension array for the detection and discrimination of pospiviroid plant pathogens. PLoS One 9, e84743.

Verhoeven, J.Th.J., Jansen, C.C.C., Willemen, T.M., Kox, L.F.F., Owens, R.A., et al., 2004. Natural infections of tomato by citrus exocortis viroid, columnea latent viroid, potato spindle tuber viroid and tomato chlorotic dwarf viroid. Eur. J. Plant Pathol. 110, 823–831.

Wang, X.F., Zhou, C.Y., Tang, K.Z., Zhou, Y., Li, Z.G., 2009. A rapid one-step multiplex RT-PCR assay for the simultaneous detection of five citrus viroids in China. Eur. J. Plant Pathol. 124, 175–180.

Wittwer, C.T., Herrmann, M.G., Moss, A.A., Rasmussen, R.P., 1997. Continuous fluorescence monitoring of rapid cycle DNA amplification. BioTechniques 22, 130–138.

CHAPTER

37

Detection and Identification of Viroids by Microarrays

Shuifang Zhu[1], Yongjiang Zhang[1], Antonio Tiberini[2] and Marina Barba[3]

[1]Chinese Academy of Inspection and Quarantine, Beijing, China
[2]Mediterranea University of Reggio Calabria, Reggio Calabria, Italy
[3]CREA-Research Centre for Plant Protection and Certification, Rome, Italy

INTRODUCTION

The microarray technology was initially developed to investigate changes in messenger RNA accumulation (Schena et al., 1995). The principle is the base-pairing of complementary sequences by hybridization, as with reverse dot-blot hybridization where a DNA or RNA target (previously treated and labeled) hybridizes to a specific complementary DNA fragment (probe) immobilized and spatially ordered onto a solid support. Prior to hybridization, the DNA/RNA samples to be tested are usually labeled by an enzymatic reaction with either fluorescent or chemoreactive compounds. This reaction allows them to be identified after hybridization has occurred.

Microarray applications cover broad genomic areas such as sequence analysis, gene expression studies, gene typing, and large-scale polymorphism screening. Due to the explosion of information arising from sequencing of plant pathogens, it became intuitive to focus the attention for possible applications of microarrays in multiple and simultaneous detection of plant pathogens. The inherently parallel nature of a microarray is ideal for both multilocus detection of a single agent as well as for simultaneous analysis of a nearly unlimited number of pathogens.

Thirty-one plant viroid species in eight genera and one unclassified viroid species were listed in the last report of the International

Viroids and Satellites.
DOI: http://dx.doi.org/10.1016/B978-0-12-801498-1.00037-1

Committee on Taxonomy of Viruses (Owens et al., 2012; Chapter 13: Viroid Taxonomy) infecting economically important horticultural, field, and ornamental crops. Theoretically, microarrays provide an ideal format for the development of multiviroid diagnostics. A microarray may be designed and developed using the nucleotide sequence data of viroids from public databases, thus allowing the simultaneous detection of all these viroids. Related viroids may be distinguished by their unique pattern of hybridization. Moreover, by selecting microarray elements derived from highly conserved regions within viroid families, genera, or species, individual viroids that are not explicitly represented on the microarray can still be detected, raising the possibility that this approach may be used for viroid discovery, especially those involved in plant diseases of uncertain etiology. Viroids that have been detected by microarray are listed in Table 37.1.

MICROARRAY CONSTRUCTION

The rigid support most commonly used for microarray construction is the classical microscope slide, which has two advantages: the intrinsic very low fluorescence and the ability to deposit the probes in spots of very small size allowing a high density. Generally, the glass is treated before use with poly-L-lysine(or 3-aminopropyltrimethoxysilane, 3-glycidoxypropyltrimethoxysilane, and an aldehyde or carboxylic acid or any other reagent that binds covalently to the probe DNA). A further development of this strategy consists of treating the glass surface with an aldehyde or epoxysilane and immobilizing the probes, previously modified by incorporation of an amino group at their ends (Zammatteo et al., 2000); in this way not only are the probes fixed to a rigid support, but they are also oriented in space.

The oligonucleotide arrays are deposited on 75 × 25 mm glass slides using an Arrayer spot system. All probes are spotted three times on the microarray. Several control probes are used to monitor the performance of the microarray. (1) The 5′ amino-linker oligonucleotide (5′-$(T)_{15}$-CTCATGCCCATGCCGATGC-3′) has a sequence that does not potentially hybridize with any viroid sequence. The targeted sequence of this probe is mixed with the samples before hybridization and used as positive control (PC) for monitoring the probe hybridization efficiency. (2) The internal control (HEX) is spotted on the microarray to monitor the probes linking to the chips. It is the 5′ Hex-linker oligonucleotide (5′-GTCACATGCGATGGATCGAGCTCCTTTATCATCGTTCCCACCTTAA-TGCA-3′). (3) Probes designed to match the highly conserved regions in plant 18S rRNAs are used as PCs to detect host plant rRNAs (Abdullahi et al., 2005). (4) The spotting buffer is used as the negative control (NC).

TABLE 37.1 Viroids Detected by Microarrays

Family/ genus	Species	Host	Reference
POSPIVIROIDAE			
Pospiviroid	Tomato apical stunt viroid (TASVd)	Tomato	Tiberini and Barba (2012), Zhang et al. (2013)
	Potato spindle tuber viroid (PSTVd)	Potato	Agindotan and Perry (2008), Zhang et al. (2013)
		Tomato	Tiberini and Barba (2012)
		Orobanche ramosa	Ivanova et al. (2010)
	Tomato chlorotic dwarf viroid (TCDVd)	Tomato	Tiberini and Barba (2012)
	Citrus exocortis viroid (CEVd)	Tomato	Tiberini and Barba (2012)
		Citrus	Zhang et al. 2013
	Tomato planta macho viroid (TPMVd)	Tomato	Tiberini and Barba (2012), Zhang et al. (2013)
	Columnea latent viroid (CLVd)	Tomato	Tiberini and Barba (2012)
		Columnea	Zhang et al. (2013)
	Chrysanthemum stunt viroid (CSVd)	Chrysanthemum	Zhang et al. (2013)
Apscaviroid	Apple scar skin viroid (ASSVd)	Apple	Zhang et al. (2013)
	Citrus dwarfing viroid (CDVd)	Citrus	Zhang et al. (2013)
Cocadviroid	Hop latent viroid (HLVd)	Hop	Zhang et al. (2013)
Coleviroid	Coleus blumei viroid 1 (CbVd-1)	Coleus	Zhang et al. (2013)
Hostuviroid	Hop stunt viroid (HSVd)	Hop	Zhang et al. (2013)
AVSUNVIROIDAE			
Avsunviroid	Avocado sunblotch viroid (ASBVd)	Avocado	Zhang et al. (2013)
Pelamoviroid	Chrysanthemum chlorotic mottle viroid (CChMVd)	Chrysanthemum	Zhang et al. (2013)
	Peach latent mosaic viroid (PLMVd)	Peach	Zhang et al. (2013)

In a standard hybridization assay, the PC and HEX probes should generate high signal intensities, while NC probes should react with signal intensities similar to the background level.

Spotted slides are air-dried for 10 minutes and then fixed by UV exposure at 240 mJ/cm^2. After being incubated in a 0.5% sodium dodecyl sulfate (SDS) solution at 60°C for 1 hour, the slides are rinsed in 100 mM Tris (pH 8.0) for 5 minutes, and kept at 4°C for use.

MICROARRAY PROBE DESIGN

A major focus in microarray design is the selection of probes. Initially, the probes mostly consisted of PCR products, from 200 to 1,000 base pairs (bp), obtained by gene amplification starting from cDNA libraries or genomic DNA (Schena et al., 1995). The production of cDNA probes, however, is very laborious, time-consuming, and requires the amplification and purification of each single probe. This manufacturing is further subject to errors and cross-contamination (Auburn et al., 2005), especially with the increasing number of probes included in the array. In addition, the cDNA probes tolerate some mismatches during hybridization, thus showing a low specificity. An alternative approach is the use of synthetic oligonucleotide probes, which are a source of high-quality DNA with uniform concentration and purity. In addition, if compared to double stranded cDNA probes, they do not require a denaturation step before hybridization. The oligonucleotide probes also have the advantage of being easily modified to increase the binding capacity and orientation to the solid support, either by the addition of a terminal reactive group or by the incorporation of a spacer, usually a dT tail, allowing a better exposure of the probe during hybridization.

The probes based on synthetic oligonucleotides generally vary in length from 20 to 70 nt. Shorter oligonucleotide (15–25 nt) probes allow a greater discrimination, but reduce sensitivity (Call, 2005); long oligonucleotide probes (longer than 70 nt) may provide more consistent hybridization signals, but the small viroid genomes limit the choice of probe selection, because only three to four nonoverlapping 70-nt sequences can be selected. Thus, a probe with 40–50 nt is a better choice for viroid detection by microarrays.

In the Combimatrix platform (Tiberini and Barba, 2012), 40-nt probes for pospiviroid detection were obtained according to the nucleotide sequences retrieved from the NCBI GenBank. In particular, the following viroids were included: tomato apical stunt, potato spindle tuber, tomato chlorotic dwarf, tomato planta macho, citrus exocortis, and

Columnea latent. The sequences were aligned with VectorNTI 10 (Invitrogen, Carlsbad, CA, USA) and used to design specific 40-nt oligonucleotides for each viroid with EXIQON "Design oligonucleotides for expression arrays" software (http://oligo.lnatools.com) and Oligoarray 2.1 (Rouillard et al., 2003). The 40-nt oligonucleotides were designed to meet the following criteria: (1) melting temperature range, 80–85°C; (2) secondary structure temperature threshold, 70°C; (3) cross-hybridization temperature threshold, 65°C; and (4) G + C content, 40%–60% (Tiberini and Barba, 2012). To confirm the specificity of the probes designed by both programs, each probe was assessed by the nucleotide identity score after BLASTN alignment. The oligonucleotide sequence was compared with all known sequences of plant viruses and viroids to select only those with no significant or very low identity, to other viruses and viroids excepting the specific targets (Tiberini and Barba, 2012).

In another microarray platform, an automated probe design protocol was applied to generate a minimal number of genus-level probes as described in Zhang et al. (2013). Nonredundant viroid nucleotide sequences within the same genus were aligned using BLASTN, and conserved regions were identified from the alignment and searched for 40-nt probes. The candidate probes were required to have: (1) a G + C content between 40% and 60%; (2) fewer than four tandem identical nucleotides; and (3) fewer than six nucleotides in the inner hairpin structure. Probes showing potential off-target hybridization to viroids of other genera, plant viruses and plant mRNAs were removed. Probes that passed the above filters were aligned to all sequences in the genus using BLASTN. A probe sequence was requested to match at least 80% of the viroid sequence. Final microarray probes were selected by maximizing the coverage of viroid genera using a minimal number of probes. Selection of probe combination was repeated three times, thus three probe sets were generated to target most viroid sequences.

TARGET PREPARATION

In general, the total RNA from infected and healthy plants was extracted using the TRIzol reagent or the RNeasy Plant Mini kit according to the manufacturer's instructions. And then the different RT-PCR strategies were used for target preparation. Two labeling strategies have been reported for successful target preparation. (1) Using random hexamer primers and oligo-dT to synthesize viroid cDNA, the universal reverse primer specific for the central conserved region of members of the genus pospiviroid was utilized for PCR incorporating Cy3 or Cy5 dye (Tiberini and Barba, 2012). (2) Using viroid species degenerate primers for the RT-PCR reaction and PCR products being labeled with nonamer random primers and Cy3-dCTP (Zhang et al., 2013).

Notably, the typical approach for the detection of viruses in plant material (incorporation of amino-allyl modified nucleotides into cDNA during reverse transcription, followed by postlabeling with a Cy dye) resulted in no hybridization of labeled viroid cDNA to the array, despite strong signals for plant controls. This result was presumably due to the high secondary structure of viroids; however, the hybridization was not improved by fragmentation of viroid RNA or cDNA (after reverse transcription).

HYBRIDIZATION, WASHING, SCANNING, AND ANALYSIS OF MICROARRAYS

Hybridization, washing, and scanning steps are basically the same in various viroid microarray platforms. In the universal oligonucleotide microarray platform (Zhang et al., 2013), the process is as follows: Cy3-labeled DNA fragments, preheated for 10 minutes at 80°C, are added to the microarray and hybridized for 16 hours at 42°C. Slides are washed for 4 minutes at 42°C with 2× SSC (1× SSC is 150 mM NaCl, 17 mM sodium citrate, pH 7.0) containing 0.2% SDS, and then washed for 4 minutes at 42°C with 0.2× SSC.

Taking into consideration dye swapping, at least two independent hybridization events need to be performed, using all the targets considered in the study to balance any interference due to the fluorophore.

After being dried by centrifugation, the microarrays are scanned using an array scanner, equipped with a 532 nm laser for Cy3 and a 635 nm laser for Cy5 fluorescence measurements. Laser power is fixed at 50% of its potential in both cases, while the photomultiplier tube power ranges from 45% to 65%, depending on the signal intensity.

Microarray image acquisition and signal analyses are performed using the scanner software. The background is subtracted as specified in the program by applying the following default local subtraction method. The background for a spot is calculated from a region near the feature (three times around and excluding the neighboring spots) and its intensity is subtracted from the mean intensity of the spot analyzed. Initially, all signals below 300 relative fluorescence units were considered negative. The threshold to consider a reaction positive was set at 3× (the mean spot intensity plus the standard deviation of the signal from all probes initially considered as negative). Microarray hybridization pseudocolor images have been generated for hop stunt viroid using the universal oligonucleotide microarray platform (Fig. 37.1) (Zhang et al., 2013).

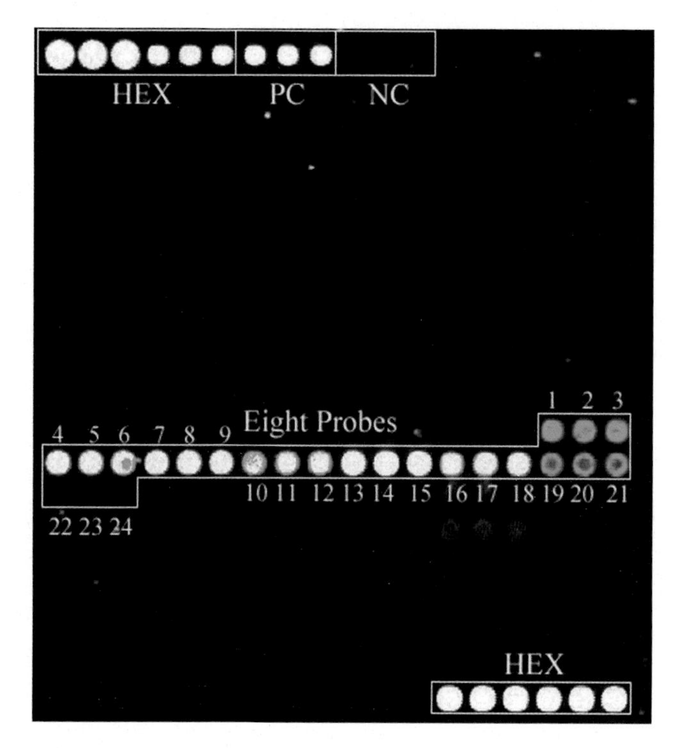

FIGURE 37.1 Microarray hybridization pseudocolor image of a hop stunt viroid (HSVd)-infected citrus sample. Eight probes were designed for HSVd and each probe was spotted on the microarray in triplicate. Seven out of the eight HSVd probes (probes 1–7, spots 1–21) were positive and one probe (probe 8, spots 22–24) was negative. The white and green dots represent probes with different intensities of fluorescent signal. Each probe was 40 nt with the following sequence:
Probe 1-TGTCTATCTGAGCCTCTGCCGCGGATCCTCTCTTGAGCCC
Probe 2-AGCTCTGCCTTCGAAACACCATCGATCGTGCCTTCTTCTT
Probe 3-TAGGAGAGAGGGCCGCGGTGCTCTGGAGTAGAGGCTTCTT
Probe 4-TTTACCTTCTTCTGGCTCTTCGAGTGAGACGCGACCGGTG
Probe 5-GGGCAACTCTTCTCAGAATCCAGCGAGAGGCGTAGGAGAG
Probe 6-TCGATCGTCCCTTCTTCTTTACCTTCTCCTGGCTCCTGGT
Probe 7-ACAGATAGACTCGGAGACGGCGCCTTGGAGAGAACTCGGG
Probe 8-GCGCCACGAGACCTCATCTCCGAGACGGAAGCTTTGTGGT
HEX (the microarray internal control), positive control (PC), and negative control (NC) are marked on the image. In a standard hybridization assay, HEX and PC probes should generate high signal intensities, while NC probes should react with signal intensities similar to the background level.
HEX with random sequence was used to monitor the probes linking to the chips. It was the 5′ Hex-linker oligonucleotide (5′-GTCACATGCGATGGATCGAGCTCCTTTATCATCG TTCCCACCTTAATGCA -3).
PC, the random sequence of this probe was mixed with the samples before hybridization for monitoring the probe hybridization efficiency. The 5′ amino-linker oligonucleotide (5′-$(T)_{15}$-CTCATGCCCATGCCGATGC-3′) had a sequence that did not potentially hybridize with any viroid sequence.

CONCLUSIONS

The microarray methodology has the following advantages. It can detect a range of viroids and/or their strains using a single test and the ease of the in silico design and the high specificity of the oligonucleotides enable discrimination of mixed infections by different pathogens and even similar strains. Its major disadvantage is that the detection process is complex and costly.

References

Abdullahi, I., Koerbler, M., Stachewicz, H., Winter, S., 2005. The 18S rDNA sequence of *Synchytrium endobioticum* and its utility in microarrays for the simultaneous detection of fungal and viral pathogens of potato. Appl. Microbiol. Biotechnol. 68, 368–375.

Agindotan, B., Perry, K.L., 2008. Macroarray detection of eleven potato-infecting viruses and potato spindle tuber viroid. Plant Dis. 92, 730–740.

Auburn, R.P., Kreil, D.P., Meadows, L.A., Fischer, B., Matilla, S.S., Russell, S., 2005. Robotic spotting of cDNA and oligonucleotide microarrays. Trends Biotechnol. 23, 374–379.

Call, D.R., 2005. Challenges and opportunities for pathogen detection using DNA microarrays. Crit. Rev. Microbiol. 31, 91–99.

Ivanova, D., Vachev, T., Baev, V., Minkov, I., Gozmanova, M., 2010. Identification of potato spindle tuber viroid small RNA in *Orobanche ramosa* by microarray. Biotechnol. Biotechnol. Equip. 24 (Suppl. 1), 144–146.

Owens, R.A., Flores, R., Di Serio, F., Li, S.-F., Pallás, V., Randles, J.W., et al., 2012. Viroids. In: King, A.M.Q., Adams, M.J., Carstens, E.B., Lefkowitz, E.,J. (Eds.), Virus Taxonomy: Ninth Report of the International Committee on Taxonomy of Viruses. Elsevier/ Academic Press, London, pp. 1221–1234.

Rouillard, J.M., Zuker, M., Gularil, E., 2003. OligoArray 2.0: design of oligonucleotide probes for DNA microarrays using a thermodynamic approach. Nucleic Acids Res. 31, 3057–3062.

Schena, M., Shalon, D., Davis, R.W., Brown, P.O., 1995. Quantitative monitoring of gene expression patterns with a complementary DNA microarray. Science 270, 467–470.

Tiberini, A., Barba, M., 2012. Optimization and improvement of oligonucleotide microarray-based detection of tomato viruses and pospiviroids. J. Virol. Methods 185, 43–51.

Zammatteo, N., Jeanmart, L., Hamels, S., Courtois, S., Louette, P., Hevesi, L., et al., 2000. Comparison between different strategies of covalent attachment of DNA to glass surfaces to build DNA microarrays. Anal. Biochem. 280, 143–150.

Zhang, Y.J., Yin, J., Jiang, D.M., Xin, Y.Y., Ding, F., Deng, Z.N., et al., 2013. A universal oligonucleotide microarray with a minimal number of probes for the detection and identification of viroids at the genus level. PLoS One 8, e64474.

CHAPTER

38

Application of Next-Generation Sequencing Technologies to Viroids

Marina Barba[1] *and Ahmed Hadidi*[2]

[1]CREA-Research Centre for Plant Protection and Certification, Rome, Italy
[2]U.S. Department of Agriculture, Beltsville, MD, United States

INTRODUCTION

The standard dye-terminator method (Sanger et al., 1977), which was then automated, is considered as a "first-generation" technology and has been used successfully in many major sequencing projects, including the completion of the human genome. However, there was a need for new and improved technologies for sequencing large numbers of human, animal, plant, and microorganism genomes. In the late 20th and early 21st century newer sequencing methods have been developed. The newer methods are referred to as next-generation (high throughput, deep) sequencing (NGS) and their use since 2004 has changed the approach to both basic and applied research in many biological disciplines, including plant virology. These technologies have lowered the costs of DNA sequencing. NGS describes platforms that produce large amounts (typically millions to billions) of DNA reads with lengths between 25 and 400 bp, which are shorter than the traditional Sanger sequence reads (300–750 bp). Recently, however, NGS technologies have advanced to produce DNA reads longer than 750 bp. The major advance offered by NGS is the ability to generate an enormous volume of data, in several cases in excess of one billion short reads per instrument run, as well as its ability to deliver fast, cost-effective, and accurate genome information.

Viroids and Satellites.
DOI: http://dx.doi.org/10.1016/B978-0-12-801498-1.00038-3

DEVELOPMENT OF NGS PLATFORMS

Several commercial NGS platforms for DNA or RNA sequencing became available after 2004 and are continuously being improved to be easier to use and more cost effective. The DNA sequencing technologies present in 2016 and both NGS systems plus their applications are shown in Tables 38.1 and 38.2, respectively.

SEQUENCING PLATFORM SELECTION

Among the important factors in selecting a NGS platform is the size of the genome being studied, its complexity, and the depth of coverage and accuracy needed. It is advisable to contact the providers of NGS services for guidance. For RNA sequencing and those projects that require high depths of coverage, Illumina and SOLiD platforms may offer the best all-round value for money, accuracy, and throughput (Radford et al., 2012). Illumina HiSeq 2500 features the largest output and lowest sequencing cost, and SOLiD 5500xl has the highest accuracy (Liu et al., 2012). Fig. 38.1 shows Illumina NGS platforms HiSeq Series sequencer (left) and NextSeq Series desktop sequencer (right). For de novo genome sequencing where longer reads may be appropriate, Roche 454 has the longest read length. For fast turnover times and limited throughput, smaller laboratory benchtop platforms (Table 38.2) may offer greater flexibility (Loman et al., 2012).

BIOINFORMATIC SOFTWARE TOOLS FOR DATA ANALYSIS

A full review of software tools available for quality control assembly and quantitative analysis of NGS is beyond the scope of this chapter, but these topics have been the subject of several review articles (Barba et al., 2014; Horner et al., 2009; Metzker, 2010; Miller et al., 2010; Pabinger et al., 2013) and books (e.g., Pevzner and Shamir, 2011). Moreover, the journal *Bioinformatics* publishes tools and algorithms that have been developed for NGS data analysis. Recently, bioinformatic tools for detecting and identifying known and unknown viroids and viroid-like (circular) RNAs in small RNA (sRNA) libraries have been reported (Wu et al., 2014; Zhang et al., 2014). Viroid sRNAs are mainly 21–24 nt in length (Barba and Hadidi, 2009; Sano et al., 2010). sRNAs with this size are retrieved, assembled with the "Velvet" program (Zerbino and Birney, 2008) and the resulting contigs are then processed

TABLE 38.1 Current Next-Generation Sequencing Technologies

Sequencing platform	Amplification method	Sequencing chemistry	Read length (bp)	Reads/run (single reads)	Error type	Accuracy (%)
454 (Life Sciences, a Roche Co.)	Emulsion PCR	Pyrosequencing	400–700 and read of 1 kb is also offered	400–600 Mb	Indel[a]	99.9
Illumina (Illumina)	Bridge PCR	Synthesis technology	50–300	25 Mb and up to 6 Gb (True Seq paired-end)	Substitution	99.9
SOLiD (Applied Biosystems)	Emulsion PCR	Ligation	85–100	400 Mb-2.4 Gb	A/T Bias	99.9
PacBio (Pacific Biosciences)	No amplification Single molecule real-time (or SMRT)	Fluorescently labeled nucleotides	10,000–15,000 or longer	0.5–0.55 M per SMR cell	Indel	88–99
Ion Torrent (Applied Biosystems)	Emulsion PCR	Detection of released protons	200–400	100 Mb-1Gb	Indel	98–99
Nanopore (Oxford Technologies)	No amplification Single molecule	Change in ionic current through the nanopores reads the DNA sequence	6000–14,000 with potential up to 98,000	Devices monitor their own data output and look for key application-specific results	Indel	65–96

[a] *Insertion or deletion of bases.*

TABLE 38.2 Current NGS Platform Systems and Their Applications

Platform systems	Applications
454 Life Sciences, a Roche company, GS FLX + System,	DNA sequencing: whole genome sequencing, de novo and resequencing of large genes in a single run, exome and amplicon sequencing and other targeted sequencing. Epigenetics sequencing: chromatin immunoprecipitation sequencing (ChIP-Seq), methylation analysis by sequencing. RNA sequencing (RNA-Seq): transcriptome sequencing which is comprised of all RNA molecules, including mRNA, rRNA, tRNA and non-coding RNA. Sequencing SNPs and indels throughout the transcriptome, sequencing capture metagenomics. Run time: 10–23 h; large reads up to 1000 bp
[a]GS Junior + System (benchtop)	[a]Microbial genome sequencing, amplicon sequencing, metagenomic analysis, transcriptome sequencing. Fast sequencing run with read length up to 800 bp
Illumina HiSeq series (2500/2000/1500/1000) and HiSeq X series (five/ten), HiScan SQ system, Genome Analyzer IIx	DNA sequencing including candidate region targeted sequencing such as exome, amplicon, gene expression genes as genes of rRNA, tRNA or small RNA, gene typing as that of human leukocyte antigen and others. Epigenetic sequencing: ChIP-Seq and methylation analysis by sequencing. RNA sequencing: transcriptome analysis, small RNA and mRNA sequencing. Gene expression profiling by sequencing, sequencing capture metagenomics. Run time: 12–30 h; output range: 30–120 Gb and up to 900–1800 Gb; maximum read length 2 × 150 bp
[b]MiSeq desktop series,	[b]Small genome sequencing, amplicon and targeted gene panel sequencing. Run time 5–55 h; output range 0.3–15 Gb; maximum read length 2 × 300 bp
[c]NextSeq desktop series	[c]Genome sequencing, targeted sequencing such as exome and amplicon, transcriptome sequencing and others. Run time 12–30 h; output range 30–120 Gb; maximum read length 2 × 150 bp

(*Continued*)

TABLE 38.2 (Continued)

Platform systems	Applications
Applied Biosystems SOLiD 5500 W Series Genetic Analyzers	DNA sequencing: whole genome and exome sequencing. RNA sequencing: whole transcriptome analysis and small RNA sequencing and gene expression profiling by Sequencing. Epigenetic sequencing: ChIP-Seq and methylation analysis by sequencing
Pacific Biosciences [d]PACBIO RSII [e]Sequal System	The Pac Bio sequencing systems are built on single molecule, real-time sequencing technology (SMRT). It has the longest average read lengths of any sequencing technology currently on the market [d]Whole genome sequencing of small genomes, targeted sequencing, complex population analysis, RNA sequencing of targeted transcripts and microbial epigenetics. Run time: from 60 min to 6 h per SMRT cell; run size: from 1 to 16 SMRT cells per run [e]Whole genome sequencing, sequencing large populations for structural variants, sequencing full-length transcripts or targeted transcripts, sequencing genomic variation in a complex population, sequencing epigenetic modificationa. Run time: 30 min to 6 h per SMRT cell; run size: from 1 to 16 SMRT cells per run
Applied Biosystems Ion Torrent [f]Ion PGM System (benchtop) [g]Ion Proton System (benchtop)	[f]Ideal for sequencing small genes and genomes. Amplicons, exomes and transcriptome sequencing. Targeted DNA or RNA sequencing; microbial genomic sequencing; viral and other microbial typing. Run time: as little as 2 h; output: from 30 Mb up to 2 Gb [g]For exome and transcriptome sequencing, whole microbial genome sequencing. Run time: as little as 2 h; output: up to 10 Gb
Oxford Nanopore Technologies PromethlON System (benchtop) GridlON System (benchtop) MinlON System (a portable device)	All systems analyze DNA or RNA. They are used for DNA sequencing; epigenetic sequencing; characterization of genetic variation; RNA and miRNA sequencing either directly or by conversion to cDNA. They offer scalable real-time, long read DNA sequencing. They run for as long as it takes to gather the needed data

FIGURE 38.1 Illumina HiSeq Series sequencer (left) and NextSeq Series desktop sequencer (right) .

to search for overlaps using the "CAP3" program. Subsequently, the circularized and non-circularized sequences are analyzed with the "PatSearch" program for the presence of conserved sequence/structural motifs in viroids. As a result, sets of circular and linear sequences with or without viroid-related structures are identified, therefore providing a catalog of potential viroids and viroid-like RNAs.

COST OF DNA SEQUENCING BY NGS

Currently, the costs per raw megabase of DNA and the cost per genome similar in size to the human genome (3000 Mb) are less than $0.02 and about $2000, respectively (Wetterstrand, 2016). Thus, NGS of viroid samples should be significantly less than the estimated cost for the human genome. In the next few years, it is expected that the third generation of DNA sequencing platforms will increase sequencing capacity and speed while reducing cost. A genome may be sequenced for $500–1000 or less in about 15 minutes.

CURRENT APPLICATION OF NGS TECHNOLOGIES TO VIROIDS

NGS technologies combined with bioinformatics were first applied to viroids in 2009. Their applications include genome and viroid sRNAs (vd-sRNAs) NGS for viroid discovery, detection, and identification (diagnostics), and viroid–host interactions (Hadidi et al., 2016).

NGS can provide thousands to millions of vd-sRNA sequences from viroid infected plant materials. When vd-sRNAs are abundant enough, viroid molecular fragments can be assembled. Since vd-sRNAs are 21–24 nt in length, their sequences can be employed directly as primer sequences to amplify viroid fragments by RT-PCR. Table 38.3 shows the use of NGS in various viroid projects.

If diseases of unknown etiology affecting fruit crops, grapevine, and hop are graft-transmissible, systemic pathogens, i.e., viroids, viruses, or phytoplasmas, are most likely to be the cause. The utilization of NGS in viroid research discovered two novel viroids, grapevine latent viroid and persimmon viroid 2, and also discovered that citrus bark cracking viroid and apple dimple fruit viroid infect hops and fig trees, respectively, thus, extending their host range (Table 38.3). Moreover, a circular RNA was discovered in Pinot noir grapevine with hammerheads in both strands and multibranched conformation stabilized by a kissing-loop interaction in the plus strand (Wu et al., 2012). The same argument applies to another viroid-like circular RNA with hammerheads in both strands and a multibranched conformation that has been reported in apple (Zhang et al., 2014). It is yet to be shown that this RNA is indeed a viroid by establishing its infectivity in grapevine or any other host. NGS has recently revealed that "severe hop stunt disease" is not caused by hop stunt viroid but by citrus bark cracking viroid, which also causes the production of severely affected hop cones. It is a new emerging and economically important disease in hops and currently has been described only from Slovenia (Jakse et al., 2015).

Conventional molecular detection and identification methods of viroids depend on prior knowledge of sequence of the viroid of interest. NGS has provided a very powerful alternative for detection and identification of viroids without a priori knowledge. The utilization of NGS in viroid research and diagnosis is sensitive, accurate, and fast using different viroid host plant species, including woody perennial crops which have low titers of these pathogens. Moreover, NGS increased the number of novel viroids and strains being discovered and characterized in different plant hosts, including extending the host range of known viroids. Thus, NGS has the potential to be used in viroid diagnostics in quarantine and certification programs as well as in programs aimed at viroid elimination from vegetatively propagated material (Barba and Hadidi, 2015; Barba et al., 2014; Hadidi and Barba, 2012; Hadidi et al., 2016), which suggests that NGS will be a significant and powerful tool in controlling viroid diseases. Follow-up studies may be required to confirm that sequences assembled by NGS can be detected in infected plants by other means such as RT-PCR.

Genome engineering editing methods such as CRISPR/Cas 9 systems have the capabilities to work on DNA or RNA sequences (Abudayyeh et al., 2016; Doudna and Charpentier, 2014; O'Connell et al., 2014). The

TABLE 38.3 Utilization of NGS in Various Studies of Viroids

Viroid	Remarks	Target RNA	Sequencing platform	Reference
DISCOVERY				
Grapevine latent viroid (GLVd)	Discovery of a novel apscaviroid species, GLVd	Total RNA	Illumina HiSeq-2000	Zhang et al. (2014)
Persimmon viroid 2 (PVd-2)	Discovery of a novel apscaviroid species, PVd-2. It shares 72%–73% sequence identity with citrus viroid VI	dsRNAs	Illumina Genome Analyzer II	Ito et al. (2013)
Citrus bark cracking viroid (CBCVd)	Discovery of a novel CBCVd variant that naturally infects hops. This study extended the host range of CBCVd to hop in addition to citrus	vdsRNAs or Total RNA	Illumina	Jakse et al. (2015)
Apple dimple fruit viroid (ADFVd)	Discovery of a novel ADFVd variant that naturally infects fig trees at a very low titer. This study extended the host range of ADFVd to fig in addition to apple	vdsRNAs	Illumina Hi Scan SQ	Chiumenti et al. (2014)
DETECTION AND IDENTIFICATION				
Citrus exocortis viroid (CEVd), Hop stunt viroid (HSVd), Grapevine yellow speckle viroid 1 (GYSVd-1)	Detection and identification of CEVd, HSVd and GYSVd-1 from both symptomatic and nonsymptomatic samples of grapevine red leaf disease	Total RNA treated with DNase	Illumina Genome Analyzer IIx	Poojari et al. (2013)
GYSVd-1, HSVd	Detection and identification of GYSVd-1 and HSVd	vdsRNAs and dsRNAs	Illumina	Chiumenti et al. (2012)
GYSVd-1, HSVd	Detection and identification of GYSVd-1 and HSVd	vdsRNAs	Illumina	Giampetruzzi et al. (2012)
Potato spindle tuber viroid (PSTVd)	Detection and identification of PSTVd in tomato	vdsRNAs	Illumina Genome Analyzer IIx	Li et al. (2012)
Australian grapevine viroid (AGVd), GYSVd, HSVd	Detection and identification of AGVd, GYSVd and HSVd in grapevine	Total RNA or dsRNA	Roche 454	Al Rwahnih et al. (2009)

(*Continued*)

TABLE 38.3 (Continued)

Viroid	Remarks	Target RNA	Sequencing platform	Reference
VDSRNAS: CHARACTERIZATION, BIOGENESIS PATHWAY, TARGETING HOST MRNAS				
HSVd, GYSVd-1, Grapevine yellow speckle viroid 2 (GYSVd-2)	Characterization of vdsRNAs of HSVd, GYSVd-1 and GYSVd-2 in grapevine	vdsRNAs	Illumina	Alabi et al. (2012)
HSVd	Study HSVd pathway involved in the biogenesis of the viroid vdsRNAs in cucumber	vdsRNAs	Illumina	Martínez et al. (2010)
Peach latent mosaic viroid (PLMVd)	To study the genesis of PLMVd vdsRNAs and viroid pathogenesis in peach	vdsRNAs	Illumina	Di Serio et al. (2009)
PSTVd	RNA-dependent RNA polymerase 6 of *Nicotiana benthamiana* restricts accumulation and precludes meristem invasion of PSTVd which replicates in nuclei	Plant sRNAs and vdsRNAs	Illumina EAS269 GAII	Di Serio et al. (2010)
GYSVd, HSVd	Different dicer-like enzymes target RNAs of GYSVd and HSVd in grapevine. Also, study suggested that the viroid RNAs may interact with host enzymes involved in the RNA-directed DNA methylation pathway	vdsRNAs	Solexa, Illumina	Navarro et al. (2009)
PLMVd	vdsRNAs of PLMVd containing the pathogenic determinant of the viroid guide the degradation of a host mRNA as predicted by RNA silencing, thus leading to symptom expression in infected plants	vdsRNAs	Illumina Genome Analyzer	Navarro et al. (2012)

(*Continued*)

TABLE 38.3 (Continued)

Viroid	Remarks	Target RNA	Sequencing platform	Reference
PSTVd	vdsRNAs derived from the virulence modulating region of two PSTVd variants target callose synthase mRNAs which greatly affect the accumulation of viroid molecules and the severity of disease symptoms	vdsRNAs	Illumina Genome Analyzer IIx	Adkar-Purushothama et al. (2015)
VARIANT MUTATION OR QUASISPECIES				
PLMVd	A single variant of PLMVd is prone to high mutation rates within a single infected GS305 peach plant. The most distant variants possessed a 17% variation level when compared to the parent sequence	DNase-treated total RNA	Roche 454	Glouzon et al. (2014)
Chrysanthemum stunt viroid (CSVd)	NGS of CSVd cDNA clones of two isolates from the US and Australia were both quasi species, while the cDNA clones of an isolate from China contained only one variant	—	Roche 454	Yoon and Palukaitis (2013)

potential to use genome engineering in association with NGS may open the door for controlling viroids (see Chapter 49: Genome Editing by CRISPR-Based Technology: Potential Applications for Viroids).

References

Abudayyeh, O.O., Gootenberg, J.S., Konermann, S., Joung, J., Slaymaker, I.M., Cox, D.B.T., et al., 2016. C2c2 is a single-component programmable RNA-guided RNA-targeting CRISPR effector. Science 353, aaf5573. Available from: http://dx.doi.org/10.1126/science.aaf5573.

Adkar-Purushothama, C.R., Brosseau, C., Giguère, T., Sano, T., Moffett, P., Perreault, J.-P., 2015. Small RNA derived from the virulence modulating region of the potato spindle tuber viroid silences callose synthase genes of tomato plants. Plant Cell. 27, 2178–2194.

Al Rwahnih, M., Daubert, S., Golino, D., Rowhani, A., 2009. Deep sequencing analysis of RNAs from a grapevine showing Syrah decline symptoms reveals a multiple virus infection that includes a novel virus. Virology 387, 395–401.

Alabi, O.J., Zheng, Y., Jagadeeswaran, G., Sunkar, R., Naidu, R.A., 2012. High-throughput sequence analysis of small RNAs in grapevine (*Vitis vinifera* L.) affected by grapevine leafroll disease. Mol. Plant Pathol. 13, 1060–1076.

Barba, M., Czosnek, H., Hadidi, A., 2014. Historical perspective, development and applications of next-generation sequencing in plant virology. Viruses 6, 106–136.

Barba, M., Hadidi, A., 2009. RNA silencing and viroids. J. Plant Pathol. 91, 243–247.

Barba, M., Hadidi, A., 2015. An overview of plant pathology and application of next-generation sequencing technologies. CAB Rev. 10, 1–21.

Chiumenti, M., Giampetruzzi, A., Pirolo, C., Morelli, M., Saldarelli, P., Minafra, A., et al., 2012. Approaches of next generation sequencing to investigate grapevine diseases of unknown etiology. Abstracts of the 17th Congress of ICVG. Davis, CA, pp. 116–117.

Chiumenti, M., Torchetti, E.M., Di Serio, F., Minafra, A., 2014. Identification and characterization of a viroid resembling apple dimple fruit viroid in fig (*Ficus carica* L.) by next generation sequencing of small RNAs. Virus Res. 188, 54–59.

Di Serio, F., Gisel, A., Navarro, B., Delgado, S., Martínez de Alba, A.E., Donvito, G., et al., 2009. Deep sequencing of the small RNAs derived from two symptomatic variants of a chloroplast viroid: implications for their genesis and for pathogenesis. PLoS One 4, e7539.

Di Serio, F., Martínez de Alba, A.E., Navarro, B.A., Gisel, A., Flores, R., 2010. RNA-dependent RNA polymerase 6 delays accumulation and precludes meristem invasion of a viroid that replicates in the nucleus. J. Virol. 84, 2477–2489.

Doudna, J.A., Charpentier, E., 2014. The new frontier of genome engineering with CRISPR-Cas9. Science 346, 1258096.

Giampetruzzi, A., Roumi, V., Roberto, R., Malossini, U., Yoshikawa, N., La Notte, P., et al., 2012. A new grapevine virus discovered by deep sequencing of virus-and viroid-derived small RNAs in cv Pinot gris. Virus Res. 163, 262–268.

Glouzon, J.P., Bolduc, F., Wang, S., Najmanovich, R.J., Perreault, J.-P., 2014. Deep-sequencing of the peach latent mosaic viroid reveals new aspects of population heterogeneity. PLoS One 9, e87297.

Hadidi, A., Barba, M., 2012. Next-generation sequencing: historical perspective and current applications in plant virology. Petria 22, 262–277.

Hadidi, A., Flores, R., Candresse, T., Barba, M., 2016. Next-generation sequencing and genome editing in plant virology. Front. Microbiol. 7, 1325.

Horner, D.S., Pavesi, G., Castrignano, T., D'Onorio De Meo, P., Liuni, S., Sammeth, M., et al., 2009. Bioinformatics approaches for genomics and post genomics applications of next generation sequencing. Brief. Bioinform. 11, 181–197.

Ito, T., Suzaki, K., Nakano, M., Sato, A., 2013. Characterization of a new apscaviroid from American persimmon. Arch. Virol. 158, 2629–2631.

Jakse, J., Radisek, S., Pokorn, T., Matousek, J., Javornik, B., 2015. Deep-sequencing revealed citrus bark cracking viroid (CBCVd) as a highly aggressive pathogen on hop. Plant Pathol. 64, 831–842.

Li, R., Gao, S., Hernandez, A.G., Wechter, W.P., Fei, Z., Ling, K.S., 2012. Deep sequencing of small RNAs in tomato for virus and viroid identification and strain differentiation. PLoS One 7, e37127.

Liu, L., Li, Y., Li, S., Hu, N., He, Y., Pong, R., et al., 2012. Comparison of next-generation sequencing systems. J. Biomed. Biotechnol.11, Article ID 251364.

Loman, N.J., Misra, R.V., Dallman, T.J., Constantinidou, C., Gharbia, S.E., Wain, J., et al., 2012. Performance comparison of benchtop high-throughput sequencing platforms. Nat. Biotechnol. 30, 434–439.

Martínez, G., Donaire, L., Llave, C., Pallás, V., Gómez, G., 2010. High-throughput sequencing of hop stunt viroid-derived small RNAs from cucumber leaves and phloem. Mol. Plant Pathol. 11, 347–359.

Metzker, M.L., 2010. Sequencing technologies-the next generation. Nat. Rev. Genet. 11, 31–46.

Miller, J.R., Koren, S., Sutton, G., 2010. Assembly algorithms for next generation sequencing data. Genetics 95, 315–327.

Navarro, B., Gisel, A., Rodio, M.E., Delgado, S., Flores, R., Di Serio, F., 2012. Small RNAs containing the pathogenic determinant of a chloroplast-replicating viroid guide the degradation of a host mRNA as predicted by RNA silencing. Plant J. 70, 991–1003.

Navarro, B., Pantaleo, V., Gisel, A., Moxon, S., Dalmay, T., Bistray, G., et al., 2009. Deep sequencing of viroid-derived small RNAs from grapevine provides new insight on the role of RNA silencing in plant-viroid interaction. PLoS One 4, e7686.

O'Connell, M.R., Oakes, B.L., Sternberg, S.H., East-Saletsky, A., Kaplan, M., Doudna, J.A., 2014. Programmable RNA recognition and cleavage by CRISPR/Cas9. Nature 516, 263–266.

Pabinger, S., Dander, A., Fischer, M., Snajder, R., Sperk, M., Efremova, M., et al., 2013. A survey of tools for variant analysis of next-generation genome sequencing data. Brief. Bioinform. 15, 256–278.

Pevzner, P., Shamir, R. (Eds.), 2011. Bioinformatics for Biologists. Cambridge University Press, Cambridge.

Poojari, S., Alabi, O.J., Fofanov, V.Y., Naidu, R.A., 2013. A leafhopper-transmissible DNA virus with novel evolutionary lineage in the family *Geminiviridae* implicated in grapevine redleaf disease by next-generation sequencing. PLoS One 8, e64194.

Radford, A.D., Chapman, D., Dixon, L., Chantrey, J., Darby, A.C., Hall, N., 2012. Application of next-generation sequencing technologies in virology. J. Gen. Virol. 93, 1853–1868.

Sanger, F., Nicklen, S., Coulsen, A.R., 1977. DNA sequencing with chain-terminator inhibitors. Proc. Natl. Acad. Sci. USA 74, 5463–5467.

Sano, T., Barba, M., Li, S.-F., Hadidi, A., 2010. Viroids and RNA silencing: mechanism, role in viroid pathogenicity and development of viroid-resistant plants. GM Crops 1, 1–7.

Wetterstrand, K.A., 2016. DNA sequencing costs: data from the NHGRI Genome Sequencing Program (GSP). Available at: www.genome.gov/sequencingcostsdata (accessed 13.09.16.).

Wu, Q., Wang, Y., Cao, M., Pantaleo, V., Burgyan, J., Li, W.-X., et al., 2012. Homology-independent discovery of replicating pathogenic circular RNAs by deep sequencing and a new computational algorithm. Proc. Natl. Acad. Sci. USA 109, 3938–3943.

Yoon, J.Y., Palukaitis, P., 2013. Sequence comparisons of global chrysanthemum stunt viroid variants: multiple polymorphic positions scattered through the viroid genome. Virus Genes. 46, 97. Available from: http://dx.doi.org/10.1007/s11262-012-0811-0.

Zerbino, D.R., Birney, E., 2008. Velvet: algorithms for de novo short read assembly using de Bruijn graphs. Genome Res. 18, 821–829.

Zhang, Z., Qi, S., Tang, N., Zhang, X., Chen, S., Zhu, P., et al., 2014. Discovery of replicating circular RNAs by RNA-seq and computational algorithms. PLoS Pathog. 10, e1004553.

PART V

CONTROL MEASURES FOR VIROIDS AND VIROID DISEASES

CHAPTER

39

Quarantine and Certification for Viroids and Viroid Diseases

Marina Barba[1] *and Delano James*[2]

[1]CREA-Research Centre for Plant Protection and Certification, Rome, Italy
[2]Centre for Plant Health-Canadian Food Inspection Agency, North Saanich, BC, Canada

INTRODUCTION

The movement of infected vegetatively propagated germplasm has contributed significantly to the worldwide distribution of viroids. Some viroids such as potato spindle tuber viroid (PSTVd) are transmitted through seeds (Lebas et al., 2005; van Brunschot et al., 2014) and pollen (Barba et al., 2007; Singh et al., 1992). The frequency of their occurrence in a large number of vegetable crops and the harm they cause have forced plant health regulatory agencies to define and apply specific phytosanitary requirements for movement of known hosts. These actions are aimed at reducing or preventing the inadvertent movement or introduction of these pathogens and the consequent economic losses due to their harmful effects on the yield and quality of crops. In recognition of the potential harm caused by PSTVd to the global trade and cultivation of host crops such as potato and tomatoes, the International Plant Protection Convention (IPPC) has developed and adopted an internationally accepted and standardized diagnostic protocol for this viroid (IPPC, 2015).

When a pest, including some viroids, is introduced and disease outbreaks occur in a country, the disruption and costs are enormous for governments, farmers, and consumers. With the increase of international travel and trade and the recognition of the threat that this poses, the IPPC came into force in 1952. It is one of three international standard setting bodies of the World Trade Organization's (WTO's) Agreement on Sanitary and Phytosanitary Measures (SPS Agreement).

Viroids and Satellites.
DOI: http://dx.doi.org/10.1016/B978-0-12-801498-1.00039-5

The aims of the IPPC are to promote fair and safe trade, and protect the health of cultivated and wild plants by preventing the introduction and spread of pests. A country that is a contracting party signatory to the IPPC is expected to form a National Plant Protection Organization (NPPO) with an Official IPPC contact point. NPPOs are responsible for certifying exports and regulating imports. The NPPOs are coordinated at the regional level by Regional Plant Protection Organizations (RPPOs). The RPPOs are responsible for gathering and disseminating information especially as it relates to the IPPC, and assist where possible with the development and implementation of international standards. There are currently nine RPPOs:

- Asia and Pacific Plant Protection Commission (APPPC)
- Comunidad Andina (CA)
- Comite de Sanidad Vegetal del Cono Sur (COSAVE)
- European and Mediterranean Plant Protection Organization (EPPO)
- Inter-African Phytosanitary Council (IAPSC)
- Near East Plant Protection Organization (NEPPO)
- North American Plant Protection Organization (NAPPO)
- Organismo Internacional Regional de Sanidad Agropecuaria (OIRSA)
- Pacific Plant Protection Organization (PPPO)

IPPC publishes, periodically, a list of quarantine pests present in different regions of the world. Table 39.1 shows viroids considered damaging in different countries and, as a consequence, covered by quarantine regulations.

The necessity to include viroids is due essentially to three major factors: (1) the severity and degree of damage that could be caused by the specific agent; (2) the economic importance of the crop that could be affected by a single viroid or a group of viroids; and (3) the wide global distribution and cultivation of crops that are known to be hosts.

PSTVd and coconut cadang-cadang viroid (CCCVd) meet all three criteria described above. They are very damaging and harmful pests, and their known hosts are of great importance and widely cultivated in many countries around the world. Hence, as indicated above, PSTVd was targeted for the development of an official Diagnostic Protocol (DP) that has been adopted by the IPPC (ISPM 27; Annex DP 7). The DP contains published and validated methods for the detection and identification of PSTVd, to facilitate safe trade among member countries. Severe PSTVd strains can reduce potato yield by up to 64%, and losses can be close to 100% when subsequent generations of crops are grown from infected seed potatoes (Singh et al., 2003). On the other hand, from the time it was first recognized to the year 1980, cadang-cadang disease killed over 30 million palm trees (Randles and Rodriguez, 2003). Economic impact includes lost production and the costs of implementing quarantine protocols

TABLE 39.1 Quarantine Viroids in Different Countries

Algeria, Uruguay, Vietnam	PSTVd
Antigua and Barbuda, Trinidad and Tobago	ASBVd
Argentina	ADFVd, CCCVd, PSTVd
Australia	All viroids
Brazil, Mauritania, Paraguay	CCCVd, PSTVd
Canada	ASSVd, PBCVd, PSTVd
Chile	ASBVd, CCCVd, PSTVd, TASVd, TCDVd
China	HSVd (cachexia isolate), CEVd, CCCVd
Costa Rica	ASBVd, CCCVd
Cuba	ASBVd, PSTVd
Ecuador	CCCVd, CSVd, PLMVd
Egypt	CChMVd, CSVd, PSTVd
EPPO/UE	CCCVd, CSVd, PSTVd
Georgia	CSVd, PLMVd, PSTVd
Iran	PLMVd, PSTVd
Jamaica	ASBVd, CCCVd, CTiVd CEVd, HSVd (cachexia isolate)
Japan	CLVd, HSVd, PSTVd, TASVd, TCDVd, TPMVd
Jordan	ADFVd, ASBVd, ASSVd, CBLVd, CEVd, CVd III, CVd IV, GYSVd, HSVd, PLMVd, PSTVd, TASVd
Lao PDR	CCCVd, CEVd, HSVd (cachexia isolate), PSTVd
Madagascar	ASBVd, CCCVd, PLMVd, PSTVd
Malaysia	ASBVd, CCCVd, PSTVd
Mauritius	ABSVd, CCCVd, CEVd, HSVd (cachexia isolate), PSTVd
Mexico	CCCVd, CChMVd, CEVd, CSVd, ELVd, HSVd, PBCVd, PLMVd, PSTVd
Mozambique	CSVd, PSTVd
Nepal	CEVd, PSTVd
Panama	CEVd, HSVd (cachexia isolate), PSTVd
Republic of Korea	ADFVd, AFCVd, PLMVd, TCDVd, PSTVd
Seychelles	CEVd, CCCVd, PSTVd

(Continued)

TABLE 39.1 (Continued)

Thailand	ASBVd, CCCVd, CTiVd, CChMVd, CEVd, CLVd, CSVd, HSVd, PLMVd, PSTVd, TASVd, TCDVd, TPMVd
Ukraine	CSVd
United States	HSVd, PBCVd, PSTVd

PSTVd, potato spindle tuber viroid; ASBVd, avocado sunblotch viroid; ADFVd, apple dimple fruit viroid; CCCVd, coconut cadang-cadang viroid; ASSVd, apple scar skin viroid; PBCVd, pear blister canker viroid; TASVd, tomato apical stunt viroid; TCDVd, tomato chlorotic dwarf viroid; HSVd, hop stunt viroid; CTiVd, coconut tinangaja viroid; CEVd, citrus exocortis viroid; CSVd, chrysanthemum stunt viroid; PLMVd, peach latent mosaic viroid; CChMVd, chrysanthemum chlorotic mottle viroid; CLVd, columnea latent viroid; TPMVd, tomato planta macho viroid; CBLVd, citrus bent leaf viroid; CVd III, citrus viroid III; CVd IV, citrus viroid IV; GYSVd, grapevine yellow speckle viroid 1 and 2; ELVd, eggplant latent viroid; AFCVd, apple fruit crinkle viroid.
Data from IPPC website.

(see Chapter 4: Economic Significance of Palm Tree Viroids and Chapter 25: Coconut Cadang-Cadang Viroid and Coconut Tinangaja Viroid).

Viroids that infect citrus are also good examples of why viroids are listed as quarantine agents by countries that are vulnerable to infections. Citrus are considered one of the most economically important fruit crops in the world and are cultivated worldwide in temperate, tropical, and subtropical regions in approximately 140 countries (Su, 2008; Chapter 2: Economic Significance of Fruit Tree and Grapevine Viroids). Citrus exocortis viroid and hop stunt viroid—cachexia strain, infect a broad range of citrus species as well as some citrus relatives and noncitrus hosts. Citrus exocortis viroid is symptomless in most citrus species, but is graft-transmitted and can be also transmitted mechanically by contaminated tools (Su, 2008; Chapter 16: Citrus Exocortis Viroid). Symptoms induced may consist of bark splitting, stunting, and a general declining of the trees that may contribute to loss of fruit production with considerable negative economic consequences.

One of the main strategies for controlling viroid diseases is to prevent the introduction of the pathogens via propagation and planting material (bud wood, plantlets, seeds, pollen) used in international trade. Prevention is a much better approach to limiting spread of these pathogens, for three main reasons: (1) viroids often show high levels of seed transmission; (2) infection rates may be high; and (3) elimination treatments are not always reliable. Furthermore, there are no therapeutic agents able to directly control viroids and their associated diseases (Barba et al., 2003b). RPPOs and NPPOs have developed acts and regulations to prohibit importation of plant materials without phytosanitary certification and to deny entry of contaminated articles (Ebbels, 2003). The plant health-related quarantine laws and regulations target viroids and plant pathogens considered a threat to local agriculture and trade (Foster and Hadidi, 1998). Strict application of quarantine regulations could involve different facilities: inspection areas, glass- and/or screenhouses, as well as field plots and laboratories used to stock, grow, and process plants. The costs of creating, maintaining, and implementing these facilities and services are considerable, but this cost is much less than any potential costs/losses that may occur as the result of any inadvertent introduction of a new pest (Barba et al., 2003a).

A second strategy used to reduce the spread of diseases among countries and the deleterious effect of domestic viroids is the adoption of certification programs. Specifically, the use of certified seed/planting material is necessary to restrict or prevent any initial introduction and or build up of infection in the field/orchard. Certification usually results in the use of healthier plants that generally give higher yields. In fact, by decreasing the initial inoculum load in the field, the spread of viroids is restricted and the damages are reduced or prevented. Seed/planting material production and certification schemes operate mainly through legislation, with the cooperation and support of the industry.

PHYTOSANITARY REGULATION IN THE EUROPEAN UNION

A policy of free circulation of plants within all European Countries has been adopted. As a consequence the EU should be considered a unique "country"/region with many different entry points (generally airports, harbors, borders). This means that the phytosanitary status of plants and plant products circulating in Europe must fit sanitary requirements that are identical for each sovereign country. Additionally, when any propagative germplasm is introduced into Europe from a country outside the Union, the inspection service present at each entry point is responsible for checking for the presence of harmful organisms and eventually preventing any circulation of infected germplasm.

All aspects covered by European Union legislation are discussed, ratified, and issued in Brussels (Belgium) where the headquarters of the European Union is located. All phytosanitary Directives, approved by EU in the frame of the Phytosanitary Committee, must be acknowledged and applied at the local level.

Quarantine and Certification are the main activities that help to guarantee the absence of viroids in germplasm produced or circulating in Europe.

PHYTOSANITARY REGULATION IN NORTH AMERICA

The NAPPO Regional Standard for Phytosanitary Measures (RSPM) provides guidance to reduce phytosanitary risk associated with the movement of host plants and products into member countries. NAPPO's RSPM 3 (2011) addresses specifically the movement of potatoes into a member country. As indicated in Table 39.1 PSTVd is regulated in all three NAPPO member countries (Canada, Mexico, and the United States).

QUARANTINE

Quarantine is a set of legislative and regulatory measures and associated activities designed to prevent or minimize the introduction, transport, and spread of harmful organisms, including viroids, by means of human activities (Foster and Hadidi, 1998). In the EU the Directive 2000/29/EC lists the pests in two Annexes (I and II), two Parts (A and B), and two Sections (I and II). Three of the viroids considered in the Directive are as follows: (1) PSTVd present in Annex IAI (Harmful organisms whose introduction into, and spread within, all member states shall be banned and not known to occur in any part of the community and relevant for the entire community); (2) CCCVd present in Annex IIAI for plants of palm (Harmful organisms whose introduction into, and spread within, all member states shall be banned if they are present on certain plants or plants products and not known to occur in any part of the community and relevant for the entire community); and (3) chrysanthemum stunt viroid present in Annex IIAII for plants of *Dendranthema* spp. (harmful organisms whose introduction into, and spread within, all member states shall be banned if they are present on certain plants or plants products and known to occur in any part of the community and relevant for the entire community). Subsequently, due to the spread of PSTVd within the EU, mainly in symptomless ornamental plants, the EPPO listed PSTVd and chrysanthemum stunt viroid in the A2 list (pests are locally present in the EPPO region) and CCCVd in A1 list (pests are absent from the EPPO region). In the NAPPO region, PSTVd is listed in Canada and the United States as an Ab2 pest (Absent, pest eradicated), and in Mexico as an Ab1 pest (Absent, no pest records) (NAPPO's RSPM 3, 2011). Table 39.2 shows viroids that are listed as regulated/quarantine pests in NAPPO member countries.

CERTIFICATION

Certification represents an effective way to guarantee that plants are maintained under defined and accepted conditions throughout their

TABLE 39.2 Viroids That Are Listed as Regulated/Quarantine Pests in NAPPO Member Countries

Plant genus or species	Viroids[a]
Chrysanthemum indicum	[c]CChMVd, [c]CSVd
Citrus L.	[c]CEVd, [c,d]HSVd (cachexia strain)
Cocos nucifera	[c]CCCVd
Malus Mills	[b]ASSVd
Pyrus L.	[b,c,d]PBCVd
Prunus persica	[c]PLMVd
Solanum melongena	[c]ELVd
Solanum tuberosoum	[b,c,d]PSTVd

[a] *Abbreviations as in Table 39.1.*
[b] *Viroids regulated by Canada.*
[c] *By Mexico.*
[d] *By the United States.*

production, and that the phytosanitary requirements (free of specified pests) necessary for the movement of propagative material are met. The material must be maintained throughout the different steps of production in ways that ensure trueness to type and preservation of the desired phytosanitary status (Barba, 1998). Certification is based on collaborations between industry, and relevant scientific and technical organizations. It involves—at different levels and with different responsibilities—more than one component. Certification safeguards the nurseryman, who imports and/or sells vegetatively propagated plant material, the country from which the products may be exported (from possible trade disputes or litigation), the country or region to which the material is exported (preventing pest introductions), and the grower who buys the nurseryman's products.

A model for a certification scheme has been produced in Europe for fruit trees (directive 2014/98/UE). This scheme follows EPPO recommendations (EPPO standard PM 4—Production of healthy plants for planting) and refers to operative schemes previously published by this intergovernmental organization. It promotes meetings among specialized researchers from member states with the purpose of integrating several procedures into a common certification scheme.

The scheme defines all the steps necessary for obtaining healthy plants, it lists the damaging pathogens to be excluded, and it suggests for this purpose the most reliable detection methods. Certified plants

FIGURE 39.1 Insect-proof screen-houses used to maintain nuclear stock free from pathogens of concern and to prevent further infections.

FIGURE 39.2 Peach nuclear stock inside screen-house maintained in suitable growing conditions. Plants are grown in pots, isolated from the soil to prevent any contamination.

are produced by direct filiations starting with nuclear stock that consists of mother plants, phenotypically confirmed well identified, free from pathogens of concern, and grown within approved facilities (preferably inside screen-houses) to prevent further infections (Fig. 39.1). All plants belonging to the certification scheme should be maintained in suitable growing conditions and should be periodically and regularly screened to confirm their phytosanitary status (Fig. 39.2).

Table 39.3 lists the viroids considered in the European certification scheme. All of them, even if to different extents, are present in EU countries.

TABLE 39.3 List of Viroids That Must Be Absent in Certified Propagative Material in the EU

Plant genus or species	Viroids[a]
Citrus L.	CEVd, HSVd (cachexia strain)
Cydonia oblonga	PBCVd
Malus Mills	ASSVd, ADFVd
Pyrus L.	PBCVd
Prunus persica	PLMVd

[a] *Abbreviations as in Table 39.1.*

CONCLUSIONS

Quarantine and phytosanitary certification programs are essential for preventing the inadvertent introduction of pests into a country or region, or export of pests to trading partners. Maintaining and cultivating plants free of harmful pests benefit both domestic agriculture and trading partners, and could help prevent trade conflict. These prevention programs are often perceived as being costly. The reality though is that if harmful pests of concern are introduced in regions where they do not exist, the economic impact and hardships that may result from crop loss or plant mortality and the costs to eradicate or manage these harmful pests usually far exceed those of prevention. Everyone benefits from effective pest management and prevention: (1) plants are more productive and the industry remains economically viable and attractive; (2) the various commodities produced are often of better quality and less expensive for the consumer; and (3) countries and governments are able to engage in safe, fair, and uninterrupted trade. Member countries that are signatories to the IPPC have an obligation to ensure that they take all necessary steps to prevent pest movement and introduction.

References

Barba, M., Gumpf, D.J., Hadidi, A., 2003a. Quarantine of imported germplasm. In: Hadidi, A., Flores, R., Randles, J.W., Semancik, J.S. (Eds.), Viroids. CSIRO Publishing, Collingwood, VIC, pp. 303–311.

Barba, M., 1998. Virus certification of fruit tree propagative material in Western Europe. In: Hadidi, A., Khetarpal, R.K., Koganezawa, H. (Eds.), Plant Virus Disease Control. APS Press, St. Paul, MN, pp. 288–293.

Barba, M., Ragozzino, E., Faggioli, F., 2007. Pollen transmission of peach latent mosaic viroid. J. Plant Pathol. 89, 287–289.

Barba, M., Ragozzino, E., Navarro, L., 2003b. Viroid elimination by thermotherapy and tissue culture. In: Hadidi, A., Flores, R., Randles, J.W., Semancik, J.S. (Eds.), Viroids. CSIRO Publishing, Collingwood, VIC, pp. 318–323.

Ebbels, D.L., 2003. Principles of plant health and quarantine. CABI Publishing, Wallingford, Oxon, p. 302.

Foster, J.A., Hadidi, A., 1998. Exclusion of plant viruses. In: Hadidi, A., Khetarpal, R.K., Koganezawa, H. (Eds.), Plant Virus Disease Control. APS Press Publishing, St. Paul, MN, pp. 208–229.

IPPC (International Plant Protection Convention), 2015. Potato spindle tuber viroid. International Standards for Phytosanitary Measures (ISPM) 27–Annex DP7.

Lebas, B.S.M., Clover, G.R.G., Ochoa-Corona, F.M., Elliott, D.R., Tang, Z., Alexander, B.J. R., 2005. Distribution of *Potato spindle tuber* viroid in New Zealand glasshouse crops of capsicum and tomato. Australian J. Plant Path. 34, 129–133.

NAPPO (North American Plant Protection Organization) Regional Standard for Phytosanitary Measures (RSPM) 3, 2011. Movement of potatoes into a NAPPO member country, 51 pp.

Randles, J.W., Rodriguez, M.J.B., 2003. Coconut cadang-cadang viroid. In: Hadidi, A., Flores, R., Randles, J.W., Semancik, J.S. (Eds.), Viroids. CSIRO Publishing, Collingwood, VIC, pp. 233–241.

Singh, R.P., Boucher, A., Somerville, T.H., 1992. Detection of potato spindle tuber viroid in the pollen and various parts of potato plants pollinated with viroid-infected pollen. Plant Dis. 76, 951–953.

Singh, R.P., Ready, K.F.M., Nie, X., 2003. Viroids of solanaceous species. In: Hadidi, A., Flores, R., Randles, J.W., Semancik, J.S. (Eds.), Viroids. CSIRO Publishing, Collingwood, VIC, pp. 125–133.

Su, H.J., 2008. Production and cultivation of virus-free citrus saplings for citrus rehabilitation in Taiwan. Asia-Pacific Consortium on Agricultural Biotechnology. New Delhi and Asia-Pacific Association of Agricultural Research Institutions, Bangkok, p. 51.

van Brunschot, S.L., Verhoeven, J.Th.J., Persley, D.M., Geering, A.D.W., Drenth, A., et al., 2014. An outbreak of potato spindle tuber viroid in tomato is linked to imported seed. Eur. J. Plant Pathol. 139, 1–7.

CHAPTER

40

Viroid Elimination by Thermotherapy, Cold Therapy, Tissue Culture, In Vitro Micrografting, or Cryotherapy

Marina Barba[1], *Munetaka Hosakawa*[2], *Qiao-Chun Wang*[3], *Anna Taglienti*[1] *and Zhibo Zhang*[3]

[1]CREA-Research Centre for Plant Protection and Certification, Rome, Italy [2]Kyoto University, Kyoto, Japan [3]Northwest A&F University, Xianyang, China

INTRODUCTION

Because viroid infections usually cause systemic disease, viroids cannot be controlled by therapeutic treatments in fields, orchards, vineyards, or palm plantations. Moreover, acquired resistance of plants to viroid infections is difficult to obtain (Matsushita et al., 2012; Omori et al., 2009; Chapter 42: Strategies to Introduce Resistance to Viroids). Hence, elimination of viroids from infected plants has been a challenge for the last four decades and different approaches have been tested (Barba et al., 2003; Kovalskaya and Hammond, 2014; Laimer and Barba, 2011). This chapter describes the use of thermotherapy, cold therapy, tissue culture, in vitro micrografting, and cryotherapy, in attempts to cure plant tissues from viroid infection.

THERMOTHERAPY

High temperature treatment was the first approach used for viroid elimination, as previously applied to plant viruses. The infected

Viroids and Satellites.
DOI: http://dx.doi.org/10.1016/B978-0-12-801498-1.00040-1

material, in vivo or in vitro, is exposed to higher temperatures for different times (weeks or months) depending on the viroid or variant, and host variety.

Apple scar skin viroid (ASSVd) was successfully eliminated from apple plants by 70 days exposure at 38°C followed by shoot-tip grafting. The eradication rate was different according to the variety considered and symptoms shown by the treated plant, ranging from 20% to 60% as confirmed by biological tests on woody indicator hosts followed by hybridization assays and RT-PCR. No symptoms of dapple apple and scar skin were observed on woody indicator plants after 3 years of field tests (Howell et al., 1998).

Hop plants infected by hop latent viroid (HLVd) were found to be viroid-free after 2 weeks at 35°C in vitro (Matousek et al., 1995). The success of this treatment was correlated with cleavage of HLVd double-stranded RNA by a nuclease, the activity of which increased significantly during thermotherapy; moreover, the distribution of the viroid was tissue-specific, being lower in the stem apex and higher in roots, and showed a negative correlation with the nuclease activity (Matousek et al., 1995).

In contrast, thermotherapy was largely ineffective to eliminate potato spindle tuber viroid (PSTVd) from infected potato plants grown in a chamber whose air temperature alternated daily between 33 and 36°C. Shoots were removed at different intervals (2–12 weeks) during treatment and axillary buds were cultured in vitro (Stace-Smith and Mellor, 1970). Only a small number of viroid-free plantlets (2.4%–6%) were obtained for all durations of the heat treatment. In fact, in tomato PSTVd actually accumulates at higher titer at growing temperatures above 30°C (Sänger et al., 1975). Similar results were obtained for chrysanthemum stunt viroid (CSVd), chrysanthemum chlorotic mottle viroid (CChMVd) (Hollings and Stone, 1970; Paludan, 1980), and hop stunt viroid (HSVd) (El-Dougdoug et al., 2010). Moreover, thermotherapy for 3 weeks in vitro at 37°C was not sufficient to eradicate HSVd from peach and pear plants (El-Dougdoug et al., 2010).

The uneven distribution of viroids throughout plant tissues mentioned above (i.e., low titer in stem apex, high titer in roots) is very common, and also was observed in PSTVd-infected *Nicotiana benthamiana* and tomato (Zhu et al., 2001); for this reason, thermotherapy has been often combined with meristem culture for viroid eradication. In pear, meristem culture applied after in vitro thermotherapy succeeded in eradicating ASSVd (as revealed by RT-PCR assay) from infected plantlets regrown from treated meristems (Postman and Hadidi, 1995). CSVd also was eliminated from chrysanthemum whole plants after 14–37 weeks at 35°C followed by meristem culture (Hollings and Stone, 1970).

Thermotherapy was also combined with other treatments, such as cold treatment or in vitro micrografting, and in some cases was more effective in eradicating viroids than thermotherapy alone. In apple plants infected by ASSVd, a dormant period of 3 months at 4°C followed by thermotherapy provided 90% elimination of the viroid, assayed on whole plants by biological indexing and molecular hybridization (Desvignes et al., 1999).

COLD THERAPY

Cold therapy consists of culturing plantlets in vitro at low temperatures for some weeks or months. This treatment is effective depending on the host–pathogen system, especially with viroids accumulating rapidly at high temperatures and which are not eliminated by thermotherapy. For example, a severe strain of PSTVd was eradicated from potato at a rate of c. 50% by growing plantlets in vitro at 5–6°C for 6 months, and then excising their meristems; the viroid eradication was then assessed on whole plantlets after regrowth (Lizárraga et al., 1980). In contrast, meristems taken from untreated plantlets were all infected. Similar results were obtained by Mahfouze et al. (2010) using dot-blot hybridization diagnostic techniques.

HSVd, which also accumulates rapidly at high temperatures, was eliminated from peach and pear plants by cold therapy at 4°C for 3 weeks with an eradication rate of 18% (El-Dougdoug et al., 2010). The whole plantlets were grown in vitro at 4°C; then the apical part about 1–2 mm in length was transferred to a fresh medium and returned to the 25°C growth chamber.

Chrysanthemum plants of different cultivars infected by CSVd, CChMVd, or cucumber pale fruit viroid (a variant of HSVd) were cold-treated at 5°C for 6 months; the meristems obtained therefrom were found to be viroid-free, with elimination rates varying according to the plant–viroid combination: 18% for CSVd on cv. "Bonnie Jean," 70% for CChMVd on cv. "Deep Ridge," and 80% for HSVd on cv. "Mistletoe" (Paduch-Cichal and Kryczynski, 1987). Similar results were obtained by Savitri et al. (2013) with CSVd.

HLVd-infected hops could also be cured by cold treatment: meristem culture of plantlets grown at 4°C for several months produced 36% viroid-free material consisting of whole plants of six cultivars and male pollinator clones (Adams et al., 1996).

High temperature-sensitive viroids also could be successfully eliminated by cold therapy: 86% ASSVd-free pear plants (as revealed by RT-PCR assay) were obtained from meristems excised after 55 days treatment at 4°C, an eradication rate nearly identical to that obtained by thermotherapy (Postman and Hadidi, 1995).

MERISTEM TISSUE CULTURE

Vascular plants have shoot apical meristems (SAMs), which are composed of undifferentiated cells. All aerial organs are differentiated from SAMs. Because lateral buds also develop from SAMs, cutting nodes with lateral buds enables clonal propagation of crops. Viroid invasion into SAMs or the tip of SAMs is not observed in many infected plants. Therefore, regeneration from leaf primordium (LP)-free (the noninfected portions of) SAMs results in viroid-free plants (Hosokawa et al., 2004a,b).

The eradication rate depends on the size of the shoot, the viroid or its variant, and the host and its variety. Dissection of a 0.1–0.3 mm shoot tip followed by regeneration of the whole plant was effective in eliminating citrus exocortis viroid from tomato (Duran-Vila and Semancik, 1986), grapevine yellow speckle viroid and HSVd from grapevine (Duran-Vila et al., 1988), and HSVd from hop (Momma and Takahashi, 1983).

In meristem-based methods for CSVd eradication, it is common that the size of the shoot tip required to obtain CSVd-free plants differs among the genotypes of the same species (Chung et al., 2001; Hosokawa et al., 2004a; Kim et al., 2012; Kovalskaya and Hammond, 2014). These data suggest that invasion of SAMs by CSVd might differ among plant cultivars. The size of the shoot tip of CSVd-infected chrysanthemum is particularly crucial to obtain viroid-free plants (Kim et al., 2012), as it affects both the meristem survival and the viroid elimination rate. Meristems of 0.5 mm or larger were more viable than smaller ones (0.2–0.4 mm with one or two LP) but no CSVd-free plants were produced. However, smaller meristems were c. 25% CSVd-free, but their survival rate was lower. Hence, a balance of these two requirements should be met. Hosokawa et al. (2005, 2006) examined CSVd and/or CChMVd, neither of which was detected in SAMs of infected chrysanthemum cultivars. Omori et al. (2009) identified chrysanthemum cultivars resistant to CSVd in Japan. When resistant plants were infected with CSVd by grafting, the viroid was not present in the newly expanded leaves generated from SAMs after the removal of the infected rootstocks. This observation suggests that CSVd-resistant cultivars have viroid-free SAMs. Nabeshima et al. (2012) evaluated many commercial Japanese cultivars by inoculating their SAMs with CSVd and identified some resistant cultivars. Zhang et al. (2015) reported that the presence of CSVd in SAMs of *Argyranthemum* depends on the cultivar.

Accordingly, work should focus on susceptible cultivars with respect to the invasion of CSVd or other viroids into SAMs.

Additional viroids were also reported to be eliminated by shoot-tip culture regardless of the viroid family, as SAMs seem to be protected from viroid invasion. Rodio et al. (2007) did not detect peach latent mosaic viroid (PLMVd) at the tips of SAMs of infected peach plants by in situ hybridization. Plus and minus strands of PLMVd were detected below SAM tips, suggesting that PLMVd not only accumulates in but also proliferates below the tip. Zhu et al. (2001) did not detect PSTVd in SAMs of *N. benthamiana* and tomato using in situ hybridization. Furthermore, Di Serio et al. (2010) did not detect PSTVd in SAMs of *N. benthamiana*, and Matsushita et al. (2011) reported that tomato chlorotic dwarf viroid (TCDVd) could not be detected in tomato SAMs.

Few studies have examined the reasons why viroids are not present in SAMs of infected plants. Zhang et al. (2015) indicated that β-1,3-glucan deposited in the plasmodesmata between cells located below SAMs could explain the protection from viroid invasion to SAMs observed in *Argyranthemum* (see below). Di Serio et al. (2010) revealed a role of RNA-directed RNA polymerase 6 (RDR6) in the resistance of PSTVd invasion into SAMs using *N. benthamiana* plants in which RDR6 was silenced.

Localization of viroids in SAMs has helped the development of various approaches for viroid eradication. There have been several studies on viroid localization in SAMs using in situ hybridization, including PSTVd in *N. benthamiana* (Di Serio et al., 2010; Zhu et al., 2001), and PLMVd in peach (Rodio et al., 2007). In hop, Momma and Takahashi (1983) did not detect HSVd in 0.2 mm SAMs containing apical domes (ADs) and the first two LPs in the infected plants. In *N. benthamiana*, Zhu et al. (2001) found that PSTVd was absent in the ADs, but present in tissues containing prophloem right below the ADs. Di Serio et al. (2010) found that PSTVd could infect SAMs, including uppermost cell layers of ADs of RDR6-silenced *N. benthamiana* plants, but not the SAMs of wild-type *N. benthamiana*. Rodio et al. (2007) reported PLMVd in the ADs and LPs of peach, leaving only a few uppermost cell layers of SAMs free of viroid. The distribution patterns of different viroids in the same host have also been studied. In tomato, although both PSTVd (Qi and Ding, 2003) and TCDVd (Matsushita et al., 2011) could not invade the apical meristems, the size of viroid-free regions of SAMs differed from each other: about 200 μm (in length) for PSTVd and 50 μm for TCDVd. These data indicate that the ability of viroids to invade SAMs differs among viroids, and explain why viroid eradication frequency varies with viroid-host combinations (El-Dougdoug et al., 2010; Grudzińska and Solarska, 2005; Hollings and Stone, 1970; Hosokawa et al., 2004a,b; Paduch-Cichal and Kryczynski, 1987; Paludan, 1985;

Savitri et al., 2013). Moreover, these studies provide a better understanding of why viroids are much more difficult to eradicate than other pathogens, such as viruses and phytoplasmas.

Recently, Zhang et al. (2015) used in situ hybridization to locate CSVd in stems, flower organ and SAMs of four diseased *Argyranthemum* cultivars. In situ hybridization with strand-specific DIG-labeled CSVd antisense-probes resulted in purple-blue signals in the CSVd-infected tissues (Fig. 40.1A) that were not seen in the healthy samples (Fig. 40.1B). CSVd was readily detected across all tissues in stems, including epidermal cells, phloem and pith in cultivars "Yellow Empire" (Fig. 40.1C) and "Border Dark Red" (Fig. 40.1D). CSVd was also detected in flower organs, including sepals, petals and ovaries in "Yellow Empire" (Fig. 40.1E) and "Border Dark Red" (Fig. 40.1F), but not in ovules (Fig. 40.1E and F). There were no differences in the ability of CSVd to infect stems and flower organs between the two cultivars. Strong viroid signals were revealed throughout SAMs (Fig. 40.1G), including ADs, and LPs 1 and 2 (Fig. 40.1H) in "Yellow Empire." Similar localization patterns were observed in "Butterfly" (Fig. 40.1I). In the diseased "Border Dark Red," CSVd was easily detected in the lower parts of ADs, in LP 3 and older tissues of SAMs (Fig. 40.1J). However, no viroid was detected in the uppermost section of ADs, and LPs 1 and 2 (Fig. 40.1K). The viroid-free area of ADs in "Border Dark Red" contained about 20 layers of cells, being approximately 0.2 mm in size. Similar patterns of CSVd distribution were observed in "Border Pink" (Fig. 40.1L). These data clearly demonstrate that the ability of CSVd to invade SAMs differs among *Argyranthemum* cultivars, thus explaining why using meristems of the same size resulted in different CSVd-free frequencies (Zhang et al., 2015).

IN VITRO MICROGRAFTING

This technique, consisting of grafting a dissected shoot tip onto a rootstock grown in vitro, may be helpful when the excised shoot tip is not able to regenerate by itself. For example, most tree species behave differently from herbaceous plants and do not regenerate to whole plants by shoot-tip culture. Similarly, LP-free shoot tips, excised in order to obtain a viroid-free tissue, need plant growth regulators to regenerate, the use of which in the culture media is strongly discouraged because of the risk of mutations. In these cases, in vitro micrografting provides a useful tool for favoring regrowth from a very small shoot tip. The elimination rate also depends on the size of the excised tip, the viroid, and the host plant and variety. The main drawback of

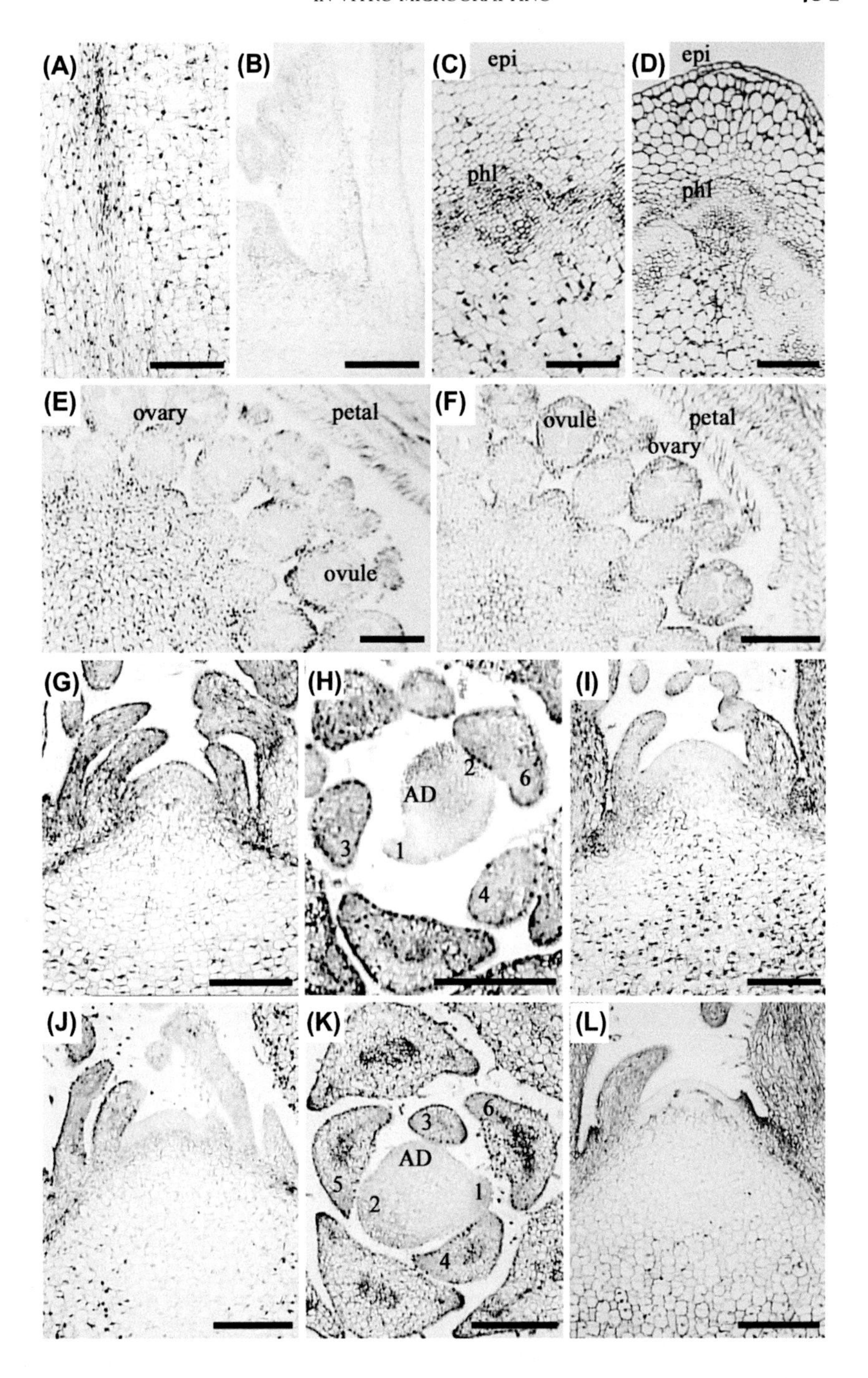
(A)
(B)
(C)
epi
phl
(D)
epi
phl
(E)
ovary
petal
ovule
(F)
ovule
petal
ovary
(G)
(H)
AD
1
2
3
4
6
(I)
(J)
(K)
AD
1
2
3
4
5
6
(L)

the method is that it requires a high level of expertise due to difficulties of technical manipulation.

PLMVd was successfully eradicated from peach using Nemaguard seedlings as rootstock, with the elimination rate depending significantly on the cultivar (Barba et al., 1995). CSVd and CChMVd were eliminated from chrysanthemum cultivar "Piato," which is one of the most recalcitrant for viroid eradication. A grafting method developed by Hosokawa et al. (2004a,b) allowed LP-free SAMs to be attached to CSVd-free chrysanthemum or cabbage root tips, obtaining high survival and viroid elimination rates. The method is based on the observation that root tips have a high rate of cell division and enable chrysanthemum LP-free SAMs to remain viable. Cabbage root tips were already observed to be suitable for rescuing the LP-free SAMs of other plant species such as *Fragaria* (strawberry) or *Dianthus*, hence the method is likely to be useful for the production of virus or viroid-free germplasm of different species.

HSVd and citrus exocortis viroid were eliminated from citrus trees by shoot-tip grafting (Navarro et al., 1975), allowing 9.5% recovery of viroid-free plants (Kapari-Isaia et al., 2002).

CRYOTHERAPY

Cryotherapy of shoot tips is an innovative technique for pathogen eradication based on cryopreservation. The principle of the method lies in the different water contents of healthy and infected cells in the apical tissue. Infected cells have larger vacuoles, higher water content, and lower nucleus/cytoplasm ratio with respect to healthy ones, hence they are more sensitive to ice damage and death during cryotherapy, which in turn operates a sort of "selection" toward healthy cells. This approach has already been successfully applied to germplasm of potato, sweet potato, grapevine, banana, raspberry, plums, and prunes

FIGURE 40.1 In situ hybridization detection of CSVd in stems, flower organs and shoot apical meristems (SAM) of *Argyranthemum* cultivars: (A) CSVd-infected tissue (positive control). (B) Healthy tissue (negative control). (C, D) Stems of CSVd-infected cv. "Yellow Empire" and "Border Dark Red"; epi and phl indicate epidermal cells and phloem, respectively. (E, F) Cross section of flower organs of CSVd-infected cv. "Yellow Empire" and "Border Dark Red." (G, H) Longitudinal and cross sections of CSVd-infected cv. "Yellow Empire" SAM. (I) Longitudinal section of CSVd-infected cv. "Butterfly" SAM. (J) and (K) longitudinal and cross sections of CSVd-infected cv. "Border Dark Red" SAM. (L) Longitudinal section of CSVd-infected cv. "Border Pink" SAM; AD indicates apical dome, numbers 1–6 indicate leaf primordia 1–6. Scale bars are 200 μm (Zhang et al, 2015). The probe is DIG-labeled and infected cells are stained indigo in color.

to eliminate pathogens such as viruses, phytoplasmas, and bacteria. Some preliminary experiments have been reported on CSVd in *Argyranthemum maderense*, but the results of in situ hybridization and histological analysis on cryotreated samples showed that CSVd invaded meristematic cells and that these cells were able to survive cryotherapy. Hence, further experiments on other host–viroid systems and on combination of cold therapy and cryotherapy are under evaluation (Zhang et al., 2014).

References

Adams, A.N., Barbara, G.J., Morton, A., Darby, P., 1996. The experimental transmission of hop latent viroid and its elimination by low temperature treatment and meristem culture. Ann. Appl. Biol. 128, 37–44.

Barba, M., Cupidi, A., Loreti, S., Faggioli, F., Martino, L., 1995. *In vitro* micrografting: a technique to eliminate peach latent mosaic viroid from peach. Acta Hortic. 386, 531–535.

Barba, M., Ragozzino, E., Navarro, L., 2003. Viroid elimination by thermotherapy and tissue cultures. In: Hadidi, A., Flores, R., Randles, J.W., Semancik, J.S. (Eds.), Viroids. CSIRO Publishing, Collingwood, VIC, pp. 318–323.

Chung, B.N., Choi, G.S., Kim, H.R., Kim, J.S., 2001. Chrysanthemum stunt viroid in *Dendranthema grandiflorum*. Plant Pathol. J. 17, 194–200.

Desvignes, J.C., Grasseau, N., Boyé, R., Cornaggia, D., Aparacio, F., Di Serio, F., et al., 1999. Biological properties of apple scar skin viroid: isolates, host range, different sensitivity of apple cultivars, elimination and natural transmission. Plant Dis. 83, 768–772.

Di Serio, F., Martínez de Alba, A.E., Navarro, B., Gisel, A., Flores, R., 2010. RNA-dependent RNA polymerase 6 delays accumulation and precludes meristem invasion of a nuclear-replicating viroid. J. Virol. 84, 2477–2489.

Duran-Vila, N., Juàrez, J., Arregui, J.M., 1988. Production of viroid-free grapevines by shoot tip culture. Am. J. Enol. Viticult. 39, 217–220.

Duran-Vila, N., Semancik, J.S., 1986. Shoot-tip culture and eradication of viroid-RNA. Sci. Hortic. 29, 199–203.

El-Dougdoug, Kh.A., Osman, M.E., Abdelkader, H.S., Dawoud, R.A., Elbaz, R.M., 2010. Elimination of hop stunt viroid (HSVd) from infected peach and pear plants using cold therapy and chemotherapy. Aust. J. Basic Appl. Sci. 4, 54–60.

Grudzińska, M., Solarska, E., 2005. The elimination of virus and hop latent viroid from hop (*Humulus lupulus* L.) in Poland. Acta Hortic. 668, 149–152.

Hollings, M., Stone, O.M., 1970. Attempts to eliminate chrysanthemum stunt from chrysanthemum by meristem-tip culture after heat treatment. Ann. Appl. Biol. 65, 311–315.

Hosokawa, M., Matsushita, Y., Ohishi, K., Yazawa, S., 2005. Elimination of chrysanthemum chlorotic mottle viroid (CChMVd) recently detected in Japan by leaf primordia-free shoot apical meristem culture from infected cultivars. J. Japan. Soc. Hort. Sci. 74, 386–391.

Hosokawa, M., Matsushita, Y., Uchida, H., Yazawa, S., 2006. Direct RT-PCR method for detecting two chrysanthemum viroids using minimal amounts of plant tissue. J. Virol. Methods 131, 28–33.

Hosokawa, M., Otake, A., Ohishi, K., Ueda, E., Hayashi, T., Yazawa, S., 2004b. Elimination of chrysanthemum stunt viroid from an infected chrysanthemum cultivar by shoot regeneration from a leaf primordium-free shoot apical meristem dome attached to a root tip. Plant Cell Rep. 22, 859–863.

Hosokawa, M., Otake, A., Sugawara, Y., Hayashi, T., Yazawa, S., 2004a. Rescue of shoot apical meristems of chrysanthemum by culturing on root tips. Plant Cell Rep. 22, 443–448.

Howell, W.E., Burgess, J., Mink, G.I., Skrzeczkowski, L.J., Zhang, Y.P., 1998. Elimination of apple fruit and bark deforming agents by heat therapy. Acta Hortic. 472, 641–646.

Kapari-Isaia, Th., Minas, G.J., Polykarpou, D., Iosephidou, E., Arseni, Sp., Kyriakou, A., 2002. Shoot-tip grafting *in vitro* for elimination of viroids and citrus psorosis virus in the local Arakapas mandarin in Cyprus. In: Proc. 15th IOCV Confer, Riverside, CA, USA, pp. 417–420.

Kim, C.-K., Jeon, S.-M., Savitri, W.D., Park, K.I., Chung, M.-Y., 2012. Elimination of chrysanthemum stunt viroid (CSVd) from a viroid-infected chrysanthemum through shoot tip culture. Flower Res. J. 20, 218–222.

Kovalskaya, N., Hammond, R.W., 2014. Molecular biology of viroid-host interactions and disease control strategies. Plant Sci. 228, 48–60.

Laimer, M., Barba, M., 2011. Elimination of systemic pathogens by thermotherapy, tissue culture, or *in vitro* micrografting. In: Hadidi, A., Barba, M., Candresse, T., Jelkmann, J. (Eds.), Virus and Virus-Like Diseases of Pome and Stone Fruits. APS Press, St. Paul, MN, pp. 389–393.

Lizárraga, R.E., Salazar, S.F., Roca, W.M., Schilde-Rentschler, L., 1980. Elimination of potato spindle tuber viroid by low temperature and meristem culture. Phytopathology 70, 754–755.

Mahfouze, S.A., El-Dougdoug, Kh.A., Allam, E.K., 2010. Production of potato spindle tuber viroid-free potato plant materials *in vitro*. N.Y. Sci. J. 3, 60–67.

Matousek, J., Trnená, L., Svoboda, P., Oriniaková, P., Lichtenstein, C.P., 1995. The gradual reduction of viroid levels in hop mericlones following heat therapy: a possible role for a nuclease degrading dsRNA. Biol. Chem. 376, 715–721.

Matsushita, Y., Aoki, K., Sumitomo, K., 2012. Selection and inheritance of resistance to chrysanthemum stunt viroid. Crop Prot. 35, 1–4.

Matsushita, Y., Usugi, T., Tsuda, S., 2011. Distribution of tomato chlorotic dwarf viroid in floral organs of tomato. Eur. J. Plant. Pathol. 130, 441–447.

Momma, T., Takahashi, T., 1983. Cytopathology of shoot apical meristem of hop plants infected with hop stunt viroid. J. Phytopathol. 106, 272–289.

Nabeshima, T., Hosokawa, M., Yano, S., Ohishi, K., Doi, M., 2012. Screening of chrysanthemum cultivars with resistance to chrysanthemum stunt viroid. J. Jpn. Soc. Hort. Sci. 81, 285–294.

Navarro, L., Roistacher, C.N., Murashige, T., 1975. Improvement of shoot-tip grafting *in vitro* for virus-free citrus. J. Amer. Soc. Hort. Sci. 100, 471–479.

Omori, H., Hosokawa, M., Shiba, H., Shitsukawa, N., Murai, K., Yazawa, S., 2009. Screening of chrysanthemum plants with strong resistance to chrysanthemum stunt viroid. J. Japan. Soc. Hort. Sci. 78, 350–355.

Paduch-Cichal, E., Kryczynski, S., 1987. A low temperature therapy and meristem-tip culture for eliminating four viroids from infected plants. J. Phytopathol. 118, 341–346.

Paludan, N., 1980. Chrysanthemum stunt and chlorotic mottle. Establishment of healthy chrysanthemum plants and storage at low temperature of chrysanthemum, carnation, campanula and pelargonium in tubes. Acta Hortic. 110, 303–315.

Paludan, N., 1985. Inactivation of viroids in *Chrysanthemum* by low-temperature treatment and meristem-tip culture. Acta Hortic. 164, 181–186.

Postman, J.D., Hadidi, A., 1995. Elimination of apple scar skin viroid from pears by *in vitro* thermotherapy and apical meristem culture. Acta Hortic. 386, 536–543.

Qi, Y., Ding, B., 2003. Inhibition of cell growth and shoot development by a specific nucleotide sequence in a noncoding viroid RNA. Plant Cell. 15, 1360–1374.

Rodio, M.E., Delgado, S., De Stradis, A., Gomez, M.D., Flores, R., Di Serio, F., 2007. A viroid RNA with a specific structural motif inhibits chloroplast development. Plant Cell. 19, 3610–3626.

Sänger, H.L., Ramm, K., 1975. Radioactive labeling of viroid RNA. In: Markham, R., et al., (Eds.), Modification of the Information Content in Plant Cells. Proc. 2nd John Innes Symp., July 1974. Norwich, England.

Savitri, W.D., Park, K.I., Jeon, S.M., Chung, M.Y., Han, J.-S., Kim, C.K., 2013. Elimination of chrysanthemum stunt viroid (CSVd) from meristem tip culture combined with prolonged cold treatment. Hort. Environ. Biotechnol. 54, 177–182.

Stace-Smith, R., Mellor, F.C., 1970. Eradication of potato spindle tuber virus by thermotherapy and axillary bud culture. Phytopathology 60, 1857–1858.

Zhang, Z., Lee, Y.K., Spetz, C., Clarke, J.L., Wang, Q.C., Blystad, D.-R., 2015. Invasion of shoot apical meristems by chrysanthemum stunt viroid differs among *Argyranthemum* cultivars. Front. Plant Sci. 6, 53.

Zhang, Z.B., Haugslien, S., Clark, J.H.L., Spetz, C., Blystad, D.-R., Wang, Q.C., et al., 2014. Cryotherapy could not eradicate chrysanthemum stunt viroid from infected *Argyranthemum maderense* "Yellow Empire." Acta Hortic. 1039, 201–208.

Zhu, Y., Green, L., Woo, Y.M., Owens, R., Ding, B., 2001. Cellular basis of potato spindle tuber viroid systemic movement. Virology 279, 66–77.

CHAPTER

41

Decontamination Measures to Prevent Mechanical Transmission of Viroids

Kai-Shu Ling

U.S. Department of Agriculture, Charleston, SC, United States

INTRODUCTION

In recent years, a significant increase in the number of reported viroid disease outbreaks on tomatoes and ornamentals has concerned vegetable growers and seed producers. These viroid disease outbreaks have occurred worldwide, including Asia (Antignus et al., 2002; Cui et al., 1992; Matsushita et al., 2008; Mishra et al., 1991), Africa (Batuman and Gilbertson, 2013; Candresse et al., 2007; Verhoeven et al., 2006; Walter, 1981), Australia (Hailstones et al., 2003; Mackie et al., 2002), New Zealand (Elliott et al., 2001; Lebas et al., 2005), Europe (Candresse et al., 2010; Navarro et al., 2009; Puchta et al., 1990; Verhoeven et al., 2004, 2008, 2012), North America (Galindo et al., 1982; Ling et al., 2009, 2013; Ling and Bledsoe, 2009; Ling and Sfetcu, 2010; Ling and Zhang, 2009; Martínez-Soriano et al., 1996; Singh et al., 1999), and South America (Ling et al., 2014; Orozco-Vargas, 1983). This sudden surge in viroid epidemics coincides with the advent of increasingly popular intensive cultural practices used in protective crop production. Those cultural practices, including grafting, deleafing, twisting, fruit picking, and long production cycles, have created ample opportunities for mechanical transmission of viroids.

A diverse array of distinct pospiviroids, including potato spindle tuber viroid (PSTVd), tomato chlorotic dwarf viroid, citrus exocortis viroid, Columnea latent viroid, tomato apical stunt viroid, tomato planta macho viroid (including Mexican papita viroid), and pepper chat

Viroids and Satellites.
DOI: http://dx.doi.org/10.1016/B978-0-12-801498-1.00041-3

fruit viroid, have been identified as causal agents of viroid disease in tomato. Since there are no cultivars displaying viroid resistance, significant yield losses have occurred. The frequent hands-on activities in greenhouse tomato production could quickly spread a viroid disease to a larger area if a proper hygiene and decontamination measure is not implemented.

UNDERSTANDING THE MEANS OF POSSIBLE VIROID INTRODUCTION

Since quite a few viroids are seed-borne and seed-transmitted to germinating seedlings, a primary source of initial viroid inoculum is through the use of contaminated seeds. It is therefore important to use a seed lot that has been tested as viroid-free. Currently, there are no effective seed treatment methods available for viroid decontamination. With localization of viroids in embryos, chemotherapy is unlikely to work as a seed treatment either. Therefore, selection of seed lots that have been certified as viroid-free is very important. For plants used for greenhouse tomato production, which are typically grafted, both scion and rootstock seeds should be tested and certified as viroid-free. Strict hygienic best practice and the use of an appropriate disinfectant are also important measures to prevent secondary viroid transmission from grafting operation and plant handling. Growers should be encouraged to demand that seed suppliers provide evidence of seed freedom from viroids. Seed companies should implement a rigorous seed certification program and employ the most sensitive and reliable seed health assays that are available.

Viroids can also be introduced through infected vegetative propagules. Seed tubers that are used for potato production should come from a potato certification program and maintained as viroid-free throughout the nursery multiplication process. For many fruit trees, such as avocado, citrus, pome, and stone fruits, as well as grapevines, plants and plant parts (like buds and cuttings) are used for propagation and planting. Fruit tree and grapevine productions generally involve grafting of a scion material to a rootstock. Pruning and fruit picking activities can also increase chances of viroid transmission. Therefore, it is necessary to adopt best hygienic practices and the use of effective disinfectants to prevent accidental viroid introduction.

Viroids may also be introduced through worker's hands from eating fresh fruits and vegetables or by handling of contaminated ornamentals. Although the chance of introducing a viroid to a production facility by workers who have come into contact with a viroid-contaminated plant product in food preparation or consumption is low, prevention measures should also include not bringing in outside vegetables or fruits to the premises during crop production. Because a number of ornamental plants have been shown to harbor PSTVd and other pospiviroids, they could serve as a source of initial inoculum during tomato production (Singh and Dilworth, 2008; Verhoeven et al., 2012).

Most if not all viroids are mechanically transmissible. For many viroids, where no definite insect vectors are known for field transmission, human cultural activities, such as plant handling, pruning of leaves, and fruit picking, serve as the major means of viroid spread. Measures for decontamination and proper cultural practices should be implemented to prevent viroid spread. In addition to frequent handwashing, disinfectants have also been shown to be effective in deactivating viroid infectivity, thereby preventing their spread.

DISINFECTANTS AGAINST MECHANICAL TRANSMISSION OF VIROIDS

Mechanical transmission is the main means of viroid spread. Therefore, decontamination of tools and hands are important measures in preventing spread of a viroid disease. Since viroid RNA is unprotected, its infectivity could be sensitive to direct contact with an effective disinfectant. Viroid infectivity tolerates heat treatment, and even disinfection of knives and pruning tools after dipping in 70% or 95% alcohol followed by flaming of blades could not completely inactivate viroid infectivity (Roistacher et al., 1969). Over the years, numerous other chemical disinfectants have been evaluated and several of them are effective in disinfecting cutting tools against transmission of viroids (Table 41.1). Among them, household bleach at a 10%–20% dilution (0.5%–1% sodium hypochlorite active ingredient) was consistently demonstrated to be effective against multiple viroids (Desjardins et al., 1987; Garnsey and Whidden, 1971; Li et al., 2015; Mackie et al., 2015; Matsuura et al., 2010; Olivier et al., 2015; Roistacher et al., 1969; Singh et al., 1989). The corrosive action of sodium hypochlorite on metal cutting tools and greenhouse structures may be neutralized post-treatment with an emulsion of vinegar–oil–water (25:2:73) (Roistacher et al., 1969). Another effective disinfection formulation is 2% formaldehyde plus 2% sodium hydroxide (Desjardins et al., 1987; Garnsey and Whidden, 1971; Matsuura et al., 2010; Roistacher et al., 1969). Recently, a new disinfectant named Virkon S (DuPont Chemical Solutions, Wilmington, DE) at 2% was effective against PSTVd and three

TABLE 41.1 List of Disinfectants and Their Effectiveness in Decontamination of Viroids

Viroid[a]	Most effective	Partially effective	Poor to no effect	Reference[b]
ASBVd	20% household bleach (1% sodium hypochlorite) 2% formaldehyde + 2% sodium hydroxide 6% hydrogen peroxide	–	–	Desjardins, et al. (1987)
CEVd	5%–10% household bleach (0.25%–0.5% sodium hypochlorite) 2% formaldehyde + 2% sodium hydroxide	2% formaldehyde 100% milk, fresh or powder 5%–10% Bromodine 1% detergent, rinsed	1% and 10% Triodine 1% Roccal 2% sodium hydroxide 10% trisodium hypochlorite 95% ethyl alcohol/flamed Diesel fuel	Garnsey and Whidden (1971)
CEVd	5%, 10%, 20% household bleach (0.25%, 0.5%, 1%, sodium hypochlorite) 2% sodium hydroxide + 2% Formalin (37% formaldehyde)	20% Phisohex 3% Formalin (37% formaldehyde)	2% trisodium phosphate 10% Borax 1% Lysol 95% ethyl alcohol Vinegar–oil–water (4:1/3:12) Alcohol and flame	Roistacher et al. (1969)
CSVd	5 μg/mL RNase in 0.005 M phosphate buffer	20% diethyl ether	2% trisodium orthophosphate 2% formaldehyde Ultraviolet radiation Sonication	Hollings and Stone (1973)
PSTVd	2%–3% sodium hypochlorite	1% sodium hypochlorite Incyte	Alcide LD Exspor	Singh et al. (1989)
PSTVd	2% Menno Florades (9% (w/v) Benzoic acid) 3% Menno Florades (9% (w/v) Benzoic acid)	–	–	Timmermann et al. (2001)

(Continued)

TABLE 41.1 (Continued)

Viroid[a]	Most effective	Partially effective	Poor to no effect	Reference[b]
PSTVd	10% Clorox regular bleach (0.5% sodium hypochlorite) 2% Virkon S (20.4% potassium peroxymonosulfate, 1.5% sodium chloride)	20% nonfat dried milk 50% lysol cleaner 0.781% Octave 1% MENNO TER forte 0.52% GREENSHIELD 0.977% StorOx	0.5% Virkon S 1.0% Virkon S 1% Menno Florades 0.4% KleenGrow 0.11% Greenhouse Guardian 0.195% Vortexx 0.078% BioSide 0.382% SaniDate 10% trisodium phosphate 0.1% DES-O-GERM	Li et al. (2015)
PSTVd	20% nonfat dried skim milk (34.78% protein) 25% household bleach (1.26% sodium hypochlorite)	–	0.5% Virkon S 1.0% Virkon S 10% Farmcleanse	Mackie et al. (2015)
PSTVd	0.5% Virocid 0.75% Hyprelva SL 1% Virkon S 1% Jet 5 16% household bleach (0.8% sodium hypochlorite)	–	1% MENNO CLEAN foam detergent	Olivier et al. (2015)
TCDVd	3% sodium hypochlorite 20% household bleach	2% sodium hydroxide + 2% formaldehyde mixture 5% trisodium phosphate	2.5% trisodium phosphate Electrolyzed acid water N hydrochloric acid 70% isopropanol	Matsuura et al. (2010)

[a] ASBVd, avocado sunblotch viroid; CEVd, citrus exocortis viroid; CSVd, chrysanthemum stunt viroid; PSTVd, potato spindle tuber viroid; TCDVd, tomato chlorotic dwarf viroid.
[b] *For the active ingredients to each disinfectant and its respective manufacturer's information, please refer to each cited reference for details.*

viruses (Li et al., 2015). Results, however, were inconsistent when a lower concentration (<1%) of Virkon S was used (Li et al., 2015; Mackie et al., 2015; Olivier et al., 2015). The active ingredient for Virkon S is potassium peroxymonosulfate. Another interesting finding was that milk is effective against viroid transmission (Li et al., 2015; Mackie et al., 2015), which offers an alternative to corrosive disinfectants. Even though milk has been shown years ago to prevent mechanical transmission of viruses, its mode of action against viruses and viroids is still unknown. On the other hand, Menno Florades, which was effective against PSTVd transmission at two higher concentrations (2% and 3%) (Timmermann et al., 2001), had only a poor effect at the recommended concentration (1%) in two independent cases (Li et al., 2015; Olivier et al., 2015). Conversely, trisodium phosphate, which is commonly used for virus decontamination, was ineffective against viroids (Garnsey and Whidden, 1971; Hollings and Stone, 1973; Li et al., 2015; Matsuura et al., 2010; Roistacher et al., 1969).

INTEGRATED DISEASE MANAGEMENT AGAINST VIROID DISEASES

Although decontamination is an important measure in viroid disease management, it is necessary to take an integrated approach for effective viroid control. The primary strategy in managing viroid diseases is to prevent introduction of an inoculum by using viroid-free seed and propagating materials. Production of viroid-free materials from viroid-infected tissues could be achieved through cold or heat treatment followed by meristem shoot-tip culture (see Chapter 40: Viroid Elimination by Thermotherapy, Cold Therapy, Tissue Culture, In Vitro Micrografting, or Cryotherapy) or other methods. Prevention could also be achieved through quarantine and importation from a country or region where the viroid has not been identified. Quarantine enforcement can only be made meaningful with simple and sensitive detection. With increasing activities in offshore hybrid seed production and a rather broad distribution of some viroids on crop plants and possibly other reservoir plant hosts, sensitive detection is necessary. Once a diseased plant is confirmed through the test, it should be removed promptly in a containment plastic bag. For those facilities where a viroid outbreak has become endemic, crop rotation to a nonhost crop should be considered, thus to break the disease cycle.

ESTABLISHING GENERAL RULES OF CONDUCT IN CROP PRODUCTION TO MINIMIZE VIROID SPREAD

Cultural activities by workers are considered the main means of viroid transmission during crop production. To minimize the chance of spreading a viroid to another area in a greenhouse/field, managers need to clearly mark the disease zone and assign selected crop workers to a designated area of work, thus minimizing further spread. It is also necessary to educate and ensure that workers use a proper disinfectant solution when touching plants. Although disinfectant efficacy may be stable in clean solutions for an extended period (within a month) (Li et al., 2015), disinfectant potency could be diminished quickly once it is contaminated with tissue debris. Therefore maintaining a fresh disinfectant solution is necessary.

In addition to disinfection of cutting tools, it is also necessary to clean and disinfect equipment. Viroids can be spread through movement and touching of plants from operating and handling of contaminated dirty equipment, such as trolleys, picking carts, and holding bins. Therefore, when moving a piece of equipment to a new area, between rows and at the end of a work day, it is necessary to clean and disinfect such equipment to prevent viroid spread to neighboring healthy plants. To avoid stepping on tissue debris and spreading it to other areas, it is necessary to frequently sweep and clean tissue debris in walkways. At the conclusion of each working day, any reusable equipment (knives, pruners, and clips) should be thoroughly cleaned with soapy water and sanitized with proper disinfectant before it is used again. To remove any potential viroid inoculum, it is also advantageous to promptly bag, remove, and dispose of leaves and tissue debris by burning or deep burial away from the production facility.

DISINFECTION OF GREENHOUSE BETWEEN CROPS

At the conclusion of each crop cycle, a total cleanup of an entire greenhouse to break the disease cycle is an effective strategy to control a viroid outbreak. The workers should wear clean protective clothing to defend themselves from splashing of detergent and disinfectants. The greenhouse concrete floor, walkways, and walls should be swept clean, power-washed, and disinfected with appropriate disinfectants as shown in Table 41.1.

Acknowledgments

This work is supported in part by grants (SCRI- 2010–600–25320) and (SCRI-2012–01507–229756) from the USDA, National Institute of Food and Agriculture, Specialty Crop Research Initiative Program to KSL.

References

Antignus, Y., Lachman, O., Pearlsman, M., Gofman, R., Bar-Joseph, M., 2002. A new disease of greenhouse tomatoes in Israel caused by a distinct strain of tomato apical stunt viroid (TASVd). Phytoparasitica 30, 502–510.

Batuman, O., Gilbertson, R.L., 2013. First report of columnea latent viroid (CLVd) in tomato in Mali. Plant Dis. 97, 692–693.

Candresse, T., Marais, A., Ollivier, F., Verdin, E., Blancard, D., 2007. First report of the presence of tomato apical stunt viroid on tomato in Senegal. Plant Dis. 91, 330.

Candresse, T., Marais, A., Tassus, X., Suhard, P., Renaudin, I., Leguay, A., et al., 2010. First report of tomato chlorotic dwarf viroid in tomato in France. Plant Dis. 94, 633.

Cui, R.C., Li, Z.F., Li, X.L., Wang, G.X., 1992. Identification of potato spindle tuber viroid (PSTVd) and its control. Acta Phytophyl. 19, 263–269.

Desjardins, P.R., Saski, P.J., Drake, R.J., 1987. Chemical inactivation of avocado sunblotch viroid on pruning and propagation tools. Calif. Avocado Soc. 1987 Yearbook. 71, 259–262.

Elliott, D.R., Alexander, B.J.R., Smales, T.E., Tang, Z., Clover, G.R.G., 2001. First report of potato spindle tuber viroid in tomato in New Zealand. Plant Dis. 85, 1027.2.

Galindo, A., Smith, D.R., Diener, T.O., 1982. Etiology of planta macho, a viroid disease of tomato. Phytopathology 72, 49–54.

Garnsey, S.M., Whidden, R., 1971. Decontamination treatments to reduce the spread of citrus exocortis virus (CEV) by contaminated tools. Proc. Fla. State Hort. Soc. 84, 63–67.

Hailstones, D.L., Tesoriero, L.A., Terras, M.A., Dephoff, C., 2003. Detection and eradication of potato spindle tuber viroid in tomatoes in commercial production in New South Wales, Australia. Australas. Plant Pathol. 32, 317–318.

Hollings, M., Stone, O.M., 1973. Some properties of chrysanthemum stunt, a virus with the characterization of an uncoated ribonucleic acid. Ann. Appl. Biol. 74, 333–348.

Lebas, B.S.M., Clover, G.R.G., Ochoa-Corona, F.M., Elliott, D.R., Tang, Z., Alexander, B.J.R., 2005. Distribution of potato spindle tuber viroid in New Zealand glasshouse crops of Capsicum and tomato. Australas. Plant Pathol. 34, 129–133.

Li, R., Baysal-Gurel, F., Abdo, Z., Miller, S.A., Ling, K.S., 2015. Evaluation of disinfectants to prevent mechanical transmission of viruses and a viroid in greenhouse tomato production. Virology J. 12, 5.

Ling, K.S., Bledsoe, M.E., 2009. First report of Mexican papita viroid infecting greenhouse tomato in Canada. Plant Dis. 93, 839.

Ling, K.S., Li, R., Groth-Helms, D., Assis-Filho, F.M., 2014. First report of potato spindle tuber viroid naturally infecting field tomatoes in the Dominican Republic. Plant Dis. 98, 701.

Ling, K.S., Li, R., Panthee, D.R., Gardner, R.G., 2013. First report of potato spindle tuber viroid naturally infecting greenhouse tomatoes in North Carolina. Plant Dis. 97, 148.

Ling, K.S., Sfetcu, D., 2010. First report of natural infection of greenhouse tomatoes by potato spindle tuber viroid in the United States. Plant Dis. 94, 1376.

Ling, K.S., Verhoeven, J.Th.J., Singh, R.P., Brown, J.K., 2009. First report of tomato chlorotic dwarf viroid in greenhouse tomatoes in Arizona. Plant Dis. 93, 1075.

Ling, K.S., Zhang, W., 2009. First report of a natural infection by Mexican papita viroid and tomato chlorotic dwarf viroid on greenhouse tomatoes in Mexico. Plant Dis. 93, 1216.

Mackie, A.E., Coutts, B.A., Barbetti, M.J., Rodoni, B.C., McKirdy, S.J., Jones, R.A.C., 2015. Potato spindle tuber viroid: stability on common surfaces and inactivation with disinfectants. Plant Dis. 99, 770–775.

Mackie, A.E., McKirdy, S.J., Rodoni, B., Kumar, S., 2002. Potato spindle tuber viroid eradicated in Western Australia. Australas. Plant Pathol. 31, 311–312.
Martínez-Soriano, J.P., Galindo-Alonso, J., Maroon, C.J.M., Yucel, I., Smith, D.R., Diener, T.O., 1996. Mexican papita viroid: putative ancestor of crop viroids. Proc. Natl. Acad. Sci. USA 93, 9397–9401.
Matsushita, Y., Kanda, A., Usugi, T., Tsuda, S., 2008. First report of a tomato chlorotic dwarf viroid disease on tomato plants in Japan. J. Gen. Plant Pathol. 74, 182–184.
Matsuura, S., Matsushita, Y., Usugi, T., Tsuda, S., 2010. Disinfection of tomato chlorotic dwarf viroid by chemical and biological agents. Crop Prot. 29, 1157–1161.
Mishra, M.D., Hammond, R.W., Owens, R.A., Smith, D.R., Diener, T.O., 1991. Indian bunchy top disease of tomato plants is caused by a distinct strain of citrus exocortis viroid. J. Gen. Virol. 72, 1781–1785.
Navarro, B., Silletti, M.R., Trisciuzzi, V.N., Di Serio, F., 2009. Identification and characterization of potato spindle tuber viroid infecting tomato in Italy. J. Plant Pathol. 91, 723–726.
Olivier, T., Sveikauskas, V., Grausgruber-Groger, S., Virscek Marn, M., Faggioli, F., Luigi, M., et al., 2015. Efficacy of five disinfectants against potato spindle tuber viroid. Crop Protect. 67, 257–260.
Orozco-Vargas, G., 1983. Thesis: Aspectos ecológicos del viroide "planta macho del jitomate." Thesis, Colegio de Postgraduados, Chapingo, Mexico, pp. 1–58.
Puchta, H., Herold, T., Verhoeven, K., Roenhorst, A., Ramm, K., Schmid-Puchta, W., et al., 1990. A new strain of potato spindle tuber viroid (PSTVd-N) exhibits major sequence differences as compared to all other PSTVd strains sequenced so far. Plant Mol. Biol. 15, 509–511.
Roistacher, C.N., Calavan, E.C., Blue, R.L., 1969. Citrus exocortis virus - chemical inactivation on tools, tolerance to heat and separation of isolates. Plant Dis. Reptr. 53, 333–336.
Singh, R.P., Boucher, A., Somerville, T.H., 1989. Evaluation of chemicals for disinfection of laboratory equipment exposed to potato spindle tuber viroid. Am. Potato J. 66, 239–245.
Singh, R.P., Dilworth, A.D., 2008. Tomato chlorotic dwarf viroid in the ornamental plant *Vinca minor* and its transmission through tomato seed. Eur. J. Plant Pathol. 123, 111–116.
Singh, R.P., Nie, X., Singh, M., 1999. Tomato chlorotic dwarf viroid: an evolutionary link in the origin of pospiviroids. J. Gen. Virol. 80, 2823–2828.
Timmermann, C., Mühlbach, H.P., Bandte, M., Büttner, C., 2001. Control of mechanical viroid transmission by the disinfection of tables and tools. Meded Rijksuniv Gent Fak Landbouwkd Toegep Biol. Wet. 66, 151–156.
Verhoeven, J.Th.J., Jansen, C.C.C., Roenhorst, J.W., 2006. First report of tomato apical stunt viroid in tomato in Tunisia. Plant Dis. 90, 528.
Verhoeven, J.Th.J., Jansen, C.C.C., Roenhorst, J.W., Steyer, S., Schwind, N., et al., 2008. First report of *Solanum jasminoides* infected by citrus exocortis viroid in Germany and the Netherlands and tomato apical stunt viroid in Belgium, and Germany. Plant Dis. 92, 973.
Verhoeven, J.Th.J., Jansen, C.C.C., Willemen, T.M., Kox, L.F.F., Owens, R.A., et al., 2004. Natural infections of tomato by citrus exocortis viroid, Columnea latent viroid, potato spindle tuber viroid and tomato chlorotic dwarf viroid. Eur. J. Plant Pathol. 110, 823–831.
Verhoeven, J.Th.J., Botermans, M., Meekes, E.T.M., Roenhorst, J.W., 2012. Tomato apical stunt viroid in the Netherlands: most prevalent pospiviroid in ornamentals and first outbreak in tomatoes. Eur. J. Plant Pathol. 133, 803–810.
Walter, B., 1981. Un viroide de la tomate en Afrique de I'Ouest: Identite avec le viroide du "potato spindle tuber"? C.R. Acad. Sci. Paris. 292, 537–542.

CHAPTER

42

Strategies to Introduce Resistance to Viroids

Rosemarie W. Hammond and Natalia Kovalskaya
U.S. Department of Agriculture, Beltsville, MD, United States

INTRODUCTION

Plants have evolved a number of different mechanisms to resist pathogen infection, such as cell death and systemic acquired resistance (Hammond-Kosack and Jones, 1996), expression of resistance genes, and RNA silencing defense responses (Ratcliff et al., 1999). Although viroid infection can result in the activation of many of these resistance mechanisms in host plants (Ding, 2009), these mechanisms do not appear to be successful in containing viroid infection. Understandings of the plant response to viroid infection, the mechanisms that determine host susceptibility, and the molecular features of viroid structure and biology, are keys to the rational design of strategies to control viroid diseases using biotechnology. In this chapter we summarize the molecular approaches that have been explored to control viroid infections. Additional sources for information on engineered resistance are, among others, Dalakouras et al. (2015), Kovalskaya and Hammond (2014), Prins et al. (2008), Sano et al. (2003), and Sano et al. (2010).

EARLY MOLECULAR METHODS—CROSS-PROTECTION

Early molecular methods to control viroid disease were based on observations that inoculation with mild strains of a virus/viroid can limit a secondary infection when a plant is challenged with a more severe strain, a phenomenon known as cross-protection (Fernow, 1967;

Viroids and Satellites.
DOI: http://dx.doi.org/10.1016/B978-0-12-801498-1.00042-5

McKinney, 1929). Niblett et al. (1978) demonstrated that preinfection of plants with chrysanthemum stunt viroid (CSVd) or mild and severe strains of potato spindle tuber viroid (PSTVd) (members of the family *Pospiviroidae*) protected chrysanthemum against a related viroid, citrus exocortis viroid (CEVd), whereas the mild strain of PSTVd protected tomato plants against expression of the symptoms caused by CEVd and a severe strain of PSTVd. In both cases, the protecting and challenging viroids were detected in plants.

Singh et al. (1990) showed that, in the highly susceptible potato cultivar Russet Burbank preinfected with a mild PSTVd strain and challenged with a severe PSTVd strain, the latter was not detected in any of the inoculated plants, whereas in the tolerant BelRus potato cultivar preinfected with the mild PSTVd strain, cross-protection was not complete; i.e., the severe PSTVd strain was detected in two out of 10 plants. In the same study, second generation potato plants grown from tubers preinfected with mild/severe PSTVd strains in the previous generation were completely protected against challenge infection after mechanical inoculation with severe/mild PSTVd strains, respectively. Graft inoculation of these same mild-strain protected plants with severe strain-infected scions led to infected plants most likely due to the continuous source of challenge inoculum (Singh et al., 1990).

The mechanism of cross-protection is not fully understood. Ratcliff et al. (1999) postulated that RNA-mediated cross-protection is functionally equivalent to posttranscriptional gene silencing (PTGS). Papaefthimiou et al. (2001) reported the accumulation of short RNA fragments characteristic of PTGS in PSTVd-infected plants. In cross-protection, the mild ("protecting") strain serves as a "primer" for PTGS initiation and the nucleotide sequences of the protecting and challenging strains must be similar (Lin et al., 2007). This view is consistent with the results obtained by Niblett et al. (1978) where chrysanthemum chlorotic mottle viroid (CChMVd), a member of the family *Avsunviroidae*, did not protect chrysanthemum plants against PSTVd, CEVd, and CSVd (family *Pospiviroidae*) infections, and PSTVd failed to protect these plants against CChMVd. To our knowledge, cross-protection has not been widely adopted commercially as a control for viroid diseases.

DOUBLE-STRANDED RNA INTERFERENCE

In nontransgenic silencing assays, Carbonell et al. (2008) studied the effect of viroid-derived dsRNAs and small RNAs (vd-sRNAs) on viroid infectivity. Experiments were conducted on tomato plants (with PSTVd and CEVd), gynura (with CEVd), and chrysanthemum (with CChMVd) where an excess of the homologous dsRNA or vd-sRNAs were coinoculated. A sequence specific and temperature dependent effect was observed where homologous RNAs reduced infectivity of the corresponding viroid RNA. Coinoculation of PSTVd-dsRNA and PSTVd on tomato was not as effective and all plants eventually became infected, although symptom appearance was delayed and less severe. The use of transgenic plants expressing viroid small RNAs to trigger the silencing pathway prior to inoculation was demonstrated to be effective in later studies (see below).

TRANSGENIC METHODS FOR CONTROLLING VIROID DISEASES

Most transgenic strategies for viroid control have been developed to mediate the degradation of single- and double-stranded (ds) viroid RNA in the cytoplasm. While viroids replicate and generate dsRNA replication intermediates in the nucleus (*Pospiviroidae*) or the chloroplast (*Avsunviroida*e), they activate the cytoplasmic RNA-mediated plant defense system with the resulting accumulation of viroid-specific small interfering RNAs (siRNAs) (Itaya et al., 2001); yet, the mature circular viroid genomes display some resistance to RNA-induced silencing complex-mediated RNA degradation (Carbonell et al., 2008; Gómez and Pallás, 2007; Itaya et al., 2007).

RNA Silencing, Antisense RNAs, and Hairpin RNAs

RNA silencing (PTGS) is a conserved, nucleotide sequence-specific process that induces inactivation of RNAs by a specific pathway, and is a natural defense mechanism. To harness the defense pathway for control of viroids, antisense RNAs and hairpin RNAs have been employed.

The first antisense RNA strategy directed against viroids was described by Matoušek et al. (1994). A short (18 nt) antisense RNA was directed against the upper central conserved region of the plus-strand PSTVd RNA, and a longer (173 nt) antisense RNA was directed against the left half of the rod-like secondary structure of the minus-strand replication intermediate. The antisense RNAs formed complexes with the corresponding target RNAs in vitro. When the antisense RNAs were expressed in transgenic potato plants, significant inhibition of PSTVd accumulation occurred at 4 weeks postinoculation. Although infection was delayed, the plants were not immune, and severely infected plants were observed in all transgenic lines at 6–8 weeks postchallenge inoculation.

Transgenic tomato lines expressing a noninfectious, near full-length hairpin RNA (hpRNA) of PSTVd displayed symptoms of viroid infection in T1-generation plants, and PSTVd-specific siRNAs were present in these seedlings even though no replicating PSTVd was generated (Wang et al., 2004). These experiments provided the first evidence that RNA silencing may play an important role in viroid pathogenesis and was confirmed in later studies (Di Serio et al., 2010; Gómez et al., 2008). Schwind et al. (2009) demonstrated that two of three transgenic tomato lines expressing the same hpRNA derived from PSTVd exhibited resistance to PSTVd infection in the T3 generation. This resistance was correlated with high-level accumulation of hpRNA-derived siRNAs in plant tissues. Although small RNAs produced by the infecting viroid did not silence viroid RNAs sufficiently to prevent their replication, the results of this work showed that hpRNA-derived siRNAs effectively targeted the PSTVd RNA.

The demonstration that siRNAs transported through grafts from rootstocks to scions induce silencing of an endogenous gene in the scion (Kasai et al., 2011), led to the exploration of the ability of PSTVd siRNAs, generated in transgenic *Nicotiana benthamiana* rootstocks, to reduce accumulation of viroid RNAs in challenge-inoculated scions (Kasai et al., 2013). A noninfectious near full-length PSTVd hpRNA was expressed from a strong companion cell-specific transcriptional promoter to increase the potential siRNAs in phloem tissue. Suppression of viroid accumulation occurred in the early stages of infection of PSTVd-challenged transgenic plants, although all plants were infected at a later stage. When wild-type scions were grafted onto the transgenic rootstocks, PSTVd accumulation was reduced in the early stage of infection (12 days postinfection, dpi), but all plants became infected at a later stage. Although the approach is theoretically promising, further improvements are needed to increase its efficacy in controlling viroid infection.

Hammerhead Ribozymes

Ribozymes are small RNA molecules that catalyze the specific cleavage of target RNAs. Transgenic tomato seedlings expressing truncated antisense RNAs targeting the minus-strand of CEVd resulted in a reduction in the accumulation of CEVd RNA in plant tissues upon challenge inoculation (Atkins et al., 1995). In contrast, transgenic plants expressing truncated antisense RNAs targeting the plus-strand of CEVd resulted in an increase in the rate of CEVd RNA accumulation when compared with control plants. Incorporation of three modified hammerhead ribozyme motifs, derived from the satellite RNA of tobacco ringspot virus, into either of the antisense RNAs did not enhance suppression of viroid replication in plants, in spite of the presence of catalytic activity of the ribozyme constructs in vitro, albeit with higher activity at 50°C than at 37°C, suggesting that RNA secondary structure affected the efficiency of cleavage in vivo.

When a hammerhead ribozyme (49 nt), designed to target the putative binding site of DNA-dependent RNA polymerase II on the minus-strand RNA of PSTVd, was expressed in potato plants, 23 out of 34 transgenic plant lines (about 68%) showed a high level of resistance and little to no viroid accumulation after plants were challenge-inoculated with PSTVd (Yang et al., 1997). The remaining lines showed a weaker level of resistance to PSTVd with low levels of PSTVd accumulation. The resistance against PSTVd replication was maintained in the vegetatively propagated progeny derived from tubers of transgenic lines. There was no report of stable transmission of the resistance trait through true seed.

Carbonell et al. (2011) tested the ability of a *trans*-cleaving extended hammerhead ribozyme derived from peach latent mosaic viroid (PLMVd) to control PSTVd infection in *N. benthamiana*. During replication of PLMVd and other members of the family *Avsunviroidae*, hammerhead ribozymes act in *cis* (self-cleaving the RNA in which they are embedded) (Hernandez and Flores, 1992). However, the ribozyme can be manipulated to act in *trans* by splitting it into the ribozyme itself and the substrate. For experiments in vivo, two cultures of *Agrobacterium tumefaciens*, transformed with constructs expressing either a *trans*-cleaving PLMVd-derived hammerhead ribozyme (HHe-PLMVd) or the substrate dPSTVd (−) which generates a head-to-tail dimeric PSTVd (−) RNA, were co-infiltrated in *N. benthamiana* leaves. The HHe-PLMVd ribozyme interfered with systemic PSTVd infection indicating that it may target the primary dimeric transcript and perhaps also the oligomeric (−) replicative intermediates. Constitutive expression in transgenic plants of a modified ribozyme like HHe-PLMVd may lead to efficient control of PSTVd infections.

Enzymes With Nuclease Activity

Another approach involved expression of the yeast dsRNA-specific ribonuclease *pac*1 gene, encoding the Pac1 enzyme that digests long dsRNAs into short oligonucleotides and cleaves small hpRNA substrates (Rotondo and Frendewey, 1996). Viroid replication involves dsRNA intermediates, and the mature, circular viroid RNA is partially double-stranded. Sano et al. (1997) demonstrated the susceptibility of in vitro PSTVd transcripts to Pac1 digestion. They then expressed the *pac*1 gene in transgenic potato, and five potato lines challenge-inoculated with PSTVd exhibited resistance (33%–90%) to PSTVd infection; resistance was maintained during vegetative propagation through seed potatoes. Resistance assays were conducted at temperature ranges of 25–32°C and 20–28°C; in both cases, none of the viroid-challenged Pac1 plants developed disease symptoms.

Three transgenic lines of chrysanthemum plants stably producing Pac1 protein were challenge-inoculated with CSVd (Ogawa et al., 2005). The infection frequency for two of three transgenic lines was less than 50% in contrast to that in control plants (100% or 78%). One of nine (11%) and two of nine plants (22%) from the different transgenic lines were infected with CSVd 60 dpi, whereas approximately 70% of control plants were infected at 30 dpi. The infection frequency in control plants reached 100% at 40 dpi, while only about 50% of transgenic plants were infected at the same time. When transgenic and control plants were grafted onto CSVd-infected plants, the transgenic plants showed better growth, and one line exhibited the least growth retardation. Analysis of CSVd accumulation in transgenic plants revealed that about half of the transgenic plants were not infected with CSVd, while in other plants the accumulation of CSVd was suppressed.

Ishida et al. (2002) reported the efficiency of the Pac1 protein to protect transgenic potato and chrysanthemum plants against PSTVd and CSVd infections, respectively. Pac1-transgenic potato plants inoculated with PSTVd did not develop disease symptoms, whereas nontransgenic control plants showed symptoms at 30–40 dpi. In transgenic potato lines, viroid RNA was detected in 25%–50% of plants, and where viroid was not detected in leaves following inoculation with PSTVd, viroid-free tubers were produced. Pac1-expressing chrysanthemum plants did not express disease symptoms after challenge inoculation with CSVd, and CSVd was detected only in 20% of the highest Pac1-expressing plants. In all of these reports, healthy transgenic plants grew without abnormal phenotypes, suggesting that there was no adverse effect of Pac1 overexpression on the physiology of the plant. The Pac1 approach avoids species-specificity and appears to be an efficient method to combat viroid infection.

Recently, a catalytic single-chain antibody, 3D8 scFv, which possesses RNase activity and catalyzes hydrolysis of nucleic acids in a sequence-independent manner, was optimized for transgenic expression in chrysanthemum plants under control of the CaMV 35S transcription promoter (Tran et al., 2016). Several of the transgenic lines expressing either the original or codon-optimized antibody accumulated lower levels of CSVd in noninoculated leaves upon challenge inoculation, indicating that systemic infection was inhibited (Tran et al., 2016). The mechanism of resistance in these plants is currently unknown.

FUTURE OUTLOOK

Efforts to produce transgenic plants with durable resistance to viroids using the various biotechnological approaches discussed above have met with some success and have resulted in variable levels of resistance. A combination of factors, such as the resistance to degradation resulting from the secondary structures of mature viroid molecules, subcellular compartmentation of viroid RNAs, and association with host proteins, may help viroids elude the host RNA silencing machinery, thereby reducing the efficacy of RNA silencing-based methods of control.

Resistant hosts have been shown to have extremely low levels of viroid replication and the inability of the viroid to move from the initial site of infection. Therefore, new approaches for molecular control of viroid diseases might include limitation of pathogen spread; i.e., interference with cell-to-cell and long-distance movement, by disrupting viroid interaction with host proteins such as phloem protein 2 or Virp1 (Dalakouras et al., 2015). Alternative strategies utilizing artificial microRNAs (amiRNAs) for plant disease resistance have been described by Dalakouras et al. (2015) and Duan et al. (2012). Micro RNAs (miRNAs) are processed from specific pre-miRNA transcripts, and amiRNAs have been used to silence plant genes and confer resistance to viruses (Niu et al., 2006). By mimicking the structure of endogenous miRNA precursors, amiRNAs can be designed to specifically target a viroid region of low secondary structure. In a recent report, tobacco plants were engineered to express a 21-nt amiRNA corresponding to positions 46–66 of the PSTVd genome (virulence modulating region) (Eamens et al., 2014). The plants exhibited symptoms resembling those of viroid infection, although no symptoms were associated with amiRNAs flanking this region. There was no report of challenge inoculation of these plants with PSTVd. Therefore, it remains to be seen if the amiRNA approach will result in viroid resistance.

References

Atkins, D., Young, M., Uzzell, S., Kelly, L., Fillati, J., Gerlach, W.L., 1995. The expression of antisense and ribozyme genes targeting citrus exocortis viroid in transgenic plants. J. Gen. Virol. 76, 1781–1790.

Carbonell, A., Flores, R., Gago, S., 2011. Trans-cleaving hammerhead ribozymes with tertiary stabilizing motifs: in vitro and in vivo activity against a structured viroid RNA. Nucleic Acids Res. 39, 2432–2444.

Carbonell, A., Martinez de Alba, A.E., Flores, R., Gago, S., 2008. Double-stranded RNA interferes in a sequence-specific manner with the infection of representative members of the two viroid families. Virology 371, 44–53.

Dalakouras, A., Dadami, E., Wassenegger, M., 2015. Engineering viroid resistance. Viruses 7, 634–646.

Di Serio, F., Martinez de Alba, A.E., Navarro, B., Gisel, A., Flores, R., 2010. RNA-dependent RNA polymerase 6 delays accumulation and precludes meristem invasion of a viroid that replicates in the nucleus. J. Virol. 84, 2477–2489.

Ding, B., 2009. The biology of viroid-host interactions. Annu. Rev. Phytopathol. 47, 105–131.

Duan, C.G., Wang, C.H., Guo, H.S., 2012. Application of RNA silencing to plant disease resistance. Silence 3, 5.

Eamens, A.L., Smith, N.A., Dennis, E.S., Wassenegger, M., Wang, M.-B., 2014. *Nicotiana* species, an artificial microRNA corresponding to the virulence modulating region of potato spindle tuber viroid directs RNA silencing of a *soluble inorganic pyrophosphatase* gene and the development of abnormal phenotypes. Virology 450-451, 266–277.

Fernow, K.H., 1967. Tomato as a test plant for detecting mild strains of potato spindle tuber virus. Phytopathology 57, 1347–1352.

Gómez, G., Martinez, G., Pallás, V., 2008. Viroid-induced symptoms in *Nicotiana benthamiana* plants are dependent on RDR6 activity. Plant Physiol. 148, 414–423.

Gómez, G., Pallás, V., 2007. Mature monomeric forms of hop stunt viroid resist RNA silencing in transgenic plants. Plant J. 51, 1041–1049.

Hammond-Kosack, K.E., Jones, J., 1996. Resistance-gene-dependent plant defense responses. Plant Cell. 8, 1773–1791.

Hernandez, C., Flores, R., 1992. Plus and minus RNAs of peach latent mosaic viroid self-cleave in vitro via hammerhead structures. Proc. Natl. Acad. Sci. USA 89, 3711–3715.

Ishida, I., Tukahara, M., Yoshioka, M., Ogawa, T., Kakitani, M., Toguri, T., 2002. Production of anti-virus, viroid plants by genetic manipulations. Pest Manage. Sci. 58, 1132–1136.

Itaya, A., Folimonov, A., Matsuda, Y., Nelson, R.S., Ding, B., 2001. Potato spindle tuber viroid as inducer of RNA silencing in infected tomato. Mol. Plant Microbe Interact. 14, 1332–1334.

Itaya, A., Zhong, X., Bundschuh, R., Qi, Y., Wang, Y., Takeda, R., et al., 2007. A structured viroid RNA serves as a substrate for dicer-like cleavage to produce biologically active small RNAs but is resistant to RNA-induced silencing complex-mediated degradation. J. Virol. 81, 2980–2994.

Kasai, A., Bau, S., Li, T., Harada, T., 2011. Graft-transmitted siRNA signal from the root induces visual manifestation of endogenous post-transcriptional gene silencing in the scion. PLoS One 6, e16895.

Kasai, A., Sano, T., Harada, T., 2013. Scion on a stock producing siRNAs of potato spindle tuber viroid (PSTVd) attenuates accumulation of the viroid. PLoS One 8, e57736.

Kovalskaya, N., Hammond, R.W., 2014. Molecular biology of viroid-host interactions and disease control strategies. Plant Sci. 228, 48–60.

Lin, S.S., Henriques, R., Wu, H.W., Niu, Q.W., Yeh, S.D., Chua, N.H., 2007. Strategies and mechanisms of plant virus resistance. Plant Biotechnol. Rep. 1, 125–134.

Matoušek, J., Schroder, A.R., Trnena, L., Reimers, M., Baumstark, T., Dedic, P., et al., 1994. Inhibition of viroid infection by antisense RNA expression in transgenic plants. Biol. Chem. Hoppe Seyler 375, 765–777.

McKinney, H.H., 1929. Mosaic diseases in the Canary Islands, West Africa and Gibraltar. J. Agri. Res. 39, 557–578.

Niblett, C.L., Dickson, E., Fernow, K.H., Horst, R.K., Zaitlin, M., 1978. Cross protection among four viroids. Virology 91, 198–203.

Niu, Q.W., Lin, S.S., Reyes, J.L., Chen, K.C., Wu, H.W., Yeh, S.D., et al., 2006. Expression of artificial microRNAs in transgenic *Arabidopsis thaliana* confers virus resistance. Nat. Biotechnol. 24, 1420–1428.

Ogawa, T., Toguri, T., Kudoh, H., Okamura, M., Momma, T., Yoshioka, M., et al., 2005. Double-stranded RNA-specific ribonuclease confers tolerance against chrysanthemum stunt viroid and tomato spotted wilt virus in transgenic chrysanthemum plants. Breed. Sci. 55, 49–55.

Papaefthimiou, I., Hamilton, A., Denti, M., Baulcombe, D., Tsagris, M., Tabler, M., 2001. Replicating potato spindle tuber viroid is accompanied by short RNA fragments that are characteristic of post-transcriptional gene silencing. Nucleic Acids Res. 29, 2395–2400.

Prins, M., Laimer, M., Noris, E., Schubert, J., Wassenegger, M., Tepfer, M., 2008. Strategies for antiviral resistance in transgenic plants. Mol. Plant Pathol. 9, 73–83.

Ratcliff, F.G., MacFarlane, S.A., Baulcombe, D.C., 1999. Gene silencing without DNA: RNA-mediated cross-protection between viruses. Plant Cell. 11, 1207–1215.

Rotondo, G., Frendewey, D., 1996. Purification and characterization of the Pac1 ribonuclease of *Schizosaccharomyces pombe*. Nucleic Acids Res. 24, 2377–2386.

Sano, T., Barba, M., Li, S.-F., Hadidi, A., 2010. Viroids and RNA silencing-mechanism, role in viroid pathogenicity and development of viroid-resistant plants. GM Crops 1, 80–86.

Sano, T., Hammond, R.W., Owens, R.A., 2003. Biotechnological approaches for controlling viroid diseases. In: Hadidi, A., Flores, R., Randles, J.W., Semancik, J.S. (Eds.), Viroids. CSIRO Publishing, Collingwood, VIC, pp. 343–349.

Sano, T., Nagayama, A., Ogawa, T., Ishida, I., Okada, Y., 1997. Transgenic potato expressing double-stranded RNA-specific ribonuclease is resistant to potato spindle tuber viroid. Nat. Biotechnol. 15, 1290–1294.

Schwind, N., Zwiebel, M., Itaya, A., Ding, B., Wang, M.-B., Krczal, G., et al., 2009. RNAi-mediated resistance to potato spindle tuber viroid in transgenic tomato expressing a viroid hairpin RNA construct. Mol. Plant Pathol. 10, 459–469.

Singh, R.P., Boucher, A., Sommerville, T.H., 1990. Cross-protection with strains of potato spindle tuber viroid in the potato plant and other solanaceous hosts. Phytopathology 80, 246–250.

Tran, D.T., Cho, S., Hoang, P.M., Kim, J., Kil, E.-J., Lee, T.-K., et al., 2016. A codon-optimized nucleic acid hydrolyzing single-chain antibody confers resistance to chrysanthemums against chrysanthemum stunt viroid infection. Plant Mol. Biol. Rep. 34, 221–232.

Wang, M.B., Bian, X.Y., Wu, L.M., Liu, L.X., Smith, N.A., Isenegger, D., et al., 2004. On the role of RNA silencing in the pathogenicity and evolution of viroids and viral satellites. Proc. Natl. Acad. Sci. USA 101, 3275–3280.

Yang, X., Yie, Y., Zhu, Y., Liu, Y., Kang, L., Wang, X., et al., 1997. Ribozyme-mediated high resistance against potato spindle tuber viroid in transgenic potatoes. Proc. Natl. Acad. Sci. USA 94, 4861–4865.

PART VI

GEOGRAPHICAL DISTRIBUTION OF VIROIDS AND VIROID DISEASES

CHAPTER

43

Geographical Distribution of Viroids in the Americas

Edward V. Podleckis

U.S. Department of Agriculture, Riverdale, MD, United States

FRUIT TREE VIROIDS

Eleven viroids have been found in fruit trees in the Americas: two, avocado sunblotch viroid (ASBVd) and peach latent mosaic viroid (PLMVd), are members of the family *Avsunviroidae*; and nine, apple scar skin viroid (ASSVd), pear blister canker viroid (PBCVd), citrus bark cracking viroid (CBCVd), citrus bent leaf viroid (CBLVd), citrus dwarf viroid (CDVd), citrus exocortis viroid (CEVd), citrus viroid V (CVd-V), hop stunt viroid (HSVd), and potato spindle tuber viroid (PSTVd), are members of the family *Pospiviroidae.*

Avocado Viroids

Semancik (2003) reported that ASBVd occurs in the United States and Central America (Guatemala and Costa Rica). The presence of ASBVd in Costa Rica has not, however, been officially confirmed. Almost 19% of accessions at the National Avocado Germplasm Repository in Florida were found infected with ASBVd (Running et al., 1996). CABI (CABI/EPPO, 2013) lists ASBVd as present in the United States, Mexico, Peru, and Venezuela. Avocado (*Persea americana*) is a host for two viroids in Peru: ASBVd (Vargas et al., 1991) and PSTVd (Querci et al., 1995), which were found infecting avocado singly or in combination.

Viroids and Satellites.
DOI: http://dx.doi.org/10.1016/B978-0-12-801498-1.00043-7

Pome Fruit Viroids

Two apple blemishing diseases are associated with variants of ASSVd in North America (Hadidi et al., 1990). Dapple apple disease was reported from apple orchards in New Hampshire, USA (Smith et al., 1956) while apple scar skin symptoms were reported in the Missouri, USA (Millikan and Martin, 1956) and in British Columbia, Canada (Welsh and Keene, 1961). According to Hadidi et al. (1991), these diseases rarely occur in North America. ASSVd was detected by dot-blot hybridization during screening for viroids in apple orchards in Rio Negro, Argentina. The identity of the viroid was confirmed by RT-PCR amplification (Nome et al., 2012).

In a survey for viroids in pears, 18 of 96 trees tested from Rio Negro, Argentina were found infected with pear blister canker viroid (Nome et al., 2011).

Stone Fruit Viroids

PLMVd has been reported globally and occurs at high rates in peach germplasm (Flores et al., 1990; Hadidi et al., 1997). Torres et al. (2004) used tissue printing and non-radioactive molecular probes to detect PLMVd in 25% of the Canadian *Prunus* trees sampled. A survey of the U.S. Department of Agriculture National Clonal Germplasm Repository at the University of California, Davis, determined that HSVd or PLMVd could be detected in about 15% of *Prunus* plants assayed (Osman et al., 2012). A survey of commercial peach cultivars (including nectarine) growing in Virginia, Colorado, Oregon, California, and Washington, USA, found over 50% of the trees were infected with PLMVd (Skrzeczkowski et al., 1996). Symptomatic commercial peach trees from the Mexican states of Puebla, Morelos, and Mexico were positive for PLMVd using molecular hybridization and RT-PCR (De La Torre-Almaráz et al., 2015). Commercial peach trees in Córdoba, Argentina, showing symptoms of yellowing were confirmed as infected with PLMVd (Nieto et al., 2008). Hadidi et al. (1997) detected PLMVd in peach trees held in quarantine after importation into the USA from Brazil. PLMVd was detected in 43% of the Chilean peach trees sampled (Torres et al., 2004). Fiore et al. (2003) also reported the occurrence of PLMVd in plum and peach trees surveyed in Chile. PLMVd was detected by molecular hybridization during a survey of peach orchards in the Canelones Department of Uruguay (Herranz et al., 2002).

HSVd was reported infecting apricots in Canada during a survey of *Prunus* in the Canadian Clonal Genebank (Michelutti et al., 2004).

Citrus Viroids

Guerrero-Gámez et al. (2013) sampled citrus trees from several groves in Nueva Leon and Tamualipas, Mexico. Using RT-PCR assays, they were able to detect CDVd, HSVd, and CEVd in 49%, 42%, and 41% of the trees sampled, respectively. Earlier, Almeyda-León et al. (2002) used RT-PCR to detect citrus viroids in commercial lime groves in Yucatan, Veracruz, Tabasco, and Colima, Mexico. That study also found CEVd and HSVd and citrus viroid III (synonym of CDVd) singly and in multiple infections. Kunta et al. (2007) used RT-PCR to detect and characterize citrus viroids in Texas. They confirmed the presence of CEVd, HSVd, and CDVd commonly in all the trees tested. CBCVd was found in one tree in that study. CBCVd was reported from a single citrus source in California (Duran-Vila et al., 1988). CVd-V was recovered from Etrog citron at the University of California, Riverside (Serra et al., 2008). CABI reports that CEVd is present in Arizona, California, Florida, Louisiana, and Texas (CABI, 2015a). Four citrus viroids, CEVd, CDVd, CBCVd (synonym of CVd IV), and HSVd (synonym of CVd II), were reported from citrus groves in Cuba (Pérez et al., 2001; Velásquez et al., 2001). A survey for citrus viroids across six major citrus producing areas in Jamaica found four citrus viroids, CEVd, CBCVd, CBLVd, and HSVd, infecting citrus trees singly and in mixed infections (Fisher et al., 2011). Roistacher et al. (1996) noted severe losses to citrus growers in Belize caused by CEVd-infected rootstocks used to control citrus tristeza virus. Surveys for citrus viroids in *Citrus* spp. from six provinces of Uruguay found 62% of the surveyed trees were either infected with a single viroid or more commonly multiple viroids (Pagliano et al., 2013). HSVd showed the highest prevalence (92%) followed by CDVd (50%), CEVd (23%), and CBLVd (19%). The authors reported that their results indicated "that CEVd, CBLVd, HSVd and CDVd are widespread throughout citrus orchards in Uruguay and highlight the presence of mixed viroid infections." Isolates of CEVd and HSVd have been detected in citrus in Brazil (Targon et al., 2003).

GRAPEVINE VIROIDS

Five viroids, all members of the family *Pospiviroidae,* Australian grapevine viroid, CEVd, grapevine yellow speckle viroid-1 (GYSVd-1),

GYSVd-2, and HSVd, are reported to infect grapevines (Adkar-Purushothama et al., 2014; Rezaian et al., 1992). Semancik et al. (1987) reported the widespread occurrence of GVYSVd-1, GYSVd-2, and HSVd in Californian grapevines. That same study also established the presence of Australian grapevine viroid in California. Isolates of CEVd and HSVd found doubly infecting grapevines were characterized in Brazil (Eiras et al., 2006).

VIROIDS OF ORNAMENTAL PLANTS

Eight viroids have been found to be associated with ornamental plants: one, chrysanthemum chlorotic mottle viroid, is a member of the family *Avsunviroidae;* and seven viroids, chrysanthemum stunt viroid (CSVd), CEVd, coleus viroid 1, Columnea latent viroid (CLVd), Iresine viroid, PSTVd, and tomato chlorotic dwarf viroid (TCDVd), are members of the family *Pospiviroidae*.

CSVd has been detected in chrysanthemum (*Dendranthema* spp.) in the United States (Lawson, 1987), as well as in chrysanthemum, *Vinca major*, and *Verbena* spp. in Canada (Bostan et al., 2004). A viroid detected in commercial chrysanthemums in Brazil was confirmed to be CSVd (Gobatto et al., 2014).

Chrysanthemum chlorotic mottle viroid was found in the United States, but is not known to exist in any other country in the Americas (Flores et al., 2003).

Surveys of commercial nurseries in New Brunswick, Canada, found CEVd widely distributed in *Verbena* and *Impatiens* plants and seeds (Singh et al., 2009). The study also found CEVd transmission rates as high as 66% and 45% in *Impatiens walleriana* and *Verbena* × *hybrida,* respectively.

Fonseca et al. (1989, 1990) reported a new viroid, now designated coleus viroid 1, in symptomless field-grown *Coleus* plants in Brazil. Shortly thereafter the viroid was detected in *Coleus* plants in Canada and found to be highly seed-transmissible (Singh and Boucher, 1991).

CLVd was isolated from symptomless *Columnea erythrophae* purchased from a commercial nursery in the US state of Maryland (Owens et al., 1978). CLVd was isolated from symptomless commercial *Nematanthus wettsteinii* plants in Canada (Singh et al., 1992).

Iresine viroid was detected in *Verbina* and *Vinca* spp., in a survey of local nurseries in New Brunswick, Canada (Bostan et al., 2004).

PSTVd was detected in three greenhouse-grown *Cestrum* samples from California (Chitambar, 2015), but the infestations have been eradicated.

TCDVd was found in US *Petunia hybrida* plants imported into The Netherlands (Verhoeven et al., 2007), and from *Vinca minor* plants originating from US nurseries and planted in Canada (Singh and Dilworth, 2009).

HOP VIROIDS

Surveys of hop gardens in major hop production areas of Washington, USA, confirmed earlier reports of the presence of hop latent viroid (Eastwell and Nelson, 2007; Nelson et al., 1997), also a member of the family *Pospiviroidae*. Those same surveys found HSVd present in all major hop growing areas of Washington.

VIROIDS OF SOLANACEOUS CROPS AND WILD SPECIES

These four viroids, PSTVd, TCDVd, tomato planta macho viroid (TPMVd), and Mexican papita viroid (MPVd; now considered a strain of TPMVd), are all members of the family *Pospiviroidae*.

Seed certification efforts eradicated PSTVd from seed potato production in western Canada (De Boer et al., 2002). PSTVd was reported to be eradicated from US seed potato producing areas based on searching for the occurrence of PSTVd in state certification records from 1990 through 2000 and a field survey testing selected crops for PSTVd infection from 1999 through 2001 (Sun et al., 2004). Ling and Sfetcu (2010) reported natural PSTVd infection of commercial greenhouse tomatoes in California. Sporadic reports of PSTVd in greenhouse-grown tomatoes elsewhere in the United States have occurred (Ling et al., 2012), but the infestations have been eradicated. CABI (CABI, 2015b; CABI/EPPO, 2012) lists PSTVd as present in Mexico but with no details as to distribution or hosts. Martinez-Soriano et al. (1996) noted surveys of potato crops in Mexico failed to find PSTVd but speculated that the viroid might be present in wild solanaceous hosts. In 2013, a severe outbreak of chlorosis and stunting was observed in a tomato field in the Dominican Republic and RT-PCR analysis determined the causal agent was PSTVd (Ling et al., 2013). PSTVd was reported from symptomatic potato fields around the potato growing area of Cartago, Costa Rica in 1999 (Badilla et al., 1999). PSTVd was reported in Argentina since the 1960s (Fernandez-Valiela, 1969), but official surveys determined that the viroid was absent from commercial potato crops and was restricted to breeding material that was subsequently destroyed. In 2009, the National Plant Protection Organization declared PSTVd eradicated and absent from Argentina (EPPO, 2009). A PSTVd variant was described by RT-PCR from symptomless leaves of field-grown pepino plants (*Solanum muricatum*) originating in the IV region of Chile (Shamloul et al., 1997). CABI removed Chile from its most recent PSTVd global distribution list because the isolate differed in five positions from the type strain of PSTVd in potato and there are no records of PSTVd detection in potato in Chile during official surveys, or in the national seed potato certification program (CABI, 2015b). PSTVd was also found in potato clones and true potato seed maintained at the International Potato Center in Lima, Peru (Peiman and Xie, 2006; Salazar et al., 1983). Several databases and reviews have reported PSTVd in Venezuela (CABI, 2015b; CABI/EPPO, 2012; Singh, 1983).

TCDVd was reported from greenhouse-grown tomatoes in Canada (Singh et al., 1999) and there have been isolated reports of TCDVd in greenhouse tomatoes in Arizona and Colorado (Ling et al., 2009; Verhoeven et al., 2004). Ling and Zhang (2009) reported the first record of commercial tomatoes in Mexico naturally infected with TCDVd.

TPMVd has only been found in Mexico (Belalcazar and Galindo, 1974). Orozco and Galindo (1986) reported seven solanaceous species as natural hosts of TPMVd. They also suggested that tomato plants in the field become infected with TPMVd by spread from wild solanaceous species. The opposite movement, however, cannot be ruled out.

MPVd was reported from Mexico infecting papita (*Solanum cardiophyllum*) (Martinez-Soriano et al., 1996). In British Columbia, Canada, Ling and Bledsoe (2009) identified MPVd infecting greenhouse-grown tomatoes, a new host for MPVd. Ling and Zhang (2009) reported commercial tomatoes in Mexico naturally infected with MPVd from a greenhouse facility outside Mexico City. The authors speculated that infected seed may have been the initial inoculum.

A country-by-country summary of the geographical distribution of viroids in the Americas is presented in Table 43.1.

TABLE 43.1 Geographical Distribution of Viroids Reported in the Americas

Viroid (Abbr.)	Country	Host	Reference
Apple scar skin viroid (ASSVd)[a]	United States	Apple	Millikan and Martin (1956), Smith et al. (1956)
	Canada	Apple	Welsh and Keene (1961)
	Argentina	Apple	Nome et al. (2012)
Australian grapevine viroid (AGVd)	United States (California)	Grapevine	Rezaian et al. (1992)
Avocado sunblotch viroid (ASBVd)	Mexico	Avocado	De La Torre-A et al. (2009)
	United States	Avocado	Running et al. (1996), Semancik (2003)
	Peru	Avocado	Vargas et al. (1991)
	Venezuela	Avocado	Rondón and Figueroa (1976)
Chrysanthemum chlorotic mottle viroid (CChMVd)	United States	*Chrysanthemum moriflora*	Flores et al. (2003)
Chrysanthemum stunt viroid (CSVd)	United States	*Dendranthema* sp.	Lawson (1987)
	Brazil	Chrysanthemum	Gobatto et al. (2014)
Citrus bark cracking viroid (CBCVd)	United States[b]	*Citrus* spp.	Kunta et al. (2007) Duran-Vila et al. (1988)
	Cuba	*Citrus* spp.	Pérez et al. (2001), Velásquez et al. (2001)
	Jamaica	*Citrus* spp.	Fisher et al. (2011)
Citrus bent leaf viroid (CBLVd)	Jamaica	*Citrus* spp.	Fisher et al. (2011)
	Uruguay	*Citrus* spp.	Pagliano et al. (2013)
Citrus dwarfing viroid (CDVd)	Mexico	*Citrus* spp.	Almeyda-León et al. (2002), Guerrero-Gámez et al. (2013)
	United States	*Citrus* spp.	Kunta et al. (2007)
	Cuba	*Citrus* spp.	Pérez et al. (2001), Velásquez et al. (2001)
	Uruguay	*Citrus* spp.	Pagliano et al. (2013)

(*Continued*)

TABLE 43.1 (Continued)

Viroid (Abbr.)	Country	Host	Reference
Citrus exocortis viroid (CEVd)	Canada	*Impatiens, Verbena*	Singh et al. (2009)
	Mexico	*Citrus* spp.	Guerrero-Gámez et al. (2013)
	United States	*Citrus* spp.	Kunta et al. (2007)
	Cuba	*Citrus* spp.	Pérez et al. (2001), Velásquez et al. (2001)
	Jamaica	*Citrus* spp.	Fisher et al. (2011)
	Brazil	Citrus	Targon et al. (2003)
		Grapevine	Eiras et al. (2006)
	Uruguay	*Citrus* spp.	Pagliano et al. (2013)
Citrus viroid V (CVd-V)	United States	Etrog citron	Serra et al. (2008)
Coleus blumei viroid-1 (CbVd-1)	Canada	*Coleus blumei*	Singh and Boucher (1991)
	Brazil	*Coleus blumei*	Fonseca et al. (1989, 1990)
Columnea latent viroid (CLVd)	Canada	*Nematanthus wettsteinii*	Singh et al. (1992)
	United States	*Columnea erythrophae*	Owens et al. (1978)
Grapevine yellow speckle viroids 1, 2 (GYSVd-1, -2)	United States (California)	Grapevine	Rezaian et al. (1992)
Hop latent viroid (HLVd)	United States	Hops	Eastwell and Nelson (2007); Nelson et al. (1997)
Hop stunt viroid (HSVd)	Canada	Apricot	Michelutti et al. (2004)
	Mexico	*Citrus* spp.	Guerrero-Gámez et al. (2013)
	United States	*Citrus* spp.	Kunta et al. (2007)
		Vitis	Sano et al. (1986)
		Hops	Eastwell and Nelson (2007), Nelson et al. (1997)
		Prunus spp.	Osman et al. (2012)
	Cuba	*Citrus* spp.	Pérez et al. (2001), Velásquez et al. (2001)

(*Continued*)

TABLE 43.1 (Continued)

Viroid (Abbr.)	Country	Host	Reference
	Jamaica	*Citrus* spp.	Bennett et al. (2010): Fisher et al. (2011)
	Brazil	*Citrus* spp.	Targon et al. (2003)
		Grapevine	Eiras et al. (2006)
	Uruguay	*Citrus* spp.	Pagliano et al. (2013)
Iresine viroid (IrVd)	Canada	*Vinca major*	Bostan et al. (2004) Ling
Mexican papita viroid (MPVd)	Canada	Tomato	and Bledsoe (2009)
	Mexico	Papita	Martinez-Soriano et al. (1996)
Peach latent mosaic viroid (PLMVd)	Canada	*Prunus* spp.	Michelutti et al. (2004)
	United States	*Prunus* spp.	Osman et al. (2012)
	Mexico	Peach	De La Torre-Almaráz et al. (2015)
	Argentina	Peach	Nieto et al. (2008)
	Brazil[c]	Peach	
	Chile	Peach	
	Uruguay	Plum	
		Peach	
Pear blister canker viroid (PBCVd)	Argentina	Pear	Nome et al. (2011)
Potato spindle tuber viroid (PSTVd)	Canada[d]	Potato	Singh (2014)
	United States[d]	Potato	Diener (1987)
		Tomato	Ling et al. (2012), Ling and Sfetcu (2010)
		Cestrum sp.	Chitambar (2015)
	Dominican Republic	Tomato	Ling et al. (2013)
	Costa Rica	Potato	Badilla et al. (1999)
	Peru	Avocado	Querci et al. (1995)
		Potato[e]	Peiman and Xie (2006), Salazar et al. (1983)
	Venezuela	Potato	CABI (2015b), CABI/EPPO (2012), Singh (1983)

(Continued)

TABLE 43.1 (Continued)

Viroid (Abbr.)	Country	Host	Reference
Tomato chlorotic dwarf viroid (TCDVd)	Canada	Tomato	Singh et al. (1999)
		Vinca minor	Singh and Dilworth (2009)
	Mexico	Tomato	Ling and Zhang (2009)
	United States	*Petunia hybrida*[f]	Verhoeven et al. (2007)
		Tomato	Ling et al. (2009)
Tomato planta macho viroid (TPMVd)	Mexico	Tomato	Belalcazar and Galindo (1974), Orozco and Vargas (1986)
		Various wild solanaceae	

[a] *Rarely occurs in North America (Hadidi et al., 1991).*
[b] *Only isolated from single trees in Texas and California.*
[c] *Detected in the United States on plants imported from Brazil.*
[d] *Not present in potato seed producing areas (De Boer et al., 2002; Singh, 2014; Sun et al., 2004); tomato and* Cestrum *sp. outbreaks eradicated.*
[e] *Reported from breeding material at International Potato Center, not from commercial fields.*
[f] *Detected in The Netherlands on plants imported from the United States.*

Acknowledgments

I would like to acknowledge Ms. Lucy Reid, Librarian USDA-APHIS Plant Epidemiology and Risk Analysis Laboratory, for her invaluable assistance in literature research and the editors for their insightful review and comments.

References

Adkar-Purushothama, C.R., Kanchepalli, P.R., Yanjarappa, S.M., Zhang, Z., Sano, T., 2014. Detection, distribution, and genetic diversity of Australian grapevine viroid in grapevines in India. Virus Genes 49, 304–311.

Almeyda-León, I.H., Iracheta-Cárdenas, M.M., Jasso-Argumedo, J., Curti-Díaz, S.A., Ruiz-Beltrán, P., Rocha-Peña, M.A., 2002. Re-examination of citrus viroids in Tahiti Lime in Mexico. Rev. Mex. Fitopatol. 20, 152–160.

Badilla, R., Hammond, R., Rivera, C., 1999. First report of potato spindle tuber viroid in Costa Rica. Plant Dis. 83, 1072.

Belalcazar, C.S., Galindo-A., J., 1974. Estudio sobre el virus de la "Planta Macho" del jitomate. Agrociencia 18, 79–88.

Bennett, S., Tennant, P., McLaughlin, W., 2010. First report of hop stunt viroid infecting citrus orchards in Jamaica. Plant Pathol. 59, 393.

Bostan, H., Nie, X., Singh, R.P., 2004. An RT-PCR primer pair for the detection of *Pospiviroid* and its application in surveying ornamental plants for viroids. J. Virol. Methods 116, 189–193.

CABI, 2015a. Datasheet: citrus exocortis viroid. Crop Protection Compendium. <www.cabi.org/cpc> (accessed 23.11.15.).

CABI, 2015b. Datasheet: potato spindle tuber viroid. Crop Protection Compendium. <http://www.cabi.org/cpc/datasheet/43659> (accessed 15.11.15.).

CABI/EPPO, 2012. Potato spindle tuber viroid. Distribution Maps of Plant Diseases No. 729. CAB International, Wallingford, UK.
CABI/EPPO, 2013. Avocado sunblotch viroid. Distribution Maps of Plant Diseases No. 1145. CAB International, Wallingford, UK.
Chitambar, J., 2015. California plant pest rating, plant pathogens, viruses and viroids, potato spindle tuber viroid. Calif. Depart. Food Agric. <http://blogs.cdfa.ca.gov/Section3162/?p = 384> (accessed 03.03.16.).
De Boer, S.H., Xu, H., DeHaan, T.L., 2002. Potato spindle tuber viroid not found in western Canadian provinces. Can. J. Plant Pathol. 24, 372–375.
De La Torre-A, R., Téliz-Ortiz, D., Pallás, V., Sánchez-Navarro, J.A., 2009. First Report of avocado sunblotch viroid in avocado from Michoacán, México. Plant Dis. 93, 202–302.
De La Torre-Almaráz, R., Pallás, V., Sánchez-Navarro, J.A., 2015. First report of peach latent mosaic viroid in peach trees from Mexico. Plant Dis. 99, 899–999.
Diener, T.O., 1987. Potato spindle tuber. In: Diener, T.O. (Ed.), The Viroids. Plenum, New York, pp. 221–233.
Duran-Vila, N., Roistacher, C.N., Rivera-Bustamante, R., Semancik, J.S., 1988. A definition of citrus viroid groups and their relationship to the exocortis disease. J. Gen. Virol. 69, 3069–3080.
Eastwell, K.C., Nelson, M.E., 2007. Occurrence of viroids in commercial hop (Humulus lupulus l.) production areas of Washington State. Plant Management Network. <http://www.plantmanagementnetwork.org/pub/php/research/2007/hop/> (accessed 17.05.15.).
Eiras, M., Targon, M.L.P.N., Fajardo, T.V.M., Flores, R., Kitajima, E.W., 2006. Citrus exocortis viroid and hop stunt viroid doubly infecting grapevines in Brazil. Fitopatol. Brasil. 31, 440–446.
EPPO, 2009. Potato spindle tuber viroid no longer occurs in Argentina. EPPO Rep. Ser. 2009 (02), 14.
Fernandez-Valiela, M.V., 1969. Introducción a la Fitopatología. Vol 1 Virus. 3 edición. INTA.
Fiore, N., Ghanem-Sabanadzovic, N., Abou, Infante, R., Myrta, A., Pallás, V., 2003. Detection of peach latent mosaic viroid in stone fruits from Chile. In: Myrta, A., Terlizzi, B.D., Savino, V. (Eds.), Virus and Virus-Like Diseases of Stone Fruits, With Particular Reference to the Mediterranean Region. CIHEAM, Bari, Italy, pp. 143–145.
Fisher, L., Bennett, S., Tennanta, P., Laughlin, W., 2011. Detection of citrus tristeza virus and citrus viroid species in Jamaica. Acta Hortic. 894, 117–121.
Flores, R., la Peña, De, Navarro, B., M., 2003. Chrysanthemum chlorotic viroid. In: Hadidi, A., Flores, R., Randles, J.W., Semancik, J.S. (Eds.), Viroids. CSIRO Publishing, Collingwood, VIC, pp. 224–227.
Flores, R., Hernández, C., Desvignes, J.C., Llácer, G., 1990. Some properties of the viroid inducing peach latent mosaic disease. Res. Virol. 141, 109–118.
Fonseca, M.E.N., Boiteux, L.S., Singh, R.P., Kitajima, E.W., 1989. A small viroid in Coleus species from Brazil. Fitopatol. Bras. 14, 94–96.
Fonseca, M.E.N., Boiteux, L.S., Singh, R.P., Kitajima, E.W., 1990. A viroid from Coleus species in Brazil. Plant Dis. 74, 80.
Gobatto, D., Chaves, A.L.R., Harakava, R., Marque, J.M., Daròs, J.A., Eiras, M., 2014. Chrysanthemum stunt viroid in Brazil: survey, identification, biological and molecular characterization and detection methods. J. Plant Pathol. 96, 111–119.
Guerrero-Gámez, C.E., Alvarado-Gómez, O.G., Gutiérrez-Mauleón, H., González-Garza, R., Álvarez-Ojeda, M.G., Luna-Rodríguez, M., 2013. Detección por RT-PCR punto final y tiempo real de tres especies de viroides en cítricos de Nuevo León y Tamaulipas, México. Rev. Mex. Fitopatol. 31, 20–28.
Hadidi, A., Giunchedi, L., Shamloul, A.M., Poggi-Pollini, C., Amer, M.A., 1997. Occurrence of peach latent mosaic viroid in stone fruits and its transmission with contaminated blades. Plant Dis. 81, 154–158.

Hadidi, A., Hansen, A.J., Parish, C.L., Yang, X., 1991. Scar skin and dapple apple viroids are seed-borne and persistent in infected apple trees. Res. Virol. 142, 289–296.

Hadidi, A., Huang, C., Hammond, R.W., Hashimoto, J., 1990. Homology of the agent associated with dapple apple disease to apple scar skin viroid and molecular detection of these viroids. Phytopathology 80, 263–268.

Herranz, M.C., Maeso, D., Soria, J., Pallás, V., 2002. First report of peach latent mosaic viroid on peach in Uruguay. Plant Dis. 86, 1405.

Kunta, M., da Gracxa, J.V., Skaria, M., 2007. Molecular detection and prevalence of citrus viroids in Texas. HortScience 42, 600–604.

Lawson, R.H., 1987. Chrysanthemum stunt. In: Diener, T.O. (Ed.), The Viroids. Plenum Press, New York, pp. 277–359.

Ling, K.S., Bledsoe, M.E., 2009. First report of Mexican papita viroid infecting greenhouse tomato in Canada. Plant Dis. 93, 839–939.

Ling, K.S., Sfetcu, D., 2010. First report of natural infection of greenhouse tomatoes by potato spindle tuber viroid in the United States. Plant Dis. 94, 1376–2376.

Ling, K.S., Zhang, W., 2009. First report of a natural infection by Mexican papita viroid and tomato chlorotic dwarf viroid on greenhouse tomatoes in Mexico. Plant Dis. 93, 1216–2216.

Ling, K.S., Li, R., Groth-Helms, D., Assis-Filho, F.M., 2013. First report of potato spindle tuber viroid naturally infecting field tomatoes in the Dominican Republic. Plant Dis. 98, 701–801.

Ling, K.S., Li, R., Panthee, D.R., Gardner, R.G., 2012. First report of potato spindle tuber viroid naturally infecting greenhouse tomatoes in North Carolina. Plant Dis. 97, 148–248.

Ling, K.S., Verhoeven, J., ThJ., Singh, R.P., Brown, J.K., 2009. First report of tomato chlorotic dwarf viroid in greenhouse tomatoes in Arizona. Plant Dis. 93, 1075–2075.

Martinez-Soriano, J.P., Galindo-Alonso, J., Maroon, C.J.M., Yucel, I., Smith, D.R., Diener, T.O., 1996. Mexican papita viroid: putative ancestor of crop viroids. Proc. Natl. Acad. Sci. USA 93, 9397–9401.

Michelutti, R., Al Rwahnih, M., Torres, H., Gómez, G., Luffman, M., Myrta, A., et al., 2004. First record of hop stunt viroid in Canada. Plant Dis. 88, 1162–2162.

Millikan, D.F., Martin, W.R., 1956. An unusual fruit symptom in apple. Plant Dis. Rep. 40, 229–230.

Nelson, M.E., Klein, R.E., Skrzeczkowski, L.J., 1997. Occurrence of hop latent viroid (HLVd) in hops in Washington State. Phytopathology 88, S108.

Nieto, A.M., Di Feo, L., Nome, C.F., 2008. First report of peach latent mosaic viroid in peach trees in Argentina. Plant Dis. 92, 1137–2137.

Nome, C., Giagetto, A., Rossini, M., Feo, L., Di Nieto, A., 2012. First report of apple scar skin viroid (ASSVd) in apple trees in Argentina. New Dis. Rep. 25, 3.

Nome, C.F., Difeo, L.V., Giayetto, A., Rossini, M., Frayssinet, S., Nieto, A., 2011. First report of pear blister canker viroid in pear trees in Argentina. Plant Dis. 95, 882–982.

Orozco, V.G., Galindo, A.J., 1986. Ecology of the tomato planta macho viroid, I. Natural host plants, agroecosystem effect on viroid incidence and influence of temperature on viroid. Rev. Mex. Fitopatol. 4, 19–28.

Osman, F., Rwahnih, M.A., Golino, D., Pitman, T., Cordero, F., Preece, J.E., et al., 2012. Evaluation of the phytosanitary status of the Prunus species in the national clonal germplasm repository in California: survey of viruses and viroids. J. Plant. Pathol. 94, 249–253.

Owens, R.A., Smith, D.R., Diener, T.O., 1978. Measurement of viroid sequence homology by hybridization with complementary DNA prepared in vitro. Virology 89, 388–394.

Pagliano, G., Umana, R., Pritsch, C., Rivas, F., Duran-Vila, N., 2013. Occurrence, prevalence and distribution of citrus viroids in Uruguay. J Plant Pathol. 95, 631–635.

Peiman, M., Xie, C., 2006. Development and evaluation of a multiplex RT-PCR for detecting main viruses and a viroid of potato. Acta Virol. 50, 129–133.

Pérez, J.M., Pérez, R., Velásquez, K., Peña, I., 2001. Diagnóstico, caracterización y manejo del complejo de viroides que afectan el cultivo de los cítricos en Cuba. Pages 268 *in* Resúmenes IV Seminario Científico Internacional de Sanidad Vegetal Cuba. Taller de Cítricos.

Querci, M., Owens, R.A., Vargas, C., Salazar, L.F., 1995. Detection of potato spindle tuber viroid in avocado growing in Peru. Plant Dis. 79, 196–202.

Rezaian, M.A., Krake, L.R., Golino, D.A., 1992. Common identity of grapevine viroids from USA and Australia revealed by PCR analysis. Intervirology 34, 38–43.

Roistacher, C.N., Canton, H. Reddy, P.S., 1996. The economics of living with citrus viroids in Belize. In: Thirteenth IOCV Conference, Pages 370–375.

Rondón, A., Figueroa, M., 1976. Mancha de sol (Sunblotch) de los aguacates (*Persea americana*) en Venezuela. Agron. Trop. 26, 463–466.

Running, C.M., Schnell, R.J., Kuhn, D.N., 1996. Detection of avocado sunblotch viroid and estimation of infection among accessions in the national germplasm collection for avocado. Proc. Fla. State Hort. Soc. 109, 235–237.

Salazar, L.F., Owens, R.A., Smith, D.R., Diener, T.O., 1983. Detection of potato spindle tuber viroid by nucleic acid spot hybridization: evaluation with tuber sprouts and true potato seed. Amer. Potato J. 60, 587–597.

Sano, T., Ohshima, K., Hataya, T., Uyeda, I., Shikata, E., Chou, T.-G., et al., 1986. A viroid resembling hop stunt viroid in grapevines from Europe, the United States and Japan. J. Gen. Virol. 67, 1673–1678.

Semancik, J.S., 2003. Avocado sunblotch viroid. In: Hadidi, A., Flores, R., Randles, J.W., Semancik, J.S. (Eds.), Viroids. CSIRO Publishing, Collingwood, VIC, pp. 171–177.

Semancik, J.S., Rivera-Bustamante, R., Goheen, A.C., 1987. Widespread occurrence of viroid-like RNAs in grapevines. Am. J. Enol. Vitic. 38, 35–40.

Serra, P., Eiras, M., Bani-Hashemian, S.M., Murcia, N., Kitajima, E.W., Daròs, J.A., et al., 2008. Citrus viroid V: occurrence, host range, diagnosis, and identification of new variants. Phytopathology 98, 1199–1204.

Shamloul, A.M., Hadidi, A., Zhu, S.F., Singh, R.P., Sagredo, B., 1997. Sensitive detection of potato spindle tuber viroid using RT-PCR and identification of a viroid variant naturally infecting pepino plants. Can. J. Plant Pathol. 19, 89–96.

Singh, R.P., Boucher, A., 1991. High incidence of transmission and occurrence of a viroid in commercial seeds of coleus in Canada. Plant Dis. 75, 184–187.

Singh, R.P., Dilworth, A.D., 2009. Tomato chlorotic dwarf viroid in the ornamental plant *Vinca minor* and its transmission through tomato seed. Eur. J. Plant. Pathol. 123, 111–116.

Singh, R.P., 1983. Viroids and their potential danger to potatoes in hot climate. Can. Plant Dis. Surv. 63, 13–18.

Singh, R.P., 2014. The discovery and eradication of potato spindle tuber viroid in Canada. Virus Dis. 25, 415–424.

Singh, R.P., Dilworth, A.D., Ao, X., Singh, M., Baranwal, V.K., 2009. Citrus exocortis viroid transmission through commercially-distributed seeds of *Impatiens* and *Verbena* plants. Eur. J. Plant Pathol. 124, 691–694.

Singh, R.P., Lakshman, D.K., Boucher, A., Tavantzis, S.M., 1992. A viroid from *Nematanthus wettsteinii* plants closely related to the columnea latent viroid. J. Gen. Virol. 73, 2769–2774.

Singh, R.P., Nie, X., Singh, M., 1999. Tomato chlorotic dwarf viroid: an evolutionary link in the origin of pospiviroids. J. Gen. Virol. 80, 2823–2828.

Skrzeczkowski, L.J., Howell, W.E., Mink, G.I., 1996. Occurrence of peach latent mosaic viroid in commercial peach and nectarine cultivars in the U.S. Plant Dis. 80, 823.

Smith, W.W., Barratt, J.G., Rich, A.E., 1956. Dapple apple, an unusual fruit symptom of apples in New Hampshire. Plant Dis. Rep. 40, 756–766.

Sun, M., Siemsen, S., Campbell, W., Guzman, P., Davidson, R., Whitworth, J.L., et al., 2004. Survey of potato spindle tuber viroid in seed potato growing areas of the United States. Amer. J. Potato Res. 81, 227–231.

Targon, M.L.P.N., Muller, G.W., Carvalho, S.A., Souza, J.M., De Machado, M.A., 2003. Detecção de viróides em citros através de hibridização de impressões de tecidos [Abstract]. Summa Phytopathol. 29, 60.

Torres, H., Gómez, G., Pallás, V., Stamo, B., Shalaby, A., Aouane, B., et al., 2004. Detection by tissue printing of stone fruit viroids, from Europe, the Mediterranean and North and South America. Acta Hortic. 657, 379–383.

Vargas, C.O., Querci, M., Salazar, L.F., 1991. Identificación y estado de diseminación del viroide del manchado solar del palto (Persea amaricana L.) en el Perú y la existencia de otros viroides en palto. Fitopatología. 26, 23–27.

Velásquez, K., Soto, M., Pérez, R., Pérez, J.M., Rodríguez, D. Durán-Vila, N., 2001. Biological and molecular characterization of two isolates of citrus viroids recovered from Cuban plantations. In: Duran-Vila, N., Milne, R. G., Graça, J.V. (Eds.), Fifteenth Conference of the International Organization of Citrus Virologists. IOCV.

Verhoeven, J.Th.J., Jansen, C.C.C., Werkman, A.W., Roenhorst, J.W., 2007. First report of tomato chlorotic dwarf viroid in *Petunia hybrida* from the United States of America. Plant Dis. 91, 324–424.

Verhoeven, J.Th.J., Jansen, C.C.C., Willemen, T.M., Kox, L.F.F., Owens, R.A., et al., 2004. Natural infections of tomato by citrus exocortis viroid, columnea latent viroid, potato spindle tuber viroid and tomato chlorotic dwarf viroid. Eur. J. Plant Pathol. 110, 823–831.

Welsh, M.F., Keene, F.W.L., 1961. Diseases of apple that are caused by viruses or have characteristics of virus diseases. Can. Plant Dis. Surv. 41, 123–147.

CHAPTER

44

Geographical Distribution of Viroids in Europe

Francesco Faggioli[1], Nuria Duran-Vila[2], Mina Tsagris[3] and Vicente Pallás[4]

[1]CREA-Research Centre for Plant Protection and Certification, Rome, Italy [2]Valencian Institute of Agricutural Research-IVIA, Valencia, Spain [3]University of Crete, Crete, Greece [4]Polytechnic University of Valencia-CSIC, Valencia, Spain

FRUIT TREE VIROIDS

Six viroids have been reported in fruit trees in Europe. They come from four of the eight viroid genera: *Pelamoviroid* (peach latent mosaic viroid, PLMVd), *Avsunviroid* (avocado sunblotch viroid, ASBVd), *Hostuviroid* (hop stunt viroid, HSVd) and *Apscaviroid* (apple dimple fruit viroid, ADFVd, apple scar skin viroid, ASSVd, and pear blister canker viroid, PBCVd). ASBVd has been detected only in its natural host, avocado, in Spain by Pallás et al. (1988).

Stone Fruit Viroids

HSVd is the viroid with the widest host range. Although its infection is usually latent in stone fruits, it has been associated with serious disorders of economic importance such as the dapple fruit disease of plum and peach (Sano et al., 1989) and with an apricot fruit disorder known as "fruit degeneration" characterized by fruit rugosity and the loss of organoleptic properties (Amari et al., 2007). HSVd has been detected in France, Spain (81% in the surveyed trees), Greece (5%), Cyprus (10%), Italy (37%), Albania (7%, 7%, and 30% for plum, peach, and apricot, respectively), and Turkey (20%, 61%, and 86%, for the same fruits,

Viroids and Satellites.
DOI: http://dx.doi.org/10.1016/B978-0-12-801498-1.00044-9

respectively) (Pallás et al., 2003; Torres et al., 2004). It is worth noting the high incidence in apricot in Spain, Italy, and Turkey, the three main producers in Europe. More recently, it has been detected in the Czech Republic (Hassan et al., 2009), Bosnia and Herzegovina (Matić et al., 2005), Serbia (Mandic et al., 2008), and Greece (Kaponi and Kyriakopoulou, 2013).

PLMVd was found in 80%–85% of peach trees in the Valencia region (Spain) (Badenes and Llácer, 1998). Remarkably, the Spanish native cultivars examined were not infected, whereas up to 85% of foreign cultivars (most from North America) were. In Italy, about 50% of peach, nectarine (a group of peach cultivars), and clingstone germplasm was infected (Loreti et al., 1998), and a similar fraction (54%) was found in peach germplasm in Zaragoza (Spain) (Rubio-Cabetas et al., 2012). More recently it was reported in the Czech Republic (Hassan et al., 2009), Bosnia and Herzegovina (Matić et al., 2005), Serbia (Mandic et al., 2008), Croatia in imported and native nectarine (94%) and peach (42%) (Škorić et al., 2008), and in peach germplasm from Romania, Austria, and the former Yugoslavia (Shamloul et al., 1995). PLMVd was occasionally detected in cherry, plum, and apricot germplasm from European countries (Hadidi et al., 1997), and in sweet cherry in Italy (Crescenzi et al., 2002). Kyriakopoulou et al. (2001) detected PLMVd in cultivated (Greece and Italy) and wild pear (Greece) (37% of the tested samples), together with HSVd, the second viroid naturally infecting stone and pome fruits.

Recently ASSVd was detected in sweet cherry in Greece (Kaponi et al., 2013).

Pome Fruit Viroids

ASSVd has been detected only rarely in European pome fruit trees despite reaching an incidence of 35% in cultivated and wild pears in Greece and Italy, and 46% in apple orchards in Eastern Anatolia, Turkey (Kyriakopoulou et al., 2003; Sipahioglu et al., 2009).

The disease caused by ADFVd was first observed in commercial orchards in Southern Italy (Di Serio et al., 1998). Recently, a viroid resembling ADFVd has been characterized in fig by next-generation sequencing of small RNAs in Italy (Chiumenti et al., 2014). However, studies on the incidence and geographical distribution of the disease in Europe are scarce.

PBCVd has been detected in France, Spain (Ambrós et al., 1995; Hernández et al., 1992), Italy (Loreti et al., 1998), and Greece (Kyriakopoulou et al., 2001), and more recently in Bosnia and Herzegovina (Lolic et al., 2007), Turkey (Yesilcollou et al., 2010), and Albania (Navarro et al., 2011).

Thus, the apscaviroids are prevalent in pome fruits whereas members of three other genera are mainly present in stone fruits

Citrus Viroids

The identification of citrus exocortis viroid (CEVd) as the first citrus viroid in Europe is linked to the outbreak of the disease caused by citrus tristeza virus. The control of the tristeza disease was based on the substitution of the widespread sour orange rootstock by other species, most of which were sensitive to CEVd. Further investigations have shown the widespread occurrence of citrus viroids, different from CEVd, in all citrus growing areas (Pallás et al., 2003). These studies showed that all tested citrus cultivars in Europe were infected with viroid mixtures, with HSVd and citrus dwarfing viroid being the most frequent. Citrus bent leaf viroid is found less frequently, and citrus bark cracking viroid very rarely. The most recently described citrus viroid V (Serra et al., 2008) has only been identified in Spain.

GRAPEVINE VIROIDS

Grapevine yellow speckle viroid 1 (GSYVd-1) and HSVd (very often in association) are the most common and widespread grapevine viroids in Italy, France, Spain, Greece, Germany, the Czech Republic, and Hungary (Gambino et al., 2014; Matousek et al., 2003; Pallás et al., 2003). These viroids have also been detected in germplasm from Austria (Sano et al., 1986), Albania, Bulgaria, Cyprus, and the former USSR (Minafra et al., 1990). Grapevine yellow speckle viroid-2 and CEVd have been reported only sporadically in Spain, Italy, and Turkey (Flores et al., 1985; Gazel and Önelge, 2003), whereas Australian grapevine viroid has been found only in Italy (Gambino et al., 2014). Using deep sequencing, GYSVd-1 and HSVd together with a badnavirus were identified in a Greek vineyard showing leaf discoloration symptoms (Maliogka et al., 2015).

POSPIVIROIDS ON ORNAMENTAL AND HERBACEOUS CROP SPECIES

Pospiviroids have been reported worldwide, including Europe, where only Mexican papita viroid and tomato planta macho viroid (now considered a single species) have never been reported. This section treats also dahlia latent viroid, which actually belongs to the genus *Hostuviroid* (Di Serio et al., 2014).

Potato spindle tuber viroid (PSTVd) is a quarantine pest detected in seeds and tubers of potato in some East European countries (Belarus, Russia, and Ukraine; EPPO/CABI, 1997). It has also been detected in the United Kingdom, and eliminated from potato accessions in the Commonwealth Potato Collection (Scottish Plant Breeding Station, 1976). Sporadic PSTVd outbreaks in tomato have been reported in Belgium, Italy, The Netherlands, and the United Kingdom (Mumford et al., 2004; Navarro et al., 2009; Verhoeven et al., 2004, 2007). Furthermore, in many European countries, PSTVd infections have been found in several ornamentals: *Brugmansia* spp., *Chrysanthemum* spp., *Petunia* spp., *Physalis peruviana, Solanum pseudocapsicum, Streptosolen jamesonii, Calibrachoa* spp., *Lycianthes rantonnetii, Datura* spp., *Solanum jasminoide* (EFSA Panel on Plant Health, 2011), and *Cestrum* spp. (Luigi et al., 2011).

Chrysanthemum stunt viroid, like PSTVd, is a quarantine pest that has been reported in several European countries (Austria, Finland, Latvia, Slovenia, and Spain, where it is now eradicated), mainly on *Dendranthema* spp. but also in other ornamentals. It is currently present, with restricted distribution, in Belgium, the Czech Republic, France, Italy, The Netherlands, Germany and the United Kingdom in different hosts: chrysanthemum, *Argyranthemum frutescens, Pericallis hybrida, Vinca* spp., *Petunia* spp., *Callibrachoa* spp., and *S. jasminoides* (EFSA Panel on Plant Health, 2012).

Tomato apical stunt viroid has been found in Austria, Belgium, Croatia, the Czech Republic, France, Germany, Italy, Finland, The Netherlands, Poland, and Slovenia, mainly in symptomless ornamentals (*Brugmansia* spp., *S. jasminoides, L. rantonnetii, Cestrum* spp., *S. jamesonii*) but also in tomato (France, Italy, and The Netherlands; now eradicated) (EFSA Panel on Plant Health, 2011).

CEVd, apart from citrus and grapevine, infects certain ornamental and herbaceous crops. In Europe, CEVd is widespread in *S. jasminoides*, whereas it has been sporadically identified in *Brassica napus, Cestrum* spp., *Citrus* spp., *Daucus carota, Vicia faba, L. rantonnetii*, tomato, eggplant, and *Verbena* spp. in Austria, Belgium, the Czech Republic, Germany, Italy, Montenegro, The Netherlands, Russia, Serbia, and Slovenia (EFSA Panel on Plant Health, 2011).

Columnea latent viroid has been detected in symptomless *Brunfelsia undulata* grown in Germany (Spieker, 1996a) and then in tomato in The Netherlands, Belgium (Verhoeven et al., 2004), France (Steyer et al., 2010), the United Kingdom (Nixon et al., 2010), and Italy (Parrella et al., 2010). In 2010, this viroid was also reported in Denmark on *Gloxinia* spp. (Nielsen and Nicolaisen, 2010).

Tomato chlorotic dwarf viroid has been reported in symptomatic tomato plants in The Netherlands (Verhoeven et al., 2004), France (Candresse et al., 2010), and recently in Norway (Fox et al., 2013). This viroid has also been reported on symptomless *P. hybrida* in the United Kingdom (James et al., 2008), Slovenia (Marn Viršček and Pleško Mavrič, 2010), and Finland (Lemmetty et al., 2011), as well as in Belgium, the Czech Republic, Germany, and Portugal (EFSA Panel on Plant Health, 2011).

Pepper chat fruit viroid has been identified in The Netherlands in greenhouse-grown sweet pepper (*Capsicum annuum* L.) showing a reduction in fruit size of up to 50% (Verhoeven et al., 2009).

Iresine viroid 1 was reported in Germany in symptomless *Iresine herbstii* (Spieker, 1996b), in *Celosia plumosa* in The Netherlands (Verhoeven et al., 2010), *Portulaca* spp. in Slovenia (Marn Viršček and Pleško Mavrič, 2012), and in *Portulaca* spp., *C. plumosa*, and *C. cristata* in Italy (Sorrentino et al., 2015).

HOP VIROIDS

HSVd has never been reported in Europe in hop (*Humulus lupulus* L), whereas hop latent viroid, genus *Cocadviroid*, was first identified in hop in Spain (Pallás et al., 1987) and Germany (Puchta et al., 1988). Despite its name, it was also found in the United Kingdom in hop plants that showed low cone yield and reduced quality for the beer industry (Barbara et al., 1990). Subsequently, hop latent viroid has been widely detected (up to 90%–100% of the tested hop germplasm) in Belgium, the Czech Republic, France, Hungary, Poland, Portugal, Russia, Spain, and Serbia (Pallás et al., 2003).

Citrus bark cracking viroid has been found very recently to cause severe stunting of hops in Slovenia; it is currently on the EPPO alert list (Jakse et al., 2015).

COLEUS VIROIDS

Coleus blumei viroid 1, 2, and 3 (CbVd-1, -2, -3), genus C*oleviroid*, have been reported in coleus (*Plectranthus scutellarioides* synonym *Coleus blumei)* in Germany (Spieker, 1996c; Spieker et al., 1990), and CbV-1 and CbVd-3 in Croatia (Jezernik, 2014). CbV-1 has also been found in Spain in several coleus cultivars (N. Duran-Vila unpublished results).

MEMBERS OF THE FAMILY *AVSUNVIROIDAE* IN HERBACEOUS CROPS

Eggplant latent viroid was first identified in Spain in symptomless eggplant (Fagoaga et al., 1994), but its incidence has not yet been evaluated.

Chrysanthemum chlorotic mottle viroid can cause potential damage to the chrysanthemum industry. Although its incidence is significant in Asia, it has only been sporadically reported in Europe, specifically in Denmark, France (Horst, 1987), and The Netherlands (Hosokawa et al., 2005).

Table 44.1 shows the reported distribution of viroids in Europe, with their hosts.

TABLE 44.1 Geographical Distribution of Viroids in Europe

Viroid	Countries	Hosts
ADFVd	Italy	Apple, fig
AGVd	Italy	Grapevine
ASBVd	Spain	Avocado
ASSVd	Greece	Sweet cherry, wild pear
	Greece, Italy	Pear
	Turkey, United Kingdom	Apple
CBCVd	Present but rare in Europe	Citrus
CBLVd	Present but not frequent in Europe	Citrus
	Slovenia	Hop
CbVd-1	Germany, Croatia, Spain	Coleus
CbVd-2	Germany	Coleus
CbVd-3	Germany, Croatia	Coleus
CChMVd	Denmark, France, The Netherlands	Chrysanthemum
CDVd	All over Europe	Citrus
CEVd	All over Europe	*S. jasminoides*, citrus
	Austria, Belgium, Czech Republic, Germany, Italy, Montenegro, The Netherlands, Russian Federation, Serbia, Slovenia	*Brassica napus*, *Cestrum* spp, *Daucus carota*, *Vicia faba*, *Lycianthes rantonnetii*, tomato, eggplant, *Verbena* spp.
	Spain, Italy, Turkey	Grapevine

(*Continued*)

TABLE 44.1 (Continued)

Viroid	Countries	Hosts
CLVd	Germany	*Brunfelsia undulata*
	Belgium, France, Italy, The Netherlands, United Kingdom	Tomato
	Denmark	*Gloxinia* spp.
CSVd	Austria, Finland, Latvia, Slovenia, Spain	*Dendranthema* spp.
	Belgium, Czech Republic, France, Germany, Italy, The Netherlands, United Kingdom	Chrysanthemum, *Argyranthemum frutescens*, *Pericallis hybrida*, *Vinca* spp., *Petunia* spp, *Callibrachoa* spp., *S. jasminoides*
CVd-V	Spain	Citrus
DLVd	The Netherlands	Dahlia
ELVd	Spain	Eggplant
GYSVd-1	All over Europe	Grapevine
GYSVd-2	Spain, Italy, Turkey	Grapevine
HLVd	Belgium, Czech Republic, France, Germany, Hungary, Poland, Portugal, Russian Federation, Spain, Serbia, United Kingdom	Hop
HSVd	All over Europe	Citrus, grapevine
	Albania, Cyprus, France, Greece, Italy, Spain, Turkey	Plum, peach, apricot
	Czech Republic	Peach, apricot
	Bosnia and Herzegovina	Apricot, plum
	Greece	Plum, peach
IrVd-1	Germany	*Iresine herbstii*
	Italy, The Netherlands	*Celosia plumosa*
	Italy, Slovenia	*Portulaca* spp.
	Italy	*Celosia cristata*
PBCVd	Albania, Bosnia and Herzegovina, France, Greece, Italy, Spain, Turkey	Pear
PCFVd	The Netherlands	Sweet pepper
PLMVd	All over Europe	Peach

(*Continued*)

TABLE 44.1 (Continued)

Viroid	Countries	Hosts
	France, Italy, former Yugoslavia, Romania	Plum, cherry, apricot
	Greece, Italy	Pear
	Greece	Wild pear
PSTVd	Belarus, Russian Federation, United Kingdom, Ukraine,	Potato
	Belgium, Italy, The Netherlands, United Kingdom	Tomato
	All over Europe	Ornamentals
TASVd	Austria, Belgium, Croatia, Czech Republic, France, Germany, Italy, Finland, The Netherlands, Poland, Slovenia	*Brugmansia* spp., *S. jasminoides*, *Lycianthes rantonnetii*, *Cestrum* spp., *Streptosolen jamesonii*
	France, Israel, Italy, The Netherlands	Tomato
TCDVd	France, The Netherlands, Norway	Tomato
	Belgium, the Czech Republic, Germany, Finland, Portugal, Slovenia, United Kingdom	*Petunia hibrida*

ADFVd, apple dimple fruit viroid; AGVd, Australian grapevine viroid; ASBVd, avocado sunblotch viroid; ASSVd, apple scar skin viroid; CVCVd, citrus bark cracking viroid; CBLVd, citrus bent leaf viroid; CbVd-1, Coleus blumei viroid 1; CbVd-2, Coleus blumei viroid 2; CbVd-3, Coleus blumei viroid 3; CChMVd, chrysanthemum chlorotic mottle viroid; CDVd, citrus dwarf viroid; CEVd, citrus exocortis viroid; CLVd, Columnea latent viroid; CSVd, chrysanthemum stunt viroid; CVd-V, citrus viroid V; DLVd, dahlia latent viroid; ELVd, eggplant latent viroid; GYSVd-1, grapevine yellow speckle viroid 1; GYSVd-2, grapevine yellow speckle viroid 2; HLVd, hop latent viroid; HSVd, hop stunt viroid; IrVd-1, Iresine viroid 1; PBCVd, pear blister canker viroid; PCFVd, pepper chat fruit viroid; PLMVd, peach latent mosaic viroid; PSTVd, potato spindle tuber viroid; TASVd, tomato apical stunt viuroid; TCDVd, tomato chlorotic dwarf viroid.

References

Amari, K., Ruiz, D., Gómez, G., Sánchez-Pina, M.A., Pallás, V., Egea, J., 2007. An important new apricot disease in Spain is associated with hop stunt viroid infection. Eur. J. Plant Pathol. 118, 173–181.

Ambrós, S., Desvignes, J.C., Llácer, G., Flores, R., 1995. Peach latent mosaic viroid and pear blister canker viroids: detection by molecular hybridization and relationships with specifics maladies affecting peach and pear trees. Acta Hortic. 386, 515–521.

Badenes, M.L., Llácer, G., 1998. Occurrence of peach latent mosaic viroid in American peach and nectarine cultivars. Acta Hortic. 472, 565–570.

Barbara, D.J., Morton, A., Adams, A.N., Green, C.P., 1990. Some effects of hop latent viroid on two cultivars of hop (*Humulus lupulus*) in the UK. Ann. Appl. Biol. 117, 359–366.

Candresse, T., Marais, A., Tassus, X., Suhard, P., Renaudin, I., Leguay, A., et al., 2010. First report of tomato chlorotic dwarf viroid in tomato in France. Plant Dis. 94, 633.

Chiumenti, M., Torchetti, E.M., Di Serio, F., Minafra, A., 2014. Identification and characterization of a viroid resembling apple dimple fruit viroid in fig (*Ficus carica* L.) by next generation sequencing of small RNAs. Virus Res. 188, 54–59.

Crescenzi, A., Piazzolla, P., Hadidi, A., 2002. First report of peach latent mosaic viroid in sweet cherry in Italy. J. Plant Pathol. 84, 168.

Di Serio, F., Flores, R., Verhoeven, J.Th.J., Li, S.-F., Pallàs, V., et al., 2014. Current status of viroid taxonomy. Arch Virol. 159, 3467–3478.

Di Serio, F., Alioto, D., Ragozzino, A., Giunchedi, L., Flores, R., 1998. Identification of apple dimple fruit viroid in different commercial varieties of apple grown in Italy. Acta Hortic. 472, 595–602.

EFSA Panel on Plant Health, 2011. Scientific opinion on the assessment of the risk of solanaceous pospiviroids for the EU territory and the identification and evaluation of risk management options. EFSA J. 9 (8), 2330; 133 pp.

EFSA Panel on Plant Health, 2012. Scientific opinion on the risk to plant health posed by chrysanthemum stunt viroid for the EU territory, with identification and evaluation of risk reduction options. EFSA J. 10 (12), 302; 87 pp.

EPPO/CABI, 1997. Potato spindle tuber viroid. In: Smith, I.M., McNamara, D.G., Scott, P.R., Holderness, M. (Eds.), Quarantine Pests for Europe, second ed. CAB International, Wallingford (GB), pp. 1305–1310.

Fagoaga, C., Pina, J.A., Duran-Vila, N., 1994. Occurrence of small RNAs in severely diseased vegetable crops. Plant Dis. 78, 749–753.

Flores, R., Duran-Vila, N., Pallás, V., Semancik, J.S., 1985. Detection of viroid and viroid-like RNAs from grapevine. J. Gen. Virol. 66, 2095–2102.

Fox, A., Daly, M., Nixon, T., Brurberg, M.B., Blystad, D., Harju, V., et al., 2013. First report of tomato chlorotic dwarf viroid (TCDVd) in tomato in Norway and subsequent eradication. New Dis. Rep. 27, 8.

Gambino, G., Navarro, B., Torchetti, E.M., La Notte, P., Schneider, A., Mannini, F., et al., 2014. Survey on viroids infecting grapevine in Italy: identification and characterization of Australian grapevine viroid and grapevine yellow speckle viroid 2. Eur. J. Plant Pathol. 140, 199–205.

Gazel, M., Önelge, N., 2003. First report of grapevine viroids in the east Mediterranean region of Turkey. Plant Pathol. 52, 405.

Hadidi, A., Giunchedi, L., Shamloul, A.M., Poggi-Pollini, G., Amer, M.A., 1997. Occurrence of peach latent mosaic viroid in stone fruits and its transmission with contaminated blades. Plant Dis. 81, 154–158.

Hassan, M., Gómez, G., Pallás, V., Myrta, A., Rysanek, P., 2009. Simultaneous detection and genetic variability of stone fruit viroids in Czech Republic. Eur. J. Plant Pathol. 124, 363–368.

Hernández, C., Elena, S.F., Moga, A., Flores, R., 1992. Pear blister canker viroid is a member of ASSVd subgroup and also has sequence homology with viroids from other subgroups. J. Gen. Virol. 73, 2503–2507.

Horst, R.K., 1987. Chrysanthemum chlorotic mottle. In: Diener, T.O. (Ed.), The Viroids. Plenum Press, New York, pp. 291–295.

Hosokawa, M., Matsushita, Y., Ohishi, K., Yazawa, S., 2005. Elimination of chrysanthemum chlorotic mottle viroid (CChMVd) recently detected in Japan by leaf-primordia free shoot apical meristem culture from infected cultivars. J. Jap. Soc. Hortic. Sci. 74, 386–391.

Jakse, J., Radisekb, S., Pokorna, T., Matousek, J., Javornik, B., 2015. Deep-sequencing revealed citrus bark cracking viroid (CBCVd) as a highly aggressive pathogen on hop. Plant Pathol. 64, 831–842.

James, T., Mulholland, V., Jeffries, C., Chard, J., 2008. First report of tomato chlorotic dwarf viroid infecting commercial petunia stocks in the United Kingdom. Plant Pathol. 57, 400.

Jezernik, K., 2014. Molecular characterization of coleus blumei viroids from coleus (*Solenostemon scutellarioides* (L.) Codd). Diploma thesis, Faculty of Science repository, Department of Biology, University of Zagreb. http://digre.pmf.unizg.hr/3348/.

Kaponi, M.S., Kyriakopoulou, P.E., 2013. First report of hop stunt viroid infecting Japanese plum, cherry plum, and peach in Greece. Plant Dis. 97, 1662.

Kaponi, M.S., Sano, T., Kyriakopoulou., P.E., 2013. Natural infection of sweet cherry trees with apple scar skin viroid. J. Plant Pathol. 95, 429–433.

Kyriakopoulou, P.E., Giunchedi, L., Hadidi, A., 2001. Peach latent mosaic and pome fruit viroids in naturally infected cultivated (*Pyrus communis)* and wild pear: implications on possible origin of these viroids in the Mediterranean region. J. Plant Pathol. 83, 51–62.

Kyriakopoulou, P.E., Osaki, H., Zhu, S.F., Hadidi, A., 2003. Apple scar skin viroid in pear. In: Hadidi, A., Flores, R., Randles, J.W., Semancik, J.S. (Eds.), Viroids. CSIRO Publishing, Collingwood, VIC, pp. 142–145.

Lemmetty, A., Laamanen, J., Soukained, M., Tegel, J., 2011. Emerging virus and viroid pathogen species identified for the first time in horticultural plants in Finland in 1997-2010. Agr. Food Sci. 20, 29–41.

Lolic, B., Afechtal, M., Matić, S., Myrta, A., Di Serio, F., 2007. Detection by tissue-printing of pome fruit viroids and characterization of pear blister canker viroid in Bosnia and Herzegovina. J. Plant Pathol. 89, 369–375.

Loreti, S., Faggioli, F., Barrale, R., Barba, M., 1998. Occurrence of viroids in temperature fruit trees in Italy. Acta Hortic. 472, 555–560.

Luigi, M., Luison, D., Tomassoli, L., Faggioli, F., 2011. First report of potato spindle tuber and citrus exocortis viroids in *Cestrum* spp. in Italy. New Dis. Rep. 23, 4.

Maliogka, V.I., Olmos, A., Pappi, P.G., Lotos, L., Efthimiou, K., Grammatikaki, G., et al., 2015. A novel grapevine badnavirus is associated with the Roditis leaf discoloration disease. Virus Res. 203, 47–55.

Mandic, B., Al Rwahnih, M., Myrta, A., Gomez, G., Pallás, V., 2008. Incidence and genetic diversity of peach latent mosaic viroid and hop stunt viroid in stone fruits in Serbia. Eur. J. Plant Pathol. 120, 167–176.

Marn Virščeк, M., Pleško Mavrič, I., 2010. First report of tomato chlorotic dwarf viroid in *Petunia* sp. in Slovenia. Plant Dis. 94, 1171.

Marn Virščeк, M., Pleško Mavrič, I., 2012. First report of iresine viroid 1 in *Portulaca* sp in Slovenia. J. Plant Pathol. 94, S4.85–S4.105.

Matić, S., Al-Rwahnih, M., Myrta, A., 2005. Occurrence of stone fruit viroids in Bosnia and Herzegovina. Phytopathol. Mediterr. 44, 285–290.

Matousek, J., Orctova, L., Patzak, J., Svoboda, P., Ludvikova, I., 2003. Molecular sampling of hop stunt viroid (HSVd) from grapevines in hop production areas in the Czech Republic and hop protection. Plant Soil Envir. 49, 168–175.

Minafra, A., Martelli, G.P., Savino, V., 1990. Viroids of grapevines in Italy. Vitis 29, 173–182.

Mumford, R.A., Jarvis, B., Skelton, A., 2004. The first report of potato spindle tuber viroid (PSTVd) in commercial tomatoes in the UK. Plant Pathol. 53, 242.

Navarro, B., Bacu, A., Torchetti, M., Kongjika, E., Susuri, L., Di Serio, F., et al., 2011. First record of pear blister canker viroid on pear in Albania. J. Plant Pathol. 93, S4.70.

Navarro, B., Silletti, M.R., Trisciuzzi, V.N., Di Serio, F., 2009. Identification and characterization of potato spindle tuber viroid infecting tomato in Italy. J. Plant Pathol. 91, 723–726.

Nielsen, S.L., Nicolaisen, M., 2010. First report of columnea latent viroid (CLVd) in *Gloxinia gymnostoma*, *G. nematanthodes* and *G. purpurascens* in a botanical garden in Denmark. New Dis Rep. 22, 4.

Nixon, T., Glover, R., Mathews-Berry, S., Daly, M., Hobden, E., Lambourne, C., et al., 2010. Columnea latent viroid (CLVd) in tomato: the first report in the United Kingdom. Plant Pathol. 59, 392.

Pallás, V., García-Luque, I., Domingo, E., Flores, R., 1988. Sequence variability in avocado viroid (ASBVd). Nucleic Acids Res. 16, 9864.

Pallás, V., Gómez, G., Duran-Vila, N., 2003. Viroids in Europe. In: Hadidi, A., Flores, R., Randles, J.W., Semancik, J.S. (Eds.), Viroids. CSIRO Publishing, Collingwood, VIC, pp. 268–274.

Pallás, V., Navarro, A., Flores, R., 1987. Isolation of a viroid-like RNA from hop different from hop stunt viroid. J. Gen. Virol. 68, 2095–2102.

Parrella, G., Pacella, R., Crescenzi, A., 2010. First record of columnea latent viroid (CLVd) in tomato in Italy. In: Abstract of Proc. III International Symposium on Tomato Diseases July 25–30, 2010, Ischia (NA) (Italy), p. 126.

Puchta, H., Ramm, K., Sänger, H.L., 1988. The molecular structure of hop latent viroid (HLV), a new viroid occurring worldwide in hops. Nucleic Acids Res. 16, 4197–4216.

Rubio-Cabetas, M.J., Montañés, M., Alonso, J.M., Pallás, V., Martínez, G., Gómez, G., 2012. Incidence of peach latent mosaic viroid (PLMVd) in a *Prunus persica* bastch germplasm collection in Spain. Acta Hortic. 940, 687–692.

Sano, T., Hataya, T., Terai, Y., Shikata, E., 1989. Hop stunt viroid strains from dapple fruit disease of plum and peach in Japan. J. Gen. Virol. 70, 1311–1319.

Sano, T., Ohshima, K., Hataya, T., 1986. A viroid resembling hop stunt viroid in grapevines from Europe, the United States and Japan. J. Gen. Virol. 67, 1673–1678.

Scottish Plant Breeding Station, 1976. Report of the Scottish Plant Breeding Station for 1975-1976. Pentlandfield, Roslin, Midlothian, UK, p. 86.

Serra, P., Eiras, M., Bani-Hashemian, S.M., Murcia, N., Kitajima, E.W., Daròs, J.A., et al., 2008. Citrus viroid V: occurrence, host range, diagnosis, and identification of new variants. Phytopathology 98, 1199–1204.

Shamloul, A., Minafra, A., Hadidi, A., Giunchedi, L., Waterworth, H., Allam, E., 1995. PLMVd: Nucleotide sequence of an Italian isolate, sensitive detection using RT-PCR and geographic distribution. Acta Hortic. 386, 522–530.

Sipahioglu, H.M., Usta, M., Ocak, M., 2009. Development of a rapid enzymatic cDNA amplification test for the detection of apple scar skin viroid (ASSVd) in apple trees from eastern Anatolia. Turkey. Arch. Phytopathol. Plant Protect. 42, 352–360.

Škorić, D., Al Rwahnih, M., Myrta, A., 2008. First record of peach latent mosaic viroid in Croatia. Acta Hortic. 781, 535–539.

Sorrentino, R., Torchetti, E.M., Minutolo, M., Di Serio, F., Alioto, D., 2015. First report of iresine viroid 1 in ornamental plants in Italy and of *Celosia cristata* as a novel natural host. Plant Dis. 99, 1655.

Spieker, R.L., 1996a. A viroid from *Brunfelsia undulata* closely related to the columnea latent viroid. Arch. Virol. 141, 1823–1832.

Spieker, R.L., 1996b. The molecular structure of iresine viroid, a new viroid species from *Iresine herbstii* (beefsteak plant). J. Gen. Virol. 77, 2631.

Spieker, R.L., 1996c. A new sequence variant of coleus blumei viroid-1 from the *Coleus blumei* cultivar "Rainbow Gold." Arch. Virol. 141, 2153–2161.

Spieker, R.L., Haas, B., Charng, Y.C., Freimuller, K., Sänger, H.L., 1990. Primary and secondary structure of a new viroid "species" (CbVd1) present in the *Coleus blumei* cultivar "Bienvenue." Nucleic Acids Res. 18, 3998.

Steyer, S., Olivier, T., Skelton, A., Nixon, T., Hobden, E., 2010. Columnea latent viroid (CLVd): first report in tomato in France. Plant Pathol. 59, 794.

Torres, H., Gomez, G., Stamo, B., Shalaby, A., Aouane, B., Gavriel, I., et al., 2004. Detection by tissue printing of stone fruit viroids, from Europe, the Mediterranean and North and South America. Acta Hortic. 657, 379–383.

Verhoeven, J.Th.J., Jansen, C.C.C., Botermans, M., Roenhorst, J.W., 2010. First report of iresine viroid-1 in *Celosia plumosa* in The Netherlands. Plant Dis. 94, 920.

Verhoeven, J.Th.J., Jansen, C.C.C., Roenhorst, J.W., Flores, R., de la Peña, M., 2009. Pepper chat fruit viroid: biological and molecular properties of a proposed new species of the genus *Pospiviroid*. Virus Res. 144, 209–214.

Verhoeven, J.Th.J., Jansen, C.C.C., Roenhorst, J.W., Steyer, S., Michelante, D., et al., 2007. First report of potato spindle tuber viroid in tomato in Belgium. Plant Dis. 91, 1055.

Verhoeven, J.Th.J., Jansen, C.C.C., Willemen, T.M., Kox, L.F.F., Owens, R.A., et al., 2004. Natural infections of tomato by citrus exocortis viroid, columnea latent viroid, potato spindle tuber viroid and tomato chlorotic dwarf viroid. Eur. J. Plant Pathol. 110, 823–831.

Yesilcollou, S., Minoia, S., Torchetti, E.M., Kaymak, S., Gumus, M., Myrta, A., et al., 2010. Molecular characterization of Turkish isolates of pear blister canker viroid and assessment of the sequence variability of this viroid. J. Plant Pathol. 92, 813–819.

CHAPTER

45

Geographical Distribution of Viroids in Africa and the Middle East

Khaled A. El-Dougdoug[1], *Kadriye Çağlayan*[2], *Amine Elleuch*[3], *Hani Z. Al-Tuwariqi*[4], *Ebenezer A. Gyamera*[5] *and Ahmed Hadidi*[6]

[1]Ain Shams University, Cairo, Egypt [2]Mustafa Kemal University, Antakya, Turkey [3]University of Sfax, Sfax, Tunisia [4]Ministry of Agriculture, Riyadh, Saudi Arabia [5]University of Cambridge, Cambridge, United Kingdom [6]U.S. Department of Agriculture, Beltsville, MD, United States

Viroids and viroid diseases, especially those of vegetable and fruit crops occur widely in African and Middle Eastern countries. The hot or warm climate of this region possibly enhances the expression of viroid disease symptoms. Natural viroid infections involve various members of the families *Avsunviroidae* and *Pospiviroidae*.

AVSUNVIROIDAE; *AVSUNVIROID*; *AVOCADO SUNBLOTCH VIROID*

Avocado sunblotch disease, now known to be caused by avocado sunblotch viroid, was recorded in Palestine in 1934 (Oppenheimer, 1955) and in South Africa in 1954 (Da Graca et al., 2003). The disease has been under control in Israel since 1964 by using certified disease-tested avocado mother trees (Spiegel et al., 1984). The disease was also reported from Turkey (Önelge and Ertuğrul, 1997) and Ghana (Acheampong et al., 2008) but at low incidence. In South Africa, fruit yield from viroid infection was reduced by up to 82% in symptomless

Viroids and Satellites.
DOI: http://dx.doi.org/10.1016/B978-0-12-801498-1.00045-0

trees but by only about 14% in trees with symptoms (Da Graca et al., 2003).

AVSUNVIROIDAE; PELAMOVIROID; PEACH LATENT MOSAIC VIROID

Peach latent mosaic viroid (PLMVd) incidence in peach trees was about 40% in Syria (Ismaeil et al., 2001), 34% in Lebanon (Choueiri et al., 2002), and 29% in Jordan (Al-Rwahnih et al., 2001). The viroid is quite common in peach in Tunisia but it is not as common in pear (Fekih Hassen et al., 2006).The incidence of PLMVd infection in Egypt on different hosts was 45% in peach, 35% in plum, 23% in pear, 18% in mango, 5% in apple, and 2% in apricot (El-Dougdoug et al., 2012a). The viroid was detected in peach, nectarine (peach cultivars with smooth fruits), and apricot in Turkey (Gümüs et al., 2007; Sipahioğlu et al., 2006) with an infection rate of 16% and 60% in peach and nectarine, respectively. The Turkish viroid isolates belong to PLMVd Group III (Gazel et al., 2008a). More than 100 new PLMVd variants were identified in Tunisian peach, pear, and almond trees, and clustered into groups and subgroups (Fekih Hassen et al., 2007). New PLMVd variants in Iranian peach and plum trees were described and distributed among three of the five groups in the phylogenetic tree of PLMVd variants reported from several hosts and countries (Yazarlou et al., 2012b).

POSPIVIROIDAE; POSPIVIROID; POTATO SPINDLE TUBER VIROID

Potato spindle tuber viroid (PSTVd) was detected in potato in Egypt (El-Dougdoug, 1988; EPPO, 2014), Nigeria (EPPO, 2014), Saudi Arabia (H.Z. Al-Tuwariqi and A. Hadidi, unpublished), Israel, and Turkey (CPC-CABI, 2010; EPPO, 2014), in tomato in Ghana (Batuman et al., 2013), and in ornamental plants in Tunisia (A. Elleuch, unpublished). PSTVd was detected in potato at low incidence in Egypt (2.5%) (El-Dougdoug, 1988) and in Turkey (0.45%–1.8%) (Bostan et al., 2010). The Turkish PSTVd has the highest sequence identity (99%) with an Italian potato isolate (Güner et al., 2012). All the above countries import potato seed mainly from Europe, which is probably the main source of PSTVd infection (Bostan et al., 2010; Hadidi et al., 2003).

POSPIVIROIDAE; POSPIVIROID; TOMATO CHLOROTIC DWARF VIROID

Tomato chlorotic dwarf viroid is seed-transmitted in tomato (Singh and Dilworth, 2009) and this finding was confirmed in Israel (Anonymous, 2012). In Tunisia the viroid was detected in the ornamental *Pittosporum tobira*; its nucleotide sequence has high identity with variants from Canada and The Netherlands (A. Elleuch, unpublished).

POSPIVIROIDAE; POSPIVIROID; TOMATO APICAL STUNT VIROID

Tomato apical stunt viroid (TASVd) was found consistently in greenhouse-grown tomato crops in Israel, causing severe symptoms and losses (Antignus et al., 2002, 2007). This concern has led EPPO to list the viroid on their "EPPO Alert List" (EPPO, 2011). TASVd-Is shares 92%, 99%, and 87% sequence identity with the Ivory Coast, Indonesian, and TASVd-S strains, respectively. TASV-S was reported from symptomless ornamental plants of *Solanum pseudocapsicum* grown in a greenhouse in Germany (Spieker et al., 1996). The experimental host range of TASVd-Is differs significantly from that of the type strain TASVd-Ivory Coast. TASVd in tomato was also reported from Tunisia (Verhoeven et al., 2006), Senegal (Candresse et al., 2007), and Ghana (Batuman et al., 2013).

POSPIVIROIDAE; POSPIVIROID; COLUMNEA LATENT VIROID

Columnea latent viroid was detected in tomato in Mali (Batuman and Gilbertson, 2012).

POSPIVIROIDAE; POSPIVIROID; CHRYSANTHEMUM STUNT VIROID

Chrysanthemum stunt viroid in chrysanthemum was reported from South Africa (Watermeyer, 1984), Egypt (El-Dougdoug et al., 2012b), and Turkey with an infection rate of about 1.3% (Bostan et al., 2010).

POSPIVIROIDAE; POSPIVIROID; CITRUS EXOCORTIS VIROID AND CITRUS VIROID COMPLEXES

Citrus exocortis viroid (CEVd) was reported from South Africa (Da Graca et al., 2003), Ghana (Opoku, 1972), Sudan, Morocco, Algeria, Tunisia, Libya, Egypt, Israel, Turkey, and Cyprus (Hadidi et al., 2003). It is also found in Saudi Arabia (H.Z. Al-Tuwariqi and A. Hadidi, unpublished) and Oman (Al-Harthi et al., 2013). In Egypt, the infection rate of CEVd was about 15% in sweet orange, 4%−12% in grapevine, and 1% in mango (El-Dougdoug et al., 2012a; Nasr-Eldin et al., 2012). In Lebanon CEVd and hop stunt viroid (HSVd) were found in 33.3% of citrus samples with 66.7% infection in Valencia orange; CEVd was detected in 50% of Washington navel orange (Saade et al., 2000). CEVd variants in Israel are divided into two classes that differ in their pathogenicity on tomato plants as well as in the number of nucleotide changes as compared with CEVd type strains (Ben Shaul et al., 1995). In Tunisia CEVd is the prevalent viroid in citrus (Najar and Duran-Vila, 2004) and its variants belong to class B (mild in tomato) (Elleuch et al., 2006). CEVd was also found to infect fig in Tunisia (Yakoubi et al., 2007).

CEVd may form complexes with other viroids in citrus trees (Duran-Vila et al., 2003). In Turkey, these complexes consist of CEVd, citrus bent leaf viroid (CBLVd), HSVd, citrus dwarfing viroid (CDVd), and citrus bark cracking viroid (Önelge and Çınar, 2010). CBLVd, HSVd, and CDVd are synonymous with citrus viroid I (CVd-I), citrus viroid II (CVd-II), and citrus viroid III (CVd-III), respectively (Lee, 2015). Citrus budlings originating from Oman, Lebanon, Syria, Egypt, and Jordan contain citrus viroid complexes consisting of CBCVd (79%), CDVd (68%), HSVd (54%), CEVd (29%), and CBLVd (28%) (Al-Harthi et al., 2013). These findings illustrate the high rate of vegetative transmission of citrus viroids by budlings in the Middle East (Al-Harthi et al., 2013).

POSPIVIROIDAE; HOSTUVIROID; HOP STUNT VIROID

HSVd in Citrus

The history of the discovery and spread of citrus cachexia (xyloporosis) disease in the Middle East has been reviewed (Hadidi et al., 2003; Bar-Joseph, 2015). The disease, which is caused by variants of HSVd (Reanwarakorn and Semancik, 1999), was reported from Cyprus, Israel, Egypt, Libya, Tunisia, Algeria, Morocco, and Sudan (Hadidi et al., 2003). In addition, it is also widespread in mandarin orchards in Turkey (Moreira, 1965).

The history and properties of citrus gummy bark (CGB) disease were described by Onelge and Semancik (2003). A HSVd etiology for CGB disease of sweet orange is supported by the similarity of symptom expression to cachexia disease of mandarins and tangelos caused by HSVd, as well as by the detection of variants thereof in CGB-infected Washington navel and the Turkish orange cultivar "Dortyol" (Önelge et al., 2004). In Egypt, CGB was shown to be induced by a variant of HSVd that was characterized at the molecular level (Sofy and El-Dougdoug, 2014; Sofy et al., 2010, 2012b). In addition, the association of a HSVd variant with gumming and stem pitting on *Citrus volkameriana* rootstock was described, and it shared 100% identity with HSVd variants CVd-IIc or Ca-905 (Sofy and El-Dougdoug, 2014). CGB disease was found in Iraq, Oman, Saudi Arabia, Syria, Sudan, United Arab Emirates, Yemen, Turkey, and Egypt (Sofy et al., 2012a).

HSVd in Other Fruit Trees and in Grapevine

The incidence of HSVd in apricot trees was 19% in Jordan (Al-Rwahnih et al., 2001), 62% in Syria (Ismaeil et al., 2001), 28% in Lebanon (Choueiri et al., 2002), and 0.1% in Turkey (Sipahioğlu et al., 2006). The viroid was reported in pear, peach, and almond in Tunisia (Fekih Hassen et al., 2006). HSVd variants from *Prunus* sources in Turkey were shown to cluster either with the recombinant P-H/cit3 group (one variant), or with the Hop-group (most variants) (Gazel et al., 2008b), which are more related to the geographical origin of the variants than to their hosts. The infection rate of HSVd in peach and pear samples in Egypt was about 93% (El-Dougdoug et al., 2010). In a subsequent study, HSVd infection in several fruit species was about 65% in peach, 40% in pear, 25% in sweet orange, 16% in mango, 10% in mandarin, 7% in apricot, 6% in plum, and 2% in apple (El-Dougdoug et al., 2012a); in grapevine the infection rate was 8%–12% (El-Dougdoug et al., 2012a; Nasr-Eldin et al., 2012). HSVd infection of grapevine in Turkey was about 6% (Gazel and Önelge, 2003); the viroid was also found in grapevine rootstocks (Gazel and Önelge, 2002).

HSVd was detected in symptomless mulberry trees from Lebanon with an infection rate of 10% (Elbeaino et al., 2012a). HSVd mulberry isolates shared 95%–96% identity with other known HSVd isolates and clustered with HSVd-citrus, regardless of their geographical origin.

The first report of HSVd in fig trees was from Tunisia (Yakoubi et al., 2007). Subsequently, it was reported from Syria and Lebanon at an infection rate of 13.3% and 12.0%, respectively; most HSVd-fig isolates from these countries and the Lebanese mulberry HSVd isolates form a distinct HSVd clade (M-group) (Elbeaino et al., 2012b, 2013).

HSVd was detected in pomegranate in Turkey (Önelge, 2000) and in Tunisia (Gorsane et al., 2010). The viroid was also detected in pistachio in Tunisia (Elleuch et al., 2013); this is the first report of a viroid infecting this plant species.

POSPIVIROIDAE; APSCAVIROID; PEAR BLISTER CANKER VIROID, APPLE DIMPLE FRUIT VIROID, APPLE SCAR SKIN VIROID

Pear blister canker viroid was detected in 5.4% of pear and quince trees in Turkey and apple dimple fruit viroid was only found in symptomatic apple trees in Lebanon (Di Serio et al., 2010). Pear blister canker viroid was also rare in Tunisia as it was detected in 2.7% of pear trees; it was only detected in the Sahel Region (Fekih Hassen et al., 2006). Apple scar skin viroid was reported in Turkish apple (Sipahioğlu et al., 2009) and Iranian apple and pear trees (Yazarlou et al., 2012a).

POSPIVIROIDAE; APSCAVIROID; AUSTRALIAN GRAPEVINE VIROID, GRAPEVINE YELLOW SPECKLE VIROID 1, GRAPEVINE YELLOW SPECKLE VIROID 2

Australian grapevine viroid and grapevine yellow speckle viroid 1 (GYSVd-1) were found in Tunisia and their genomic nucleotide sequences were determined (Elleuch et al., 2002, 2003). Sequence variability of Australian grapevine viroid was not clustered in any specific domain or region of the genome (Elleuch et al., 2003).

The infection rates for GYSVd-1 and grapevine yellow speckle viroid 2 (GYSVd-2) in the Eastern Mediterranean region of Turkey were 12.5% and 2.17%, respectively (Gazel and Önelge, 2003; Önelge and Gazel, 2001), whereas it was very low (0.09%) in Eastern Anatolia for the two viroids (Gökçek and Önelge, 2007). The two viroids also infect grapevine rootstocks (Gazel and Önelge, 2002).

POSPIVIROIDAE; APSCAVIROID; CITRUS BENT LEAF VIROID, CITRUS DWARFING VIROID

An extensive study on CDVd from Israel revealed that both point mutation and RNA recombination contribute to viroid sequence diversity (Owens et al., 2000). The natural variability of CDVd from Tunisia was also reported (Elleuch et al., 2006). Dwarfed "Meyer" lemon infected with CDVd was reported from Turkey (Önelge et al., 2000). CBLVd and CDVd were detected in citrus budlings originating in Oman, Lebanon, Syria, Egypt, and Jordan (Al-Harthi et al., 2013). The two viroids from Oman share 94%–100% sequence identity with viroids from other parts of the world (Al-Harthi et al., 2013).

POSPIVIROIDAE; APSCAVIROID; CITRUS VIROID V

Citrus viroid V (CVd-V) was found in six sweet orange sources from Oman and the viroid isolates were identical to the reference CVd-V variant (Serra et al., 2008). The viroid was also detected in sweet orange trees in Turkey (Önelge and Yurtmen, 2012). The Turkish viroid isolate shares 96% and 98% identity with two different CVd-V reference isolates deposited in GenBank. CVd V was also detected in Tunisia in the majority of citrus varieties (Hamdi et al., 2015). The Tunisian variants showed 80%–91% nucleotide identity with other known variants; two main CVd-V groups were identified.

POSPIVIROIDAE; COCADVIROID; CITRUS BARK CRACKING VIROID

CBCVd in *Citrus lemon* was reported from Turkey in Meyer lemon or Meyer lemon on sour orange rootstock (Önelge et al., 2000). The viroid was also reported from Sudan (Mohamed et al., 2009) and South Africa (Cook et al., 2012). Nucleotide sequences of the South African isolates were identical to reference sequences from the Middle East.

Table 45.1 summarizes the geographical distribution of viroids in Africa and the Middle East.

TABLE 45.1 Geographical Distribution of Viroids in Africa and the Middle East

Viroid	Country	Host
Apple dimple fruit viroid	Lebanon	Apple
Apple scar skin viroid	Iran, Turkey	Apple, pear
Australian grapevine viroid	Tunisia	Grapevine
Avocado sunblotch viroid	Ghana, Israel, South Africa, Turkey	Avocado
Chrysanthemum stunt viroid	Egypt, South Africa, Turkey	Chrysanthemum
Citrus bark cracking viroid	Sudan, South Africa, Turkey	Citrus
Citrus bent leaf viroid	Egypt, Jordan, Lebanon, Oman, Syria	Citrus

(*Continued*)

TABLE 45.1 (Continued)

Viroid	Country	Host
Citrus dwarfing viroid	Egypt, Israel, Jordan, Lebanon, Oman, Syria, Tunisia, Turkey	Citrus
Citrus exocortis viroid	Algeria, Cyprus, Egypt, Ghana, Israel, Lebanon, Libya, Morocco, Oman, Saudi, Arabia, South Africa, Sudan, Tunisia, Turkey	Citrus, tomato, fig
Citrus viroid V	Oman, Tunisia, Turkey	Citrus
Columnea latent viroid	Mali	Columnea
Grapevine yellow speckle viroid 1	Tunisia, Turkey	Grapevine
Grapevine yellow speckle viroid 2	Turkey	Grapevine
Hop stunt viroid	Algeria, Cyprus, Egypt, Iraq, Israel, Jordan, Lebanon, Libya, Morocco, Oman, Saudi Arabia, Sudan, Syria, Tunisia, Turkey, UAE Yemen	Citrus, several other fruit trees, grapevine
Peach latent mosaic viroid	Egypt, Iran, Jordan, Lebanon, Syria, Tunisia, Turkey	Peach, several other fruit trees
Pear blister canker viroid	Turkey, Tunisia	Pear, quince
Potato spindle tuber viroid	Egypt, Ghana, Israel, Nigeria, Saudi Arabia, Tunisia, Turkey	Potato, tomato, ornamental plants
Tomato apical stunt viroid	Ghana, Israel, the Ivory Coast, Senegal, Tunisia	Tomato, ornamental plants
Tomato chlorotic dwarf viroid	Israel, Tunisia	Tomato, ornamental plants

Acknowledgments

This chapter is dedicated to the memory of Prof. Esmat K. Allam who introduced K.A.E.D. to viroid research in the mid-1980s.

References

Acheampong, A.K., Akromah, R., Ofori, F.A., Takrama, J.F., Zeidan, M., 2008. Is there avocado sunblotch viroid in Ghana? Afr. J. Biotechnol. 7, 3540–3545.

Al-Harthi, A., Al-Sadi, A.M., Al-Saady, A.A., 2013. Potential of citrus budlings originating in the Middle East as sources of citrus viroids. Crop Prot. 48, 13–15.

Al-Rwahnih, M., Myrta, A., Abou-Ghanem, N., Di Terlizzi, B., Savino, V., 2001. Viruses and viroids of stone fruits in Jordan. EPPO Bull. 31, 95–98.

Anonymous, 2012. Risk management proposal: *Solanum lycopersicum* (tomato) seed for sowing, from all countries. Ministry for Primary Industries, New Zealand, 17 pp.

Antignus, Y., Lachman, O., Pearlsman, M., 2007. Spread of tomato apical stunt viroid (TASVd) in greenhouse tomato crops is associated with seed transmission and bumble bee activity. Plant Dis. 91, 47–50.

Antignus, Y., Lachman, O., Pearlsman, M., Gofman, R., Bar-Joseph, M., 2002. A new disease of greenhouse tomatoes in Israel caused by a distinct strain of tomato apical stunt viroid (TASVd). Phytoparasitica 30, 502–510.

Bar-Joseph, M., 2015. Xyloporosis: A history of the emergence and eradication of a citrus viroid disease. IOCV J. Cit. Pathol. 2, 27202.

Batuman, O., Gilbertson, R.L., 2012. First report of columnea latent viroid (CLVd) in tomato in Mali. Plant Dis. 97, 692.

Batuman, O., Osei, M.K., Mochiah, M.B., Lamptey, J.N., Miller, S., Gilbertson, R.L., 2013. The first report of tomato apical stunt viroid (TASVd) and potato spindle tuber viroid (PSTVd) in tomatoes in Ghana. Phytopathology 103 (Suppl. 2), S2.12.

Ben Shaul, A., Guan, Y., Mogilner, N., Hadas, R., Mawassi, M., Gafny, R., et al., 1995. Genomic diversity among populations of two citrus viroids from different graft-transmissible dwarfing complexes in Israel. Phytopathology 85, 359–364.

Bostan, H., Gazel, M., Elibüyük, İ.Ö., Çağlayan, K., 2010. Occurrence of pospiviroid in potato, tomato and some ornamental plants in Turkey. Afr. J. Biotechnol. 9, 2613–2617.

Candresse, T., Marais, A., Ollivier, F., Verdin, E., Blancard, D., 2007. First report of the presence of tomato apical stunt viroid on tomato in Senegal. Plant Dis. 91, 330.

Choueiri, E., Abou Ghanem-Sabanadzovic, N., Khazzaka, K., Sabanadzovic, S., Di Terlizzi, B., Myrta, A., et al., 2002. First record of hop stunt viroid on apricot in Lebanon. J. Plant Pathol. 84, 69.

Cook, G., van Vuuren, S.P., Breytenbach, J.H.J., Manicom, B.Q., 2012. Citrus viroid IV detected in *Citrus sinensis* and *C. reticulata* in South Africa. Plant Dis. 96, 772.

CPC-CABI: Crop Protection Compendium CAB International, 2010.

Da Graca, J.V., van Vuuren, S.P., 2003. Viroids in Africa. In: Hadidi, A., Flores, R., Randles, J.W., Semancik, J.S. (Eds.), Viroids. CSIRO Publishing, Collingwood, VIC, pp. 290–292.

Di Serio, F., Afechtal, M., Attard, D., Choueiri, E., Gumus, M., Kaymak, S., et al., 2010. Detection by tissue printing hybridization of pome fruit viroids in the Mediterranean basin. Julius-Kühn Archiv. 427, 357–360.

Duran-Vila, N., Semancik, J.S., 2003. Citrus viroids. In: Hadidi, A., Flores, R., Randles, J.W., Semancik, J.S. (Eds.), Viroids. CSIRO Publishing, Collingwood, VIC, 178–194.

Elbeaino, T., Abou Kubaa, R., Choueiri, E., Digiaro, M., Navarro, B., 2012a. Occurrence of hop stunt viroid in Mulberry (*Morus alba*) in Lebanon and Italy. J. Phytopathol. 160, 48–51.

Elbeaino, T., Abou Kubaa, R., Ismaeil, E., Mando, J., Digiaro, M., 2012b. Viruses and hop stunt viroid of fig trees in Syria. J. Plant Pathol. 94, 687–691.

Elbeaino, T., Choueiri, E., Digiaro, M., 2013. First report of hop stunt viroid in Lebanese fig trees. J. Plant Pathol. 95, 218.

El–Dougdoug, K.A., 1988. Studies on some viroids in Egypt. PhD Thesis, Ain Shams University, Cairo, Egypt, 185 pp.

El-Dougdoug, K.A., Osman, M.E., Hayam, S.A., Rehab, D.A., Elbaz, R.M., 2010. Biological and molecular detection of HSVd-infecting peach and pear trees in Egypt. Austr. J. Basic Appl. Sci. 4, 19–26.

El-Dougdoug, K.A., Rehab, D.A., Rezk, A.A., Sofy, A.R., 2012a. Incidence of fruit trees viroid diseases by tissue print hybridization in Egypt. Int. J. Virol. 8, 114–120.

El-Dougdoug, K.A., Rezk, A.A., Rehab, D.A., Sofy, A.R., 2012b. Partial nucleotide sequence and secondary structure of chrysanthemum stunt viroid Egyptian isolate from infected-chrysanthemum plants. Int. J. Virol. 8, 191–202.

Elleuch, A., Djilani Khouaja, F., Hamdi, I., Bsais, N., Perreault, J.P., Marrakchi, M., et al., 2006. Sequence analysis of three citrus viroids infecting a single Tunisian citrus tree (*Citrus reticulata*, Clementine). Gen. Mol. Biol. 29, 705–710.

Elleuch, A., Fakhfakh, H., Pelchat, M., Landry, P., Marrakchi, M., Perreault, J.-P., 2002. Sequencing of Australian grapevine viroid and yellow speckle viroid I isolated from a Tunisian grapevine without passage in an indicator plant. Eur. J. Plant Pathol. 108, 815–820.

Elleuch, A., Hamdi, I., Ellouze, O., Ghrab, M., Fakhfakh, H., Drira, N., 2013. Pistachio (*Pistacia vera* L.) is a new natural host of hop stunt viroid. Virus Genes 47, 330–337.

Elleuch, A., Perreault, J.-P., Fakhfakh, H., 2003. First report of Australian grapevine viroid from the Mediterranean Region. J. Plant Pathol. 85, 53–57.

EPPO, 2014. PQR database. European and Mediterranean Plant Protection Organization, Paris, France. http://www.eppo.int/DATABASES/pqr/pqr.htm.

EPPO. 2011. Alert List: Tomato apical stunt pospiviroid. Updated 26 August 2011. Available at: http://www.eppo.org/QUARANTINE/Alert_List/viruses/TASVD0.htm.

Fekih Hassen, I., Massart, S., Motard, J., Roussel, S., Parisi, O., Kummert, J., et al., 2007. Molecular features of new peach latent mosaic viroid variants suggest that recombination may have contributed to the evolution of this infectious RNA. Virology 360, 50–57.

Fekih Hassen, I., Roussel, S., Kummert, J., Fakhfakh, H., Marrakchi, M., Jijakli, M.H., 2006. Development of a rapid RT-PCR test for the detection of peach latent mosaic viroid, pear blister canker viroid, hop stunt viroid and apple scar skin viroid in fruit trees from Tunisia. J. Phytopathol. 154, 217–223.

Gazel, M., Önelge, N., 2002. Detection of viroids in grapevine rootstocks by polyacrylamide gel electrophoresis. J. Turk. Phytopathol. 31, 167.

Gazel, M., Önelge, N., 2003. First report of grapevine viroids in the east Mediterranean region of Turkey. Turk. J. Plant Pathol. 52, 405.

Gazel, M., Ulubaş Serçe, Ç., Çağlayan, K., Luigi, M., Faggioli, F., 2008. Incidence and genetic diversity of peach latent mosaic viroid ısolates in Turkey. Turk. J. Plant Pathol. 90, 495–503.

Gökçek, B., Önelge, B., 2007. Gaziantep ili Bağlarında bağ sarı benek (GYSVd-1 ve 2) hastalığının araştırılması. Türkiye II. Bitki Koruma Kongresi 27-29 Ağustos 2007, Isparta. p. 118.

Gorsane, F., Elleuch, A., Hamdi., I., Salhi-Hannachi, A., Fakhfakh, H., 2010. Molecular detection and characterization of hop stunt viroid sequence variants from naturally infected pomegranate (*Punica granatum* L.) in Tunisia. Phytopathol. Mediterr. 49, 152–162.

Gümüs, M., Paylan, I.C., Matic, S., Myrta, A., Sipahioglu, H.M., Erkan, S., 2007. Occurrence and distribution of stone fruit viruses and viroids in commercial plantings of *Prunus* species in Western Anatolia. Turk. J. Plant Pathol. 89, 265–268.

Güner, Ü., Sipahioğlu, H.M., Usta, M., 2012. Incidence and genetic stability of potato spindle tuber pospiviroid in potato in Turkey. Turk. J. Agric. For. 36, 353–363.

Hadidi, A., Mazyad, H.H., Madkour, M.A., Bar-Joseph, M., 2003. Viroids in the middle east. In: Hadidi, A., Flores, R., Randles, J.W., Semancik, J.S. (Eds.), Viroids. CSIRO Publishing, Collingwood, VIC, pp. 275–278.

Hamdi, I., Elleuch, A., Bessaies, N., Grubb, C.D., Fakhfakh, H., 2015. First report of citrus viroid V in North Africa. J. Gen. Plant Pathol. 81, 87.

Ismaeil, F., Abou Ghanem-Sabanadzovic, N., Myrta, A., Di Terlizzi, B., Savino, V., 2001. First record of peach latent mosaic viroid and hop stunt viroid in Syria. J. Plant Pathol. 82, 227.

Lee, R.F., 2015. Control of virus disease of citrus. Adv. Virus Res. 91, 143–173.

Mohamed, M.E., Hashemian, S.M.B., Dafalla, G., Bove, Duran-Vila, N., 2009. Occurrence and identification of citrus viroids from Sudan. J. Plant Pathol. 91, 185–190.

Moreira, S., 1965. Report to government of Turkey on virus diseases of citrus. FAO/UNDP Report, Rome, No:1982, 19 pp.

Najar, A., Duran-Vila, N., 2004. Viroid prevalence in Tunisian citrus. Plant Dis. 88, 1286.

Nasr-Eldin, M.A., El-Dougdoug, K.A., Othman, B.A., Ahmed, S.A., Abdel-Aziz, S.H., 2012. Three viroids frequency naturally infecting grapevine in Egypt. Int. J. Virol. 8, 1–13.

Önelge, N., 2000. Occurrence of hop stunt viroid (HSVd) on pomegranate (*Punica granatum*) trees in Turkey. J. Turk. Phytopathol. 29, 49–52.

Önelge, N., Çınar, A., 2010. Virus and virus-like diseases in Turkey citriculture. In: Proceedings of 17th Conference, IOCV, 233–236.

Önelge, N., Çinar, A., Szychowski, J.S., Vidalakis, G., Semancik, J.S., 2004. Citrus viroid II variants associated with "Gummy Bark" disease. Eur. J. Plant Pathol. 110, 1047–1052.

Önelge, N., Ertuğrul, B., 1997. Detection of avocado sunblotch viroid (ASBVd) in Turkish avocado introduction material by polyacrylamide gel electrophoresis. J. Turk. Phytopathol. 26, 97–101.

Önelge, N., Gazel, M., 2001. Detection of the viroids of grapevine in East Mediterranean Region. J. Turk. Phytopathol. 30, 70.

Önelge, N., Kersting, U., Guang, Y., Bar Joseph, M., Bozan, O., 2000. Nucleotide sequences of citrus viroids CVd-IIIa and CVd-IV obtained from dwarfed Meyer lemon trees grafted on sour orange. J. Plant Dis. Prot. 107, 387–391.

Onelge N and Semancik J.S., Citrus gummy bark, In: Hadidi A., Flores R., Randles J.W. and Semancik J.S., (Eds.). *Viroids*, 2003, CSIRO Publishing; Collingwood, VIC, 330–332.

Önelge, N., Yurtmen, M., 2012. First report of citrus viroid V in Turkey. J. Plant Pathol. 94 (Suppl. 4), 88.

Opoku, A.A., 1972. Incidence of exocortis virus disease of citrus in Ghana. Ghana J. Agric. Sci. 5, 65–71.

Oppenheimer, C., 1955. Growing new subtropical fruit trees in Israel Hassadeh Library. Tel Aviv.110–113.

Owens, R.A., Yang, G., Gundersen-Rindal, D., Hammond, R.W., Candresse, T., Bar Joseph, M., 2000. Both point mutation and RNA recombination contribute to citrus viroid III sequence diversity. Virus Genes 20, 243–252.

Reanwarakorn, K., Semancik, J.S., 1999. Correlation of hop stunt viroid variants to cachexia and xyloporrosis diseases of citrus. Phytopathology 89, 568–574.

Saade, P., D'Onghia, A.M., Khoury, W., Turturo, C., Savino. V., 2000. Surveys and certification sanitary status of citrus in Lebanon, Proc. 14th Conf. IOCV, 326–331.

Serra, P., Eiras, M., Bani-Hashemian, S.M., Murcia, N., Kitajima, E.W., Daros, J.A., et al., 2008. Citrus viroid V: occurrence, host range, diagnosis and identification of new variants. Phytopathology 98, 1199–1204.

Singh, R.P., Dilworth, A.D., 2009. Tomato chlorotic dwarf viroid in the ornamental plant *Vinca minor* and its transmission through tomato seed. Eur. J. Plant Pathol. 123, 111–116.

Sipahioğlu, H.M., Demir, S., Myrta, A., Al Rwahnih, M., Polat, B., Schena, L., et al., 2006. Viroid, phytoplasma and fungal diseases of stone fruit in eastern Anatolia, Turkey. New Zeal. J. Crop Hort. 34, 1–6.

Sipahioğlu, H.M., Usta, M., Ocak, M., 2009. Development of a rapid enzymatic cDNA amplification test for the detection of apple scar skin viroid (ASSVd) in apple trees from eastern Anatolia. Turkey. Arch. Phytopathol. and Plant Prot. 42, 352–360.

Sofy, A.R., El-Dougdoug, K.A., 2014. First record of a hop stunt viroid variant associated with gumming and stem pitting on *Citrus volkameriana* trunk rootstock in Egypt. New Dis. Rep. 30, 11.

Sofy, A.R., Mousa, A.A., Soliman, A.M., El-Dougdoug, K.A., 2012a. The limiting of climatic factors and predicting of suitable habitat for citrus gummy bark disease occurrence using GIS. Int. J. Virol. 8, 165–177.

Sofy, A.R., Soliman, A.M., Mousa, A.A., El-Dougdoug, K.A., 2012b. Molecular characterization and bioinformatics analysis of viroid isolate associated with citrus gummy bark disease in Egypt. Int. J. Virol. 8, 133–150.

Sofy, A.R., Soliman, A.M., Mousa, A.A., Ghazal, S.A., El-Dougdoug, K.A., 2010. First record of citrus viroid II (CVd-II) associated with gummy bark disease in sweet orange (*Citrus sinensis*) in Egypt. New Dis. Rep. 21, 24.

Spiegel, S., Alper, M., Allen, R.N., 1984. Evaluation of biochemical methods for the diagnosis of the avocado sunblotch viroid in Israel. Phytoparasitica 12, 37–43.

Spieker, R.L., Marinkovic, S., Sänger, H.L., 1996. A viroid from *Solanum pseudocapsicum* closely related to the apical stunt viroid. Arch. Virol. 141, 1387–1395.

Verhoeven, J.T., Jansen, C.C., Roenhorst, J.W., 2006. First report of tomato apical stunt viroid in tomato in Tunisia. Plant Dis. 90, 528.

Watermeyer, S.R., 1984. Detection of chrysanthemum stunt viroid in South Africa by polyacrylamide gel electrophoresis and bioassays. Plant Dis. 68, 485–488.

Yakoubi, S., Elleuch, A., Besaies, N., Marrakchi, M., Fakhfakh, H., 2007. First report of hop stunt viroid and citrus exocortis viroid on fig with symptoms of fig mosaic disease. J. Phytopathol. 155, 125–128.

Yazarlou, A., Jafarpour, B., Tarıghı, S., Habili, N., Randles, J.W., 2012b. New Iranian and Australian peach latent mosaic viroid variants and evidence for rapid sequence evolution. Arch. Virol. 157, 343–347.

Yazarlou, A., Jafarpour, B., Habili, N., Randles, J.W., 2012a. First detection and molecular characterization of new apple scar skin viroid variants from apple and pear in Iran. Australasian Plant Dis. Notes 7, 99–102.

CHAPTER

46

Geographical Distribution of Viroids in Oceania

Andrew D.W. Geering

The University of Queensland, St Lucia, QLD, Australia

AVOCADO SUNBLOTCH VIROID (ASBVD)

In Australia, commercial avocado production began in the Sunraysia region and in subtropical Queensland/NSW in the 1920/30s, using varieties imported from California (Anderson, 1977; Cadman and El-Zeftawi, 1977). Nearly all Australian records of ASBVd derive from these regions with the oldest avocado industries (Allen and Dale, 1981; Palukaitis et al., 1979, 1981; Trochoulias and Allen, 1970) and it is therefore likely that the viroid was brought into the country from the USA in infected seed or budwood.

In 1978, the Avocado Nursery Voluntary Accreditation Scheme (ANVAS) commenced with aims to control the spread of ASBVd and *Phytophthora cinnamomi*. Under ANVAS, nuclear and multiplication trees are tested for ASBVd every five and 20 years, respectively. The success of the scheme can be gauged by the fact that there have only been two confirmed records of ASBVd in Australia in the last decade, one each from Tamborine Mountain, Queensland and Red Cliffs, Victoria (A.D.W. Geering, unpublished).

The only other country in Oceania with a major avocado industry is New Zealand, and it declared area freedom from ASBVd in 2009 (Pugh and Thomson, 2009).

CITRUS VIROIDS

Citrus exocortis viroid (CEVd), hop stunt viroid (HSVd), and citrus dwarfing viroid (CDVd) all occur in Australian citrus, but have limited

Viroids and Satellites.
DOI: http://dx.doi.org/10.1016/B978-0-12-801498-1.00046-2

distributions (Gillings et al., 1991). A clean budwood and rootstock seed scheme has been operated by the Australian Citrus Propagation Association (Auscitrus), which has provided high health status planting material to the industry since 1967 (T. Herrmann, pers. comm.).

Three strains of HSVd have been found in Australian citrus, labeled IIa, b, and c (Gillings et al., 1991). Only strain IIb, which is the rarest of the three, is regarded as pathogenic and causes cachexia disease in mandarin and tangelo (Gillings et al., 1991). Strain IIa has been found in association with CDVd strain IIIb in dwarfed sweet orange but whether it contributes to the phenotype is unclear (Gillings et al., 1991; Hutton et al., 2000).

Strain variation has also been observed for CEVd (Gillings et al., 1991; Visvader and Symons, 1983, 1985). Visvader and Symons (1985) divided CEVd isolates from Australian citrus into two groups according to reactions in tomato, which were either severe or asymptomatic. However, symptoms in tomato and citrus do not correlate (Murcia et al., 2011) and all Australian isolates, even those that do not produce scaling on trifoliate orange rootstocks, cause severe, persistent epinasty, and stunting when indexed on Etrog citron (Gillings et al., 1991; Schwinghamer and Broadbent, 1987).

In Australia, CDVd strain IIIb has been used as a dwarfing agent for high density plantings of sweet orange on trifoliate orange rootstock (Hutton et al., 2000). Infected bud sticks are sold out of Yanco Agricultural Institute to enable patch-graft inoculation of young trees but industry uptake of the technology has been low (N.J. Donovan, pers. comm.).

New Zealand has only a small citrus industry and CDVd has been detected in lemon and sweet orange, and CEVd in a range of citrus (Quemin et al., 2011; Veerakone et al., 2015). In other parts of Oceania, CEVd is recorded from lime, lemon, and bitter orange in Fiji, from the former two species in the Cook Islands, and from lemon in Samoa (Davis and Ruabete, 2010). Additionally, HSVd is recorded from mandarin in French Polynesia (Davis and Ruabete, 2010).

GRAPEVINE VIROIDS

Five viroids infect grapevines in Australia, namely grapevine yellow speckle viroid 1 (GYSVd-1), grapevine yellow speckle viroid 2 (GYSVd-2), Australian grapevine viroid, CEVd, and HSVd. Cultivars such as Sultana are uniformly infected with one or more viroids, suggesting introduction with the first cuttings (Taylor and Woodham, 1972).

Grapevine yellow speckle disease is caused by GYSVd-1 and GYSVd-2, which are found in table and wine grapes throughout Australia

(Koltunow et al., 1989; Salman et al., 2014; Wan Chow Wah and Symons, 1997). Disease symptoms are induced by high temperatures and bright sunlight (Salman et al., 2014). GYSVd-2 has always been found together with GYSVd-1, although single infections of GYSVd-1 are known (Koltunow et al., 1989; Salman et al., 2014; Wan Chow Wah and Symons, 1997).

Australian grapevine viroid, CEVd and HSVd are widely distributed as asymptomatic infections in Australian vineyards (Koltunow et al., 1988; Rezaian et al., 1988; Salman et al., 2014; Wan Chow Wah and Symons, 1997). RT-PCR testing of grapevine planting material for viroids is not routinely done, as they are not considered to be economically important (N. Habili, pers. comm.).

In New Zealand, GYSVd-1 and HSVd were detected in a commercial vineyard near Auckland during February 2009 (Ward et al., 2011). One would expect that the viroid status in New Zealand vineyards would mirror that of Australia, as many varieties have the same origins. However, the cooler, cloudier, wetter growing conditions in summer would not favor yellow speckle disease symptom expression.

HOP VIROIDS

There are no records of HSVd from Australian hop gardens, despite the fact that it is very common in vineyards in the same general vicinities (Koltunow et al., 1988). Hop latent viroid (HLVd) is, however, ubiquitous and asymptomatic in the hop gardens of Myrtleford, Victoria and the Derwent Valley, Tasmania (Crowle, 2010). HLVd is also ubiquitous in the hop gardens of New Zealand and HSVd has also been recorded (Hay et al., 1992; Veerakone et al., 2015).

CHRYSANTHEMUM VIROIDS

Chrysanthemum stunt disease was first observed in Australia in 1951, although details of the locality were not provided (Brierley, 1953). In 1987, chrysanthemum stunt viroid (CSVd) was discovered in the nuclear stocks of a specialist chrysanthemum propagator in Victoria, and this detection led the grower to destroy all infected nuclear and foundation stocks, along with all summer flowering mother stocks, actions that cost AUD 3 million (Hill et al., 1996; Moran and Bate, 1988). In 1993, the first nationwide survey of chrysanthemum was done and CSVd was detected only in greenhouse chrysanthemums from Queensland at an incidence of 0.7% (Hill et al., 1996). In 2003, a survey of 42 pyrethrum crops in northwest Tasmania was done and CSVd not found (Pethybridge and Wilson, 2004).

There is one old record of CSVd from New Zealand, which is based on field symptoms and biological indexing (Veerakone et al., 2015).

Knowledge of the status of chrysanthemum chlorotic mottle viroid in Australia is poor, other than it is has been detected several times at a wholesale chrysanthemum nursery in Brisbane and cuttings from this nursery have been sent interstate (A.D.W. Geering, unpublished). There are no records of this viroid from New Zealand (Veerakone et al., 2015).

PALM VIROIDS

Coconut cadang-cadang viroid related sequences have been detected in coconut palms with abnormalities such as chlorosis, stunting, sterility, or misshapen nuts in Indonesia, Papua New Guinea, the Solomon Islands, and Vanuatu, and in oil palms with foliar orange-spotting disease in Indonesia and the Solomon Islands (Hanold and Randles, 1989, 1991, 1998; Vadamalai et al., 2006; J.W. Randles, pers. comm.). Within the southwest Pacific region, the host range of the viroid may be broader, as viroid-like molecules that cross-hybridize with a coconut cadang-cadang viroid probe under high stringency conditions have been detected in *Areca catechu*, *Pandanus* spp., *Alpinia* spp., *Zingiber* sp., and unidentified species in the *Commelinaceae* and *Marantaceae* (Hanold and Randles, 1991; 1998).

The only other palm-infecting viroid, coconut tinangaja viroid, has a very localized distribution, restricted to Guam and probably also Anatahan in the Marianas Islands (Hodgson et al., 1998; Wall and Randles, 2003). In the late 1990s, the incidence of coconut tinangaja disease on Guam was about 30%, with highest incidences along the public beaches (Wall and Wiecko, 1997).

PEAR BLISTER CANKER VIROID (PBCVD)

PBCVd is broadly distributed in all pear-growing states of Australia except Tasmania, where only limited surveying has been done (Joyce et al., 2006). Hosts include pear cvs "Red Princess," "Red Sensation," "Beurre Bosc," "Baron De Mello," "Packham's Triumph," "Williams Bon Chretien," and "Winter Nellis," as well as nashi cvs "Red Danjou" and "Ko Sui" and quince cv. "Champion" (Joyce et al., 2006). All infections have been asymptomatic and judging by the age of some of the trees, PBCVd is likely to have been present in Australia for at least 50 years.

There are no records of viroids infecting pome fruit in New Zealand (Veerakone et al., 2015).

PEACH LATENT MOSAIC VIROID (PLMVD)

PLMVd was detected in a single block of nectarine cv. "Maygrand" in the Riverland area of South Australia in 1998, which was planted in about 1985 (Di Serio et al., 1999; Randles et al., 2003). Although 68% of the tested trees were infected, there was no evidence of spread to the neighboring block. It is likely that PLMVd entered Australia in imported budwood of the cultivar in question (Randles et al., 2003).

A single specimen of peach with calico disease symptoms from Central Otago, New Zealand, was lodged in the University of Otago Herbarium in 1956. After nearly 50 years of storage at ambient temperature and humidity, PLMVd was still able to be detected in this specimen by RT-PCR (Guy, 2013). Whether PLMVd has persisted in New Zealand is unknown (P.L. Guy, pers. comm.).

SOLANACEOUS PLANT VIROIDS

Potato spindle tuber viroid (PSTVd) is present in some areas of Australia but is subject to official control. PSTVd was first detected in Australia in 1981 in the potato breeders' collection at Glen Innes, NSW (Schwinghamer and Conroy, 1983), and then at Toolangi, Victoria and near Adelaide, South Australia, where germplasm from Glen Innes had been sent (Cartwright, 1984; Mason and Heath, 1984). To achieve eradication, all breeding material in the National Breeding Program was destroyed except for 100 single tested tubers (Moran, 1997). There have been no recent detections in potato, and the National Seed Potato Certification Scheme requires seed producers to source minitubers from virus and PSTVd-tested mother stocks maintained in tissue culture.

Most recent outbreaks of PSTVd have been in greenhouse tomato crops, in southwest Western Australia, South Australia, Victoria, New South Wales, and Queensland (Hailstones et al., 2003; Mackie et al., 2002; van Brunschot et al., 2014b; M.J. Gibbs, pers. comm.). PSTVd has also been detected in a weedy *Solanum* sp. from the Northern Territory by graft-inoculation to cherry tomato (Behjatnia et al., 1996), but as testing was not done on the original host plant material, and PSTVd is seed-borne in tomato, there remains some doubt about this pathogen record. Each time there has been an outbreak in a greenhouse crop, a biosecurity incursion response has been mounted and the viroid locally eradicated.

The majority of records in tomato have been of the Naaldwijk strain (Puchta et al., 1990), which elsewhere in the world has been isolated frequently from *Physalis peruviana* (Verhoeven et al., 2010). As *P. peruviana* is a common weed in Australia, it was speculated that this species was a source of infection for the PSTVd outbreaks in tomato (Verhoeven et al., 2010), but this hypothesis has not been supported by field evidence (Mackie et al., 2016). Almost all commercial quantities of tomato seed in Australia are imported, and most outbreaks of PSTVd have been attributed to seed-borne infection.

The Naaldwijk strain of PSTVd has become established in the field in the Gascoyne Horticultural District of central coastal Western Australia, in tomato, capsicum, and chilli crops, as well as in weedy and indigenous plant species in the *Solanaceae, Amaranthaceae* (syn. *Chenopodiaceae*), *Asteraceae*, and *Malvaceae* (Mackie et al., 2016). Patterns of spread suggest mechanical transmission on contaminated farming equipment such as slashers or tractors (A.E. Mackie, pers. comm.). Detections of PSTVd increased after the record-breaking Gascoyne River flood of December 2010, and PSTVd was found in native saltbushes (*Atriplex semilunaris* and *Rhagodia eremaea*) that were remote from any farm (Mackie et al., 2016). In this area, farmers have not noticed any yield impacts from infection, but the Naaldwijk strain is a threat to the Australian potato industry, as spindle tuber symptoms have been observed following experimental inoculations (A.E. Mackie, pers. comm.).

In 2012, a nationwide survey for PSTVd was done, covering 90 tomato farms and 45 ornamental nurseries (M.J. Gibbs, pers. comm.). PSTVd and pepper chat fruit viroid were detected in a greenhouse tomato crop at Virginia, South Australia, but both have now been eradicated (Anonymous, 2013). Additionally, the S1 genotype of PSTVd was detected in *Solanum laxum* (syn. *Solanum jasminoides*) plants sold by retail and wholesale nurseries in south-east Queensland (van Brunschot et al., 2014a; Verhoeven et al., 2010). Other than PSTVd, CEVd was detected in a small number of ornamental nursery plants (*Calibrachoa, Petunia, Petchoa*) in four Australian states (M.J. Gibbs, pers. comm.; van Brunschot et al., 2014a) and also tomato chlorotic dwarf viroid was detected in a single *Calibrachoa* plant from South Australia—this plant was subsequently destroyed (M.J. Gibbs, pers. comm.).

There have been transient occurrences of the Naaldwijk strain of PSTVd in New Zealand, and each time, all infected material has been destroyed. The first detection was in greenhouse-grown tomatoes in South Auckland in 2000, and pathogen delimitation surveys revealed four more affected tomato sites, three close to the original outbreak site and one in Nelson on the South Island (Elliott et al., 2001; Lebas et al., 2005). Surveys of surrounding potato crops suggested that PSTVd was confined to the greenhouses (Elliott et al., 2001).

In 2001, PSTVd was detected for the first time in capsicum in New Zealand, this time from a greenhouse in North Auckland (Lebas et al., 2005). Additional delimitation surveys were done and PSTVd detected in two other properties near the original outbreak site where capsicums were grown (Lebas et al., 2005). The tomato and capsicum isolates of PSTVd from New Zealand were identical in sequence, suggesting a common origin (Lebas et al., 2005).

The only other record of PSTVd from New Zealand was in 2009 from *P. peruviana* in a home garden in Christchurch. Contaminated seed that was imported from Germany was considered the cause of the outbreak (Ward et al., 2010).

Acknowledgments

I thank M.J. Gibbs, A.E Mackie, J.W. Randles, J.E. Thomas, D. Persley, B.C. Rodoni, B.S.M Lebas, P.L. Guy, N. Habili, N.J. Donovan, D.L. Hailstones, P. Barkley, and K. Pegg for very useful discussions, and in some cases, providing unpublished information.

References

Allen, R., Dale, J.L., 1981. Application of rapid biochemical methods for detecting avocado sunblotch disease. Ann. Appl. Biol. 98, 451–461.

Anderson, K., 1977. Brief review of avocado growing in sub-tropical Australia. Australian Avocado Research Workshop. Binna Burra Lodge, Lamington National Park, pp. 4–6.

Anonymous, 2013. Exotic plant pest alert: pepper chat fruit viroid. Biosecurity SA, Adelaide, South Australia.

Behjatnia, S.A.A., Dry, I.B., Krake, L.R., Condé, B.D., Connelly, M.I., Randles, J.W., et al., 1996. New potato spindle tuber viroid and tomato leaf curl geminivirus strains from a wild *Solanum* sp. Phytopathology 86, 880–886.

Brierley, P., 1953. Virus diseases of the chrysanthemum. In: Plant Diseases: The Yearbook of Agriculture, 1953. United States Department of Agriculture, pp. 596–601.

Cadman, R., El-Zeftawi, B.M., 1977. Avocado plantings in Sunraysia, Australia. California Avocado Society 1977 Yearbook. 61, 29–36.

Cartwright, D.N., 1984. Potato spindle tuber viroid survey – South Australia. Australas. Plant Pathol. 13, 4–5.

Crowle, D.R., 2010. Molecular variation of viruses infecting hops in Australia and associated studies. PhD Thesis, The University of Tasmania, Hobart.

Davis, R.I., Ruabete, T.K., 2010. Records of plant pathogenic viruses and virus-like agents from 22 Pacific island countries and territories: a review and an update. Australas. Plant Pathol. 39, 265–291.

Di Serio, F., Malfitano, M., Flores, R., Randles, J.W., 1999. Detection of peach latent mosaic viroid in Australia. Australas. Plant Pathol. 28, 80–81.

Elliott, D.R., Alexander, B.J.R., Smales, T.E., Tang, Z., Clover, G.R.G., 2001. First report of potato spindle tuber viroid in tomato in New Zealand. Plant Dis. 85, 1027–2027.

Gillings, M.R., Broadbent, P., Gollnow, B.I., 1991. Viroids in Australian citrus: relationship to exocortis, cachexia and citrus dwarfing. Funct. Plant Biol. 18, 559–570.

Guy, P.L., 2013. Ancient RNA? RT-PCR of 50-year-old RNA identifies peach latent mosaic viroid. Arch. Virol. 158, 691–694.

Hailstones, D.L., Tesoriero, L.A., Terras, M.A., Dephoff, C., 2003. Detection and eradication of potato spindle tuber viroid in tomatoes in commercial production in New South Wales, Australia. Australas. Plant Pathol. 32, 317–318.

Hanold, D., Randles, J.W., 1989. Cadang-cadang-like viroid in oil palm in the Solomon Islands. Plant Dis. 73, 183.

Hanold, D., Randles, J.W., 1991. Detection of coconut cadang-cadang viroid-like sequences in oil and coconut palm and other monocotyledons in the south-west Pacific. Ann. Appl. Biol. 118, 139–151.

Hanold D. and Randles J.W. 1998. Report on ACIAR-Funded Research on Viroids and Viruses of Coconut Palm and Other Tropical Monocotyledons. ACIAR Working Paper No. 51. ACIAR, Canberra, 222 pp.

Hay, F.S., Close, R.C., Fletcher, J.D., Ashby, J.W., 1992. Incidence and spread of viruses in hop (*Humulus lupulus* L.) in New Zealand. New Zeal. J. Crop Hort. 20, 319–327.

Hill, M.F., Giles, R.J., Moran, J.R., Hepworth, G., 1996. The incidence of chrysanthemum stunt viroid, chrysanthemum B carlavirus, tomato aspermy cucumovirus and tomato spotted wilt tospovirus in Australian chrysanthemum crops. Australas. Plant Pathol. 25, 174–178.

Hodgson, R.A.J., Wall, G.C., Randles, J.W., 1998. Specific identification of coconut tinangaja viroid for differential field diagnosis of viroids in coconut palm. Phytopathology 88, 774–781.

Hutton, R.J., Broadbent, P., Bevington, K.B., 2000. Viroid dwarfing for high density citrus plantings. In: Janick, J. (Ed.), Horticultural Reviews, vol. 24. John Wiley & Sons, Inc, New York, pp. 277–317.

Joyce, P.A., Constable, F.E., Crosslin, J., Eastwell, K., Howell, W.E., Rodoni, B.C., 2006. Characterisation of pear blister canker viroid isolates from Australian pome fruit orchards. Australas. Plant Pathol. 35, 465–471.

Koltunow, A.M., Krake, L.R., Johnson, S.D., Rezaian, M.A., 1989. Two related viroids cause grapevine yellow speckle disease independently. J. Gen. Virol. 70, 3411–3419.

Koltunow, A.M., Krake, L.R., Rezaian, M.A., 1988. Hop stunt viroid in Australian grapevine cultivars: potential for hop infection. Australas. Plant Pathol. 17, 7–10.

Lebas, B.S.M., Clover, G.R.G., Ochoa-Corona, F.M., Elliott, D.R., Tang, Z., Alexander, B.J.R., 2005. Distribution of potato spindle tuber viroid in New Zealand glasshouse crops of capsicum and tomato. Australas. Plant Pathol. 34, 129–133.

Mackie, A.E., Rodoni, B.C., Barbetti, M.J., McKirdy, S.J., Jones, R.A.C., 2016. Potato spindle tuber viroid: alternative host reservoirs and strain found in a remote subtropical irrigation area. Eur. J. Plant Pathol. 145, 433–446.

Mackie, A.E., McKirdy, S.J., Rodoni, B., Kumar, S., 2002. Potato spindle tuber viroid eradicated in Western Australia. Australas. Plant Pathol. 31, 311–312.

Mason, A., Heath, R.R., 1984. A survey of potatoes in the Victorian potato breeding program for the presence of potato spindle tuber viroid. Australas. Plant Pathol. 13, 20–21.

Moran, J., 1997. Final project report for HAL Project PT410. PCR protocols for the detection of chrysanthemum stunt and potato spindle tuber viroids. Horticulture Research and Development Corporation, Gordon, NSW, 52 pp.

Moran, J., Bate, C., Chrysanthemum stunt viroid. The Flower Link, June 1988, 15–19.

Murcia, N., Bernad, L., Duran-Vila, N., Serra, P., 2011. Two nucleotide positions in the citrus exocortis viroid RNA associated with symptom expression in Etrog citron but not in experimental herbaceous hosts. Mol. Plant Pathol. 12, 203–208.

Palukaitis, P., Hatta, T., Alexander, D.M., Symons, R.H., 1979. Characterization of a viroid associated with avocado sunblotch disease. Virology 99, 145–151.

Palukaitis, P., Rakowski, A.G., Alexander, D.M., Symons, R.H., 1981. Rapid indexing of the sunblotch disease of avocados using a complementary DNA probe to avocado sunblotch viroid. Ann. Appl. Biol. 98, 439–449.

Pethybridge, S.J., Wilson, C.R., 2004. A survey for viruses and a viroid in Tasmanian pyrethrum crops. Australas. Plant Pathol. 33, 301–303.

Puchta, H., Herold, T., Verhoeven, K., Roenhorst, A., Ramm, K., Schmidt-Puchta, W., et al., 1990. A new strain of potato spindle tuber viroid (PSTVd-N) exhibits major sequence differences as compared to all other PSTVd strains sequenced so far. Plant Mol. Biol. 15, 509–511.

Pugh, K., Thomson, V., 2009. Protecting our avocados. Biosecurity Magazine 93, 16–17.

Quemin, M.F., Lebas, B.S.M., Veerakone, S., Harper, S.J., Clover, G.R.G., Dawson, T.E., 2011. First molecular evidence of citrus psorosis virus and citrus viroid III from *Citrus* spp. in New Zealand. Plant Dis. 95, 775–875.

Randles, J.W., Rezaian, M.A., Hanold, D., Harding, R.M., Skrzeckzkowski, L., Whattam, M., 2003. Viroids in Australasia. In: Hadidi, A., Flores, R., Randles, J.W., Semancik, J.S. (Eds.), Viroids. CSIRO Publishing, Collingwood, VIC, pp. 279–282.

Rezaian, M.A., Koltunow, A.M., Krake, L.R., 1988. Isolation of three viroids and a circular RNA from grapevines. J. Gen. Virol. 69, 413–422.

Salman, T.M., Habili, N., Shi, B., 2014. Effect of temperature on symptom expression and sequence polymorphism of grapevine yellow speckle viroid 1 in grapevine. Virus Res. 189, 243–247.

Schwinghamer, M.J., Broadbent, P., 1987. Association of viroids with a graft-transmissible dwarfing symptom in Australian orange trees. Phytopathology 77, 205–209.

Schwinghamer, M.W., Conroy, R.J., 1983. A viroid similar to potato spindle tuber viroid in the New South Wales potato breeders' collection. Australas. Plant Pathol. 12, 4–6.

Taylor, R., Woodham, R., 1972. Grapevine yellow speckle disease; a newly recognized graft-transmissible disease of *Vitis*. Aust. J. Agr. Res. 23, 447–452.

Trochoulias, T., Allen, R., 1970. Sun-blotch disease of avocado in N.S.W. Agr. Gazette NSW. 81, 167.

Vadamalai, G., Hanold, D., Rezaian, A.M., Randles, J.W., 2006. Variants of the coconut cadang-cadang viroid isolated from an African oil palm (*Elaies guineensis* Jacq.) in Malaysia. Arch. Virol. 151, 1447–1456.

van Brunschot, S.L., Persley, D.M., Roberts, A., Thomas, J.E., 2014a. First report of pospiviroids infecting ornamental plants in Australia: potato spindle tuber viroid in *Solanum laxum* (synonym *S. jasminoides*) and citrus exocortis viroid in *Petunia* spp. New Dis. Rep. 29, 3.

van Brunschot, S.L., Verhoeven, J.T.J., Persley, D.M., Geering, A.D.W., Drenth, A., Thomas, J.E., 2014b. An outbreak of potato spindle tuber viroid in tomato is linked to imported seed. Eur. J. Plant Pathol. 139, 1–7.

Veerakone, S., Tang, J.Z., Ward, L.I., Liefting, L.W., Perez-Egusquiza, Z., Lebas, B.S.M., et al., 2015. A review of the plant virus, viroid, phytoplasma and liberibacter records for New Zealand. Australas. Plant Pathol. 44, 463–514.

Verhoeven, J.T.J., Jansen, C.C.C., Botermans, M., Roenhorst, J.W., 2010. Epidemiological evidence that vegetatively propagated, solanaceous plant species act as sources of potato spindle tuber viroid inoculum for tomato. Plant Pathol. 59, 3–12.

Visvader, J.E., Symons, R.H., 1983. Comparative sequence and structure of different isolates of citrus exocortis viroid. Virology 130, 232–237.

Visvader, J.E., Symons, R.H., 1985. Eleven new sequence variants of citrus exocortis viroid and the correlation of sequence with pathogenicity. Nucleic Acids Res. 13, 2907–2920.

Wall, G.C., Randles, J.W., 2003. Coconut tinangaja viroid. In: Hadidi, A., Flores, R., Randles, J.W., Semancik, J.S. (Eds.), Viroids. CSIRO Publishing, Collingwood, VIC, pp. 242–245.

Wall, G.C., Wiecko, A.T., 1997. Survey of Tinangaja on Guam's coconut populations. Phytopathology 87, S101.

Wan Chow Wah, Y.F., Symons, R.H., 1997. A high sensitivity RT-PCR assay for the diagnosis of grapevine viroids in field and tissue culture samples. J. Virol. Methods 63, 57–69.

Ward, L.I., Burnip, G.M., Liefting, L.W., Harper, S.J., Clover, G.R.G., 2011. First report of grapevine yellow speckle viroid 1 and hop stunt viroid in grapevine (*Vitis vinifera*) in New Zealand. Plant Dis. 95, 617–717.

Ward, L.I., Tang, J., Veerakone, S., Quinn, B.D., Harper, S.J., Delmiglio, C., et al., 2010. First report of potato spindle tuber viroid in Cape Gooseberry (*Physalis peruviana*) in New Zealand. Plant Dis. 94, 479–579.

CHAPTER

47

Geographical Distribution of Viroids in South, Southeast, and East Asia

Dattaraj B. Parakh[1], *Shuifang Zhu*[2] *and Teruo Sano*[3]

[1]ICAR-National Bureau of Plant Genetic Resources, New Delhi, India
[2]Chinese Academy of Inspection and Quarantine, Beijing, China
[3]Hirosaki University, Hirosaki, Japan

In this chapter, the status of over 30 viroids in South, Southeast, and East Asia is documented with special emphasis on China, India, and Japan. Importance of viroid diseases and the economic losses they cause are discussed.

FRUIT TREE VIROIDS

This section includes viroids infecting jujube, persimmon, citrus fruit trees, pome fruit trees, and stone fruit trees.

Hop stunt viroid (HSVd) has been found on jujube in China (Zhang et al., 2009).

Persimmon latent viroid has been recorded on Japanese persimmon in Japan (Nakaune and Nakano, 2008), while persimmon viroid 2 has been recorded in Japan on American persimmon grafted onto a Japanese persimmon rootstock that showed poor growth (Ito et al., 2013).

An isolate of citrus viroid VI (CVd-VI) and a distinct isolate of apple fruit crinkle viroid (AFCVd) were found in Japanese persimmon tress (Nakaune and Nakano, 2008).

Viroids and Satellites.
DOI: http://dx.doi.org/10.1016/B978-0-12-801498-1.00047-4

Citrus Fruit Viroids

Citrus dwarfing viroid (CDVd) has been recorded in citrus from Japan. This viroid, in mixed infections with citrus bent leaf viroid (CBLVd), HSVd, and citrus viroid IV (CVd-IV), produces the severe bark scaling characteristic of exocortis disease in trifoliate orange rootstocks (Ito et al., 2002).

Citrus exocortis viroid (CEVd) has been recorded on citrus from China, India, Indonesia, Japan, South Korea, Malaysia, Pakistan, the Philippines, Taiwan, Thailand, and Vietnam (CABI/EPPO, 2014a). Roy and Ramchandran (2003) described a variant of CEVd (370 nt) from yellow corky vein diseased *Citrus aurantifolia*.

Citrus viroid V (CVd-V) has been recorded from China, Japan, Nepal, and Pakistan (Cao et al., 2013). A total of 44 citrus samples of 275 examined were positive in Japan. Two variants were detected which shared 91%–96% nucleotide identity with other reported sequences. The isolates from Pakistan consisted of 292–295 nt and formed two main groups, suggesting that this country might be one of the geographic origins of the viroid (Ito et al., 2001; Serra et al., 2008a, b; Ito and Ohta, 2010).

CVd-VI has been recorded in Japan (Ito et al., 2001). The pathogen causes mild petiole necrosis and very mild leaf bending in Etrog citron cv. Arizona 861-S1.

Citrus bark cracking viroid (CBCVd) has been recorded in citrus from China (Cao et al., 2010) and Japan (Hataya, 1997).

CBLVd has been recorded on citrus from China, Japan, Pakistan, and the Philippines. The isolates from the Philippines consist of 328–329 nt (Hataya et al., 1998), whereas a China isolate consisted of 318 nt. In addition, a distinct isolate named CVd-I-LSS with 82%–85% sequence identity to CBLVd was found in citrus from Japan.

HSVd has been recorded on citrus in China, India (Ramchandran et al., 2005; Roy and Ramchandran, 2003), Japan (Sano et al., 1986a), and Taiwan. Citrus cachexia-inducing variants of HSVd have been recorded in China and Japan. In India, a variant of HSVd (295 nt), showing 100% sequence identity with six citrus isolates was detected in cv. "Kagzi lime" (*C. aurantifolia*) leaves showing yellow corky vein disease. In Central India, infected citrus trees of "Nagpur" mandarin (*Citrus reticulata*) grafted on lemon and sweet orange (*Citrus sinensis*) showed bark scaling, splitting, and yellowing of leaves.

Pome Fruit Viroids

AFCVd has been reported in apple in Japan (Koganezawa et al., 1989).

Apple dimple fruit viroid has been recorded on apple from China and Japan. In China, symptoms on cv. "Fuji" were slightly deformed fruit with significantly crinkled skin, while cv. "Gala" showed discolored spots with a varying number of dimples similar to those described on cv. "Pink Lady" and cv. "Braeburn." In Japan, symptoms on cv. "Fuji" appear as dimpled and yellowish fruit with scattered yellow spots on the calyx and on cv. "Jonagold" as mild dappling similar to that produced by AFCVd or apple scar skin viroid (ASSVd) (He et al., 2010).

ASSVd has been recorded from China, India, Japan, and South Korea (CABI/EPPO, 2012). The hosts include apple and pear in China, apple in India, and apple and Japanese pear in Japan. In China, ASSVd occurs in eight provinces, infecting more than 0.37 million trees (Li, 2008). Zhu et al. (1995) reported variation in nucleotide sequence. Most of the commercial varieties of pear in China are symptomless carriers but some sensitive cultivars like "Dargazi" or specific ASSVd variants are associated with rusty skin or fruit crinkle symptoms. Japanese pear fruit dimple disease, also associated with ASSVd, was reported in pear cvs. "Niitaka" and "Yoshino" in Japan. In India, Behl et al. (1998) reported a variant of ASSVd in apple cv. "Golden Delicious" showing "bumpy fruit" symptoms. Walia et al. (2009) have reported seven new variants and Kumar et al. (2012) recorded the highest ASSVd infection rate (27.6%) in apple.

Pear blister canker viroid has been reported in Japan on the European pear cv. "La France" (Sano et al., 1997).

HSVd has been found on pear in China (Yang et al., 2007).

Stone Fruit Viroids

ASSVd has been found on apricot and peach in China and Himalayan wild cherry (*Prunus cerasoides*) in India.

HSVd has been found on almond and apricot in China (Yang et al., 2006), peach (dapple fruit) in China and Japan (Sano et al., 1989), and plum in China (Yang et al., 2007) and Japan (Sano et al., 1989). In China, HSVd was detected from latent and symptomatic apricot cv. "Yin Bai" with yellow spots having an irregular border scattered over the leaf surface. In Japan, HSVd caused dapple fruit on sensitive plum cvs. "Taiyo," "Oishiwase," "Beauty," and "Santa Rosa," and yellow fruit on cv. "Soldam." In South Korea, HSVd caused dapple fruit disease on plum cv. "Oishiwase" at Gyeonggi-do and Gyeongsangbukdo.

Peach latent mosaic viroid (PLMVd) has been recorded in peach in China, Japan, Nepal, and Pakistan. In China, symptoms of PLMVd on leaves are rare; however, Xu et al. (2010) reported chlorosis along leaf margins including yellow leaf and mosaic that affects fruit quality. A high infection rate of 64.3% for PLMVd in peach has made the disease a major problem in China (Zhang et al., 2000). In Japan, the infection rate was more than 90% in peach and 5.3%–3.7% in Japanese plum, Japanese apricot, and cherry trees (Osaki et al., 1999).

GRAPEVINE VIROIDS

Australian grapevine viroid has been recorded on grapevines from China and India. The isolates from China showed regional disparity and cultivar-specificity (Jiang et al., 2008) and those from India consisted of two distinct clusters in a phylogenetic tree (Adkar-Purushothama et al., 2014).

Grapevine latent viroid has been reported in China in grapevine (Zhang et al., 2014).

Grapevine yellow speckle viroid 1 has been recorded in China, India, and Japan (Li et al., 2007; Sahana et al., 2013; Sano et al., 2000), while grapevine yellow speckle viroid 2 has been recorded in China (Li et al., 2007; Jiang et al., 2012).

HSVd has been found in grapevine in China (Li et al., 2006), India, and Japan (Sano et al., 1986b).

ORNAMENTAL VIROIDS

CEVd infects asymptomatically moss verbena and trailing verbena and the latter may become doubly infected with tomato chlorotic dwarf viroid (TCDVd) (Singh et al., 2006).

Chrysanthemum stunt viroid (CSVd) has been recorded in chrysanthemum from China, India, Japan, and South Korea (CABI/EPPO, 2007). In China, 1.2%–11.5% infection was observed in wild and commercial cultivars showing chlorosis and stunting, and in symptomless cultivars (Zhang et al., 2011a; Zhu and Su, 1991). In India, 50%–70% infection was observed in plants showing mild chlorosis, stunting and delayed flowering (Mathur et al., 2002). In Japan, CSVd was observed in 77.2% of *Dahlia* spp. tested and eight wild chrysanthemums (Nakashima et al., 2007). In South Korea, the incidence of CSVd infection in chrysanthemum was up to 100% (Chung et al., 2005).

Chrysanthemum chlorotic mottle viroid (CChMVd) has been recorded in China, Japan, and South Korea. In China, CChMVd was detected from chrysanthemum plants showing mild mottle or mild chlorotic spots on leaves (Zhang et al., 2011b). In Japan, out of 223 plants sampled, 46 from 21 cultivars, mostly asymptomatic, were found positive for CChMVd (Yamamoto and Sano, 2005). In South Korea, certain cultivars showed reduced plant growth while others were not affected (Chung et al., 2006).

Coleus blumei viroid 1 to 6 (CbVd-1 to 6): CbVd-1 has been recorded from China, India, Indonesia, Japan, South Korea, and Taiwan. In India, coleus leaf samples showed irregular chlorotic spots and patches (Adkar-Purushothama et al., 2012). In Japan, sequence and host range was reported by Ishiguro et al. (1996). In South Korea, the viroid causes discoloration and growth retardation in some cultivars but is symptomless in others. CbVd-2 has been recorded in China (Fu et al., 2011). CbVd-3 has been recorded in China (Jiang et al., 2011). CbVd-4 is an in vitro generated inverse chimera (Hou et al., 2009b). CbVd-5 (tentative species) has been recorded in China, India, Indonesia, and Japan (Hou et al., 2009a; Tsushima and Sano, 2015). CbVd-6 (tentative species) has been reported in China and Japan (Hou et al., 2009a,b).

Dahlia latent viroid has been recorded in Japan in 38 of 78 symptomless dahlias examined (Tsushima et al., 2015).

Iresine viroid 1 has been recorded in India on *Alternanthera sessilis* (Singh et al., 2006).

Potato spindle tuber viroid (PSTVd) has been recorded in dahlia in Japan (CABI/EPPO, 2014b). Dahlia is a symptomless carrier of PSTVd in Japan (Tsushima et al., 2015).

TCDVd was detected from symptomless petunia in Japan (Shiraishi et al., 2013).

PALM TREE VIROIDS

Coconut cadang-cadang viroid occurs in the Philippines (Randles and Rodriguez, 2003) and Malaysia (Wu et al., 2013). Natural hosts include "buri" palm, coconut palm, and African oil palm. The viroid causes cadang-cadang disease on coconut palm in the Philippines and orange leaf spotting disorder in African oil palm.

Coconut tinangaja viroid has been recorded only on coconut palm in Guam (Wall and Randles, 2003).

HOP VIROIDS

Hop latent viroid has been recorded in China and Japan. In a survey in China, 96 symptomless hop leaf samples, collected from seven districts of Xingiang, revealed the presence of hop latent viroid in all areas, with incidence ranging between 56% and 88% (Liu et al., 2008).

HSVd has been recorded in hop in China (Guo et al., 2008), Japan (Sasaki and Shikata, 1977), and South Korea (Lee et al., 1988).

AFCVd has been reported in hops in Japan (Sano et al., 2004).

SOLANACEOUS CROP VIROIDS

A distinct CEVd strain (372 nt) causes Indian bunchy top disease of tomato (Mishra et al., 1991).

Pepper chat fruit viroid has been recorded in Thailand. Tomato plants showed stunting, leaf necrosis, distortion, and discoloration (Reanwarakorn et al., 2011).

PSTVd has been recorded on potato in China and India; in tomato in Afghanistan, China, India, and Japan; and in wild *Solanum* spp. in India (CABI/EPPO, 2014b). In China, PSTVd was widespread in the early 1990s in a Helongjiang province, as it is seed-transmissible in true potato seeds and tubers, and efforts were made to eradicate it (Singh et al., 1991). In India, PSTVd was documented by Khurana et al. (1997). Tuber losses up to 59%–100% in Punjab state were reported by Singh and Kaur (2014). In Japan, PSTVd-infected greenhouse-grown tomato plants exhibited purple colored leaves in the tips. A variant named PSTVd-Rubber-KER having 97.5% nucleotide identity with other PSTVd variants was characterized associated with the tapping panel dryness syndrome of *Hevea brasiliensis* (rubber tree) (Kumar et al., 2015).

Tomato apical stunt viroid has been recorded in Indonesia on tomato.

TCDVd has been recorded in India and Japan on tomato (CABI/EPPO, 2009). The infected greenhouse-grown tomato plants exhibited leaf chlorosis and yellowing, severe stunting, and reduced and cracked fruit symptoms.

Table 47.1 summarizes the geographical distribution of viroids in South, Southeast, and East Asia.

TABLE 47.1 Geographical Distribution of Viroids in South, Southeast, and East Asia

Viroid	Host	Country
Apple dimple fruit viroid	Apple	China, Japan
Apple fruit crinkle viroid[a]	Apple, Japanese persimmon, hop	Japan
Apple scar skin viroid	Apple, pear, apricot, peach, wild cherry	China, India, Japan, South Korea
Australian grapevine viroid	Grapevine	China, India
Chrysanthemum chlorotic mottle viroid	Chrysanthemum	China, Japan, South Korea

(*Continued*)

TABLE 47.1 (Continued)

Viroid	Host	Country
Chrysanthemum stunt viroid	Cultivated and wild Chrysanthemum, *Dahlia* spp.	China, India, Japan, South Korea
Citrus bark cracking viroid	Citrus	China, Japan
Citrus bent leaf viroid	Citrus	China, Japan, Pakistan, the Philippines
Citrus dwarfing viroid	Citrus	Japan
Citrus exocortis viroid	Citrus, moss verbena and trailing verbena, tomato	China, India, Indonesia, Japan, South Korea, Malysia, Pakistan, the Philippines, Taiwan, Thailand, Vietnam
Citrus viroid IV	Citrus	Japan
Citrus viroid V	Citrus	China, Japan, Nepal, Pakistan
Citrus viroid VI	Citrus, persimmon	Japan
Coconut cadang-cadang viroid	"Buri" palm, coconut palm, African oil palm	the Philippines, Malaysia
Coconut tinangaja viroid	Coconut palm	Guam
Coleus blumei viroid 1 to 6[b]	Coleus	China, India, Indonesia, Japan, South Korea, Taiwan
Dahlia latent viroid	Dahlia	Japan
Grapevine latent viroid[a]	Grapevine	China
Grapevine yellow speckle viroid 1	Grapevine	China, India, Japan
Grapevine yellow speckle viroid 2	Grapevine	China
Hop latent viroid	Hop	China, Japan
Hop stunt viroid	Almond, apricot, citrus, grapevine, hop, jujube, peach, pear, plum	China, India, Japan, South Korea, Taiwan
Iresine viroid 1	*Alternanthera sessilis*	India
Peach latent mosaic viroid	Peach, cherry, Japanese apricot, Japanese plum,	China, Japan, Nepal, Pakistan

(*Continued*)

TABLE 47.1 (Continued)

Viroid	Host	Country
Pear blister canker viroid	Pear	Japan
Pepper chat fruit viroid	Tomato	Thailand
Persimmon latent viroid[a]	Persimmon	Japan
Persimmon viroid 2	Persimmon	Japan
Potato spindle tuber viroid	Dahlia, potato, rubber tree, tomato, wild *Solanum* spp.	Afghanistan, China, India, Japan
Tomato apical stunt viroid	Tomato	Indonesia
Tomato chlorotic dwarf viroid	Petunia, tomato	India, Japan

[a] *Not yet accepted as a species.*
[b] *Coleus blumei viroids 4–6 not yet accepted as species.*

Acknowledgments

D.B.P. would like to acknowledge the full support of the Director, ICAR-National Bureau of Plant Genetic Resources, New Delhi, India.

References

Adkar-Purushothama, C.R., Nagaraja, H., Sreenivasa, M.Y., Sano, T., 2012. First report of Coleus blumei viroid infecting Coleus in India. Plant Dis. 97, 149.

Adkar-Purushothama, C.R., Poornachandra Rao, K., Yanjarappa, S.M., Zhang, Z., Sano, T., 2014. Detection, distribution, and genetic diversity of Australian grapevine viroid in grapevines in India. Virus Genes 49, 304–311.

Behl, M.K., Khurana, S.M.P., Parakh, D.B., Hadidi, A., 1998. Bumpy fruit and other viroid and viroid-like diseases of apple in H. P. India. Acta Hortic. 472, 627–630.

CABI/EPPO, 2007. Chrysanthemun stunt viroid. [Distribution map]. Distribution Maps of Plant Diseases 2007 No. April pp. Map 730 (Edition 2). Wallingford, UK.

CABI/EPPO, 2009. Tomato chlorotic dwarf viroid. [Distribution Map]. Distribution Maps of Plant Diseases 2009 No. October pp. Map 1071 (Edition 1). Wallingford, UK.

CABI/EPPO, 2012. Apple scar skin viroid. [Distribution map]. Distribution Maps of Plant Diseases 2012 No. October pp. Map 1127 (Edition 1). Wallingford, UK.

CABI/EPPO, 2014a. Citrus exocortis viroid. [Distribution map]. Distribution Maps of Plant Diseases 2014 No. April pp. Map 291 (Edition 5). Wallingford, UK.

CABI/EPPO, 2014b. Potato spindle tuber viroid. [Distribution map]. Distribution Maps of Plant Diseases 2014 No. April pp. Map 729 (Edition 3). Wallingford, UK.

Cao, M., Atta, S., Su, H., Wang, X., Wu, Q., Li, Z., et al., 2013. Molecular characterization and phylogenetic analysis of citrus viroid V isolates from Pakistan. Eur. J. Plant Pathol. 135, 11–21.

Cao, M.J., Liu, Y.Q., Wang, X.F., Yang, F.Y., Zhou, C.Y., 2010. First report of citrus bark cracking viroid and citrus viroid V infecting citrus in China. Plant Dis. 94, 922.

Chung, B.N., Kim, D.C., Kim, J.S., Cho, J.D., 2006. Occurrence of chrysanthemum chlorotic mottle viroid in chrysanthemum *(Dendranthema grandiflorum)* in Korea. Plant Pathol. J. 22, 334–338.

Chung, B.N., Lim, J.H., Choi, S.Y., Kim, J.S., Lee, E.J., 2005. Occurrence of chrysanthemum stunt viroid in Korea. Plant Pathol. J. 21, 377–382.

Fu, F.H., Li, S.F., Jiang, D.M., Wang, H.Q., Liu, A.Q., Sang, L.W., 2011. First report of Coleus blumei viroid 2 from commercial coleus in China. Plant Dis. 95, 494.

Guo, L., Liu, S., Wu, Z., Mu, L., Xiang, B., Li, S., 2008. Hop stunt viroid (HSVd) newly reported from hop in Xinjiang, China. Plant Pathol. 57, 764.

Hadidi, A., Barba, M., 2011. Economic impact of pome and stone fruits viruses and viroids. In: Hadidi, A., Barba, M., Candresse, T., Jelkmann, W. (Eds.), Viruses and Virus-Like Diseases of Pome and Stone Fruits. APS, St. Paul, MN, pp. 1–7.

Hataya, T., 1997. Characteristics and detection methods of viroids detected from citrus in Japan. Shokubutsu Boeki 51, 163–167.

Hataya, T., Nakahara, K., Ohara, T., Leki, H., Kano, T., 1998. Citrus viroid Ia is a derivative of citrus bent leaf viroid (CVd-Ib) by partial sequence duplications in the right terminal region. Arch. Virol. 143, 971–980.

He, Y.-H., Isono, S., Kawaguchi-Ito, Y., Taneda, A., Kondo, K., Iijima, A., et al., 2010. Characterization of a new apple dimple fruit viroid variant that causes yellow dimple fruit formation in "Fuji" apple trees. J. Gen. Plant Pathol. 76, 324–330.

Hou, W.Y., Li, S.F., Wu, Z.J., Jiang, D.M., Sano, T., 2009b. Coleus blumei viroid 6: a new tentative member of the genus *Coleviroid* derived from natural genome shuffling. Arch. Virol. 154, 993–997.

Hou, W.Y., Sano, T., Li, F., Wu, Z.J., Li, L., Li, S.F., 2009a. Identification and characterization of a new coleviroid (CbVd-5). Arch. Virol. 154, 315–320.

Ishiguro, A., Sano, T., Harada, Y., 1996. Nucleotide sequence and host range of Coleus viroid isolated from Coleus (*Coleus blumei* Benth.) in Japan. Ann. Phytopathol. Soc. Japan 62, 84–86.

Ito, T., Ieki, H., Ozaki, H., 2001. Characterization of a new citrus viroid species tentatively termed citrus viroid OS. Arch. Virol. 146, 975–982.

Ito, T., Ieki, H., Ozaki, H., Iwanami, T., Nakahara, K., Hataya, T., et al., 2002. Multiple citrus viroids in citrus from Japan and their ability to produce exocortis-like symptoms in citron. Phytopathology 92, 542–547.

Ito, T., Ohta, S., 2010. First report of citrus viroid V in Japan. J. Gen. Plant Pathol. 76, 348–350.

Ito, T., Suzaki, K., Nakano, M., Sato, A., 2013. Characterization of a new apscaviroid from American persimmon. Arch. Virol. 158, 262–263.

Jiang, D.M., Li, S.F., Fu, F.H., Wu, Z.J., Xie, L.H., 2011. First report of occurrence of Coleus blumei viroid 3 from *Coleus blumei* in China. J. Pl. Pathol. 4, Supplement S4, 82.

Jiang, D.M., Peng, S., Wu., Z.J., Cheng, Z.M., Li, S.F., 2008. Genetic diversity and phylogenetic analysis of Australian grapevine viroid (AGVd) isolated from different grapevines in China. Virus Genes 38, 178–183.

Jiang, D.M., Sano, T., Tsuji, M., Araki, H., Sagawa, K., Purushothama, C.R., et al., 2012. Comprehensive diversity analysis of viroids infecting grapevine in China and Japan. Virus Res. 169, 237–245.

Khurana, S.M.P., Singh, R.A., Sati, S.C., Saxena, J., Dubey, R.C., 1997. Virus, viroid and mycoplasma diseases of potato in India and their control. Himalayan Microbial Diversity. Part I. Today and Tomorrow's Printers and Publishers, New Delhi, pp. 219–247.

Koganezawa, H., Ohnuma, Y., Sakuma, T., Yanse, H., 1989. 'Apple fruit crinkle', a new graft-transmitted fruit disorder of apple. Bull. Fruit Tree Res. Stn. MAFF Ser.c. (Morioka) 16, 57–62.

Kumar, A., Pandey, D.M., Abraham, T., Mathews, J., Jyothsna, P., Ramchandran, P., et al., 2015. Molecular characterization of viroid associated with tapping panel dryness syndrome of *Hevea brasiliensis* from India. Curr. Sci. 108 (8), 1520–1527.

Kumar, S., Singh, R.M., Ram, R., Badyal, J., Hallan, V., Zaidi, A.A., et al., 2012. Determination of major viral and sub viral pathogens incidence in apple orchards in Himachal Pradesh. Indian J. Virol. 23, 75–79.

Lee, J.Y., Puchta, H., Ramm, K., Sänger, H.L., 1988. Nucleotide sequence of the Korean strain of hop stunt viroid (HSV). Nucl Acids Res. 16, 8708.

Li, J., 2008. Strengthening the control of apple scar skin disease. J. Anhui Agric. Sci. 36, 40–43.

Li, S.F., Guo, R., Peng, S., Sano, T., 2007. Grapevine yellow speckle viroid 1 and grapevine yellow speckle viroid 2 isolates from China. J. Plant Pathol. 89, 572.

Li, S.F., Guo, R., Tsuji, M., Sano, T., 2006. Two grapevine viroids in China and the possible detection of a third. Plant Pathol. 55, 564.

Liu, S., Li, S., Zhu, J., Xiang, B., Cao, L., 2008. First report of hop latent viroid (HLVd) in China. Plant Pathol. 57, 400.

Mathur, S., Garg, A.P., Ramchandran, P., 2002. Detection of viroid infecting chrysanthemum in India. Indian Phytopathol. 55, 479–482.

Mishra, M.D., Hammond, R.W., Owens, R.A., Smith, D.R., Diener, T.O., 1991. Indian bunchy top disease of tomato plants is caused by a distinct strain of citrus exocortis viroid. J. Gen. Virol. 72, 1781–1785.

Nakashima, A., Hosokawa, M., Maeda, S., Yazawa, S., 2007. Natural infection of chrysanthemum stunt viroid in dahlia plants. J. Gen. Plant Pathol. 73, 225–227.

Nakaune, R., Nakano, M., 2008. Identification of a new Apscaviroid from Japanese persimmon. Arch. Virol. 153, 969–972.

Osaki, K., Yamaguchi, M., Sato, Y., Tomita, Y., Kawai, Y., Miyamoto, Y., et al., 1999. Peach latent mosaic viroid isolated from stone fruits in Japan. Ann. Phytopathol. Soc. Jpn. 65, 3–8.

Ramchandran, P., Agarwal, J., Roy, A., Ghosh, D.K., Das, D.R., Ahlwat, Y.S., 2005. First record of a hop stunt viroid variant on Nagpur mandarin and Mosambi sweet orange trees on rough lemon and Rangpur lime rootstocks. Plant Pathol. 54, 571.

Randles, J.W., 2003. Economic impact of viroid diseases. In: Hadidi, A., Flores, R., Randles, J.W., Semancik, J.S. (Eds.), Viroids. CSIRO Publishing, Collingwood, VIC, pp. 3–11.

Randles, J.W., Rodriguez, M.J.B., 2003. Coconut cadang-cadang viroid. In: Hadidi, A., Flores, R., Randles, J.W., Semancik, J.S. (Eds.), Viroids. CSIRO Publishing, Collingwood, VIC, pp. 233–241.

Reanwarakorn, K., Klinkong, S., Porsoongnurn, J., 2011. First report of natural infection of pepper chat fruit viroid in tomato plants in Thailand. New Dis. Rep. 24, 6.

Roy, A., Ramchandran, P., 2003. Occurrence of a hop stunt viroid (HSVd) variant in yellow corky vein disease of citrus in India. Curr. Sci. 85, 1608–1612.

Sahana, A.B., Adkar-Purushothama, C.R., Chennappa, G., Zhang, Z.X., Sreenivasa, M.Y., Sano, T., 2013. First report of grapevine yellow speckle viroid-1 and hop stunt viroid infecting grapevines (*Vitis vinifera*) in India. Plant Dis. 97, 1517.

Sano, T., Hataya, T., Sasaki, A., Shikata, E., 1986a. Etrog citron is latently infected with hop stunt viroid-like RNA. Proc. Japan Acad. 62 (B), 325–328.

Sano, T., Hataya, T., Terai, Y., Shikata, E., 1989. Hop stunt viroid strains from dapple fruit disease of plum, peach in Japan. J. Gen. Virol. 70, 1311–1319.

Sano, T., Kobayashi, T., Ishiguro, A., Motomura, Y., 2000. Two types of grapevine yellow speckle viroid isolated from commercial grapevine had the nucleotide sequence of yellow speckle symptom-inducing type. J. Gen. Plant Pathol. 66, 68–70.

Sano, T., Li, S., Ogata, T., Ochiai, N., Suzuki, C., Ohnuma, S., et al., 1997. Pear blister canker viroid isolated from European pear in Japan. Ann. Phytopathol. Soc. Jpn. 63, 89–94.

Sano, T., Ohshima, K., Hataya, T., Uyeda, I., Shikata, E., Chou, T.G., et al., 1986b. A viroid resembling hop stunt viroid in grapevines from Europe, the United States and Japan. J. Gen. Virol. 67, 1673–1678.

Sano, T., Yoshida, H., Goshono, M., Monma, T., Kawasaki, H., Ishizaki, K., 2004. Characterization of a new viroid strain from hops: evidence for viroid speciation by isolation in different host species. J. Gen. Plant Pathol. 70, 181–187.

Sasaki, M., Shikata, E., 1977. On some properties of hop stunt disease agent, a viroid. Proc. Jpn. Acad. Ser. B. 53, 109–112.

Serra, P., Barbosa, C.J., Daros, J.A., Flores, R., Duran-Vila, N., 2008a. Citrus viroid V: molecular characterization and synergistic interactions with other members of the genus Apscaviroid. Virology 370, 102–112.

Serra, P., Eiras, M., Bani-Hashemian, S.M., Murcia, N., Kitajima, E.W., Daros, J.A., et al., 2008b. Citrus viroid V: occurrence, host range, diagnosis, and identification of new variants. Phytopathology 98, 1199–1204.

Shiraishi, T., Maejima, K., Komatsu, K., Hashimoto, M., Okano, Y., Kitazawa, Y., et al., 2013. First report of tomato chlorotic dwarf viroid isolated from symptomless petunia plants (*Petunia* spp.) in Japan. J. Gen. Plant Pathol. 79, 214–216.

Singh, B., Kaur, A., 2014. Incidence of potato spindle tuber viroid (PSTVd) in potato growing areas of Punjab. Vegetos 27, 96–100.

Singh, R.P., Boucher, A., Wang, R.G., 1991. Detection, distribution and long-term persistence of potato spindle tuber viroid in true potato seed from Heilongjiang. China. Am. Potato J. 68, 65–74.

Singh, R.P., Dilworth, A.D., Baranwal, V.K., Gupta, K.N., 2006. Detection of citrus exocortis viroid, Iresine viroid and tomato chlorotic dwarf viroid in new ornamental host plants in India. Plant Dis. 90, 1457.

Tsushima, T., Matsushita, Y., Fuji, S., Sano, T., 2015. First report of dahlia latent viroid and potato spindle tuber viroid mixed-infection in commercial ornamental dahlia in Japan. New Dis. Rep. 31, 11.

Tsushima, T., Sano, T., 2015. First report of *Coleus blumei* viroid 5 infection in vegetatively propagated clonal Coleus cv. 'Aurora black cherry' in Japan. New Dis. Rep. 32, 7.

Walia, Y., Kumar, Y., Rana, T., Bharadwaj, P., Ram, R., Thakur, P.D., et al., 2009. Molecular characterization and variability analysis of apple scar skin viroid in India. J. Gen. Pl. Pathol. 75, 307–311.

Wall, G.C., Randles, J.W., 2003. Coconut tinangaja viroid. In: Hadidi, A., Flores, R., Randles, J.W., Semancik, J.S. (Eds.), Viroids. CSIRO Publication, Collingwood, VIC, pp. 242–245.

Wu, Y.H., Cheong, L.C., Meon, S., Lau, W.H., Kong, L.L., Joseph, H., et al., 2013. Characterization of coconut cadang-cadang viroid variants from oil palm affected by orange spotting disease in Malaysia. Arch. Virol. 158, 1407–1410.

Xu, W.X., Dou, R.M., Wang, G.P., Hong, N., Wang, Z.H., 2010. Population structure and sequence analysis of a peach latent mosaic viroid isolate from a peach tree showing discoloration along leaf veins. Acta Phytopathol. Sinica. 40, 317–321.

Yamamoto, H., Sano, T., 2005. Occurrence of chrysanthemum chlorotic mottle viroid in Japan. J. Gen. Plant. Pathol. 71, 156–157.

Yang, Y.A., Wang, H.Q., Cheng, Z.M., Sano, T., Li, S.F., 2007. First report of hop stunt viroid from plum in China. Plant Pathol. 56, 339.

Yang, Y.A., Wang, H.Q., Guo, R., Cheng, Z.M., Li, S.F., Sano, T., 2006. First report of hop stunt viroid in apricot in China. Plant Dis. 90, 828.

Zhang, B.L., Liu, G.Y., Liu, C.Q., Wu, Z.J., Jiang, D.M., Li, S.H., 2009. Characterization of hop stunt viroid (HSVd) isolates from jujube trees (*Ziziphus jujuba*). Eur. J. Plant Pathol. 125, 665–669.

Zhang, S.Y., Zhang, Z.P., Hong, N., Wang, G.P., 2000. Identification of peach latent mosaic viroid. China Fruits 1, 30–31.

Zhang, Z., Qi, S., Tang, N., Zhang, X., Chen, S., Zhu, P., et al., 2014. Discovery of replicating circular RNAs by RNA-seq and computational algorithms. PLoS Pathog. 10, e1004553.

Zhang, Z.X., Ge, B.B., Pan, S., Zhao, Z., Wang, H.Q., Li, S.F., 2011a. Molecular detection and sequence analysis of chrysanthemum stunt viroid. Acta Hortic. Sin. 38, 2349–2356.

Zhang, Z.Z., Pan, S., Li, S.F., 2011b. First report of chrysanthemum chlorotic mottle viroid in chrysanthemum in China. Plant Dis. 95, 1320.

Zhu, S.F., Hadidi, A., Hammond, R.W., Yang, X., Hansen, A.J., 1995. Nucleotide sequence and secondary structure of pome fruit viroids from dapple apple diseased apples, pear rusty skin diseased pears and apple scar skin symptomless pears. Acta Hortic. 386, 554–559.

Zhu, S.F., Su, X.Z., 1991. A two-dimensional electrophoretic detection of chrysanthemum stunt viroid. Plant Quarant. 5, 173–174 (in Chinese).

PART VII

SPECIAL TOPICS

CHAPTER

48

Seed, Pollen, and Insect Transmission of Viroids

Rosemarie W. Hammond

U.S. Department of Agriculture, Beltsville, MD, United States

SEED AND POLLEN TRANSMISSION IN MEMBERS OF THE FAMILY *POSPIVIROIDAE*

Potato spindle tuber viroid (PSTVd) has been reported to be present in, or on, true seeds and pollen of potato (Singh et al., 1992) and could be transmitted to tomato at a rate of 0%–100% (averaging 31% depending on the cultivar) (Fernow et al., 1970). This was not correlated with age of the seed or the cultivar. After crossing, Hunter et al. (1969) and Singh (1970) obtained 87%–100% and 12%, respectively, transmission through potato seed when both parents were infected.

Benson and Singh (1964) and Singh (1970) reported a tomato seed transmission rate of 11% for PSTVd, when both parents used for a cross were infected, while Kryczynski et al. (1988) reported seed transmission in 18 of 20 combined seed samples from PSTVd-infected fruits. The frequency of seed transmission of PSTVd in tomato was critically evaluated by Simmons et al. (2015) as ranging from 62.3% to 69%, while the frequency of PSTVd transmission to germinated seedlings was 50.9%. Germination efficiency of PSTVd-infected tomato seed was significantly lower (53%) than noninfected seed (98%) and viroid titers in infected seed were significantly lower in the former (Simmons et al., 2015). In addition, 60% of infected seedlings did not display symptoms until they were 2–3 weeks old. The negative effect of PSTVd on sexual reproduction and viroid transmission in true potato seed was observed by Grasmick and Slack (1986). Hooker et al. (1978) reported a reduction in PSTVd-infected tomato pollen viability by 50% in synthetic media, and

Viroids and Satellites.
DOI: http://dx.doi.org/10.1016/B978-0-12-801498-1.00048-6

germination and growth on stigma and style, respectively, was severely impaired. Aberrant pollen mother cells were due to nondisjunction at anaphase I. These studies suggest that the lower infection incidence in seedlings may be due to the deleterious effects of viroids on seed viability.

PSTVd is also seed-transmitted in petunia by invasion of the embryo through infected pollen grains or through the infected embryo sac (Matsushita and Tsuda, 2014). The distribution of PSTVd and tomato chlorotic dwarf viroid (TCDVd) in ovary and ovules of infected tomato revealed that TCDVd was absent in the shoot apical meristem and from ovules of tomato, and was not seed-transmitted; however, seed transmission of TCDVd was reported by Singh and Dilworth (2009). There are conflicting reports of the presence of TCDVd on the surface of seed coat (Matsushita et al., 2011) or systemically in seed (Singh and Dilworth, 2009), and TCDVd was detected in at least one commercial seed lot (Candresse et al., 2010). PSTVd was present in ovules and seed-transmitted, and was not a contaminant of seed surfaces in tomato seeds (Matsushita et al., 2011). Seed transmission rate is the rate at which infected seeds germinate and lead to an infected plant; however, some seeds may not germinate if infected, so may lead to a lower estimation of infection. PSTVd can accumulate in all floral organs in *Nicotiana benthamiana* and tomato; however, it was found in sepals, but its absence in petals, ovary, and stamens suggested restricted movement into these organs and not suppression of replication (Zhu et al., 2002).

Pepper chat fruit viroid (PCFVd) was seed-transmitted in pepper (*Capsicum annuum* L.) at a rate of 19% (Verhoeven et al., 2009). The presence of grapevine yellow speckle viroid 1 and hop stunt viroid was confirmed in seedlings from eight grapevine (*Vitis vinifera*) varieties in Australia (Wan Chow Wah and Symons, 1999). The viroid profile in infected seedlings revealed a differential transmission of sequence variants through seed. Chrysanthemum stunt viroid was reported to be seed-transmitted in chrysanthemum (Chung and Pak, 2008). Chrysanthemum stunt viroid and cucumber pale fruit viroid (an isolate of hop stunt viroid) were found to be transmitted through seed and pollen in tomato (Kryczynski et al., 1988). Coleus blumei viroid was detected in commercial coleus seed in Canada, at a rate of 16%–68% depending on the cultivar and seed supplier (Singh et al., 1991), and in India (Ramachandran et al., 1992).

Tomato apical stunt viroid was transmitted at a rate of 80% through tomato seed (Antignus et al., 2007). Disinfestation of the seed did not prevent viroid transmission to germinated seedlings. Hop latent viroid is not transmissible through hop seeds and Matousek et al. (2008)

attributed elimination of the viroid to pollen nucleases during development of pollen.

Seed transmission of PSTVd was variable, at 0%–90.2% in tomato, 0.3% in *C. annuum* var. "Grossum," 0.5% in *C. annuum* var. "Angulosum." 1.2% in *Glebionis coronaria*, and 81% in *Petunia* × *hybrida*. For TCDVd and Columnea latent viroid (CLVd), transmission rates were 25% in *Petunia* × *hybrida* and 5.3%–100% in tomato (Matsushita and Tsuda, 2016). No transmission of PSTVd, citrus exocortis viroid, tomato apical stunt viroid, or CLVd was found in tomato inoculated when flowering (Faggioli et al., 2015), in stark contrast to other studies. Seed transmission efficiency may be affected if plants are infected after the onset of flowering, when the connection from the embryogenic suspensor to the developing embryo degenerates, reducing the opportunity for direct invasion of the embryo (Maule and Wang, 1996).

Apple scar skin viroid (ASSVd) was reported as not seed-transmitted or transmitted at a low rate in Oriental pear (Hurtt and Podleckis, 1995). In apple, Howell et al. (1998) reported that, while seed-borne, ASSVd was not seed-transmitted; however, in two separate reports, ASSVd and dapple apple viroid (a strain of ASSVd) were found to be seed-borne (Hadidi et al., 1991), and Kim et al. (2006) showed a 7.7% infection rate in seedlings germinated from ASSVd-positive apple seeds. Seed transmission of citrus exocortis viroid in citrus does not occur (Semancik, 1980; Duran-Vila and Semancik, 2003), however, it does occur in tomato (Semancik, 1980) and through commercially distributed seeds of *Impatiens* and *Verbena* (Singh et al., 2009). Pacumbaba et al. (1994) demonstrated seed and pollen transmission of coconut cadang-cadang viroid and Manalo et al. (2000) reported seed transmission at a rate of 4% of F1 progeny when healthy palms were pollinated with fresh pollen from infected palms.

SEED AND POLLEN TRANSMISSION IN MEMBERS OF THE FAMILY *AVSUNVIROIDAE*

Avocado sunblotch viroid is transmitted from symptomless infected mother trees to progeny seed (Wallace and Drake, 1962), with a rate from symptomatic trees of 5% (Wallace and Drake, 1953). Experimental transmission through infected pollen by bees (*Apis mellifera*) was recorded at 1%–4% (Desjardins et al., 1979) and symptomatic and asymptomatic infected trees can serve as pollen donors (Desjardins et al., 1984).

Peach latent mosaic viroid is not seed-transmitted (Howell et al., 1998; Flores et al., 1992). It was, however, shown to be pollen-transmitted in peach trees (Barba et al., 2007).

Eggplant latent viroid (ELVd) is seed-transmitted in eggplant with an efficiency of 20% (Fadda et al., 2003; Fagoaga and Duran-Vila, 2003). Analysis of two lots of 50 seedlings from germinated seeds of two ELVd-infected eggplants revealed ELVd infection ratios of 26% and 16%.

DETECTION METHODS AND DECONTAMINATION OF SEED

Large scale testing for PSTVd in potato seed showed a detection sensitivity of one contaminated seed in 80–150 noncontaminated seeds (Borkhardt et al., 1994; Salazar et al., 1983). One infected in 2000 noninfected tomato seeds could be detected using RT-PCR (Hoshino et al., 2006) and Bakker et al. (2015) reported that a high throughput, multiplex TaqMan real-time RT-PCR could detect one tomato seed infected with PSTVd and TCDVd in 1000 noninfected seeds. Other amplification methods have been developed, including reverse transcription loop-mediated isothermal amplification, which could detect one infected in 400 noninfected tomato seeds (Tsutsumi et al., 2010), and a rapid diagnostic assay based on isothermal reverse transcription recombinase polymerase amplification and a lateral flow immunoassay for the detection of TCDVd in tomato seed (Hammond and Zhang, 2016).

Decontamination of seed with 1% sodium hypochlorite did not reduce seed transmission (Fadda et al., 2003; Simmons et al., 2015; Wan Chow Wah and Symons, 1999).

SEED HEALTH AND THE IMPACT OF VIROID SEED TRANSMISSION ON GROWERS AND INTERNATIONAL SEED TRADE

Agricultural effects of seed transmission of pathogens include (1) direct or indirect injury to the seed; (2) infected plants from germinating seedlings; (3) survival of the inoculum from one crop season to the next; (4) dissemination of the disease worldwide through exchange of seeds; and (5) complex regulatory issues for seed exchange. PSTVd has been eradicated in potato in Canada because a multifaceted eradication program was employed that included strict seed potato certification programs (Singh, 2014). Current phytosanitary requirements are not consistent among countries and the methods have not yet been harmonized (see Chapter 39: Quarantine and Certification for Viroids and Viroid Diseases).

The potential role of seed (and vegetative propagule) transmission in the spread of viroid diseases exists. For example, before 2007, CLVd was only described in ornamentals. In the United Kingdom and France, it was found in tomato in 2007. CLVd was discovered in a tomato nursery in S.E. England in 2009, with 20%–60% of plants infected. In The Netherlands, tomato samples collected from 1988–2002 were evaluated in 2004 (Verhoeven et al., 2004). Sequence analysis of RT-PCR products revealed two lineages of CLVd—one from Columnea also found in other ornamentals, and one that is present in tomato; the latter is severe on potato. In 2011, CLVd discovered in Mali in tomatoes (Batuman and Gilbertson, 2013) had a 99% identity to the tomato lineage isolate reported in The Netherlands. As a source of viroid infection could not be associated with any particular seed lot, ornamentals may have been the source of CLVd infection. A similar analysis revealed that a tomato isolate found in The Netherlands and United Kingdom (2003) was very similar to isolates from Australia and New Zealand, raising speculation that those isolates (Verhoeven et al., 2004) may have originated in Oceania (through seed or ornamentals).

In 2011, a severe disease infected 3% of 80,000 cherry tomato plants (cv. Perino) in a commercial greenhouse in Queensland, Australia. A variant of PSTVd was discovered with 96.8% sequence identity to a variant found in greenhouse tomatoes in North Carolina, USA (Ling et al., 2013) in the *Physalis peruviana* cluster (Oceania, North America, some European countries). The source of the outbreak was an imported seed lot of planted gourmet tomato (cv. Tiger) (Van Brunschot et al., 2014). Spread to cherry tomato was most likely by mechanical transmission. Under government-assisted direction, all plants in the greenhouse were destroyed, and the greenhouse was decontaminated with 1% sodium hypochlorite, and was left empty for 6 weeks prior to replanting.

A new disease of greenhouse-grown sweet pepper (*C. annuum* L.) in The Netherlands in 2006 was caused by a new pospiviroid, PCFVd (Verhoeven et al., 2009). In pepper, fruit size was reduced by 50% and plant growth was slightly reduced. PCFVd is transmitted mechanically and through seed in tomato and pepper (19% in pepper) and, in 2012, was intercepted in Australia in shipments of traded tomato seed (from Israel and Thailand) (Chambers et al., 2013). PCFVd was previously reported in The Netherlands, Thailand, and Canada; however a direct link to seed transmission in outbreaks in Canada and The Netherlands

(Verhoeven et al., 2009, 2011) was not obvious as different cultivars from different seed companies were infected.

INSECT TRANSMISSION

Six insect pests were evaluated for their ability to transmit PSTVd to healthy potato; however, they were relatively insignificant as only the tarnished plant bug (*Lygus lineolaris*) transmitted PSTVd to 2 of 183 plants (Schumann et al., 1980).

The green peach aphid (*Myzus persicae*) transmitted PSTVd from plants coinfected with potato leafroll virus (PLRV) to potato, *Physalis floridana*, and *Datura stramonium* (Querci et al., 1997; Salazar et al., 1995; Syller et al., 1997) although this species was an insignificant vector of PSTVd alone. The coat protein of PLRV transencapsidates PSTVd, allowing cotransmission of both pathogens by the insect vector (Querci et al., 1997; Syller et al., 1997). This cotransmission by aphid vectors has implications for epidemiology, transmission, and control of PSTVd in potato fields. The efficiency of aphid transmission of PSTVd when cotransmitted with PLRV was variable, ranging from 0% to 55%, depending on the potato cultivar used as the inoculum source or test plant (Syller and Marczewski, 2001). In potato plants highly resistant to PLRV, only PSTVd was in 23% of plants indicating that PLRV can serve as a carrier for PSTVd (Syller and Marczewski, 2001). Additional studies revealed that PSTVd is encapsidated in vivo at a low frequency by velvet tobacco mottle virus (Francki et al., 1986).

Tomato planta macho viroid could be transmitted at high efficiency by *M. persicae* to *Senecio nigrescens* and *Psychotria foetens* (Galindo et al., 1986); however, it is not known if this was mechanical transmission by the insects or feeding-associated transmission.

Other aspects of pospiviroid transmission by bees, aphids, or whiteflies are described in Chapter 5, Viroid Biology and Chapter 15, Other Pospiviroids Infecting Solanaceous Plants.

Recently, ASSVd was found to be transmitted by the whitefly *Trialeurodes vaporariorum* from viroid-infected cucumber and bean to cucumber, bean, tomato, and pea plants (Walia et al., 2015). The transmission frequency was enhanced by the *Cucumus sativus* phloem protein 2, which formed a complex with ASSVd RNA.

In fewer reports, members of the *Avsunviroidae* have been reported to be insect-transmitted. Peach latent mosaic viroid was experimentally transmitted at a low rate by *M. persicae* (Flores et al., 1992).

A summary of seed, pollen, and insect transmission of viroids is presented in Table 48.1.

TABLE 48.1 Seed, Pollen, and Insect Transmission of Viroids

Viroid[a]	Seed	Pollen	Insect	References[b]
ASBVd	+ (Avocado)	+ (Avocado)	+	1–4
ASSVd	+ (Apple, pear)		+	5–9
CbVd	+ (*Coleus*)			10, 11
CCCVd	+ (Coconut)	+ (Coconut)		12, 13
CEVd	+ (Tomato, *Impatiens, Verbena*)			14, 15
CLVd	+ (Tomato, *Petunia hybrida*)			16
CPFVd	+ (Tomato)	+ (Tomato)		17
CSVd	+ (Tomato)	+ (Tomato)		17, 18
ELVd	+ (Eggplant)			19, 20
HSVd	+ (Grape)			21
GYSVd1	+ (Grape)			21
PCFVd	+ (Tomato, pepper)	+		22
PLMVd		+ (Peach)		23
PSTVd	+ (Potato, tomato, *Capsicum annuum, Glebionis coronaria, Petunia hybrida*)	+ (Potato, tomato)	+	17, 24–36
TASVd	+ (Tomato)		+	37, 38
TCDVd	+ (Tomato, *Petunia hybrida*)		+	16, 39, 40
TPMVd			+	41

[a] ASBVd, *avocado sunblotch viroid;* ASSVd, *apple scar skin viroid;* CbVd, *coleus blumei viroid;* CCCVd, *coconut cadang-cadang viroid;* CEVd, *citrus exocortis viroid;* CLVd, *columnea latent viroid;* CPFVd, *cucumber pale fruit viroid;* CSVd, *chrysanthemum stunt viroid;* ELVd, *eggplant latent viroid;* HSVd, *hop stunt viroid;* GYSVd1, *grapevine yellow speckle viroid 1;* PCFVd, *pepper chat fruit viroid;* PLMVd, *peach latent mosaic viroid;* PSTVd, *potato spindle tuber viroid;* TASVd, *tomato apical stunt viroid;* TCDVd, *tomato chlorotic dwarf viroid;* TPMVd, *tomato planta macho viroid.*

[b] *1, Wallace and Drake (1962); 2, Wallace and Drake (1953); 4, Desjardins et al. (1979); 5, Hadidi et al. (1991); 6, Howell et al. (1998); 7, Hurtt and Podleckis (1995); 8, Kim et al. (2006); 9, Walia et al. (2015); 10, Ramachandran et al. (1992); 11, Singh et al. (1991); 12, Manalo et al. (2000); 13, Pacumbaba et al. (1994); 14, Semancik (1980); 15, Singh et al. (2009); 16, Matsushita and Tsuda (2016); 17, Kryczynski et al. (1988); 18, Chung and Pak (2008); 19, Fadda et al. (2003); 20, Fagoaga and Duran-Vila (2003); 21, Wan Chow Wah and Symons (1999); 22, Verhoeven et al. (2009); 23, Barba et al. (2007); 24, Fernow et al. (1970); 25, Hunter et al. (1969); 26, Singh (1970); 27, Benson and Singh (1964); 28, Simmons et al. (2015); 29, Matsushita and Tsuda (2014); 30, Matsushita et al. (2011); 31, Matsushita and Tsuda (2016); 32, Salazar et al. (1995); 33, Querci et al. (1997); 34, Syller et al. (1997); 35, Syller and Marczewski (2001); 36, Schumann et al. (1980); 37, Antignus et al. (2007); 38, Van Bogaert et al. (2015); 39, Singh and Dilworth (2009); 40, Matsuura et al. (2010); 41, Galindo et al. (1986).*

References

Antignus, Y.O., Lachman, O., Pearlsman, M., 2007. Spread of tomato apical stunt viroid (TASVd) in greenhouse tomato crops is associated with seed transmission and bumble bee activity. Plant Dis. 91, 47–50.

Bakker, D., Bruinsma, M., Dekter, R.W., Toonen, M.A.J., Verhoeven, J.Th.J., et al., 2015. Detection of PSTVd and TCDVd in seeds of tomato using real-time RT-PCR. Bull. OEPP/EPPO 45, 14–21.

Barba, M., Ragozzino, E., Faggioli, F., 2007. Pollen transmission of peach latent mosaic viroid. J. Plant Pathol. 89, 287–289.

Batuman, O., Gilbertson, R.L., 2013. First report of columnea latent viroid (CLVd) in tomato in Mali. Plant Dis. 97, 692.

Benson, A.P., Singh, R.P., 1964. Seed transmission of potato spindle tuber virus in tomato. Am. Potato J. 41, 294.

Borkhardt, B., Vongsasitorn, D., Albrechtsen, S.E., 1994. Chemiluminescent detection of potato spindle tuber viroid in true potato seed using a digoxigenin labeled DNA probe. Potato Res. 37, 249–255.

Candresse, T., Marais, A., Tassus, X., Suhard, P., Renaudin, I., Leguay, A., et al., 2010. First report of tomato chlorotic dwarf viroid in tomato in France. Plant Dis. 94, 633.

Chambers, G.A., Seyb, A.M., Mackie, J., Constable, F.E., Rodoni, B.C., Letham, D., et al., 2013. First report of pepper chat fruit viroid in traded tomato seed, an interception by Australian biosecurity. Plant Dis. 97, 1386.

Chung, B.N., Pak, H.S., 2008. Seed transmission of chrysanthemum stunt viroid in chrysanthemum (*Dendranthema grandiflorum*) in Korea. Plant Pathol. J. 24, 31–35.

Desjardins, P.R., Drake, R.J., Atkins, E.L., Bergh, O., 1979. Pollen transmission of avocado sunblotch virus experimentally demonstrated. California Agr. 33, 14–15.

Desjardins, P.R., Drake, R.J., Sasaki, P.J., Atkins, E.L., Bergh, O., 1984. Pollen transmission of avocado sunblotch viroid and the fate of the pollen recipient tree. Phytopathology 74, 845.

Duran-Vila, N., Semancik, J.S., 2003. Citrus viroids. In: Hadidi, A., Flores, R., Randles, J.W., Semancik, J.S. (Eds.), Viroids. CSIRO Publishing, Collingwood, VIC, pp. 178–194.

Fadda, Z., Daros, J.A., Fagoaga, C., Flores, R., Duran-Vila, N., 2003. Eggplant latent viroid, the candidate type species for a new genus within the family *Asvunviroidae* (hammerhead viroids). J. Virol. 77, 6528–6532.

Faggioli, F., Luigi, M., Sveikauskas, V., Olivier, T., Virscek Marn, M., Mavric Plesko, I., et al., 2015. An assessment of the transmission rate of four pospiviroid species through tomato seeds. Eur. J. Plant Pathol. 143, 613–617.

Fagoaga, C., Duran-Vila, N., 2003. Eggplant latent. In: Hadidi, A., Flores, R., Randles, J.W., Semancik, J.S. (Eds.), Viroids. CSIRO Publishing, Collingwood, VIC, p. 333.

Fernow, K.H., Peterson, L.C., Plaisted, R.L., 1970. Spindle tuber virus in seeds and pollen of infected potato plants. Am. Potato J. 47, 75–80.

Flores, R., Hernández, C., Avinent, L., Hermoso, A., Llácer, G., Juárez, J., et al., 1992. Studies on the detection, transmission and distribution of peach latent mosaic viroid in peach trees. Acta. Hortic. 309, 325–330.

Francki, R.I.B., Zaitlin, M., Palukaitis, P., 1986. In vivo encapsidation of potato spindle tuber viroid by velvet tobacco mottle virus particles. Virology 155, 469–473.

Galindo, J., López, M., Aguilar, T., 1986. Significance of *Myzus persicae* in the spread of tomato planta macho viroid. Fitopatol. Bras. 11, 400–410.

Grasmick, M.E., Slack, S.A., 1986. Effect of potato spindle tuber viroid on sexual reproduction and viroid transmission in true potato seed. Can. J. Bot. 64, 336–340.

Hadidi, A., Hansen, A.J., Parish, C.L., Yang, X., 1991. Scar skin and dapple apple viroids are seed-borne and persistent in infected apple trees. Res. Virol. 142, 289–296.

Hammond, R.W., Zhang, S., 2016. Development of a rapid diagnostic assay for the detection of tomato chlorotic dwarf viroid based on isothermal reverse-transcription-recombinase polymerase amplification. J. Virol. Methods 236, 62–67.

Hooker, W.J., Nimnoi, P.N., Tai, W., Young, T.C., 1978. Germination reduction in PSTV-infected tomato pollen. Am. Potato J. 55, 378.

Hoshino, S., Okuta, T., Isaka, M., Tsutsumi, N., Miyai, N., Ikeshiro, T., et al., 2006. Detection of potato spindle tuber viroid (PSTVd) in tomato and potato seeds. Res. Bull. Pl. Prot. Japan. 42, 75–79.

Howell, W.E., Skrzeczkowski, L.J., Mink, G.I., Nunez, A., Wessels, T., 1998. Non-transmission of apple scar skin viroid and peach latent mosaic viroid through seed. Acta Hortic. 472, 635–639.

Hunter, D.E., Darling, H.M., Beale, W.L., 1969. Seed transmission of potato spindle tuber virus. Am. Potato. J. 46, 247–250.

Hurtt, S.S., Podleckis, E.V., 1995. Apple scar skin viroid is not seed transmitted or transmitted at a low rate in Oriental pear. Acta Hortic. 386, 544–630.

Kim, H.-R., Lee, S.-H., Lee, D.-H., Kim, J.-S., Park, J.-W., 2006. Transmission of apple scar skin viroid by grafting, using contaminated pruning equipment, and planting infected seeds. Plant Pathol. J. 22, 63–67.

Kryczynski, S., Paduch-Cichal, E., Skrzeczkowski, L.J., 1988. Transmission of three viroids through seed and pollen of tomato plants. J. Phytopathol. 121, 51–57.

Ling, K.S., Li, R., Panthee, D.R., Gardner, R.G., 2013. First report of potato spindle tuber viroid naturally infecting greenhouse tomatoes in North Carolina. Plant Dis. 97, 148.

Manalo, G.G., Estioko, L.P., Rodriguez, M.J.B., 2000. Studies on the transmission of the coconut cadang-cadang viroid. Report Phil. Coconut Auth. Quezon City, Philippines.

Matousek, J., Orctova, L., Skopek, J., Pesina, K., Steger, G., 2008. Elimination of hop latent viroid upon developmental activation of pollen nucleases. Biol. Chem. 389, 905–918.

Matsushita, Y., Tsuda, S., 2014. Distribution of potato spindle tuber viroid in reproductive organs of petunia during its developmental stages. Phytopathology. 104, 964–969.

Matsushita, Y., Tsuda, S., 2016. Seed transmission of potato spindle tuber viroid, tomato chlorotic dwarf viroid, tomato apical stunt viroid, and columnea latent viroid in horticultural crops. Eur. J. Plant Pathol. 145, 1007–1011.

Matsushita, Y., Usugi, T., Tsuda, S., 2011. Distribution of tomato chlorotic dwarf viroid in floral organs of tomato. Eur. J. Plant Pathol. 130, 441–447.

Matsuura, S., Matsushita, Y., Kozuka, R., Shimizu, S., Tsuda, S., 2010. Transmission of tomato chlorotic dwarf viroid by bumblebees (*Bombus ignites*) in tomato plants. Eur. J. Plant Pathol. 126, 111–115.

Maule, A.J., Wang, D., 1996. Seed transmission of plant viruses: a lesson in biological complexity. Trends Microbiol. 4, 153–158.

Pacumbaba, E.P., Zelazny, B., Orense, J.C., Rillo, E.P., 1994. Evidence for pollen and seed transmission of the coconut cadang-cadang viroid in *Cocos nucifera*. J. Phytopathol. 142, 37–42.

Querci, M., Owens, R.A., Bartolini, I., Lazarte, V., Salazar, L.F., 1997. Evidence for heterologous encapsidation of potato spindle tuber viroid in particles of potato leafroll virus. J. Gen. Virol. 18, 1207–1211.

Ramachandran, P., Kumar, D., Varma, A., Pandey, P.K., Singh, R.P., 1992. *Coleus* viroid in India. Current Sci. 62, 271–272.

Salazar, L.F., Owens, R.A., Smith, D.R., Diener, T.O., 1983. Detection of potato spindle tuber viroid by nucleic acid spot hybridization: evaluation with tuber sprouts and true potato seed. Am. Potato J. 60, 587–597.

Salazar, L.F., Querci, M., Bartolini, I., Lazarte, V., 1995. Aphid transmission of potato spindle tuber viroid assisted by potato leaf roll virus. Fitopatologia 30, 56–58.

Schumann, G.L., Tingey, W.M., Thurston, H.D., 1980. Evaluation of six insect pests for transmission of potato spindle tuber viroid. Am. Potato J. 57, 205–211.

Semancik, J.S., 1980. Citrus exocortis viroid. CMI/AAB Descr. Pl. Viruses. No. 226, 4.

Simmons, H.E., Ruchti, T.B., Munkvold, G.P., 2015. Frequencies of seed infection and transmission to seedlings by potato spindle tuber viroid (a pospiviroid) in tomato. J. Plant Pathol. Microbiol. 6, 275. Available from: http://dx.doi.org/10.4172/2157-7471.1000275.

Singh, R.P., 1970. Seed transmission of potato spindle tuber virus in tomato and potato. Am. Potato J. 47, 225–227.

Singh, R.P., 2014. The discovery and eradication of potato spindle tuber viroid in Canada. Virus Dis. 25, 415–424.

Singh, R.P., Dilworth, A.D., 2009. Tomato chlorotic dwarf viroid in the ornamental plant *Vinca minor* and its transmission through tomato seed. Eur. J. Plant Pathol. 123, 111–116.

Singh, R.P., Boucher, A., Singh, A., 1991. High incidence of transmission and occurrence of a viroid in commercial seeds of *Coleus* in Canada. Plant Dis. 75, 184–187.

Singh, R.P., Boucher, A., Somerville, T.H., 1992. Detection of potato spindle tuber viroid in the pollen and various parts of potato plant pollinated with viroid-infected pollen. Plant Dis. 76, 951–953.

Singh, R.P., Dilworth, A.D., Ao, X., Singh, M., Baranwal, V.K., 2009. Citrus exocortis viroid transmission through commercially-distributed seeds of *Impatiens* and *Verbena* plants. Eur. J. Plant Pathol. 124, 691–694.

Syller, J., Marczewski, W., 2001. Potato leafroll virus-assisted aphid transmission of potato spindle tuber viroid to potato leafroll virus-resistant potato. J. Phytopathol. 149, 195–201.

Syller, J., Marczewski, W., Pawlowicz, J., 1997. Transmission by aphids of potato spindle tuber viroid encapsidated by potato leafroll luteovirus particles. Eur. J. Plant Pathol. 103, 285–289.

Tsutsumi, N., Yanagisawa, H., Fujiwara, Y., Ohara, T., 2010. Detection of potato spindle tuber viroid by reverse transcription loop-mediated isothermal amplification. Res. Bull. Pl. Prot. Japan 46, 61–67.

Van Bogaert, N., De Jonghe, K., Van Damme, E.J.M., Maes, M., Smagghe, G., 2015. Quantitation and localization of pospiviroids in aphids. J. Virol. Methods 211, 51–54.

Van Brunschot, S.L., Verhoeven, J.Th.J., Persley, D.M., Geering, A.D.W., Drenth, A., et al., 2014. An outbreak of potato spindle tuber viroid in tomato is linked to imported seed. Eur. J. Plant Pathol. 139, 1–7.

Verhoeven, J.Th.J., Jansen, C.C.C., Willemen, T.M., Kox, L.F.F., Owens, R.A., et al., 2004. Natural infection of tomato by citrus exocortis viroid, columnea latent viroid, potato spindle tuber viroid and tomato chlorotic dwarf viroid. Eur. J. Plant Pathol. 110, 823–831.

Verhoeven, J.Th.J., Jansen, C.C.C., Roenhorst, J.W., Flores, R., de la Peña, M., 2009. Pepper chat fruit viroid: biological and molecular properties of a proposed new species of the genus *Pospiviroid*. Virus Res. 144, 209–214.

Verhoeven, J.Th.J., Botermans, M., Jansen, C.C.C., Roenhorst, J.W., 2011. First report of pepper chat fruit viroid in capsicum pepper in Canada. New Disease Rep. 23, 15.

Walia, Y., Dhir, S., Zaidi, A.A., Hallan, V., 2015. Apple scar skin viroid naked RNA is actively transmitted by the whitefly *Trialeurodes vaporariorum*. RNA Biol. 12, 1131–1138.

Wallace, J.M., Drake, R.J., 1953. Seed transmission of the avocado sunblotch virus. Citrus Leaves 33, 18.

Wallace, J.M., Drake, R.J., 1962. A high rate of seed transmission of avocado sun-blotch virus from symptomless trees and the origin of such trees. Phytopathology 52, 237–241.

Wan Chow Wah, Y.F., Symons, R.H., 1999. Transmission of viroids via grape seeds. J. Phytopathol. 147, 285–291.

Zhu, Y., Qi, Y., Xun, Y., Owens, R., Ding, B., 2002. Movement of potato spindle tuber viroid reveals regulatory points of phloem-mediated RNA traffic. Plant Physiol. 130, 138–146.

CHAPTER

49

Genome Editing by CRISPR-Based Technology: Potential Applications for Viroids

Ahmed Hadidi[1] *and Ricardo Flores*[2]

[1]U.S. Department of Agriculture, Beltsville, MD, United States
[2]Polytechnic University of Valencia-CSIC, Valencia, Spain

INTRODUCTION

Genome editing is an approach in which a genome sequence is directly changed by adding, replacing, or removing DNA bases. A nuclease is directed to the desired genome location where a DNA break is introduced and for this purpose different types of nucleases have been developed. All nucleases are composed of two components: the major component is the nuclease itself (the catalytic domain), which cleaves the DNA, while the secondary component recognizes and binds DNA in a sequence-specific or nonspecific manner. There are three main classes of nucleases engineered for genome editing purposes that can be used to knock out genes or to introduce designed sequences into the genome. Two of the three classes are zinc-finger nucleases (ZFNs) and transcription activator-like effector nucleases (TALENs) (Gaj et al., 2013). ZFNs and TALENs induce DNA double-strand breaks that stimulate nonhomologous end joining or homology-directed repair at targeted genomic sites, and they require protein engineering to bind to the desired DNA sequence (Gaj et al., 2013). The third class is the clustered, regularly interspaced, short palindromic repeats (CRISPR)/Cas system, which employs a nuclease called Cas 9 to introduce a dsDNA cleavage (Doudna and Charpentier, 2014). Unlike ZFNs or TALENs,

Viroids and Satellites.
DOI: http://dx.doi.org/10.1016/B978-0-12-801498-1.00049-8

this system does not use a protein-based DNA recognition domain. The Cas 9 nuclease is guided to the target DNA binding site by an RNA sequence that is designed to precisely bind to a complementary DNA sequence, allowing for the Cas9 nuclease to make a specific cut. Thus, with CRISPR-Cas, a synthesized guide RNA is needed. Additional advantages of the CRISPR-Cas system over the other genome editing systems are: (1) multiple genomic sites may be edited at once; (2) only a few manipulating tools are required, making it faster and easier to use; (3) it is not species-specific as some other genome engineering systems are; and (4) it can be used to manipulate a number of different genes at a time. Moreover, the ability to redirect the dsDNA targeting capability of CRISPR-Cas9 for RNA-guided single-strand RNA binding and/or cleavage (which is denoted RCas9, an RNA targeting Cas9) has been shown (Abudayyeh et al., 2016; East-Seletsky et al., 2016; O'Connell et al., 2014) and the successful application of this system for targeting human hepatitis C virus RNA in eukaryotic cells has been accomplished (Price et al., 2015). Recently, the nuclease Cpf1 has been reported as an alternative to the Cas9 enzyme (Zetsche et al., 2015). CRISPR-Cpf1 is smaller in size than CRISPR/Cas9 and it was documented to make genome editing easier and more precise.

CRISPR/CAS9 EDITING CONFERS RESISTANCE TO PLANT VIRUSES

DNA Geminiviruses

Current conventional and molecular strategies of controlling DNA geminiviruses have met with marginal success (Aragao and Faria, 2009; Lapidot et al., 2015; Pilartz and Jeske, 1992; Reyes et al., 2013; Vanderschuren et al., 2007; Yang et al., 2014). Recently, the application of the CRISPR-Cas systems to geminiviruses has been shown to enhance resistance to tomato yellow leaf curl virus (genus *Begomovirus*) in *Nicotiana benthamiana* (Ali et al., 2015), bean yellow dwarf virus (genus *Mastrevirus*) in *N. benthamiana* (Baltes et al., 2015), and beet severe curly top virus (genus *Curtovirus*) in *N. benthamiana* and in *Arabidopsis* (Ji et al., 2015). Ali et al. (2015) were also successful in enhancing *N. benthamiana* resistance to three different geminiviruses simultaneously.

RNA Viruses

Recessive plant genes, such as those coding for translation initiation factors, have been reported to confer host resistance to infection by

RNA viruses (Truniger and Aranda, 2009; Sanfaçon, 2015). The translation initiation factor eIF4E and its isoform have been shown to interact with the small VPg protein of members of the families *Potyviridae* and *Secoviridae* and of the genus *Sobemovirus* in infected plant cells. Disrupting the covalent linkage of VPg to the viral RNA 5′-terminus by mutagenesis or silencing interferes with virus infectivity (Sanfaçon, 2015).

The CRISPR-Cas9 system has been used successfully recently to target two sites of eIF4E gene function in cucumber to develop plants resistance to infection by three positive strand RNA viruses, namely cucumber vein yellowing virus (family: *Potyviridae,* genus: *Ipomovirus*), and papaya ringspot virus-w and zucchini yellow mosaic virus (family: *Potyviridae,* genus: *Potyvirus*) (Chandrasekaran et al., 2016). Complete resistance in *Arabidopsis* to turnip mosaic virus (genus: *Potyvirus*) using a CRISPR-Cas9 system targeting the plant eIF (iso) 4E gene has also been reported (Pyott et al., 2016).

INTRODUCING VIROID RESISTANCE IN PLANTS

No plant genes have been reported to confer host resistance to infection by viroids. Thus, conventional plant breeding has not been used for producing varieties resistant to viroids. Moreover, since stable and durable resistance to viroid infection has not been found in most viroid host species, different molecular strategies have been applied to introduce viroid resistance in these plants (Dalakouras et al., 2015; Kovalskaya and Hammond, 2014; Chapter 42: Strategies to Introduce Resistance to Viroids). These strategies have been partially successful and new molecular approaches for controlling viroid diseases by genome editing of viroid RNAs using CRISPR/Cas9 or CRISPR/Cpf1 systems could be useful, as they have been in controlling plant viruses (Hadidi et al., 2016) as well as DNA and RNA human viruses (Price et al., 2015; Zhen et al., 2015).

TARGETING VIROID SPECIFIC NUCLEOTIDE SITES FOR GENOME EDITING

Because the genome information in viroids is very much compressed as a consequence of their small size, most artificial mutations are expected to be deleterious, particularly when affecting a critical function. By the way of example we highlight below in three representative viroids some functionally relevant motifs that could be specifically targeted by CRISPR-based technologies.

Potato Spindle Tuber Viroid

Five sequence/structural domains were recognized in 1985 in potato spindle tuber viroid (PSTVd) and related viroids that are responsible for specific functions presumably through interaction with host factors (Keese and Symons, 1985). They are: the terminal left (TL), pathogenicity (P), variable (V), central (C) and its innermost central conserved region (CCR), and the terminal right (TR) domains.

PSTVd TL domain

Initiation Site of PSTVd (−) RNA Synthesis

The initiation of de novo-synthesized PSTVd (−) RNA in vitro has been mapped at nucleotide position U359 or C1 at the hairpin loop of the TL domain of PSTVd (+) RNA (Kolonko et al., 2006). The proposed initiation site is also supported by infectivity and replication ability of site-directed PSTVd (+) RNA mutants.

Bulges Associated With PSTVd (+) RNA Replication

Three bulges in the TL domain of PSTVd (+) strand are required for interaction in vitro with the host RNA polymerase II involved in viroid replication (Bojić et al., 2012). Sequence changes of these bulges reduce or prevent binding with the same polymerase, which was not found to interact with PSTVd (−) strand (Bojić et al., 2012).

PSTVd CCR

Hairpin I

This motif (HP I), located between positions 79 and 110, is formed by the nucleotides of the upper CCR strand and their flanking inverted repeats (Riesner et al., 1979). In oligomeric PSTVd (+) strands, HP I potentially facilitates the adoption of a conserved double-stranded structure that is the substrate for their in vivo cleavage (Gas et al., 2007).

Hairpin II

The hairpin II (HP II) is located between positions 227 and 328 on the lower strand of PSTVd (Riesner et al., 1979). It is critical for viroid infectivity (Loss et al., 1991), and restoring the correct base-pairing of the G-C-rich stem of HP II in a non-infectious recombinant PSTVd led to recovery of infectivity (Candresse et al., 2001). HP II has been directly detected in vitro and in vivo, and suggested to be an essential element for PSTVd replicative intermediates (Schroder and Riesner, 2002).

Loop E

Transcription factor for polymerase IIIA (TFIIIA) is a zinc-finger protein that specifically binds to a sequence about 55 bp in the eukaryotic 5S DNA (Brown and Flint, 2005; Ryan and Darby, 1998). It is localized in the nucleolus where rRNA genes are transcribed (Brown and Flint, 2005). PSTVd, which also accumulates preferentially in the nucleolus (Harders et al., 1989), contains in vivo a loop E motif (Eiras et al., 2007; Wang et al., 2007) that is also present in 5S rRNA (Branch et al., 1985). Loop E is located at the CCR between positions 5′-G97 to C103–3′ and 5′-G255 to C262–3′ (Branch et al., 1985; Zhong et al., 2006). PSTVd (+) RNA has been reported to interact in vitro with TFIIIA from *Arabidopsis thaliana* (Eiras et al., 2011) and in vitro and in vivo with the same transcription factor from *N. benthamiana* (Wang et al., 2016). Its downregulation and overexpression result in reduced and increased accumulation of PSTVd, respectively (Wang et al., 2016). Loop E has been involved in multiple functional roles including replication (Zhong et al., 2006, 2008) and, particularly, ligation (Gas et al., 2007).

PSTVd TR Domain

PSTVd RY Motif

Tomato plant nuclei contain the so-called viroid-binding protein 1 (VirP1). It was isolated by its specific interaction with PSTVd (+) RNA in vitro and in vivo and it has been proposed to play a role in PSTVd infection by transferring the viroid molecules to host nuclei and bringing them into contact with chromatin (Martínez de Alba et al., 2003). VirP1 binding to PSTVd requires the presence of an "RY motif" located in the TR domain close to the TR hairpin of the rod-like conformation proposed for the viroid (Gozmanova et al., 2003). Two asymmetric internal PSTVd loops, with sequence elements 5′-ACAGG-3′ (positions 173–177) in the upper strand and 3′-CUCUUCC-5′ (positions 190–184) in the lower strand, form the RY motif and sequence alteration in this motif abolishes viroid binding to VirP1 (Gozmanova et al., 2003). For instance, PSTVd with A173 changed to G173 was not infectious (Gora-Sochacka et al., 1997) and a PSTVd mutant defective for the VirP1 binding motif remained localized and did not spread systemically in the host plant (Hammond, 1994). Moreover, protoplast transfection has revealed that VirP1-suppressed cells cannot sustain viroid replication (Kalantidis et al., 2007).

Peach Latent Mosaic Viroid

Initiation Sites for PLMVd (+) and (−) RNA Strands

The in vivo initiation site for peach latent mosaic viroid (PLMVd) (+) RNA is at C51 and at U286 for the viroid (−) RNA, with these sites mapping at similar double-stranded motifs of 6−7 bp that also contain the highly conserved GUC triplet preceding the self-cleavage site in both polarity strands. Within the branched secondary structures predicted for the two PLMVd strands, these motifs are located at the base of a similar long hairpin that presumably contains the promoters for a chloroplastic RNA polymerase (Delgado et al., 2005). Similar positions for the viroid initiation sites were also reported: A50/C51 for PLMVd (+) strand and U284 for PLMVd (−) strand (Motard et al., 2008). The minor differences in location of the initiation sites could be due to primary and secondary structural differences among PLMVd variants near the sites, which are located within, or very close to, a highly conserved CAGACG box that may contribute to their positioning (Motard et al., 2008). Since the in vivo initiation of PLMVd RNAs occurs near the self-cleavage (and self-ligation) sites, any CRISPR/Cas editing in this region would attenuate both the transcription and processing of viroid RNAs.

PLMVd Interacting Proteins

PLMVd has been reported to interact in vitro with six proteins isolated from peach leaf tissue. The eukaryotic translation elongation factor eEF1A is one of these proteins, which also interacts in vitro with both PSTVd and citrus exocortis viroid (Dubé et al., 2009). Further investigation is needed to identify the viroid sequences that interact with eEF1A and the role in viroid infection of eEF1A. This protein also interacts with the 3′-terminus of the genomic RNA of tobacco mosaic virus and with the RNA-dependent RNA polymerase mediating its replication. Moreover, downregulation of eEF1A mRNA levels by virus-induced gene silencing reduced accumulation of viral RNA and virus spread, suggesting a fundamental role of eEF1A in infection by tobacco mosaic virus (Yamaji et al., 2010).

Avocado Sunblotch Viroid (ASBVd)

A single initiation site of ASBVd (+) strand has been mapped in vivo at position U121 in the A + U-rich right terminal loop of the viroid (+) RNA secondary structure, while U119 is the single initiation site of the ASBVd (−) strand located in an A + U-rich terminal loop structurally similar to that containing the initiation site of the ASBVd (+) strand (Navarro and Flores, 2000). The initiation sites of the viroid (+) and (−)

strands are only two residues apart and each site starts with the same sequence, UAAAA, which suggests that the viroid promoters are formed, at least in part, by the sequences flanking the two initiation sites (Navarro and Flores, 2000).

FINAL REMARKS

CRISPR-Cas9 genome editing technology for developing resistance to viroids or other plant pathogens of plants producing food or any other valuable items, does not incorporate foreign DNA from plant pathogens into the host genome as in transgenic plants. Because of this feature, the U.S. Department of Agriculture (USDA) has recently announced that it would not regulate foods or plants developed by CRISPR-Cas9 genome editing (Anonymous, 2016; Waltz, 2016), thus adding extra value to modified plants generated by this technology.

Acknowledgments

Research in R.F. laboratory is currently funded by grant BFU2014–56812-P from the Spanish Ministerio de Economía y Competitividad (MINECO).

References

Abudayyeh, O.O., Gootenberg, J.S., Konermann, S., Joung, J., Slaymaker, I.M., Cox, D.B.T., et al., 2016. C2c2 is a single-component programmable RNA-guided RNA-targeting CRISPR effector. Science 353, aaf5573. Available from: http://dx.doi.org/10.1126/science.aaf5573.

Ali, Z., Abulfaraj, A., Idris, A., Ali, S., Tashkandi, M., Mahfouz, M.M., 2015. CRISPR/Cas9-mediated viral interference in plants. Genome Biol. 16, 238.

Anonymous, 2016. CRISPR foods dodge USDA. Science 352, 388.

Aragao, F.J.L., Faria, J.C., 2009. First transgenic geminivirus-resistant plant in the field. Nat. Biotechnol. 27, 1086–1088.

Baltes, N.J., Hummel, A.W., Konecna, E., Cegan, R., Bruns, A.N., Bisaro, D.M., et al., 2015. Conferring resistance to geminiviruses with the CRISPR-Cas prokaryotic immune system. Nat. Plants. 1, 15145.

Bojić, T., Beeharry, Y., Zhang, D.J., Pelchat, M., 2012. Tomato RNA polymerase II interacts with the rod-like conformation of the left terminal domain of the potato spindle tuber viroid positive RNA genome. J. Gen. Virol. 93, 1591–1600.

Branch, A.D., Benenfeld, B.J., Robertson, H.D., 1985. Ultraviolet light-induced crosslinking reveals a unique region of local tertiary structure in potato spindle tuber viroid and HeLa 5S RNA. Proc. Natl. Acad. Sci. USA 82, 6590–6594.

Brown, R.S., Flint, J., 2005. TFIIIA: a sophisticated zinc finger protein. In: Luchi, S., Kuldell, N. (Eds.), Zinc Finger Proteins. Springer, USA, pp. 14–19.

Candresse, T., Gora-Sochacka, A., Zagorcki, W., 2001. Restoration of secondary hairpin II is associated with restoration of infectivity of non-viable recombinant viroid. Virus Res. 75, 29–34.

Chandrasekaran, J., Brumin, M., Wolf, D., Leibman, D., Klap, C., Pearlsman, M., et al., 2016. Development of broad virus resistance in non-transgenic cucumber using CRISPR/Cas9 technology. Mol. Plant Pathol. 17, 1140–1153.

Dalakouras, A., Dadami, E., Wassenegger, M., 2015. Engineering viroid resistance. Viruses 7, 634–646.

Delgado, S., Martínez de Alba, E., Hernández, C., Flores, R., 2005. A short double-stranded RNA motif of peach latent mosaic viroid contains the initiation and the self-cleavage sites of both polarity strands. J. Virol. 79, 12934–12943.

Doudna, J.A., Charpentier, E., 2014. The new frontier of genome engineering with CRISPR-Cas9. Science 346, 1258096.

Dubé, A., Bisaillon, M., Perreault, J.P., 2009. Identification of proteins from *Prunus persica* that interact with peach latent mosaic viroid. J. Virol. 83, 12057–12067.

East-Seletsky, A., O'Connell, M.R., Knight, S.C., Burstein, D., Cate, J.H.D., Tjian, R., et al., 2016. Two distinct RNase activities of CRISPR-C2c2 enable guide-RNA processing and RNA detection. Nature 538, 270–273. Available from: http://dx.doi.org/10.1038/nature, 19802.

Eiras, M., Kitajima, E.W., Flores, R., Daròs, J.A., 2007. Existence in vivo of the loop E motif in potato spindle tuber viroid RNA. Arch. Virol. 152, 1389–1393.

Eiras, M., Nohales, M.A., Kitajima, E.W., Flores, R., Daròs, J.A., 2011. Ribosomal protein L5 and transcription factor IIIA from *Arabidopsis thaliana* bind *in vitro* specifically potato spindle tuber viroid RNA. Arch. Virol. 156, 529–533.

Gaj, T., Gersbach, C.A., Barbas III, C.F., 2013. ZFN, TALEN, and CRISPR/Cas-based methods for genome engineering. Trends Biotechnol. 31, 397–405.

Gas, M.E., Hernández, C., Flores, R., Daròs, J.A., 2007. Processing of nuclear viroids in vivo: an interplay between RNA conformations. PLoS Pathog. 3, 1813–1826.

Gora-Sochacka, A., Kierzek, A., Candresse, T., Zagorski, W., 1997. The genetic stability of potato spindle tuber viroid (PSTVd) molecular variants. RNA. 3, 68–74.

Gozmanova, M., Denti, M.A., Minkov, I.N., Tsagris, M., Tabler, M., 2003. Characterization of the RNA motif responsible for the specific interaction of potato spindle tuber viroid RNA (PSTVd) and the tomato protein Virp1. Nucleic Acids Res. 31, 5534–5543.

Hadidi, A., Flores, R., Candresse, T., Barba, M., 2016. Next-generation sequencing and genome editing in plant virology. Front. Microbiol. 7, 1325.

Hammond, R.W., 1994. *Agrobacterium*-mediated inoculation of PSTVd cDNAs onto tomato reveals the biological effect of apparently lethal mutations. Virology 201, 36–45.

Harders, J., Lukács, N., Robert-Nicoud, M., Riesner, D., 1989. Imaging of viroids in nuclei from tomato leaf tissue by in situ hybridization and confocal laser scanning microscopy. EMBO J. 8, 3941–3949.

Ji, X., Zhang, H., Zhang, Y., Wang, Y., Gao, C., 2015. Establishing a CRISPR-Cas-like immune system conferring DNA virus resistance in plants. Nat. Plants. 1, 15144.

Kalantidis, K., Denti, M.A., Tzortzakaki, S., Marinou, E., Tabler, M., Tsagris, M., 2007. Virp1 is a host protein with a major role in potato spindle tuber viroid infection in Nicotiana plants. J. Virol. 81, 12872–12880.

Keese, P., Symon, R.H., 1985. Domains in viroids: evidence of intermolecular RNA rearrangements and their contribution to viroid evolution. Proc. Natl. Acad. Sci. USA 82, 4582–4586.

Kolonko, N., Bannach, O., Aschermann, K., Hu, K.H., Moors, M., Schmitz, M., et al., 2006. Transcription of potato spindle tuber viroid by RNA polymerase II starts in the left terminal loop. Virology 347, 392–404.

Kovalskaya, N., Hammond, R.W., 2014. Molecular biology of viroid-host interactions and disease control strategies. Plant Sci. 228, 48–60.

Lapidot, M., Karniel, U., Gelbart, D., Fogel, D., Evenor, D., Kutsher, Y., et al., 2015. A novel route controlling begomovirus resistance by the messenger RNA surveillance factor pelota. PLoS Genet. 11, e1005538.

Loss, P., Schmitz, M., Steger, G., Riesner, D., 1991. Formation of a thermodynamically metastable structure containing hairpin II is critical for infectivity of potato spindle tuber viroid RNA. EMBO J. 10, 719–727.

Martínez de Alba, A.E., Sägesser, R., Tabler, M., Tsagris, M., 2003. A bromodomain-containing protein from tomato specifically binds potato spindle tuber viroid RNA *in vitro* and *in vivo*. J. Virol. 77, 9685–9694.

Motard, J., Boldue, F., Thompson, D., Perreault, J.-P., 2008. The peach latent mosaic viroid replication initiation site is located at a universal position that appears to be defined by a conserved sequence. Virology 373, 362–375.

Navarro, J.A., Flores, R., 2000. Characterization of the initiation sites of both polarity strands of a viroid RNA reveals a motif conserved in sequence and structure. EMBO J. 19, 2662–2670.

O'Connell, M.R., Oakes, B.L., Sternberg, S.H., East-Saletsky, A., Kaplan, M., Doudna, J.A., 2014. Programmable RNA recognition and cleavage by CRISPR/Cas9. Nature 516, 263–266.

Pilartz, M., Jeske, H., 1992. Abutilon mosaic geminivirus double-stranded DNA is packed into minichromosomes. Virology 189, 800–802.

Price, A.A., Sampson, T.R., Ratner, H.K., Grakoui, A., Weiss, D.S., 2015. Cas9-mediated targeting of viral RNA in eukaryotic cells. Proc. Natl. Acad. Sci. USA 112, 6164–6169.

Pyott, D.E., Sheehan, E., Molnar, A., 2016. Engineering of CRISPR/Cas9-mediated potyvirus resistance in transgene-free *Arabidopsis* plants. Mol. Plant Pathol. 17, 1276–1288.

Reyes, M.I., Nash, T.E., Dallas, M.M., Ascencio-Ibanez, J.T., Hanley-Bowdoin, L., 2013. Peptide aptamers that bind to geminivirus replication proteins confer a resistance phenotype to tomato yellow leaf curl virus and tomato mottle virus infection in tomato. J. Virol. 87, 9691–9706.

Riesner, D., Henco, K., Rokohl, U., Klotz, G., Kleinschmidt, A.K., Domdey, H., et al., 1979. Structure and structure formation of viroids. J. Mol. Biol. 133, 85–115.

Ryan, R.F., Darby, M.K., 1998. The role of zinc finger linkers in p43 and TFIIIA binding to 5S rRNA and DNA. Nucleic Acids Res. 26, 703–709.

Sanfaçon, H., 2015. Plant translation factors and virus resistance. Viruses 7, 3392–3419.

Schroder, A.R.W., Riesner, D., 2002. Detection and analysis of hairpin II, an essential metastable structural element in viroid replication intermediates. Nucleic Acids Res. 30, 3349–3359.

Truniger, V., Aranda, M.A., 2009. Recessive resistance to plant viruses Adv. Virus Res. 75, 119–159.

Vanderschuren, H., Stupak, M., Futterer, J., Gruissem, W., Zhang, P., 2007. Engineering resistance to geminiviruses—review and perspectives. Plant Biotechnol. J. 5, 207–220.

Waltz, E., 2016. CRISPR-edited crops free to enter market, skip regulation. Nat. Biotech. 34, 582.

Wang, Y., Qu, J., Ji, S., Wallace, A.J., Wu, J., Li, Y., et al., 2016. A land plant-specific transcription factor directly enhances transcription of a pathogenic noncoding RNA template by DNA-dependent RNA polymerase II. Plant Cell. 28, 1094–1107.

Wang, Y., Zhong, X., Itaya, A., Ding, B., 2007. Evidence for the existence of the loop E motif of potato spindle tuber in vivo. J. Virol. 81, 2074–2077.

Yamaji, Y., Sakurai, K., Hamada, K., Komatsu, K., Ozeki, J., Yoshida, A., et al., 2010. Significance of eukaryotic translation elongation factor 1A in tobacco mosaic virus infection. Arch. Virol. 155, 263–268.

Yang, C.-F., Chen, K.-C., Cheng, Y.-H., Raja, J.A.J., Huang, Y.-L., Chien, W.-C., et al., 2014. Generation of marker-free transgenic plants concurrently resistant to a DNA geminivirus and a RNA tospovirus. Sci. Rep. 4, 5717.

Zetsche, B., Gootenberg, J.S., Abudayyeh, O.O., Slaymaker, I.M., Makarova, K.S., Essletzbichler, P., et al., 2015. Cpf1 is a single RNA-guided endonuclease of a class 2 CRISPR-Cas system. Cell 163, 759–771.

Zhen, S., Hua, L., Liu, Y.H., Gao, L.C., Fu, J., Wan, D.Y., et al., 2015. Harnessing the clustered regularly interspaced short palindromic repeat (CRISPR)/CRISPR-associated Cas9 system to disrupt the hepatitis B virus. Gene Ther. 22, 404–412.
Zhong, X., Archual, A.J., Amin, A.A., Ding, B.A., 2008. Genomic map of viroid RNA motifs critical for replication and systemic trafficking. Plant Cell. 20, 35–47.
Zhong, X., Leontis, N., Qiang, S., Itaya, A., Qi, Y., Boris-Lawrie, K., et al., 2006. Tertiary structural and functional analysis of a viroid RNA motif by isostericity matrix and mutagenesis reveal its essential role in replication. J. Virol. 80, 8566–8581.

SECTION II

SATELLITES

PART VIII

INTRODUCTION

CHAPTER

50

Satellite Viruses and Satellite Nucleic Acids

Peter Palukaitis

Seoul Women's University, Seoul, South Korea

INTRODUCTION

What constitutes a viral satellite? The definition has changed somewhat since the first use of the term "satellite virus" to describe an extra RNA encapsidated by a novel capsid protein into a smaller virus particle associated with and dependent upon another virus for replication and probably movement, but which was not required for the infection of the helper virus (tobacco necrosis virus; TNV) (Kassanis, 1962). Some years later, a RNA molecule dependent upon a helper virus for replication and movement, as well as encapsidation, was described in association with tobacco ringspot virus (TRSV) and was given the name "satellite RNA" (Schneider et al., 1972). (Note: Two papers first describing properties of unusual smaller-sized virions associated with TNV were published earlier by Kassanis and Nixon (1960, 1961), as were two papers describing a satellite-like agent associated with TRSV by Schneider (1969, 1971). However, in both cases, these agents were not given these current names until the corresponding third papers were published, at which point sufficient evidence had accumulated as to their general nature.) DNA satellites of geminiviruses also were discovered, although the first one identified (Dry et al., 1997) was later described as being a defective form of a larger satellite DNA designated DNAβ (Saunders et al., 2000), one of many since described and now referred to as betasatellites (see Chapter 62: Betasatellites of Begomoviruses).

As more satellites were described, they were considered unique and different from either subgenomic RNAs or defective (interfering)

Viroids and Satellites.
DOI: http://dx.doi.org/10.1016/B978-0-12-801498-1.00050-4

RNAs/DNAs of viruses, which had sequences identical to and derived from part(s) of the viral genome (see Fig. 1 of Hu et al., 2009). This still holds true for most satellites; however, there are exceptions, since some satellites also contain additional sequence similarity to their helper viruses. In particular, this applies to two satellites: (1) a satellite-like chimera called satC of turnip crinkle virus (TCV), consisting of a 5′ half derived from a satellite RNA (similar to satD of TCV), fused to sequences derived from the 3′ nontranslated region of the helper virus genomic RNA (Simon and Howell, 1986) and (2) the satellite RNA of cymbidium ringspot virus, which contains five stretches of 8–10 nt and one stretch of 48 out of 51 nt identical to the helper virus (Rubino et al., 1990).

In addition, the original definition of satellites, based on the limited examples known at the time, implied that they were not necessary for the infection cycle of the helper virus and thus they were completely supernumerary (Francki, 1985; Fritsch and Mayo, 1989; Murant and Mayo, 1982). This view also changed to some extent when some satellite-like RNAs were shown to be required for either specific movement within the plant or plant-to-plant transmission, such as RNAs 3 and 4 of beet necrotic yellow vein virus (BNYVV; Peltier et al., 2008) or the satellite-like RNAs of groundnut rosette virus (GRV; Taliansky et al., 2000). Thus, BNYVV RNAs 3 and 4 are no longer considered satellite-like, but the various GRV satellite-like RNAs are still considered satellite-like, even though they are required for aphid transmission (Simon et al., 2004; Taliansky et al., 2000), an essential function of the infection cycle. BNYVV RNA 5 is still considered to be a satellite-like RNA despite it being unnecessary for the infection cycle of BNYVV, and many isolates do not have this RNA (Chiba et al., 2011). Its satellite-like status is due to the observation that the BNYVV RNA 5-encoded protein (p26) is more similar (43%) to the p29 protein encoded by RNA 3 of the related benyvirus, beet soil-borne mosaic virus (BSBMV), than the BSBMV-encoded p29 is to the p25 protein encoded by the corresponding BNYVV RNA 3 (23%) (Ratti et al., 2009). Since the BSBMV RNA 3 can substitute for BNYVV RNA 3 in promoting systemic infection by BNYVV RNAs 1 and 2 (Ratti et al., 2009), it appears as if BNYVV RNA 5 is in evolutionary transition from having been a genomic component of one virus to becoming a satellite RNA of another virus!

The term virusoid has been used to describe small, circular satellite RNAs (reviewed in Symons and Randles, 1999). These circular satellite RNAs, all associated with viruses that are members of the genus *Sobemovirus*, have viroid-like structures and contain ribozymes involved in the self-cleavage of the concatemeric RNAs produced during their rolling-circle replication. The term virusoid was used first (Haseloff et al., 1982) to describe highly-structured circular RNAs similar to viroids, packaged by viruses in vivo, but not capable of independent replication, although whether or not these RNAs were satellite RNAs was not known conclusively at that time. Later, when this distinction was clear, the term virusoid was used to discriminate the sobemovirus satellite RNAs, all of which adopted a circular form of the satellite RNA, from the other structurally similar satellite RNAs that adopted a linear form of the satellite RNA, such as the satellite RNA of TRSV (as described in Symons and Randles, 1999). This distinction is not considered in the taxonomic characterization of different classes of satellite RNAs, where all satellite RNAs containing a circular state at some point in their infection cycles are grouped together (Briddon et al., 2012). Nevertheless, the term virusoid still appears occasionally in the literature.

Based on observations of many satellites, but in a limited number of hosts, different satellites were found to have one of three biological effects: (1) amelioration of the disease induced by the helper virus, which usually, but not always, coincided with a reduction in the titer of the helper virus (e.g., Gallitelli and Hull, 1985; Rasochová and Miller, 1996; Schneider, 1969, 1971; Waterworth et al., 1979); (2) exacerbation of the disease induced by the helper virus (e.g., Davies and Clark, 1983; Gonsalves et al., 1982; Kaper and Waterworth, 1977; Takanami, 1981); or (3) none (reviewed in Roossinck et al., 1992). These effects also could be host-specific (see Chapter 52: Biology of Satellites). Thus, satellites were generally thought of as parasitic and sometimes disease inducing. More recently, however, a third relationship has been considered—mutualism, in which the satellite has some effect to the benefit of the helper virus (Simon et al., 2004). This will be considered further in this book (see Chapter 52: Biology of Satellites and Chapter 62: Betasatellites of Begomoviruses).

Other plant viral agents that have been described as satellite-like, such as the so-called alphasatellites of both geminiviruses (Zhou, 2013) and nanoviruses (Mandal, 2010), are small circular DNAs that are dependent upon their associated virus for encapsidation and movement/transmission functions, but are capable of autonomous replication, and therefore are not true satellites. Rather, these "alphasatellites" are similar to the polerovirus-associated RNAs, beet western yellows virus-associated RNA, carrot red leaf virus-associated RNA, and tobacco vein-distorting virus-associated RNA, all of which also are dependent upon a helper virus for their encapsidation and transmission, but replicate independently (Briddon et al., 2012). Hence, these various other subviral, dependent, infectious agents will not be covered in this book.

DETECTION AND IDENTIFICATION OF SATELLITES

How do you know when a viral satellite is present? It depends on whether you are looking for it or not! If you are not expecting satellites to be present, then your first clue may be some change in pathogenicity (see Chapter 54: Satellite RNAs: Their Involvement in Pathogenesis and RNA Silencing), although this could also simply be due to some mutation in the virus itself, affecting pathogenicity, host range, or tropism. However, some satellites do not affect the pathogenicity of the helper virus. In the case of a satellite virus, there may be smaller particles present [observed by either electron microscopy (Kassanis and Nixon, 1961; Kassanis et al., 1970) or sedimentation in gradients (Kassanis and Nixon, 1960, 1961)], which are absent from other sources of the same virus and do not react with antiserum to the helper virus (Kassanis and Nixon, 1961). For well-known satellites, detection can be done rapidly by the reverse transcription polymerase chain reaction (Celix et al., 1997; Nolasco et al., 1993); however, this method cannot detect novel satellites present in infected plants, and thus classical, or more general, methods are needed in the first instance. Extra bands may be observed after gel electrophoresis of either nucleic acids extracted from virus particles (Gould et al., 1978), or double-stranded RNAs extracted from infected leaves (Dodds et al., 1984). This was the most common procedure for detecting satellite RNAs, especially when the RNA was a new satellite. Such bands also may be defective nucleic acids, subgenomic RNAs, degradation products of genomic RNAs, or nicked circular satellites RNAs, but usually one is comparing samples to the same virus without the extra RNA present and this should eliminate the question of subgenomic RNAs (Dodds et al., 1984; Gould et al., 1978). Establishing that the extra nucleic acid is either a satellite or a defective nucleic acid can be done biologically, by purifying the nucleic acid, mixing it with a helper virus isolate that does not contain this extra material, inoculating to plants, and recovering the extra nucleic acid only if the helper virus is present. By contrast, a coinfecting viroid would be able to propagate without the associated virus. Establishing that the extra nucleic acid is a defective form of the virus requires hybridizing a northern/Southern blot with a probe to either the 3′ terminal region of a viral genomic RNA (Hillman et al., 1987) or a common region of a geminiviral DNA (Saunders et al., 2000), since defective nucleic acids always contain such sequences. However, this would also eliminate hybrid molecules, such as satC of TCV, unless the novel satellite RNA sequence was identified as well (Simon and Howell, 1986). Therefore, only by sequencing will the supernumerary nucleic acids be identified as satellite-like, or a defective form of a virus nucleic acid. However, none of the tests listed above would determine whether the extra nucleic acid is a true satellite or a virus-associated nucleic acid, such as the so-called alphasatellite DNAs (Gronenborn, 2004; Mansoor et al., 1999; Saunders and Stanley, 1999) and polerovirus-associated RNAs (Mo et al., 2011; Passmore et al., 1993; Watson et al., 1998). In such cases, replication in the absence of the associated virus, either in agroinfiltrated cells but without a progressive systemic infection, or in protoplasts, would differentiate a satellite from a virus-associated nucleic acid. More recently, deep sequencing of small RNAs (20–30 nt in size) and assembly of overlapping sequences has been used to identify satellite RNAs of bamboo mosaic virus (Lin et al., 2010), a betasatellite DNA of a geminivirus (Yang et al., 2011), and a new satellite-like RNA in grapevines (Wu et al., 2012). In the last case, a program was developed to identify viroids or circular satellite RNAs without prior sequence knowledge, but it could not discriminate between these two subviral pathogens. Another deep sequencing approach, but using enriched double-stranded RNAs as the starting material, was used to identify the presence of a large satellite-like RNA in nucleic acids extracted from grapevines (Al Rwahnih et al., 2013). These methods all require evidence of helper virus-dependent replication before the nucleic acid can be proven to be a satellite. Finally, if the suspected nucleic acid has some role beyond replication in the infection cycle of the virus, such as systemic movement, or transmission, then it may be considered an essential genomic component rather than only satellite-like.

OVERVIEW

This section of the book dealing with plant viral satellites is not organized by particular hosts, individual satellites, or the nature of the helper virus. Rather, it is organized by (Part IX) the impact and economic significance of satellites (see Chapter 51: Economic Significance of Satellites); (Part X) the characteristics of satellites, viz., biology (see Chapter 52: Biology of Plant Viral Satellites), replication (see Chapter 53: Replication of Plant Viral Satellites), pathogenesis and RNA silencing (see Chapter 54:: Plant Viral Satellites: Their Involvement in Pathogenesis and RNA Silencing), satellite RNA-based vectors (see Chapter 55: Development and Application of Satellite-Based Vectors), possible origin and evolution of satellites (see Chapter 56: Origin and Evolution of Satellites), and taxonomy (see Chapter 57: Satellite Taxonomy); (Part XI) the physical types of viral satellites, viz., satellite viruses (see Chapter 58: Biology and Pathogenesis of Plant Satellite Viruses), large satellite RNAs (see Chapter 59: Large Satellite

RNAs), small linear satellite RNAs (see Chapter 60: Small Linear Satellite RNAs), small circular satellite RNAs (see Chapter 61: Small Circular Satellite RNAs), and betasatellite DNAs (see Chapter 62: Betasatellites of Begomoviruses); and (Part XII) the use of satellites as biocontrol agents (see Chapter 63: Viral Satellites as Viral Biocontrol Agents). This organizational approach allows comparisons and contrasts to be made between the numerous satellites described in the literature and should provide the reader with an understanding of the similarities and differences between satellites at both conceptual and specific levels.

Acknowledgments

This work was supported by a special grant from Seoul Women's University.

References

Al Rwahnih, M., Daubert, S., Sudarshana, M.R., Rowhani, A., 2013. Gene for a novel plant virus satellite from grapevine identifies a viral satellite lineage. Virus Genes 47, 114–118.

Briddon, R.W., Ghabrial, S., Lin, N.-S., Palukaitis, P., Scholthof, K.-B.G., Vetten, H.-J., 2012. Satellite and other virus-dependent nucleic acids. In: King, A.M.Q., Adams, M.J., Carstens, E.B., Lefkowitz, E.J. (Eds.), Virus Taxonomy. Classification and Nomenclature of Viruses. Ninth Report of the International Committee on Taxonomy of Viruses. Elsevier Academic Press, San Diego, CA, pp. 1211–1219.

Celix, A., Rodriguez-Cerezo, E., Garcia-Arenal, F., 1997. New satellite RNAs, but no DI RNAs, are found in natural populations of tomato bushy stunt tombusvirus. Virology 239, 277–284.

Chiba, S., Kondo, H., Miyanishi, M., Andika, I.B., Han, C., Tamada, T., 2011. The evolutionary history of beet necrotic yellow vein virus deduced from genetic variation, geographical origin and spread, and the breaking of host resistance. Mol. Plant Microbe Interact. 24, 207–218.

Davies, D.L., Clark, M.F., 1983. A satellite-like nucleic acid of arabis mosaic virus associated with hop nettlehead disease. Ann. Appl. Biol. 103, 439–448.

Dodds, J.A., Morris, T.J., Jordan, R.L., 1984. Plant viral double-stranded RNA. Annu. Rev. Phytopathol. 22, 151–168.

Dry, I.B., Krake, L.R., Rigden, J.E., Rezaian, M.A., 1997. A novel subviral agent associated with a geminivirus: the first report of a DNA satellite. Proc. Natl. Acad. Sci. USA 94, 7088–7093.

Francki, R.I.B., 1985. Plant virus satellites. Annu. Rev. Microbiol. 39, 151–174.

Fritsch, C., Mayo, M.A., 1989. Satellites of plant viruses. In: Mandahar, C.L. (Ed.), Plant Viruses, vol. I. Structure and Replication. CRC Press, Inc, Boca Raton, FL, pp. 289–321.

Gallitelli, D., Hull, R., 1985. Characterization of satellite RNAs associated with tomato bushy stunt virus and five other definitive tombusviruses. J. Gen. Virol. 66, 1533–1543.

Gonsalves, D., Provvidenti, R., Edwards, M.C., 1982. Tomato white leaf: the relation of an apparent satellite RNA and cucumber mosaic virus. Phytopathology 72, 1533–1538.

Gould, A.R., Palukaitis, P., Symons, R.H., Mossop, D.W., 1978. Characterization of a satellite RNA associated with cucumber mosaic virus. Virology 84, 443–455.

Gronenborn, B., 2004. Nanoviruses: genome organisation and protein function. Vet. Microbiol. 98, 103–109.

Haseloff, J., Mohamed, N.A., Symons, R.H., 1982. Viroid RNAs of the cadang-cadang disease of coconuts. Nature 229, 316–321.

Hillman, B.I., Carrington, J.C., Morris, T.J., 1987. A defective interfering RNA that contains a mosaic of a plant virus genome. Cell 51, 427–433.

Hu, C.-C., Hsu, Y.-H., Lin, N.-S., 2009. Satellite RNAs and satellite viruses of plants. Viruses 1, 1325–1350.

Kaper, J.M., Waterworth, H.E., 1977. Cucumber mosaic virus associated RNA 5: causal agent for tomato necrosis. Science 196, 429–431.

Kassanis, B., 1962. Properties and behaviour of a virus depending for its multiplication on another. J. Gen. Microbiol. 27, 477–488.

Kassanis, B., Nixon, H.L., 1960. Activation of one plant virus by another. Nature 187, 713–714.

Kassanis, B., Nixon, H.L., 1961. Activation of one tobacco necrosis virus by another. J. Gen. Microbiol. 25, 459–471.

Kassanis, B., Vince, D.A., Woods, R.D., 1970. Light and electron microscopy of cells infected with tobacco necrosis and satellite viruses. J. Gen. Virol. 7, 143–151.

Lin, K.Y., Cheng, C.-P., Chang, B.C.-H., Wang, W.-C., Huang, Y.-W., Lee, Y.-S., et al., 2010. Global analysis of small interfering RNAs derived from bamboo mosaic virus and its associated satellite RNAs in different plants. PLoS One 5, e11928.

Mandal, B., 2010. Advances in small isometric multicomponent ssDNA viruses infecting plants. Indian J. Virol. 21, 18–30.

Mansoor, S., Khan, S.H., Bashir, A., Saeed, M., Zafar, Y., Malik, K.A., et al., 1999. Identification of a novel circular single-stranded DNA associated with cotton leaf curl disease in Pakistan. Virology 259, 190–199.

Mo, X.-H., Chen, Z.-B., Chen, J.-P., 2011. Molecular identification and phylogenetic analysis of a viral RNA associated with the Chinese tobacco top disease complex. Ann. Appl. Biol. 158, 188–193.

Murant, A.F., Mayo, M.A., 1982. Satellites of plant viruses. Annu. Rev. Phytopathol. 20, 49–70.

Nolasco, G., de Blas, C., Torres, V., Ponz, F., 1993. A method combining immunocapture and PCR amplification in a microtiter plate for the detection of plant viruses and subviral pathogens. J. Virol. Methods 45, 201–218.

Passmore, B.K., Sanger, M., Chin, A.-H., Falk, B.W., Bruening, G., 1993. Beet western yellows virus-associated RNA: an independently replicating RNA that stimulates virus accumulation. Proc. Natl. Acad. Sci. USA 90, 10168–10172.

Peltier, C., Hleibieh, K., Thiel, H., Klein, E., Bragard, C., Gilmer, D., 2008. Molecular biology of beet necrotic yellow vein virus. Plant Viruses 2, 14–24.

Rasochová, L., Miller, W.A., 1996. Satellite RNA of barley yellow dwarf-RPV virus reduces accumulation of RPV helper virus RNA and attenuates RPV symptoms in oats. Mol. Plant Microbe Interact. 9, 646–650.

Ratti, C., Hleibieh, K., Bianchi, L., Schirmer, A., Autonell, C.R., Gilmer, D., 2009. Beet soil-borne mosaic virus RNA-3 is replicated and encapsidated in the presence of BNYVV RNA-1 and -2 and allows long distance movement in *Beta macrocarpa*. Virology 385, 392–399.

Roossinck, M.J., Sleat, D., Palukaitis, P., 1992. Satellite RNAs of plant viruses: structures and biological effects. Microbiol. Rev. 56, 265–279.

Rubino, L., Burgyan, J., Grieco, F., Russo, M., 1990. Sequence analysis of cymbidium ringspot virus satellite and defective interfering RNAs. J. Gen. Virol. 71, 1655–1660.

Saunders, K., Bedford, I.D., Briddon, R.W., Markham, P.G., Wong, S.M., Stanley, J., 2000. A unique virus complex causes *Ageratum* yellow vein disease. Proc. Natl. Acad. Sci. USA 97, 6890–6895.

Saunders, K., Stanley, J., 1999. A nanovirus-like DNA component associated with yellow vein disease of *Ageratum conyzoides*: evidence for interfamilial recombination between plant DNA viruses. Virology 264, 142–152.

Schneider, I.R., 1969. Satellite-like particle of tobacco ringspot virus that resembles tobacco ringspot virus. Science 166, 1627–1629.

Schneider, I.R., 1971. Characteristics of a satellite-like virus of tobacco ringspot virus. Virology 45, 108–122.

Schneider, I.R., Hull, R., Markham, R., 1972. Multidense satellite of tobacco ringspot virus: a regular series of components of different densities. Virology 47, 320–330.

Simon, A.E., Howell, S.H., 1986. The virulent satellite RNA of turnip crinkle virus has a major domain homologous to the 3′ end of the helper virus genome. EMBO J. 5, 3423–3428.

Simon, A.E., Roossinck, M.J., Halveda, Z., 2004. Plant virus satellites and defective interfering RNAs: new paradigms for a new century. Annu. Rev. Phytopathol. 42, 415–437.

Symons, R.H., Randles, J.W., 1999. Encapsidated circular viroid-like satellite RNA (virusoids) of plants. Curr. Topics Microbiol. Immunol. 239, 81–105.

Takanami, Y., 1981. A striking change in symptoms on cucumber mosaic-virus infected tobacco plants induced by a satellite RNA. Virology 109, 120–126.

Taliansky, M.E., Robinson, D.J., Murant, A.F., 2000. Groundnut rosette disease virus complex: biology and molecular biology. Adv. Virus Res. 55, 357–400.

Waterworth, H.E., Kaper, J.M., Tousignant, M.E., 1979. CARNA 5, the small cucumber mosaic virus–dependent replicating RNA, regulates disease expression. Science 204, 845–847.

Watson, M.T., Tian, T., Estabrook, E., Falk, B.W., 1998. A small RNA resembling the beet western yellows luteovirus ST9-associated RNA is a component of the California carrot motley dwarf complex. Phytopathology 88, 164–170.

Wu, Q., Wang, Y., Cao, M., Pantaleo, V., Burgyan, J., Li, W.-X., et al., 2012. Homology-independent discovery of replicating pathogenic circular RNAs by deep sequencing and a new computational algorithm. Proc. Natl. Acad. Sci. USA 109, 3938–3943.

Yang, X., Wang, Y., Guo, W., Xie, Y., Xie, Q., Fan, L., et al., 2011. Characterization of small interfering RNAs derived from the geminivirus/betasatellite complex using deep sequencing. PLoS One 6, e16928.

Zhou, X., 2013. Advances in understanding begomovirus satellites. Annu. Rev. Phytopathol. 51, 357–381.

PART IX

IMPACT OF SATELLITES

CHAPTER

51

Economic Significance of Satellites

Tiziana Mascia and Donato Gallitelli

University of Bari Aldo Moro, Bari, Italy

INTRODUCTION

Plant diseases account for at least 10% loss of global food production and, more importantly, they impact on food security in certain countries (Strange and Scott, 2005). Crop losses due to fungi and bacteria are generally well described, while current information on losses caused by viruses is limited (Nicaise, 2014; Waterworth and Hadidi, 1998), scarce, or not documented adequately. It is even more difficult to quantify how much of the damage is attributable to the presence of viral satellites. Most estimates of damage were provided well before the discovery of their disease etiology and there are very few examples in which a real intensification of the disease symptoms has been attributed unequivocally to the presence of a satellite. Host genetic determinants for symptoms induced by either DNA or RNA viral satellites are beginning to be disclosed (Ding et al., 2009; Smith et al., 2011; Shimura et al., 2011). This progress offers new opportunities to dissect the contribution of the virus infection to the disease from that of its satellite, such as where the combined infection induces symptoms distinct from those of the helper virus alone.

For the purpose of this chapter we will focus first on the impact on the economy and food security of infections supported by some begomovirus–betasatellite complexes in certain crops of Africa, Asia, and the Indian subcontinent. Then, for examples of economic effects of RNA viruses, we will report on the impact of outbreaks of groundnut rosette virus (GRV) and cucumber mosaic virus (CMV) associated with harmful variants of their satellite RNAs.

Viroids and Satellites.
DOI: http://dx.doi.org/10.1016/B978-0-12-801498-1.00051-6

BEGOMOVIRUS BETASATELLITES

Begomoviruses, members of the family *Geminiviridae*, are transmitted by the whitefly *Bemisia tabaci*, have genomes that consist of either one (DNA A) or two (DNA A and DNA B) circular single-stranded DNA components and are frequently associated with either betasatellites (see Chapter 62: Betasatellites of Begomoviruses) or alphasatellites. Many begomovirus–betasatellite complexes are responsible for economically important diseases in different plant species in Africa and Asia (Zhou, 2013). Besides high economic losses, their most negative impact is on food security, because large areas planted for food crops have been destroyed. Such catastrophic plant diseases exacerbate the current deficit of food supply in which at least 800 million people are inadequately fed. Fourteen crop plants provide the bulk of food for human consumption and among them is cassava, which is listed as one of the six stable carbohydrates sources in general botany text books. In addition to cassava, there are numerous plants that are less intensively grown yet fulfill important nutritional requirements, like okra. If affected by severe viral epidemics, losses in these crops may contribute to destabilizing food security in restricted areas. An exhaustive list of other begomovirus–betasatellite complexes with high negative impact on staple crops in Africa, Asia, and the Indian subcontinent is reported in a number of reviews (Briddon and Stanley, 2006; Patil and Fauquet, 2009; Sattar et al., 2013; Kenyon et al., 2014; Leke et al., 2015; Legg et al., 2006, 2015). Details of selected examples are given below.

THE COTTON AND OKRA LEAF CURL DISEASES

Globally about 32.6 million ha are devoted to cotton cultivation with production estimated at 27.6 million tons for 2011/12 (Sattar et al., 2013). Between 1992 and 1997, cotton production from Pakistan was seriously compromised by an epidemic disease denoted cotton leaf curl disease complex. Plants displayed very characteristic symptoms, consisting of severe leaf curling, vein swelling, vein darkening, and enations that produced leaf-like organs as large as the leaf from which they emerged. Symptoms differed with cotton variety and the age of the plant at infection. Late season infection frequently led to mild symptoms, while early infected plants were severely stunted with tightly rolled leaves and produced no harvestable lint (Sattar et al., 2013). However, cotton leaf curl disease complex was not a limitation to cotton cultivation in Pakistan until 1988, when the highly susceptible American variety S12 replaced the local varieties (Briddon and Markham, 2000). The disease was first recorded in small plots of the S12 cotton variety grown at Moza Khokran near Multan, but between 1992 and 1994 the epidemic expanded to affect from 120,000 to 400,000 ha with a cost to the Pakistan economy estimated at US$5 billion (Briddon, 2003). The disease was particularly severe in the Indian subcontinent and its etiology was attributed to the three whitefly-transmitted begomoviruses, cotton leaf curl Kokhran virus, cotton leaf curl Multan virus (CLCuMuV) and papaya leaf curl virus, which in all instances required the presence of cotton leaf curl Multan betasatellite to induce the specific symptoms in cotton (Briddon et al., 2001; Mansoor et al., 2003; Sattar et al., 2013). Thus the 1990s epidemic in cotton involved multiple monopartite begomoviruses, with many plants containing more than one virus, and a single betasatellite (Sattar et al., 2013). The large number of whiteflies present was the key component for the very rapid spread of the epidemic, probably also as a consequence of the effect of betasatellite on jasmonic acid responsive genes in infected plants (Zhang et al., 2012). Subsequent to the epidemic, efforts made through conventional selection/breeding yielded varieties with excellent resistance to CLCuMuV, and losses due to the disease diminished, thus restoring cotton production in Pakistan to preepidemic levels (Rahman et al., 2005; Briddon et al., 2014). These varieties remained in use until the resistance was broken by a second epidemic that started in 2001 near the town of Burewala and spread rapidly to most of the cotton growing areas of Pakistan and northwestern India. Molecular analyses demonstrated that a begomovirus–betasatellite complex was also involved in this instance, but unlike in the first epidemic, both the virus and the satellite were recombinants. The recombinant betasatellite, referred to as "Burewala strain" mostly derived from the cotton leaf curl Multan betasatellite associated with the first epidemic and contained about 100 nt of sequence from the tomato leaf curl betasatellite. The begomovirus associated with the resistance-breaking epidemic was a recombinant between CLCuMuV and cotton leaf curl Kohkran virus, both involved in the 1990s epidemic (Sattar et al., 2013 and references therein; Briddon et al., 2014). Postepidemic analyses revealed that these complexes were not uniformly widespread like those responsible for the first epidemic, as regional genetic variations in their composition were detected. The mechanism for resistance breaking is still unknown (Akhtar et al., 2014).

Okra is a malvaceous crop grown for its edible seed pods, which contain 20%–24% protein and 13%–22% oil. The crop is commonly known in India as bhendi, whose production is around 3.8 million tons constituting 70% of the total fresh vegetable income excluding onion (Sheikh Safiuddin et al., 2013). Bhendi yellow vein mosaic disease (BYVMD) is the major limitation to bhendi production in India accounting for 96% of losses (Pun and Doraiswamy, 1999). The disease is characterized by chlorosis and yellowing of veins and veinlets, small leaves, few and small fruits, and stunting (Sheikh Safiuddin et al., 2013). Detailed analyses demonstrated that BYVMD is caused by bhendi yellow vein mosaic virus and an associated betasatellite, as bhendi yellow vein mosaic virus can systemically infect bhendi but produces only mild leaf curling in this host. Severe symptoms of the BYVMD are elicited only when the betasatellite is coinoculated with the begomovirus to bhendi, thus accounting for most of the losses. BYVMD has been described recently from Thailand where the BYVMD-betasatellite complex has been found associated with 50%−100% of symptomatic plants (Tsai et al., 2013).

THE AFRICAN CASSAVA MOSAIC DISEASE PANDEMIC

Cassava (*Manihot esculenta*), family *Euphorbiaceae*, is a source of food for more than 700 million people in developing countries and is cultivated in an estimated global area of 18.6 million ha with total annual production of 238 million tons (Patil and Fauquet, 2009). The African cassava mosaic disease (CMD) pandemic devastated Uganda's cassava production between 1992 and 1997, causing losses valued in excess of US$60 million annually (Legg et al., 2006). The impact was not only economic, as food security was also compromised. Farmers abandoned the crop in large parts of the country, and in eastern districts widespread food shortages led to some famine-related deaths. The key characteristic of the pandemic was its very rapid spread—estimated at 20−50 km/year—vectored by super abundant *B. tabaci* populations, so that soon the epidemic involved the neighboring countries of Sudan, Kenya, Tanzania, and the eastern part of the Democratic Republic of Congo, with a similar impact on cassava cultivation (Legg and Fauquet, 2004; Legg et al., 2006, 2015; Alabi et al., 2011). Severely affected plants were found infected with the begomoviruses African cassava mosaic virus (ACMV), East African cassava mosaic virus (EACMV), and a novel recombinant virus denoted Uganda strain of EACMV (EACMV-UG) (Zhou et al., 1997; Deng et al., 1997). Mixed infections of ACMV and EACMV were an important feature of the severe CMD, but plants infected with EACMV-UG expressed more severe symptoms than those infected with ACMV. The main strategy to control the pandemic was the development and implementation of CMD-resistant cassava varieties by national and international breeding programs. During the pandemic no DNA satellites were detected, but a number of cassava plants in the disease-affected area of northwestern Tanzania showed unusually severe virus-like symptoms (Ndunguru et al., 2005; Legg et al., 2006; Alabi et al., 2011). Detailed laboratory investigation revealed the presence of two novel small DNA molecules, denoted DNA II and III, dependent on geminiviruses for replication and movement within the plant, thus confirming their status as satellite DNAs (Ndunguru et al., 2005; Briddon and Stanley, 2006). When these satellite molecules occurred in coinfections with geminiviruses, they caused increased viral accumulation and novel, severe disease symptoms. Moreover, the satellite DNAs broke the high resistance to geminiviruses in the West African cassava landrace TME 3, which had become an important component of the CMD pandemic mitigation effort. This observation has raised concern about the impact of these satellites on cassava production and their possible role in the current CMD pandemic is a question currently under investigation. Legg et al. (2015) listed 10 begomoviruses involved in the CMD pandemic in Africa and two in the Indian subcontinent. Patil and Fauquet (2010), studying the interaction between different betasatellites and begomoviruses in the etiology of CMD, came to the conclusion that an encounter of geminiviruses associated with CMD with betasatellites normally associated with other geminiviruses could have created new disease complexes with resistance-breaking properties, posing new serious threats to cassava production (Alabi et al., 2011).

SATELLITES ASSOCIATED WITH RNA VIRUSES

Despite the large number of satellite RNAs, some of which exacerbate symptoms induced by the helper virus, there are no firm data available for most on their economic relevance in agriculture. Notable exceptions are from satellites associated with GRV and CMV. Data on the ability of other viral RNA satellites to increase disease severity are provided mostly from laboratory experiments, although negative impacts on real agriculture cannot be excluded (Roossinck et al., 1992 and references therein).

Rosette is the most destructive virus disease of groundnut (peanut) (*Arachis hypogaea*) in Africa, where it is an important source of protein and cooking oil. In 1975, rosette disease epidemics affected about 0.7 million ha of groundnut in Nigeria and caused yield losses estimated at over 0.5 million tons, with a value estimated at US$250 million. In 1995, about 43,000 ha were affected in eastern Zambia, with losses amounting to US$5 million, and following the epidemic in Malawi, many farmers abandoned the crop causing a 23% decrease of the area under groundnut (Murant et al., 1998). Rosette disease is caused by GRV, a satellite RNA, and groundnut rosette assistor virus, on which the other two components depend for transmission by *Aphis craccivora* (Murant, 1990). Neither GRV nor groundnut rosette assistor virus cause obvious symptoms in groundnut, while different variants of the satellite RNA are responsible for the different major forms of rosette disease (Murant and Kumar, 1990).

Lethal necrosis, top stunting (synonym curl stunt), and internal fruit browning (synonym internal fruit necrosis) are three destructive diseases of tomato codetermined by specific satellite RNA (satRNA) variants associated with CMV. After a first appearance in 1972, in the Alsace region of France, tomato lethal necrosis recurred in 60%–80% of tomato fields of Italy and Spain between 1987 and 1990 (Fig. 51.1 panels A–C) (Jordá et al., 1992; Gallitelli, 2000), as well as in other countries of the Mediterranean basin and in Japan. Replacement of local landraces and old tomato varieties with new genetically homogeneous F1 types, importation of the new, so-called, Asian strains of CMV (IB strains) by unknown means, and movement of extremely high numbers of aphids due to favorable climatic conditions, determined what became an ecological disaster. Epidemics were particularly detrimental to commercial fields of canning tomato, because two new or unusual disease phenotypes emerged, in addition to the well-known filimorphism and lethal necrosis. One phenotype, referred to as tomato top stunting or curl stunt was

FIGURE 51.1 Progression of lethal necrosis (panels A–C) and internal fruit necrosis (panels D–F) on canning tomato fruits during CMV epidemics in Italy. Lethal necrosis was induced by CMV strains carrying a necrogenic satellite RNA while internal fruit necrosis was caused by CMV plus an ameliorative satellite RNA, as plants were vigorous with no disease symptoms on leaves.

observed in Italy and Spain and was codetermined by a variant of CMV satRNA, as in the absence of the associated satRNA the helper CMV strain induced the classical filimorphism. The other phenotype, denoted tomato internal fruit necrosis or internal fruit browning, had a different etiology depending whether or not an ameliorative satRNA was present in the inoculum. In Italy, the disease was characterized by extended necrosis of the mesocarp and of tissues surrounding the pedicel (Fig. 51.1 panels D–F) while plants were vigorous and did not show foliar symptoms (Gallitelli, 2000). This phenotype correlated positively with the presence of an ameliorative satRNA denoted Tfn-satRNA, which was responsible for the absence of leaf symptoms (Crescenzi et al., 1993). In the Spanish experience, plants with fruit internal browning showed also the filimorphism condition typically induced by ordinary CMV strains (Jordá et al., 1992) and did not contain a satRNA. Economically, internal fruit necrosis with Tfn-satRNA involved was even more damaging than lethal necrosis, which may cause death of all plants in a field. Unlike lethal necrosis, foliage and plant growth seemed normal and farmers were unaware that fruits had internal necrosis and were unmarketable until they ripened, by which time most of the production costs had already been incurred. In Italy, losses in total tomato production have been estimated at 300,000 tons, which is equivalent to 12,500 ha of tomato production fields. In Spain the disease was first detected north of Valencia in 1986 and affected most areas of tomato crops in eastern Spain by 1990, with disease incidence close to 100%. Abandonment of traditional tomato-growing areas, severe crop losses, low product quality, and a huge increase in, and often indiscriminate, use of insecticides and herbicides were other common features of the epidemic.

References

Akhtar, S., Tahir, M.N., Baloch, G.R., Javaid, S., Khan, A.Q., Amin, I., et al., 2014. Regional changes in the sequence of cotton leaf curl Multan betasatellite. Viruses 6, 2186–2203.

Alabi, O.J., Kumar, P.L., Naidu, R.A., 2011. Cassava mosaic disease: a curse to food security in Sub-Saharan Africa. Online. APSnet Features. Available from: http://dx.doi.org/10.1094/APSnetFeature-2011-0701.

Briddon, R.W., 2003. Cotton leaf curl disease a multicomponent begomovirus complex. Mol. Plant Pathol. 4, 427–434.

Briddon, R.W., Akbar, F., Iqbal, Z., Amrao, L., Amin, I., Saeed, M., et al., 2014. Effects of genetic changes to the begomovirus/betasatellite complex causing cotton leaf curl disease in South Asia post-resistance breaking. Virus Res. 186, 114–119.

Briddon, R.W., Mansoor, S., Bedford, I.D., Pinner, M.S., Saunders, K., Stanley, J., et al., 2001. Identification of DNA components required for induction of cotton leaf curl disease. Virology 285, 234–243.

Briddon, R.W., Markham, P.G., 2000. Cotton leaf curl virus disease. Virus Res. 71, 151–159.

Briddon, R.W., Stanley, J., 2006. Subviral agents associated with plant single-stranded DNA viruses. Virology 344, 198–210.

Crescenzi, A., Barbarossa, L., Cillo, F., Di Franco, A., Vovlas, N., Gallitelli, D., 1993. Role of cucumber mosaic virus and its satellite RNA in the etiology of tomato fruit necrosis in Italy. Arch. Virol. 131, 321–333.

Deng, D., Otim-Nape, G.W., Sangare, A., Ogwal, S., Beachy, R.N., Fauquet, C.M., 1997. Presence of a new virus closely associated with cassava mosaic outbreak in Uganda. Afr. J. Root Tuber Crops 2, 23–28.

Ding, C., Qing, L., Li, Z., Liu, Y., Qian, Y., Zhou, X., 2009. Genetic determinants of symptoms on viral DNA satellites. App. Environ. Microbiol. 75, 5380–5389.

Gallitelli, D., 2000. The ecology of cucumber mosaic virus and sustainable agriculture. Virus Res. 71, 9–21.

Jordá, C., Alfaro, A., Aranda, M.A., Moriones, E., García-Arenal, F., 1992. Epidemic of cucumber mosaic virus plus satellite RNA in tomatoes in Eastern Spain. Plant Dis. 76, 363–366.

Kenyon, L., Tsai, W.-S., Shih, S.-L., Lee, L.-M., 2014. Emergence and diversity of begomoviruses infecting solanaceous crops in East and Southeast Asia. Virus Res. 186, 104–113.

Legg, J.P., Kumar, P.L., Makeshkumar, T., Tripathi, L., Ferguson, M., Kanju, E., et al., 2015. Cassava virus diseases: biology, epidemiology, and management. Adv. Virus Res. 91, 85–142.

Legg, J.P., Fauquet, C.M., 2004. Cassava mosaic geminiviruses in Africa. Plant Mol. Biol. 56, 585–599.

Legg, J.P., Owor, B., Seruwagi, P., Ndunguru, J., 2006. Cassava mosaic virus disease in east and central Africa: epidemiology and management of a regional pandemic. Adv. Virus Res. 67, 355–418.

Leke, W.N., Mignouna, D.,B., Brown, J.K., Kvamheden, A., 2015. Begomovirus disease complex: emerging threat to vegetable production systems of West and Central Africa. Agr. Food Secur. 4, 1.

Mansoor, S., Briddon, R.W., Bull, S.E., Bedford, I.D., Bashir, A., Hussain, M., et al., 2003. Cotton leaf curl disease is associated with multiple monopartite begomoviruses supported by single DNA β. Arch. Virol. 148, 1969–1986.

Murant, A.F., 1990. Dependence of groundnut rosette virus on its satellite RNA as well as on groundnut rosette assistor luteovirus for transmission by *Aphis craccivora*. J. Gen. Virol. 71, 2163–2166.

Murant, A.F., Kumar, J.K., 1990. Different variants of the satellite RNA of groundnut rosette virus are responsible for the chlorotic and green form of groundnut rosette disease. Ann. Appl. Biol. 117, 85–92.

Murant, A.F., Robinson, D.J., Taliansky, M.E., 1998. Groundnut rosette virus. CMI/AAB Descr. Plant Viruses No. 355. www.dpvweb.net/dpv/showdpv.php?dpvno=355.

Ndunguru, J., Legg, J.P., Aveling, T.A.S., Thompson, G., Fauquet, C.M., 2005. Molecular biodiversity of cassava begomoviruses in Tanzania: evolution of cassava geminiviruses in Africa and evidence for East Africa being a center of diversity of cassava geminiviruses. Virol. J. 2, 21.

Nicaise, V., 2014. Crop immunity against viruses: outcomes and future challenges. Front. Plant Sci. 5, 660.

Patil, B.L., Fauquet, C.M., 2009. Cassava mosaic geminiviruses: actual knowledge and perspectives. Mol. Plant Pathol. 10, 685–701.

Patil, B.L., Fauquet, C.M., 2010. Differential interaction between cassava mosaic geminiviruses and geminivirus satellites. J. Gen. Virol. 91, 1871–1882.

Pun, K.B., Doraiswamy, S., 1999. Effect of age of okra plants on susceptibility to Okra yellow vein mosaic virus. Indian J. Virol. 15, 57–58.

Rahman, M., Hussain, D., Malik, T.A., Zafar, Y., 2005. Genetics of resistance against cotton leaf curl disease in *Gossypium hirsutum*. Plant Pathol. 54, 764–772.

Roossinck, M.J., Sleat, D., Palukaitis, P., 1992. Satellite RNAs of plant viruses: structures and biological effects. Microbiol. Rev. 56, 265–279.

Sattar, M.N., Kvarnheden, A., Saeed, M., Briddon, R.W., 2013. Cotton leaf curl disease—An emerging threat to cotton production worldwide. J. Gen. Virol. 94, 695–710.

Sheikh Safiuddin, M.A., Khan, Z., Mahmood, I., 2013. Effect of bhendi yellow vein mosaic virus on yield components of okra plants. J. Plant Pathol. 95, 391–393.

Shimura, H., Pantaleo, V., Ishihara, T., Myojo, N., Inaba, J.-I., Sueda, K., et al., 2011. A viral satellite RNA induces yellow symptoms on tobacco by targeting a gene involved in chlorophyll biosynthesis using the RNA silencing machinery. PLoS Pathog. 7, e1002021.

Smith, N.A., Eamens, A.L., Wang, M.-B., 2011. Viral small interfering RNAs target host genes to mediate disease symptoms in plants. PLoS Pathog. 7, e1002022.

Strange, R.N., Scott, P.R., 2005. Plant disease: a threat to global food security. Annu. Rev. Phytopathol. 43, 83–116.

Tsai, W.S., Shih, S.L., Lee, L.M., Wang, J.T., Duangsong, U., Kenyon, L., 2013. First report of bhendi yellow vein mosaic virus associated with yellow vein mosaic of okra (*Abelmoschus esculentus*) in Thailand. Plant Dis. 97, 291.

Waterworth, H., Hadidi, A., 1998. Economic losses due to plant viruses. In: Hadidi, A., Khetarpal, R.K., Koganezawa, H. (Eds.), Plant Virus Disease Control. APS Press, St. Paul, MN, pp. 1–13.

Zhang, T., Luan, J.-B., Qi, J.-F., Huang, C.-J., Li, M., Zhou, X.-P., et al., 2012. Begomovirus-whitefly mutualism is achieved through repression of plant defences by a virus pathogenicity factor. Mol. Ecol. 21, 1294–1304.

Zhou, X., 2013. Advances in understanding begomovirus betasatellites. Annu. Rev. Phytopathol. 51, 357–381.

Zhou, X., Liu, Y., Calvert, L., Munoz, C., Otim-Nape, G.W., Robinson, D.J., et al., 1997. Evidence that DNA-A of a geminivirus associated with severe cassava mosaic disease in Uganda has arisen by interspecific recombination. J. Gen. Virol. 78, 2101–2111.

PART X

SATELLITE CHARACTERISTICS

CHAPTER

52

Biology of Satellites

Luisa Rubino

National Research Council, Bari, Italy

INTRODUCTION

Satellites are subviral agents incapable of autonomous replication, which are dependent on the presence of a helper virus for their multiplication, but are generally dispensable for the helper infection cycle. Satellites share the requirement for virus-encoded replication proteins with defective interfering (DI) RNAs, but differ from these in sequence composition: DI RNAs are entirely derived from the helper virus genome, whereas satellites share little or no significant sequence similarity with the helper genome.

Satellites have limited or no coding capability and in addition to replication, they depend on the cognate virus for encapsidation, movement, and transmission. This concept applies to satellite nucleic acids, which do not have any coding ability or may seldom encode nonstructural proteins, and are always encapsidated by the helper virus coat protein (CP). In contrast, satellites that encode their own structural protein for encapsidation are denoted satellite viruses; in addition, a satellite RNA (satRNA) may be associated with some helper virus–satellite virus combinations. Extensive information on satellite viruses and nucleic acids has been described in several comprehensive reviews (Hu et al., 2009; Roossinck et al., 1992; Simon et al., 2004; Vogt and Jackson, 1999; Zhou, 2013) and will be dealt with separately in this book. This chapter will focus on some biological features of plant satellite viruses and nucleic acids.

GEOGRAPHICAL DISTRIBUTION, HOST RANGE, AND TRANSMISSION

Due to the strict association with their helper virus, satellite transmission, host range, and distribution coincide with those of the cognate

Viroids and Satellites.
DOI: http://dx.doi.org/10.1016/B978-0-12-801498-1.00052-8

virus. However, under natural field conditions, the environment may influence satRNA populations in distinct geographic areas, as shown for the satRNA of panicum mosaic virus (PMV) (Cabrera et al., 2000). Satellites may have a role in the determination of the host range or in the efficiency of transmission of the helper virus. For instance, in coinfection with their cognate virus, betasatellites, a class of circular, single-stranded DNAs, contribute to determine the virus host range and are essential for eliciting disease symptoms, such as enations, leaf curling, and yellow veins (Zhou, 2013), and are responsible for many economically important diseases in Africa and Asia. In association with their helper viruses, betasatellites can infect a wide host range (at least 37 genera in 17 families), including vegetables, ornamentals, and fiber plants. Satellites share the same vector with the helper virus and some of them are needed for the helper transmission by vectors, as for the satRNA associated with the umbravirus groundnut rosette virus (GRV). GRV satRNA, which is responsible for the rosette symptoms in groundnut (peanut), is essential for the encapsidation of its helper virus by the luteovirus groundnut rosette assistor virus and to mediate the aphid transmission of GRV by *Aphis craccivora* (Robinson et al., 1999).

No biological vector has been identified for tobacco mosaic satellite virus (STMV) or panicum mosaic satellite virus (SPMV) so far, and the plasmodiophorid fungus *Polymyxa graminis* has been only suspected to transmit maize white line mosaic virus (MWLMV), but no data are available on the transmission of its satellite SMWLMV (De Zoeten, 2004). Tobacco necrosis satellite virus (STNV) is transmitted by the chytrid fungus *Olpidium brassicae* with the same mechanism as TNV; i.e., adsorption to the plasmalemma and the flagellum and transport inside the zoospore cytoplasm, when the flagellum is retracted prior to encystment and penetration into the host (Rubino and Martelli, 2008).

SATELLITE AFFINITY FOR THE HELPER VIRUS

Satellites have been mostly found in association with plant rather than animal viruses. With the exception of STMV, which is associated with a rod-shaped helper virus, other satellite viruses are associated with spherical viruses. High resolution structural analysis showed that they have icosahedral particles about 17 nm in diameter, comprising 60 subunits of a 17–24 kDa CP, arranged in a $T = 1$ lattice, and containing the ~1 kb single-stranded positive-sense RNA genome (Scholthof et al., 1999). In natural infections, STMV was detected in *Nicotiana glauca* only in association with the rod-shaped tobacco mild green mosaic virus, but in experimental conditions its replication can be supported by other tobamoviruses on a wide host range of experimental and crop plants, like tomato and pepper (Valverde et al., 1991). Interestingly, specific sequence changes in the 5′ region of the STMV RNA are required for adaptation to TMV (Yassi and Dodds, 1998). The association between satellite virus and its cognate virus is more specific for SPMV, STNV, and SMWLMV, for which no heterologous helpers have been reported.

The replication of satRNA associated with members of the genus *Tombusvirus* can be supported by different helpers. The determinants for efficient satRNA replication reside in the sequence of open reading frame 1 product (Celix et al., 1999; Rubino and Russo, 2012), but the expression of the replicase proteins alone is not sufficient to replicate cymbidium ringspot virus (CymRSV) satRNA in plant and yeast cells (Rubino et al., 2004). The cucumber mosaic virus (CMV) satRNA replication efficiency may vary depending on the helper strain, and it can be supported by the related tomato aspermy virus (Moriones et al., 1992; Palukaitis and García-Arenal, 2003).

Alphasatellites and betasatellites are generally associated with monopartite begomoviruses of the Old World (Zhou, 2013), and rarely with bipartite begomoviruses (Paprotka et al., 2010; Romay et al., 2010; Sivalingam and Varma, 2012). *Trans*-replication of a betasatellite by different begomoviruses is a common event in nature (Nawaz-ul-Rehman et al., 2012), whereas the concomitant replication of two different betasatellites by one helper virus occurs rarely (Qing and Zhou, 2009). Because of adaptation of betasatellites to their helper virus, cognate helper viruses *trans*-replicate betasatellites more efficiently than noncognate helpers (Qing and Zhou, 2009; Zhou et al., 2003). The betasatellite capability of being replicated by different helpers and the reassortment in mixed infections make the occurrence of new devastating disease complexes very likely, as in cotton leaf curl disease in Pakistan (Nawaz-ul-Rehman et al., 2012) or a severe disease of tomato in Mali (Chen et al., 2009).

EFFECTS ON SYMPTOMS

Satellites behave as molecular parasites of their helper viruses. They may have a neutral effect, or may interfere with the helper virus replication, affecting its accumulation. Some satellites modulate the helper virus-induced symptoms, either attenuating or exacerbating them, depending on the virus–satellite combination (Collmer and Howell, 1992; Hu et al., 2009; Roossinck et al., 1992). Finally, new symptoms, different from those exerted by the helper virus alone, may be produced in association with satellites (Takanami, 1981). The capability of attenuating virus symptoms or lowering the helper titer qualifies satellites as possible agents for virus control and plant resistance (Cillo and Palukaitis, 2014; Tien and Gusui, 1991).

Symptoms induced by the helper virus are not influenced by coinfection with STMV, with the exception of pepper, where symptoms are different and attenuated in the presence of STMV, compared to those exerted by the helper only (Dodds, 1998). No differences have been reported in MWLMV symptoms on sweet corn in coinfections with SMWLMV (Gingery and Louie, 1985), whereas a slight reduction in virus titer and in the size, but not in the number of local lesions, has been reported in TNV-STNV coinfected plants (Rubino and Martelli, 2008).

The synergistic coinfection of PMV and SPMV in the natural host St. Augustine grass exacerbates PMV symptoms, due to increased virus concentration and more rapid spread, leading to St. Augustine decline, characterized by severe chlorotic mottling, leaf mosaic, and stunting (Scholthof, 1999). The involvement of SPMV CP in symptom exacerbation was demonstrated by its ability to elicit symptoms in nonhost plants after expression from a potato virus X-based vector (Qiu and Scholthof, 2004). Alphasatellites associate with begomo virus–betasatellite complexes, without contributing to the pathogenesis of the disease (Nawaz-Ul-Rehman et al., 2010), whereas betasatellites dramatically influence the helper virus symptom elicitation through the multifunctional protein βC1 (Cui et al., 2004; Saunders et al., 2004). Protein βC1 is able to induce virus-like symptoms when expressed transiently or constitutively (Saced et al., 2005; Saunders et al., 2004).

Large linear single-stranded satRNA replication or protein expression generally exerts no, or a limited, effect on symptom expression, as in the nepovirus large satRNAs (Gottula et al., 2013; Saldarelli et al., 1993). An exception is represented by the satRNA associated with the lilac isolate of arabis mosaic virus, having an important host-specific effect (Liu et al., 1991). In the potexvirus bamboo mosaic virus, the competition for replication between the satRNA and its helper genome may or not result in symptom attenuation (Chen et al., 2012). Small nepovirus satRNAs have a protective effect in the helper-induced symptom expression, as in the tobacco ringspot virus satRNA, for which reduction in virus accumulation and in the size and number of local lesions have been reported (Fritsch and Mayo, 1989), or in chicory yellow mottle virus small satRNA (Piazzolla et al., 1986). By contrast, arabis mosaic virus symptoms are exacerbated by its small satRNA (Davies and Clark, 1983).

Four tombusvirus-associated small linear satRNAs have been characterized so far, namely CymRSV satRNA, and tomato bushy stunt virus (TBSV) satRNAs B1, B10, and L (Celix et al., 1997; Gallitelli and Hull,

1985; Rubino and Russo, 2010; Rubino et al., 1990). CymRSV, TBSV B1, and L satRNAs do not interfere with symptom expression, whereas satRNA B10 prevents systemic necrosis and the death of infected plants (Celix et al., 1997, 1999; Rubino and Russo, 2010; Rubino et al., 1992) (Fig. 52.1, upper panel).

The best characterized small linear satRNAs are those associated with CMV. These satRNAs can modulate CMV symptom expression in a complex relationship with the cognate virus and the host plant, ranging from attenuation to exacerbation. Benign CMV satRNAs mostly attenuate virus symptoms, often but not always reducing virus accumulation and infectivity (Kaper and Collmer, 1988; Mossop and Francki, 1979; Palukaitis, 2016). In contrast, necrogenic CMV satRNAs dramatically increase symptom severity, inducing lethal necrosis on tomato and chlorosis on tobacco and tomato, due to the pathogenicity determinants mapping to one or few nucleotides (Sleat et al., 1994; Palukaitis, 2016). The coinfection of a CMV satRNA with the white leaf CMV strain

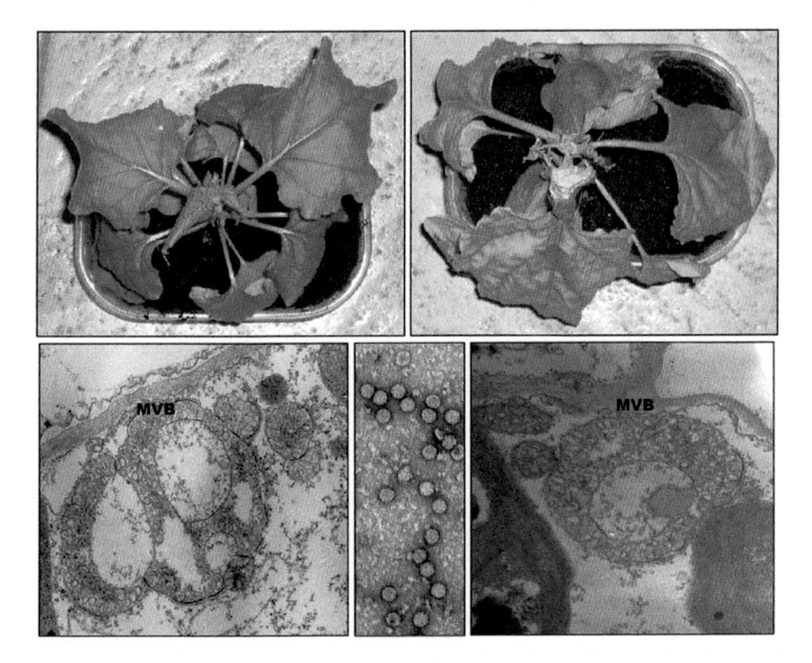

FIGURE 52.1 Upper panel: *Nicotiana benthamiana* plants coinfected by TBSV and B10 (left) or L (right) satRNAs, with symptom attenuation induced by B10 satRNA, and not by satRNA L. Lower panel: *N. benthamiana* mesophyll cells coinfected by TBSV and B10 (left) or L (right) satRNAs, showing deranged peroxisomes (multivesicular bodies; MVB). Middle, purified TBSV particles. *Source: Upper panel, right, from Rubino, L., Russo, M., 2010. Properties of a novel satellite RNA associated with tomato bushy stunt virus infections. J. Gen. Virol. 91, 2393–2401.*

induces the tomato white leaf disease, whereas the same helper–satRNA combination is symptomless on tobacco (Gonsalves et al., 1982). More recently, compelling evidence shows that another CMV satRNA (Y) incites a characteristic yellowing in tobacco by targeting a gene involved in chlorophyll biosynthesis using the RNA silencing machinery (Shimura et al., 2011; Smith et al., 2011). Altogether these data confirm the importance of the satRNA species, the helper virus strain and the host plant in the resulting symptomatology.

CYTOPATHOLOGY

The association between some satellites and their helper viruses was analyzed at the ultrastructural level. STMV infected cells showed the accumulation of virus crystals and proteins within membrane-bound structures. The vesiculation of these membranous structures indicates that they are sites for genome replication. In cells doubly infected by STMV and tobacco mild green mosaic virus, the presence of characteristic features of both infecting agents suggests that their replication is separately compartmentalized in the cell (Dodds, 1998). Ultrastructural analysis of TBSV satRNA L- and B10-infected cells showed cytopathological features typical of tombusviruses; i.e., multivesicular bodies derived from deranged peroxisomes, thus suggesting that satRNA replication occurs in the same cell compartments as the helper virus (Rubino and Russo, 2010 and unpublished data) (Fig. 52.1, lower panel). The subcellular localization of GFP-fused SPMV CP to cell wall, nucleus and nucleolus, cell membranes, and cytoplasm was determined by fluorescence microscopy (Qi et al., 2008). When CMV satRNA-Q transcription is induced independently from the helper virus, it localizes to the nucleus (Choi et al., 2012), supporting the hypothesis of a link between satRNAs and viroids.

References

Cabrera, O., Roossinck, M.J., Scholthof, K.-B.G., 2000. Genetic diversity of panicum mosaic virus satellite RNAs in St. Augustine grass. Phytopathology. 90, 977–980.

Celix, A., Burgyan, J., Rodríguez-Cerezo, E., 1999. Interactions between tombusviruses and satellite RNAs of tomato bushy stunt virus: a defect in satRNA B1 replication maps to ORF1 of a helper virus. Virology 262, 129–138.

Celix, A., Rodríguez-Cerezo, E., García-Arenal, F., 1997. New satellite RNAs, but no DI RNAs, are found in natural populations of tomato bushy stunt tombusvirus. Virology 239, 277–284.

Chen, H.C., Kong, L.R., Yeh, T.Y., Cheng, C.P., Hsu, Y.H., Lin, N.S., 2012. The conserved 5′ apical hairpin stem loops of bamboo mosaic virus and its satellite RNA contribute to replication competence. Nucleic Acids Res. 40, 4641–4652.

Chen, L.F., Rojas, M., Kon, T., Gamby, K., Xoconostle-Cazares, B., Gilbertson, R.L., 2009. A severe symptom phenotype in tomato in Mali is caused by a reassortant between a novel recombinant begomovirus (tomato yellow leaf curl Mali virus) and a betasatellite. Mol. Plant Pathol. 10, 415–430.

Choi, S.H., Seo, J.K., Kwon, S.J., Rao, A.L., 2012. Helper virus-independent transcription and multimerization of a satellite RNA associated with cucumber mosaic virus. J. Virol. 86, 4823–4832.

Cillo, F., Palukaitis, P., 2014. Transgenic resistance. Adv. Virus Res. 90, 35–146.

Collmer, C.W., Howell, S.H., 1992. Role of satellite RNA in the expression of symptoms caused by plant viruses. Annu. Rev. Phytopathol. 30, 419–442.

Cui, X.F., Tao, X.R., Xie, Y., Fauquet, C.M., Zhou, X.P., 2004. A DNA β associated with tomato yellow leaf curl China virus is required for symptom induction. J. Virol. 78, 13966–13974.

Davies, D.L., Clark, M.F., 1983. A satellite-like nucleic acid of arabis mosaic virus associated with hop nettlehead disease. Ann. Appl. Biol. 103, 439–448.

De Zoeten, G.A., 2004. Maize white line mosaic. In: Lapierre, H., Signoret, P.A. (Eds.), Viruses and Virus Diseases of *Poaceae* (*Gramineae*). INRA Editions, Paris, France, pp. 678–682.

Dodds, J.A., 1998. Satellite tobacco mosaic virus. Annu. Rev. Phytopathol. 36, 295–310.

Fritsch, C., Mayo, M.A., 1989. Satellites of plant viruses. In: Mandahar, C.L. (Ed.), Plant Viruses, Structure and Replication. CRC Press, Boca Raton, FL, pp. 290–315.

Gallitelli, D., Hull, R., 1985. Characterization of satellite RNAs associated with tomato bushy stunt virus and five other definitive tombusviruses. J. Gen. Virol. 66, 1533–1543.

Gingery, R.E., Louie, R., 1985. A satellite-like virus particle associated with maize white line mosaic virus. Phytopathology 70, 1019–1022.

Gonsalves, D., Provvidenti, R., Edwards, M.C., 1982. Tomato white leaf: the relation of an apparent satellite RNA and cucumber mosaic virus. Phytopathology 72, 1533–1538.

Gottula, J., Lapato, D., Cantilina, K., Saito, S., Bartlett, B., Fuchs, M., 2013. Genetic variability, evolution and biological effects of grapevine fanleaf virus satellite RNAs. Phytopathology 103, 1180–1187.

Hu, C.-C., Hsu, Y.-H., Lin, N.-S., 2009. Satellite RNAs and satellite viruses of plants. Viruses 1, 1325–1350.

Kaper, J.M., Collmer, C.W., 1988. Modulation of viral plant diseases by secondary RNA agents. In: Domingo, E., Holland, J.J., Ahlquist, P. (Eds.), RNA Genetics. Vol. 3. Variability of RNA Genomes. CRC Press, Boca Raton, FL, pp. 171–194.

Liu, Y.Y., Cooper, J.I., Edwards, M.L., Hellen, C.U.T., 1991. A satellite RNA of arabis mosaic nepovirus and its pathological impact. Ann. Appl. Biol. 118, 577–587.

Moriones, E., Díaz, I., Rodríguez-Cerezo, E., Fraile, A., García-Arenal, F., 1992. Differential interactions among strains of tomato aspermy virus and satellite RNA of cucumber mosaic virus. Virology 186, 475–480.

Mossop, D.W., Francki, R.I.B., 1979. Comparative studies on two satellite RNAs of cucumber mosaic virus. Virology 95, 395–404.

Nawaz-ul-Rehman, M.S., Briddon, R.W., Fauquet, C.M., 2012. A melting pot of Old World begomoviruses and their satellites infecting a collection of Gossypium species in Pakistan. PLoS One 7, e40050.

Nawaz-Ul-Rehman, M.S., Nahid, N., Mansoor, S., Briddon, R.W., Fauquet, C.M., 2010. Post-transcriptional gene silencing suppressor activity of two non-pathogenic alphasatellites associated with a begomovirus. Virology 405, 300–308.

Palukaitis, P., 2016. Satellite RNAs and satellite viruses. Mol. Plant Microbe Interact. 29, 181–186.

Palukaitis, P., García-Arenal, F., 2003. Cucumoviruses. Adv. Virus Res. 62, 241–323.

Paprotka, T., Metzler, V., Jeske, H., 2010. The first DNA 1-like alpha satellites in association with New World begomoviruses in natural infections. Virology. 404, 148–157.

Piazzolla, P., Vovlas, C., Rubino, L., 1986. Symptom regulation induced by chicory yellow mottle virus satellite-like RNA. J. Phytopathol. 115, 124–129.

Qi, D., Omarov, R.T., Scholthof, K.B., 2008. The complex subcellular distribution of satellite panicum mosaic virus capsid protein reflects its multifunctional role during infection. Virology 376, 154–164.

Qing, L., Zhou, X.P., 2009. Trans-replication of, and competition between, DNA β satellites in plants inoculated with tomato yellow leaf curl China virus and tobacco curly shoot virus. Phytopathology 99, 716–720.

Qiu, W., Scholthof, K.B., 2004. Satellite panicum mosaic virus capsid protein elicits symptoms on a nonhost plant and interferes with a suppressor of virus-induced gene silencing. Mol. Plant Microbe Interact. 17, 263–271.

Robinson, D.J., Ryabov, E.V., Raja, S.K., Roberts, I.M., Taliansky, M.E., 1999. Satellite RNA is essential for encapsidation of groundnut rosette umbravirus RNA by groundnut rosette assistor luteovirus coat protein. Virology 254, 105–114.

Romay, G., Chirinos, D., Geraud-Pouey, F., Desbiez, C., 2010. Association of an atypical alphasatellite with a bipartite New World begomovirus. Arch. Virol. 155, 1843–1847.

Roossinck, M.J., Sleat, D., Palukaitis, P., 1992. Satellite RNAs of plant viruses: structures and biological effects. Microbiol. Rev. 56, 265–279.

Rubino, L., Burgyan, J., Grieco, F., Russo, M., 1990. Sequence analysis of cymbidium ringspot virus satellite and defective interfering RNAs. J. Gen. Virol. 71, 1655–1660.

Rubino, L., Carrington, J.C., Russo, M., 1992. Biologically active cymbidium ringspot virus satellite RNA in transgenic plants suppresses accumulation of DI RNAs. Virology. 188, 429–437.

Rubino, L., Martelli, G.P., 2008. Necrovirus. In: Mahy, B.W.H., Van Regenmortel, M.H.V. (Eds.), Encyclop. Virol., 5 Vols. Elsevier, Oxford, pp. 403–405.

Rubino, L., Pantaleo, V., Navarro, B., Russo, M., 2004. Expression of tombusvirus open reading frames 1 and 2 is sufficient for the replication of defective interfering, but not satellite, RNA. J. Gen. Virol. 85, 3115–3122.

Rubino, L., Russo, M., 2010. Properties of a novel satellite RNA associated with tomato bushy stunt virus infections. J. Gen. Virol. 91, 2393–2401.

Rubino, L., Russo, M., 2012. A single amino acid substitution in the ORF1 of cymbidium ringspot virus determines the accumulation of two satellite RNAs. Virus Res. 168, 84–87.

Saced, M., Behjatnia, S.A., Mansoor, S., Zafar, Y., Hasnain, S., Rezaian, M.A., 2005. A single complementary sense transcript of a geminiviral DNA β satellite is determinant of pathogenicity. Mol. Plant Microbe Interact. 18, 7–14.

Saldarelli, P., Minafra, A., Walter, B., 1993. A survey of grapevine fanleaf nepovirus isolates for the presence of satRNA. Vitis 32, 99–102.

Saunders, K., Norman, A., Gucciardo, S., Stanley, J., 2004. The DNA β satellite component associated with Ageratum yellow vein disease encodes an essential pathogenicity protein (βC1). Virology 324, 37–47.

Scholthof, K.-B.G., 1999. A synergism induced by satellite panicum mosaic virus. Mol. Plant Microbe Interact. 12, 163–166.

Scholthof, K.-B.G., Jones, R.W., Jackson, A.O., 1999. Biology and structure of plant satellite viruses activated by icosahedral helper viruses. In: Vogt, P.K., Jackson, A.O.E. (Eds.), Satellites and Defective Viral RNAs. Springer-Verlach, Berlin, pp. 123–143.

Shimura, H., Pantaleo, V., Ishihara, T., Myojo, N., Inaba, J.I., Sueda, K., et al., 2011. A viral satellite RNA induces yellow symptoms on tobacco by targeting a gene involved in chlorophyll biosynthesis using the RNA silencing machinery. PLoS Pathog. 7, e1002021.

Simon, A.E., Roossinck, M.J., Havelda, Z., 2004. Plant virus satellite and defective interfering RNAs: new paradigms for a new century. Annu. Rev. Phytopathol. 42, 415–437.

Sivalingam, P.N., Varma, A., 2012. Role of betasatellite in the pathogenesis of a bipartite begomovirus affecting tomato in India. Arch. Virol. 157, 1081–1092.
Sleat, D.E., Zhang, L., Palukaitis, P., 1994. Mapping determinants within cucumber mosaic virus and its satellite RNA for the induction of necrosis in tomato plants. Mol. Plant Microbe Interact. 7, 189–195.
Smith, N.A., Eamens, A.L., Wang, M.B., 2011. Viral small interfering RNAs target host genes to mediate disease symptoms in plants. PLoS Pathog. 7, e1002022.
Takanami, Y., 1981. A striking change in symptoms on cucumber mosaic virus-infected tobacco plants induced by a satellite RNA. Virology 109, 120–126.
Tien, P., Gusui, W., 1991. Satellite RNA for the biocontrol of plant disease. Adv. Virus Res. 39, 321–339.
Valverde, R.A., Heick, J.A., Dodds, J.A., 1991. Interactions between satellite tobacco mosaic virus, helper tobamoviruses, and their hosts. Phytopathology 81, 99–104.
Vogt, P.K., Jackson, A.O.E. (Eds.), 1999. Satellites and Defective Viral RNAs. Springer-Verlag, Berlin.
Yassi, M.N.A., Dodds, J.A., 1998. Specific sequence changes in the 5′-terminal region of the genome of satellite tobacco mosaic virus are required for adaptation to tobacco mosaic virus. J. Gen. Virol. 79, 905–913.
Zhou, X., 2013. Advances in understanding begomovirus satellites. Annu. Rev. Phytopathol. 51, 357–381.
Zhou, X.P., Xie, Y., Tao, X.R., Zhang, K.Z., Li, Z.H., Fauquet, C.M., 2003. Characterization of DNA β associated with begomoviruses in China and evidence for co-evolution with their cognate viral DNA-A. J. Gen. Virol. 84, 237–247.

CHAPTER

53

Replication of Satellites

Ying-Wen Huang[1], Chung-Chi Hu[1], Yau-Heiu Hsu[1] and Na-Sheng Lin[2]

[1]National Chung Hsing University, Taichung, Taiwan [2]Institute of Plant and Microbial Biology, Academia Sinica, Taipei, Taiwan

INTRODUCTION

Plant viral satellites are completely dependent on the replication enzymes and cofactors encoded by their helper viruses (HVs) and host plants. Although the satellites are not phylogenetically related to their cognate HVs, these molecular parasites are able to efficiently utilize the replication machineries of their HVs. Studies of the "*trans*-replication" mechanisms of satellites by HVs have provided many insights into the interactions between HV replication machineries and the *cis*-elements of satellites. Several comprehensive reviews have described in detail the replication of plant viral satellites, including satellite viruses, satellite RNAs (satRNAs), and satellite DNAs (satDNAs) (Dodds, 1998; Hu et al., 2009; Huang et al., 2010; Simon et al., 2004; Zhou, 2013). Here, we summarize recent advances in the understanding of *cis*-acting elements required for satRNA replication, focusing on the similarities and differences as compared with those of the respective HVs, and the underlying mechanisms (Table 53.1). These studies have revealed the strategies satellites may employ to maintain a delicate balance with the HV, and have provided essential information for the development of satellite-based applications.

SATELLITE VIRUS REPLICATION

Satellite viruses depend on their cognate HVs for replication, movement, and transmission, but encode their own coat protein (CP) for

Viroids and Satellites.
DOI: http://dx.doi.org/10.1016/B978-0-12-801498-1.00053-X

TABLE 53.1 Characteristics of Representative Plant Viral Satellites

	Helper virus	Satellite				
				Similarities[a]		
Satellite type	**Genus/ species**	**Abbreviation**	**Size (nt)**	**Terminal sequence identity**	**Terminal structural motif**	**Differences[b]**
Sat virus	*Necrovirus*/ TNV-D	STNV-C	1221	5′-UTR: 47% 3′-UTR: 36%	3′-TLS	ND
	Tobamovirus/ TMV	STMV	1059	3′-terminus (250 nt): 65% 3′-termini: GGCCCA	3′-TLS aminoacylation with histidine	ND
Large satRNA	*Potexvirus*/ BaMV	satBaMV	836	5′-terminus (94 nt): 63%	3′-UTR: 3 stem loops	No 3′ pseudoknot for RdRp binding
				5′-termini: GAAAAC 3′-conserved ACCUAA	5′-UTR: AHSL	Noninterchangeable with BaMV 5′- and 3′-UTRs; specific determinant in AHSL for satBaMV-mediated downregulation of BaMV replication
Small linear satRNA	*Carmovirus*/ TCV	satC	356	3′-terminus: 90%	3′-terminus: a set of hairpins and pseudoknots	Noninterchangeable with TCV 3′-UTR
		satD	194	3′-termini: CCUGCCC	3′-terminus: H5- and H4b-like hairpins	ND

	Cucumovirus/ CMV-Q	Q-satRNA	336	3′-terminus sequence similar with that of CMV TLS	3′-TLS	No 3′-terminal aminoacylation Transcription in the nucleus
Small circular satRNA	*Polerovirus*/ CYDV-RPV	satRPV	322	ND	ND	Rolling-circle replication Y-shaped RdRp binding motif

[a] *Similarities: the features of satellites that mimic those of the respective HVs.*
[b] *Differences: the differences between the satellites and their HVs.*
TNV-D, tobacco necrosis virus strain D; *TMV*, tobacco mosaic virus; *BaMV*, bamboo mosaic virus; *TCV*, turnip crinkle virus; *CMV-Q*, cucumber mosaic virus strain Q; *CYDV-RPV*, cereal yellow dwarf virus serotype RPV; *AHSL*, apical hairpin stem loop; *RdRp*, RNA-dependent RNA polymerase; *TLS*, tRNA like structure; *UTR*, untranslated region; *ND*, not determined by experiments.

encapsidation. Identified satellite virus-encoded CPs are not required for replication of the respective satellite virus. Replication has been extensively studied for two model satellite viruses: satellite tobacco necrosis virus (STNV) and satellite tobacco mosaic virus (STMV). STNV strain C (STNV-C) is well characterized (Bringloe et al., 1999). The 5′- and 3′-untranslated regions (UTRs) of STNV-C RNA share only 47% and 36% sequence identity, respectively, with the corresponding regions of its HV, TNV strain D (TNV-D), but are interchangeable with the counterpart sequences of TNV-D for replication, suggesting that STNV-C and its HV use similar mechanisms for replication. Competition for similar replication machineries may explain why coinfection of STNV-C caused substantial suppression of TNV replication (Bringloe et al., 1999). In contrast, for STMV, a spherical virus associated with rod-shaped tobamoviruses, only the 3′-terminus exhibits substantial identity with the HVs (65% with tobacco mosaic virus, TMV) (Mirkov et al., 1989). However, the STMV 3′-terminus, which is predicted to fold into a tRNA-like structure (TLS) and is aminoacylated with histidine, similar to that of TMV (Felden et al., 1994; Gultyaev et al., 1994), is also interchangeable with the counterpart sequence of TMV, suggesting that TMV replicase recognizes a similar STMV 3′-terminal sequence and/or structure to initiate RNA replication (Dodds, 1998).

LARGE SATRNA REPLICATION

Large satRNAs, with a length of 0.8–1.5 kb, encode at least one nonstructural protein. Most known large satRNAs are associated with nepoviruses, but one is associated with a potexvirus, bamboo mosaic virus (BaMV).

SatRNAs of Nepoviruses

The satRNA-encoded nonstructural proteins are essential for the replication of a few nepovirus satRNAs; for instance, satRNAs of tomato black ring virus and arabis mosaic virus (Hemmer et al., 1993; Liu and Cooper, 1993).

SatRNA of BaMV

Although the P20 protein, encoded by the satRNA associated with BaMV (satBaMV), is involved in systemic movement (Palani et al., 2006; Vijayapalani et al., 2012), it is dispensable for replication (Lin et al.,

1996). The replication of satBaMV RNA has been well studied, and can be used as a model system for large satRNAs.

SatBaMV RNA is a single-stranded (ss) positive-sense (+) RNA of 836 nt, encapsidated by BaMV CPs into a rod-shaped particle (Lin and Hsu, 1994). Similar to other satRNAs of plant viruses, satBaMV RNA shares no overall similarity with BaMV RNA, but mimicries have been identified at both the 5′- and 3′-termini (Table 53.1).

The 5′-UTR of satBaMV contains one hypervariable region that folds into a conserved apical hairpin stem loop (AHSL) comprising an apical loop and two internal loops (IL-I and IL-II) (Chen et al., 2007). The AHSL of satBaMV is not only important for satBaMV replication but also shares remarkable similarities with that of BaMV in terms of structure and function (Chen et al., 2010). Mutational studies indicated that preservation of AHSL structure and two specific nucleotides (C_{83} in the IL-I and C_{60} in the IL-II) in interfering satBaMV isolate BSL6 (Hsu et al., 1998) play a crucial role in downregulating BaMV replication (Chen et al., 2007; Hsu et al., 2006). Similarly, BaMV 5′-AHSL also contains C_{86} in IL-I and C_{64} in IL-II, with a mutation in C_{86} or C_{64} completely abolishing or reducing BaMV replication, respectively (Chen et al., 2012). These results imply that the AHSL in satBaMV can tolerate more sequence variations than the AHSL in BaMV (Chen et al., 2007, 2010, 2012). Moreover, BSL6-mediated downregulation of BaMV replication is dose-dependent, suggesting that the downregulation arises from competition for viral and/or host factors (Chen et al., 2012). Besides the integral structure of the AHSL of satBaMV, other elements or secondary structures relevant for replication have been mapped in the 5′-UTR of satBaMVs (Annamalai et al., 2003; Chen et al., 2010).

For the similarities between the 3′-UTRs of satBaMV and BaMV, they include: (1) two small SLs and one large SL in satBaMV 3′-UTR resembling those found in the BaMV 3′-UTR; (2) the conserved hexanucleotide ACCUAA involved in accumulation of all known potexviral RNAs (Bancroft et al., 1991); and (3) the putative polyadenylation signal AAUAAA. The maintenance of intact folded secondary structures and conserved sequences are important for efficient replication of satBaMV (Huang et al., 2009). These similarities imply that the satBaMV 3′-UTR contains sufficient sequence and/or structural elements to be recognized by the BaMV replicase and efficiently initiate (−)-satBaMV RNA synthesis (Huang et al., 2010). However, some remarkable discrepancies were also identified between BaMV and satBaMV 3′-UTRs (Huang et al., 2009): (1) unlike BaMV, satBaMV 3′-UTR does not form a 3′-terminal pseudoknot required for RNA-dependent RNA polymerase binding and BaMV replication and (2) the 5′- and 3′-UTRs of BaMV and satBaMV are not interchangeable, despite their high degree of structural similarity. Thus, BaMV and satBaMV may exploit different sets of replication complexes containing different host factors, a view further supported by the identification of a host factor, heat shock protein 90; this protein can bind to the pseudoknot region of BaMV 3′-UTR and is required for the replication of BaMV, but not of satBaMV (Huang et al., 2012).

SMALL LINEAR SATRNA REPLICATION

Unlike the large satRNAs, small satRNAs (usually <700 nucleotides in length) do not harbor any functional open reading frames, and therefore the only *cis*-acting elements for their replication are contained in their highly structured genomes. Recently, extensive information on their replication has been obtained from satRNAs associated with turnip crinkle virus (TCV) and cucumber mosaic virus (CMV).

satRNA of TCV

Several small linear satRNAs are associated with TCV. Of these, satD is a typical satRNA that shares no sequence identity with TCV, but contains 3′ structural elements similar to the H5 and H4b hairpins of TCV involved in the replication of TCV and another TCV satRNA, satC (Zhang et al., 2006c). In contrast, satC is a nontypical satRNA consisting of satD and two fragments from the 3′ region of TCV RNA that account for the 90% sequence identity between the 3′ regions of satC and its HV (Simon and Howell, 1986). The 3′ portion of satC contains a set of structures, designated Pr, H5, H4b, and H4a from 3′ to 5′, which are also present in the 3′-UTR of TCV RNA. The hairpin Pr is the promoter region for (−) RNA synthesis of satC and TCV; this hairpin is also found in other carmoviruses (Simon, 2015). Despite the high similarities in sequence and structure, the Pr hairpins of satC and TCV are not interchangeable, and the promoter activity for (−) RNA synthesis is much more efficient in satC than in TCV, suggesting that upstream sequences are involved in efficient replication of satC (Zhang et al., 2006b). Hairpin H5 contains a symmetrical internal loop, with its 3′-side (GGGC) interacting with 3′-terminal nucleotides (GCCC) to form a pseudoknot (Ψ1) that is critical for accumulation of both satC and TCV (Zhang et al., 2006b). The loop region of hairpin H4b interacts with the sequence flanking the 3′-side of H5 to form Ψ2, which is important for efficient accumulation of satC and TCV (Guo et al., 2009; Zhang et al., 2006a). The H4a hairpin, together with H4b, Ψ2, and one derepressor, are required for the conformational changes of satC that activate (−) RNA synthesis. Ψ2 stabilizes the preactive structure, and then triggers a structural switch to generate the active form for stimulating (−) RNA synthesis (Zhang et al., 2006a,c). The stabilization of the preactive structure of progeny RNA is thought to restrict the access of the viral replicase, thus blocking RNA synthesis from (−) satC RNA intermediates (Simon and Gehrke, 2009). A motif 1-hairpin, located in the region joining satD and TCV on the (−) strand of satC, serves as an enhancer for (+) satC RNA accumulation (Nagy et al., 1999), and reduces TCV virion formation to increase the level of free CP, which in turn suppresses the host RNA silencing defense system (Zhang and Simon, 2003). These findings reveal that satC utilizes multiple strategies to promote its own survival and survival of the HV.

satRNA of CMV-Q (Q-satRNA)

The sequence and structure of the Q-satRNA 3′-terminus resembles that of CMV-Q (Gordon and Symons, 1983). The 3′-terminus of CMV folds into a TLS that can be aminoacylated with tyrosine, and is important for the initiation of CMV replication. However, the 3′-terminus of Q-satRNA cannot be aminoacylated, indicating that Q-satRNA may recruit the assembled replication complexes from its HV. A recent study revealed that, in the presence or absence of HV, the (−) strand of Q-satRNA can be transcribed, enabling the formation of double-stranded (ds-) and multimeric intermediates in the nucleus analogous to those associated with certain viroids (Choi et al., 2012). Therefore, Q-satRNA is hypothesized to have two replication phases: viroid-like and virus-like (Rao and Kalantidis, 2015). In the viroid-like phase, Q-satRNA is imported into the nucleus through an interaction with a bromodomain-containing host protein (Chaturvedi et al., 2014). Q-satRNA multimers are generated through repetitive additions of a nontemplate heptanucleotide motif (HNM) to the 3′-end of the (+)-monomer by an unknown host mechanism, and the ligation of the 5′-end of the second monomeric unit. Subsequently, (−) multimers are transcribed from the (+) multimers by a host RNA polymerase; the resulting multimers, which contain the complementary HNM (cHNM) at their junction, are designated as cHNM-(−) multimers. Following export into the cytoplasm, the cHNM-(−) multimers are believed to be the templates for HV-driven synthesis to begin the virus-like replication phase; such multimers have been demonstrated to be the preferred templates for HV-dependent replication. During the virus-like phase, Q-satRNA (−) multimers are transported by HV replication complexes to the site of HV replication, presumably in the tonoplast. The HV replication complexes synthesize (+) Q-satRNA progeny from cHNM-(−) multimers, using the cHNM for termination and reinitiation of RNA synthesis, to generate monomeric progeny (Seo et al., 2013). This model highlights a mechanism of Q-satRNA replication that is distinct from that of its HV.

SMALL CIRCULAR SATRNA REPLICATION

Small circular satRNAs were also known as viroid-like satRNAs because they replicate through a rolling-circle replication mechanism, in a similar manner to viroids (Symons, 1997). Despite the different replication mechanisms employed by viroid-like satRNAs and their HVs, the circular satRNAs have still evolved ways of utilizing the replication machineries provided by their HVs. The best studied example via a rolling-circle replication mechanism is satRNA of cereal yellow dwarf virus-RPV (Song and Miller, 2004).

SATDNA REPLICATION

For replication of satDNAs, the readers are referred to Chapter 62, Betasatellites of Begomoviruses.

Acknowledgments

This research was supported by the Ministry of Science and Technology of the Republic of China grants (MOST-104-2313-B-005 -023 -MY3, MOST-103-2321-B-005-002, NSC-102-2321-B-005-002, NSC-101-2321-B-005-006, NSC-101-2321-B-005-015-MY2, NSC-101-2313-B-005-036-MY3).

References

Annamalai, P., Hsu, Y.H., Liu, Y.P., Tsai, C.H., Lin, N.S., 2003. Structural and mutational analyses of cis-acting sequences in the 5′-untranslated region of satellite RNA of bamboo mosaic potexvirus. Virology 311, 229–239.

Bancroft, J.B., Rouleau, M., Johnston, R., Prins, L., Mackie, G.A., 1991. The entire nucleotide sequence of foxtail mosaic virus RNA. J. Gen. Virol. 72, 2173–2181.

Bringloe, D.H., Pleij, C.W., Coutts, R.H., 1999. Mutation analysis of cis-elements in the 3′- and 5′-untranslated regions of satellite tobacco necrosis virus strain C RNA. Virology 264, 76–84.

Chaturvedi, S., Kalantidis, K., Rao, A.L.N., 2014. A bromodomain-containing host protein mediates the nuclear importation of a satellite RNA of cucumber mosaic virus. J. Virol. 88, 1890–1896.

Chen, H.C., Hsu, Y.H., Lin, N.S., 2007. Downregulation of bamboo mosaic virus replication requires the 5′ apical hairpin stem loop structure and sequence of satellite RNA. Virology 365, 271–284.

Chen, H.C., Kong, L.R., Yeh, T.Y., Cheng, C.P., Hsu, Y.H., Lin, N.S., 2012. The conserved 5′ apical hairpin stem loops of bamboo mosaic virus and its satellite RNA contribute to replication competence. Nucleic Acids Res. 40, 4641–4652.

Chen, S.C., Desprez, A., Olsthoorn, R., 2010. Structural homology between bamboo mosaic virus and its satellite RNAs in the 5′ untranslated region. J. Gen. Virol. 91, 782–787.

Choi, S.H., Seo, J.K., Kwon, S.J., Rao, A.L.N., 2012. Helper virus-independent transcription and multimerization of a satellite RNA associated with cucumber mosaic virus. J. Virol. 86, 4823–4832.

Dodds, J.A., 1998. Satellite tobacco mosaic virus. Annu. Rev. Phytopathol. 36, 295–310.
Felden, B., Florentz, C., McPherson, A., Giege, R., 1994. A histidine accepting tRNA-like fold at the 3′-end of satellite tobacco mosaic virus RNA. Nucleic Acids Res. 22, 2882–2886.
Gordon, K.H., Symons, R.H., 1983. Satellite RNA of cucumber mosaic virus forms a secondary structure with partial 3′-terminal homology to genomal RNAs. Nucleic Acids Res. 11, 947–960.
Gultyaev, A.P., van Batenburg, E., Pleij, C.W., 1994. Similarities between the secondary structure of satellite tobacco mosaic virus and tobamovirus RNAs. J. Gen. Virol. 75, 2851–2856.
Guo, R., Lin, W., Zhang, J.C., Simon, A.E., Kushner, D.B., 2009. Structural plasticity and rapid evolution in a viral RNA revealed by in vivo genetic selection. J. Virol. 83, 927–939.
Hemmer, O., Oncino, C., Fritsch, C., 1993. Efficient replication of the in vitro transcripts from cloned cDNA of tomato black ring virus satellite RNA requires the 48K satellite RNA-encoded protein. Virology 194, 800–806.
Hsu, Y.H., Chen, H.C., Cheng, J., Annamalai, P., Lin, B.Y., Wu, C.T., et al., 2006. Crucial role of the 5′ conserved structure of bamboo mosaic virus satellite RNA in downregulation of helper viral RNA replication. J. Virol. 80, 2566–2574.
Hsu, Y.H., Lee, Y.S., Liu, J.S., Lin, N.S., 1998. Differential interactions of bamboo mosaic potexvirus satellite RNAs, helper virus, and host plants. Mol. Plant Microbe Interact. 11, 1207–1213.
Hu, C.C., Hsu, Y.H., Lin, N.S., 2009. Satellite RNAs and satellite viruses of plants. Viruses 1, 1325–1350.
Huang, Y.W., Hu, C.C., Lin, C.A., Liu, Y.P., Tsai, C.H., Lin, N.S., et al., 2009. Structural and functional analyses of the 3′ untranslated region of bamboo mosaic virus satellite RNA. Virology 386, 139–153.
Huang, Y.W., Hu, C.C., Lin, N.S., Hsu, Y.H., 2010. Mimicry of molecular pretenders: the terminal structures of satellites associated with plant RNA viruses. RNA Biol. 7, 162–171.
Huang, Y.W., Hu, C.C., Liou, M.R., Chang, B.Y., Tsai, C.H., Meng, M.H., et al., 2012. Hsp90 interacts specifically with viral RNA and differentially regulates replication initiation of bamboo mosaic virus and associated satellite RNA. PLoS Pathog. 8, e1002726.
Lin, N.S., Hsu, Y.H., 1994. A satellite RNA associated with bamboo mosaic potexvirus. Virology 202, 707–714.
Lin, N.S., Lee, Y.S., Lin, B.Y., Lee, C.W., Hsu, Y.H., 1996. The open reading frame of bamboo mosaic potexvirus satellite RNA is not essential for its replication and can be replaced with a bacterial gene. Proc. Natl. Acad. Sci. USA 93, 3138–3142.
Liu, Y.Y., Cooper, J.I., 1993. The multiplication in plants of arabis mosaic virus satellite RNA requires the encoded protein. J. Gen. Virol. 74, 1471–1474.
Mirkov, T.E., Mathews, D.M., Duplessis, D.H., Dodds, J.A., 1989. Nucleotide-sequence and translation of satellite tobacco mosaic-virus RNA. Virology 170, 139–146.
Nagy, P.D., Pogany, J., Simon, A.E., 1999. RNA elements required for RNA recombination function as replication enhancers in vitro and in vivo in a plus-strand RNA virus. EMBO J. 18, 5653–5665.
Palani, V.P., Kasiviswanathan, V., Chen, J.C., Chen, W., Hsu, Y.H., Lin, N.S., 2006. The arginine-rich motif of bamboo mosaic virus satellite RNA-encoded P20 mediates self-interaction, intracellular targeting, and cell-to-cell movement. Mol. Plant Microbe Interact. 19, 758–767.
Rao, A.L.N., Kalantidis, K., 2015. Virus-associated small satellite RNAs and viroids display similarities in their replication strategies. Virology 479–480, 627–636.
Seo, J.K., Kwon, S.J., Chaturvedi, S., Choi, S.H., Rao, A.L.N., 2013. Functional significance of a hepta nucleotide motif present at the junction of cucumber mosaic virus satellite RNA multimers in helper-virus dependent replication. Virology 435, 214–219.

Simon, A.E., 2015. 3'UTRs of carmoviruses. Virus Res. 206, 27–36.
Simon, A.E., Gehrke, L., 2009. RNA conformational changes in the life cycles of RNA viruses, viroids, and virus-associated RNAs. Biochim. Biophys. Acta. 1789, 571–583.
Simon, A.E., Howell, S.H., 1986. The virulent satellite RNA of turnip crinkle virus has a major domain homologous to the 3' end of the helper virus genome. EMBO J. 5, 3423–3428.
Simon, A.E., Roossinck, M.J., Havelda, Z., 2004. Plant virus satellite and defective interfering RNAs: new paradigms for a new century. Annu. Rev. Phytopathol. 42, 415–437.
Song, S.I., Miller, W.A., 2004. *cis* and *trans* requirements for rolling circle replication of a satellite RNA. J. Virol. 78, 3072–3082.
Symons, R.H., 1997. Plant pathogenic RNAs and RNA catalysis. Nucleic Acids Res. 25, 2683–2689.
Vijayapalani, P., Chen, C.F., Liou, M.R., Hsu, Y.H., Lin, N.S., 2012. Phosphorylation of bamboo mosaic virus satellite RNA (satBaMV)-encoded protein P20 downregulates the formation of the satBaMV-P20 ribonucleoprotein complex. Nucleic Acids Res. 40, 638–649.
Zhang, G., Simon, A.E., 2003. A multifunctional turnip crinkle virus replication enhancer revealed by in vivo functional SELEX. J. Mol. Biol. 326, 35–48.
Zhang, G., Zhang, J., George, A.T., Baumstark, T., Simon, A.E., 2006a. Conformational changes involved in initiation of minus-strand synthesis of a virus-associated RNA. RNA. 12, 147–162.
Zhang, J., Zhang, G., Guo, R., Shapiro, B.A., Simon, A.E., 2006c. A pseudoknot in a preactive form of a viral RNA is part of a structural switch activating minus-strand synthesis. J. Virol. 80, 9181–9191.
Zhang, J., Zhang, G., McCormack, J.C., Simon, A.E., 2006b. Evolution of virus-derived sequences for high-level replication of a subviral RNA. Virology 351, 476–488.
Zhou, X.P., 2013. Advances in understanding begomovirus satellites. Annu. Rev. Phytopathol. 51, 357–381.

CHAPTER

54

Satellite RNAs: Their Involvement in Pathogenesis and RNA Silencing

Chikara Masuta and Hanako Shimura
Hokkaido University, Sapporo, Japan

INTRODUCTION

"Satellites" comprise satellite viruses and satellite nucleic acids, and the satellite nucleic acids are further divided into satellite DNAs and satellite RNAs (satRNAs) (Briddon et al., 2012). In plants, circular single-stranded (ss) DNA satellites (satDNA) are often associated with ssDNA viruses, geminiviruses (Nawaz-ul-Rehman and Fauquet, 2009). Plant satRNAs are apparently molecular parasites depending for their replication and encapsidation on their helper viruses (Hu et al., 2009; Roossinck et al., 1992; Simon et al., 2004). SatRNAs often attenuate the viral symptoms but some cause more severe symptoms. Here we overview the molecular mechanisms for the symptom attenuation/exacerbation induced by several satellite nucleic acids, paying special attention to the role played by the host RNA silencing.

SYMPTOM ALTERATIONS BY SATELLITE NUCLEIC ACIDS

satDNA

Alphasatellites and betasatellites are associated with DNA viruses, and those associated with plant viruses in the genus *Begomovirus* (family *Geminiviridae*) have been most extensively studied at the molecular level. To be precise, alphasatellites are not true satellites because they

Viroids and Satellites.
DOI: http://dx.doi.org/10.1016/B978-0-12-801498-1.00054-1

can replicate independently. Betasatellites enhance the pathogenicity of the helper begomovirus by modulating host defenses (Idris et al., 2005; Jose and Usha, 2003; Saeed et al., 2005; Zhou et al., 2003). Although alphasatellites do not contribute to disease development (Briddon and Stanley, 2006; Saunders et al., 2000), recent studies have shown that they can attenuate symptoms caused by the helper virus (Idris et al., 2011; Nawaz-ul-Rehman et al., 2010).

satRNA

A general explanation on satRNAs has been given in Chapter 50, Satellite Viruses and Satellite Nucleic Acids.

Large Linear satRNAs

Bamboo mosaic virus (BaMV) satRNA has been studied extensively (see Chapter 53: Replication of Satellites; Wang et al., 2014). BaMV satRNA has been reported to reduce BaMV RNA replication and suppress the symptoms induced by BaMV; this attenuation may be due to competition with the helper for viral polymerase (Hsu et al., 2006).

Small Linear satRNAs

Among this group, satRNAs of turnip crinkle virus (TCV) and cucumber mosaic virus (CMV) are the best characterized. The 356-nt satRNA C of TCV has been reported to enhance TCV-induced symptoms by interfering with virion formation (Sun and Simon, 2003). CMV satRNAs comprise 330–400 nt and mostly attenuate CMV-induced symptoms, but some enhance the symptoms.

Circular satRNAs

These satRNAs often attenuate the symptoms induced by their helper viruses. For example, stunting of cowpea plants infected with tobacco ringspot virus is significantly reduced by a satRNA (Gerlach et al., 1986). However, arabis mosaic virus satRNA can enhance symptom severity on hops and *Chenopodium quinoa* (Davies and Clark, 1983).

PATHOGENICITY INVOLVING RNA SILENCING MACHINERY IN SATELLITE NUCLEIC ACIDS

A Case Example of satDNAs: Begomovirus Betasatellite

The βC1 protein encoded by a betasatellite of tomato yellow leaf curl China virus, which contains a monopartite genome, can interfere with both posttranscriptional gene silencing (PTGS) and transcriptional gene silencing (Cui et al., 2005; Li et al., 2014; Yang et al., 2011). The interference by βC1 with PTGS occurs by two distinct mechanisms. First, βC1 is an RNA silencing suppressor (RSS) that directly inhibits PTGS (Cui et al., 2005). Second, this protein induces the expression of the *Nb-rgsCam* gene, which encodes an endogenous RSS that inhibits expression of host *RDR6* (Li et al., 2014). In addition, βC1 can interfere with *S*-adenosyl homocysteine hydrolase, resulting in a decrease in methylation-mediated transcriptional gene silencing (Yang et al., 2011). Because the helper virus also encodes a RSS such as AC2/AL2 and C2/L2, the coinfection of βC1 and the helper virus would further enhance the ability to suppress RNA silencing and thus confer great benefit to the begomovirus/betasatellite association.

A Case Example of Small Linear satRNAs: CMV satRNA

At the molecular level, the most thoroughly studied satRNAs are those associated with CMV (García-Arenal and Palukaitis, 1999; Kouadio et al., 2013). Early molecular work on CMV satRNAs included studies on their secondary structure. Rodriguez-Alvarado and Roossinck (1997) analyzed the structure of a CMV satRNA (D4-satRNA) not only in vitro but also in vivo, and identified the particular base-paired region responsible for inducing the lethal necrosis in tomato. These secondary structures contribute to the generation of satRNA-derived small interfering RNAs (sat-siRNAs). Du et al. (2007) showed that dicer-like 4 (DCL4) can recognize the secondary structure of CMV satRNA Shandong strain (SD-satRNA) and produce 21-nt sat-siRNAs from an ss SD-satRNA in *Arabidopsis* plants.

SAT-SIRNAS

Unlike endogenous siRNAs and miRNAs, sat-siRNAs are not modified at the 3′ ends with a methyl group in infected cells, especially when there is a strong RSS such as HC-Pro; they are thus more vulnerable to degradation by exonucleases (Ebhardt et al., 2005). Although Ebhardt et al. (2005) argued that the unmodified sat-siRNAs may be used for priming and activating the RDR6-dependent secondary siRNA pathway to make the host plant survive longer, the biological meaning of this phenomenon needs more consideration since Wang et al. (2010) showed that production of sat-siRNAs is RDR-independent and thus not amplified by either RDR1 or RDR6. The observation that host RDRs do not target satRNAs would partially explain why satRNAs are tolerant to RNA silencing in the presence of abundant sat-siRNAs.

MECHANISMS BY WHICH SATRNAS REDUCE CMV GENOMIC RNAS AND ATTENUATE SYMPTOMS

The classical explanation originally proposed by Kaper (1982), that satRNAs compete with their helper virus for the viral replicase in infected cells and thus reduce helper virus accumulation (Gal-On et al., 1995; Simon et al., 2004), is likely, but some observations cannot be explained by this "competition" mechanism. For example, a satRNA whose replication is supported by both CMV and tomato aspermy virus (TAV) causes the decrease of CMV accumulation, but not of TAV, indicating that the satRNA did not compete with TAV in the infected cells although the satRNA was multiplied by the TAV replicase (McGarvey et al., 1994). Attenuation of symptoms by satRNAs is generally associated with reduced CMV accumulation (Liao et al., 2007; McGarvey et al., 1994). However, there are many exceptions to this observation, showing that symptom severity is not necessarily correlated with CMV accumulation (Escriu et al., 2000; Masuta et al., 1989; Moriones et al., 1992; Wang et al., 2002, 2004).

After the discovery of RNA silencing, siRNAs and miRNAs were identified as key modulators playing crucial roles in plant development, differentiation, and stress responses (Martínez de Alba et al., 2013). Because viral RSSs can interfere with the RNA silencing pathways, virus-induced symptoms can be attributed to a disturbance in siRNA and/or miRNA pathways. Hou et al. (2011) found that a CMV satRNA (SD-satRNA) could reduce the accumulation of the CMV subgenomic RNA (RNA 4A) that codes for the 2b protein (2b). Feng et al. (2012) reported that a benign CMV satRNA (Yn12-satRNA), which attenuates CMV symptoms in tomato, greatly reduced RNA 4A and 2b and that symptom severity in the presence of the satRNA was correlated with the extent of perturbation of a miRNA pathway. But how can a satRNA reduce RNA 4A and 2b? One possible answer suggested by Zhu et al. (2011) is that the 21-nt siRNA derived from the particular region of SD-satRNA (sat-siR-12) could target the 3′ untranslated regions of all three genomic RNAs of CMV. In the absence of 2b, sat-siR-12 efficiently directed cleavage of CMV RNAs using the RDR6-dependent secondary siRNA pathway. In the presence of 2b, the sat-siR-12-mediated inactivation of CMV RNAs was greatly suppressed in a transgenic plant expressing an artificial sat-siR-12. Alternatively, Shen et al. (2015) recently proposed a simpler explanation. Even though satRNAs cannot reduce 2b accumulation, high levels of sat-siRNAs could saturate 2b, thus reducing its capacity to bind host siRNAs and miRNAs.

MOLECULAR MECHANISMS UNDERLYING THE DISTINCT SYMPTOMS INDUCED BY CMV SATRNA

Aside from symptom attenuation, some CMV satRNAs can enhance symptoms induced by CMV. Among such distinct symptoms, lethal systemic necrosis of tomato and yellow chlorosis on tobacco or tomato, have been intensively studied. Soon after sequence information of CMV satRNAs became available, the nucleotide sequence domains responsible for the two symptoms were determined (Devic et al., 1990; Jaegle et al., 1990; Kuwata et al., 1991; Masuta and Takanami, 1989; Sleat and Palukaitis, 1990, 1992; Sleat et al., 1994; Zhang et al., 1994). The satRNA determinant responsible for systemic necrosis on tomato was mapped to just a few nucleotides at the 3′ end, while that for yellowing on tobacco was located in the central region; other CMV satRNAs do not have the latter determinant.

For systemic necrosis on tomato, D-satRNA appears to act via programmed cell death (PCD), but not through RNA silencing, on the basis of nuclear DNA fragmentation and chromatin condensation (Xu and Roossinck, 2000). Xu et al. (2004) further showed that D-satRNA failed to induce cell death in transgenic tomato plants that express animal antiapoptotic genes (*bcl-XL* and *ced-9*), supporting systemic necrosis as a type of PCD. In addition, this PCD requires the plant hormone ethylene (Irian et al., 2007), not involved in animal apoptosis.

For yellowing on tobacco, the molecular mechanism was elucidated by two simultaneously published studies on CMV Y satRNA (Y-sat) (Shimura et al., 2011; Smith et al., 2011). Both groups verified that Y-sat-induced bright yellow symptoms were caused by Y-sat-derived siRNAs, leading to specific downregulation of the magnesium protoporphyrin chelatase subunit I (*ChlI*) gene, with a key role in chlorophyll biosynthesis. A unique 22-nt sequence region, designated SYR, in the Y-sat was found to be complementary to a sequence located at the 3′ half of *ChlI*. The SYR-derived siRNA thus targets *ChlI* mRNA via RNA silencing, resulting in a decrease of chlorophyll and eventually yellowing on tobacco (Fig. 54.1).

OTHER EXAMPLES OF SMALL LINEAR SATRNAS: TCV SATC AND CYMBIDIUM RINGSPOT VIRUS (CYMRSV) SATRNA

TCV satC enhances TCV-induced symptom severity although it reduces TCV accumulation (Zhang and Simon, 2003). The RSS of TCV is the coat protein (CP), which is supposed to inhibit DCL2/DCL4 in the

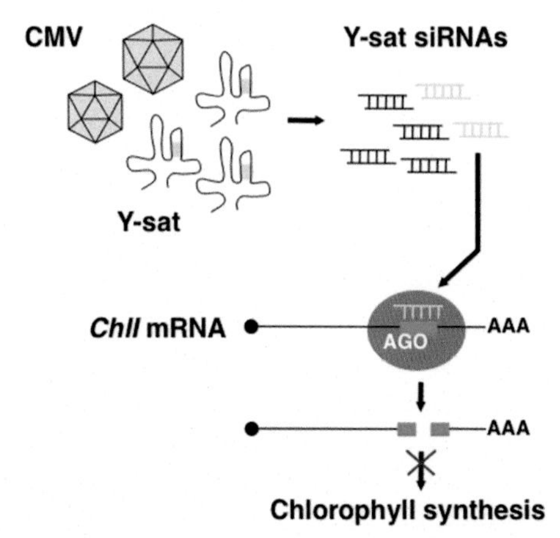

FIGURE 54.1 Yellow symptoms induced by Y-satellite RNA (Y-sat) of cucumber mosaic virus (CMV) through the host RNA silencing machinery. Y-sat is dependent on CMV for its replication. Replication of Y-sat generates a large amount of Y-sat-derived siRNAs. Y-sat has a 22-nt long sequence that is complementary to the magnesium protoporphyrin chelatase subunit I gene (*ChlI*) mRNA. The Y-sat siRNAs are incorporated into AGO and then guide the AGO complex to cleavage of *ChlI* mRNA, which results in a decrease in *ChlI* mRNA level, impairing the chlorophyll biosynthesis pathway. The photograph is a representative yellow symptom on *N. benthamiana* infected with CMV + Y-sat (left). CMV alone causes green mosaic symptoms (right).

silencing pathway. Levels of a mutant TCV (M1) with a CP deficient in RSS activity, were reduced in comparison with those of wild-type TCV in *Arabidopsis* Col-0 plants. When the plants were coinoculated with satC and TCV, TCV levels were reduced in both Col-0 and dcl2/dcl4 (silencing deficient) protoplasts, while addition of satC to M1 did not affect TCV levels in either type of protoplast. These results indicate that satC-mediated reduction of TCV is not attributable to host RNA silencing (Manfre and Simon, 2008; Sun and Simon, 2003). In addition, host

RNA silencing was necessary to enhance satC accumulation only when CP lacked RSS activity (Manfre and Simon, 2008), suggesting that TCV CP functions to decrease satRNA through its RSS activity.

Although satRNAs generally reduce the levels of their helper viruses, CymRSV-derived siRNAs that target CymRSV satRNAs reduce satRNA accumulation (Pantaleo and Burgyán, 2008). The viral siRNAs are generated from the sequence common to the virus and its satRNA and seem to preferentially target its satRNA rather than the original helper RNA thus functioning in favor of the helper virus; the authors concluded that CymRSV used the host antiviral silencing to control its satRNA accumulation.

CONCLUDING REMARKS

SatRNAs accumulate large amounts of their double-stranded RNAs during their replication, and thus large amounts of sat-siRNAs are also generated by host DCLs. In such satRNA-infected tissues, host silencing to maintain cellular homeostasis including miRNA-mediated gene regulation and heterochromatin silencing, must be deeply affected. As we have here illustrated with some examples, it appears reasonable to assume that satRNA pathogenesis primarily results from the interactions between the host RNA silencing and the satRNAs.

Acknowldgments

C.M. would like to dedicate this chapter to the memory of Dr. Yoichi Takanami, who directed him to work on CMV satRNAs.

References

Briddon, R.W., Ghabrial, S., Lin, N.-S., Palukaitis, P., Scholthof, K.-B.G., Vetten, H.-J., 2012. Satellite and other virus-dependent nucleic acids. In: King, A.M.Q., Adams, M.J., Carstens, E.B., Lefkowitz, E.J. (Eds.), Virus taxonomy: Classification and Nomenclature of Viruses, Ninth Report of the International Committee on Taxonomy of Viruses. Academic Press, San Diego, CA, pp. 1211–1219.

Briddon, R.W., Stanley, J., 2006. Subviral agents associated with plant single-stranded DNA viruses. Virology 344, 198–210.

Cui, X., Li, G., Wang, D., Hu, D., Zhou, X., 2005. A begomovirus DNA β-encoded protein binds DNA, functions as a suppressor of RNA silencing, and targets the cell nucleus. J. Virol. 79, 10764–10775.

Davies, D.L., Clark, M.F., 1983. A satellite-like nucleic acid of arabis mosaic virus associated with hop nettlehead disease. Ann. Appl. Biol. 103, 439–448.

Devic, M., Jaegle, M., Baulcombe, D., 1990. Cucumber mosaic virus satellite RNA (strain Y): analysis of sequences which affect systemic necrosis on tomato. J. Gen. Virol. 71, 1443–1449.

Du, Q.-S., Duan, C.-G., Zhang, Z.-H., Fang, Y.-Y., Fang, R.-X., Xie, Q., et al., 2007. DCL4 targets cucumber mosaic virus satellite RNA at novel secondary structures. J. Virol. 81, 9142–9151.

Ebhardt, H.A., Thi, E.P., Wang, M.-B., Unrau, P.J., 2005. Extensive 3′modification of plant small RNAs is modulated by helper component-proteinase expression. Proc. Natl. Acad. Sci. USA 102, 13398–13403.

Escriu, F., Perry, K.L., García-Arenal, F., 2000. Transmissibility of cucumber mosaic virus by *Aphis gossypii* correlates with viral accumulation and is affected by the presence of its satellite RNA. Phytopathology 90, 1068–1072.

Feng, J., Lai, L., Lin, R., Jin, C., Chen, J., 2012. Differential effects of cucumber mosaic virus satellite RNAs in the perturbation of microRNA-regulated gene expression in tomato. Mol. Biol. Rep. 39, 775–784.

Gal-On, A., Kaplan, I., Palukaitis, P., 1995. Differential effects of satellite RNA on the accumulation of cucumber mosaic virus RNAs and their encoded proteins in tobacco vs zucchini squash with two strains of CMV helper virus. Virology 208, 58–66.

García-Arenal, F., Palukaitis, P., 1999. Structure and functional relationships of satellite RNAs of cucumber mosaic virus. In: Vogt, P.K., Jackson, A.O. (Eds.), Satellites and Defective Viral RNAs. Springer-Verlag, Berlin, Heidelberg, pp. 37–63.

Gerlach, W.L., Buzayan, J.M., Schneider, I.R., Bruening, G., 1986. Satellite tobacco ringspot virus RNA: biological activity of DNA clones and their in vitro transcripts. Virology 151, 172–185.

Hou, W.-N., Duan, C.-G., Fang, R.-X., Zhou, X.-Y., Guo, H.-S., 2011. Satellite RNA reduces expression of the 2b suppressor protein resulting in the attenuation of symptoms caused by cucumber mosaic virus infection. Mol. Plant Pathol. 12, 595–605.

Hsu, Y.-H., Chen, H.-C., Cheng, J., Annamali, P., Lin, B.-Y., Wu, C.-T., et al., 2006. Crucial role of the 5′conserved structure of bamboo mosaic virus satellite RNA in downregulation of helper viral RNA replication. J. Virol. 80, 2566–2574.

Hu, C.-C., Hsu, Y.-H., Lin, N.-S., 2009. Satellite RNAs and satellite viruses of plants. Viruses 1, 1325–1350.

Idris, A.M., Briddon, R.W., Bull, S.E., Brown, J.K., 2005. Cotton leaf curl Gezira virus-satellite DNAs represent a divergent, geographically isolated Nile basin lineage: predictive identification of a satDNA Rep-binding motif. Virus Res. 109, 19–32.

Idris, A.M., Shahid, M.S., Briddon, R.W., Khan, A.J., Zhu, J.K., Brown, J.K., 2011. An unusual alphasatellite associated with monopartite begomoviruses attenuates symptoms and reduces betasatellite accumulation. J. Gen. Virol. 92, 706–717.

Irian, S., Xu, P., Dai, X., Zhao, P.X., Roossinck, M.J., 2007. Regulation of a virus-induced lethal disease in tomato revealed by LongSAGE analysis. Mol. Plant Microbe Interact. 20, 1477–1488.

Jaegle, M., Devic, M., Longstaff, M., Baulcombe, D., 1990. Cucumber mosaic virus satellite RNA (Y strain): analysis of sequences which affect yellow mosaic symptoms on tobacco. J. Gen. Virol. 71, 1905–1912.

Jose, J., Usha, R., 2003. Bhendi yellow vein mosaic disease in India is caused by association of a DNA β satellite with a begomovirus. Virology 305, 310–317.

Kaper, J.M., 1982. Rapid synthesis of double-stranded cucumber mosaic virus-associated RNA 5: mechanism controlling viral pathogenesis? Biochem. Biophys. Res. Commun. 105, 1014–1022.

Kouadio, K.T., De Clerck, C., Agneroh, T.A., Parisi, O., Lepoivre, P., Jijakli, H., 2013. Role of satellite RNAs in cucumber mosaic virus-host plant interactions. A review. Biotechnol. Agron. Soc. Environ. 17, 644–650.

Kuwata, S., Masuta, C., Takanami, Y., 1991. Reciprocal phenotype alterations between two satellite RNAs of cucumber mosaic virus. J. Gen. Virol. 72, 2385–2389.

Li, F., Huang, C., Li, Z., Zhou, X., 2014. Suppression of RNA silencing by a plant DNA virus satellite requires a host calmodulin-like protein to repress RDR6 expression. PLoS Pathog. 10, e21003921.

Liao, Q., Zhu, L., Du, Z., Zeng, R., Feng, J., Chen, J., 2007. Satellite RNA-mediated reduction of cucumber mosaic virus genomic RNAs accumulation in *Nicotiana tabacum*. Acta Biochim. Biophys. Sinica. 39, 217–223.

Manfre, A.J., Simon, A.E., 2008. Importance of coat protein and RNA silencing in satellite RNA/virus interactions. Virology 379, 161–167.

Martínez de Alba, A.E., Elvira-Matelot, E., Vaucheret, H., 2013. Gene silencing in plants: a diversity of pathways. Biochim. Biophys. Acta. 1829, 1300–1308.

Masuta, C., Takanami, Y., 1989. Determination of sequence and structural requirements for pathogenicity of a cucumber mosaic virus satellite RNA (Y-satRNA). Plant Cell. 1, 1165–1173.

Masuta, C., Komari, T., Takanami, Y., 1989. Expression of cucumber mosaic virus satellite RNA from cDNA copies in transgenic tobacco plants. Ann. Phytopathol. Soc. Japan 55, 49–55.

McGarvey, P.B., Montasser, M.S., Kaper, J.M., 1994. Transgenic tomato plants expressing satellite RNA are tolerant to some strains of cucumber mosaic virus. J. Am. Soc. Hortic. Sci. 119, 642–647.

Moriones, E., Diaz, I., Rodriguez-Cerezo, E., Fraile, A., Garcia-Arenal, F., 1992. Differential interactions among strains of tomato aspermy virus and satellite RNAs of cucumber mosaic virus. Virology 186, 475–480.

Nawaz-ul-Rehman, M.S., Fauquet, C.M., 2009. Evolution of geminiviruses and their satellites. FEBS Lett. 583, 1825–1832.

Nawaz-ul-Rehman, M.S., Nahid, N., Mansoor, S., Briddon, R.W., Fauquet, C.M., 2010. Post-transcriptional gene silencing suppressor activity of two non-pathogenic alphasatellites associated with a begomovirus. Virology 405, 300–308.

Pantaleo, V., Burgyán, J., 2008. Cymbidium ringspot virus harnesses RNA silencing to control the accumulation of virus parasite satellite RNA. J. Virol. 82, 11851–11858.

Rodriguez-Alvarado, G., Roossinck, M.J., 1997. Structural analysis of a necrogenic strain of cucumber mosaic cucumovirus satellite RNA in planta. Virology 236, 155–166.

Roossinck, M.J., Sleat, D., Palukaitis, P., 1992. Satellite RNAs of plant viruses: structures and biological effects. Microbiol. Rev. 56, 265–279.

Saeed, M., Behjatnia, S.A.A., Mansoor, S., Zafar, Y., Hasnain, S., Rezaian, M.A., 2005. A single complementary-sense transcript of a geminiviral DNA β satellite is determinant of pathogenicity. Mol Plant Microbe Interact. 18, 7–14.

Saunders, K., Bedford, I.D., Briddon, R.W., Markham, P.G., Wong, S.M., Stanley, J., 2000. A unique virus complex causes Ageratum yellow vein disease. Proc. Natl. Acad. Sci. USA 97, 6890–6895.

Shen, W., Au, P.C.K., Shi, B.J., Smith, N.A., Dennis, E.S., Guo, H.S., et al., 2015. Satellite RNAs interfere with the function of viral RNA silencing suppressors. Front. Plant Sci. 6, 281.

Shimura, H., Pantaleo, V., Ishihara, T., Myojo, N., Inaba, J., Sueda, K., et al., 2011. A viral satellite RNA induces yellow symptoms on tobacco by targeting a gene involved in chlorophyll biosynthesis using the RNA silencing machinery. PLoS Pathog. 7, e1002021.

Simon, A.E., Roossinck, M.J., Havelda, Z., 2004. Plant virus satellite and defective interfering RNAs: new paradigms for a new century. Annu. Rev. Phytopathol. 42, 415–437.

Sleat, D.E., Palukaitis, P., 1990. Site-directed mutagenesis of a plant viral satellite RNA changes its phenotype from ameliorative to necrogenic. Proc. Natl. Acad. Sci. 87, 2946–2950.

Sleat, D.E., Palukaitis, P., 1992. A single nucleotide change within a plant virus satellite RNA alters the host specificity of disease induction. Plant J. 2, 43–49.

Sleat, D.E., Sleat, D.E., Zhang, L., Palukaitis, P., 1994. Mapping determinants within cucumber mosaic virus and its satellite RNA for the induction of necrosis in tomato-plants. Mol. Plant Microbe Interact. 7, 189–195.

Smith, N.A., Eamens, A.L., Wang, M.B., 2011. Viral small interfering RNAs target host genes to mediate disease symptoms in plants. PLoS Pathog. 7, e1002022.

Sun, X., Simon, A.E., 2003. Fitness of a turnip crinkle virus satellite RNA correlates with a sequence-nonspecific hairpin and flanking sequences that enhance replication and repress the accumulation of virions. J. Virol. 77, 7880–7889.

Wang, I.-N., Hu, C.-C., Lee, C.-W., Yen, S.-M., Yeh, W.-B., Hsu, Y.-H., et al., 2014. Genetic diversity and evolution of satellite RNAs associated with Bamboo mosaic virus. PLoS One 9, e108015.

Wang, X.-B., Wu, Q., Ito, T., Cillo, F., Li, W.-X., Chen, X., et al., 2010. RNAi-mediated viral immunity requires amplification of virus-derived siRNAs in *Arabidopsis thaliana*. Proc. Natl. Acad. Sci. USA 107, 484–489.

Wang, Y., Gaba, V., Yang, J., Palukaitis, P., Gal-On, A., 2002. Characterization of synergy between cucumber mosaic virus and potyviruses in cucurbit hosts. Phytopathology 92, 51–58.

Wang, Y., Lee, K.C., Gaba, V., Wong, S.M., Palukaitis, P., Gal-On, A., 2004. Breakage of resistance to cucumber mosaic virus by co-infection with zucchini yellow mosaic virus: enhancement of CMV accumulation independent of symptom expression. Arch. Virol. 149, 379–396.

Xu, P., Rogers, S.J., Roossinck, M.J., 2004. Expression of antiapoptotic genes bcl-xL and ced-9 in tomato enhances tolerance to viral-induced necrosis and abiotic stress. Proc. Natl. Acad. Sci. USA 101, 15805–15810.

Xu, P., Roossinck, M.J., 2000. Cucumber mosaic virus D satellite RNA-induced programmed cell death in tomato. Plant Cell. 12, 1079–1092.

Yang, X., Xie, Y., Raja, P., Li, S., Wolf, J.N., Shen, Q., et al., 2011. Suppression of methylation-mediated transcriptional gene silencing by βC1-SAHH protein interaction during geminivirus-betasatellite infection. PLoS Pathog. 7, e1002329.

Zhang, F., Simon, A.E., 2003. Enhanced viral pathogenesis associated with a virulent mutant virus or a virulent satellite RNA correlates with reduced virion accumulation and abundance of free coat protein. Virology 312, 8–13.

Zhang, L., Kim, C.H., Palukaitis, P., 1994. The chlorosis-induction domain of the satellite RNA of cucumber mosaic virus: identifying sequences that affect accumulation and the degree of chlorosis. Mol. Plant Microbe Interact. 7, 208–213.

Zhou, X., Xei, Y., Tao, X., Zhang, Z., Li, Z., Fauquet, C., 2003. Characterization of DNAβ associated with begomoviruses in China and evidence for co-evolution with their cognate viral DNA-A. J. Gen. Virol. 84, 237–247.

Zhu, H., Duan, C.-G., Hou, W.-N., Du, Q.-S., Lv, D.-Q., Fang, R.-X., et al., 2011. Satellite RNA-derived small interfering RNA satsiR-12 targeting the 3′ untranslated region of cucumber mosaic virus triggers viral RNAs for degradation. J. Virol. 85, 13384–13397.

CHAPTER

55

Development and Application of Satellite-Based Vectors

Ming-Ru Liou[1], *Chung-Chi Hu*[1], *Na-Sheng Lin*[2] *and Yau-Heiu Hsu*[1]

[1]**National Chung Hsing University, Taichung, Taiwan** [2]**Institute of Plant and Microbial Biology, Academia Sinica, Taipei, Taiwan**

INTRODUCTION

Plant viral satellites are molecular parasites associated with helper viruses, as discussed in detail in Chapter 50, Satellite Viruses and Satellite Nucleic Acids. Studies on plant viral satellites have contributed greatly to our understanding of molecular virology, and also revealed the potential of these satellites as vectors for various applications (Huang et al., 2010; Roossinck et al., 1992; Simon et al., 2004). Certain features of satellites make them especially suitable for the development of vector systems: (1) plant viruses and associated satellites constitute two natural component systems that uncouple the biological functions of the helper viruses from those of the foreign genes on the satellites, which may enhance system stability (Gosselé et al., 2002); (2) in their host plants satellites can replicate efficiently and accumulate to high levels, usually higher than those of their helper viruses; and (3) genetic manipulations of satellites are relatively easy, due to their smaller genome size as compared to their helper viruses. Thus, several satellite-based vector systems have been established (Hu et al., 2009; Xu and Roossinck, 2011; Zhou, 2013; Zhou and Huang, 2012). Two major biotechnological applications are the use of satellites for ectopic expression of foreign proteins and silencing of endogenous or foreign genes. In the following sections, examples of satellite-based vectors for these two applications are presented, as well as current challenges and possible future trends of satellite-based vector systems.

Viroids and Satellites.
DOI: http://dx.doi.org/10.1016/B978-0-12-801498-1.00055-3

IN PLANTA EXPRESSION OF FOREIGN GENE

Despite the aforementioned potential advantages, the use of satellite-based vectors for ectopic expression of foreign genes is limited. This is probably due to constraints on the sizes of satellite genomes for encapsidation and/or the strict requirement for the maintenance of certain secondary structures, which severely limit sequence alterations. Bamboo mosaic virus satellite RNA (satBaMV) was used as a satRNA-based foreign gene expression vector *in planta* (Lin et al., 1996), and remains a unique example of such application. SatBaMV is a linear RNA with an open reading frame (ORF) that encodes a 20-kDa P20 protein (Lin and Hsu, 1994), which is not required for satBaMV replication. Thus, the ORF of P20 can be replaced with a foreign gene for ectopic expression in infected *Chenopodium quinoa*. The high yield (2 μg/g of leaf) of chloramphenicol acetyltransferase protein in plants demonstrated that satBaMV can be used as an expression vector (Lin et al., 1996). The ectopic expression of endogenous or foreign genes has also been used in functional genomics studies. For example, a satBaMV-based vector has served to express triple gene block proteins (TGBps) of various potexviruses in a complementary experiment to study cell-to-cell movement of BaMV (Lin et al., 2006), which also is a potexvirus. This study revealed the requirement for species-specific interactions among TGBps of BaMV for cell-to-cell movement, which may also be the case for the TGBps of other potexviruses. The satBaMV-based vector system has also been used to analyze promoters that control synthesis of potexvirus subgenomic RNAs (Lee et al., 2000). Insertion of subgenomic promoter-like sequences upstream of the P20 ORF led to the synthesis of satBaMV subgenomic RNA in infected cells coinoculated with BaMV. The satBaMV-based vector was used to reveal one core promoter-like sequence, two upstream enhancers, and one downstream enhancer in the BaMV subgenomic prometor-like sequence in vivo (Lee et al., 2000).

The satBaMV-based expression system has several advantages over alternatives. The accumulation level of satRNAs is usually higher than those of the helper viruses, which may result in higher expression levels of foreign proteins. In addition, the ORF of the target gene is not inserted into the viral genomic RNA, and thus does not directly interfere with the normal biological functions of the helper virus. Such a two-component system is relatively easy to construct, maintain, and manipulate to suit the experimental requirements. Studies of satBaMV-based vectors have demonstrated that the satellite-based system is suitable for mutational studies of essential genes or vital regulatory sequences of helper viruses, since satRNAs are not required in the infection cycle of the helper virus. Such satellite-based vector systems are also appropriate for combined studies of multiple genes, as two or more satRNA-based vectors harboring different genes may coexist with the same helper virus in a single plant (Hu et al., 2009). However, it is possible that the target genes may not be expressed at the same level.

It should be noted that such applications are currently limited to satRNAs that harbor nonessential ORFs. However, it is expected that a deeper understanding of satRNA biology will enable satRNA-based expression vectors to be used in the production of vaccines or biopharmaceutical proteins *in planta* in the future.

GENE SILENCING VECTOR

Plant virus-based gene silencing vectors have become common genetic tools for plant functional genomic studies. Expression of a plant virus-based vector carrying an endogenous or foreign gene fragment may trigger a sequence-specific RNA inactivation process in the inoculated plant, which is referred to as virus-induced gene silencing (VIGS) (Baulcombe, 1999). To date, more than 30 plant viruses have been modified to serve as VIGS vectors. Similarly, plant viral satellites can also be used as vectors to trigger such "induced gene silencing" effects. In this section, the use of various types of satellite-based vectors in induced gene silencing studies (as summarized in Table 55.1) is discussed.

Satellite Virus-Based Vector: Satellite Tobacco Mosaic Virus (STMV)

STMV was the first subviral agent to be developed as a VIGS vector, to generate a satellite virus-induced silencing system (SVISS) (Gosselé et al., 2002). The potential of STMV as a satellite virus-based vector for gene silencing was demonstrated by knocking out the expression of 14 variably expressed endogenous genes (e.g., phytoene desaturase, glutamine synthetase, chalcone synthase, cellulose synthase, acetolactate synthase, and transketolase) involved in different biochemical pathways in *Nicotiana tabacum*. The efficiency of SVISS was supported by the observation that most of the relevant phenotypes were silenced at 10–12 days postinoculation in various plant tissues (Gosselé and Metzlaff, 2005; Gosselé et al., 2002). The authors also demonstrated the ability of STMV-based SVISS to induce simultaneous knockout of an endogenous gene and a transgene by using a tandem-repeat construct (Gosselé et al., 2002). However, this vector is based on a frame-shift mutant of STMV, which does not produce a functional CP. All other deletion mutants of STMV, which provide spaces for the insertion of

TABLE 55.1 Plant Virus Satellites That were Developed as Vectors for Practical Applications

Satellite-based vectors	Practical applications	References
Satellite virus		
Satellite tobacco mosaic virus	Gene expression vector	Gosselé et al. (2002), Gosselé and Metzlaff (2005)
Satellite RNA		
Bamboo mosaic virus satellite RNA	Gene expression vector, gene silencing vector	Lin et al. (1996), Liou et al. (2014)
Satellite DNA		
Alphasatellite		
Tobacco curly shoot virus DNA1	Gene silencing vector	Huang et al. (2009, 2011)
Betasatellites		
Bhendi yellow vein mosaic virus β-satellite	Gene silencing vector	Jeyabharathy et al. (2015)
Cotton leaf curl Multan betasatellite	Gene silencing vector	Kharazmi et al. (2012), Kumar et al. (2014)
Tomato leaf curl virus satellite DNA	Gene silencing vector	Li et al. (2008)
Tomato yellow leaf curl China virus DNAβ	Gene silencing vector	Cai et al. (2007), He et al. (2008), Qian et al. (2006), Tao and Zhou (2004), Tao et al. (2006), Zhou and Huang (2012)

target gene fragments, exhibited poor accumulation in plants, and did not induce silencing phenotypes. Thus, the functional SVISS vector is not encapsidated, raising major concerns regarding the stability of this system and the persistence of silencing effects.

SatRNA-Based Vector: satBaMV

SatBaMV has not only been developed as a foreign protein expression vector (Lin et al., 1996), but also modified for use as a dual gene silencing vector system together with its helper virus, BaMV (Liou et al., 2014). It has recently been shown that both satBaMV and BaMV vectors can effectively silence endogenous genes in dicotyledonous (*Nicotiana benthamiana*) and monocotyledonous (*Brachypodium distachyon*) plants. Such a dual gene silencing vector has several advantages: (1) double-silenced plants can be easily obtained without tedious traditional genetic manipulations, such as chemical mutagenesis or either T-DNA or transposon insertions; (2) the vectors can be used to investigate the interactions between different genes or pathways (such as those involved in metabolism or disease resistance); and (3) the silencing phenotype of one target gene induced by a satRNA-based vector can serve as an indicator for the silencing of the other target gene induced by the helper virus-based vector, since the satRNA is dependent on the helper virus (Liou et al., 2014), thus enhancing the selection efficiency of simultaneous cosuppression of two genes. Furthermore, the efficiency of gene silencing was enhanced by satBaMV-, but not BaMV-based vectors, in newly emerging leaves of *N. benthamiana* deficient for RNA-dependent RNA polymerase 6, suggesting that gene silencing mechanisms induced by satRNAs and RNA viruses may differ.

SatDNA-Based Vector: Begomovirus Satellites

Two types of subviral DNAs, alphasatellite (formerly known as DNA1, but not true satellites) and betasatellite (DNAβ), have been reported to be associated with begomoviruses (Zhou, 2013). Alphasatellites encode their own replication initiator protein (Rep) and depend on the helper virus for encapsidation, movement, and transmission, but not for replication. In contrast, betasatellites encode a βC1 protein involved in symptom development and suppression of gene silencing. To date, one alphasatellite and four betasatellites have been successfully developed as gene silencing vectors. These satDNAs are the DNA1 associated with tobacco curly shoot virus (designated as 2mDNA1, an alphasatellite) (Huang et al., 2009, 2011), and the betasatellites associated with tomato yellow leaf curl China virus (Cai et al., 2007; He et al., 2008; Qian et al., 2006; Tao and Zhou, 2004; Tao et al., 2006; Zhou and Huang, 2012), tomato leaf curl virus (Li et al., 2007, 2008), cotton leaf curl Multan virus (Kharazmi et al., 2012; Kumar et al., 2014), and bhendi yellow vein mosaic virus (Jeyabharathy et al., 2015). The βC1 ORF was replaced with a multiple cloning site for insertion of target gene fragments, to enhance the efficiency of gene silencing. In the alphasatellite-based vector, the multiple cloning site was inserted between the Rep ORF and the A-rich motif. Begomovirus satDNA-based vectors exhibit several advantages over other gene silencing systems. These satDNA-based gene silencing vectors can: (1) infect a broad range of plants; (2) replicate in plant nuclei; (3) be trans-replicated by different noncognate helper begomovirus species and induce gene silencing in the new host plants (Huang et al., 2009; Kharazmi et al., 2012; Qian et al., 2006); (4) silence genes in different tissues (including root, shoot apex, flower, and fruit); (5) tolerate high-temperature conditions (Cai et al., 2007); and (6) maintain a persistent silencing effect until flowering and fruiting (Huang et al., 2009; Zhou, 2013). In addition, these vectors do not induce severe symptoms in plants, avoiding the possibility of mistaking viral symptoms for gene silencing phenotypes. Furthermore, the coexistence of alpha- and betasatellite (Xie et al., 2010) or association of two betasatellites (Qing and Zhou, 2009; Zhou et al., 2003) with a single helper begomovirus in the same plant may enable simultaneous silencing of two or more genes using satDNA-based vector systems. However, competition between different satDNAs within the plant is a major concern.

CONCLUDING REMARKS: CURRENT CHALLENGES AND FUTURE PERSPECTIVES OF SATELLITE-BASED VECTORS

Despite the many advantages of satellite-based systems as ectopic expression or induced gene silencing vectors, two major drawbacks still limit the development of such vector systems: (1) the strict requirement of certain nucleotide sequences for the maintenance of higher-order structures for replication and/or encapsidation may restrict engineering of the satellite genome and (2) the requirement for encapsidation also limits the size of genes (fragments) that can be inserted into the vector. In addition, some satellites are known to differentially affect the symptoms caused by their helper viruses upon inoculation of different host plants (Roossinck et al., 1992; Shen et al., 2015; Shimura et al., 2011; Zhou, 2013). Further studies on the molecular bases of the trilateral interactions among satellites, helper viruses, and the host plants are required to overcome these challenges.

As more satellites associated with plant viruses are identified, it is expected that more versatile satellite-based systems may be developed. Future trends for the development and applications of satellite-based vector systems may include: (1) vectors with easily adjustable expression levels or gene silencing efficiencies, possibly by using different combinations of satellites/vectors; (2) vectors that enable simultaneous expression and/or silencing of multiple genes, possibly through the use of different, compatible, satellite-based vectors in the same plant; and (3) vectors with applications in the biological control of viral diseases, possibly through the expression of certain host defense genes or the silencing of host factors involved in viral infection cycles. Future studies and developments in this field will lay the foundation for novel applications of satellite-based vectors.

Acknowledgments

This research was supported by the Ministry of Science and Technology of the Republic of China grants (NSC-98-2321-B-005-005-MY3 and NSC-97-2752-B-005-005-PAE) and Council of Agriculture (97AS-1.2.1-ST-a2), Taiwan, Republic of China.

References

Baulcombe, D.C., 1999. Gene silencing: RNA makes RNA makes no protein. Curr. Biol. 9, R599–R601.

Cai, X., Wang, C., Xu, Y., Xu, Q., Zheng, Z., Zhou, X., 2007. Efficient gene silencing induction in tomato by a viral satellite DNA vector. Virus Res. 125, 169–175.

Gosselé, V., Metzlaff, M., 2005. Using satellite tobacco mosaic virus vectors for gene silencing. Curr. Protoc. Microbiol. Chapter 16, Unit 16I.5.

Gosselé, V., Faché, I., Meulewaeter, F., Cornelissen, M., Metzlaff, M., 2002. SVISS – a novel transient gene silencing system for gene function discovery and validation in tobacco plants. Plant J. 32, 859–866.

He, X., Jin, C., Li, G., You, G., Zhou, X., Zheng, S., 2008. Use of the modified viral satellite DNA vector to silence mineral nutrition-related genes in plants: silencing of the tomato ferric chelate reductase gene, *FRO1*, as an example. Sci. China Ser. C Life Sci. 51, 402–409.

Hu, C.C., Hsu, Y.H., Lin, N.S., 2009. Satellite RNAs and satellite viruses of plants. Viruses 1, 1325–1350.

Huang, C., Xie, Y., Zhou, X., 2009. Efficient virus-induced gene silencing in plants using a modified geminivirus DNA1 component. Plant Biotechnol. J. 7, 254–265.

Huang, C.J., Zhang, T., Li, F.F., Zhang, X.Y., Zhou, X.P., 2011. Development and application of an efficient virus-induced gene silencing system in *Nicotiana tabacum* using geminivirus alphasatellite. J. Zhejiang Univ. Sci. B. 12, 83–92.

Huang, Y.W., Hu, C.C., Lin, N.S., Hsu, Y.H., 2010. Mimicry of molecular pretenders: the terminal structures of satellites associated with plant RNA viruses. RNA Biol. 7, 162–171.

Jeyabharathy, C., Shakila, H., Usha, R., 2015. Development of a VIGS vector based on the β-satellite DNA associated with bhendi yellow vein mosaic virus. Virus Res. 195, 73–78.

Kharazmi, S., Behjatnia, S.A., Hamzehzarghani, H., Niazi, A., 2012. Cotton leaf curl Multan betasatellite as a plant gene delivery vector trans-activated by taxonomically diverse geminiviruses. Arch. Virol. 157, 1269–1279.

Kumar, J., Gunapati, S., Kumar, J., Kumari, A., Kumar, A., Tuli, R., et al., 2014. Virus-induced gene silencing using a modified betasatellite: a potential candidate for functional genomics of crops. Arch. Virol. 159, 2109–2113.

Lee, Y.S., Hsu, Y.H., Lin, N.S., 2000. Generation of subgenomic RNA directed by a satellite RNA associated with bamboo mosaic potexvirus: analyses of potexvirus subgenomic RNA promoter. J. Virol. 74, 10341–10348.

Li, D., Behjatnia, S.A.A., Dry, I.B., Randles, J.W., Eini, O., Rezaian, M.A., 2007. Genomic regions of tomato leaf curl virus DNA satellite required for replication and for satellite-mediated delivery of heterologous DNAs. J. Gen. Virol. 88, 2073–2077.

Li, D., Behjatnia, S.A.A., Dry, I.B., Walker, A.R., Randles, J.W., Rezaian, A.M., 2008. Tomato leaf curl virus satellite DNA as a gene silencing vector activated by helper virus infection. Virus Res. 136, 30–34.

Lin, M.K., Hu, C.C., Lin, N.S., Chang, B.Y., Hsu, Y.H., 2006. Movement of potexviruses requires species-specific interactions among the cognate triple gene block proteins, as revealed by a trans-complementation assay based on the bamboo mosaic virus satellite RNA-mediated expression system. J. Gen. Virol. 87, 1357–1367.

Lin, N.S., Hsu, Y.H., 1994. A satellite RNA associated with bamboo mosaic potexvirus. Virology 202, 707–714.

Lin, N.S., Lee, Y.S., Lin, B.Y., Lee, C.W., Hsu, Y.H., 1996. The open reading frame of bamboo mosaic potexvirus satellite RNA is not essential for its replication and can be replaced with a bacterial gene. Proc. Natl. Acad. Sci. USA 93, 3138–3142.

Liou, M.R., Huang, Y.W., Hu, C.C., Lin, N.S., Hsu, Y.H., 2014. A dual gene-silencing vector system for monocot and dicot plants. Plant Biotechnol. J. 12, 330–343.

Qian, Y., Mugiira, R.B., Zhou, X., 2006. A modified viral satellite DNA-based gene silencing vector is effective in association with heterologous begomoviruses. Virus Res. 118, 136–142.

Qing, L., Zhou, X., 2009. Trans-replication of, and competition between, DNAβ satellites in plants inoculated with tomato yellow leaf curl China virus and tobacco curly shoot virus. Phytopathology 99, 716–720.

Roossinck, M.J., Sleat, D., Palukaitis, P., 1992. Satellite RNAs of plant viruses: structures and biological effects. Microbiol. Rev. 56, 265–279.

Shen, W.X., Au, P.C., Shi, B.J., Smith, N.A., Dennis, E.S., Guo, H.S., et al., 2015. Satellite RNAs interfere with the function of viral RNA silencing suppressors. Front Plant Sci. 6, 281.

Shimura, H., Pantaleo, V., Ishihara, T., Myojo, N., Inaba, J., Sueda, K., et al., 2011. A viral satellite RNA induces yellow symptoms on tobacco by targeting a gene involved in chlorophyll biosynthesis using the RNA silencing machinery. PLoS Pathog. 7, e1002021.

Simon, A.E., Roossinck, M.J., Havelda, Z., 2004. Plant virus satellite and defective interfering RNAs: new paradigms for a new century. Annu. Rev. Phytopathol. 42, 415–437.

Tao, X., Qian, Y., Zhou, X., 2006. Modification of a viral satellite DNA-based gene silencing vector and its application for leaf or flower color change in *Petunia hybrida*. Chin. Sci. Bull. 51, 2208–2213.

Tao, X., Zhou, X., 2004. A modified viral satellite DNA that suppresses gene expression in plants. Plant J. 38, 850–860.

Xie, Y., Wu, P., Liu, P., Gong, H., Zhou, X., 2010. Characterization of alphasatellites associated with monopartite begomovirus/betasatellite complexes in Yunnan, China. Virol. J. 7, 178.

Xu, P., Roossinck, M.J., 2011. Plant virus satellites. Encyclopedia of Life Sciences (ELS). John Wiley & Sons Ltd, Chichester. Available from: http://dx.doi.org/10.1002/9780470015902.a0000771, pub2.

Zhou, X., Huang, C., 2012. Virus-induced gene silencing using begomovirus satellite molecules. Methods Mol. Biol. 894, 57–67.

Zhou, X., 2013. Advances in understanding begomovirus satellites. Annu. Rev. Phytopathol. 51, 357–381.

Zhou, X., Xie, Y., Tao, X., Zhang, K., Li, Z., Fauquet, C.M., 2003. Characterization of DNAβ associated with begomoviruses in China and evidence for co-evolution with their cognate viral DNA-A. J. Gen. Virol. 84, 237–247.

CHAPTER

56

Origin and Evolution of Satellites

Fernando García-Arenal and Aurora Fraile

Center for Plant Biotechnology and Genomics, Polytechnic University of Madrid-INIA, Madrid, Spain

INTRODUCTION

Satellites, like their helper viruses (HV), vary genetically because, during replication, mutations (*sensu lato*, i.e., base substitutions, insertions, deletions, inversions, recombinations) are introduced in their nucleic acids thus producing genetic variants. The frequency distribution of genetic variants in the population of an organism defines its genetic structure, which may change with time in a process called evolution. Evolution may lead eventually to the rise of new taxa, such as virus or satellite species, and one aspect of the study of evolution is to clarify the origin and history of organisms and their resulting taxonomic relationships. For viruses, and satellites, this paleontological side of evolutionary studies relies almost entirely on phylogenetic reconstructions based on nucleotide or amino acid sequences of extant individuals of different taxa. Interesting as this intellectual exercise might be, and in spite of large methodological advances in phylogenetics, it abounds with pitfalls, which deepen the farther back in time reconstructions attempt to dig (e.g., Duchêne et al., 2014).

The other important goal of the study of evolution is understanding its mechanisms and how they shape the genetic structure of populations. The study of the genetic variation of virus and satellite populations, and of changes in their genetic structure, can be addressed with higher confidence from sequence analyses of extant individuals and, moreover, is amenable to experimentation. The study of satellite evolution under this perspective has addressed central questions

Viroids and Satellites.
DOI: http://dx.doi.org/10.1016/B978-0-12-801498-1.00056-5

of evolutionary biology, and is highly relevant for understanding the emergence of new viral diseases and for developing strategies for disease control. In consequence, much effort has been made to understand the evolution of those satellites that modulate the pathogenicity of their helper viruses.

In this chapter we will review knowledge on the evolution of satellite nucleic acids, not considering satellite viruses, for which less information is available, and we will not discuss satellites other than those associated with plant-infecting viruses.

ORIGIN OF SATELLITE NUCLEIC ACIDS

As with viruses, the origin of satellites is unknown and the subject of much speculation. An evolutionary line could link nondependent viruses to satellite viruses and to large coding satellites, such as bamboo mosaic virus satellite RNA, by size and information content reduction (Palukaitis et al., 2008). On the basis of sequence analyses it has been proposed that alphasatellites of begomoviruses derive from nanoviruses (Zhou, 2013), that betasatellites derive from preexisting DNA elements of unknown origin after donation by the helper virus (HV) DNA of the conserved stem-loop motif required for replication, that they acquired the A-rich conserved region to reach a size allowing encapsidation (Mansoor et al., 2003), and that viroids and small circular satellite RNAs are phylogenetically related (Elena et al., 2001). None of these hypotheses addresses the origin of satellites.

Satellite nucleic acids do not share with their HV sequences other than those involved in recognition by the HV replication machinery, with the notable exception of the virulent satellite RNA (satRNA) of turnip crinkle virus (TCV) (Simon and Howell, 1986). The analysis of the repair of 3′ terminal deletions of TCV satRNA D has suggested that it could have originated by repeated recombination events linking short stretches of the TCV genomic plus and minus strands (Simon et al., 2004). Alternatively, satRNAs could derive from the genome of the host plant considering the similarities found between structural features of introns and the satRNA of peanut stunt virus, or between sequence stretches of the *Arabidopsis thaliana* genome (the only plant genome available at the time) and cucumber mosaic virus (CMV) satRNA (Collmer et al., 1985; Simon et al., 2004). De novo emergence of CMV-satRNA was reported after eight serial passages in tobacco plants of initially satRNA-free Fny-CMV, but not LS-CMV derived from cDNA clones (Hajimorad et al., 2009). This report is consistent with an origin of satRNA in the host plant genome, but also with a role of the HV in its generation. Recently, Zahid et al. (2015) found that the CMV Y-satRNA sequence in a 35S-GUS-Sat transgene was specifically methylated in tobacco plants in the absence of Y-satRNA replication, suggesting that 24-nt repeat-associated siRNAs homologous to Y-satRNA exist in tobacco inducing methylation of its cDNA. Further analyses revealed the existence of such 24-nt repeat associated siRNAs and of multiple DNA fragments in tobacco plants with sequence similarity to Y-satRNA. These results are compatible with an origin of CMV-satRNAs from repetitive DNA elements in *Nicotiana* genomes. Still, the widespread occurrence of highly similar satRNAs in CMV isolates from different hosts and different regions of the world, as discussed below and shown in Fig. 56.1, needs to be explained.

Thus, the origin of satellite nucleic acids is still far from being understood.

EXPERIMENTAL EVOLUTION OF SATELLITE NUCLEIC ACIDS

The evolution of satellite nucleic acids has been explored experimentally since early in the study of these subviral entities. Most work was done with satRNAs, specifically those associated with CMV and TCV.

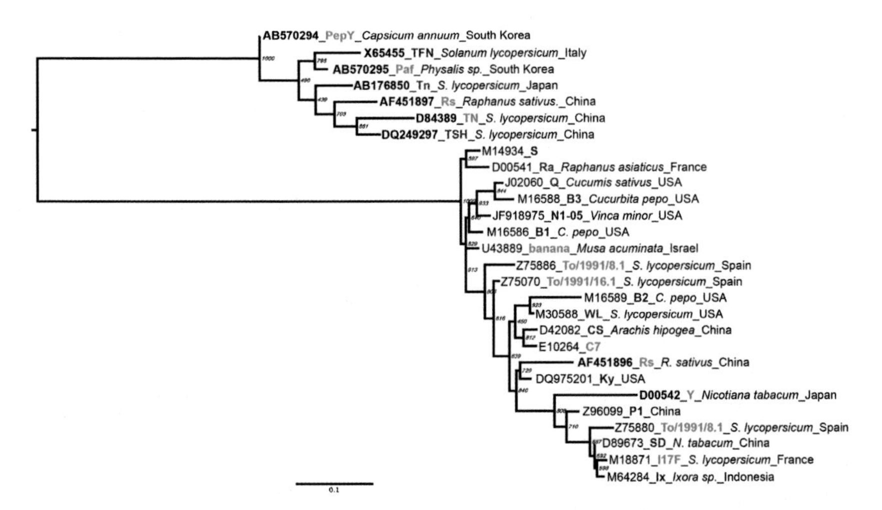

FIGURE 56.1 Maximum likelihood phylogenetic tree of 29 CMV-satRNA variants randomly sampled from databases. Accession numbers are indicated, those of the class of large satRNAs are in bold. The associated CMV isolate is indicated in green, blue, and red for subgroup IA, IB, and II isolates, respectively. When known, host plant and country of isolation are also indicated.

CMV-satRNA was used as a model system for the study of genetic variation of plant-pathogenic RNAs and the evolutionary factors that shape their populations, and much experimental work was published in the 1980s and 1990s (reviewed in García-Arenal and Palukaitis, 1999). Since the first report of a CMV-satRNA sequence, it was apparent that isolates were heterogeneous populations of sequence variants (Richards et al., 1978), and that the most frequently occurring sequence could change upon passage (Kurath and Palukaitis, 1990). These findings prompted studies to identify factors that select sequence variants upon passage. It was found that CMV-satRNA populations responded to selection exerted by the host plant species (García-Luque et al., 1984; Kurath and Palukaitis, 1990; Moriones et al., 1991); the nature of the HV (Palukaitis and Roossinck, 1995; Roossinck and Palukaitis, 1995); the environmental conditions (Kaper et al., 1995; White et al., 1995); and the nature of the satellite RNA sequence itself (Palukaitis and Roossinck, 1995). It was shown that selection could result in important phenotypic responses of the satRNA, including changes from a benign, attenuating, phenotype to a pathogenic one (Palukaitis and Roossinck, 1996), which showed the need for understanding CMV-satRNA evolution, under natural, field conditions.

Experiments on the evolution of TCV satRNAs have significantly contributed to understanding RNA recombination as a driver of virus evolution, as well as to identify evolutionary constraints related to the maintenance of RNA functional structures. TCV is the HV of two satRNAs: satD (194–230 nt), with no substantial sequence identity with TCV RNA except in the 3′ terminal 7 nt, and satC (356 nt), which is a recombinant between satD and two regions of the 3′ end of TCV RNA (Simon and Howell, 1986). Recombination between satD and TCV RNA is frequent in coinfected turnip protoplasts, generating large populations of similar, but not identical recombinants. Analyses of these recombinant populations showed that crossovers in TCV RNA located at structural elements similar to those found in promoters, and a model was proposed in which the RNA-dependent RNA polymerase generated recombinant molecules by reinitiating RNA at these structures during plus-strand synthesis (Cascone et al., 1993; Nagy et al., 1998, 1999; Zhang et al., 1991). Recombinant satC has replication enhancer motifs derived from TCV RNA as well as new motifs generated in the recombinant molecule (Nagy et al., 1998, 1999). In a series of elegant experiments, sequences in the junction region were randomized, and scoring for fitness in protoplasts or plants by in vivo genetic selection allowed identification of the optimal product of satC evolution. Interestingly, the evolutionary outcome can vary, giving structures such as those in the original recombinants, or alternative ones with replication enhancer capacity (e.g., Guo et al., 2009; Murawaski et al., 2015).

Thus, experimental evolution of CMV and TCV satRNAs has significantly contributed to answer fundamental questions on viral RNA evolution and on the underlying molecular mechanisms.

GENETIC VARIATION AND STRUCTURE OF NATURAL POPULATIONS OF SATELLITE NUCLEIC ACIDS

Studies on the genetic variation and population structure have been reported for many satellite RNAs, either large, encoding, or small, non-coding ones. Their frequency in the HV population varies from very high (up to 100% in West Africa) in rice yellow mottle virus satRNA (Pinel et al., 2003) to very low, as in CMV-satRNA (Palukaitis and García-Arenal, 2003). Genetic variation also is high as shown for the satRNAs of rice yellow mottle virus, bamboo mosaic virus, or grapevine fanleaf virus (Gottula et al., 2013; Pinel et al., 2003; Yeh et al., 2004). Common features, though, seem to be a lack of specific association between strains of HV and satRNA (Gottula et al., 2013; Yeh et al., 2004) and restrictions to genetic variation due to the preservation of RNA structures (Fraile and García-Arenal, 1991; Yeh et al., 2004).

The evolution of satRNAs under natural conditions has been analyzed in most detail for CMV-satRNA. Databases contain the sequence of more than 180 CMV satRNA variants, associated with over 65 CMV isolates of subgroups I and II. Most variants from all over the world contain 332–342 nt, and variation is not associated with HV, host plant species or geographic origin (Fig. 56.1). Larger variants, of 386–405 nt, are mostly associated with subgroup IB isolates from East Asia, or if from other regions, always in association with Asian CMV strains (Fig. 56.1) (García-Arenal and Palukaitis, 1999). The larger satRNAs are similar in sequence to the smaller ones, except for insertions. Sequence identity among CMV-satRNA variants, excluding the insertions in the larger ones, range between 73% and 99%, this range being similar to that reported for the genome of the HV CMV (Palukaitis and García-Arenal, 2003).

CMV satRNA evolution in the field was analyzed primarily in association with epidemics of tomato necrosis in the Mediterranean Basin. As in passage experiments, evolution was due primarily to mutation accumulation along major evolutionary lines corresponding to necrogenic or nonnecrogenic variants in tomato (Aranda et al., 1993; Grieco et al., 1997). Recombination between these lines also occurred (Aranda et al., 1997). In field conditions, however, no correlation was found between satRNA variants and host plant or HV species

(Alonso-Prados et al., 1998; Grieco et al., 1997), suggesting that other factors may overshadow host or HV-associated selection. One factor would be random genetic drift: CMV-satRNA is transmitted by aphids in CMV particles, and transmission results in severe population bottlenecks, with effective founder population sizes of about 1–2, for CMV isolates with or without satRNAs (Betancourt et al., 2008), which underscores the relevance of genetic drift. Also, because satRNA depresses CMV accumulation, the efficiency of transmission of isolates supporting a satRNA may be much less than for those without one (Escriu et al., 2000b; Jacquemond, 1982), increasing drift. Another factor overshadowing host- and HV-associated selection can be the highly efficient transmission of the satRNA to the CMV progeny (Escriu et al., 2000a,b). Thus, CMV satRNA can easily be transmitted by aphids to different plant species previously infected by diverse satRNA-free CMV isolates (Alonso-Prados et al., 1998; Escriu et al., 2000b), cross-transmission preventing host or HV-associated selection. Accordingly, it was shown that the population dynamics and genetics of CMV-satRNA and its HV differ, and that the satRNA spreads epidemically as a parasite on the CMV population (Alonso-Prados et al., 1998).

Since CMV isolates supporting a satRNA would be transmitted less efficiently than satRNA-free ones, the conditions for the emergence of CMV isolates with satRNAs, e.g., those causing tomato necrosis, are not obvious. This important subject was addressed by Escriu et al. (2003) by experimental and model analyses. Values for accumulation, virulence, and transmission in tomato were experimentally estimated for CMV isolates free of satRNA (Y isolates) or supporting attenuating (A isolates) or necrogenic (N isolates) satRNA variants. These parameters were introduced into a model describing the population dynamics of uninfected or Y-, A-, and N-infected tomato plants, and allowing for mixed infections and competition between the different types of CMV isolates. N isolates could only invade the CMV population in coinfection with A isolates, and invasion required high density of vector populations, in agreement with field observations during tomato necrosis epidemics. Since the pathogenicity of CMV-satRNA and its effect on HV multiplication depends on the host plant species, analyses were extended to include melon, a host in which N isolates attenuate symptoms. Again it was found that mixed infections and high aphid densities were required for N -isolate emergence, but also that a low frequency of the most competent host, tomato, was sufficient to maintain satRNA-supporting CMV isolates in the less competent host, melon, while the reverse did not occur (Betancourt et al., 2013). These analyses help to understand the evolution and emergence of CMV isolates supporting satRNAs, and also why satRNAs are highly infrequent, most of the time, in CMV populations (Palukaitis and García-Arenal, 2003).

The genetic diversity of betasatellites associated with different begomovirus species has also been analyzed extensively. Different species of betasatellites have been reported in association with Old World (mostly) monopartite begomoviruses. Diversity within each species can be high (e.g., 28% for tomato yellow leaf curl China virus betasatellite; Zhou et al., 2003), and is generated both by mutation accumulation in noncoding regions outside of the conserved and A-rich regions (Guo et al., 2008; Zhou et al., 2003) and by recombination (Briddon and Stanley, 2006). At odds with most satRNAs, phylogenetic compatibility analyses indicate coevolution between betasatellite and HV variants (Zhou et al., 2003), and show that their genetic variation is spatially structured (Guo et al., 2008). These results suggest limited interisolate or interregion transmission, which may be surprising considering the polyphagy of the begomovirus whitefly vectors and remains an unexplained feature of betasatellite evolution. Sequence diversity among betasatellite species is 40%–60%, and there is a significant association between the betasatellite and the begomovirus species. Notably, betasatellites from viruses infecting the *Malvaceae* are phylogenetically distinct from those of other host plants (mostly in the *Solanaceae*), strongly suggesting a role of the HV and/or host plant in their evolution (Zhou, 2013). However, betasatellites can be transreplicated by heterologous begomoviruses or even by geminiviruses from other genera, as first reported for tomato leaf curl virus satellite (Dry et al., 1997). Heterologous transreplication can be initially inefficient, as shown for cotton leaf curl Multan betasatellite replicated by the New World cabbage leaf curl virus, but is quickly optimized by betasatellite evolution (Nawaz-ul-Rehman et al., 2009). These results are also in apparent contradiction to the host-associated genetic structure of betasatellites but, whatever the explanation, set a caveat about the possibility of betasatellites spreading epidemically into new virus/host combinations or into new geographical regions.

CONCLUDING REMARKS

The analysis of plant virus satellite evolution has been an active field of research for more than 25 years. Research has focused on relatively few systems that have been analyzed extensively, and which are taken as exemplars in this review. Results have significantly contributed to understanding central questions regarding the processes and mechanisms of virus evolution, as well as the role of regulatory structures in RNA recombination and replication. In more recent times, an effort has been made to understand the factors that shape the dynamics and genetic structure of populations of those satellites that determine severe host syndromes. These studies contribute to a more precise understanding of the emergence of new viral diseases, which is presently a particularly hot topic in virology.

Acknowledgments

This work was partially funded by grant CGL2013–44952-R (Plan Estatal de I + D + i, Spain).

References

Alonso-Prados, J.L., Aranda, M.A., Malpica, J.M., García-Arenal, F., Fraile, A., 1998. Satellite RNA of cucumber mosaic cucumovirus spreads epidemically in natural populations of its helper virus. Phytopathology 88, 520–524.

Aranda, M.A., Fraile, A., Dopazo, J., Malpica, J.M., García-Arenal, F., 1997. Contribution of mutation and RNA recombination to the evolution of a plant pathogenic RNA. J. Mol. Evol. 44, 81–88.

Aranda, M.A., Fraile, A., García-Arenal, F., 1993. Genetic variability and evolution of the satellite RNA of cucumber mosaic virus during natural epidemics. J. Virol. 67, 5896–5901.

Betancourt, M., Escriu, F., Fraile, A., García-Arenal, F., 2013. Virulence evolution of a generalist plant virus in a heterogeneous host system. Evol. Appl. 6, 875–890.

Betancourt, M., Fereres, A., Fraile, A., García-Arenal, F., 2008. Estimation of the effective number of founders that initiate an infection after aphid transmission of a multipartite plant virus. J. Virol. 82, 12416–12421.

Briddon, R.W., Stanley, J., 2006. Subviral agents associated with plant single-stranded DNA viruses. Virology 344, 198–210.

Cascone, P.J., Haydar, T., Simon, A.E., 1993. Sequences and structures required for RNA recombination between virus-associated RNAs. Science 260, 801–805.

Collmer, C.W., Hadidi, A., Kaper, J.M., 1985. Nucleotide-sequence of the satellite of peanut stunt virus reveals structural homologies with viroids and certain nuclear and mitochondrial introns. Proc. Natl. Acad. Sci. USA 82, 3110–3114.

Dry, I.B., Krake, L.R., Rigden, J.E., Rezaian, M.A., 1997. A novel subviral agent associated with a geminivirus: the first report of a DNA satellite. Proc. Natl. Acad. Sci. USA 94, 7088–7093.

Duchêne, S., Holmes, E.C., Ho, S.Y.W., 2014. Analyses of evolutionary dynamics in viruses are hindered by a time-dependent bias in rate estimates. Proc. Roy. Soc. B-Biol. Sci. 281, 20140732.

Elena, S.F., Dopazo, J., de la Pena, M., Flores, R., Diener, T.O., Moya, A., 2001. Phylogenetic analysis of viroid and viroid-like satellite RNAs from plants: a reassessment. J. Mol. Evol. 53, 155–159.

Escriu, F., Fraile, A., García-Arenal, F., 2000a. Evolution of virulence in natural populations of the satellite RNA of cucumber mosaic virus. Phytopathology 90, 480–485.

Escriu, F., Fraile, A., García-Arenal, F., 2003. The evolution of virulence in a plant virus. Evolution 57, 755–765.

Escriu, F., Perry, K.L., García-Arenal, F., 2000b. Transmissibility of cucumber mosaic virus by *Aphis gossypii* correlates with viral accumulation and is affected by the presence of its satellite RNA. Phytopathology 90, 1068–1072.

Fraile, A., García-Arenal, F., 1991. Secondary structure as a constraint on the evolution of a plant viral satellite RNA. J. Mol. Biol. 221, 1065–1069.

García-Arenal, F., Palukaitis, P., 1999. Structure and functional relationships of satellite RNAs of cucumber mosaic virus. Curr. Top. Microbiol. 239, 37–63.
García-Luque, I., Kaper, J.M., Díaz-Ruiz, J.R., Rubio-Huertos, M., 1984. Emergence and characterization of satellite RNAs associated with Spanish cucumber mosaic virus isolates. J. Gen. Virol. 65, 539–547.
Gottula, J., Lapato, D., Cantilina, K., Saito, S., Barlett, B., Fuchs, M., 2013. Genetic variability, evolution and biological effects of grapevine fanleaf virus satellite RNAs. Phytopathology 103, 1180–1187.
Grieco, F., Lanave, C., Gallitelli, D., 1997. Evolutionary dynamics of cucumber mosaic virus satellite RNA during natural epidemics in Italy. Virology 229, 166–174.
Guo, R., Lin, W., Zhang, J., Simon, A.E., Kushner, D.B., 2009. Structural plasticity and rapid evolution in a viral RNA revealed by in vivo genetic selection. J. Virol. 83, 927–939.
Guo, W., Jiang, T., Zhang, X., Li, G.X., Zhou, X.P., 2008. Molecular variation of satellite DNA β molecules associated with malvastrum yellow vein virus and their role in pathogenicity. Appl. Environ. Microb. 74, 1909–1913.
Hajimorad, M.R., Ghabrial, S.A., Roossinck, M.J., 2009. De novo emergence of a novel satellite RNA of cucumber mosaic virus following serial passages of the virus derived from RNA transcripts. Arch. Virol. 154, 137–140.
Jacquemond, M., 1982. Cucumber mosaic virus associated RNA-5. Experimental transmission of the tomato necrosis by aphids. Agronomie 2, 641–646.
Kaper, J.M., Geletka, L.M., Wu, G.S., Tousignant, M.E., 1995. Effect of temperature on cucumber mosaic virus satellite-induced lethal tomato necrosis is helper virus strain dependent. Arch. Virol. 140, 65–74.
Kurath, G., Palukaitis, P., 1990. Serial passage of infectious transcripts of a cucumber mosaic-virus satellite RNA clone results in sequence heterogeneity. Virology 176, 8–15.
Mansoor, S., Briddon, R.W., Zafar, Y., Stanley, J., 2003. Geminivirus disease complexes: an emerging threat. Trends Plant Sci. 8, 128–134.
Moriones, E., Fraile, A., García-Arenal, F., 1991. Host-associated selection of sequence variants from a satellite RNA of cucumber mosaic virus. Virology 184, 465–468.
Murawaski, A.M., Nieves, J.L., Chattopadhyay, M., Young, M.Y., Szarko, C., Tajalli, H.F., et al., 2015. Rapid evolution on in vivo-selected sequences and structures replacing 20% of a subviral RNA. Virology 483, 149–162.
Nagy, P.E., Pogany, J., Simon, A.E., 1999. RNA elements required for RNA recombination function as replication enhancers in vitro and in vivo in a plus-strand RNA virus. EMBO J. 18, 5653–5665.
Nagy, P.E., Zhang, C., Simon, A.E., 1998. Dissecting RNA recombination in vitro: role of RNA sequences and the viral replicase. EMBO J. 17, 2392–2403.
Nawaz-ul-Rehman, M.S., Mansoor, S., Briddon, R.W., Fauquet, C.M., 2009. Maintenance of an Old World betasatellite by a New World helper begomovirus and possible rapid adaptation of the betasatellite. J. Virol. 83, 9347–9355.
Palukaitis, P., García-Arenal, F., 2003. Cucumoviruses. Adv. Virus Res. 62, 241–323.
Palukaitis, P., Rezaian, A., Garcia-Arenal, F., 2008. Satellite Nucleic Acids and Viruses. In: Mahy, B.W.J., Van Regenmortel, M.H.V. (Eds.), Encyclopedia of Virology, third ed Elsevier, Oxford, pp. 526–535.
Palukaitis, P., Roossinck, M.J., 1995. Variation in the hypervariable region of cucumber mosaic virus satellite RNAs is affected by the helper virus and the initial sequence context. Virology 206, 765–768.
Palukaitis, P., Roossinck, M.J., 1996. Spontaneous change of a benign satellite RNA of cucumber mosaic virus to a pathogenic variant. Nat. Biotechnol. 14, 1264–1268.
Pinel, A., Abubakar, Z., Traoré, O., Konaté, G., Fargette, D., 2003. Molecular epidemiology of the RNA satellite of rice yellow mottle virus in Africa. Arch. Virol. 148, 1721–1733.

Richards, K.E., Jonard, G., Jacquemond, M., Lot, H., 1978. Nucleotide-sequence of cucumber mosaic virus-associated RNA-5. Virology 89, 395–408.

Roossinck, M.J., Palukaitis, P., 1995. Genetic analysis of helper virus-specific selective amplification of cucumber mosaic virus satellite RNAs. J. Mol. Evol. 40, 25–29.

Simon, A.E., Howell, S.H., 1986. The virulent satellite RNA of turnip crinkle virus has a major domain homologous to the 3′ end of the helper virus genome. EMBO J. 5, 3423–3428.

Simon, A.E., Roossinck, M.J., Havelda, Z., 2004. Plant virus satellite and defective interfering RNAs: new paradigms for a new century. Annu. Rev. Phytopathol. 42, 415–437.

White, J.L., Tousignant, M.E., Geletka, L.M., Kaper, J.M., 1995. The replication of a necrogenic cucumber mosaic virus satellite is temperature-sensitive in tomato. Arch. Virol. 140, 53–63.

Yeh, W.B., Hsu, Y.H., Chen, H.C., Lin, N.S., 2004. A conserved secondary structure in the hypervariable region at the 5 end of bamboo mosaic virus satellite RNA is functionally interchangeable. Virology 330, 105–115.

Zahid, K., Zhao, J.H., Smith, N.A., Schumann, U., Fang, Y.Y., Dennis, E.S., et al., 2015. *Nicotiana* small RNA sequences support a host genome origin of cucumber mosaic virus satellite RNA. PLoS Genet. 11, e1004906.

Zhang, C., Cascone, P.J., Simon, A.E., 1991. Recombination between satellite and genomic RNAs of turnip crinkle virus. Virology 184, 791–794.

Zhou, X.P., 2013. Advances in understanding begomovirus satellites. Annu. Rev. Phytopathol. 51, 357–381.

Zhou, X.P., Xie, Y., Tao, X.R., Zhang, K.Z., Li, Z.H., Fauquet, C.M., 2003. Characterization of DNAβ associated with begomoviruses in China and evidence for co-evolution with their cognate viral DNA-A. J. Gen. Virol. 84, 237–247.

CHAPTER

57

Satellite Taxonomy

Peter Palukaitis

Seoul Women's University, Seoul, South Korea

INTRODUCTION

Unlike viruses that are grouped together into various taxa, viral satellites do not share a common phylogeny. Therefore, viral satellites are not grouped into species, genera, families, and orders. Rather, viral satellites are grouped on common physical/genetic characteristics (Briddon et al., 2012). If viral satellites encode their own capsid protein they are described as satellite viruses; all other viral satellites are considered to be satellite nucleic acids, containing single-stranded (ss) DNA genomes, double-stranded (ds) RNA genomes, or ss RNA genomes. The last group is further divided into three types of viral satellite RNAs, consisting of large linear ss satellite RNAs, small linear ss satellite RNAs, or small circular ss satellite RNAs; the last are circular at some point in their replication cycle (Briddon et al., 2012). Beyond this level are individual viral satellite RNAs, including several satellite-like RNAs. There are no definitive plant viral satellites with ds RNA genomes (Briddon et al., 2012), although there are several tentative members that have been found in infected plants, but probably are satellites of fungal viruses associated with specific plant diseases (Di Serio et al., 2006; Minoia et al., 2014).

SATELLITE VIRUSES

All satellite viruses encode their own capsid proteins. At this time, there are only four known plant satellite viruses: tobacco necrosis satellite virus, panicum mosaic satellite virus, maize white line satellite virus, and tobacco mosaic satellite virus. These satellite viruses are named after their associated helper viruses: tobacco necrosis virus

Viroids and Satellites.
DOI: http://dx.doi.org/10.1016/B978-0-12-801498-1.00057-7

(species *Tobacco necrosis virus*, genus *Necrovirus*, family *Tombusviridae*), panicum mosaic virus (species *Panicum mosaic virus*, genus *Panicovirus*, family *Tombusviridae*), maize white line mosaic virus (species *Maize white line mosaic virus*, genus *Aureusvirus*, family *Tombusviridae*), and tobacco mosaic virus (species *Tobacco mosaic virus*, genus *Tobamovirus*, family *Virgaviridae*), respectively (Briddon et al., 2012) (see Chapter 58: Biology and Pathogenesis of Satellite Viruses).

SATELLITE DNAS

Plant viral satellite DNAs, usually consisting of ~1.3 kb circular ss DNAs, are found in association with virus members of the genus *Begomovirus* in the family *Geminiviridae*. There are at least 61 identified plant viral satellite DNAs (Briddon et al., 2012), now collectively called betasatellite DNAs, encoding a protein designated βC1, which is an RNA silencing suppressor and pathogenicity enhancer. There are also deletion forms of some betasatellite DNAs, and forms that have recombined with the helper virus, complicating the relationships of various betasatellites to each other (see Chapter 62: Betasatellites of Begomoviruses).

LARGE LINEAR SS SATELLITE RNAS

These viral satellite RNAs consist of linear ss RNAs, 0.8–1.5 kb in size, and each encoding a single nonstructural protein of 20–45 kDa (Briddon et al., 2012). In most cases this protein is required for replication of the satellite RNA. In some cases, the satellite RNA may be replicated by other viruses related to the helper virus found associated with the satellite RNA. The satellite RNAs are named after the associated helper virus. There are 10 known large linear ss satellite RNAs (Table 57.1) and several satellite-like RNAs (see below for the latter). Nine of the large linear ss satellite RNA are supported by helper viruses that are members of the family *Secoviridae*, while one such satellite RNA, bamboo mosaic virus satellite RNA, is supported by a member of the genus *Potexvirus*, family *Alphaflexiviridae*. Two of the above satellite RNAs have "large" as part of their name, since there are also small linear satellite ss RNAs associated with the same helper viruses. The helper virus, strawberry latent ringspot virus, is an unassigned member of the family *Secoviridae*, while the other eight helper virus members of this family are all in the genus *Nepovirus*, subfamily *Comovirinae*, family *Secoviridae*. The helper viruses beet ringspot virus and tomato black ring virus (TBRV) were known previously as TBRV-S serotype and TBRV-G

TABLE 57.1 Large Linear Satellite RNAs of Plant Viruses

Satellite RNA	Associated virus	Genus	Family
Arabis mosaic virus large satellite RNA	Arabis mosaic virus	*Nepovirus*	*Secoviridae*
Beet ringspot virus satellite RNA	Beet ringspot virus	*Nepovirus*	*Secoviridae*
Black current reversion virus satellite RNA	Black current reversion virus	*Nepovirus*	*Secoviridae*
Chicory yellow mottle virus large satellite RNA	Chicory yellow mottle virus	*Nepovirus*	*Secoviridae*
Grapevine Bulgarian latent virus satellite RNA	Grapevine Bulgarian latent virus	*Nepovirus*	*Secoviridae*
Grapevine fanleaf virus satellite RNA	Grapevine fanleaf virus	*Nepovirus*	*Secoviridae*
Myrobalan latent ringspot virus satellite RNA	Myrobalan latent ringspot virus	*Nepovirus*	*Secoviridae*
Tomato black ring virus satellite RNA	Tomato black ring virus	*Nepovirus*	*Secoviridae*
Strawberry latent ringspot virus satellite RNA	Strawberry latent ringspot virus	Unassigned	*Secoviridae*
Bamboo mosaic virus satellite RNA	Bamboo mosaic virus	*Potexvirus*	*Alphaflexiviridae*

serotype, respectively (Briddon et al., 2012) (see Chapter 59: Large Satellite RNAs).

SMALL LINEAR SS SATELLITES RNAS

These viral satellite RNAs consist of linear ss RNAs, although often with a high degree of secondary structure, less than 0.7 kb in size, and do not encode any functional proteins (Briddon et al., 2012). These 13 satellite RNAs are named after their associated helper viruses, which are distributed in four genera in the family *Tombusviridae*, one genus in the family *Bromoviridae*, and one unassigned genus (*Umbravirus*) (Table 57.2). In some cases, the satellite RNAs consist of several different sized RNAs, such as several cucumber mosaic virus satellite RNAs, tomato bushy stunt virus satellite RNAs, and turnip crinkle virus (TCV) satellite RNAs D and F; TCV satellite RNA C is a hybrid molecule with the 5′ half derived from an RNA similar to TCV satellite RNA D fused to sequences derived from the 3′ nontranslated region of the helper

TABLE 57.2 Small Linear Satellite RNAs of Plant Viruses

Satellite RNA	Associated virus	Genus	Family
Artichoke mottled crinkle virus satellite RNA	Artichoke mottled crinkle virus	*Tombusvirus*	*Tombusviridae*
Black beet scorch virus satellite RNA	Black beet scorch virus	*Necrovirus*	*Tombusviridae*
Carnation Italian ringspot virus satellite RNA	Carnation Italian ringspot virus	*Tombusvirus*	*Tombusviridae*
Cymbidium ringspot virus satellite RNA	Cymbidium ringspot virus	*Tombusvirus*	*Tombusviridae*
Panicum mosaic virus satellite RNA	Panicum mosaic virus	*Panicovirus*	*Tombusviridae*
Pelargonium leaf curl virus satellite RNA	Pelargonium leaf curl virus	*Tombusvirus*	*Tombusviridae*
Petunia asteroid mosaic virus satellite RNA	Petunia asteroid mosaic virus	*Tombusvirus*	*Tombusviridae*
Tomato bushy stunt virus satellite RNA	Tomato bushy stunt virus	*Tombusvirus*	*Tombusviridae*
Turnip crinkle virus satellite RNA D and F	Turnip crinkle virus	*Carmovirus*	*Tombusviridae*
Cucumber mosaic virus satellite RNA	Cucumber mosaic virus	*Cucumovirus*	*Bromoviridae*
Peanut stunt virus satellite RNA	Peanut stunt virus	*Cucumovirus*	*Bromoviridae*
Carrot mottle mimic virus satellite RNA	Carrot mottle mimic virus	*Umbravirus*	Unassigned
Pea enation mosaic virus satellite RNA	Pea enation mosaic virus	*Umbravirus*	Unassigned

virus genomic RNA (Briddon et al., 2012; Simon and Howell, 1986) (see Chapter 60: Small Linear Satellite RNAs).

SMALL CIRCULAR SS SATELLITE RNAS

These small ss satellite RNAs have a circular form at some stage of their infection cycle; either the stable encapsidated form, or a form present only during replication to generate concatemer linear forms of the satellite RNA (in the case of the satellite RNAs associated with viruses that are members of the families *Secoviridae* and *Luteoviridae*) (see Chapter 53: Replication of Satellites). There are 10 recognized small circular satellite RNAs; two associated with subterranean clover mottle virus. They vary in size from 220 to 457 nt. These satellite RNAs and their associated helper viruses are listed in Table 57.3. With one possible exception, which needs to be confirmed, they do not appear to be translated (AbouHaidar et al., 2014). These small circular RNAs have a high degree of secondary structure, usually with short branches off a rod-like central structure. This physical relationship to viroids led to some laboratories referring to the small circular satellite RNAs associated with virus members of the (unassigned) genus *Sobemovirus* as virusoids (see Chapter 50: Satellite Viruses and Satellite Nucleic Acids). Early work on the small circular satellite RNAs associated with velvet tobacco mottle virus and Solanum nodiflorum mottle virus (SNMV) indicated that these circular RNAs were essential for the infection of the associated viruses and vice versa (Gould et al., 1981), and thus these RNAs were neither satellite RNAs nor viroids. However, this conclusion was not upheld by subsequent work showing that the small circular RNA of

TABLE 57.3 Small Circular Satellite RNAs of Plant Viruses

Satellite RNA	Associated virus	Genus	Family	Size (nt)
Arabis mosaic virus small satellite RNA	Arabis mosaic virus	*Nepovirus*	*Secoviridae*	300
Chicory yellow mottle virus small satellite RNA	Chicory yellow mottle virus	*Nepovirus*	*Secoviridae*	457
Tobacco ringspot virus satellite RNA	Tobacco ringspot virus	*Nepovirus*	*Secoviridae*	354
Cereal yellow dwarf virus-RPV satellite RNA	Cereal yellow dwarf virus-RPV	*Polerovirus*	*Luteoviridae*	322
Lucerne transient streak virus satellite RNA	Lucerne transient streak virus	*Sobemovirus*	Unassigned	324
Rice yellow mottle virus satellite RNA	Rice yellow mottle virus	*Sobemovirus*	Unassigned	220
Solanum nodiflorum mottle virus satellite RNA	Solanum nodiflorum mottle virus	*Sobemovirus*	Unassigned	377
Subterranean clover mottle virus satellite RNA 1	Subterranean clover mottle virus	*Sobemovirus*	Unassigned	388
Subterranean clover mottle virus satellite RNA 2	Subterranean clover mottle virus	*Sobemovirus*	Unassigned	332
Velvet tobacco mottle virus satellite RNA	Velvet tobacco mottle virus	*Sobemovirus*	Unassigned	365

SNMV was not required for the replication of the genomic RNA of SNMV (Jones and Mayo, 1984) and that lucerne transient streak virus genomic RNA could replicate without its associated small circular RNA (Jones et al., 1983), as well as amplify the small circular RNA of SNMV (Jones and Mayo, 1983). Thus, the nature of these small circular RNAs was again restored to the status of small circular satellite RNAs. Interestingly, work showing that a velvet tobacco mottle virus isolate could be obtained without its small circular RNA, as determined by polyacrylamide gel electrophoresis (Francki et al., 1986), has not been supported by analysis of the same isolate using reverse-transcription polymerase chain reaction (Arthur et al., 2014), raising the possibility again that one or more of these small circular satellite RNAs are only satellite-like. Nevertheless, in describing the smallest circular satellite RNA, that of rice yellow mottle virus, the authors have again used the term virusoid (AbouHaidar et al., 2014) (see Chapter 61: Small Circular Satellite RNAs).

SATELLITE-LIKE RNAS

Several satellite RNAs are required for the complete infectious cycle of the associated virus. These include the various satellite RNAs of groundnut rosette virus and the satellite RNA of tobacco bushy top virus, and hence are referred to as satellite-like RNAs (Briddon et al., 2012). The helper viruses of these satellite-like RNAs are both members of the unassigned genus *Umbravirus* and require a luteovirus for plant-to-plant transmission (Liu et al., 2014; Simon et al., 2004; Taliansky et al., 2000). This transmission is facilitated by an unknown mechanism involving the satellite-like RNAs, which does not require the production of putative satellite-like RNA-encoded proteins (Robinson et al., 1999; Taliansky et al., 2000).

The hybrid TCV satellite RNA C also qualifies as being satellite-like. It is dependent upon the helper virus for both replication and movement, and its presence alters the accumulation and pathogenicity of the helper virus (Simon and Howell, 1986; Simon et al., 2004). However, as it contains appreciable sequence identity to its helper virus, it is not a canonical satellite.

Recently, using deep sequencing of ds RNAs extracted from grapevines, Al Rwahnih et al. (2013) described the presence of a new putative satellite, which may be a satellite RNA or a satellite virus; however, as it has not been studied biologically and a possible helper virus has not been identified, its exact nature cannot be ascertained at this time. It was suggested as being a satellite since the larger (151 amino acid) putative protein encoded by this RNA showed 21% sequence identity to the proteins encoded by both panicum mosaic satellite virus (159 amino acids) and bamboo mosaic virus satellite RNA (183 amino acids), while proteins encoded by the latter two RNAs showed 39% sequence identity (Al Rwahnih et al., 2013). Thus, satellite viruses may have evolved from large linear ss satellite RNAs.

A viroid-like RNA was also identified by deep sequencing of small RNAs from grapevines. A dimer transcript of this RNA was not transmissible back to the grapevines, suggesting it may be a small circular satellite-like RNA rather than a viroid (Wu et al., 2012), but further experimentation is needed to establish its actual nature.

Acknowledgments

This work was supported by a special grant from Seoul Women's University.

References

AbouHaidar, M.G., Venkataraman, S., Golshani, A., Liu, B., Ahmad, T., 2014. Novel coding, translation, and gene expression of a replicating covalently closed circular RNA of 220 nt. Proc. Natl. Acad. Sci. USA 111, 14542–14547.

Al Rwahnih, M., Daubert, S., Sudarshana, M.R., Rowhani, A., 2013. Gene for a novel plant virus satellite from grapevine identifies a viral satellite lineage. Virus Genes 47, 114–118.

Arthur, K., Collins, N.C., Yazarlou, A., Randles, J.W., 2014. Nucleotide sequence diversity in velvet tobacco mottle virus: a virus with a unique Australian pathosystem. Virus Genes 48, 168–173.

Briddon, R.W., Ghabrial, S., Lin, N.-S., Palukaitis, P., Scholthof, K.-B.G., Vetten, H.-J., 2012. Satellite and other virus-dependent nucleic acids. In: King, A.M.Q., Adams, M.J., Carstens, E.B., Lefkowitz, E.J. (Eds.), Virus Taxonomy. Classification and Nomenclature of Viruses. Ninth Report of the International Committee on Taxonomy of Viruses. Elsevier Academic Press, San Diego, CA, pp. 1211–1219.

Di Serio, F., Daròs, J.A., Ragozzino, A., Flores, R., 2006. Close structural relationship between two hammerhead viroid-like RNAs associated with cherry chlorotic rusty spot disease. Arch. Virol. 151, 1539–1549.

Francki, R.I.B., Grivell, C.J., Gibb, K.S., 1986. Isolation of velvet tobacco mottle virus capable of replication with and without a viroid-like RNA. Virology 148, 381–384.

Gould, A.R., Francki, R.I.B., Randles, J.W., 1981. Studies on encapsidated viroid-like RNA. IV. Requirement for infectivity and specificity of two RNA components from velvet tobacco mottle virus. Virology 110, 420–426.

Jones, A.T., Mayo, M.A., 1983. Interaction of lucerne transient streak virus and the viroid-like RNA-2 of *Solanum nodiflorum* mottle virus. J. Gen. Virol. 64, 1771–1774.

Jones, A.T., Mayo, M.A., 1984. Satellite nature of the viroid-like RNA-2 of solanum nodiflorum mottle virus and the ability of other plant viruses to support the replication of viroid-like RNA molecules. J. Gen. Virol. 65, 1713–1731.

Jones, A.T., Mayo, M.A., Duncan, G.H., 1983. Satellite-like properties of small circular RNA molecules in particles of lucerne transient streak virus. J. Gen. Virol. 64, 1167–1173.

Liu, F., Tan, G., Li, X., Chen, H., Li, R., Li, F., 2014. Simultaneous detection of four causal agents of tobacco bushy top disease by a multiple one-step RT-PCR. J. Virol. Methods 205, 99–103.

Minoia, S., Navarro, B., Covelli, L., Barone, M., García-Becedas, M.T., Ragozzino, A., et al., 2014. Viroid-like RNAs from cherry trees affected by leaf scorch disease: further data supporting their association with mycoviral double-stranded RNAs. Arch. Virol. 159, 589–593.

Robinson, D.J., Ryabov, E.V., Raj, S.K., Roberts, I.M., Taliansky, M.E., 1999. Satellite RNA is essential for encapsidation of groundnut rosette umbravirus RNA by groundnut rosette assistor luteovirus coat protein. Virology 254, 105–114.

Simon, A.E., Howell, S.H., 1986. The virulent satellite RNA of turnip crinkle virus has a major domain homologous to the 3′ end of the helper virus genome. EMBO J. 5, 3423–3428.

Simon, A.E., Roossinck, M.J., Halveda, Z., 2004. Plant virus satellites and defective interfering RNAs: new paradigms for a new century. Annu. Rev. Phytopathol. 42, 415–437.

Taliansky, M.E., Robinson, D.J., Murant, A.F., 2000. Groundnut rosette disease virus complex: biology and molecular biology. Adv. Virus Res. 55, 357–400.

Wu, Q., Wang, Y., Cao, M., Pantaleo, V., Burgyan, J., Li, W.-X., et al., 2012. Homology-independent discovery of replicating pathogenic circular RNAs by deep sequencing and a new computational algorithm. Proc. Natl. Acad. Sci. USA 109, 3938–3943.

PART XI

TYPES OF SATELLITES

A. SATELLITE VIRUSES

CHAPTER

58

Biology and Pathogenesis of Satellite Viruses

Jesse D. Pyle[1] *and Karen-Beth G. Scholthof*[2]

[1]**Harvard Medical School, Boston, MA, United States** [2]**Texas A&M University, College Station, TX, United States**

INTRODUCTION

Four plant satellite viruses (SVs) have been described: satellite tobacco necrosis virus (STNV), satellite panicum mosaic virus (SPMV), satellite maize white line mosaic virus (SMWLMV), and satellite tobacco mosaic virus (STMV) (Dodds, 1998; Krupovic et al., 2015; Scholthof et al., 1999). Plant SVs have positive-sense single-stranded (ss) RNAs that encode a single capsid protein (CP) for the assembly of icosahedral virions (Fig. 58.1). The crystal structures of STMV, STNV, and SPMV have been determined (Ban and McPherson, 1995; Jones and Liljas, 1984; Larson et al., 2014). SVs are viable agents for study of cap-independent translation, RNA:CP interactions driving virion assembly, and molecular plant–virus interactions.

BIOLOGICAL AND MOLECULAR PROPERTIES OF STNV

STNV-1 and STNV-2 are supported by tobacco necrosis virus A (TNV-A), while TNV-D supports STNV-C (Table 58.1). The TNV-STNV complex is primarily transmitted plant-to-plant through root systems by the obligate fungus *Olpidium brassicae*. Differences in CP sequence likely influence packaging, virion structure, and efficiency of transmission by *Olpidium* spp. (Bringloe et al., 1998). Similarly, SV RNA sequence differences are predicted to affect formation of distinct hairpin and

Viroids and Satellites.
DOI: http://dx.doi.org/10.1016/B978-0-12-801498-1.00058-9

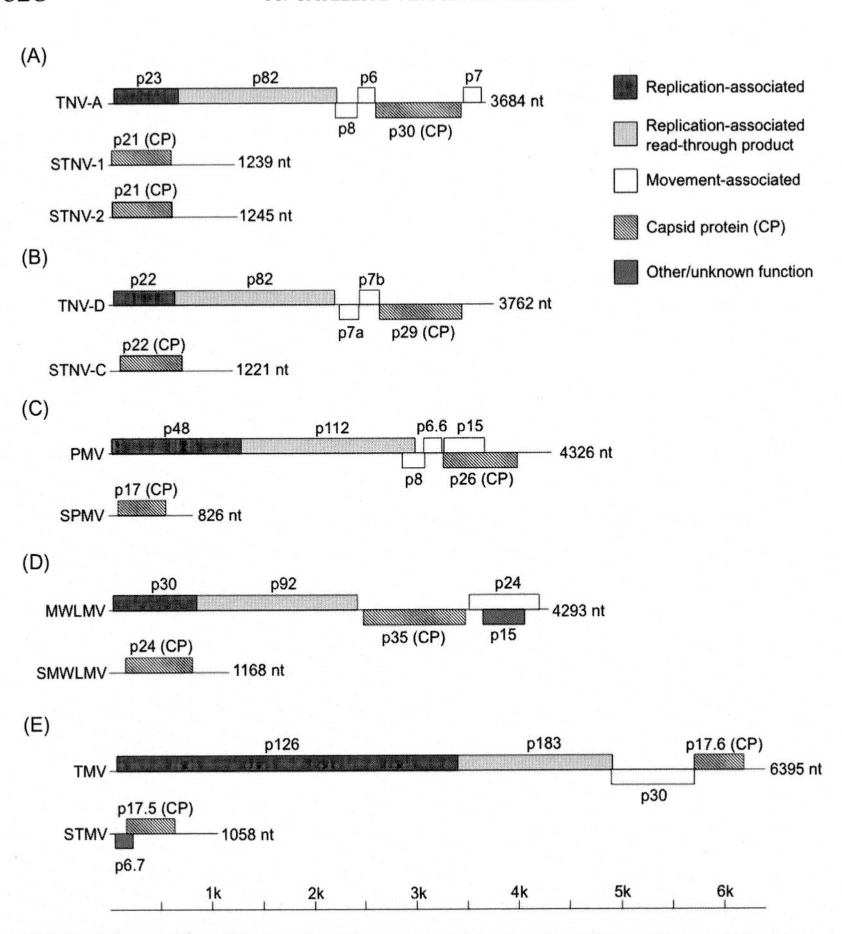

FIGURE 58.1 Representative genomes of the plant satellite viruses and their helper viruses. Tobacco necrosis virus A (TNV-A) with STNV-1 and STNV-2 (A), tobacco necrosis virus D (TNV-D) and STNV-C (B), panicum mosaic virus (PMV) and SPMV (C), maize white line mosaic virus (MWLMV) and SMWLMV (D), and tobacco mosaic virus (TMV) and STMV (E). Shaded boxes represent the known or unknown functions of the viral and subviral open reading frames. Scale bar is in thousands of bases.

pseudoknot secondary RNA structures to determine strain-specific replication by their helper viruses (HVs) (Bringloe et al., 1998). Additionally, STNV CP monomers assemble in vitro only in the presence of the STNV genomic RNA, suggesting a significant role for the viral nucleic acid in virion formation (Bunka et al., 2011).

One primary research focus with STNV has been the cap-independent translation-enhancing features of the 5′- and 3′-untranslated regions (UTRs). STNV strains utilize a 3′-translation enhancer domain (TED) of approximately 130 nt immediately downstream of the

TABLE 58.1 Biological and Taxonomic Characteristics of the Plant Satellite Viruses and Their Corresponding Helper Viruses

Virus/satellite virus	Genus[a]	Family	Genome	CP (kDa)	Virion
Tobacco necrosis virus (TNV)-A	*Alphanecrovirus*	*Tombusviridae*	+ssRNA (3684 nt)	30	Icosahedral ($T = 3$)
Tobacco necrosis satellite virus (STNV)-1	*Albetovirus*	na[b]	+ssRNA (1239 nt)	21	Icosahedral ($T = 1$)
Tobacco necrosis satellite virus (STNV)-2	*Albetovirus*	na	+ssRNA (1245 nt)	21	Icosahedral ($T = 1$)
Tobacco necrosis virus (TNV)-D	*Betanecrovirus*	*Tombusviridae*	+ssRNA (3762 nt)	29	Icosahedral ($T = 3$)
Tobacco necrosis satellite virus (STNV)-C	*Albetovirus*	na	+ssRNA (1221 nt)	22	Icosahedral ($T = 1$)
Panicum mosaic virus (PMV)	*Panicovirus*	*Tombusviridae*	+ssRNA (4326 nt)	26	Icosahedral ($T = 3$)
Panicum mosaic satellite virus (SPMV)	*Papanivirus*	na	+ssRNA (826 nt)	17.5	Icosahedral ($T = 1$)
Maize white line mosaic virus (MWLMV)	*Aureusvirus*	*Tombusviridae*	+ssRNA (4293 nt)	35	Icosahedral ($T = 3$)
Maize white line mosaic satellite virus (SMWLMV)	*Aumaivirus*	na	+ssRNA (1168 nt)	24	Icosahedral ($T = 1$)
Tobacco mosaic virus (TMV)	*Tobamovirus*	*Virgaviridae*	+ssRNA (6395 nt)	17.5	Rigid rod
Tobacco mosaic satellite virus (STMV)	*Virtovirus*	na	+ssRNA (1058 nt)	17	Icosahedral ($T = 1$)

[a] *The genus designations are proposed (Krupovic et al., 2015) and under consideration by the International Committee on Taxonomy of Viruses (ictvonline.org).*
[b] *No assigned family.*

CP ORF. The TED, in concert with a 38 nt region in the 5′-UTR, activates and enhances translation of both STNV RNA and heterologous RNAs in vitro and in vivo and when translocated to different regions on the SV genome (Meulewaeter et al., 1998; Timmer et al., 1993). Additionally, a 100 nt fragment of the 3′-TED directly interacts with the cap-binding subunits of the eIF4F complex to facilitate SV translation (Gazo et al., 2004), further implicating its role in translational modulation.

BIOLOGICAL AND MOLECULAR PROPERTIES OF STMV

STMV is supported by coinfection with strains of tobacco mosaic virus (TMV). STMV was first reported in *Nicotiana glauca* with TMV-U5 (now tobacco mild green mosaic virus), a tobamovirus (Dodds, 1998). STMV coinfection is limited primarily to the *Solanaceae* and does not alter the typical symptoms induced by HV alone (Table 58.2). However, coinfections of STMV and tobacco mild green mosaic virus in Jalapeño and Pimiento pepper hosts can noticeably enhance symptoms, characterized by severe chlorotic lesions on the leaf surfaces (Dodds, 1998).

STMV encodes a nonessential 6.3-kDa protein of unknown function and a 17.5-kDa CP used to assemble icosahedral virions. The 51 nt STMV 5′-UTR resembles bromovirus and cucumovirus 5′-termini and the 3′-UTR forms highly complex secondary structures that likely are important for recognition and replication by the HV machinery, encapsidation and CP translation (Dodds, 1998; Sivanandam et al., 2015). A striking feature of the 3′-UTR, that may be biologically and evolutionarily important, is a 50 nt region on STMV RNA (positions 919–960) that has >90% sequence similarity with the 3′-UTR of tomato mottle

TABLE 58.2 Host Range and Effect on Symptomatology for Each Helper-Satellite Coinfection

Coinfection	Effect on disease phenotype	Host range (coinfections)
TNV + STNV	Attenuation	*Nicotiana clevelandii, N. tabacum, Phaseolus vulgaris, Tulipa gesneriana*
PMV + SPMV	Exacerbation	*Brachypodium distachyon, Eremochloa ophiuroides, Panicum miliaceum, P. virgatum, Pennisetum glaucum, Setaria italica, S. viridis, Stenotaphrum secundatum*
MWLMV + SMWLMV	Unclear/no effect	*Zea mays*
TMV + STMV	Exacerbation[a]/ no effect	*Capsicum annum, Datura metaloides, Solanum lycopersicum, N. benthamiana, N. glauca, N. rustica, N. sylvestris, N. tabacum, Vinca rosea*

[a] *Only for Jalapeño and Pimiento* (C. annuum) *varieties*

mosaic virus, rehmannia mosaic virus, and over 50 strains of TMV (J. Pyle, unpublished data).

BIOLOGICAL AND MOLECULAR PROPERTIES OF SMWLMV

Maize is the only known natural host for SMWLMV (Table 58.2). Both the HV and SV are soil-borne, although a vector has not been identified (Russo et al., 2008). SMWLMV is the largest ssRNA genome of the plant SVs, encoding a 24-kDa CP to assemble isosahedral $T = 1$ virions (Zhang et al., 1991). SMVLMV has no significant sequence similarity with its HV. Since the initial discovery of MWLMV and SMWLMV, the two agents have rarely been detected in the field.

BIOLOGICAL AND MOLECULAR PROPERTIES SPMV

SPMV and its HV, panicum mosaic virus (PMV), are economically important pathogens of switchgrass (*Panicum virgatum*) and St. Augustine grass (*Stenotaphrum secundatum*), a turfgrass (Scholthof et al., 1999; Stewart et al., 2015). The 826-nt RNA of SPMV is the smallest known genome for a SV (Fig. 58.1). SPMV encodes a 17-kDa CP (SPCP), with evidence that a 9.5-kDa *N*-terminal-truncated form of the CP accumulates and localizes specifically to the host cell wall and membrane (Omarov et al., 2005).

Coinfections of SPMV and its HV result in greatly exacerbated symptoms on host plants to include severe chlorosis and stunting, reduced seed production, and often death of the plant (Fig. 58.2) (Mandadi et al., 2014). This synergism is also characterized by the more rapid systemic accumulation of PMV and SPMV CPs and genomic RNAs, compared to infection by PMV alone (Mandadi and Scholthof, 2012). SPCP exhibits distinct cytotoxic effects, with the induction of localized necrotic lesions when transiently expressed in *Nicotiana benthamiana* (Qiu and Scholthof, 2004).

In addition to its structural role in the virions for protection and transmission of SPMV genomic RNA, SPCP is also predicted to serve a critical function in the cell-to-cell transport of the SPMV genomic RNA, suggesting its inherent ability to stabilize diverse RNA species during systemic movement. Much of this activity involves the *N*-terminal arginine-rich motif, RNA-binding domain of SPCP (Omarov et al., 2005; Qi and Scholthof, 2008). SPCP also possesses the ability to stabilize a tomato bushy stunt virus-GFP based expression vector *in trans*,

FIGURE 58.2 Disease synergism induced during coinfections of PMV and SPMV in host grasses. Infection by PMV alone causes mild symptoms. The presence of SPMV in coinfected plants results in exacerbated symptoms including severe stunting, systemic chlorosis, and loss of biomass. Mock-, PMV-, and PMV + SPMV-infections are shown for *Setaria viridis* (A), *Brachypodium distachyon* (B), and *Panicum miliaceum* (C) at 42, 21, and 14 days postinoculation, respectively.

suggesting its potential use for enhancing transient gene expression *in planta* (Everett et al., 2010).

The PMV + SPMV disease synergism results in altered expression of multiple defense-related genes, such as PR genes, receptor-like kinases, WRKY transcription factors, and genes involved in antiviral hormone signaling pathways (Mandadi and Scholthof, 2012). Among the upregulated genes were several SPMV-specific splicing and transcription factors. PMV and SPMV have a significant impact on changes to the host spliceosome through alterations of alternative splicing of host mRNAs

(Mandadi and Scholthof, 2015) and suggest new avenues of study toward defining molecular mechanisms affecting disease progression.

PLANT SATELLITE VIRUS STRUCTURES

Owing to their small and relatively simple particles and genomes plant SVs are a subject of extensive protein and nucleic acid structural analyses. All four SVs form small (ca. 17 nm), nonenveloped $T = 1$ icosahedral virions with 60 identical CP subunits per particle. With the exception of SMWLMV, the structures of the SV CPs and virions have been solved at high resolutions using X-ray crystallography (Fig. 58.3).

Recent studies provide compelling evidence that SV RNA secondary structures have embedded information for correct packaging and organization of the virion RNA:CP interactions, in addition to the primary gene expression functionality of the RNA coding regions and 5′- and 3′-UTRs (Ban et al., 1995; Ford et al., 2013; Patel et al., 2015). Packaging

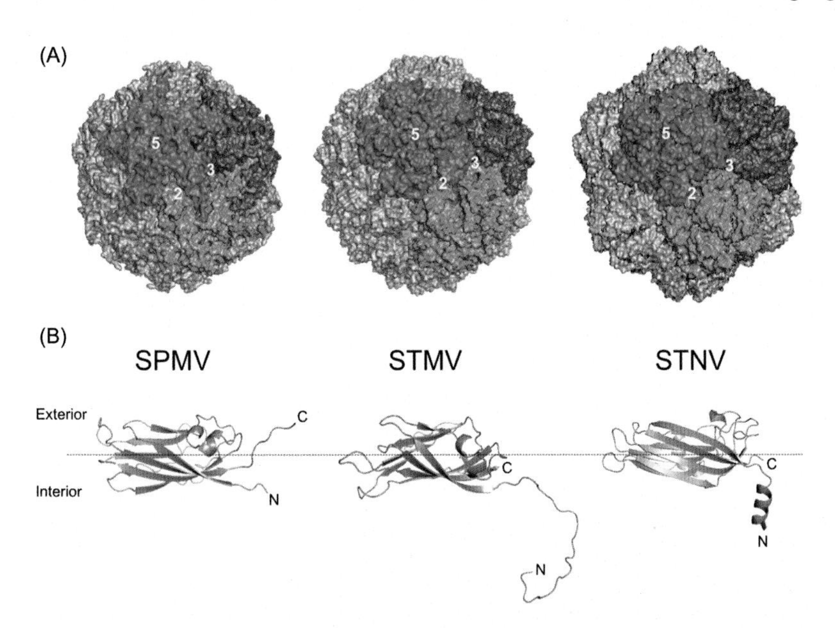

FIGURE 58.3 Structures of the SV virions and CPs. Reconstructed surface images of the SPMV (Protein data bank (PDB) ID: 1STM), STMV (PDB ID: 1A34), and STNV (PDB ID: 2BUK) $T = 1$ icosahedral virions (A). Individual colored pentamers (blue, maroon, and orange) are indicated, along with the five-, three-, and twofold axes. Cartoon images of the jellyroll β-sheet structures of the SPMV, STMV, and STNV CPs (B). A dashed line indicates orientation of the CPs within the capsid shell, with the interior and exterior faces of the virion indicated. The amino- and carboxyl-termini of the CPs are labeled as N and C, respectively. All images were generated using PyMOL (pymol.org).

signals (PSs), primarily short RNA stem-loop structures, have been identified or predicted for STNV, SPMV, and STMV.

Empty SV particles do not self-assemble, most likely due to strong electrostatic repulsions of the arginine-rich *N*-termini of the SV CPs. This effect is reduced by PS:CP interactions that seed the assembly of the virion. Thus, the combination of PSs with differing affinities for CP, and sufficient free CP, regulate the packaging kinetics to allow for efficient and specific assembly of infectious SV particles. Yet there must be some mutational flexibility, to allow for beneficial changes to the SV genome without compromising the secondary RNA PS structures or perturbing the essential biological functions of the CPs.

The PSs encoded by STMV form as short-range stem-loop structures as the RNA exits the replicase complex (Kuznetsov et al., 2010), likely in different conformations depending on its destination as translated or packaged RNA. The question of two RNA structures came about because the predicted 3′-tRNA-like end conflicted with known packaging constraints. Using SHAPE analysis, long-range secondary base-pairing was observed, with a region delimited by nt positions 169–646 being reminiscent of viroid RNA. A tRNA-like 3′-end structure was also observed, emphasizing its role in replication and RNA stability—possibly preventing degradation or abrogating silencing, akin to viroids (Athavale et al., 2013). Using cryo-electron microscopy and SHAPE analyses, the viroid-like secondary structure has been realized in solution (Garmann et al., 2015). The control of folding and parameters that determine if the RNA is packaged are key aspects of understanding the biology and infection by these SVs, as well as their comingled plant and virus origins.

Acknowledgments

This project was supported in part by an Agriculture and Food Research Initiative Competitive Grant (2016–67013–24738) from the USDA National Institute of Food and Agriculture, awarded to K.-B.G.S.

References

Athavale, S.S., Gossett, J.J., Bowman, J.C., Hud, N.V., Williams, L.D., Harvey, S.C., 2013. *In vitro* secondary structure of the genomic RNA of satellite tobacco mosaic virus. PLoS One 8, e54384.

Ban, N., Larson, S.B., McPherson, A., 1995. Structural comparison of the plant satellite viruses. Virology 214, 571–583.

Ban, N., McPherson, A., 1995. The structure of satellite panicum mosaic virus at 1.9 Å resolution. Nat. Struct. Biol. 2, 882–890.

Bringloe, D.H., Gultyaev, A.P., Pelpel, M., Pleij, C.W.A., Coutts, R.H., 1998. The nucleotide sequence of satellite tobacco necrosis virus strain C and helper-assisted replication of wild-type and mutant clones of the virus. J. Gen. Virol. 79, 1539–1546.

Bunka, D.H.J., Lane, S.W., Lane, C.L., Dykeman, E.C., Ford, R.J., Barker, A.M., et al., 2011. Degenerate RNA packaging signals in the genome of satellite tobacco necrosis virus: implications for the assembly of a T = 1 capsid. J. Mol. Biol. 413, 51–65.

Dodds, J.A., 1998. Satellite tobacco mosaic virus. Annu. Rev. Phytopathol. 36, 295–310.

Everett, A.L., Scholthof, H.B., Scholthof, K.-B.G., 2010. Satellite panicum mosaic virus coat protein enhances the performance of plant virus gene vectors. Virology 396, 37–46.

Ford, R.J., Barker, A.M., Bakker, S.E., Coutts, R.H., Ranson, N.A., Phillips, S.E.V., et al., 2013. Sequence-specific, RNA–protein interactions overcome electrostatic barriers preventing assembly of satellite tobacco necrosis virus coat protein. J. Mol. Biol. 425, 1050–1064.

Garmann, R.F., Gopal, A., Athavale, S.S., Knobler, C.M., Gelbart, W.M., Harvey, S.C., 2015. Visualizing the global secondary structure of a viral RNA genome with cryo-electron microscopy. RNA. 21, 877–886.

Gazo, B.M., Murphy, P., Gatchel, J.R., Browning, K.S., 2004. A novel interaction of cap-binding protein complexes eukaryotic initiation factor (eIF) 4F and eIF(iso)4F with a region in the 3′-untranslated region of satellite tobacco necrosis virus. J. Biol. Chem. 279, 13584–13592.

Jones, T.A., Liljas, L., 1984. Structure of satellite tobacco necrosis virus after crystallographic refinement at 2.5 Å resolution. J. Mol. Biol. 177, 735–767.

Krupovic, M., Kuhn, J.H., Fischer, M.G., 2015. A classification system for virophages and satellite viruses. Arch. Virol. 1–15.

Kuznetsov, Y.G., Dowell, J.J., Gavira, J.A., Ng, J.D., McPherson, A., 2010. Biophysical and atomic force microscopy characterization of the RNA from satellite tobacco mosaic virus. Nucleic Acids Res. 38, 8284–8294.

Larson, S.B., Day, J.S., McPherson, A., 2014. Satellite tobacco mosaic virus refined to 1.4 Å resolution. Acta Crystallographica Section D. 70, 2316–2330.

Mandadi, K.K., Pyle, J.D., Scholthof, K.-B.G., 2014. Comparative analysis of antiviral responses in *Brachypodium distachyon* and *Setaria viridis* reveals conserved and unique outcomes among C3 and C4 plant defenses. Mol. Plant Microbe Interact. 27, 1277–1290.

Mandadi, K.K., Scholthof, K.-B.G., 2012. Characterization of a viral synergism in the monocot *Brachypodium distachyon* reveals distinctly altered host molecular processes associated with disease. Plant Physiol. 160, 1432–1452.

Mandadi, K.K., Scholthof, K.-B.G., 2015. Genome-wide analysis of alternative splicing landscapes modulated during plant-virus interactions in *Brachypodium distachyon*. Plant Cell. 27, 71–85.

Meulewaeter, F., Danthinne, X., Van, M., Montagu, Cornelissen, M., 1998. 5′- and 3′-sequences of satellite tobacco necrosis virus RNA promoting translation in tobacco. Plant J. 14, 169–176.

Omarov, R.T., Qi, D., Scholthof, K.-B.G., 2005. The capsid protein of satellite panicum mosaic virus contributes to systemic invasion and interacts with its helper virus. J. Virol. 79, 9756–9764.

Patel, N., Dykeman, E.C., Coutts, R.H.A., Lomonossoff, G.P., Rowlands, D.J., Phillips, S.E.V., et al., 2015. Revealing the density of encoded functions in a viral RNA. Proc. Natl. Acad. Sci. USA 112, 2227–2232.

Qi, D., Scholthof, K.-B.G., 2008. Multiple activities associated with the capsid protein of satellite panicum mosaic virus are controlled separately by the N- and C-terminal regions. Mol. Plant Microbe Interact. 21, 613–621.

Qiu, W., Scholthof, K.-B.G., 2004. Satellite panicum mosaic virus capsid protein elicits symptoms on a nonhost plant and interferes with a suppressor of virus-induced gene silencing. Mol. Plant Microbe Interact. 17, 263–271.

Russo, M., De Stradis, A., Boscia, D., Rubino, L., Redinbaugh, M.G., Abt, J.J., et al., 2008. Molecular and ultrastructural properties of maize white line mosaic virus. J. Plant Pathol. 90, 363–369.

Scholthof, K.-B.G., Jones, R.W., Jackson, A.O., 1999. Biology and structure of plant satellite viruses activated by icosahedral helper viruses. In: Vogt, P.K., Jackson, A.O. (Eds.), Satellites and Defective Viral RNAs. Springer-Verlag, Berlin/Heidelberg, pp. 123–143.

Sivanandam, V., Mathews, D., Rao, A.L.N., 2015. Properties of satellite tobacco mosaic virus phenotypes expressed in the presence and absence of helper virus. Virology 483, 163–173.

Stewart, C.L., Pyle, J.D., Jochum, C.C., Vogel, K.P., Yuen, G.Y., Scholthof, K.-B.G., 2015. Multi-year pathogen survey of biofuel switchgrass breeding plots reveals high prevalence of infections by panicum mosaic virus and its satellite virus. Phytopathology 105, 1146–1154.

Timmer, R.T., Benkowski, L.A., Schodin, D., Lax, S.R., Metz, A.M., Ravel, J.M., et al., 1993. The 5′ and 3′ untranslated regions of satellite tobacco necrosis virus RNA affect translational efficiency and dependence on a 5′ cap structure. J. Biol. Chem. 268, 9504–9510.

Zhang, L., Zitter, T.A., Palukaitis, P., 1991. Helper virus-dependent replication, nucleotide sequence and genome organization of the satellite virus of maize white line mosaic virus. Virology 180, 467–473.

B. SATELLITE NUCLEIC ACIDS

CHAPTER

59

Large Satellite RNAs

Mazen Alazem and Na-Sheng Lin

Institute of Plant and Microbial Biology, Academia Sinica, Taipei, Taiwan

INTRODUCTION

Helper virus (HV)-dependent subviral agents can exist as satellite viruses, which encode their own structural proteins, or satellite RNAs or DNAs, some of which encode nonstructural proteins (Briddon and Stanley, 2006; Hu et al., 2009). Descriptions and taxonomic analyses of subviral agents are available on some public databases, such as DPVweb and Subviral RNA Database (Adams and Antoniw, 2006; Rocheleau and Pelchat, 2006). Satellite RNAs (satRNAs) are classified as small circular satellites (comprising ~350–400 nt), small linear satellites (<800 nt), or large linear satellites (>800 nt). While small circular and linear satRNAs do not encode any functional protein, large satRNAs usually encode nonstructural proteins (Hu et al., 2009). The HVs of most of these large satRNAs are nepoviruses, which are transmitted primarily through seed and by soil-borne nematodes (Briddon et al., 2012; Dunez and Le Gall, 2011). The genus *Nepovirus* (*Comovirinae; Secoviridae*) and the unassigned member strawberry latent ringspot virus (SLRV) (species *Strawberry latent ringspot virus,* family *Secoviridae*) consist of single-stranded positive-sense RNA viruses with icosahedral symmetry and contain bipartite RNA genomes (Dunez and Le Gall, 2011; Briddon et al., 2012). Several economically important nepoviruses including SLRV, are associated with satRNAs, and up to eight large satellites have been reported (Table 59.1). The satRNA of bamboo mosaic virus (satBaMV) is currently unique in that bamboo mosaic virus (BaMV) is the only potexvirus that has associated satRNAs (Lin and Hsu, 1994; Verchot-Lubicz and Baulcombe, 2011). The potexvirus genome is a monopartite single-stranded positive-sense RNA encapsidated in filamentous virions. Transmission of potexviruses is mechanical, with no known insect vectors (Verchot-Lubicz and Baulcombe, 2011).

Viroids and Satellites.
DOI: http://dx.doi.org/10.1016/B978-0-12-801498-1.00059-0

TABLE 59.1 Characteristics of Large satRNAs Associated With Plant Viruses

Large satellite RNAs	Accession No.	HV	Taxon	satRNA size: nt/aa	Transmission	Host range and economic importance of HV	Roles of the satellite	Ref.
satArMV	NC_003523	ArMV	*Secoviridae*	1092–1139 nt 353 aa	Soil-inhabiting nematode	Many dicot and monocot plants/Grapevine degeneration disease	Interferes with HV in model plants	1, 2
satBaMV	F4: AAP31396 L6: AY205210	BaMV	*Potexvirus*	~836 nt 182 aa	Mechanical	Affects several monocot and dicot plants/Bamboo Mosaic disease	Interferes with HV accumulation	3, 4
satBRV	NP_624321	BRV	*Secoviridae*	1432 nt 416 aa	Mites	Dicot plants/Blackcurrant reversion disease	NA	1, 5, 6
satChYMV	NP_620554	ChYMV	*Secoviridae*	~1165 nt 374 aa	Soil-inhabiting nematode	Few dicot plants	Interferes with HV in model plants	1, 7, 8
satGFLV	NP_443755	GFLV	*Secoviridae*	~1104 nt 341 aa	Soil-inhabiting nematode	Few dicot plants/ Grapevine infectious degeneration virus	Interferes with HV in model plants	9
satGBLV	No sequence available	GBLV	*Secoviridae*	N/A	Soil-inhabiting nematode	Few dicot plants	NA	10
satTBRV	X00978	TBRV	*Secoviridae*	1375 nt 424 aa	Soil-inhabiting nematode	Many dicot and monocot plants	NA	1, 11, 12

satMLRSV	No sequence available	MLRSV	*Secoviridae*	N/A	Soil-inhabiting nematode	Dicot plants	NA	13
satSLRV	NP_620833	SLRV	*Secoviridae*	1118 nt 331 aa	Soil-inhabiting nematode	Many dicot plants: berries and stone fruits	NA	1, 12, 14

satArMV, Arabis mosaic virus large satellite RNA; *satBaMV*, Bamboo mosaic virus satellite RNA; *satBRV*, Blackcurrant reversion virus satellite RNA; *satChYMV*, Chicory yellow mottle virus satellite RNA; *satGFLV*, Grapevine fanleaf virus satellite RNA; *satGBLV*, Grapevine Bulgarian latent virus satellite RNA; *satTBRV*, Tomato black ring virus (synonym: Beet ringspot virus) satellite RNA; *satMLRV*, Myrobalan latent ringspot virus satellite RNA; *satSLRV*, Strawberry latent ringspot virus satellite RNA; *HV*, Helper virus; *NA*, not available; Ref: References - 1, Dunez and Le Gall (2011); 2, Wetzel et al. (2006); 3, Lin et al. (1993); 4, Vijayapalani et al. (2012); 5, Latvala-Kilby et al. (2000); 6, Susi (2004); 7, Rubino et al. (1990); 8, Piazzolla et al. (1986); 9, Andret-Link et al. (2004); 10, Gallitelli et al. (1983); 11, Gallitelli et al. (2004); 12, Meyer et al. (1984); 13, Fritsch et al. (1984); 14, Kreiah et al. (1993). *nt*, nucleotide; *aa*, amino acid. SLRV is an unassigned species in the Secoviridae, to which the genus Nepovirus also belongs. Taxonomy of these species is as described in the Ninth Report of the International Committee on Taxonomy of Viruses (King et al., 2011).

We will review satBaMV separately from the large satRNAs of SLRV and the nepoviruses (which will be collectively referred to as secovirid-satRNAs).

ROLES AND PHYLOGENY OF THE PROTEINS ENCODED BY THE LARGE SECOVIRID-SATRNAS

Large secovirid-satRNAs usually encode one nonstructural protein of 37–45 kDa in size, although the size of the proteins encoded by all members has not been determined (Gottula et al., 2013; Latvala-Kilby et al., 2000). A few of these proteins are important for satRNA replication: such as those encoded by arabis mosaic virus large satellite RNA (satARMV) (Liu and Cooper, 1993), tomato black ring virus satellite RNA (Hemmer et al., 1993), and grapevine fanleaf virus satellite RNA (satGFLV) (Hans et al., 1993). Specifically, a cysteine- and histidine-rich region in the encoded protein of satGFLV is important for its replication (Latvala-Kilby et al., 2000). Interestingly, this region is highly conserved in proteins encoded by satArMV and chicory yellow mottle virus satellite RNA (satChYMV) (Latvala-Kilby et al., 2000), suggesting that these uncharacterized proteins may also be involved in satRNA replication. BLAST searches (https://blast.ncbi.nlm.nih.gov/Blast.cgi) did not reveal significant similarities between secovirid-satRNA proteins and other proteins within the NCBI public databases. The secovirid-satRNA proteins themselves had amino acid identities of ~70% between satArMV and satGFLV, and ~47% between blackcurrant reversion virus satellite RNA and satChYMV. These pairs clustered in phylogenetic analyses, with only satSLRV separated far from the rest of the group (Fig. 59.1).

The most studied examples from this family are the satRNAs associated with ArMV and GFLV. Both viruses are economically important to a wide range of plants, especially grapevine, in which infection causes grapevine degeneration disease (Gottula et al., 2013; Lamprecht et al., 2013; Nourinejhad Zarghani et al., 2014; Wetzel et al., 2006). Although both viruses are serologically distinct, their satRNAs exhibit high identity (up to 70% for some strains). Some isolates of satGFLV are more similar to satArMV isolates than to other satGFLV isolates (Lamprecht et al., 2013). In consonance with this similarity, some satGFLV isolates (such as F13) are able to replicate with specific ArMV strains (ArMV-S) more efficiently than with some GFLV strains (Hans et al., 1993). Such findings imply that certain strains of blackcurrant reversion virus satRNA and satChYMV may also rely on each other's helper virus (HV) for replication. However, recognition of satRNAs by

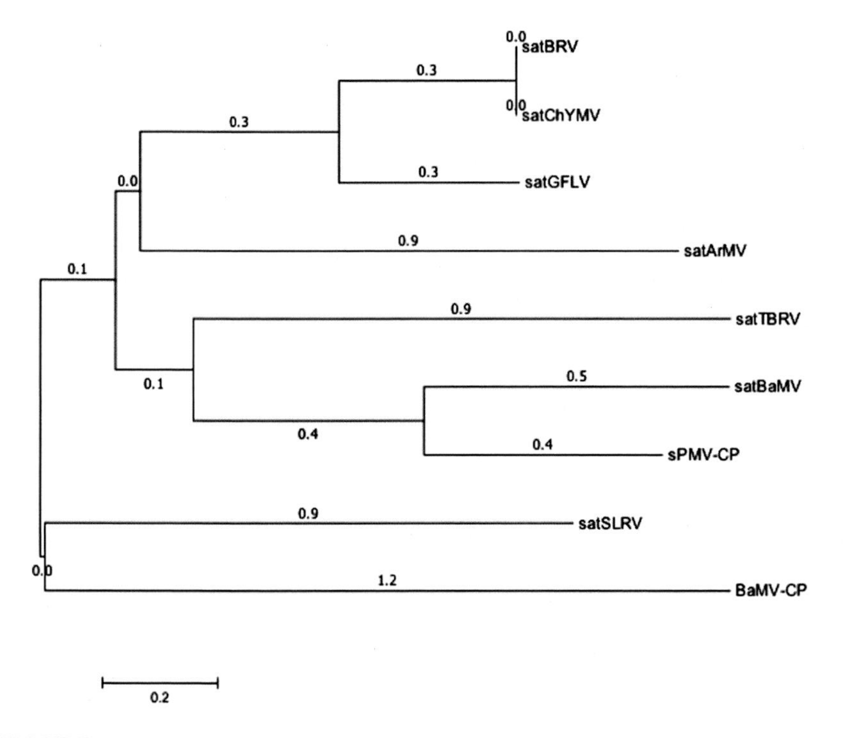

FIGURE 59.1 Phylogeny of large satRNAs generated using the Neighbor Joining method by MEGA 6.0. Accession numbers for the strains used in the analysis are indicated in Table 59.1. Numbers in the figure indicate the phylogenetic distance. Satellites are: arabis mosaic virus large satellite RNA (satArMV), bamboo mosaic virus satellite RNA (satBaMV), blackcurrant reversion virus satellite RNA (satBRV), chicory yellow mottle virus satellite RNA (satChYMV), grapevine fanleaf virus satellite RNA (satGFLV), grapevine Bulgarian latent virus satellite RNA (satGBLV), tomato black ring virus (synonym: beet ringspot virus) satellite RNA (satTBRV), myrobalan latent ringspot virus satellite RNA (satMLRV), strawberry latent ringspot virus satellite RNA (satSLRV). BaMV (capsid protein)-CP was used as an outlier, satellite panicum mosaic virus (sPMV)-CP was used here for comparison purposes.

HV-encoded replicase has not yet been studied for most secovirid-satRNAs.

TBRV has a large satellite of 1375 nucleotides (Oncino et al., 1995). This satRNA, however, is strain-specific for HV-dependency; i.e., it replicates efficiently with the G strain, but not with the L strain of TBRV. Interestingly, the 5′ and the 3′ untranslated regions (UTRs) of this satRNA are not implicated in this specificity (Oncino et al., 1995). Two other members of this group, grapevine Bulgarian latent virus satellite RNA and myrobalan latent ringspot virus satellite RNA, were previously reported, but have not been subjected to molecular characterization (Roossinck et al., 1992).

EFFECTS OF SECOVIRID-SATRNAS ON HV ACCUMULATION

Several satRNAs interfere with the accumulation or spread of their HVs. The negative effect on the accumulation of their HVs suggests that such satRNAs can be used as agents for biological control (Lin et al., 2013; Shen et al., 2015). However, the interfering roles of secovirid-satRNAs are not well understood in most cases. The reported effects vary depending on the host examined. For example, satGFLV does not seem to affect GFLV induced-symptoms in a grapevine host (Saldarelli et al., 1993), but slightly delays symptom development in *Chenopodium quinoa* (Fuchs et al., 1991). Similarly, satArMV (lilac strain) does not affect titers of its HV, but does prevent development of virus-induced tip necrosis symptoms (Liu et al., 1991). Molecular studies on the interfering roles of secovirid-satRNAs may provide valuable information on resistance for use in breeding programs.

SATBAMV

One of the best characterized large satRNAs is satBaMV, with different studies describing its structure, replication, movement, phylogeny, and effects on BaMV and hosts. The genome of satBaMV is ~830 nt (excluding the poly-A tail), and encodes a 20 kDa protein (P20). Furthermore, the satBaMV genome shares limited similarity with the BaMV genome at the 5′ UTR (Lin and Hsu, 1994). Interestingly, P20 showed ~47% amino acid identity with the structural protein of satellite panicum mosaic virus (sPMV)-CP (Liu and Lin, 1995). It is noteworthy that secovirid-satRNA proteins are clustered more closely to the proteins encoded by satBaMV and sPMV than to satSLRV-encoded protein (Fig. 59.1).

ROLES OF THE SATBAMV P20 PROTEIN

Unlike proteins encoded by secovirid-satRNAs, the satBaMV P20 protein is not required for replication, as revealed by the observation that mutations or replacement of the P20 gene did not abolish satBaMV replication in BaMV-coinfected plants (Lin et al., 1996). P20 shows high binding affinity to RNAs of BaMV and satBaMV at both 5′- and 3′-UTRs (Tsai et al., 1999). The arginine-rich motif at the *N*-terminus of P20 is an RNA binding domain required for the formation of stable satBaMV ribonucleoprotein complexes and self-interaction. Phosphorylation of

P20 regulates negatively ribonucleoprotein complex formation and P20 self-interaction (Vijayapalani et al., 2012). In addition, disruption of RNA-binding affinity by deleting the arginine-rich arm, or mimicking phosphorylation of P20, drastically limited satBaMV long-distance movement in *Nicotiana benthamiana* coinfected with BaMV (Vijayapalani et al., 2006, 2012). However, P20 protein can functionally complement *in trans* the systemic trafficking of P20-defective satBaMV in infected *N. benthamiana*, indicating a strong role for P20 in satBaMV long-distance movement. The satBaMV alone was recently found to be a mobile RNA in the absence of HV but systemically traffics via P20-fibrillarin-satBaMV ribonucleoprotein complex through the phloem (Chang et al., 2016).

EFFECTS OF SATBAMV ON HV ACCUMULATION

Structurally, satBaMV and BaMV share conserved secondary structures at their 5′- and 3′-UTRs (Huang et al., 2009). The dependence of satBaMV on HV for replication is related to the structural resemblance between satBaMV and BaMV at both UTRs. Disrupting the secondary structures through deletion or substitution had a critical effect on the replication of satBaMV (Huang et al., 2009). Moreover, some of these isolates, such as BSL6, have interfering roles, causing almost complete attenuation of BaMV symptoms and titers (Chen et al., 2007; Hsu et al., 1998, 2006). It has been shown that the apical hairpin stem loop at the 5′-UTR plays a crucial role in satBaMV-mediated downregulation of BaMV (Hsu et al., 2006). In vivo competition assays revealed that BSL6 RNA has greater affinity than BaMV RNA for the viral replicase (Chen et al., 2012). Although the replication machineries of HV and satBaMV are similar, host heat shock protein 90 is specifically required for BaMV replication only (Huang et al., 2012).

Phylogenetic analysis of satBaMVs collected from different regions and bamboo hosts showed that genetic diversity is not correlated with geographical distribution or host species. The observed clustering was different to that of BaMV, implying that satBaMV evolutionary trends are distinct from those of BaMV (Wang et al., 2014).

Such in-depth studies of satBaMV have been successfully translated into useful applications. In combination with BaMV, satBaMV has been developed as a dual gene silencing vector, which should be a useful tool for genomic studies (Liou et al., 2014). The development of satRNAs as vectors is discussed further in Chapter 55, Development and Application of Satellite-Based Vectors. The interfering BSL6 isolate has also become a powerful agent for transgenic resistance against BaMV (Lin et al., 2013).

Acknowledgments

This research was supported by grants from the Ministry of Science and Technology (NSC 1002313B001002MY3) and Academia Sinica Investigator Award.

References

Adams, M.J., Antoniw, J.F., 2006. DPVweb: a comprehensive database of plant and fungal virus genes and genomes. Nucleic Acids Res. 34, D382–D385.

Andret-Link, P., Schmitt-Keichinger, C., Demangeat, G., Komar, V., Fuchs, M., 2004. The specific transmission of grapevine fanleaf virus by its nematode vector Xiphinema index is solely determined by the viral coat protein. Virology 320, 12–22.

Briddon, R.W., Ghabrial, S., Lin, N.-S., Palukaitis, P., Scholthof, K.-B.G., Vetten, H.-J., 2012. Satellite and other virus-dependent nucleic acids. In: King, A.M.Q., Adams, M.J., Carstens, E.B., Lefkowitz, E.J. (Eds.), Virus Taxonomy. Classification and Nomenclature of Viruses. Ninth Report of the International Committee on Taxonomy of Viruses. Elsevier Academic Press, San Diego, CA, pp. 1211–1219.

Briddon, R.W., Stanley, J., 2006. Subviral agents associated with plant single-stranded DNA viruses. Virology 344, 198–210.

Chang, C.-H., Hsu, F.-C., Lee, S.-C., Lo, Y.-S., Wang, J.-D., Shaw, J., et al., 2016. The nucleolar fibrillarin protein is required for helper virus-independent long-distance trafficking of a subviral satellite RNA in plants. Plant Cell, 28, 2586–2602.

Chen, H.C., Hsu, Y.H., Lin, N.S., 2007. Downregulation of bamboo mosaic virus replication requires the 5′ apical hairpin stem loop structure and sequence of satellite RNA. Virology 365, 271–284.

Chen, H.C., Kong, L.R., Yeh, T.Y., Cheng, C.P., Hsu, Y.H., Lin, N.S., 2012. The conserved 5′ apical hairpin stem loops of bamboo mosaic virus and its satellite RNA contribute to replication competence. Nucleic Acids Res. 40, 4641–4652.

Dunez, J., Le Gall, O., 2011. Nepovirus. In: Tidona, C., Darai, G. (Eds.), The Springer Index of Viruses. Springer New York, pp. 361–369.

Fritsch, C., Koenig, I., Murant, A.F., Raschke, J.H., Mayo, M.A., 1984. Comparisons among satellite RNA species from five isolates of tomato black ring virus and one isolate of myrobalan latent ringspot virus. J. Gen. Virol. 65, 289–294.

Fuchs, M., Pinck, M., Etienne, L., Pinck, L., Walter, B., 1991. Characterization and detection of grapevine fanleaf virus by using cDNA probes. Phytopathology 81, 559–565.

Gallitelli, D., Rana, G.L., Vovlas, C., Martelli, G.P., 2004. Viruses of globe artichoke: an overview. J. Plant Pathol. 86, 267–281.

Gallitelli, D., Savino, V., De Sequeira, O., 1983. Properties of a distinctive strain of grapevine Bulgarian latent virus. Phytopathol. Mediterr. 22, 27–32.

Gottula, J., Lapato, D., Cantilina, K., Saito, S., Bartlett, B., Fuchs, M., 2013. Genetic variability, evolution, and biological effects of grapevine fanleaf virus satellite RNAs. Phytopathology 103, 1180–1187.

Hans, F., Pinck, M., Pinck, L., 1993. Location of the replication determinants of the satellite RNA associated with grapevine fanleaf nepovirus (strain F13). Biochimie 75, 597–603.

Hemmer, O., Oncino, C., Fritsch, C., 1993. Efficient replication of the in vitro transcripts from cloned cDNA of tomato black ring virus satellite RNA requires the 48K satellite RNA-encoded protein. Virology 194, 800–806.

Hsu, Y.H., Chen, H.C., Cheng, J., Annamalai, P., Lin, B.Y., Wu, C.T., et al., 2006. Crucial role of the 5′ conserved structure of bamboo mosaic virus satellite RNA in downregulation of helper viral RNA replication. J. Virol. 80, 2566–2574.

Hsu, Y.H., Lee, Y.S., Liu, J.S., Lin, N.S., 1998. Differential interactions of bamboo mosaic potexvirus satellite RNAs, helper virus, and host plants. Mol. Plant Microbe Interact. 11, 1207–1213.

Hu, C.C., Hsu, Y.H., Lin, N.S., 2009. Satellite RNAs and satellite viruses of plants. Viruses 1, 1325–1350.

Huang, Y.W., Hu, C.C., Lin, C.A., Liu, Y.P., Tsai, C.H., Lin, N.S., et al., 2009. Structural and functional analyses of the 3′ untranslated region of bamboo mosaic virus satellite RNA. Virology 386, 139–153.

Huang, Y.W., Hu, C.C., Liou, M.R., Chang, B.Y., Tsai, C.H., Meng, M., et al., 2012. Hsp90 interacts specifically with viral RNA and differentially regulates replication initiation of bamboo mosaic virus and associated satellite RNA. PLoS Pathog. 8, e1002726.

King, A.M., Adams, M.J., Carstens, E.B., Lefkowitz, E.J., 2011. Virus taxonomy: classification and nomenclature of viruses: Ninth Report of the International Committee on Taxonomy of Viruses. Elsevier Academic Press, San Diego, CA.

Kreiah, S., Cooper, J.I., Strunk, G., 1993. The nucleotide-sequence of a satellite RNA associated with strawberry latent ringspot virus. J. Gen. Virol. 74, 1163–1165.

Lamprecht, R.L., Spaltman, M., Stephan, D., Wetzel, T., Burger, J.T., 2013. Complete nucleotide sequence of a South African isolate of grapevine fanleaf virus and its associated satellite RNA. Viruses 5, 1815–1823.

Latvala-Kilby, S., Lemmetty, A., Lehto, K., 2000. Molecular characterization of a satellite RNA associated with blackcurrant reversion nepovirus. Arch. Virol. 145, 51–61.

Lin, K.Y., Hsu, Y.H., Chen, H.C., Lin, N.S., 2013. Transgenic resistance to bamboo mosaic virus by expression of interfering satellite RNA. Mol. Plant Pathol. 14, 693–707.

Lin, N.S., Chai, Y.J., Huang, T.Y., Chang, T.Y., Hsu, Y.H., 1993. Incidence of bamboo mosaic potexvirus in Taiwan. Plant Dis. 77, 448–450.

Lin, N.S., Hsu, Y.H., 1994. A satellite RNA associated with bamboo mosaic potexvirus. Virology 202, 707–714.

Lin, N.S., Lee, Y.S., Lin, B.Y., Lee, C.W., Hsu, Y.H., 1996. The open reading frame of bamboo mosaic potexvirus satellite RNA is not essential for its replication and can be replaced with a bacterial gene. Proc. Natl. Acad. Sci. USA 93, 3138–3142.

Liou, M.R., Huang, Y.W., Hu, C.C., Lin, N.S., Hsu, Y.H., 2014. A dual gene-silencing vector system for monocot and dicot plants. Plant Biotechnol. J. 12, 330–343.

Liu, J.S., Lin, N.S., 1995. Satellite RNA associated with bamboo mosaic potexvirus shares similarity with satellites associated with sobemoviruses. Arch. Virol. 140, 1511–1514.

Liu, Y.Y., Cooper, J., Edwards, M.L., Hellen, C.U., 1991. A satellite RNA of arabis mosaic nepovirus and its pathological impact. Ann. Appl. Biol. 118, 577–587.

Liu, Y.Y., Cooper, J.I., 1993. The multiplication in plants of arabis mosaic virus satellite RNA requires the encoded protein. J. Gen. Virol. 74, 1471–1474.

Meyer, M., Hemmer, O., Fritsch, C., 1984. Complete nucleotide-sequence of a satellite RNA of tomato black ring virus. J. Gen. Virol. 65, 1575–1583.

Nourinejhad Zarghani, S., Dupuis-Maguiraga, L., Bassler, A., Wetzel, T., 2014. Mapping of the exchangeable and dispensable domains of the RNA 2-encoded 2A (HP) protein of arabis mosaic nepovirus. Virology 458-459, 106–113.

Oncino, C., Hemmer, O., Fritsch, C., 1995. Specificity in the association of tomato black ring virus satellite RNA with helper virus. Virology 213, 87–96.

Piazzolla, P., Vovlas, C., Rubino, L., 1986. Symptom regulation induced by chicory yellow mottle virus satellite-like RNA. J. Phytopathol. 115, 124–129.

Rocheleau, L., Pelchat, M., 2006. The Subviral RNA Database: a toolbox for viroids, the hepatitis delta virus and satellite RNAs research. BMC Microbiol. 6, 24.

Roossinck, M.J., Sleat, D., Palukaitis, P., 1992. Satellite RNAs of plant viruses: structures and biological effects. Microbiol. Rev. 56, 265–279.

Rubino, L., Tousignant, M.E., Steger, G., Kaper, J.M., 1990. Nucleotide sequence and structural analysis of two satellite RNAs associated with chicory yellow mottle virus. J. Gen. Virol. 71, 1897–1903.

Saldarelli, P., Minafra, A., Walter, B., 1993. A survey of grapevine fanleaf nepovirus Isolates for the presence of satellite RNA. Vitis 32, 99–102.

Shen, W., Au, P.C.K., Shi, B.-J., Smith, N.A., Dennis, E.S., Guo, H.-S., et al., 2015. Satellite RNAs interfere with the function of viral RNA silencing suppressors. Front. Plant Sci. 6, 281.

Susi, P., 2004. Black currant reversion virus, a mite-transmitted nepovirus. Mol. Plant Pathol. 5, 167–173.

Tsai, M.S., Hsu, Y.H., Lin, N.S., 1999. Bamboo mosaic potexvirus satellite RNA (satBaMV RNA)-encoded P20 protein preferentially binds to satBaMV RNA. J. Virol. 73, 3032–3039.

Verchot-Lubicz, J., Baulcombe, D., 2011. Potexvirus. In: Tidona, C., Darai, G. (Eds.), The Springer Index of Viruses. Springer New York, pp. 505–515.

Vijayapalani, P., Chen, J.C., Liou, M.R., Chen, H.C., Hsu, Y.H., Lin, N.S., 2012. Phosphorylation of bamboo mosaic virus satellite RNA (satBaMV)-encoded protein P20 downregulates the formation of satBaMV-P20 ribonucleoprotein complex. Nucleic Acids Res. 40, 638–649.

Vijayapalani, P., Kasiviswanathan, V., Chen, J.C., Chen, W., Hsu, Y.H., Lin, N.S., 2006. The arginine-rich motif of bamboo mosaic virus satellite RNA-encoded P20 mediates self-interaction, intracellular targeting, and cell-to-cell movement. Mol. Plant Microbe Interact. 19, 758–767.

Wang, I.N., Hu, C.C., Lee, C.W., Yen, S.M., Yeh, W.B., Hsu, Y.H., et al., 2014. Genetic diversity and evolution of satellite RNAs associated with the bamboo mosaic virus. PLoS One 9, e108015.

Wetzel, T., Bassler, A., Amren, M.A.W., Krczal, G., 2006. A RT/PCR-partial restriction enzymatic mapping (PREM) method for the molecular characterisation of the large satellite RNAs of arabis mosaic virus isolates. J Virol. Methods 132, 97–103.

CHAPTER

60

Small Linear Satellite RNAs

Mikyeong Kim[1] *and Marilyn J. Roossinck*[2]

[1]National Institute of Agricultural Sciences, RDA, Wanju-gun, South Korea

[2]Pennsylvania State University, University Park, PA, United States

INTRODUCTION

Plant viral satellite RNAs (satRNAs) are subviral RNA agents that depend on their helper viruses for replication and encapsidation/dissemination, and can be considered molecular parasites of plant viruses. They are not required for any essential functions of their helper viruses and contain no RNA sequence homology with the helper virus. This chapter is concerned with the small linear satRNAs that do not encode any proteins, and therefore their biology is completely dictated by the primary and higher-order structures of the RNA. The small linear satRNAs may affect the host directly or indirectly through interactions with the helper virus. We summarize some of the properties of small linear satRNAs with emphasis on replication and interactions with helper viruses and hosts.

NUCLEOTIDE SEQUENCE AND CLASSIFICATION

Small linear single-strand satRNAs recognized by the International Committee on Taxonomy of Viruses include those supported by members of genera *Carmovirus, Cucumovirus, Necrovirus, Tombusvirus,* and *Umbravirus* (King et al., 2012). The small RNAs associated with groundnut rosette virus and tobacco bushy top virus are considered as satRNAs by some but are essential for the life cycle of the virus (Murant, 1990), and turnip crinkle virus (TCV) supports both satRNAs and some defective interfering (DI) RNA/satRNA chimeras that result from recombination between satRNAs and the helper virus (Simon and Howell, 1986). More than 14 species of small linear satRNAs have been

Viroids and Satellites.
DOI: http://dx.doi.org/10.1016/B978-0-12-801498-1.00060-7

reported, with the majority supported by viruses from the families *Bromoviridae* and *Tombusviridae* (Abraham et al., 2014; Hu et al., 2009; Simon et al., 2004; Xi et al., 2006; Xu and Roossinck, 2011) (Table 60.1). The most well-studied of these satRNAs are those of cucumber mosaic virus (CMV), and 158 genomes from CMV satRNAs can be found in GenBank. The nomenclature of CMV satRNAs is sometimes related to symptoms, such as WL for the satRNA that induces white leaf symptoms, but in many cases names have been chosen by researchers without clear reference to meaning.

TABLE 60.1 Small Linear Satellite RNAs

Virus family	Virus genus	Virus species	Satellite RNAs	Size	Accession No.
Bromoviridae	*Cucumovirus*	Cucumber mosaic virus	Several types	330–405	Not shown[a]
		Peanut stunt virus	Ag	393	EF469733
			P4	393	Z98198
			P	393	EF535259
			P6	393	Z98197
Tombusviridae	*Carmovirus*	Turnip crinkle virus	C	355	X12750
			D	194	Not reported
			F	230	X12749
	Necrovirus	Beet black scorch virus	Ir-Bj1sat	617	FJ176575
			Ir-Ha1	617	FN543474
			Ir-Kh1sat	617	FN543473
			Ir-Msh1sat	617	FN543472
			X	615	AY394497
			Xinjiang	616	JN635326
	Tombusvirus	Artichoke mottled crinkle virus	–	~700	Not reported
		Carnation Italian ringspot virus	–	~700	Not reported
		Cymbidium ringspot virus	–	621	NC004009
		Pelargonium leaf curl virus	–	~700	Not reported

(Continued)

TABLE 60.1 (Continued)

Virus family	Virus genus	Virus species	Satellite RNAs	Size	Accession No.
		Petunia asteroid mosaic virus	–	~700	Not reported
		Tomato bushy stunt virus	B1	822	AF022787
			B10	612	AF022788
			L	615	FJ666076
	Umbravirus	Carrot mottle mimic virus	Thessaloniki	748	EU914919
		Carrot mottle virus	Quedlinburg	748	EU914920
		Ethiopian tobacco bushy top	2–18	521	KJ918747
		Pea enation mosaic virus	WSG	714	U03564

[a] *Accession numbers for CMV satRNAs are not listed because there are too many.*

STRUCTURE AND REPLICATION

Small linear satRNAs appear to be highly structured, with between 50% and 70% of the nucleotides involved in base-pairing. This feature could explain the high stability and survivability of satRNA in plants and when purified. Replication of satRNAs includes a two-stage process: synthesis of minus-strand RNAs from plus-strand RNA templates and synthesis of progeny plus-strand RNAs from minus-strand RNA intermediates. This process requires both *cis*-acting signals located on the satRNAs and *trans*-acting factors such as helper virus-encoded RNA-dependent RNA polymerase and host factors. The satRNA structure is likely related to all aspects of the satRNA biological activities such as replication, encapsidation, pathogenesis, and interaction with the helper virus and plant hosts. The CMV satRNAs range in size from 330 to 405 nt with a high degree of secondary structure. While most analyses of RNA structure have been done with computer modeling, a few studies have examined the structure of CMV satRNAs using enzymatic or chemical probing in vitro (Bernal and García-Arenal, 1997; García-Arenal et al., 1987). One study of the CMV D-satRNA, which induces necrosis in tomato, used chemical probing *in planta*, and showed that the structure of the satRNA was significantly different in virions, as well as in purified satRNA compared with structures *in planta*; none of these structures resembled the computer models (Rodríguez-Alvarado and Roossinck, 1997).

Studies on cucumovirus satRNA replication determined that these satRNAs did not have circular replicative intermediates, although satRNAs of CMV and peanut stunt virus (PSV; genus *Cucumovirus*) produced multimeric forms in virus infected plants (Roossinck et al., 1992; Kuroda et al., 1997). CMV Q-satRNA expressed in the nucleus in the absence of helper virus was transcribed into multimeric forms (Choi et al., 2012). Comparison of the Q-satRNA multimeric forms in the presence and absence of CMV-Q helper virus suggested that the function of a hepta-nucleotide motif present at the junction of Q-satRNA multimers contributes to helper virus dependent replication (Seo et al., 2013). These models have been explained and discussed in a recent review (Rao and Kalantidis, 2015).

SatRNAs of CMV are also supported by tomato aspermy virus (TAV; genus *Cucumovirus*) (Devic et al., 1990; Gould et al., 1978; Jaegle et al., 1990; Moriones et al., 1992; Palukaitis and Roossinck, 1995), although some studies showed differential interactions among strains of TAV and CMV satRNAs (García-Arenal and Palukaitis, 1999; Moriones et al., 1992). Experiments on the 3′ end repair of CMV Q-satRNAs by helper viruses using agroinfiltration of *Nicotiana benthamiana* showed that the TAV-V strain replicase was competent to repair Q-satRNAs deletion mutants, but accumulation of Q-satRNA multimers with TAV-V as a helper virus was not observed (Sivanandam et al., 2015).

PSV can harbor its own satRNA that is not supported by CMV; nor can PSV support the replication of CMV satRNAs (Kaper and Tousignant, 1984; Kaper et al., 1978). The satRNAs of CMV and PSV are of similar size but do not have any evidence of sequence homology.

TCV, a member of the genus *Carmovirus* in the *Tombusviridae*, supports satRNAs C, D, and F. The three satRNAs have a conserved region (7 nt, CCUGCCC-OH) at the 3′ end and have different features including length and biological effects on the helper virus (Simon and Howell, 1986). SatRNA C (sat C) is a chimeric RNA and has two major domains; the 5′ portion is similar to sat D, and the 3′ region is derived from TCV. Analyses of RNA structures by computer modeling predicted hairpins and pseudoknots in the 3′ terminus of sat C and TCV, with 94% sequence identity. Two similar secondary structural elements at the 3′ terminus are important for satRNA replication. Molecular modeling also predicted the 3D structure of a tRNA-like element in TCV that is not shared with sat C. Studies of sat C have provided a model for *cis*-acting elements in the replication of satRNAs (Guo et al., 2011; Nagy et al., 1999).

The X satRNA is associated with beet black scorch virus, a member of the genus *Necrovirus* and consists of 615 nt. A highly ordered secondary structure estimated from computer analysis is approximately 70% base-paired, consisting of 26 short inverted repeats of 3–17 nt throughout the entire genome. The structure at both the 5′- and 3′-termini of the monomeric RNA, or the presence of intact juxtaposed multimeric forms, may be involved in replication (Guo et al., 2005).

In members of the genus *Tombusvirus*, three small satRNAs (B1, B10 and L) have been described as associated with tomato bushy stunt virus (TBSV), as well as one satRNA with cymbidium ringspot virus (CymRSV) (Célix et al., 1997; Rubino and Russo, 2010). Although by definition satRNAs do not share sequence identity with their helper virus genomes, analyses of tombusvirus satRNAs indicate the conservation of certain 5′- and 3′-terminal RNA sequences, or secondary and tertiary structures (Fabian et al., 2003). The 5′-terminus of CymRSV B1 and B10 satRNAs contains two RNA domains, a T-shaped domain and a downstream domain. Although limited sequence identity was observed, these corresponding higher-order RNA structures may be important for maintenance of the necessary structural features for satRNA replication and accumulation (Chernysheva and White, 2005). Helper viruses CymRSV and carnation Italian ringspot virus (genus *Tombusvirus*) support the replication of B1 and B10 satRNAs of TBSV, and the TBSV L satRNA is supported by carnation Italian ringspot virus, but not by CymRSV, implying that the tombusvirus satRNAs have different strategies for replication than their helper viruses (Rubino and Russo, 2010).

EFFECTS ON HELPER VIRUSES AND HOSTS

SatRNAs can affect the titer of the helper virus, and/or modulate its pathology, including attenuation or exacerbation of virus-induced symptoms. Attenuation of disease symptoms can be correlated with reduced accumulation of helper viruses. However, some satRNAs exacerbate the symptoms even if they also reduce helper virus accumulation. For example, the D-satRNA lowers the CMV titer and induces necrosis in infected tomato plants in the presence of CMV, but does not cause significant changes in TAV titer even though D-satRNA attenuates TAV-induced symptoms in tobacco (García-Arenal and Palukaitis, 1999). Some CMV satRNAs can induce specific yellowing symptoms in *Nicotiana* species such as the Y-satRNA of CMV (Masuta and Takanami, 1989). SatRNA-induced reduction of helper virus accumulation may be due to competition with their helper virus for replication enzymes or nucleotides (Gal-On et al., 1995; Roossinck et al., 1992).

Generally, complex interactions involving the host, the helper virus, and the satRNA are involved in symptom production. Numerous RNA transcript changes were seen in tomato plants undergoing necrosis induced by the D-satRNA of CMV (Irian et al., 2007). The genetic basis of the necrotic phenotype mapped to multiple loci in tomato (Xu et al., 2012). The satRNA C, a virulent TCV-associated satRNA, which is really a chimera of a satRNA and a DI RNA, normally intensifies symptoms but can attenuate symptoms, either if the TCV coat protein is replaced with that of the related cardamine chlorotic fleck virus, or if TCV contains an alteration in the coat protein initiation codon (Kong et al., 1995; Wang and Simon, 1999). On the other hand, the related TCV satRNAs D and F do not exacerbate symptoms (Simon et al., 2004). To understand the effect of satRNA of CMV on pathogenicity, and the possible involvement of host RNA silencing pathways in pathogenicity, Hou et al. (2011) investigated the relationship between the 2b silencing suppressor of the SD-CMV strain (ShanDong strain) and its SD-satRNA. The SD-satRNA attenuates the yellowing phenotype induced by the helper virus in *N. benthamiana* and *Arabidopsis* and the accumulation level of 2b coding subgenomic RNA 4A was greatly reduced in the presence of SD-satRNA. The satRNA serves as part of the RNA silencing target to be degraded by the host to small interfering RNA (siRNA), resulting in reduced CMV RNA-derived siRNA production. The results show that the effect of satRNAs on helper virus-induced symptoms can involve the host RNA silencing mechanism, and the plant silencing mechanism is involved in the pathogenicity of satRNAs (Hou et al., 2011; Kouadio et al., 2013). In addition, siRNAs derived from the Y-satRNA yellowing region directly targeted a host gene involved in chlorophyll synthesis in tobacco (Shimura et al., 2011).

Tombusvirus-associated satRNAs have few effects on symptoms. TBSV B10 satRNA attenuated the disease symptoms induced by the helper virus in *Nicotiana clevelandii* and *N. benthamiana*, whereas the B1 sat RNA of TBSV did not influence the symptoms (Célix et al., 1997, 1999). TBSV L satRNA and CymRSV satRNA do not interfere with symptom expression in *N. benthamiana*, regardless of the helper virus, although CymRSV satRNA may be indirectly involved in viral pathogenesis by influencing DI RNA accumulation (Rubino and Russo, 2010).

EVOLUTION AND ORIGINS

A large number of satRNAs associated with several groups of plant viruses have been reported. The small linear satRNAs are highly structured; they tend to be very stable and highly infectious. In one experiment CMV satRNA molecules were shown to be stable on plants for 25 days in the absence of helper virus (Jacquemond and Lot, 1982). These properties confound attempts to find origins of satRNAs, which have been suggested to come from the genome of the helper virus or that of the host plant. A novel CMV satRNA spontaneously appeared in tobacco plants following serial passages of CMV generated from cDNA clones that are free of satRNAs (Hajimorad et al., 2009). In addition no satRNA sequence identity with the helper virus and host plant suggested that satRNAs originate from other coinfecting entities (Hu et al., 2009). With more plant genomic information available, a recent study found multiple DNA fragments in *Nicotiana* plants related to CMV satRNAs, and suggested that the CMV satRNAs originated from the *Nicotiana* genome (Zahid et al., 2015).

CONCLUSIONS

Since the first description of small linear satellites of plant viruses extensive studies of plant viral satellites including satRNAs have characterized many new satellites, determined their pathogenicity, and examined various biological and biochemical functions. The different types of satRNA that have been characterized by sequence, structure, and effects on the symptoms induced by their helper viruses, have contributed to a body of knowledge about the relationship between RNA molecules and biological effects. Some satRNAs attenuate the disease symptoms induced by their helper viruses so that satRNAs are suggested to have viral control potential in agriculture, although significant risks may be involved. The small size of the satRNA genome makes it an attractive reporter for the study of helper virus evolutionary mechanisms (Pita et al., 2007).

Many studies have described interactions between satRNA, helper virus, and host plant including novel information on RNA biology, pathogen symptom modulation, and origins of small parasitic RNAs. However, in spite of new information on satRNA biology, including movement, replication in the host, and epidemiology, there remain many unanswered questions, and further study would undoubtedly be fruitful.

Acknowledgments

This paper was supported by grants from the cooperative research project (PJ008537) funded by the Rural developement Administration of Korea.

References

Abraham, A.D., Menzel, W., Bekele, B., Winter, S., 2014. A novel combination of a new umbravirus, a new satellite RNA and potato leafroll virus causes tobacco bushy top disease in Ethopia. Arch. Virol. 159, 3395–3399.

Bernal, J.J., García-Arenal, F., 1997. Analysis of the in vitro secondary structure of cucumber mosaic virus satellite RNA. RNA. 3, 1052–1067.

Célix, A., Burgyán, J., Rodríguez-Cerezo, E., 1999. Interactions between tombusviruses and satellite RNAs of tomato bushy stunt virus: a defect in sat RNA B1 replication maps to ORF1 of a helper virus. Virology 262, 129–138.

Célix, A., Rodriguez-Cerezo, E., Garcia-Arenal, F., 1997. New satellite RNAs, but no DI RNAs, are found in natural populations of tomato bushy stunt tombusvirus. Virology 239, 277–284.

Choi, S.H., Seo, J.K., Kwon, S.J., Rao, A.L., 2012. Helper virus-independent transcription and multimerization of a satellite RNA associated with cucumber mosaic virus. J. Virol. 86, 4823–4832.

Chernysheva, O.A., White, K.A., 2005. Modular arrangement of viral *cis*-acting RNA domains in a tombusvirus satellite RNA. Virology. 332, 640–649.

Devic, M., Jaegle, M., Baulcombe, D., 1990. Cucumber mosaic virus satellite RNA (strain Y): analysis of sequences which affect systemic necrosis on tomato. J. Gen. Virol. 71, 1443–1449.

Fabian, M.R., Na, H., Ray, D., White, K.A., 2003. 3′-terminal RNA secondary structures are important for accumulation of tomato bushy stunt DI RNAs. Virology 313, 567–580.

Gal-On, A., Kaplan, I., Palukaitis, P., 1995. Differential effects of satellite RNA on the accumulation of cucumber mosaic virus RNAs and their encoded proteins in tobacco vs zucchini squash with two strains of CMV helper virus. Virology 208, 58–66.

García-Arenal, F., Palukaitis, P., 1999. Structure and functional relationships of satellite RNAs of cucumber mosaic virus. In: Vogt, P.K., Jackson, A.O. (Eds.), Satellites and Defective Viral RNAs. Springer, Berlin, pp. 37–63.

García-Arenal, F., Zaitlin, M., Palukaitis, P., 1987. Nucleotide sequence analysis of six satellite RNAs of cucumber mosaic virus: primary sequence and secondary structure alterations do not correlate with differences in pathogenicity. Virology 158, 339–347.

Gould, A.R., Palukaitis, P., Symons, R.H., Mossop, D.W., 1978. Characterization of a satellite RNA associated with cucumber mosaic virus. Virology 84, 443–455.

Guo, L.-H., Cao, Y.-H., Li, D.-W., Niu, S.-N., Cai, Z.-N., Han, C.-G., et al., 2005. Analysis of nucleotide sequences and multimeric forms of a novel satellite RNA associated with beet black scorch virus. J. Virol. 79, 3664–3674.

Guo, R., Meskauskas, A., Dinman, J.D., Simon, A.E., 2011. Evolution of a helper virus-derived, ribosome binding translational enhancer in an untranslated satellite RNA of turnip crinkle virus. Virology 419, 10–16.

Hajimorad, M.R., Ghabrial, S.A., Roossinck, M.J., 2009. De novo emergence of a novel satellite RNA of cucumber mosaic virus following serial passages of the virus derived from RNA transcripts. Arch. Virol. 154, 137–140.

Hou, W.-N., Duan, C.-G., Fang, R.-X., Zhou, X.-Y., Guo, H.-S., 2011. Satellite RNA reduces expression of the 2b suppressor protein resulting in the attenuation of symptoms caused by cucumber mosaic virus infection. Mol. Plant Pathol. 12, 595–605.

Hu, C.-C., Hsu, Y.-H., Lin, N.-S., 2009. Satellite RNAs and satellite viruses of plants. Virology 1, 1325–1350.

Irian, S., Xu, P., Dai, X., Zhao, P.X., Roossinck, M.J., 2007. Regulation of a virus-induced lethal disease in tomato revealed by LongSAGE analysis. Mol. Plant Microbe Interact. 20, 1477–1488.

Jacquemond, M., Lot, H., 1982. L'ARN satellite du virus de la mosaïque du concombre III. -La propriété de survie *in vivo*. Agronomie 2, 533–538.

Jaegle, M., Devic, M., Longstaff, M., Baulcombe, D., 1990. Cucumber mosaic virus satellite RNA (Y strain): analysis of sequences which affect yellow mosaic symptoms on tobacco. J. Gen. Virol. 71, 1905–1912.

Kaper, J.M., Tousignant, M.E., 1984. Viral satellites: parasitic nucleic acids capable of modulating disease expression. Endeavour 8, 194–199.

Kaper, J.M., Tousignant, M.E., Díaz-Ruíz, J.R., Tolin, S.A., 1978. Peanut stunt virus-associated RNA 5: second tripartite genome virus with an associated satellite-like replicating RNA. Virology 88, 166–170.

King, A.M.Q., Adams, M.J., Carstens, E.B., Lefkowitz, E.J. (Eds.), 2012. Virus Taxonomy, Ninth Report of the International Committee on Taxonomy of Viruses. Elsevier Academic Press, San Diego, CA.

Kong, Q., Oh, J.-W., Simon, A.E., 1995. Symptom attenuation by a normally virulent satellite RNA of turnip crinkle virus is associated with the coat protein open reading frame. Plant Cell. 7, 1625–1634.

Kouadio, K.T., DeClerck, C., Agneroh, T.A., Parisi, O., Lepoivre, P., Jijakli, H., 2013. Role of satellite RNAs in cucumber mosaic virus-host plant interactions. A review. Biotechnol. Agro. Soc. Environ. 17, 644–650.

Kuroda, T., Natsuaki, T., Wang, W.Q., 1997. Formation of multimers of cucumber mosaic virus satellite RNA. J. Gen. Virol. 78, 941–946.

Masuta, C., Takanami, Y., 1989. Determination of sequence and structural requirements for pathogenicity of a cucumber mosaic virus satellite RNA (Y-satRNA). Plant Cell. 1, 1165–1173.

Moriones, E., Díaz, I., Rodríguez-Cerezo, E., Fraile, A., García-Arenal, F., 1992. Differential interactions among strains of tomato aspermy virus and satellite RNAs of cucumber mosaic virus. Virology 186, 475–480.

Murant, A.F., 1990. Dependence of groundnut rosette virus on its satellite RNA as well as on groundnut rosette assistor luteovirus for transmission by *Aphis craccivora*. J. Gen. Virol. 71, 2163–2166.

Nagy, P.D., Pogany, J., Simon, A.E., 1999. RNA elements required for RNA recombination function as replication enhancer *in vitro and in vivo* in a plus-stranded RNA virus. EMBO J. 18, 5653–5665.

Palukaitis, P., Roossinck, M.J., 1995. Variation in the hypervariable region of cucumber mosaic virus satellite RNAs is affected by the helper virus and the initial sequence context. Virology 206, 765–768.

Pita, J.S., deMiranda, J.R., Schneider, W.L., Roossinck, M.J., 2007. Environment determines fidelity for an RNA virus replicase. J. Virol. 81, 9072–9077.

Rao, A., Kalantidis, K., 2015. Virus-associated small satellite RNAs and viroids display similarities in their replication strategies. Virology 479-480, 627–636.

Rodríguez-Alvarado, G., Roossinck, M.J., 1997. Structural analysis of a necrogenic strain of cucumber mosaic cucumovirus satellite RNA *in planta*. Virology 236, 155–166.

Roossinck, M.J., Sleat, D., Palukaitis, P., 1992. Satellite RNAs of plant viruses: structures and biological effects. Microbiol. Rev. 56, 265–279.

Rubino, L., Russo, M., 2010. Properties of a novel satellite RNA associated with tomato bushy stunt virus infections. J. Gen. Virol. 91, 2393–2401.

Seo, J.K., Kwon, S.J., Chaturvedi, S., Choi, S.H., Rao, A.L.N., 2013. Functional significance of a hepta nucleotide motif present at the junction of cucumber mosaic virus satellite RNA multimers in helper-virus dependent. Virology 435, 214–219.

Shimura, H., Pantaleo, V., Ishihara, T., Myojo, N., et al., 2011. A viral satellite RNA induces yellow symptoms on tobacco by targeting a gene involved in chlorophyll biosynthesis using the RNA silencing machinery. PLoS Pathog. 7, 1–12.

Simon, A.E., Howell, S.H., 1986. The virulent satellite RNA of turnip crinkle virus has a major domain homologous to the 3′ end of the helper virus genome. EMBO J. 5, 3423–3428.

Simon, A.E., Roossinck, M.J., Havelda, Z., 2004. Plant virus satellite and defective interfering RNAs: new paradigms for a new century. Annu. Rev. Phytopathol. 42, 415–437.

Sivanandam, V., Varady, E., Rao, A.L.N., 2015. Heterologous replicase driven 3′ end repair of cucumber mosaic virus satellite RNA. Virology 478, 18–26.

Wang, J., Simon, A.E., 1999. Symptom attenuation by a satellite RNA *in vivo* is dependent on reduced levels of virus coat protein. Virology 259, 234–245.

Xi, D., Lan, L., Wang, J., Xu, W., Xiang, B., Lin, H., 2006. Variation analysis of two cucumber mosaic viruses and their associated satellite RNAs from sugar beet in China. Virus Genes 33, 293–298.

Xu, P., Roossinck, M.J., 2011. Plant Virus Satellites, Encyclopedia of Life Sciences, 2 vols. John Wiley and Sons, Ltd, Chichester, p. 11.

Xu, P., Wang, H., Coker, F., Ma, J.-Y., Tang, Y., Taylor, M., et al., 2012. Genetic loci controlling lethal cell death in tomato caused by viral satellite RNA infection. Mol. Plant Microbe Interact. 25, 1034–1044.

Zahid, K., Zhao, J.-H., Smith, N.A., Schumann, U., Fang, Y.-Y., Dennis, E.S., et al., 2015. Nicotiana small RNA sequences support a host genome origin of cucumber mosaic virus satellite RNA. PLoS Gen. 11, e10049606.

CHAPTER

61

Small Circular Satellite RNAs

Beatriz Navarro, Luisa Rubino and Francesco Di Serio

National Research Council, Bari, Italy

INTRODUCTION

The infectivity of small circular satellite RNAs (sc-satRNAs) depends on a helper virus. However, sc-satRNAs resemble viroids in several features, including the small (220–257 nt) and single-stranded genome, the accumulation in the infected host as covalently-closed circular RNAs, the replication through rolling-circle mechanisms based on RNA intermediates only, and the inability to code for proteins, with one exception. In addition, sc-satRNAs are endowed with self-cleaving (and some with self-ligation) activity mediated by ribozymes, another feature shared with some viroids, thus reinforcing the hypothesis that these two groups of subviral infectious RNAs may have a monophyletic origin (Elena et al., 2001). However, a key biological difference exists between sc-satRNAs and viroids regarding the molecular machinery needed for replication, which is supplied by the helper virus and the host in the first case, but only by the host in the second. In addition, in contrast to viroids, sc-satRNAs are encapsidated by the helper virus coat protein. This chapter focuses on the main structural and biological features of sc-satRNAs, pointing out the role of hammerhead and hairpin ribozymes in their replication. Readers looking for additional information on sc-satRNAs may check previous reviews (Bruening et al., 1991; Diener, 1991; Francki, 1987; Rao and Kalantidis, 2015; Roossinck et al., 1992; Rubino et al., 2003; Symons and Randles, 1999; Taliansky and Palukaitis, 1999).

Viroids and Satellites.
DOI: http://dx.doi.org/10.1016/B978-0-12-801498-1.00061-9

STRUCTURAL PROPERTIES

sc-satRNAs share little or no sequence similarity with the helper virus and host genomes, accumulate in vivo as circular and linear forms, and do not code for proteins. The latter feature, supported by studies with several sc-satRNAs (Kiberstis and Zimmern, 1984; Morris-Krsinich and Forster, 1983; Owens and Schneider, 1977; Rubino et al., 1990), has been recently questioned for the sc-satRNA of rice yellow mottle virus (RYMV) (see below).

sc-satRNAs are associated with members of the genera *Sobemovirus*, *Nepovirus*, and *Polerovirus* (Table 61.1). Circular and linear forms of sc-satRNAs coexist in infected tissues; the circular form is encapsidated predominantly by sobemoviruses and the linear form by nepo and poleroviruses (Table 61.1). sc-satRNAs encapsidated as circular forms are also termed virusoids.

Apart from the conserved nucleotides of the ribozymes (see below), only a few sequence elements are shared among sc-satRNAs. The sc-satRNA associated with the lucerne transient streak virus (LTSV) and RYMV share a 50-nt region with high sequence identity (89%) located in the left terminal domain of their proposed secondary structure (Collins et al., 1998). This region includes a GAUUUU motif conserved in the same position in all sc-satRNAs (Keese et al., 1983) and is suggested to play a role in replication (Davies et al., 1990).

Due to their high level of self-complementary sequences, a compact secondary structure constituted by double-stranded regions interposed between bulges and loops, has been proposed for sc-satRNAs, with some of them assuming quasi rod-like and others more branched conformations (Rubino et al., 2003). In the sc-satRNA associated with cereal yellow dwarf virus-RPV (CYDV-RPV), the proposed branched secondary structure was confirmed by in vitro probing, and structural elements that may interact with viral proteins were proposed (Song and Miller, 2004).

BIOLOGICAL PROPERTIES

Replication of sc-satRNAs depends on both the helper virus and the host, although related viruses can support replication of a different sc-satRNA as exemplified by some sobemoviruses (Roossinck et al., 1992). sc-satRNAs can attenuate or exacerbate the symptoms caused by their helper viruses and they can alter (usually reduce) viral RNA accumulation. In sc-satRNAs associated with nepoviruses, both situations have been reported: sc-satRNAs of tobacco ringspot virus (TRSV) and chicory

TABLE 61.1 Small Circular Satellite RNAs (sc-satRNAs)

Helper virus	Genus of the helper virus	Encapsidated RNA	sc-satRNA size (nt)	(+) ribozyme[a]	(–) ribozyme[a]	Reference
Lucerne transient streak virus	*Sobemovirus*	Circular	322 and 324	HH	HH	Keese et al. (1983)
Solanum nodiflorum mottle virus	*Sobemovirus*	Circular	377	HH	–	Haseloff and Symons (1982)
Subterranean clover mottle virus	*Sobemovirus*	Circular	332 and 328	HH	–	Davies et al. (1990)
Velvet tobacco mottle virus	*Sobemovirus*	Circular	366	HH	–	Haseloff and Symons (1982)
Rice yellow mottle virus	*Sobemovirus*	Circular	220	HH	–	Collins et al. (1998)
Arabis mosaic virus satellite	*Nepovirus*	Linear	300–301	HH	HP	Kaper and Collmer (1988)
Chicory yellow mottle virus	*Nepovirus*	Linear	457	HH	HP	Rubino et al. (1990)
Tobacco ringspot virus satellite	*Nepovirus*	Linear	359	HH	HP	Buzayan et al. (1986c)
Cereal yellow dwarf virus-RPV[b]	*Polerovirus*	Linear	322	HH	HH	Miller et al. (1991)

[a] *HH, hammerhead ribozyme; HP, hairpin ribozyme.*
[b] *The helper virus, formerly known as the RPV serotype of the barley yellow dwarf virus (genus* Luteovirus*), has been reclassified as cereal yellow dwarf virus-RPV (CYDV-RPV) in the new genus* Polerovirus.

yellow mottle virus reduce the accumulation of their supporting viruses and attenuate the severity of the symptoms induced by the viruses alone (Kaper and Collmer, 1988; Piazzolla et al., 1986), whereas the sc-satRNA of the arabis mosaic virus has an opposite effect in *Chenopodium quinoa* and hop (Davies and Clark, 1983). The ability of some satellites to attenuate the symptoms induced by their helper viruses suggested that satellites may act as antiviral agents (Gerlach et al., 1987).

sc-satRNAs-derived small RNAs of 21–24 nt similar to microRNAs and small interfering RNAs, the hallmarks of RNA silencing (Axtell, 2013), have been identified in plants infected by the sc-satRNA associated with CYDV-RPV and proposed to play a role in pathogenesis by downregulation of specific host mRNAs (Wang et al., 2001, 2004). This hypothesis, recently validated for a small linear satellite RNA of cucumber mosaic virus (Shimura et al., 2011; Smith et al., 2011) and a chloroplast-replicating viroid (Navarro et al., 2012), is awaiting experimental proof in sc-satRNAs. Since the replicating CYDV-RPV sc-satRNA efficiently induced de novo cytosine methylation of its homologous transgenic DNA, resembling the RNA-directed DNA methylation induced by nuclear-replicating viroids (Wassenegger et al., 1994), the possibility that sc-sRNAs could interfere with host gene expression by targeting DNA for methylation has also been suggested (Wang et al., 2001). RNA silencing could also act as a plant defense against sc-satRNAs promoting their sequence-specific degradation. The compact secondary structure may help sc-satRNAs to escape RNA silencing-mediated degradation. Moreover, viral-encoded proteins supply at least two additional barriers against RNA silencing: sc-satRNA encapsidation by viral coat protein (Wang et al., 2004) and impairment of RNA silencing by virus-encoded RNA silencing suppressor proteins.

ROLE OF SELF-CLEAVAGE IN ROLLING-CYCLE REPLICATION

Studies showing that synthesis of sc-satRNA associated with velvet tobacco mottle virus is not inhibited in vivo by the low concentrations of α-amanitin that impair the activity of host DNA-dependent RNA polymerase II, indicated that, in contrast to nuclear-replicating viroids, this enzyme is not involved in sc-satRNAs replication (Wu et al., 1986). In addition, actinomycin D, which inhibits host DNA-dependent RNA synthesis, did not affect replication of TRSV sc-satRNA in protoplasts coinfected with TRSV (Buckley and Bruening, 1990), thus suggesting the involvement of the RNA-dependent RNA-polymerase (RNA replicase) encoded (in part) by the virus. This view was further supported by more recent studies that identified the RNA replicase as the only viral component needed for CYDV-RPV sc-satRNA replication (Song and Miller, 2004). Since helper viruses replicate in the cytoplasm associated with membranous vesicles, sc-satRNA replication likely occurs in the same subcellular compartment, although no conclusive experimental evidence has been obtained yet (Flores et al., 2011; Rao and Kalantidis, 2015).

Based on the presence in infected tissues of circular and multimeric forms of sc-satRNAs and their ability to self-cleave through the ribozymes that at least one of the strands can form, a rolling-circle replication mechanism was proposed (Branch and Robertson, 1984). Similarly to viroids, symmetric and asymmetric variants were suggested depending on the presence or absence in the infected tissues of the circular RNA of minus polarity, respectively (Forster and Symons, 1987). Since sc-satRNAs lack coding capacity, the plus polarity has been arbitrarily assigned to the sc-satRNA strand that accumulates at higher level in vivo. According to the rolling-circle mechanism, the monomeric circular plus strand RNA is reiteratively transcribed to produce multimeric linear minus RNA strands that either self-cleave and ligate to form circular monomers (symmetric variant) or remain as multimers (asymmetric variant). The minus strands serve as templates for the synthesis of linear oligomeric plus strand RNAs, which are finally cleaved and circularized to generate the monomeric circular RNA progeny (Flores et al., 2011).

Cleavage of oligomeric RNA intermediates is an autolytic process mediated by *cis*-acting ribozymes. According to the active conformation assumed, the ribozymes in sc-satRNAs are termed hammerhead or hairpin ribozymes, the self-cleavage of which generate 5′ hydroxyl and 2′–3′ cyclic phosphodiester termini. The first evidence of self-cleavage in vitro was obtained with purified natural linear oligomers (Prody et al., 1986) or in vitro-generated transcripts (Buzayan et al., 1986a) of the sc-satRNA associated with TRSV, which resulted in infectious linear monomers. Based on the structural features characteristic of ribozymes, self-cleavage mediated by hammerhead ribozymes was predicted and experimentally shown in vitro for the plus polarity strand of all sc-satRNAs, whereas self-cleavage of the minus strand was reported only for some of them, including those associated with LTSV and CYDV-RPV (mediated by hammerhead ribozymes) and with nepoviruses (mediated by hairpin ribozymes) (Table 61.1). Self-cleaving activity in one or both polarity strands is indicative of the asymmetric or symmetric replication pathway followed by the sc-satRNA, respectively. Strong support for the ribozyme role in vivo was supplied by data showing reversion in vivo of mutations impairing the autolytic processing in vitro of the sc-satRNA associated with LTSV (Sheldon and Symons, 1993).

After self-cleavage, an RNA ligase that circularizes the monomeric linear RNAs is needed to complete the rolling-circle replication. This catalytic activity, likely virus-independent (Chay et al., 1997), may be mediated by the same ribozyme acting in the reverse direction. Interestingly, the RNA ligase activity of the hairpin ribozyme is significantly higher than that of the hammerhead ribozyme (Fedor, 2000). In line with this view, in vitro circularization in the absence of proteins has been shown for the minus strand of TRSV sc-satRNA, which contains a hairpin ribozyme (Buzayan et al., 1986b). In contrast, low efficiency self-ligation in vitro was observed for the plus RNA of the same satellite containing a hammerhead ribozyme (Nelson et al., 2005; Prody et al, 1986). Similarly to viroids containing hammerhead ribozymes, circularization of the monomeric plus linear sc-satRNA of TRSV most likely requires a host RNA ligase (Kiberstis et al., 1985) that must colocalize with the site of sc-satRNA and helper virus replication (Flores et al., 2011). Alternative RNA foldings have been proposed to facilitate efficient ligation of linear sc-satRNA forms (Chay et al., 1997).

STRUCTURE OF THE HAMMERHEAD AND HAIRPIN RIBOZYMES OF SC-SATRNAS

The minimal hammerhead ribozyme consists of three base-paired helices of variable composition and length surrounding a 15-nt core containing 11 conserved nt around the self-cleavage site (Flores et al., 2001). This site is preceded by a GUC motif in most natural hammerheads. Exceptions to this rule are the hammerheads of the minus strands of sc-satRNAs associated with LTSV (Keese et al., 1983) and velvet tobacco mottle virus (Haseloff and Symons, 1982), where the triplet is GUA, and the hammerhead of the plus strand of the CYDV-RPV sc-satRNA (Miller et al., 1991), in which an AUA triplet precedes the self-cleavage site (Fig. 61.1). Another peculiarity is the presence of an extra U after the conserved GA doublet in the plus hammerheads of the sc-satRNAs associated with LTSV and arabis mosaic virus (Fig. 61.1). The sc-satRNA hammerheads reported in Fig. 61.1 are presented in a Y-shaped conformation derived from crystallography studies that revealed complex noncanonical interactions between the residues forming the central core (Pley et al., 1994). Tertiary interactions outside the catalytic core, essential for in vivo self-cleavage of some viroid hammerheads (De la Peña et al., 2003; Khvorova et al., 2003), were also identified in the sc-satRNA TRSV hammerhead (Chi et al., 2008). In the CYDV-RPV-associated

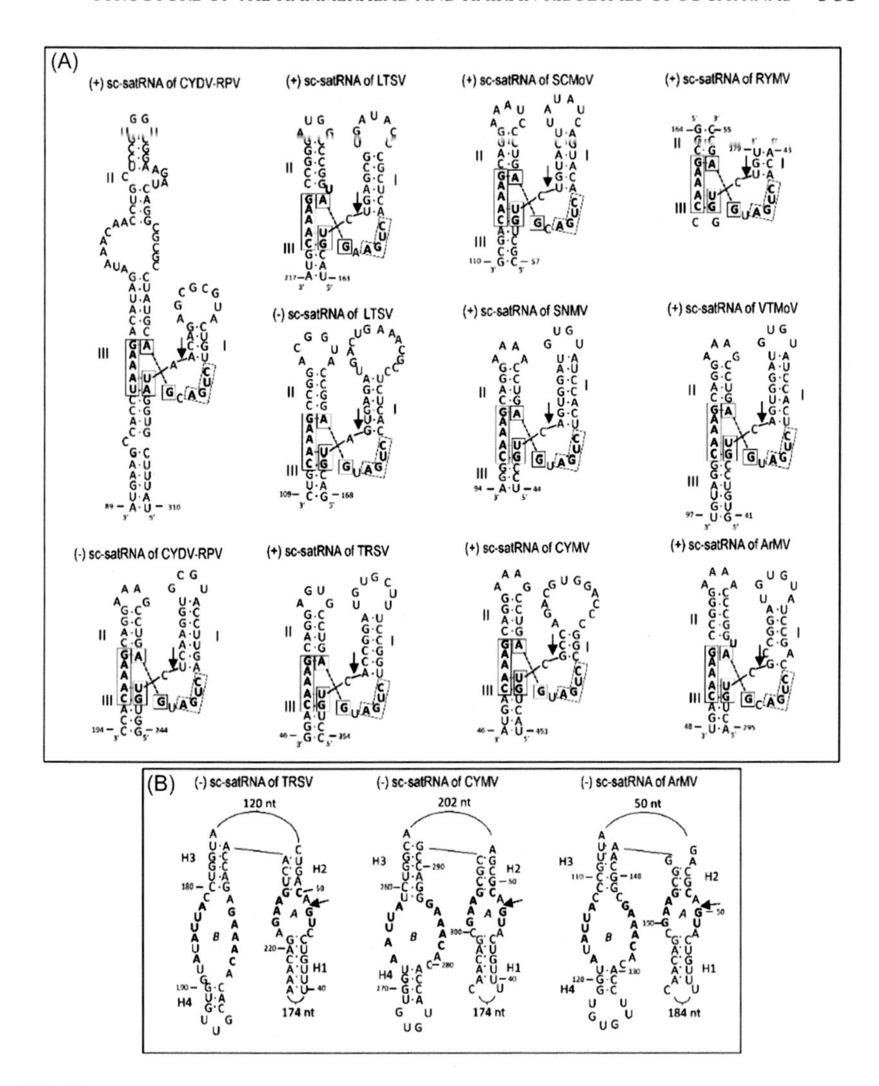

FIGURE 61.1 Sequence and secondary structure of the minimal ribozyme domains of sc-satRNAs. Panel A: Hammerhead structures are represented according to Y-shape. The nucleotides forming the conserved catalytic core are boxed and in bold; stems are numbered according to a previous convention (Forster and Symons, 1987). The extra U after the conserved GA in the plus strand hammerhead of sc-satRNAs associated with LTSV and ArMV is highlighted in bold. The plus strand of sc-satRNA associated with RYMV most likely self-cleaves in vivo by a double-hammerhead structure (not shown). Panel B: Minimal hairpin ribozyme structures are represented based on Fedor (2000). H1, H2, H3, and H4 indicate base-paired helices, including interacting loops (A and B). Conserved nucleotides in bold are required for activity and regions of RNA that are not involved in hairpin structure are drawn as continuous lines. In both panels, the self-cleavage site is indicated by an arrow and numbering refers to plus strands.

sc-satRNA, a pseudoknot inhibits the hammerhead-mediated self-cleavage of the monomeric plus RNA (Miller and Silver, 1991), while self-cleavage of the multimeric RNAs of this satellite likely occurs via an alternative double-hammerhead structure, thus suggesting that a conformational switch of sc-satRNAs associated with CYDV-RPV controls ribozyme activity during replication (Song and Miller, 2004).

Hairpin ribozymes have been identified only in the minus strand of the three sc-satRNAs associated with nepovirus (Fig. 61.1). The secondary structure proposed for the hairpin ribozyme (also called paperclip) contains a catalytic core that consists of four base-paired helices separated by two unpaired conserved loops (Berzal-Herranz et al., 1993). The crystal structure of the hairpin ribozyme has been resolved and showed the presence of tertiary interactions that facilitate the catalysis (Rupert and Ferré-D'Amaré, 2001).

DO SC-SATRNAS CODE FOR FUNCTIONAL PROTEINS?

Recently, AbouHaidar et al. (2014) showed that the sc-satRNA associated with RYMV, with the smallest size (220 nt) among sc-satRNAs and viroids, codes for a 16-kDa highly-basic polypeptide of unknown function that has been detected in infected tissues and in purified virions. The mechanism of translation initiation proposed for the RYMV sc-satRNA-encoded protein is unusual since it does not operate by the conventional ribosome-scanning pathway. The authors propose an alternative mechanism based on direct and reiterative (at least two rounds) translation of the circular RYMV sc-satRNA. Whether this is a situation restricted to this sc-satRNA is not known. However, extension of the coding properties to the other sc-satRNAs is unlikely because a similar coding potential seems absent in the other sc-satRNAs, and inconsistent with the insertion of one or two nucleotides observed in natural sequence variants of several sc-satRNAs (Table 61.1).

OTHER PUTATIVE SC-SATELLITE RNAS

A new member of the sc-satRNA group could be a small circular RNA of 365 nt identified in mulberry trees (mulberry small circular RNAs, mscRNA) in China. This RNA was initially considered a viroid associated with mulberry mosaic dwarf disease (Wang et al., 2010). However, attempts to infect healthy mulberry seedlings with purified circular forms and in vitro transcripts of the mscRNA were unsuccessful, suggesting that this RNA is not a viroid. Moreover, in vitro self-cleaving hammerhead and hairpin ribozymes have been identified in the mscRNA plus and minus strands, respectively, a feature only found in sc-satRNA associated with nepoviruses. The satellite nature of mscRNA is further supported by the finding of a nepovirus in mulberry plants coinfected by mscRNA (S. Li, B. Navarro and T. Sano, unpublished data).

Acknowledgments

Research in B.N., L.R., and F.D.S. laboratories has been partially funded by a dedicated grant of the Ministero dell'Economia e Finanze Italiano to the CNR (CISIA, Legge n. 191/ 2009) and by a joint project in the frame of scientific cooperation between Consiglio Nazionale delle Ricerche (Italy) and Chinese Academy of Agricultural Sciences (China) 2011–13 to B.N.

References

AbouHaidar, M.G., Venkataraman, S., Golshani, A., Liu, B., Ahmad, T., 2014. Novel coding, translation, and gene expression of a replicating covalently closed circular RNA of 220 nt. Proc. Natl. Acad. Sci. USA 111, 14542–14547.

Axtell, M.J., 2013. Classification and comparison of small RNAs from plants. Annu. Rev. Plant Biol. 64, 137–159.

Berzal-Herranz, A., Joseph, S., Chowrira, B.M., Butcher, S.E., Burke, J.M., 1993. Essential nucleotide sequences and secondary structure elements of the hairpin ribozyme. EMBO J. 12, 2567–2573.

Branch, A.D., Robertson, H.D., 1984. A replication cycle for viroids and other small infectious RNAs. Science 223, 450–455.

Bruening, G., Feldstein, P.A., Buzayan, J.M., van Tol, H., deBear, J., Gough, G.R., et al., 1991. Satellite tobacco ringspot virus satellite RNA: self-cleavage and ligation reactions in replication. In: Maramorosch, K. (Ed.), Viroids and Satellites: Molecular Parasites at the Frontier of Life. CRC Press, Boca Raton, FL, pp. 141–158.

Buckley, B., Bruening, G., 1990. Effect of actinomycin D on replication of satellite tobacco ringspot virus RNA in plant protoplasts. Virology 177, 298–304.

Buzayan, J.M., Gerlach, W.L., Bruening, G., 1986a. Non-enzymatic cleavage and ligation of RNAs complementary to a plant virus satellite RNA. Nature 323, 349–353.

Buzayan, J.M., Gerlach, W.L., Bruening, G., 1986b. Satellite tobacco ringspot virus RNA: a subset of the RNA sequence is sufficient for autolytic processing. Proc. Natl. Acad. Sci. USA 83, 8859–8862.

Buzayan, J.M., Gerlach, W.L., Bruening, G., Keese, P., Gould, A.R., 1986c. Nucleotide sequence of satellite tobacco ringspot virus RNA and its relationship to multimeric forms. Virology 151, 186–199.

Chay, C.A., Guan, X., Bruening, G., 1997. Formation of circular satellite tobacco ringspot virus RNA in protoplasts transiently expressing the linear RNA. Virology 239, 413–425.

Chi, Y.I., Martick, M., Lares, M., Kim, R., Scott, W.G., Kim, S.H., 2008. Capturing hammerhead ribozyme structures in action by modulating general base catalysis. PLoS Biol. 6, e234.

Collins, R.F., Gellatly, D.L., Sehgal, O.P., Abouhaidar, M.G., 1998. Self-cleaving circular RNA associated with rice yellow mottle virus is the smallest viroid-like RNA. Virology 241, 269–275.

Davies, C., Haseloff, J., Symons, R.H., 1990. Structure, self-cleavage, and replication of two viroid-like satellite RNAs (virusoids) of subterranean clover mottle virus. Virology 177, 216–224.

Davies, D.L., Clark, M.F., 1983. A satellite-like nucleic acid of arabis mosaic virus associated with hop nettled disease. Ann. Appl. Biol. 103, 439–448.

De la Peña, M., Gago, S., Flores, R., 2003. Peripheral regions of natural hammerhead ribozymes greatly increase their self-cleavage activity. EMBO J. 22, 5561–5570.

Diener, T.O., 1991. The frontiers of life: the viroids and viroid-like satellite RNAs. In: Maramorosch, K. (Ed.), Viroids and Satellites: Molecular Parasites at the Frontier of Life. CRC Press, Boca Raton, FL, pp. 1–21.

Elena, S.F., Dopazo, J., Flores, R., Diener, T.O., Moya, A., 2001. Phylogenetic analysis of viroid and viroid-like satellite RNAs from plants: a reassessment. J. Mol. Evol. 53, 155–159.

Fedor, M.J., 2000. Structure and function of the hairpin ribozyme. J. Mol. Biol. 297, 269–291.

Flores, R., Grubb, D., Elleuch, A., Nohales, M.Á., Delgado, S., Gago, S., 2011. Rolling-circle replication of viroids, viroid-like satellite RNAs and hepatitis delta virus: variations on a theme. RNA Biol. 8, 200–206.

Flores, R., Hernández, C., de la Peña, M., Vera, A., Daròs, J.A., 2001. Hammerhead ribozyme structure and function in plant RNA replication. Methods Enzymol. 341, 540–552.

Forster, A.C., Symons, R.H., 1987. Self-cleavage of plus and minus RNAs of a virusoid and a structural model for the active sites. Cell 49, 211–220.

Francki, R.I., 1987. Encapsidated viroidlike RNA. In: Diener, T.O. (Ed.), The Viroids. Plenum, New York, pp. 205–218.

Gerlach, W.L., Llewellyn, D., Haseloff, J., 1987. Construction of a plant disease resistance gene from the satellite RNA of tobacco ringspot virus. Nature 328, 802–805.

Haseloff, J., Symons, R.H., 1982. Comparative sequence and structure of viroid-like RNAs of two plant viruses. Nucleic Acids Res. 10, 3681–3691.

Kaper, J.M., Collmer, C.W., 1988. Modulation of viral plant diseases by secondary RNA agents. In: Domingo, E., Holland, J.J., Alquist, P. (Eds.), RNA Genetics. CRC Press, Boca Raton, FL, pp. 171–194.

Keese, P., Bruening, G., Symons, R.H., 1983. Comparative sequence and structure of circular RNAs from two isolates of lucerne transient streak virus. FEBS Lett. 159, 185–190.

Khvorova, A., Lescoute, A., Westhof, E., Jayasena, S.D., 2003. Sequence elements outside the hammerhead ribozyme catalytic core enable intracellular activity. Nat. Struct. Biol. 10, 708–712.

Kiberstis, P.A., Haseloff, J., Zimmern, D., 1985. 2′ Phosphomonoester, 3′-5′ phosphodiester bond at a unique site in a circular viral RNA. EMBO J. 4, 817–827.

Kiberstis, P.A., Zimmern, D., 1984. Translational strategy of solanum nodiflorum mottle virus RNA: synthesis of a coat protein precursor in vitro and in vivo. Nucleic Acids Res. 12, 933–943.

Miller, W.A., Hercus, T., Waterhouse, P.M., Gerlach, W.L., 1991. A satellite RNA of barley yellow dwarf virus contains a novel hammerhead structure in the self-cleavage domain. Virology 183, 711–720.

Miller, W.A., Silver, S.L., 1991. Alternative tertiary structure attenuates self-cleavage of the ribozyme in the satellite RNA of barley yellow dwarf virus. Nucleic Acids Res. 19, 5313–5320.

Morris-Krsinich, B.A.M., Forster, R.L.S., 1983. Lucerne transient streak virus RNA and its translation in rabbit reticulocyte lysate and wheat germ extract. Virology 128, 176–185.

Navarro, B., Gisel, A., Rodio, M.E., Delgado, S., Flores, R., Di Serio, F., 2012. Small RNAs containing the pathogenic determinant of a chloroplast-replicating viroid guide degradation of a host mRNA as predicted by RNA silencing. Plant J. 70, 991–1003.

Nelson, J.A., Shepotinovskaya, I., Uhlenbeck, O.C., 2005. Hammerheads derived from sTRSV show enhanced cleavage and ligation rate constants. Biochemistry 44, 14577–14585.

Owens, R.A., Schneider, I.R., 1977. Satellite of tobacco ringspot virus RNA lacks detectable mRNA activity. Virology 890, [illegible]–[illegible]4

Piazzolla, P., Vovlas, C., Rubino, L., 1986. Symptom regulation induced by chicory yellow mottle virus satellite-like RNA. J. Phytopathol. 115, 124–129.

Pley, H.W., Flaherty, K.M., McKay, D.B., 1994. Three-dimensional structure of a hammerhead ribozyme. Nature 372, 68–74.

Prody, G.A., Bakos, J.T., Buzayan, J.M., Schneider, I.R., Bruening, G., 1986. Autolytic processing of dimeric plant virus satellite RNA. Science 231, 1577–1580.

Rao, A.L.N., Kalantidis, K., 2015. Virus-associated small satellite RNAs and viroids display similarities in their replication strategies. Virology 479-480, 627–636.

Roossinck, M.J., Sleat, D., Palukaitis, P., 1992. Satellite RNAs of plant viruses: structures and biological effects. Microbiol. Rev. 56, 265–279.

Rubino, L., Tousignant, M.E., Steger, G., Kaper, J.M., 1990. Nucleotide sequence and structural analysis of two satellite RNAs associated with chicory yellow mottle virus. J. Gen. Virol. 71, 1897–1903.

Rubino., L., Martelli, G.P., Di Serio, F., 2003. Viroid-like satellite RNAs. In: Hadidi, A., Flores, R., Randles, J.W., Semancik, J.S. (Eds.), Viroids. CSIRO Publishing, Collingwood, VIC, pp. 76–86.

Rupert, P.B., Ferré-D'Amaré, A.R., 2001. Crystal structure of a hairpin ribozyme-inhibitor complex with implications for catalysis. Nature 410, 780–786.

Sheldon, C.C., Symons, R.H., 1993. Is hammerhead self-cleavage involved in the replication of a virusoid in vivo? Virology 194, 463–474.

Shimura, H., Pantaleo, V., Ishihara, T., Myojo, N., Inaba, J., Sueda, K., et al., 2011. A viral satellite RNA induces yellow symptoms on tobacco by targeting a gene involved in chlorophyll biosynthesis using the RNA silencing machinery. PLoS Pathog. 7, e1002021.

Smith, N.A., Eamens, A.L., Wang, M.B., 2011. Viral small interfering RNAs target host genes to mediate disease symptoms in plants. PLoS Pathog. 7, e1002022.

Song, S.I., Miller, W.A., 2004. Cis and trans requirements for rolling circle replication of a satellite RNA. J. Virol. 78, 3072–3082.

Symons, R.H., Randles, J.W., 1999. Encapsidated circular viroid-like satellite RNAs (virusoids) of plants. Curr. Top. Microbiol. Immunol. 239, 81–105.

Taliansky, M.E., Palukaitis, P.F., 1999. Satellite RNAs and satellite virus. In: Webster, B.G., Granoff, A. (Eds.), Encyclopedia of Virology. Academic Press, San Diego, CA, pp. 1607–1615.

Wang, M.B., Bian, X.Y., Wu, L.M., Liu, L.X., Smith, N.A., Isenegger, D., et al., 2004. On the role of RNA silencing in the pathogenicity and evolution of viroids and viral satellites. Proc. Natl. Acad. Sci. USA 101, 3275–3280.

Wang, M.B., Wesley, S.V., Finnegan, E.J., Smith, N.A., Waterhouse, P.M., 2001. Replicating satellite RNA induces sequence-specific DNA methylation and truncated transcripts in plants. RNA. 7, 16–28.

Wang, W.B., Fei, J.M., Wu, Y., Bai, X.C., Yu, F., Shi, G.F., et al., 2010. A new report of a mosaic dwarf viroid-like disease on mulberry trees in China. Pol. J. Microbiol. 59, 33–36.

Wassenegger, M., Heimes, S., Riedel, L., Sänger, H.L., 1994. RNA-directed de novo methylation of genomic sequences in plants. Cell 76, 567–576.

Wu, J.-G., Lu, W.-J., Tien, P., 1986. Multiplication of velvet tobacco mottle virus in Nicotiana clevelandii protoplasts is resistant to α-amanitin. J. Gen. Virol. 67, 2757–2762.

CHAPTER

62

Betasatellites of Begomoviruses

Xiuling Yang[1] *and Xueping Zhou*[1,2]

[1]Chinese Academy of Agricultural Sciences, Beijing, China
[2]Zhejiang University, Hangzhou, China

INTRODUCTION

Betasatellites, formerly designated as DNAβ, are small (~1360 nucleotides), circular single-stranded DNA molecules that are frequently associated with many monopartite begomoviruses in the family *Geminiviridae* (Mansoor et al., 2003a, 2006; Zhou, 2013) (for geminivirus classification, refer to Brown et al., 2012, 2015; Varsani et al., 2014). Since the first full-length betasatellite was identified in 1999 in an ageratum yellow vein virus-infected *Ageratum conyzoides*, more than 800 full-length betasatellite sequences have been deposited in the public database. Begomovirus-betasatellite disease complexes have been reported in at least 27 countries (Fig. 62.1) infecting a wide range of dicotyledonous host species within at least 37 different genera (Briddon and Stanley, 2006; Zhou, 2013). A betasatellite was reported to be associated with eupatorium yellow vein virus, the earliest recorded plant virus disease (Saunders et al., 2003). Symptom phenotypes induced by these disease complexes include leaf curling, leaf crumpling, yellow leaf curling, yellow mosaic, vein yellowing, vein swelling and darkening, and formation of ectopic enations. To date, betasatellites have become popular models to study the molecular biology of gene regulation and plant–virus interaction (Zhou, 2013).

GENOME ORGANIZATION OF BETASATELLITES

Sequence analysis has revealed three conserved structural features of betasatellites, including a highly conserved βC1 ORF on the complementary strand, a satellite conserved region (SCR) of about 100 nt, and an

Viroids and Satellites.
DOI: http://dx.doi.org/10.1016/B978-0-12-801498-1.00062-0

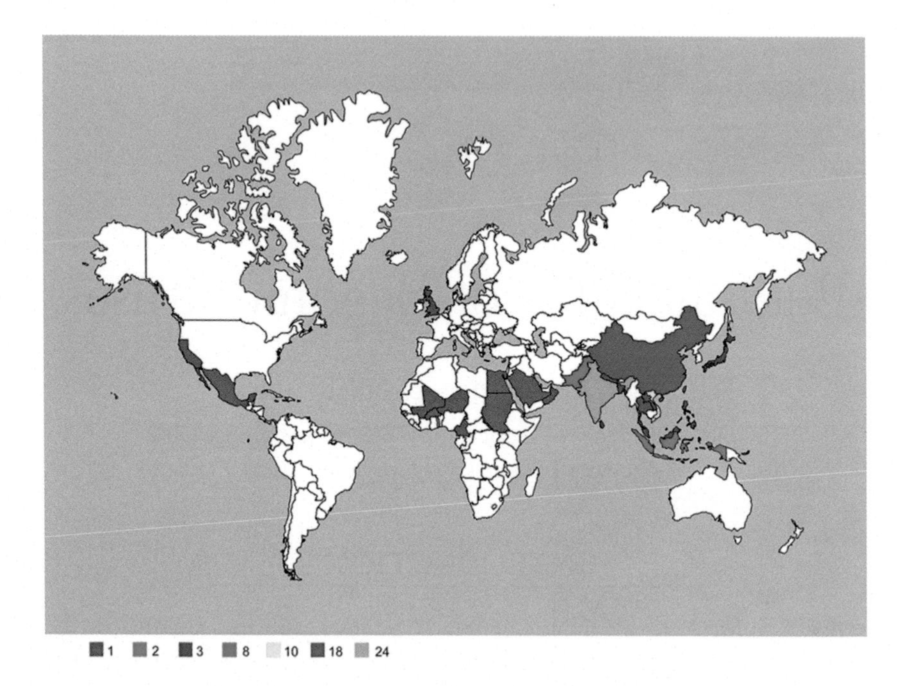

FIGURE 62.1 Geographical distributions of recognized betasatellites. Countries from which betasatellites have been isolated and published are highlighted. Different colors are used to differentiate the number of betasatellite species isolated in each country.

adenine (A)-rich region. The SCR contains a putative hairpin structure with the loop sequence TAATATTAC that might serve as origin for replication. The A-rich region, just upstream of the SCR, is typically between 160 and 280 nt in length and has an A content of 57%–65% (Briddon et al., 2003). This region was suggested to be a "stuffer" to increase the size of betasatellite and for encapsidation, and possibly has a role in complementary-strand DNA replication (Briddon et al., 2003; Saunders et al., 2000). The gene *βC1* typically encodes a protein of approximately 13.5 kDa, which is multifunctional and participates in diverse cellular processes.

BETASATELLITE–BEGOMOVIRUS INTERACTION

Betasatellites are true satellite DNAs that depend entirely on their helper viruses for replication, encapsidation, movement in plants, and insect transmission. Of the identified begomovirus/betasatellite complexes, numerous betasatellites are essential for their helper viruses to induce typical disease symptoms (Briddon et al., 2001; Cui et al., 2004;

Saunders et al., 2000). In some selected cases, the helper virus alone infects and induces disease symptoms in the presence or absence of betasatellite (Kon and Gilbertson, 2012; Li et al., 2005). Different from the interaction of DNA B components with their cognate DNA A, which is highly stringent, the interaction of betasatellites with their helper viruses is much less specific. Betasatellites are transreplicated by geographically divergent and biologically diverse geminiviruses. For instance, cotton leaf curl Multan betasatellite was found to associate with at least seven distinct begomoviruses, either as dual or multiple infectious in the field (Mansoor et al., 2003b). A betasatellite can also interact with a New World bipartite begomovirus under laboratory conditions, leading to enhanced disease symptoms in *Nicotiana benthamiana* plants and rapid sequence changes in the betasatellite (Nawaz-ul-Rehman et al., 2009). Consistent with the transreplication of betasatellites by diverse begomoviruses, betasatellites are capable of substituting for the movement function of DNA B component for systemic spread (Nawaz-ul-Rehman et al., 2009; Patil and Fauquet, 2010; Saeed et al., 2007). It is speculated that promiscuity in betasatellite replication plays important roles in adapting to new hosts, as the presence of ageratum yellow vein virus betasatellite facilitates the bipartite begomovirus Sri Lankan cassava mosaic virus to infect and produce yellow vein symptoms in ageratum plants, whereas Sri Lankan cassava mosaic virus is unable to systemically infect ageratum plants (Saunders et al., 2002).

BIOLOGICAL FUNCTIONS OF BETASATELLITES

As reviewed recently, our understanding of betasatellites has progressed at a rapid pace during the past few years. Betasatellites have been shown to be multifunctional, being involved in symptom modulation, suppression of RNA silencing-mediated plant defenses, and accelerating whitefly–virus mutualism (Zhou, 2013).

Symptom Development

Initial studies to resolve the etiology of ageratum yellow vein disease and cotton leaf curl disease clearly demonstrated the profound effects of betasatellites on symptom development and viral DNA accumulation of helper begomoviruses (Briddon et al., 2001; Saunders et al., 2000). Two distinct lines of evidence showed that βC1 is directly related to symptom induction. Firstly, coinoculation of helper begomoviruses with betasatellite mutants containing a disrupted βC1 ORF failed to induce typical symptoms in infected plants (Cui et al., 2004; Saunders et al.,

2004). Secondly, constitutive expression of βC1 in transgenic plants, or expression of βC1 from a potato virus X vector, resulted in severe developmental abnormalities that phenocopied to a large extent the disease induced by the helper virus and betasatellite (Cui et al., 2005; Qazi et al., 2007; Saeed et al., 2005; Saunders et al., 2004). Ageratum yellow leaf curl virus betasatellite was reported to be involved in pathogenicity determination not only for begomoviruses, but also for mastreviruses (Kumar et al., 2014a,b).

Suppression of RNA Silencing-Mediated Plant Defense

RNA silencing is a fundamental mechanism that regulates gene expression in most eukaryotes. Silencing can occur both at the transcriptional level by DNA methylation and at the posttranscriptional level through several pathways mediated by microRNAs or small interfering RNAs. Like many RNA viruses, geminiviruses have been reported to be inducers and targets of the RNA silencing machinery. Accordingly, both geminiviruses and betasatellites encode silencing suppressors that employ various strategies to counteract this host defense (Burgyan and Havelda, 2011; Díaz-Pendón and Ding, 2008; Guo et al., 2014; Hanley-Bowdoin et al., 2013; Pumplin and Voinnet, 2013; Raja et al., 2010).

The first demonstration of the role of a betasatellite in counteracting RNA silencing-based plant defense came from the work of Cui et al. (2005), who showed that βC1 behaves as a suppressor of RNA silencing. In plants in which a green fluorescent protein (GFP) transgene was silenced, infection of tomato yellow leaf curl China virus (TYLCCNV) and its associated betasatellite TYLCCNB, but not TYLCCNV alone, could reverse established silencing in newly emerging leaves of infected plants. Expression of βC1 was also able to prevent local silencing in transient *Agrobacterium* assays. In vitro binding assays indicated that βC1 binds single-stranded DNA and double-stranded DNA in a sequence nonspecific manner (Cui et al., 2005). Further work involving the βC1 protein of several distinct betasatellites further confirmed that the βC1 protein can function as an RNA silencing suppressor. Mutagenesis analysis showed that the nuclear localization of TYLCCNB and tomato leaf curl China betasatellite-encoded βC1 were required for the silencing suppression activity (Cui et al., 2005; Yang et al., 2011a).

In addition to suppressing RNA silencing, βC1 has been shown to counteract transcriptional gene silencing (TGS) in plants, in several ways (Yang et al., 2011b). Coinoculation of TYLCCNV with TYLCCNB was capable of reversing established TGS of a GFP transgene in *N. benthamiana* while causing substantial reduction of cytosine methylation levels in both TYLCCNV and host genomes. TYLCCNB could complement silencing suppression-defective beet curly top virus *L2*-mutants to reverse TGS, prevent host recovery, and reduce cytosine methylation in the beet curly top virus genome. Similar reductions in host genome cytosine methylation and suppression of GFP-directed TGS were also observed following expression of βC1 from a recombinant potato virus X vector. Moreover, transgenic expression of βC1 in *Arabidopsis* plants reactivated *F-box*, an endogenous locus known to be silenced by methylation, and reduced genome-wide DNA methylation (Yang et al., 2011b). In a recent study, Saeed et al. (2015) showed that cotton leaf curl Kokhran virus-encoded AC2, Rep, and βC1 proteins orchestrate suppression of TGS of a GFP transgene. It is not known whether distinct TGS suppressors in the same virus are assigned to different tasks during evasion of TGS. Nevertheless, taken together, these results showed that TGS is a significant barrier to begomovirus infection, and that βC1 is able to suppress DNA methylation-mediated TGS.

Acceleration of Whitefly–Begomovirus Mutualism

Begomoviruses are exclusively transmitted by the whitefly *Bemisia tabaci* (Gennadius) (*Hemiptera: Aleyrodidae*). Interactions between begomoviruses and whitefly via host plants are extremely complex, and can be mutualistic, neutral, or negative, depending on the species involved (Colvin et al., 2006; Luan et al., 2014). These interactions, especially mutualism between begomovirus and whitefly, exert important influences on the dynamics of whitefly populations and the epidemiology of virus diseases (Colvin et al., 2006). Study of the fecundity and longevity of both invasive and indigenous biotypes of whiteflies in plants infected by betasatellite-associated TYLCCNV or tobacco curly shoot virus showed that invasive whiteflies have developed indirect mutualism with TYLCCNV and tobacco curly shoot virus via host plants (Jiu et al., 2007). This vector–virus mutualism is achieved through repression of jasmonic acid (JA) defenses by betasatellites in plants (Zhang et al., 2012). Coinfection of TYLCCNB with TYLCCNV or stable transgene-expression of the βC1 protein reduced the transcription of some JA-biosynthesis and JA-responsive genes, as well as the JA level in tobacco plants through a salicylic acid-independent mechanism. Impairing or enhancing JA responses in tobacco plants enhances or depresses, respectively, the performance of the whitefly, confirming the importance of JA-regulated defenses in tobacco resistance to whitefly vector (Zhang et al., 2012). TYLCCNV/TYLCCNB infection also suppresses the terpenoid-mediated plant defense against whiteflies, showing the key role of plant terpenoids in shaping whitefly–begomovirus mutualism (Luan et al., 2013). Since JA mediates the production of terpenoids in

Nicotiana attenuata (Kessler and Baldwin, 2001), it is reasonable to speculate that betasatellites may suppress the synthesis of tobacco terpenoids through repressing the JA-signaling pathway. Further study showed that βC1 acts as the key viral genetic factor for the suppression of terpene synthesis, which in turn subverts plant resistance and promotes whitefly performance (Li et al., 2014).

CONCLUSIONS

Since their discovery in 1999, betasatellites have been recognized as important components of geminivirus disease complexes that have posed a serious threat to crop production throughout the Old World. As betasatellites can be transreplicated by different geminiviruses in a flexible manner, and can accelerate whitefly–begomovirus mutualism, new geminivirus–betasatellite disease complexes may emerge as severe threats for crop production in future. Continuous efforts should be made to comprehensively assess the spatial and temporal distribution and diversity of geminivirus–betasatellite disease complexes in nature.

Despite their simple genome organization, betasatellites have been shown to be involved in symptom modulation, suppression of RNA silencing-mediated plant defenses, and accelerating whitefly–virus mutualism, and the betasatellite-encoded βC1 protein plays an important role in these functions. The interaction between βC1 and host-encoded proteins is a rather complex process. Identification of several βC1-interacting partners has advanced our understanding of the interplay between βC1 and the host plant. Although we have witnessed considerable progress in understanding the diversity and biological functions of betasatellites, the interplay between betasatellites and host plants is still poorly understood and mechanisms between plant defense and betasatellite counterdefense remain to be elucidated.

Acknowledgments

This research was supported the National Natural Science Foundation of China (31390422) and the National Key Basic Research and Development Program of China (2012CB114004).

References

Briddon, R.W., Bull, S.E., Amin, I., Idris, A.M., Mansoor, S., Bedford, I.D., et al., 2003. Diversity of DNAβ, a satellite molecule associated with some monopartite begomoviruses. Virology 312, 106–121.

Briddon, R.W., Mansoor, S., Bedford, I.D., Pinner, M.S., Saunders, K., Stanley, J., et al., 2001. Identification of DNA components required for induction of cotton leaf curl disease. Virology 285, 234–243.

Briddon, R.W., Stanley, J., 2006. Subviral agents associated with plant single-stranded DNA viruses. Virology 344, 198–210.

Brown, J.K., Fauquet, C.M., Briddon, R.W., Zerbini, M., Moriones, E., Navas-Castillo, J., 2012. Geminiviridae. In: King, A.M.Q., Adams, M.J., Carstens, E.B., Lefkowitz, E.J. (Eds.), Virus Taxonomy: Ninth Report of the International Committee on Taxonomy of Viruses. Elsevier, London, pp. 351–373.

Brown, J.K., Zerbini, F.M., Navas-Castillo, J., Moriones, E., Ramos-Sobrinho, R., Silva, J.C., et al., 2015. Revision of begomovirus taxonomy based on pairwise sequence comparisons. Arch. Virol. 160, 1593–1619.

Burgyan, J., Havelda, Z., 2011. Viral suppressors of RNA silencing. Trends Plant Sci. 16, 265–272.

Colvin, J., Omongo, C.A., Govindappa, M.R., Stevenson, P.C., Maruthi, M.N., Gibson, G., et al., 2006. Host-plant viral infection effects on arthropod-vector population growth, development and behaviour: management and epidemiological implications. Adv. Virus Res. 67, 419–452.

Cui, X.F., Li, G.X., Wang, D.W., Hu, D.W., Zhou, X.P., 2005. A begomovirus DNA β-encoded protein binds DNA, functions as a suppressor of RNA silencing, and targets the cell nucleus. J. Virol. 79, 10764–10775.

Cui, X.F., Tao, X.R., Xie, Y., Fauquet, C.M., Zhou, X.P., 2004. A DNAβ associated with tomato yellow leaf curl China virus is required for symptom induction. J. Virol. 78, 13966–13974.

Díaz-Pendón, J.A., Ding, S.W., 2008. Direct and indirect roles of viral suppressors of RNA silencing in pathogenesis. Annu. Rev. Phytopathol. 46, 303–326.

Guo, W., Liew, J.Y., Yuan, Y.A., 2014. Structural insights into the arms race between host and virus along RNA silencing pathways in *Arabidopsis thaliana*. Biol. Rev. 89, 337–355.

Hanley-Bowdoin, L., Bejarano, E.R., Robertson, D., Mansoor, S., 2013. Geminiviruses: masters at redirecting and reprogramming plant processes. Nat. Rev. Microbiol. 11, 777–788.

Jiu, M., Zhou, X.P., Tong, L., Xu, J., Yang, X., Wan, F.H., et al., 2007. Vector-virus mutualism accelerates population increase of an invasive whitefly. PLoS One 2, e182.

Kessler, A., Baldwin, I.T., 2001. Defensive function of herbivore-induced plant volatile emissions in nature. Science 291, 2141–2144.

Kon, T., Gilbertson, R.L., 2012. Two genetically related begomoviruses causing tomato leaf curl disease in Togo and Nigeria differ in virulence and host range but do not require a betasatellite for induction of disease symptoms. Arch. Virol. 157, 107–120.

Kumar, J., Kumar, J., Singh, S.P., Tuli, R., 2014a. βC1 is a pathogenicity determinant: not only for begomoviruses but also for a mastrevirus. Arch. Virol. 159, 3071–3076.

Kumar, J., Kumar, J., Singh, S.P., Tuli, R., 2014b. Association of satellites with a mastrevirus in natural infection: complexity of wheat dwarf India virus disease. J. Virol. 88, 7093–7104.

Li, R., Weldegergis, B.T., Li, J., Jung, C., Qu, J., Sun, Y.W., et al., 2014. Virulence factors of geminivirus interact with MYC2 to subvert plant resistance and promote vector performance. Plant Cell. 26, 4991–5008.

Li, Z.H., Xie, Y., Zhou, X.P., 2005. Tobacco curly shoot virus DNAβ is not necessary for infection but intensifies symptoms in a host-dependent manner. Phytopathology 95, 902–908.

Luan, J.B., Wang, X.W., Colvin, J., Liu, S.S., 2014. Plant-mediated whitefly-begomovirus interactions: research progress and future prospects. Bull. Entomol. Res. 104, 267–276.

Luan, J.B., Yao, D.M., Zhang, T., Walling, L.L., Yang, M., Wang, Y.J., et al., 2013. Suppression of terpenoid synthesis in plants by a virus promotes its mutualism with vectors. Ecol. Lett. 16, 390–398.

Mansoor, S., Briddon, R.W., Bull, S.E., Bedford, I.D., Bashir, A., Hussain, M., et al., 2003b. Cotton leaf curl disease is associated with multiple monopartite begomoviruses supported by single DNAβ. Arch. Virol. 148, 1969–1986.

Mansoor, S., Briddon, R.W., Zafar, Y., Stanley, J., 2003a. Geminivirus disease complexes: an emerging threat. Trends Plant Sci. 8, 128–134.
Mansoor, S., Zafar, Y., Briddon, R.W., 2006. Geminivirus disease complexes: the threat is spreading. Trends Plant Sci. 11, 209–212.
Nawaz-ul-Rehman, M.S., Mansoor, S., Briddon, R.W., Fauquet, C.M., 2009. Maintenance of an old world betasatellite by a new world helper begomovirus and possible rapid adaptation of the betasatellite. J. Virol. 83, 9347–9355.
Patil, B.L., Fauquet, C.M., 2010. Differential interaction between cassava mosaic geminiviruses and geminivirus satellites. J. Gen. Virol. 91, 1871–1882.
Pumplin, N., Voinnet, O., 2013. RNA silencing suppression by plant pathogens: defence, counter-defence and counter-counter-defence. Nat. Rev. Microbiol. 11, 745–760.
Qazi, J., Amin, I., Mansoor, S., Iqbal, M.J., Briddon, R.W., 2007. Contribution of the satellite encoded gene βC1 to cotton leaf curl disease symptoms. Virus Res. 128, 135–139.
Raja, P., Wolf, J.N., Bisaro, D.M., 2010. RNA silencing directed against geminiviruses: post-transcriptional and epigenetic components. Biochim. Biophys. Acta. 1799, 337–351.
Saeed, M., Behjatnia, S.A., Mansoor, S., Zafar, Y., Hasnain, S., Rezaian, M.A., 2005. A single complementary-sense transcript of a geminiviral DNAβ satellite is determinant of pathogenicity. Mol. Plant Microbe Interact. 18, 7–14.
Saeed, M., Krczal, G., Wassenegger, M., 2015. Three gene products of a begomovirus-betasatellite complex restore expression of a transcriptionally silenced green fluorescent protein transgene in *Nicotiana benthamiana*. Virus Genes 50, 340–344.
Saeed, M., Zafar, Y., Randles, J.W., Rezaian, M.A., 2007. A monopartite begomovirus-associated DNAβ satellite substitutes for the DNA B of a bipartite begomovirus to permit systemic infection. J. Gen. Virol. 88, 2881–2889.
Saunders, K., Bedford, I.D., Briddon, R.W., Markham, P.G., Wong, S.M., Stanley, J., 2000. A unique virus complex causes ageratum yellow vein disease. Proc. Natl. Acad. Sci. USA 97, 6890–6895.
Saunders, K., Bedford, I.D., Yahara, T., Stanley, J., 2003. The earliest recorded plant virus disease. Nature 422, 831.
Saunders, K., Norman, A., Gucciardo, S., Stanley, J., 2004. The DNAβ satellite component associated with ageratum yellow vein disease encodes an essential pathogenicity protein (βC1). Virology 324, 37–47.
Saunders, K., Salim, N., Mali, V.R., Malathi, V.G., Briddon, R., Markham, P.G., et al., 2002. Characterisation of Sri Lankan cassava mosaic virus and Indian cassava mosaic virus: evidence for acquisition of a DNA B component by a monopartite begomovirus. Virology 293, 63–74.
Varsani, A., Navas-Castillo, J., Moriones, E., Hernndez-Zepeda, C., Idris, A., Brown, J.K., et al., 2014. Establishment of three new genera in the family *Geminiviridae*: *Becurtovirus*, *Eragrovirus* and *Turncurtovirus*. Arch. Virol. 159, 2193–2203.
Yang, X.L., Guo, W., Ma, X.Y., An, Q.L., Zhou, X.P., 2011a. Molecular characterization of tomato leaf curl China virus, infecting tomato plants in China, and functional analyses of its associated betasatellite. Appl. Environ. Microbiol. 77, 3092–3101.
Yang, X.L., Xie, Y., Raja, P., Li, S., Wolf, J.N., Shen, Q.T., et al., 2011b. Suppression of methylation-mediated transcriptional gene silencing by βC1-SAHH protein interaction during geminivirus-betasatellite infection. PLoS Pathog. 7, e1002329.
Zhang, T., Luan, J.B., Qi, J.F., Huang, C.J., Li, M., Zhou, X.P., et al., 2012. Begomovirus-whitefly mutualism is achieved through repression of plant defences by a virus pathogenicity factor. Mol. Ecol. 21, 1294–1304.
Zhou, X.P., 2013. Advances in understanding begomovirus satellites. Annu. Rev. Phytopathol. 51, 357–381.

PART XII

APPLICATION TO CONTROL OF VIRUSES

CHAPTER

63

Satellites as Viral Biocontrol Agents

Tiziana Mascia and Donato Gallitelli
University of Bari Aldo Moro, Bari, Italy

INTRODUCTION

Although some of the viral satellites do not encode any functional protein, the effects they can have on the disease phenotype that helper viruses induce in some hosts may range from exacerbation to attenuation. Symptom attenuation is the outcome observed most commonly, and in some plant species it is accompanied by a reduction in accumulation of the helper virus, which has led to the suggestion that competition for limited quantities of the replicase complex could be involved. Recent evidence from several systems also suggests a role for the host-driven antiviral RNA-interference (RNAi) response in the way certain viral satellites originate (Zahid et al., 2015), depress accumulation of the helper virus, and determine symptoms without encoding proteins (Hu et al., 2009; Shimura et al., 2011; Smith et al., 2011; Wang et al., 2004). Therefore, viral satellites that reduce symptoms are of interest for viral disease control, although attenuation of disease severity can occur in certain hosts but not in others and may depend on the strain of helper virus (Hu et al., 2009; Simon et al., 2004).

For the purpose of this chapter we will focus on the satellite RNA (satRNA) of cucumber mosaic virus (CMV, a member of the genus *Cucumovirus*, family *Bromoviridae,*), which was proposed as a biocontrol agent in commercial fields of vegetable crops to limit damage caused by the helper virus and some of its harmful satRNA variants.

CMV is an ubiquitous plant pathogen with the broadest host range described for any plant virus in nature, and has isometric particles that encapsidate three linear, plus-sense, single-stranded genomic RNAs and

Viroids and Satellites.
DOI: http://dx.doi.org/10.1016/B978-0-12-801498-1.00063-2

two subgenomic RNAs (Palukaitis and García-Arenal, 2003). The virus is transmitted efficiently by more than 80 aphid species in a nonpersistent manner and causes epidemics with important yield and qualitative losses in vegetable crops of temperate regions (Alonso-Prados et al., 1997; Gallitelli, 2000; Jordá et al., 1992; Tien and Wu, 1991). Some isolates of CMV function as helper virus for replication, encapsidation, and transmission of a single-stranded, linear, noncoding satRNA. CMV satRNAs exist as more than 100 variants, ranging from 330 to 405 nt in size, which occasionally cause hypervirulent CMV strains to emerge with devastating effects on crop production. Such epidemics are best exemplified by those that occurred in France, Japan, Italy, and other countries of the Mediterranean basin in the mid-1980s, when millions of hectares of processing tomato were struck by diseases characterized by necrosis and whole plant death (synonym: lethal necrosis), top stunting (synonym: curl stunt), and internal fruit browning, all co-determined by specific satRNA variants.

Interestingly, some CMV satRNAs attenuate such effects so that plants show mild symptoms, if any, and quality and quantity of the produce are preserved to acceptable levels. From a practical point of view it is even more appealing that symptom-attenuating satRNAs (hereafter denoted ameliorative satRNAs) protect plants not only from virulent CMV strains but also from those supporting replication of destructive satRNA variants.

STRATEGIES TO USE CMV SATRNA AS A BIOCONTROL AGENT

The ability of ameliorative CMV satRNAs to attenuate disease symptoms induced by CMV has led to the suggestion that they can be used as biocontrol agents. Two main approaches have been proposed and both of them have been tested in field conditions to protect tomato crops from hypervirulent CMV strains with necrogenic satRNAs: preinoculation of tomato seedlings with a mild strain of CMV supporting an ameliorative satellite, and expression of an ameliorative satRNA sequence in transgenic plants.

In the preinoculation strategy, tomato plants at the cotyledon or two-leaf stage are rub-inoculated or sprayed with a mild CMV strain containing an ameliorative satellite and kept in an aphid-proof greenhouse until ready for transplanting (Gallitelli et al., 1991; Montasser et al., 1991; Sayama et al., 1993). This takes 3–4 weeks, which is also the time needed to stabilize the virus and its satellite at low levels in plant tissues after a peak in replication reached within 2 weeks of inoculation.

In our experience, field infections occurred within 2–3 weeks after transplanting (Fig. 63.1, panel A) under climatic conditions that favored the emergence of alate forms of aphids from crowded colonies set on infected plant reservoirs. Thus, it is expected that CMV and the ameliorative satRNA are in a sort of "basal replication" by the time plants are exposed to natural inoculum. Besides tomato, the preinoculation approach has been tested in tobacco and pepper in China (Tien and Wu, 1991) and in pepper and melon in USA (Montasser et al., 1998).

This preinoculation strategy, also known as cross-protection, preimmunization, or vaccination, is the most immediate response to particularly dramatic situations such as sudden epidemic events. This offers the possibility to check plants for adverse effects before transplanting them to the field, and is flexible enough to be adapted to new cultivars as soon as they are released to the market. Major disadvantages are the labor needed to set up and store sufficient inoculum, to inoculate plants, and to dedicate enough greenhouse space in the nursery to keep treated plants separated from others before transplanting. However, since one of the prerequisites of successful epidemics is the genetic uniformity of the host (Thresh, 1982), an alternate strategy involves only a portion of the plants being preinoculated, but then transplanted, scattered among untreated individuals, so as to break the *continuum* required for necrogenic disease progression throughout the crop. A third major disadvantage is the yield reduction, usually in the range of 5%–15%, due to the infection of the mild CMV strain itself. To attenuate this "side effect," Dashti et al. (2012, 2014) showed that a mixture of two plant growth-promoting rhizobacteria, *Pseudomonas aeruginosa* and *Stenotrophomonas rhizophilia*, could compensate for qualitative and quantitative yield

FIGURE 63.1 Panel A: A tomato field devastated by lethal necrosis within a month after transplanting in a commercial field of the Basilicata Region (southern Italy) in 1988. Panel B: An experimental field of tomato plants transformed with an ameliorative satRNA set in the Basilicata Region in 1993 upon authorization of the relevant Italian Ministry. The field was surrounded by six rows of pepper, eggplants, melon, zucchini squash, watermelon, and common bean to attract aphids eventually moving from transgenic plants.

losses. Since such rhizobacteria are known to enhance the systemic defenses against other foliar pathogens, this complex interaction brought the fruit yield in tomato to values equivalent to that of the healthy controls, with more than 91% disease prevention (Dashti et al., 2012, 2014).

In the transgenic approach, the plant genome is engineered to express an ameliorative satRNA. No adverse effects have been observed from the accumulation of low levels of primary satRNA transcripts and, when the transformed plants were inoculated with a satellite-free CMV, they showed attenuation of symptoms in comparison to inoculated untransformed genotypes. The transgenic approach was tested first in tobacco (Baulcombe et al., 1986; Harrison et al., 1987; Jacquemond et al., 1988). Reports on field releases of tomato plants expressing constitutively a CMV satRNA were provided by Valanzuolo et al. (1994), who tested two tomato lines of the cultivar "San Marzano" transformed with the Ra variant of satRNA in a highly CMV-infected area of Southern Italy (Fig. 63.1, panel B), and by Yie and Tien (1993) and Wai et al. (1995), who tested in China and the United States, respectively, CMV satRNA transgenic lines of tomato by mechanical CMV challenge inoculation. In all instances a delay in symptom appearance and about a 50% increase in yield were observed, compared to nontransformed plants. Recourse to transgenesis has a major obvious advantage over the preinoculation approach, because the satRNA sequence is incorporated stably in the plant genome, inherited in its progeny and therefore "ready-to-use." Despite the satisfactory performance of transgenic plants expressing an ameliorative satellite tested in either laboratory or small-scale field experiments in highly infected areas, they have not yet been deployed on a large scale in agriculture. This is probably because they are not "flexible" enough for a timely response to the market request of new commercial varieties; their implementation in agriculture is still a matter of debate in many countries and there is the perception for potential risks (see below).

MECHANISMS AND POTENTIAL RISKS

CMV ameliorative satRNAs can function as biocontrol agents of their helper viruses with or without their own satRNAs, whether applied by mechanical inoculation or expressed as a transgene in the host plant. In cross-protection, ameliorative satRNAs probably attenuate disease symptoms by depressing the accumulation of all or individual CMV RNAs (Liao et al., 2007). It was suggested that in tobacco plants this attenuation was related to the CMV 2b gene, which is a suppressor of

RNAi, because the accumulation of CMV genomic RNAs was reduced by either the addition of satRNA or the deletion of the 2b gene (Liao et al., 2007). In some tomato genotypes this condition leads to latent infections (Cillo et al., 2007). Symptomless infections were observed also when an ameliorative satRNA was present in tomato plants with a mixed infection of CMV and potato virus Y. The latter observation is relevant for field infections, as a number of reports indicate that CMV is often found in solanaceous, cucurbit, and legume crops in mixed infections with potyviruses, where the two viruses are known to interact synergistically, worsening the effects of the respective single infections on the produce (Mascia and Gallitelli, 2014, and references therein). The presence of an ameliorative satRNA did perturb the interplay between CMV and potato virus Y in tomato, acting as a dominant factor in reducing selectively the accumulation of CMV genomic RNA to barely detectable levels (Mascia et al., 2010). From a practical point of view, the depressive effect of the ameliorative satRNA on CMV is beneficial for the produce and introduces also a bias against CMV for aphid transmission, as a decrease in the efficiency of CMV transmission from CMV satRNA-infected tomato plants was documented in several instances (Betancourt et al., 2011; Escriu et al., 2000).

In the transgenic approach, plants of a tomato line expressing an ameliorative CMV satRNA variant were tolerant to symptom production by CMV strains and downregulated virus accumulation (Cillo et al., 2004). The plant line was initially susceptible to necrosis elicited by a CMV strain supporting a necrogenic variant of satRNA, but by 21 days after inoculation, new leaves developed without necrosis and plants recovered from disease. Thus, it was hypothesized that two different mechanisms of resistance operated in the transgenic plant line: one against the virus through the downregulation operated by the ameliorative satRNA amplified from the primary transgene transcript during viral replication, and the other against the necrogenic satRNA mediated by an RNAi-based response. Evidence for the latter was provided from a correlation between reduced accumulation levels of transgenic primary transcript of satRNA, the accumulation of small interfering RNAs derived from satRNAs and the progressively reduced accumulation of the necrogenic satRNA during the plant transition from diseased to symptomless. Wang and associates (2004) proposed that CMV satRNAs are insensitive to RNAi because of their secondary structure unless they are replicated by CMV helper and the resulting double-stranded forms become targets for RNAi. Du et al. (2007) reported that Dicer-like 4, a key enzyme in the RNAi pathway, can target the satRNAs of CMV at specific secondary structures to produce small interfering RNAs against the satRNAs. The hypothesis of a delayed involvement of RNAi seems congruent with the initial necrosis observed in transgenic plants challenged with CMV plus a necrogenic satRNA, and with the induction of recovery at a later stage of infection. The hypothesis is congruent also with the observation that if nontransformed tomato plants were either inoculated simultaneously with an ameliorative and a necrogenic satRNA, or the necrogenic satRNA was inoculated 3 days after the first inoculation, the ratio of the two satRNAs was largely in favor of the necrogenic variant and the disease resulted in the typical lethal necrosis symptoms (Jacquemond and Lot 1981; Smith et al., 1992). An exception to this situation was shown when the concentration of the necrogenic variant was about 100 times less than its ameliorative counterpart (Cillo et al., 2004), a situation that probably occurs when cross-protected plants are exposed to field infection.

As for potential risks, plants infected intentionally with a mild CMV strain and an ameliorative satRNA variant as a protective measure against hypervirulent CMV strains carrying necrogenic satRNAs could perhaps introduce environmental risks. Two major risks envisaged are that deleterious satellite variants could arise by mutation from the ameliorative one employed in plant protection (Palukaitis and Roossinck, 1996), and that the ameliorative variant can have unknown adverse effects on hosts for which host range studies have not been done. Although both these possibilities exist and have been shown to occur under experimental conditions, surveys done in places where cross-protected or transgenic plants were exposed to field conditions did not support these hypotheses. Emergence of new variants harmful to other hosts was not documented and no significative outbreaks of CMV-carrying necrogenic variants recurred after the destructive one ended. Betancourt et al. (2011) demonstrated that necrogenic and non-necrogenic CMV satRNA variants do not differ in symptom modulation in melon, although, as in tomato, necrogenic satRNAs are favored in this host, so that it could act as a reservoir of necrogenic as well as for some nonnecrogenic satRNAs. Finally, the picture would be incomplete without considering that the depressive effect of ameliorative satRNA acts as a bias against aphid-mediated transmission of CMV.

References

Alonso-Prados, J.L., Fraile, A., García-Arenal, F., 1997. Impact of cucumber mosaic virus and watermelon mosaic virus 2 infection on melon production in central Spain. J. Plant Pathol. 79, 131–134.

Baulcombe, D.C., Saunders, G.R., Bevan, M.W., Mayo, M.A., Harrison, B.D., 1986. Expression of biologically active viral satellite RNA from the nuclear genome of transformed plants. Nature (London) 321, 446–449.

Betancourt, M., Fraile, A., García-Arenal, F., 2011. Cucumber mosaic virus satellite RNAs that induce similar symptoms in melon plants show large differences in fitness. J. Gen Virol. 92, 1930–1938.

Cillo, F., Finetti-Sialer, M.M., Papanice, M.A., Gallitelli, D., 2004. Analysis of mechanisms involved in the cucumber mosaic virus satellite RNA-mediated transgenic resistance in tomato plants. Mol. Plant Microbe Interact. 17, 98–108.

Cillo, F., Pasciuto, M.M., De Giovanni, C., Finetti-Sialer, M.M., Ricciardi, L., Gallitelli, D., 2007. Response of tomato and its wild relatives in the genus *Solanum* to cucumber mosaic virus and satellite RNA combinations. J. Gen. Virol. 88, 31663176.

Dashti, N.H., Ali, N.A., Cherian, V.M., Montasser, M.S., 2012. Application of plant growth promoting rhizobacteria (PGPR) in combination with a mild strain of cucumber mosaic virus (CMV) associated with viral satellite RNAs to enhance growth and protection against a virulent strain of CMV in tomato. Can. J. Plant. Pathol. 34, 177–186.

Dashti, N.H., Montasser, M.S., Nedaa, Y.A.A., Cherian, V.M., 2014. Influence of plant growth promoting rhizobacteria on fruit yield, pomological characteristics and chemical contents in cucumber mosaic virus-infected tomato plants. Kuwait J. Sci. 41, 205–220.

Du, Q.S., Duan, C.G., Zhang, Z.H., Fang, Y.Y., Fang, R.X., Xie, Q., et al., 2007. DCL4 targets cucumber mosaic virus satellite RNA at novel secondary structures. J. Virol. 81, 9142–9151.

Escriu, F., Perry, K.L., García-Arenal, F., 2000. Transmissibility of cucumber mosaic virus by *Aphis gossypii* correlates with viral accumulation and is affected by the presence of its satellite RNA. Phytopathology 90, 1068–1072.

Gallitelli, D., 2000. The ecology of cucumber mosaic virus and sustainable agriculture. Virus Res. 71, 9–21.

Gallitelli, D., Vovlas, C., Martelli, G., Montasser, M.S., Tousignant, M.E., Kaper, J.M., 1991. Satellite-mediated protection of tomato against cucumber mosaic virus: II. Field test under natural epidemic conditions in southern Italy. Plant Dis. 75, 93–95.

Harrison, B.D., Mayo, M.A., Baulcombe, D.C., 1987. Virus resistance in transgenic plants that express cucumber mosaic virus satellite RNA. Nature (London) 328, 799–802.

Hu, C.-C., Hsu, Y.-H., Lin, N.-S., 2009. Satellite RNAs and satellite viruses of plants. Viruses 1, 1325–1350.

Jacquemond, M., Anselem, J., Tepfer, M., 1988. A gene coding for a monomeric form of cucumber mosaic virus satellite RNA confers tolerance to CMV. Mol. Plant Microbe Interact. 1, 311–316.

Jacquemond, M., Lot, H., 1981. L'ARN satellite du virus de la mosaïque du concombre. I. Comparison de l'aptitude a induire la necrose de la tomate d'ARN satellites isolés de plusieurs souches du virus. Agronomie 1, 927–932.

Jordá, C., Alfaro, A., Aranda, M.A., Moriones, E., García-Arenal, F., 1992. Epidemic of cucumber mosaic virus plus satellite RNA in tomatoes in eastern Spain. Plant Dis. 76, 363–366.

Liao, Q., Zhu, L., Du, Z., Zeng, R., Peng, J., Chen, J., 2007. Satellite RNA-mediated reduction of cucumber mosaic virus genomic RNAs accumulation in *Nicotiana tabacum*. Acta Biochim. Biophys. Sin. (Shanghai). 39, 217–223.

Mascia, T., Cillo, F., Fanelli, V., Finetti-Sialer, M.M., De Stradis, A., Palukaitis, P., et al., 2010. Characterization of the interactions between cucumber mosaic virus and potato virus Y in mixed infections in tomato. Mol. Plant Microbe Interact. 23, 1514–1524.

Mascia, T., Gallitelli, D., 2014. Synergism in plant-virus interactions: a case study of CMV and PVY in mixed infection in tomato. In: Gaur, R.K., Hohn, T., Sharma, P. (Eds.), Plant Virus-Host Interactions. Academic Press Elsevier, Oxford, UK, pp. 195–206.

Montasser, M.S., Tousignant, M.E., Kaper, J.M., 1998. Viral satellite RNAs for the prevention of cucumber mosaic virus (CMV) disease in field-grown pepper and melon plants. Plant Dis. 82, 1298–1303.

Montasser, M., Tousignant, M., Kaper, J.M., 1991. Satellite mediated protection of tomato against cucumber mosaic virus: I. Greenhouse experiments and simulated epidemic conditions in the field. Plant Dis. 75, 86–92.

Palukaitis, P., García-Arenal, F., 2003. Cucumoviruses. Adv. Virus Res. 62, 241–323.

Palukaitis, P., Roossinck, M.J., 1996. Spontaneous change of a benign satellite RNA of cucumber mosaic virus to pathogenic variant. Nature Biotechnol. 14, 1264–1268.

Sayama, H., Sato, T., Kominato, M., Natsuaki, T., Kaper, J.M., 1993. Field testing of a satellite-containing attenuated strain of cucumber mosaic virus for tomato protection in Japan. Phytopathology 83, 405–410.

Shimura, H., Pantaleo, V., Ishihara, T., Myojo, N., Inaba, J.-i, Sueda, K., et al., 2011. A viral satellite RNA induces yellow symptoms on tobacco by targeting a gene involved in chlorophyll biosynthesis using the RNA silencing machinery. PLoS Pathog. 7, e1002021.

Simon, A.E., Roossinck, M.J., Havelda, Z., 2004. Plant virus satellite and defective interfering RNAs: new paradigms for a new century. Annu. Rev. Phytopathol. 42, 415–437.

Smith, C.R., Tousignant, M.E., Geletka, L.M., Kaper, J.M., 1992. Competition between cucumber mosaic virus RNAs in tomato seedlings and protoplast: a model for satellite-mediated control of tomato necrosis. Plant Dis. 76, 1270–1274.

Smith, N.A., Eamens, A.L., Wang, M.-B., 2011. Viral small interfering RNAs target host genes to mediate disease symptoms in plants. PLoS Pathog. 7, e1002022.

Thresh, J.M., 1982. Cropping practices and virus spread. Annu. Rev. Phytopathol. 20, 193–218.

Tien, P., Wu, G.S., 1991. Satellite RNA for the biocontrol of plant disease. Adv. Virus Res. 39, 321–339.

Valanzuolo, S., Catello, S., Colombo, M., Dani, M.M., Monti, M.M., Uncini, L., et al., 1994. Cucumber mosaic virus resistance in transgenic San Marzano tomatoes. Acta Hortic. 376, 377–386.

Wai, T., Stommel, J.R., Tousignant, M.E., Kaper, J.M., 1995. Field test of satellite-transgenic tomato resistance against CMV. Phytopathology 85, 1192.

Wang, M.B., Bian, X.Y., Wu, L.M., Liu, L.X., Smith, N.A., Isenegger, D., et al., 2004. On the role of RNA silencing in the pathogenicity and evolution of viroids and viral satellites. Proc. Natl. Acad. Sci. USA 101, 3275–3280.

Yie, Y., Tien, P., 1993. Plant virus satellite RNAs and their role in engineering resistance to virus diseases. Seminars Virol. 4, 363–368.

Zahid, K., Zhao, J.H., Smith, N.A., Schumann, U., Fang, Y.Y., Dennis, E.S., et al., 2015. Nicotiana small RNA sequences support a host genome origin of cucumber mosaic virus satellite RNA. PLoS Genet. 11, e1004906.

Index

Note: Page numbers followed by "*f*" and "*t*" refer to figures and tables, respectively.

D

E

F

H

O

S

T

W

X

Y

Z

Printed in the United States
By Bookmasters